Teach Yourself Electricity and Electronics

About the Authors

Stan Gibilisco, a full-time writer, is an electronics hobbyist and engineer. He has been a ham radio operator since 1966. Stan has authored several titles for the McGraw-Hill *Demystified* and *Know-It-All* series, along with numerous other technical books and dozens of magazine articles. His *Encyclopedia of Electronics* (TAB Books, 1985) was cited by the American Library Association as one of the "best references of the 1980s." Stan maintains a website at **www.sciencewriter.net.**

Dr. Simon Monk has a degree in Cybernetics and Computer Science and a PhD in Software Engineering. Dr. Monk spent several years as an academic before he returned to industry, co-founding the mobile software company Momote Ltd. He has been an active electronics hobbyist since his early teens and is a full-time writer on hobby electronics and open source hardware. Dr. Monk is the author of numerous electronics books, including *Programming Arduino*, *Hacking Electronics*, and *Programming the Raspberry Pi*.

Teach Yourself Electricity and Electronics

Sixth Edition

Stan Gibilisco

and

Simon Monk

New York Chicago San Francisco Athens London
Madrid Mexico City Milan New Delhi
Singapore Sydney Toronto

McGraw-Hill Education books are available at special quantity discounts to use as premiums and sales promotions or for use in corporate training programs. To contact a representative, please visit the Contact Us page at www.mhprofessional.com.

Teach Yourself Electricity and Electronics, Sixth Edition

1 2 3 4 5 6 7 8 9 DOC 21 20 19 18 17 16

ISBN 978-1-25-958553-1

MHID 1-25-958553-0

Sponsoring Editor
Michael McCabe

Editorial Supervisor
Donna M. Martone

Production Supervisor
Pamela A. Pelton

Project Manager
Nancy W. Dimitry

Art Director, Cover
Jeff Weeks

Copy Editor
Nancy W. Dimitry

Proofreader
Donald L. Dimitry

Composition
Gabriella Kadar

Indexer
WordCo Indexing Services

In Memory of Jack

Contents

Part 3 Basic Electronics

Preface

This book will help you learn the fundamentals of electricity and electronics without taking a formal course. It can serve as a do-it-yourself study guide or as a classroom text. This sixth edition contains new material about switching power supplies, class-D amplifiers, lithium-polymer batteries, microcontrollers, and Arduino.

You'll find a multiple-choice quiz at the end of every chapter. The quizzes are "open-book," meaning that you may (and should) refer to the chapter text as you work out the answers. When you have finished a chapter, take the quiz, write down your answers, and then give your list of answers to a friend. Have the friend tell you your score, but not which questions you got wrong. That way, you can take the test again without bias.

When you reach the end of each section, you'll encounter a multiple-choice test. A final exam concludes this course. The questions are a bit easier than the ones in the chapter-ending quizzes, but the tests are "closed-book." Don't refer back to the text as you take the part-ending tests or the final exam. For all 35 chapter-ending quizzes, all four tests, and the final exam, a satisfactory score is at least three-quarters of the answers correct. The answer key is in Appendix A.

If you need a mathematics or physics refresher, you can select from several of Stan Gibilisco's McGraw-Hill books dedicated to those topics. If you want to bolster your mathematics knowledge base before you start this course, study *Algebra Know-It-All* and *Pre-Calculus Know-It-All.* On the practical side, check out *Electricity Experiments You Can Do at Home.*

If you get bitten by the microcontroller bug, then you'll find Simon Monk's *Programming Arduino: Getting Started with Sketches* and *Programming Arduino Next Steps: Going Further with Sketches* useful companions to this book.

The authors welcome ideas and suggestions for future editions.

Stan Gibilisco
and
Simon Monk

Preface

Teach Yourself Electricity and Electronics

1
PART
Direct Current

1
CHAPTER
Background Physics

YOU MUST UNDERSTAND SOME PHYSICS PRINCIPLES TO GRASP THE FUNDAMENTALS OF ELECTRICITY and electronics. In science, we can talk about *qualitative* things or *quantitative* things, that is, "what" versus "how much." For now, let's focus on "what" and worry about "how much" later!

Atoms

All matter consists of countless tiny particles in constant motion. These particles have density far greater than anything we ever see. The matter we encounter in our everyday lives contains mostly space, and almost no "real stuff." Matter seems continuous to us only because of the particles' submicroscopic size and incredible speed. Each chemical *element* has its own unique type of particle called its *atom*.

Atoms of different elements always differ! The slightest change in an atom can make a tremendous difference in its behavior. You can live by breathing pure *oxygen*, but you couldn't survive in an atmosphere comprising pure *nitrogen*. Oxygen will cause metal to corrode, but nitrogen will not. Wood will burn in an atmosphere of pure oxygen, but won't even ignite in pure nitrogen. Nevertheless, both oxygen and nitrogen are *gases* at room temperature and pressure. Neither gas has any color or odor. These two substances differ because oxygen has eight *protons*, while nitrogen has only seven.

Nature provides countless situations in which a slight change in atomic structure makes a major difference in the way a sample of matter behaves. In some cases, we can force such changes on atoms (*hydrogen* into *helium*, for example, in a *nuclear fusion* reaction); in other cases, a minor change presents difficulties so great that people have never made them happen (*lead* into *gold*, for example).

Protons, Neutrons, and Atomic Numbers

The *nucleus*, or central part, of an atom gives an element its identity. An atomic nucleus contains two kinds of particles, the *proton* and the *neutron*, both of which have incredible density. A teaspoonful of protons or neutrons, packed tightly together, would weigh tons at the earth's surface. Protons and neutrons have nearly identical mass, but the proton has an electric charge while the neutron does not.

The simplest and most abundant element in the universe, hydrogen, has a nucleus containing one proton. Sometimes a nucleus of hydrogen has a neutron or two along with the proton, but not very often. The second most common element is helium. Usually, a helium atom has a nucleus with two protons and two neutrons. Inside the sun, nuclear fusion converts hydrogen into helium, generating the energy that makes the sun shine. The process is also responsible for the energy produced by a hydrogen bomb.

Every proton in the universe is identical to every other proton. Neutrons are all alike, too. The number of protons in an element's nucleus, the *atomic number,* gives that element its unique identity. With three protons in a nucleus we get *lithium*, a light metal solid at room temperature that reacts easily with gases, such as oxygen or chlorine. With four protons in the nucleus we get *beryllium*, also a light metal solid at room temperature. Add three more protons, however, and we have nitrogen, which is a gas at room temperature.

In general, as the number of protons in an element's nucleus increases, the number of neutrons also increases. Elements with high atomic numbers, such as lead, are therefore much more dense than elements with low atomic numbers, such as *carbon*. If you hold a lead shot in one hand and a similar-sized piece of charcoal in the other hand, you'll notice this difference.

Isotopes and Atomic Weights

For a given element, such as oxygen, the number of neutrons can vary. But no matter what the number of neutrons, the element keeps its identity, based on the atomic number. Differing numbers of neutrons result in various *isotopes* for a given element.

Each element has one particular isotope that occurs most often in nature, but all elements have multiple isotopes. Changing the number of neutrons in an element's nucleus results in a difference in the weight, and also a difference in the density, of the element. Chemists and physicists call hydrogen whose atoms contain a neutron or two in the nucleus (along with the lone proton) *heavy hydrogen* for good reason!

The *atomic weight* of an element approximately equals the sum of the number of protons and the number of neutrons in the nucleus. Common carbon has an atomic weight of 12. We call it *carbon 12* (symbolized C12). But a less-often-found isotope has an atomic weight very close to 14, so we call it *carbon 14* (symbolized C14).

Electrons

Surrounding the nucleus of an atom, we usually find a "swarm" of particles called *electrons*. An electron carries an electric charge that's *quantitatively* equal to, but *qualitatively* opposite from, the charge on a proton. Physicists arbitrarily call the electron charge *negative,* and the proton charge *positive.* The charge on a single electron or proton constitutes the smallest possible quantity of electric charge. All charge quantities, no matter how great, are theoretically whole-number multiples of this so-called *unit electric charge.*

One of the earliest ideas about the atom pictured the electrons embedded in the nucleus, like raisins in a cake. Later, scientists imagined the electrons as orbiting the nucleus, making the atom resemble a miniature solar system with the electrons as "planets," as shown in Fig. 1-1.

Today, we know that the electrons move so fast, with patterns of motion so complex, that we can't pinpoint any single electron at any given instant of time. We can, however, say that at any moment, a particular electron will just as likely "reside" inside a defined sphere as outside it. We call

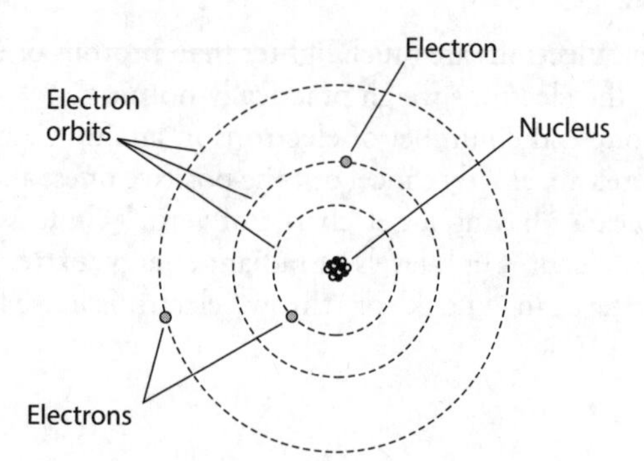

1-1 An early model of the atom, developed around the year 1900. Electrostatic attraction holds the electrons in "orbits" around the nucleus.

an imaginary sphere of this sort, centered at the nucleus of an atom, an *electron shell*. These shells have specific, predictable radii. As a shell's radius increases, the amount of energy in an electron "residing in" the shell also increases. Electrons commonly "jump" from one shell to another within an atom, thereby gaining energy, as shown in Fig. 1-2. Electrons can also "fall" from one shell to another within an atom, thereby losing energy.

Electrons can move easily from one atom to another in some materials. In other substances, it is difficult to get electrons to move. But in any case, we can move electrons a lot more easily than we can move protons. Electricity almost always results, in some way, from the motion of electrons

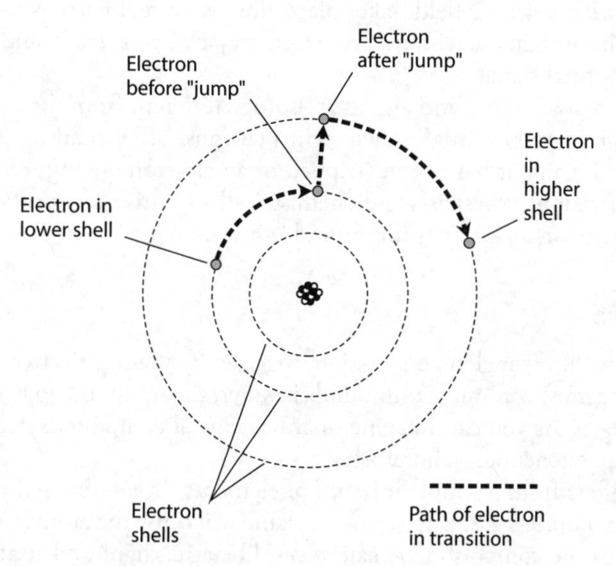

1-2 Electrons move around the nucleus of an atom at defined levels, called shells, which correspond to discrete energy states. Here, an electron gains energy within an atom.

in a material. Electrons are much lighter than protons or neutrons. In fact, compared to the nucleus of an atom, the electrons weigh practically nothing.

Quite often, the number of electrons in an atom equals the number of protons. The negative charges, therefore, exactly cancel out the positive ones, and we get an *electrically neutral* atom, where "neutral" means "having a net charge of zero." Under some conditions, an excess or shortage of electrons can occur. High levels of radiant energy, extreme heat, or the presence of an electric field (discussed later) can "knock" or "throw" electrons loose from atoms, upsetting the balance.

Ions

If an atom has more or fewer electrons than protons, then the atom carries an electrical charge. A shortage of electrons produces a positive charge; an excess of electrons produces a negative charge. The element's identity remains the same no matter how great the excess or shortage of electrons. In the extreme, all the electrons might leave the influence of an atom, leaving only the nucleus; but even then, we still have the same element. We call an electrically charged atom an *ion*. When a substance contains many ions, we say that the substance is *ionized*.

The gases in the earth's atmosphere become ionized at high altitudes, especially during the daylight hours. Radiation from the sun, as well as a constant barrage of high-speed subatomic particles from space, strips electrons from the nuclei. The ionized gases concentrate at various altitudes, sometimes returning signals from surface-based radio transmitters to the earth, allowing for long-distance broadcasting and communication.

An ionized material can conduct electricity fairly well even if, under normal conditions, it conducts poorly or not at all. Ionized air allows a *lightning stroke* (a rapid electrical *discharge* that causes a visible flash) hundreds or even thousands of meters long to occur, for example. The ionization, caused by a powerful electric field, takes place along a jagged, narrow path called the *channel*. During the stroke, the atomic nuclei quickly attract stray electrons back, and the air returns to its electrically neutral, normal state.

An element can exist as an ion and also as an isotope different from the most common isotope. For example, an atom of carbon might have eight neutrons rather than the usual six (so it's C14 rather than C12), and it might have been stripped of an electron, giving it a positive unit electric charge (so it's a positive ion). Physicists and chemists call a positive ion a *cation* (pronounced "cat-eye-on") and a negative ion an *anion* (pronounced "an-eye-on").

Compounds

Atoms of two or more different elements can join together by sharing electrons, forming a chemical *compound*. One of the most common compounds is water, the result of two hydrogen atoms joining with an atom of oxygen. As you can imagine, many chemical compounds occur in nature, and we can create many more in chemical laboratories.

A compound differs from a simple mixture of elements. If we mix hydrogen gas with oxygen gas, we get a colorless, odorless gas. But a spark or flame will cause the atoms to combine in a chemical reaction to give us the compound we call *water*, liberating light and heat energy. Under ideal conditions, a violent explosion will occur as the atoms merge almost instantly, producing a "hybrid" particle, as shown in Fig. 1-3.

Compounds often, but not always, have properties that drastically differ from either (or any) of the elements that make them up. At room temperature and pressure, both hydrogen and oxygen are gases. But under the same conditions, water exists mainly in liquid form. If the temperature falls

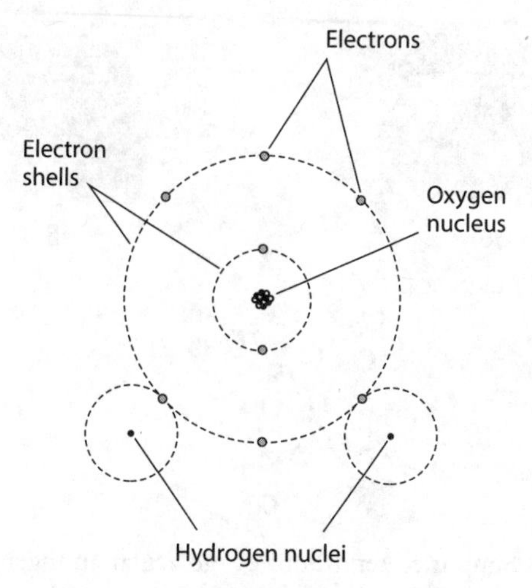

1-3 Two hydrogen atoms readily share electrons with a single atom of oxygen.

enough, water turns solid at standard pressure. If it gets hot enough, water becomes a gas, odorless and colorless, just like hydrogen or oxygen.

Another common example of a compound is rust, which forms when *iron* joins with oxygen. While iron appears to us as a dull gray solid and oxygen appears as a gas, rust shows up as a red-brown powder, completely unlike either iron or oxygen. The chemical reaction that produces rust requires a lot more time than the reaction that produces water.

Molecules

When atoms of elements join in groups of two or more, we call the resulting particles *molecules*. Figure 1-3 portrays a molecule of water. Oxygen atoms in the earth's atmosphere usually pair up to form molecules, so you'll sometimes see oxygen symbolized as O_2. The "O" represents oxygen, and the subscript 2 indicates two atoms per molecule. We symbolize water by writing H_2O to show that each molecule contains two atoms of hydrogen and one atom of oxygen.

Sometimes oxygen atoms exist all by themselves; then, we denote the basic particle as O, indicating a lone atom. Sometimes, three atoms of oxygen "stick" together to produce a molecule of *ozone*, a gas that has received attention in environmental news. We symbolize ozone by writing O_3. When an element occurs as single atoms, we call the substance *monatomic*. When an element occurs as two-atom molecules, we call the substance *diatomic*. When an element occurs as three-atom molecules, we call the substance *triatomic*.

Whether we find it in solid, liquid, or gaseous form, all matter consists of molecules or atoms that constantly move. As we increase the temperature, the particles in any given medium move faster. In a solid, we find molecules interlocked in a rigid matrix so they can't move much (Fig. 1-4A), although they vibrate continuously. In a liquid, more space exists between the molecules (Fig. 1-4B), allowing them to slide around. In a gas, still more space separates the molecules, so they can fly freely (Fig. 1-4C), sometimes crashing into each other.

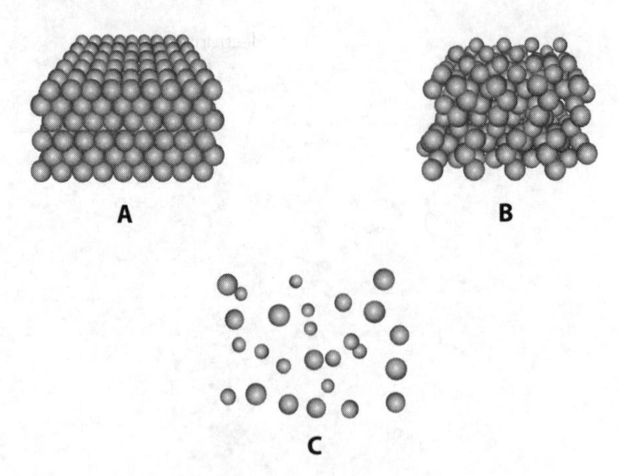

1-4 Simplified renditions of molecular arrangements in a solid (A), a liquid (B), and a gas (C).

Conductors

We define an electrical *conductor* as a substance in which the electrons can move with ease. The best known conductor at room temperature is pure elemental *silver*. *Copper* and *aluminum* also conduct electricity well at room temperature. Various other metals constitute fair to good conductors. In most electrical circuits and systems, we find copper or aluminum wire.

Some liquids conduct electricity quite well. *Mercury* provides a good example. Salt water conducts fairly well, but it depends on the concentration of dissolved salt. Gases or mixtures of gases, such as air, usually fail to conduct electricity because the large distances between the atoms or molecules prevent the free exchange of electrons. If a gas becomes ionized, however, it can conduct fairly well.

In an electrical conductor, the electrons "jump" from atom to atom (Fig. 1-5), predominantly from negatively charged locations toward positively charged locations. In a typical electrical circuit, many trillions, quadrillions, or quintillions of electrons pass a given point every second.

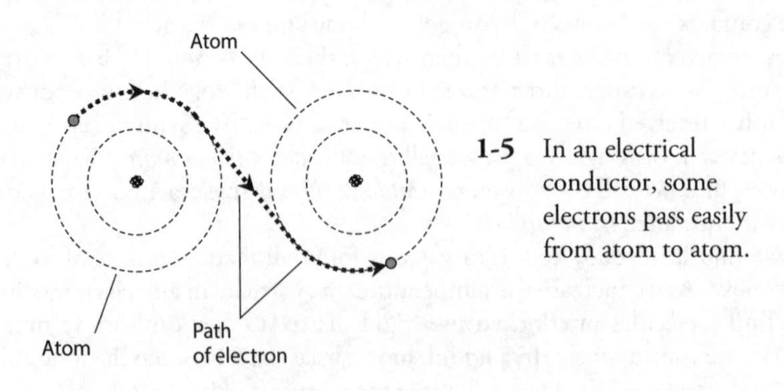

1-5 In an electrical conductor, some electrons pass easily from atom to atom.

Insulators

An electrical *insulator* prevents electron movement among atoms, except occasionally in tiny amounts. Most gases make good electrical insulators. Glass, dry wood, dry paper, and plastics also insulate well. Pure water normally insulates, although some dissolved solids can cause it to conduct. Certain metal oxides can function as good insulators, even if the metal in its pure form makes a good conductor.

Sometimes, you'll hear an insulating material called a *dielectric*. This term arises from the fact that a sample of the substance can keep electrical charges apart to form an *electric dipole*, preventing the flow of electrons that would otherwise equalize the charge difference. We encounter dielectrics in specialized components, such as *capacitors*, through which electrons *should not* directly travel.

Engineers commonly use porcelain or glass in electrical systems. These devices, called insulators in the passive rather than the active sense, are manufactured in various shapes and sizes for different applications. You can see them on utility lines that carry high *voltage*. The insulators hold the wire up without risking a *short circuit* with a metal tower or a *bleedoff* (slow discharge) through a salt-water-soaked wooden pole.

If we try hard enough, we can force almost any electrical insulator to let electrons move by forcing ionization to occur. When electrons are stripped away from their atoms, they can roam more or less freely. Sometimes a normally insulating material gets charred, or melts down, or gets perforated by a spark. Then it loses its insulating properties, and electrons can move through it.

Resistors

Some substances, such as carbon, allow electrons to move among atoms *fairly* well. We can modify the conductivity of such materials by adding impurities such as clay to a carbon paste, or by winding a long, thin strand of the material into a coil. When we manufacture a component with the intent of giving it a specific amount of conductivity, we call it a *resistor*. These components allow us to limit or control the rate of electron flow in a device or system. As the conductivity improves, the *resistance* decreases. As the conductivity goes down, the resistance goes up. Conductivity and resistance vary in *inverse proportion*.

Engineers express resistance in units called *ohms*. The higher the resistance in ohms, the more opposition a substance offers to the movement of electrons. For wires, the resistance is sometimes specified in terms of *ohms per unit length* (foot, meter, kilometer, or mile). In an electrical system, engineers strive to minimize the resistance (or *ohmic value*) because resistance converts electricity into heat, reducing the *efficiency* that the engineers want and increasing the *loss* that they don't want.

Semiconductors

In a *semiconductor*, electrons flow easily under some conditions, and with difficulty under other conditions. In their pure form, some semiconductors carry electrons almost as easily as good conductors, while other semiconductors conduct almost as poorly as insulators. But semiconductors differ fundamentally from plain conductors, insulators, or resistors. In the manufacture of a semiconductor device, chemists treat the materials so that they conduct well some of the time, and poorly some of the time—and we can control the conductivity by altering the conditions. We find semiconductors in *diodes*, *transistors*, and *integrated circuits*.

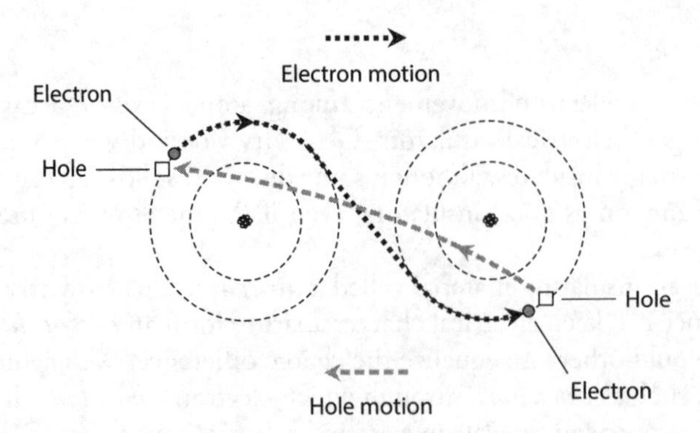

1-6 In a sample of semiconductor material, the holes travel in a direction opposite the electron motion.

Semiconductors include substances, such as *silicon, selenium,* or *gallium,* that have been "doped" by the addition of *impurities,* such as *indium* or *antimony.* Have you heard of *gallium-arsenide diodes, metal-oxide transistors,* or *silicon rectifiers*? Electrical conduction in these materials occurs as a result of the motion of electrons, but the physical details of the process are rather complicated. Sometimes engineers speak of the movement of *holes* rather than electrons. A hole is a sort of electron deficiency. You might think of it as a place where an electron normally belongs, but for some reason it's missing. Holes travel opposite to the flow of electrons, as shown in Fig. 1-6.

When electrons make up most of the *charge carriers* in a substance, we have an *N-type semiconductor.* When most of the charge carriers are holes, we have a *P-type semiconductor.* A sample of P-type material passes some electrons, and a sample of N-type material carries some holes. We call the more abundant charge carrier the *majority carrier,* and the less abundant one the *minority carrier.*

Current

Whenever charge carriers move through a substance, an electric *current* exists. We express and measure current indirectly in terms of the number of electrons or holes passing a single point in one second. Electric current flows rapidly through any conductor, resistor, or semiconductor. Nevertheless, the charge carriers actually move at only a small fraction of the speed of light in a vacuum.

A great many charge carriers go past any given point in one second, even in a system carrying relatively little current. In a household electric circuit, a 100-watt (100-W) light bulb draws about *six quintillion* (6 followed by 18 zeroes or 6×10^{18}) charge carriers per second. Even the smallest bulb carries *quadrillions* (numbers followed by 15 zeros) of charge carriers every second. Most engineers find it inconvenient to speak of current in terms of charge carriers per second, so they express current in *coulombs per second* instead. We might think of a coulomb as an "engineer's superdozen"—approximately 6,240,000,000,000,000,000 (6.24×10^{18}) electrons or holes. When 1 coulomb (1 C) of charge carriers passes a given point per second, we have an *ampere,* the standard unit of electric current. A 60-W bulb in your desk lamp draws about half an ampere (0.5 A). A typical electric utility heater draws 10 A to 12 A.

When a current flows through a resistance—always the case because even the best conductors have finite, nonzero resistance—we get heat. Sometimes we observe light as well. Old-fashioned *incandescent lamps* are deliberately designed so that the currents through their filaments produce visible light.

Static Electricity

When you walk on a carpeted floor while wearing hard-soled shoes, an excess or shortage of electrons can develop on your body, creating *static electricity*. It's called "static" because the charge carriers don't flow—until you touch a metallic object connected to earth ground or to some large fixture. Then an abrupt discharge occurs, accompanied by a spark, a snapping or popping noise, and a startling sensation.

If you acquire a much greater charge than you do under ordinary circumstances, your hair will stand on end because every strand will repel every other as they all acquire a static charge of the same polarity. When the discharge takes place, the spark might jump a centimeter or more. Then it will more than startle you; you could actually get hurt. Fortunately, charge buildups of that extent rarely, if ever, occur with ordinary carpet and shoes. However, a device called a *Van de Graaff generator* (Fig. 1-7), found in physics labs, can cause a spark several centimeters long. Use caution if you work around these things. They can be dangerous.

Lightning provides the most spectacular example of the effects of static electricity on this planet. Lightning strokes commonly occur between clouds, and between clouds and the ground. The stroke is preceded by a massive static charge buildup. Figure 1-8 illustrates *cloud-to-cloud* (A) and *cloud-to-ground* (B) electric dipoles caused by weather conditions. In the scenario shown at B, the positive charge in the earth follows along beneath a storm cloud.

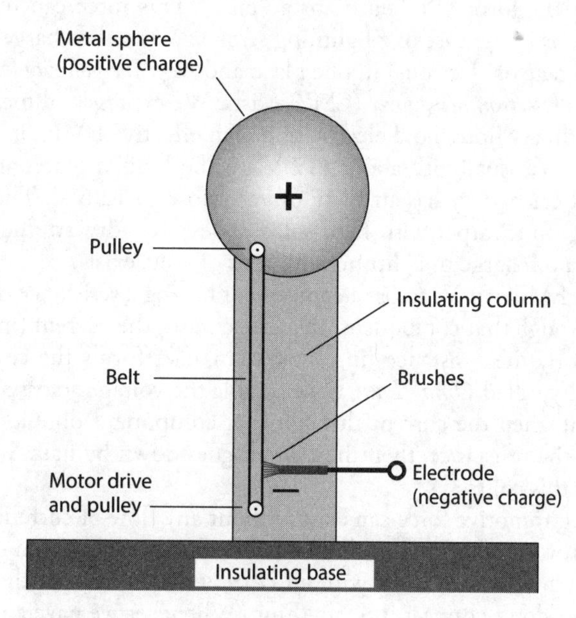

1-7 Simplified illustration of a Van de Graaff generator. This machine can create a charge sufficient to produce a spark several centimeters long.

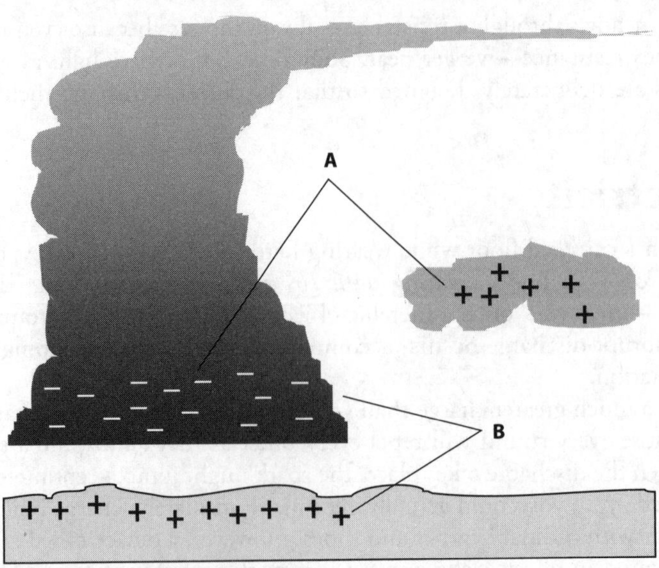

1-8 Electrostatic charges can build up between clouds (A) or between a cloud and the earth's surface (B).

Electromotive Force

Charge carriers can move in an orderly fashion only if they experience a well-defined directional force in the form of a "push" or a "pull." This force can result from a buildup of static electric charges, as in the case of a lightning stroke. When the charge builds up, attended by *positive polarity* (shortage of electrons) in one place and *negative polarity* (excess of electrons) in another place, a powerful *electromotive force* (EMF) exists. We express and measure EMF in units called *volts*.

Ordinary household electricity has an effective EMF, or *voltage*, of between 110 volts (110 V) and 130 V; usually it's about 117 V. In the United States and most other countries, a new, fully charged car battery has an EMF of very close to 12.6 V. The static charge that you acquire when walking on a carpet with hard-soled shoes on a dry afternoon can reach several thousand volts. Before a discharge of lightning, millions of volts exist.

An EMF of 1 V, across a component having a resistance of 1 ohm, will cause a current of 1 A to flow through that component. In a DC circuit, the current (in amperes) equals the voltage (in volts) divided by the resistance (in ohms). This fact forms the cornerstone for a classic relationship in electricity called *Ohm's Law*. If we double the voltage across a component whose resistance remains constant, then the current through that component doubles. If we keep the voltage constant but double the resistance, then the current goes down by half. We'll examine Ohm's law more closely later in this course.

Electromotive force can exist without any flow of current, producing static electricity, as we've seen. However, an EMF without current also exists between the two wires of an electric lamp when the switch is off. An EMF without current exists between the terminals of a common *flashlight cell* when we don't connect it to anything. Whenever we have an EMF between two points, an electric current will flow if we provide a conductive path between those points. Voltage, or EMF, is sometimes called *electric potential* or *potential difference* for this reason. An EMF has the potential (that is, the ability) to move charge carriers, given the right conditions.

A huge EMF doesn't necessarily drive a lot of current through a conductor or resistance. Think of your body after you've spent some time walking around on the carpet. Although the EMF might seem deadly in terms of sheer magnitude (thousands of volts), relatively few coulombs of charge carriers accumulate on your body. In relative terms, not that many electrons flow through your finger when you touch an external object. That's why you don't get a severe shock. However, if plenty of coulombs are available, then even a modest EMF, such as 117 V (typical of a household utility outlet), can drive a lethal current through your body. That's why it's dangerous to repair an electrical device when it's connected to a source of power. The utility plant can deliver an unlimited number of coulombs.

Non-Electrical Energy

In scientific experiments, we often observe phenomena that involve energy in non-electrical form. Visible light provides an excellent example. A light bulb converts electricity into radiant energy that we can see. This fact motivated people like Thomas Edison to work with electricity, making discoveries and refining devices that make our lives convenient today. We can also convert visible light into electricity. A *photovoltaic cell* (also called a *solar cell*) works this sort of magic.

Light bulbs always give off heat as well as light. In fact, incandescent lamps actually give off more energy as heat than as light. You've probably had experience with electric heaters, designed for the purpose of changing electrical energy into heat energy. This "heat" is actually a form of radiant energy called *infrared* (IR), which resembles visible light, except that IR has a longer *wavelength* and you can't see the rays.

We can convert electricity into *radio waves*, *ultraviolet* (UV) rays, and *X rays*. These tasks require specialized devices such as radio transmitters, *mercury-vapor lamps*, and *electron tubes*. Fast-moving protons, neutrons, electrons, and atomic nuclei also constitute non-electrical forms of energy.

When a conductor moves in a magnetic field, electric current flows in that conductor. This effect allows us to convert mechanical energy into electricity, obtaining an *electric generator*. Generators can also work backwards, in which case we have an *electric motor* that changes electricity into mechanical energy.

A magnetic field contains energy of a unique kind. The science of *magnetism* is closely related to electricity. The oldest and most universal source of magnetism is the *geomagnetic field* surrounding the earth, which arises as a result of the alignment of iron atoms in the core of the planet.

A changing magnetic field creates a fluctuating electric field, and a fluctuating electric field produces a changing magnetic field. This phenomenon, called *electromagnetism*, makes it possible to send wireless signals over long distances. The electric and magnetic fields keep producing one another over and over again through space.

Dry cells, *wet cells*, and *batteries* convert *chemical energy* into electrical energy. In an automotive battery, for example, acid reacts with metal electrodes to generate a potential difference. When we connect the poles of the battery to a component having finite resistance, current flows. Chemical reactions inside the battery keep the current going for a while, but the battery eventually runs out of energy. We can restore the chemical energy to a lead-acid automotive battery (and certain other types) by driving current through it for a period of time, but some batteries (such as most ordinary flashlight cells and lantern batteries) become useless when they run out of chemical energy.

Quiz

Refer to the text in this chapter if necessary. A good score is at least 18 correct answers out of these 20 questions. The answers are listed in the back of this book.

1. The number of protons in the nucleus of an atom *always*
   (a) equals its atomic number.
   (b) equals its atomic weight.
   (c) equals the number of electrons.
   (d) equals the number of neutrons plus the number of electrons.

2. The number of neutrons in the nucleus of an atom *sometimes*
   (a) equals its atomic number.
   (b) equals its atomic weight.
   (c) equals the number of protons.
   (d) More than one of the above

3. The atomic weight of an atom *always*
   (a) equals the number of electrons.
   (b) equals the number of protons.
   (c) equals the number of neutrons.
   (d) approximately equals the number of neutrons plus the number of protons.

4. When an atom has a net negative electric charge, we can call it
   (a) an anion.
   (b) a cation.
   (c) diatomic.
   (d) positronic.

5. An atom can have
   (a) more than one isotope.
   (b) only one isotope.
   (c) no more protons than neutrons.
   (d) no more neutrons than protons.

6. An element whose atoms can have more than one atomic weight
   (a) cannot exist.
   (b) always has an electric charge.
   (c) shares protons with surrounding atoms.
   (d) is a common occurrence in nature.

7. A compound comprising three atoms
   (a) cannot exist.
   (b) always has an electric charge.
   (c) shares protons with surrounding atoms.
   (d) is a common occurrence in nature.

8. Ionization by itself *never* causes
   (a) the conductivity of a substance to improve.
   (b) an atom to gain or lose protons.
   (c) an electrically neutral atom to become charged.
   (d) an atom to gain or lose electrons.

9. Which of the following substances is the worst electrical conductor?
   (a) Mercury
   (b) Aluminum
   (c) Glass
   (d) Silver

10. Which of the following substances allows electrons to move among its atoms with the greatest ease?
    (a) Copper
    (b) Pure water
    (c) Dry air
    (d) Porcelain

11. If we place 12 V across a component whose resistance equals 6 ohms, how much current will flow through the component?
    (a) 0.5 A
    (b) 2 A
    (c) 72 A
    (d) We need more information to say.

12. If we double the resistance in the situation of Question 11 but don't change the voltage, the current will
    (a) not change.
    (b) get cut in half.
    (c) double.
    (d) quadruple.

13. The term *static electricity* refers to
    (a) voltage with no current.
    (b) current with no voltage.
    (c) current through an infinite resistance.
    (d) voltage that never changes.

14. Which of the following general statements applies to dielectric materials?
    (a) They have extremely low resistance (practically zero).
    (b) They have extremely high resistance (practically infinite).
    (c) They have resistance that depends on the current through them.
    (d) They produce two different voltages at the same time.

15. We can express the quantity of electrons flowing past a fixed point per unit of time in
    (a) coulombs.
    (b) volts.
    (c) ohms.
    (d) amperes.

16. In a lightning stroke, the term *channel* means
    (a) a current-carrying path of ionized air.
    (b) alternating-current frequency.
    (c) a stream of moving protons and neutrons.
    (d) a flowing stream of cool gas.

17. The term *electromotive force* (EMF) is an alternative expression for
    (a) current.
    (b) charge.
    (c) voltage.
    (d) resistance.

18. When you shuffle across a carpeted floor on a dry winter afternoon, you can acquire a potential difference, with respect to ground, of
    (a) an ohm or two.
    (b) up to about 200 ohms.
    (c) millions of ohms.
    (d) None of the above

19. Which of the following devices directly converts chemical energy to electricity?
    (a) A generator
    (b) A dry cell
    (c) A motor
    (d) A photovoltaic cell

20. Which of the following devices directly converts visible light to electricity?
    (a) A generator
    (b) A dry cell
    (c) A motor
    (d) A photovoltaic cell

2
CHAPTER

Electrical Units

LET'S LEARN ABOUT THE STANDARD UNITS THAT ENGINEERS USE IN DIRECT-CURRENT (DC) CIRCUITS. Many of these principles apply to common utility alternating-current (AC) systems as well.

The Volt

In Chap. 1, you learned about the volt, the standard unit of electromotive force (EMF), or potential difference. An accumulation of electrostatic charge, such as an excess or shortage of electrons, always occurs when we have a potential difference between two points or objects. A power plant, an electrochemical reaction, light rays striking a semiconductor chip, and other phenomena can also produce voltages. We can get an EMF when we move an electrical conductor through a fixed magnetic field, or when we surround a fixed electrical conductor with a fluctuating magnetic field.

A potential difference between two points, called *poles*, invariably produces an *electric field*, represented by *electric lines of flux*, as shown in Fig. 2-1. We call such a pair of electrically charged poles an *electric dipole*. One pole carries relatively positive charge, and the other pole carries relatively negative charge. The positive pole always has fewer electrons than the negative pole. Note that the electron numbers are relative, not absolute! An electric dipole can exist even if both poles carry surplus electrons, or if both poles suffer from electron deficiencies, relative to some external point of reference having an absolutely neutral charge.

The abbreviation for volt (or volts) is V. Sometimes, engineers use smaller units. The *millivolt* (mV) equals 0.001 V. The *microvolt* (μV) equals 0.000001 V. Units larger than the volt also exist. One kilovolt (kV) represents 1000 V. One *megavolt* (MV) equals 1,000,000 V, or 1000 kV.

In an everyday dry cell, the poles maintain a potential difference somewhere between 1.2 and 1.7 V. In an automotive battery, it's in the range of 12 V to 14 V. In household AC utility wiring, the potential difference alternates polarity and maintains an effective value of approximately 117 V for electric lights and most small appliances, and 234 V for washing machines, ovens, or other large appliances. In some high-power radio transmitters, the EMF can range in the thousands of volts. The largest potential differences on our planet—upwards of 1 MV—build up in thunderstorms, sandstorms, and violent erupting volcanoes.

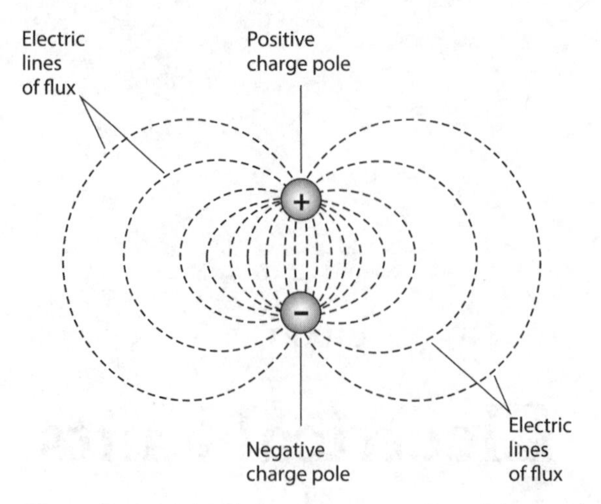

2-1 Electric lines of flux always exist near poles of electric charge.

The existence of a voltage always means that *charge carriers*, which are mostly electrons in a conventional circuit, will travel between the charge poles if we provide a decent path for them to follow. Voltage represents the driving force, or "pressure," that impels charge carriers to move. If we hold all other factors constant, a high voltage will make the charge carriers flow in greater quantity per unit of time, thereby producing a larger electrical current than a low voltage. But that statement oversimplifies the situation in most practical systems, where "all other factors" rarely "hold constant"!

Current Flow

If we provide a conducting or semiconducting path between two poles having a potential difference, charge carriers flow in an attempt to equalize the charge between the poles. This current continues for as long as the path remains intact, and as long as a charge difference exists between the poles.

Sometimes the charge difference between two electric poles decreases to zero after a current has flowed for a while. This effect takes place in a lightning stroke, or when you touch a radiator after shuffling around on a carpet. In these instances, the charge between the poles equalizes in a fraction of a second. In other cases, the charge takes longer to dissipate. If you connect a piece of wire directly between the positive and negative poles of a dry cell, the cell "runs out of juice" after a few minutes. If you connect a light bulb across the cell to make a "flashlight," the charge difference may take an hour or two to get all the way down to zero.

In household electric circuits, the charge difference never equalizes unless a power failure occurs. Of course, if you short-circuit an AC electrical outlet (don't!), the fuse or breaker will blow or trip, and the charge difference will immediately drop to zero. But if you put a standard utility light bulb at the outlet, the charge difference will continue to exist at "full force" even as the current flows. The power plant can maintain a potential difference of 117 V across a lot of light bulbs indefinitely.

Have you heard that the deadly aspect of electricity results from current, not voltage? Literally, that's true, but the statement plays on semantics. You could also say "It's the heat, not the fire, that burns you." Okay! But a deadly current can arise only in the presence of an EMF sufficient to drive a certain amount of current through your body. You don't have to worry about deadly currents flowing between your hands when you handle a 1.5-V dry cell, even though, in theory, such

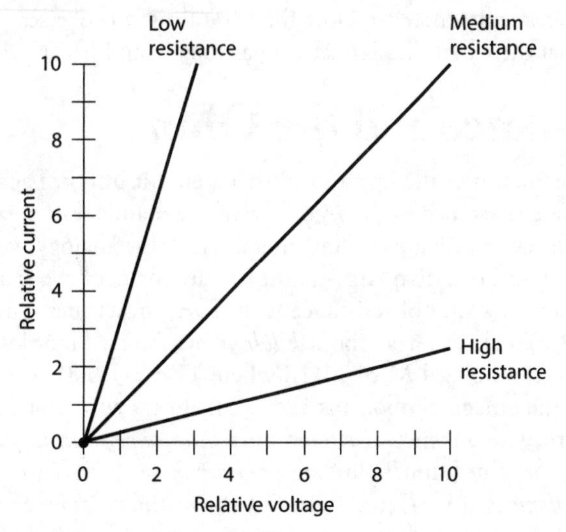

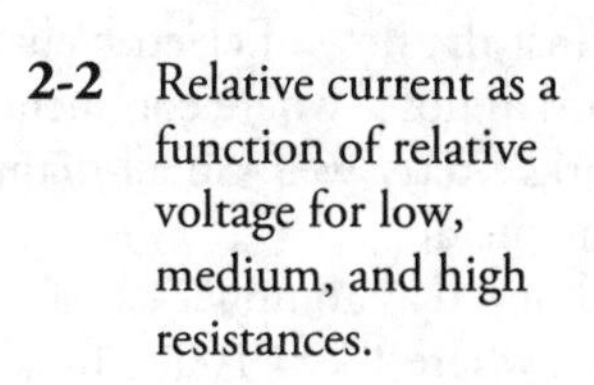
2-2 Relative current as a function of relative voltage for low, medium, and high resistances.

a cell could produce currents strong enough to kill you if your *body resistance* were much lower. You're safe when handling flashlight cells, but you've got good reason to fear for your life around household utility circuits. An EMF of 117 V can easily pump enough current through your body to electrocute you.

It all goes back to Ohm's Law. In an electric circuit whose conductance (or resistance) never varies, the current is directly proportional to the applied voltage. If you double the voltage, you double the current. If you cut the voltage in half, the current goes down by half. Figure 2-2 shows this relationship as a graph in general terms. Here, we assume that the *power supply* can always provide as many charge carriers per unit of time as we need.

The Ampere

Current expresses the rate at which charge carriers flow past a fixed point per unit of time. The standard unit of current is the *ampere,* which represents one coulomb (6,240,000,000,000,000,000, or 6.24×10^{18}) of charge carriers flowing past a given point every second.

An ampere is a comparatively large amount of current. The abbreviation is A. Often, you'll want to express current in terms of *milliamperes,* abbreviated mA, where 1 mA = 0.001 A. You'll also sometimes hear of *microamperes* (μA), where 1 μA = 0.000001 A or 0. 001 mA. You might even encounter *nanoamperes* (nA), where 1 nA = 0.000000001 A = 0.001 μA.

A current of a few milliamperes will give you a rude electrical shock. About 50 mA will jolt you severely, and 100 mA can kill you if it flows through your heart. An ordinary utility light bulb draws 0.5 A to 1 A of current in a household utility circuit. An electric iron draws approximately 10 A; an entire household normally uses between 10 A and 100 A, depending on the size of the house and the kinds of appliances it has, and also on the time of day, week, or year.

The amount of current that flows in an electrical circuit depends on the voltage, and also on the resistance. In some electrical systems, extremely large currents, say 1000 A, can flow. You'll get a current like this if you place a metal bar directly across the output terminals of a massive electric generator. The bar has an extremely low resistance, and the generator can drive many coulombs of charge carriers through the bar every second. In some semiconductor electronic devices, a few nanoamperes will suffice to allow for complicated processes. Some electronic clocks draw so little current that their batteries last as long as they would if you left them on the shelf.

Resistance and the Ohm

Resistance quantifies the opposition that a circuit imposes against the flow of electric current. You can compare resistance to the *reciprocal* of the diameter of a garden hose (where conductance compares to the actual diameter). For metal wire, this analogy works pretty well. Small-diameter wire has higher resistance than large-diameter wire made of the same metal.

The standard unit of resistance is the *ohm*, sometimes symbolized as an upper-case Greek letter omega (Ω). You'll also hear about *kilohms* (symbolized k or kΩ), where 1 k = 1000 ohms, or about 1 *megohm* (symbolized M or MΩ), where 1 M = 1,000,000 ohms or 1000 k. In this book, we'll never use the omega symbol. Instead, we'll always write out "ohm" or "ohms" in full.

Electric wire is sometimes rated for *resistance per unit length.* The standard unit for this purpose is the *ohm per foot* (ohm/ft) or the *ohm per meter* (ohm/m). You might also come across the unit *ohm per kilometer* (ohm/km). Table 2-1 shows the resistance per unit of length for various common sizes of solid copper wire at room temperature as a function of the wire size, as defined by a scheme known as the *American Wire Gauge* (AWG).

When we place a potential difference of 1 V across a component whose resistance equals 1 ohm, assuming that the power supply can deliver an unlimited number of charge carriers, we get a current of 1 A. If we double the resistance to 2 ohms, the current decreases to 0.5 A. If we cut the resistance by a factor of 5 to get only 0.2 ohms, the current increases by the same factor, from 1 A to 5 A. The current flow, for a constant voltage, varies in *inverse proportion* to the resistance. Figure 2-3 shows the current, through components of various resistances, given a constant potential difference of 1 V.

Table 2-1. Approximate resistance per unit of length in ohms per kilometer (ohms/km) at room temperature for solid copper wire as a function of the wire size in American Wire Gauge (AWG).

Wire size, AWG #	Ohms/km
2	0.52
4	0.83
6	1.3
8	2.7
10	3.3
12	5.3
14	8.4
16	13
18	21
20	34
22	54
24	86
26	140
28	220
30	350

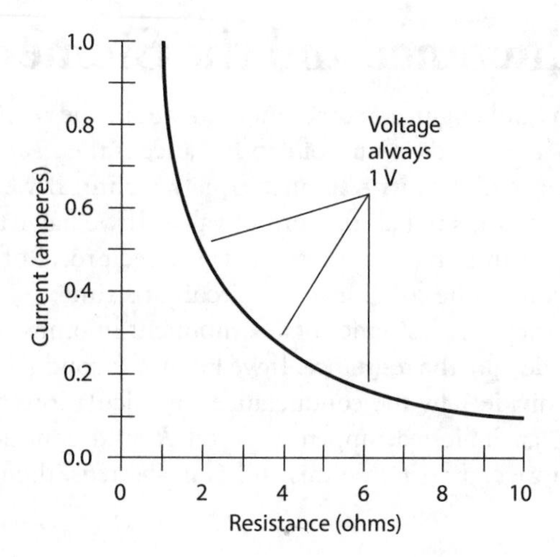

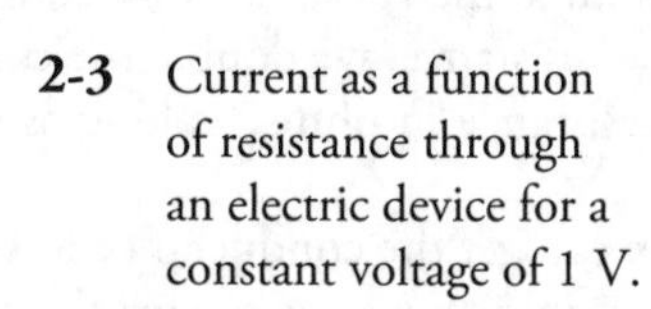

2-3 Current as a function of resistance through an electric device for a constant voltage of 1 V.

Whenever an electric current flows through a component, a potential difference appears across that component. If the component has been deliberately manufactured to exhibit a certain resistance, we call it a *resistor*. Figure 2-4 illustrates this effect. In general, the potential difference arises in direct proportion to the current through the resistance. Engineers take advantage of this effect when they design electronic circuits, as you'll learn later in this book.

Electrical circuits always have some resistance. No such thing as a perfect conductor (an object with mathematically zero resistance) exists in the "real world." When scientists cool certain metals down to temperatures near *absolute zero*, the substances lose practically all of their resistance, so that current can flow around and around for a long time. This phenomenon is called *superconductivity*. But nothing can ever become an *absolutely perfect* conductor.

Just as a *perfectly* resistance-free substance cannot exist in the real world, we'll never encounter an *absolutely infinite* resistance, either. Even dry air conducts electric current to some extent, although the effect is usually so small that scientists and engineers can ignore it. In some electronic applications, engineers select materials based on how "nearly infinite" their resistance appears; but when they say that, they exploit a figure of speech. They really mean to say that the resistance is so gigantic that we can consider it "infinite" for all practical purposes.

In electronics, the resistance of a component often varies, depending on the conditions under which that component operates. A transistor, for example, might have high resistance some of the time, and low resistance at other times. High/low resistance variations can take place thousands, millions, or billions of times each second. In this way, oscillators, amplifiers, and digital devices function in radio receivers and transmitters, telephone networks, digital computers, and satellite links (to name just a few applications).

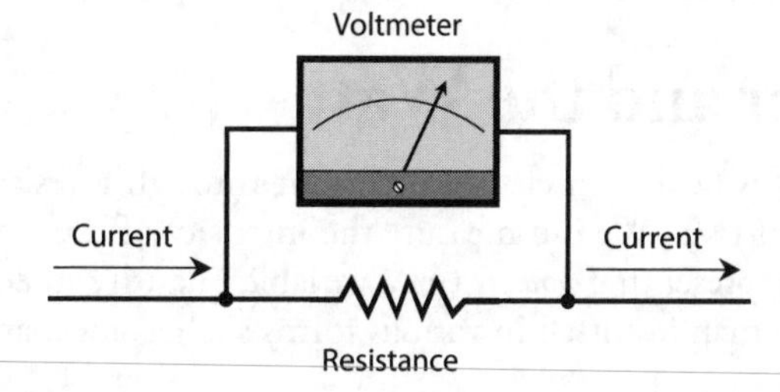

2-4 Whenever current passes through a component having resistance, a voltage exists across that component.

Conductance and the Siemens

Electricians and engineers sometimes talk about the *conductance* of a material, rather than about its resistance. The standard unit of conductance is the *siemens,* abbreviated S. When a component has a conductance of 1 S, its resistance equals 1 ohm. If we double the resistance of a component, its conductance drops to half the former value. If we halve the resistance, we double the conductance. Conductance in siemens always equals the reciprocal of resistance in ohms, as long as we confine our attention to one component or circuit at a time.

If we know the resistance of a component in ohms, we can get the conductance in siemens; we simply divide 1 by the resistance. If we know the conductance in siemens, we can get the resistance in ohms; we divide 1 by the conductance. In calculations and equations, engineers denote resistance by writing an italicized, uppercase letter R, and conductance by writing an italicized, uppercase letter G. If we express R in ohms and G in siemens, then

$$G = 1/R$$

and

$$R = 1/G$$

In "real-world" electrical and electronic circuits, you'll often use units of conductance much smaller than the siemens. A resistance of 1 k represents a conductance of one *millisiemens* (1 mS). If we encounter a component whose resistance equals 1 M, its conductance is one *microsiemens* (1 μS). You'll sometimes hear about *kilosiemens* (kS) or *megasiemens* (MS), representing resistances of 0.001 ohm and 0.000001 ohm, respectively. Short lengths of heavy wire have conductance values in the range of kilosiemens. A heavy, solid copper or silver rod might exhibit a conductance in the megasiemens range.

If a component has a resistance of 50 ohms, its conductance equals 1/50 S or 0.02 S. We can also call this quantity 20 mS. Now imagine a piece of wire with a conductance of 20 S. Its resistance equals 1/20 ohm or 0.05 ohm. You won't often hear or read the term "milliohm" in technical conversations or papers, but you might say that an 0.05-ohm length of wire has a resistance of 50 milliohms, and you'd be technically correct.

When you want to determine the *conductivity* of a component, circuit, or system, you must exercise caution or you might end up calculating the wrong value. If wire has a resistance per unit length of 10 ohms/km, you can't say that it has a conductivity of 1/10, or 0.1, S/km. A 1-km span of such wire does indeed have a conductance of 0.1 S, but a 2-km span of the same wire has a resistance of 20 ohms because you have twice as much wire. That's not twice the conductance, but half the conductance, of the 1-km span. If you say that the conductivity of the wire is 0.1 S/km, then you might be tempted to say that 2 km of the wire has 0.2 S of conductance. That would be a mistake! Conductance decreases with increasing wire length.

Figure 2-5 illustrates the resistance and conductance values for various lengths of wire having a resistance per unit length of 10 ohms/km.

Power and the Watt

Whenever we drive an electrical current through a resistive component, the temperature of that component rises. We can measure the intensity of the resulting *heat* in units called *watts* (symbolized W), representing *power*. (As a variable quantity in equations, we denote power by writing P.) Power can manifest itself in various forms such as mechanical motion, radio waves, visible light, or

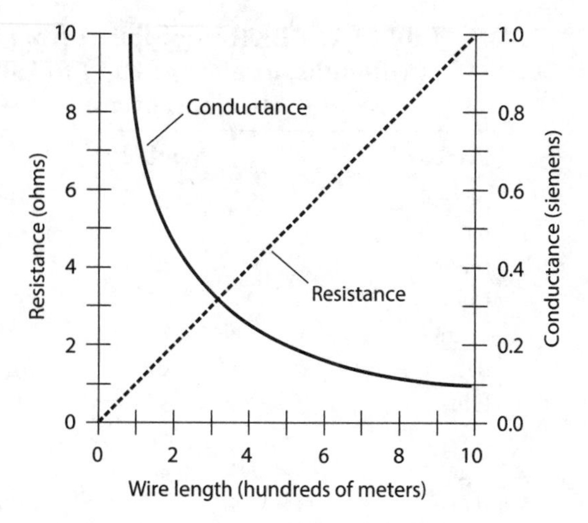

2-5 Resistance and conductance for various lengths of wire having a resistivity of 10 ohms/km.

noise. But we'll always find heat (in addition to any other form of power) in an electrical or electronic device, because no "real-world" system operates with 100-percent efficiency. Some power always goes to waste, and this waste shows up mainly as heat.

Look again at Fig. 2-4. A certain potential difference appears across the resistor, although the illustration does not reveal the actual voltage. A current flows through the resistor; again, the diagram doesn't tell us the value. Suppose that we call the voltage across the resistor E and the current through the resistor I, expressed in volts (V) and amperes (A), respectively. If we let P represent the power in watts dissipated by the resistor, then

$$P = EI$$

If the voltage E across the resistance is caused by two flashlight cells in series, giving us 3 V, and if the current I through the resistance (a flashlight bulb, perhaps) equals 0.2 A, then $E = 3$ V and $I = 0.2$ A, and we can calculate the power P as

$$P = EI = 3 \times 0.2 = 0.6 \text{ W}$$

Suppose the voltage equals 220 V, giving rise to a current of 400 mA. To calculate the power, we must convert the current into amperes: 400 mA = 400/1000 A = 0.400 A. Then we have

$$P = EI = 220 \times 0.400 = 88.0 \text{ W}$$

You will often hear about *milliwatts* (mW), *microwatts* (μW), *kilowatts* (kW), and *megawatts* (MW). By now, you should know what these units represent when you see the prefixes. Otherwise, you can refer to Table 2-2, which lists the common *prefix multipliers* for physical units.

Once in a while, you'll want to take advantage of the power equation to find a current through a component or a voltage across a component. In that case, you can use the variant

$$I = P/E$$

to find current, or

$$E = P/I$$

Table 2-2. Prefix multipliers from 0.000000000001 (trillionths, or units of 10^{-12}) to 1,000,000,000,000 (trillions, or units of 10^{12}).

Prefix	Symbol	Multiplier
pico-	p	0.000000000001 (or 10^{-12})
nano-	n	0.000000001 (or 10^{-9})
micro-	μ	0.000001 (or 10^{-6})
milli-	m	0.001 (or 10^{-3})
centi-	c	0.01 (or 10^{-2})
deci-	d	0.1 (or 10^{-1})
kilo-	k	1000 (or 10^{3})
mega-	M	1,000,000 (or 10^{6})
giga-	G	1,000,000,000 (or 10^{9})
tera-	T	1,000,000,000,000 (or 10^{12})

to find the voltage. Always convert to standard units (volts, amperes, and watts) before performing calculations with any of these formulas. Otherwise, you risk getting an answer that's too large or small by one or more *orders of magnitude* (powers of 10)!

A Word about Notation

Sometimes, symbols and abbreviations appear in italics, and sometimes they don't. We'll encounter subscripts often, and sometimes even the subscripts are italicized. Following are some rules that apply to notation in electricity and electronics.

- We never italicize the abbreviations for units such as volts (V), amperes (A), and watts (W).
- We never italicize the abbreviations for objects or components such as resistors (R), batteries (B), capacitors (C), and inductors (L).
- We never italicize the abbreviations for quantifying prefixes such as kilo- (k), micro- (μ), mega- (M), or nano- (n).
- Labeled points in drawings might or might not be italicized. It doesn't matter, as long as a diagram remains consistent within itself. We might call a point either P or *P*, for example.
- We always italicize the symbols for mathematical constants and variables such as time (*t*), the speed of light in a vacuum (*c*), velocity (*v*), and acceleration (*a*).
- We always italicize the symbols for electrical quantities such as voltage (*E* or *V*), current (*I*), resistance (*R*), and power (*P*).
- We never italicize numeric subscripts. We might denote a certain resistor as R_2, but never as R_{2}; we might denote a certain amount of current as I_4, but never as I_{4}.
- For non-numeric subscripts, the same rules apply as for general symbols.

Once in a while we'll see the same symbol italicized in one place and not in another—even within a single diagram or discussion! We might, for example, talk about "resistor number 3" (symbolized R_3), and then later in the same paragraph talk about its value as "resistance number 3" (symbolized R_3). Still later, we might talk about "the *n*th resistor in a combination of resistors" (R_n) and then "the *n*th resistance in a combination of resistances" (R_n).

Energy and the Watt-Hour

Have you heard the terms "power" and "energy" used interchangeably, as if they mean the same thing? Well, they don't! The term *energy* expresses power dissipated over a certain period of time. Conversely, the term *power* expresses the instantaneous rate at which energy is expended at a particular moment in time.

Physicists measure energy in units called *joules*. One joule (1 J) technically equals a *watt-second*, the equivalent of 1 W of power dissipated for 1 s of time (1 W · s or 1 Ws). In electricity, you'll more often encounter the *watt-hour* (symbolized W · h or Wh) or the *kilowatt-hour* (symbolized kW · h or kWh). As their names imply, a watt-hour represents the equivalent of 1 W dissipated for 1 h, and 1 kWh represents the equivalent of 1 kW of power dissipated for 1 h.

An energy quantity of 1 Wh can manifest itself in infinitely many ways. A light bulb rated at 60 W consumes 60 Wh in 1 h, the equivalent of a watt-hour per minute (1 Wh/min). A lamp rated at 100 W consumes 1 Wh in 1/100 h, or 36 s. Whenever we double the power, we halve the time required to consume 1 Wh of energy. But in "real-world" scenarios, the rate of power dissipation rarely remains constant. It can change from moment to moment in time.

Figure 2-6 illustrates two hypothetical devices that consume 1 Wh of energy. Device A uses its power at a constant rate of 60 W, so it consumes 1 Wh in 1 min. The power consumption rate of device B varies, starting at zero and ending up at more than 60 W. How do we know that this second device really consumes 1 Wh of energy? To figure that out, we must determine the *area under the graph*. In this case, the graph encloses a simple triangle. We recall from our basic geometry courses that the area of a triangle equals half the base length times the height. Device B receives power for 72 s, or 1.2 min; that's $1.2/60 = 0.02$ h. The area under the graph is therefore $1/2 \times 100 \times 0.02 = 1$ Wh.

When you calculate energy values, you must always keep in mind the units with which you work. In the example of Fig. 2-6, you would use the watt-hour, so you must multiply watts by hours. If you multiply watts by minutes or seconds, you'll get the wrong kind of unit in your answer—and sometimes a "unit" that doesn't have any technical definition at all!

Often, graphs of power versus time show up as complex curves, not "neat" figures, such as rectangles or triangles. Consider the graph of power consumption in your home, as a function of time, over the course of a hypothetical day. That graph might resemble the curve in Fig. 2-7. Obviously, you won't find a simple formula to calculate the area under this curve, but you can use another

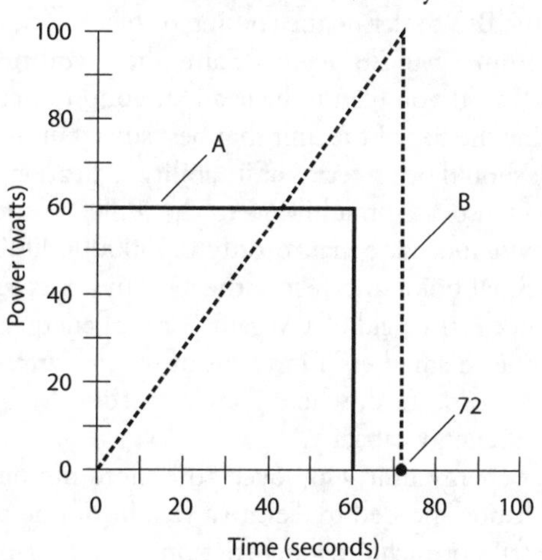

2-6 Two devices that consume 1 Wh of energy. Device A dissipates a constant amount of power as time passes. Device B dissipates an increasing amount of power as time passes.

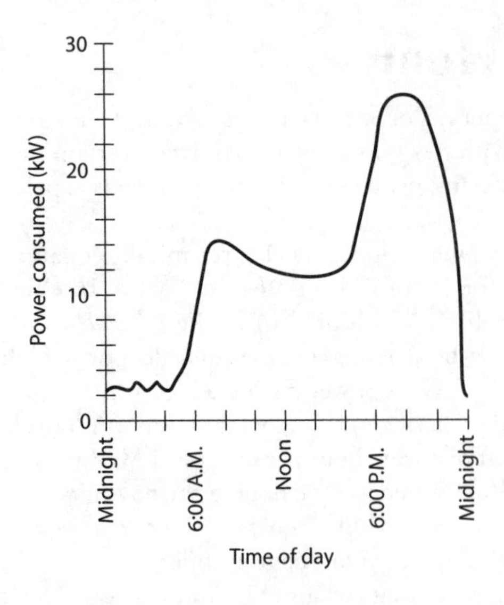

2-7 Graph showing the amount of power consumed by a hypothetical household, as a function of the time of day.

scheme to determine the total energy consumed by your household over a period of time. You can employ a special meter that measures electrical energy in kilowatt-hours (kWh).

Every month, the power company sends a representative, or employs a wireless device, to record the number of kilowatt-hours that your meter displays. The power company then subtracts the reading taken the previous month from the current value. A few days later, you get a bill for the month's energy usage. The "power meter" (a misnomer, because it's really an energy meter) automatically keeps track of total consumed energy, without anybody having to go through high-level mathematical calculations to find the areas under irregular curves, such as the graph of Fig. 2-7.

Other Energy Units

The joule, while standard among scientists, isn't the only energy unit that exists. You'll occasionally encounter the *erg*, a tiny unit equivalent to 0.0000001 of a joule. Some scientists use the erg in laboratory experiments involving small amounts of expended energy.

Most folks have heard or read about the *British thermal unit (Btu)*, equivalent to 1055 joules. People use the Btu to define the cooling or heating capacity of air-conditioning equipment. To cool your room from 85 to 78 degrees Fahrenheit, you need a certain amount of energy, perhaps best specified in Btu. If you plan to have an air conditioner or furnace installed in your home, an expert will determine the size of the unit that best suits your needs. That person might tell you how "powerful" the unit should be, in terms of its ability to heat or cool in British thermal units per hour (Btu/h).

Physicists also use, in addition to the joule, a unit of energy called the *electron-volt* (eV). It's a minuscule unit indeed, equal to only 0.00000000000000000016 joule (you can count 18 zeroes after the decimal point and before the 1). Physicists represent this number as 1.6×10^{-19}. A single electron in an electric field of 1 V gains 1 eV of energy. Atomic physicists rate *particle accelerators* (or, informally, "atom smashers") in terms of *megaelectron-volts* (MeV, where 1 MeV = 1,000,000 eV), or *gigaelectron-volts* (GeV, where 1 GeV = 1000 MeV), or *teraelectron-volts* (TeV, where 1 TeV = 1000 GeV) of energy capacity.

Another energy unit, employed to denote mechanical work, is the *foot-pound* (ft-lb). It's the amount of "labor" needed to elevate a weight of one pound (1 lb) straight upward by a distance of one foot (1 ft), not including any friction. One foot-pound equals 1.356 J.

Table 2-3. Conversion Factors between Joules and Various Other Energy Units

Unit	To convert energy in the unit at left to joules, multiply by	To convert energy in joules to the unit at left, multiply by
British thermal units (Btu)	1055	0.000948
electron-volts (eV)	1.6×10^{-19}	6.2×10^{18}
ergs	0.0000001 (or 10^{-7})	10,000,000 (or 10^{7})
foot-pounds (ft-lb)	1.356	0.738
watt-hours (Wh)	3600	0.000278
kilowatt-hours (kWh)	3,600,000 (or 3.6×10^{6})	0.000000278 (or 2.78×10^{-7})

Table 2-3 summarizes all of the energy units described here, along with conversion factors to help you change from any particular unit to joules or vice-versa. The table includes watt-hours and kilowatt-hours. In electricity and electronics, you'll rarely need to concern yourself with any energy unit other than these two.

Alternating Current and the Hertz

Direct current (DC) always flows in the same direction, but household utility current reverses direction at regular intervals. In the United States (and much of the world), the current direction reverses once every 1/120 second, completing one full cycle every 1/60 second. In some countries, the direction reverses every 1/100 second, taking 1/50 second to go through a complete cycle. When we encounter a periodically reversing current flow of this sort, we call it *alternating current* (AC).

Figure 2-8 shows a common "117-V" utility AC wave as a graph of voltage versus time. If you're astute, you'll see that the maximum positive and negative EMFs don't equal 117 V. Instead, they come close to 165 V. The *effective voltage* for an AC wave usually differs from the *instantaneous maximum*, or *peak*, voltage. For the waveform shown in Fig. 2-8, the effective value is approxi-

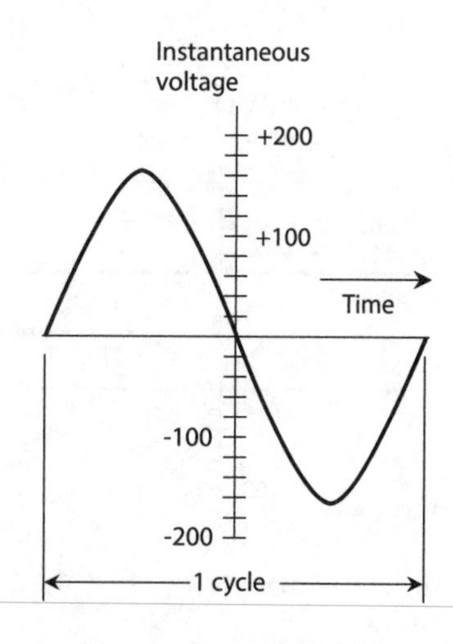

2-8 One cycle of utility alternating current (AC). The instantaneous voltage is the voltage at any particular instant in time. The peak voltages are approximately plus and minus 165 V.

mately 0.707 times the peak value (the theoretically exact multiplication factor equals the reciprocal of the square root of 2). Conversely, the peak value is approximately 1.414 times the effective value (theoretically the factor equals the square root of 2).

The *hertz* (symbolized *Hz*) is the basic unit of AC *frequency*. One hertz represents one complete cycle per second. Because a typical utility AC cycle repeats itself every 1/60 second, we say that the wave has a *frequency* of 60 Hz. In the United States, 60 Hz is the standard frequency for AC. In much of the rest of the world, however, it's 50 Hz.

In wireless communications, you'll hear about *kilohertz* (kHz), *megahertz* (MHz), and *gigahertz* (GHz). These units relate to each other as follows:

- 1 kHz = 1000 Hz = 10^3 Hz
- 1 MHz = 1000 kHz = 1,000,000 Hz = 10^6 Hz
- 1 GHz = 1000 MHz = 1,000,000 kHz = 1,000,000,000 Hz = 10^9 Hz

Usually, but not always, the waves have shapes like the one shown in Fig. 2-8. Engineers and technicians call it a *sine wave* or a *sinusoid*.

Rectification and Pulsating Direct Current

Batteries and other sources of DC produce constant voltage, which we can graphically portray by plotting a straight, horizontal line on a coordinate grid, showing voltage as a function of time. Figure 2-9 shows a representation of DC. For pure DC, the peak voltage equals the effective voltage. In some systems that derive their power from sources other than batteries, the instantaneous DC voltage fluctuates rapidly with time. This situation exists, for example, if we pass the sinusoid of Fig. 2-8 through a *rectifier* circuit, which allows the current to flow in only one direction.

Rectification changes AC into DC. To obtain rectification, we can use a device called a *diode*. When we rectify an AC wave, we can either cut off or invert one half of the AC wave to get *pulsating DC* output. Figure 2-10 illustrates two different waveforms of pulsating DC. In the waveform at A, we simply remove the negative (bottom) half of the cycle. In the situation at B, we invert the negative portion of the wave, making it positive instead—a "mirror image" of its former self. Figure 2-10A shows *half-wave rectification*; it involves only half the waveform. Figure 2-10B illustrates *full-wave rectification*, in which both halves of the waveform contribute to the output. In the output

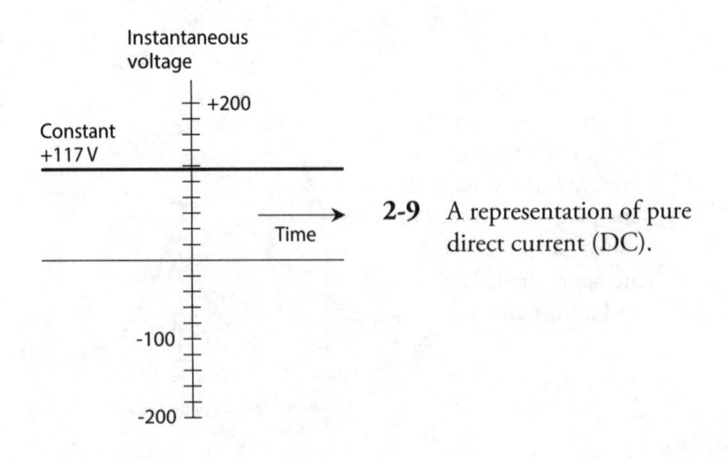

2-9 A representation of pure direct current (DC).

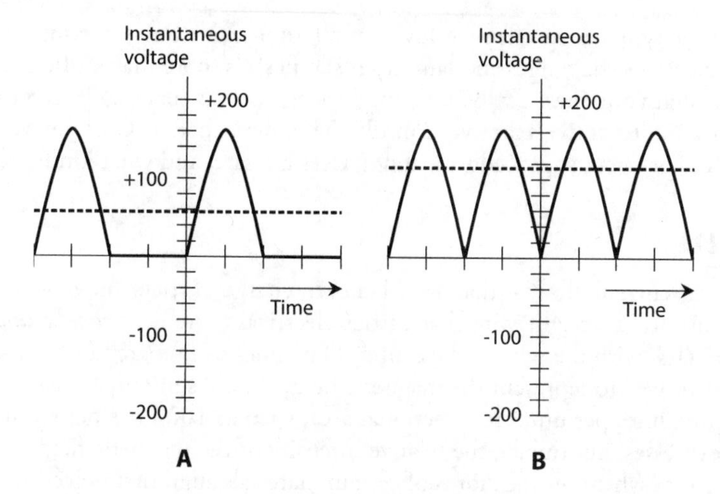

2-10 At A, half-wave rectification of common utility AC. At B, full-wave rectification of common utility AC. Effective voltages are shown by the dashed lines.

of a full-wave rectifier, all of the original current still flows, even though it doesn't alternate as the input does.

The effective value, compared with the peak value, for pulsating DC depends on whether we apply half-wave or full-wave rectification to an AC wave. In Figs. 2-10 A and B, the effective voltages appear as dashed lines, and the instantaneous voltages show up as solid curves. The instantaneous voltage changes from *instant* to *instant* in time (that's where the term comes from).

In Fig. 2-10B, the effective voltage equals $2^{-1/2}$ (roughly 0.707) times the peak voltage, just as with ordinary AC. The direction of current flow, for many kinds of devices, doesn't make any difference. But in Fig. 2-10A, half of the wave has been lost, cutting the effective value in half so that it's only $2^{-1/2}/2$ (approximately 0.354) times the peak voltage.

In household "wall-outlet" AC for powering-up conventional appliances in the United States, we observe a peak EMF of about 165 V, and an effective EMF of about 117 V. If we subject this electricity to full-wave rectification, both the peak and the effective EMFs remain at these values. If we put such a wave through a half-wave rectifier, the peak EMF remains the same, but the effective output EMF drops to about 58.5 V.

Stay Safe!

For all "intents and purposes," one rule applies concerning safety around electrical apparatus. Never forget it, even for a moment. One careless move can kill anyone.

> **Warning!**
>
> If you have any doubts about whether or not you can safely work with a device, assume that you *cannot*. In that case, have a professional electrician work on it.

Household electricity, with an effective EMF of about 117 V (but sometimes twice that for large appliances, such as electric ranges and laundry machines), is more than sufficient to kill you if it drives current through your chest cavity. Certain devices, such as spark coils, can produce lethal currents even from an automotive battery. Consult the American Red Cross or your electrician concerning what types of circuits, procedures, and devices are safe, and what kinds aren't.

Magnetism

Whenever an electric current flows—that is, whenever charge carriers move—a *magnetic field* appears in the vicinity. In a straight wire that carries electrical current, *magnetic lines of flux* surround the wire in circles, with the wire at the center. (The lines of flux aren't physical objects, but they offer a convenient way to represent the magnetic field.) You'll sometimes hear or read about a certain number of flux lines per unit cross-sectional area, such as 100 lines per square centimeter. That terminology expresses, informally, the relative intensity of the magnetic field.

Magnetic fields arise whenever the atoms of certain materials align themselves. Iron is the most familiar element with this property. The atoms of iron in the earth's core have become aligned to some extent, a complex phenomenon caused by the rotation of our planet and its motion with respect to the magnetic field of the sun. The magnetic field surrounding the earth gives rise to fascinating effects, such as the concentration of charged particles that you see as the *aurora borealis* during a "solar storm."

When you wind a piece of wire into a tight coil, the resulting magnetic flux takes a shape similar to the flux field surrounding the earth. Two well-defined *magnetic poles* develop, as shown in Fig. 2-11. You can increase the intensity of such a field by placing a special core inside the coil. Iron, steel, or some other material that can be readily magnetized works very well for this purpose. We call such substances *ferromagnetic materials*.

A ferromagnetic core doesn't increase the total *quantity* of magnetism in and around a coil, but it can produce a more *intense* field. This is the principle by which an electromagnet works. It also

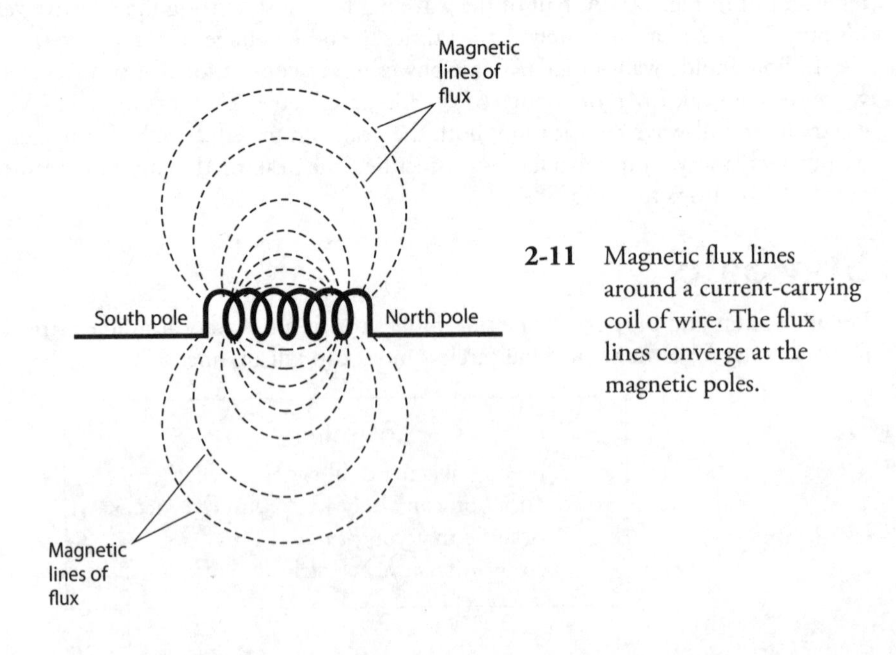

2-11 Magnetic flux lines around a current-carrying coil of wire. The flux lines converge at the magnetic poles.

facilitates the operation of electrical transformers for utility current. Technically, magnetic lines of flux emerge from north poles and converge toward south poles. Therefore, the magnetic field "flows" from the north end of a coil or bar magnet to the south end, following the lines of flux through the surrounding space.

Magnetic Units

We can express the overall quantity of a magnetic field in units called *webers*, abbreviated Wb. One weber is mathematically equivalent to one volt-second (1 V · s). For weaker magnetic fields, we can use a smaller unit called the *maxwell* (symbolized Mx). One maxwell equals 0.00000001 Wb, or 0.01 microvolt-second (0.01 μV · s).

We can express the *flux density* of a magnetic field in terms of webers or maxwells per square meter or per square centimeter. A flux density of one weber per square meter (1 Wb/m^2) represents one *tesla* (1 T). One *gauss* (1 G) equals 0.0001 T, or one maxwell per square centimeter (1 Mx/cm^2).

In general, as the electric current through a wire increases, so does the flux density near the wire. A coiled wire produces a greater flux density for a given electrical current than a single, straight wire. As we increase the number of turns in a coil of a specific diameter that carries a constant current, the flux density in and around the coil increases.

Sometimes, engineers specify magnetic field strength in *ampere-turns* (At). The ampere-turn quantifies a phenomenon called *magnetomotive force*. A one-turn wire loop, carrying 1 A of current, produces a magnetomotive force of 1 At. Doubling the number of turns with constant current doubles the magnetomotive force. Doubling the current for a constant number of turns also doubles the magnetomotive force. If you have 10 A flowing in a 10-turn coil, the magnetomotive force equals 10 × 10, or 100 At. If you have 100 mA flowing in a 200-turn coil, the magnetomotive force equals 0.1 × 200, or 20 At. (Remember that 100 mA = 0.1 A.)

Once in a while, you might hear or read about a unit of magnetomotive force called the *gilbert* (Gb). One gilbert equals approximately 0.796 At. Conversely, 1 At equals approximately 1.26 Gb.

Tip

A DC-carrying coil's magnetomotive force depends only on the current through the coil and the number of turns that it has. That's it! Nothing else makes any difference.

Quiz

Refer to the text in this chapter if necessary. A good score is at least 18 correct answers. The answers are listed in the back of this book.

1. In an electric dipole of constant polarity, the positive charge center
   (a) has more electrons than the negative charge center.
   (b) has the same number of electrons as the negative charge center.
   (c) has fewer electrons than the negative charge center.
   (d) sometimes has more electrons than the negative charge center, sometimes has the same number, and sometimes has fewer.

2. If you touch two points that have DC voltage between them, one point with your left hand and the other point with your right hand, which of the following voltages would present the greatest electrocution hazard?
   (a) 1.5 V
   (b) 15 V
   (c) 150 V
   (d) All three voltages would present equal electrocution hazards because it's the current that kills, not the voltage.

3. If you increase the DC voltage across a resistor by a factor of 100 but you also increase the resistance to keep the current constant, then (assuming the resistor doesn't burn out) the resistor will dissipate
   (a) 100 times as much power as it did before.
   (b) 10 times as much power as it did before.
   (c) the same amount of power as it did before.
   (d) 1/10 as much power as it did before.

4. If a length of wire exhibits 500 mS of conductance, then it has a resistance of
   (a) 0.02 ohm.
   (b) 0.2 ohm.
   (c) 2 ohms.
   (d) an amount that depends on how much current the wire carries.

5. A 330-ohm resistor has a conductance of
   (a) 0.303 mS.
   (b) 3.03 mS.
   (c) 30.3 mS.
   (d) 303 mS.

6. A circuit breaker is rated for 15.0 A in a 13.8-V DC automotive system (with the alternator running). This breaker should cut off the current if you connect a set of devices that demand a total of more than
   (a) 207 W.
   (b) 20.7 W.
   (c) 1.09 W.
   (d) 920 mW.

7. A heater warms a space by 1,000,000 J over a period of time. This amount of energy represents
   (a) 1055 Btu.
   (b) 948 Btu.
   (c) 10.55 Btu.
   (d) None of the above. The British thermal unit quantifies power, not energy!

8. Suppose that a 6.00-V battery delivers 4.00 W of power to a light bulb. How much current flows through the bulb?
   (a) 24.0 A
   (b) 1.50 A
   (c) 667 mA
   (d) We must know the bulb's resistance to calculate the current.

9. Imagine that a span of wire 200 m long has a conductance of 900 mS. A 600-m length of this wire would have a conductance of
   (a) 8.10 S.
   (b) 2.70 S.
   (c) 300 mS.
   (d) 100 mS.

10. Which of the following units quantifies energy?
    (a) The erg
    (b) The kilowatt-hour
    (c) The joule
    (d) All of the above

11. Suppose that an AC cycle repeats at a constant rate of one full cycle every 0.02 second. This wave has a frequency of
    (a) 500 Hz.
    (b) 200 Hz.
    (c) 50 Hz.
    (d) 20 Hz.

12. In many countries outside the United States, utility AC electricity has a frequency of
    (a) 33 Hz.
    (b) 50 Hz.
    (c) 75 Hz.
    (d) 100 Hz.

13. If we could see them, the magnetic flux contours near a straight, current-carrying wire would look like
    (a) concentric circles with the wire at their centers.
    (b) straight lines parallel to the wire.
    (c) straight lines that all pass through the wire at right angles.
    (d) spirals that originate on the wire and all lie in planes perpendicular to the wire.

14. A high DC voltage across a *load* (a component with DC resistance)
    (a) gives rise to poor conductance.
    (b) can exist even if the load has low resistance.
    (c) invariably drives a lot of current through the load.
    (d) All of the above

15. Suppose that DC flows through a wire coil. The magnetomotive force produced by this coil depends on
    (a) the number of turns in the coil.
    (b) the diameter of the coil.
    (c) the resistance of the coil.
    (d) the material around which the coil is wound.

16. Suppose that 3 A of current flows through a 100-turn, circular loop of wire wound around a powdered-iron rod. Then we remove the rod, leaving the coil with an air core. The magnetomotive force
    (a) decreases.
    (b) increases.
    (c) stays the same.
    (d) drops to zero.

17. Which, if any, of the following units can express magnetomotive force?
    (a) The ampere-turn per square meter
    (b) The weber per square meter
    (c) The maxwell per square meter
    (d) None of the above

18. Given a sine-wave AC input, the output of a *full-wave* rectifier
    (a) has an average voltage equal to the peak voltage.
    (b) comprises constant DC just like a battery produces.
    (c) is pulsating DC.
    (d) is also a sine wave.

19. Given a sine-wave AC input, the output of a *half-wave* rectifier
    (a) has an average voltage equal to the peak voltage.
    (b) comprises constant DC just like a battery produces.
    (c) is pulsating DC.
    (d) is also a sine wave.

20. Which of the following units can express overall magnetic-field quantity?
    (a) The weber
    (b) The coulomb
    (c) The volt
    (d) The watt

3

CHAPTER

Measuring Devices

LET'S LOOK AT THE INSTRUMENTS THAT ENGINEERS USE TO MEASURE ELECTRICAL QUANTITIES. Some measuring devices work because electric and magnetic fields produce forces proportional to the field intensity. Some meters determine electric current by measuring the amount of heat produced as charge carriers move through a medium with a known resistance. Some meters have small motors whose speed depends on the measured quantity of charge carriers. Still other meters count electrical pulses or cycles.

Electromagnetic Deflection

Early experimenters noticed that an electric current produces a magnetic field. They could detect this field by placing a magnetic compass near a wire. The compass pointed toward magnetic north when the wire carried no current. But when the experimenters drove DC through the wire by connecting it to the terminals of a battery, the compass needle deflected toward the east or west. The extent of the deflection depended on the distance between the compass and the wire, and also on how much current the wire carried. When scientists first observed this effect, they called it *electromagnetic deflection*, and that's still a good descriptive term today.

Experimenters tried various compass-and-wire arrangements to find out how much they could force the compass needle to rotate. The experimenters also wanted to make the device as sensitive as possible. When scientists wrapped the wire in a coil around the compass, as shown in Fig. 3-1, they got a device that could detect small currents. They called this effect *galvanism*, and they called the coil-around-a-compass device a *galvanometer*. The extent of any given galvanometer's needle displacement increased with increasing current. The experimenters had almost reached their goal of building a meter that could quantitatively measure current, but one final challenge remained: calibrate the galvanometer somehow.

You can make a galvanometer at home. Buy a cheap compass, about two feet of insulated bell wire, and a large 6-V lantern battery. Wind the wire around the compass four or five times, as shown in Fig. 3-1, and align the compass so that the needle points along the wire turns when the wire is disconnected from the battery. Make sure that the compass lies flat on a horizontal surface, such as a table or desk. Then connect one end of the wire to the negative (−) terminal of the battery. Touch

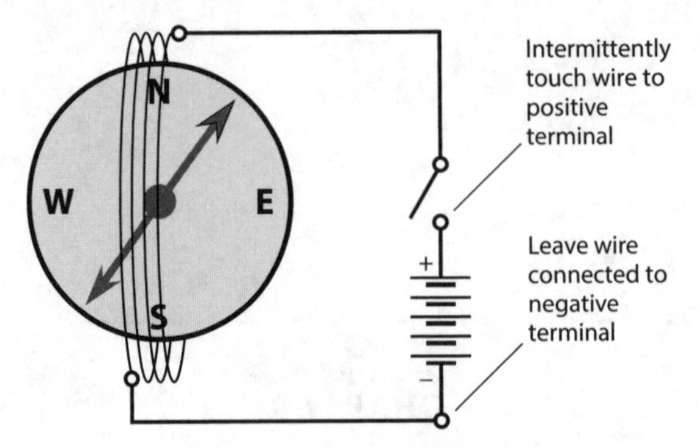

3-1 A simple galvanometer. The magnetic compass must lie flat.

the other end to the positive (+) terminal for a moment, and watch the compass needle. Don't leave the wire connected to the battery for more than a few seconds at a time.

You can buy a *resistor* and a *linear-taper potentiometer* at an electronics retail outlet and set up an experiment that shows how galvanometers measure current. For a 6-V lantern battery, the fixed resistor should have a value of at least 330 ohms and should be rated to dissipate at least ¼ W of power. The potentiometer should have a maximum value of 10 k. Connect the resistor and the potentiometer in series between one end of the bell wire and one terminal of the battery, as shown in Fig. 3-2. Short-circuit the center contact of the potentiometer to one of its end contacts. Use the resulting two terminals in the circuit.

When you adjust the potentiometer, the compass needle should deflect more or less, depending on the current through the wire. As the resistance decreases, the current increases, and so does the number of degrees by which the needle deflects. You can vary the current by changing the potentiometer setting. You can reverse the direction of needle deflection by reversing the battery polarity. Early experimenters calibrated their galvanometers by referring to the "degrees" scale around the perimeter of the compass, and generating graphs of degrees versus amperes. They calculated the theoretical currents in amperes by dividing the known voltage by the known resistance, taking advantage of Ohm's law, about which you'll learn more in the next chapter.

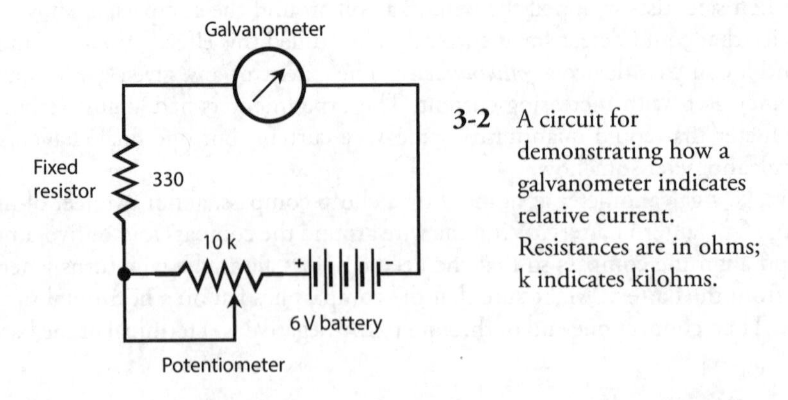

3-2 A circuit for demonstrating how a galvanometer indicates relative current. Resistances are in ohms; k indicates kilohms.

Electrostatic Deflection

Electric fields produce forces, just as magnetic fields do. Have you noticed this effect when your hair stands on end in dry, cold weather? If you live in a place where the winters get severe, shuffle around on a rug while wearing hard-soled shoes next January, and see if you can get your hair to stand up. Have you heard that people's hair stands up just before a lightning bolt hits nearby? Sometimes it does, but not always.

The *electroscope* is a common physics lab device for demonstrating electrostatic force. It consists of two foil leaves, attached to a conducting rod, and placed in a sealed container so that air currents can't disturb the leaves (Fig. 3-3). When a charged object comes near, or touches, the contact at the top of the rod, the leaves stand apart from each other because the leaves become flooded with like electric charges—either an excess or a deficiency of electrons—and "like poles always repel." The extent to which the leaves stand apart depends on the amount of electric charge. It's difficult to measure this deflection and correlate it with charge quantity; electroscopes don't make good meters. But electrostatic forces can operate against tension springs or magnets, allowing engineers to build sensitive, accurate *electrostatic meters*.

An electrostatic meter can quantify alternating (or AC) electric charges as well as direct (or DC) charges. This property gives electrostatic meters an advantage over electromagnetic meters, such as the galvanometer. If you connect a source of AC to the coil of the galvanometer device portrayed in Fig. 3-1, the current in one direction pulls the meter needle one way, the current in the other direction pushes the needle the opposite way, and the opposing forces alternate so fast that the needle doesn't have time to deflect noticeably in either direction! But if you connect a source of AC to an electrostatic meter, the plates repel whether the charge is positive or negative at any given instant in time. The alternations make no difference in the direction of the force.

Most electroscopes aren't sensitive enough to show much deflection with ordinary 117-V utility AC. Don't try connecting 117 V to an electroscope anyway, however. That electricity can present an electrocution hazard if you bring it out to points where you can come into physical contact with it.

An electrostatic meter has another useful property. The device does not draw any current, except a tiny initial current needed to put a charge on the plates. Sometimes, an engineer or experimenter doesn't want a measuring device connected to a significant amount of current because so-called *current drain* affects the behavior of the circuit under test. Galvanometers, by contrast, always need some current to produce an indication.

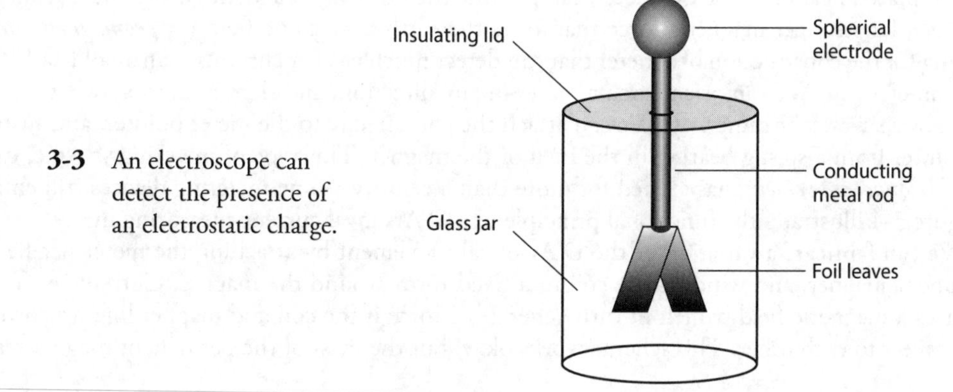

3-3 An electroscope can detect the presence of an electrostatic charge.

If you have access to a laboratory electroscope, try charging it up with a glass rod after rubbing the rod against a dry cloth. When you pull the rod away from the electroscope, the foil leaves will remain standing apart. The charge "sits there" on the foil, trapped! If the electroscope drew any current, the leaves would fall back together again, just as the galvanometer compass needle returns to magnetic north the instant you take the wire away from the battery.

Thermal Heating

Whenever current flows through a substance with a finite, nonzero resistance, the temperature of that substance rises. The extent of the temperature increase depends on the current; for any particular sample, more current generates more heat. If we choose a metal or alloy with known physical properties, tailor a wire from that alloy to a certain length and diameter, use a sensitive, accurate thermometer to measure the wire's temperature, and place the entire assembly inside a thermally insulated package, we end up with a *hot-wire meter*. This device allows us to measure AC as well as DC because the heating doesn't depend on the direction of current flow. Hot-wire meters allow for the measurement of AC at frequencies up to several gigahertz.

We can take advantage of a variation of the hot-wire principle by placing two different metals (called *dissimilar metals*) into direct contact with each other, forming a boundary called a *junction*. The junction heats up when current flows through it. Engineers call this effect the *thermocouple principle*. As with the hot-wire meter, we can use a thermometer to measure the extent of the heating. The thermocouple principle works in reverse as well. When we apply heat to a thermocouple, it generates DC, which we can measure with a galvanometer. This effect allows us to build an *electronic thermometer*.

Ammeters

A magnetic compass, surrounded by a coil of wire, makes an effective but temperamental current-measuring meter. The compass must lie flat on a horizontal surface. We must align the coil with the compass needle under no-current conditions. We must rotate the compass so that the needle points at the "N" on the scale (that is, 0° *magnetic azimuth*) under no-current conditions. All of these restrictions add up to quite an annoyance for experimenters working in labs among complex electronic systems. You'll hardly ever see a compass galvanometer in a professional engineer's or technician's workshop.

The external magnetic field for a galvanometer need not come from the earth. A permanent magnet, placed near or inside the meter, can provide the necessary magnetic field. A nearby magnet supplies a far stronger magnetic force than does the earth's magnetic field (or *geomagnetic field*), allowing for the construction of a meter that can detect much weaker currents than an old-fashioned galvanometer can. We can orient such a meter in any direction, and slant it any way we want, and it will always work the same way. We can attach the coil directly to the meter pointer, and suspend the pointer from a spring bearing in the field of the magnet. This type of metering scheme, called the *D'Arsonval movement,* has existed for more than a century. Some metering devices still employ it. Figure 3-4 illustrates the functional principle of a D'Arsonval current-measuring meter.

We can fabricate a variation of the D'Arsonval movement by attaching the meter needle to a permanent magnet, and winding the coil in a fixed form around the magnet. Current in the coil produces a magnetic field, which in turn generates a force if the coil and magnet line up correctly with respect to each other. This scheme works okay, but the mass of the permanent magnet results

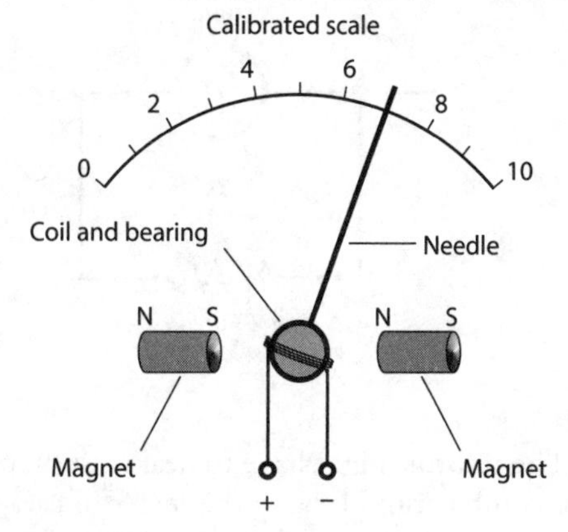

3-4 Functional drawing of a D'Arsonval movement for measuring current.

in a slower needle response, and the meter is more prone to *overshoot* than the true D'Arsonval movement. In overshoot, the inertia of the magnet's mass, once overcome by the magnetic force, causes the needle to fly past the actual point for the current reading, and then to wag back and forth a couple of times before coming to rest in the right place.

Yet another alternative avails itself: We can substitute an *electromagnet* in place of the permanent magnet in a D'Arsonval meter assembly. The electromagnet works with the same current that flows in the coil attached to the meter needle. This arrangement gets rid of the need for a massive, permanent magnet inside the meter. It also eliminates the possibility that the meter sensitivity will change over time if the strength of the permanent magnet deteriorates. Such a demise can result from exposure to heat, severe mechanical vibration, or the mere passage of years.

The sensitivity of any D'Arsonval type meter depends on the amount of current needed to produce a certain force inside the device. That force, in turn, depends on the strength of the permanent magnet (if the meter uses a permanent magnet) and the number of turns in the coil. As the strength of the magnet and/or the number of coil turns increases, the amount of current necessary to produce a given force goes down. In an electromagnet type D'Arsonval meter, the combined number of coil turns affects the sensitivity. If we hold the current constant, the force increases in direct proportion to the number of coil turns. The more magnetomotive force the coils produce, the greater the needle deflection for a given amount of current, and the less current it takes to cause a certain amount of needle movement. The most sensitive D'Arsonval current meters can detect a microampere or two. The amount of current for *full scale deflection* (the needle goes all the way up without banging against the stop pin) can be as little as about 50 μA in commonly available microammeters.

Sometimes we want an ammeter that will allow for a wide range of current measurements. We can't easily change the full-scale deflection of a meter because that task would require altering the number of coil turns and/or the strength of the magnet inside the assembly. However, all ammeters have a certain amount of *internal resistance* (even though in a well-designed ammeter, the internal resistance is extremely low; in an ideal one, it would be zero). If we connect a resistor, having the same internal resistance as the meter, in parallel with the meter, the resistor will draw half the current while the meter draws the other half. Then it will take twice the current through the assembly to deflect the meter to full scale, as compared with the meter alone. By choosing a resistor of a specific value, we can increase the full-scale deflection of any ammeter by a fixed, convenient factor, such as 10, or 100, or 1000. The resistor must be able to carry the necessary current without

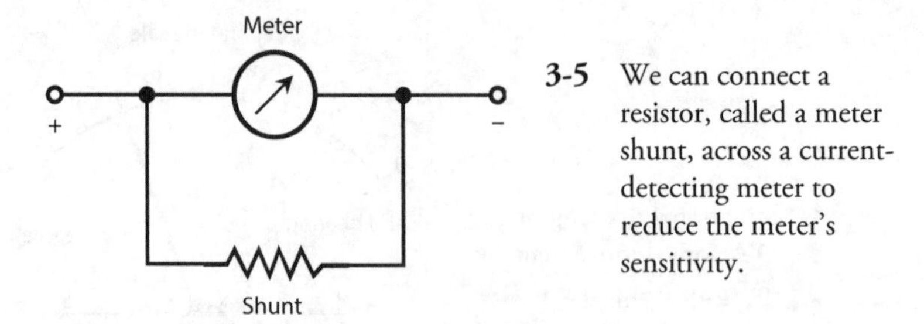

3-5 We can connect a resistor, called a meter shunt, across a current-detecting meter to reduce the meter's sensitivity.

overheating. The resistor might have to deal with practically all of the current flowing through the meter/resistor combination, leaving the meter to carry only 1/10, or 1/100, or 1/1000 of the current. We call a resistor in this application a *shunt* (Fig. 3-5).

Voltmeters

Current, as we have seen, consists of a flow of charge carriers. Voltage, also called electromotive force (EMF) or potential difference, is manifest as "electric pressure" that makes current possible. Given a circuit having a constant resistance, the current through the circuit varies in direct proportion to the voltage across the circuit.

Early experimenters saw that they could use ammeters to measure voltage indirectly. An ammeter acts as a constant-resistance circuit (although the resistance is low). If you connect an ammeter directly across a source of voltage such as a battery, the meter needle deflects. In fact, a milliammeter needle will probably "hit the pin" if you connect it right across a battery; the meter might even suffer permanent damage. (Never connect milliammeters or microammeters directly across voltage sources!) An ammeter, perhaps with a range of 0-10 A, might not deflect to full scale if you place it across a battery, but the meter coil will rapidly drain the battery. Some batteries, such as automotive lead-acid cells, can rupture or explode under these conditions.

Ammeters, as we've learned, have low internal resistance. That's because they're intended for connection in *series* with other parts of a circuit. A circuit under test should "see" an ammeter as a short circuit—ideally like a piece of copper wire—so the meter won't affect the operation of the circuit. Ammeters aren't meant to be connected directly across a source of voltage! However, if you place a large resistor in series with an ammeter, and then connect the combination across a battery or other type of power supply, you no longer have a short circuit. The ammeter will give an indication that varies in direct proportion to the source voltage. The smaller the full-scale reading of the ammeter, the larger the resistance needed to get a meaningful indication on the meter. With a microammeter and a gigantic resistance in series, you can construct a *voltmeter* that will draw only a little current from the source.

You can tailor a voltmeter to have various full-scale ranges by switching different values of resistance in series with the microammeter, as shown in Fig. 3-6. The meter exhibits high internal resistance because the resistors have large ohmic values. As the supply voltage increases, so does the meter's internal resistance because the necessary series resistance increases as the voltage increases.

A voltmeter should have high internal resistance; the higher the better. (An ideal voltmeter would have infinite internal resistance.) You don't want the meter to draw significant current from the power source; ideally it wouldn't draw any current at all. You don't want circuit behavior to

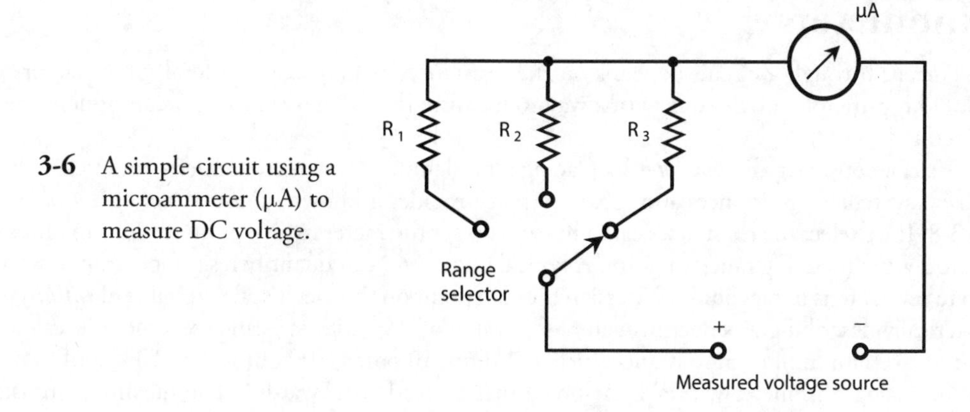

3-6 A simple circuit using a microammeter (μA) to measure DC voltage.

change, even a tiny bit, when you connect or disconnect the meter. The less current a voltmeter draws, the less it affects the behavior of anything that operates from the power supply.

An alternative type of voltmeter uses electrostatic deflection, rather than electromagnetic deflection, to produce its readings. Remember that electric fields produce forces, just as magnetic fields do. Therefore, a pair of electrically charged plates attract or repel each other. An *electrostatic voltmeter* takes advantage of the attractive force between two plates having opposite electric charge, or having a large potential difference. Figure 3-7 portrays the functional mechanics of an electrostatic voltmeter. It constitutes, in effect, a sensitive, calibrated electroscope. The device draws essentially no current from the power supply. Nothing but air exists between the plates, and air constitutes a nearly perfect electrical insulator. A properly designed electrostatic meter can indicate AC voltage as well as DC voltage. However, the construction tends to be fragile, and mechanical vibration can influence the reading.

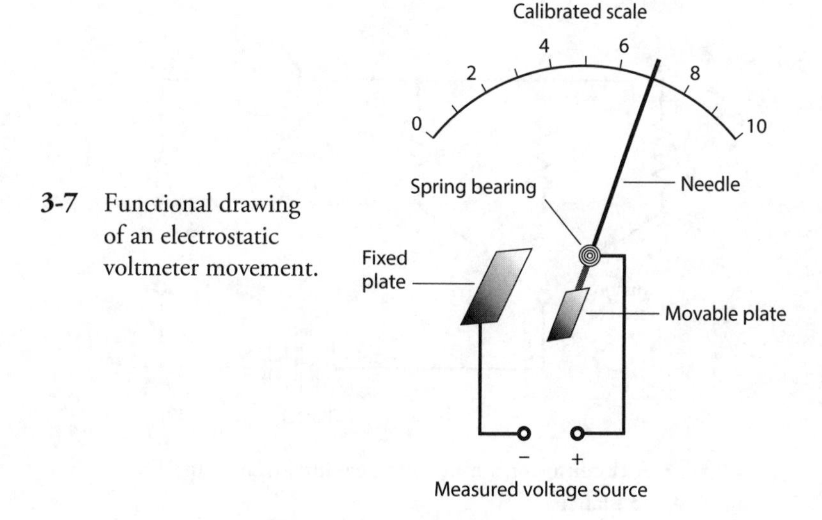

3-7 Functional drawing of an electrostatic voltmeter movement.

Ohmmeters

The current through a circuit depends on the resistance, as long as we hold all other factors constant. This principle provides us with a way to measure the DC resistance of a component, device, or circuit.

We can construct an *ohmmeter* by placing a milliammeter or microammeter in series with a set of fixed, switchable resistances and a battery that provides a known, constant voltage, as shown in Fig. 3-8. If we select the resistances carefully, we can get the meter to give us indications in ohms over practically any metering range we want. A typical ohmmeter can quantify resistances from less than 1 ohm to several tens of megohms. We assign the zero point on the meter scale the value of *infinity ohms*, theoretically describing a perfect insulator. The value of the series resistance sets the full-scale meter point to a certain minimum resistance, such as 1 ohm, 10 ohms, 100 ohms, 1 k, 10 k, and so on.

An analog ohmmeter with a D'Arsonval meter "reads backwards." The maximum resistance corresponds to the left-hand end of the meter scale, and the resistance decreases as we move toward the right on the scale. Engineers must calibrate an ohmmeter at the site of manufacture, or else in a well-equipped electronics lab. A slight error in the values of the series resistors can cause gigantic errors in measured resistance. The resistors must have precise *tolerances*. In other words, they must exhibit values close to what the manufacturer claims, to within a fraction of 1 percent if possible. For an ohmmeter to work right, its internal battery must provide a precise, constant, and predictable voltage.

An ohmmeter built from a milliammeter or microammeter always has a *nonlinear scale*. The increments vary in size, depending on where you look on the scale. In most ohmmeters, the graduations "squash together" towards the "infinity" end of the scale. Because of this nonlinearity, you might find it difficult to read the meter for high values of resistance unless you select the optimum meter range by switching the appropriate resistance in series with the meter device.

A technician will usually connect an ohmmeter in a circuit with the meter set for the highest resistance range first. Then she'll switch the range down until the meter needle comes to rest in a

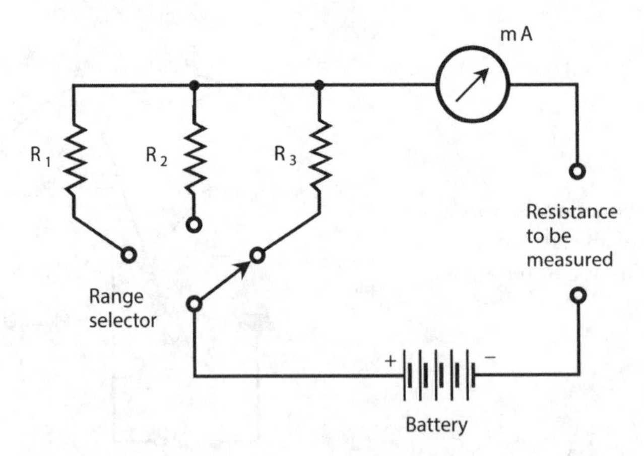

3-8 A circuit using a milliammeter (mA) to measure DC resistance.

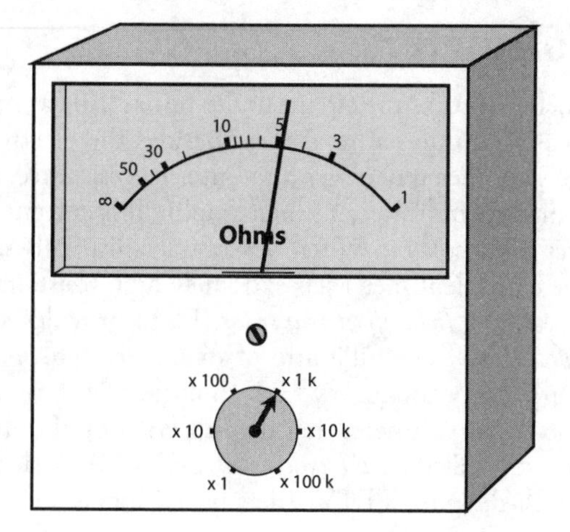

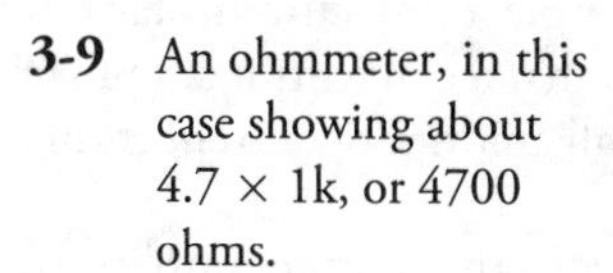
3-9 An ohmmeter, in this case showing about 4.7 × 1k, or 4700 ohms.

readable part of the scale. After taking down the actual meter reading from the scale, the technician multiplies that number by the appropriate amount, as indicated on the range switch. Figure 3-9 shows an ohmmeter reading. The meter itself indicates approximately 4.7, and the range switch says 1 k. This combination indicates a resistance of about 4.7 k, or 4700 ohms.

Ohmmeters give inaccurate readings if a potential difference exists between the points in the circuit to which the meter is connected. The external voltage either adds to, or subtracts from, the ohmmeter's internal battery voltage. Sometimes, in this type of situation, an ohmmeter might tell you that a circuit has "more than infinity" ohms! The needle will hit the pin at the left end of the scale. When you use an ohmmeter to measure DC resistance, you must always make certain that no voltage exists between the points where you intend to connect the meter terminals. You can easily check for such "distracting voltage" using a voltmeter before you use the ohmmeter. If you observe a voltage between those points, you'll probably have to power-down the entire circuit before you try to measure the resistance.

Multimeters

In every good electronics lab, you'll find a test instrument called a *multimeter*, which combines several measuring devices into a single unit. The *volt-ohm-milliammeter* (VOM) is the most often-used type of multimeter. As its name implies, the VOM combines voltage, resistance, and current measuring capabilities. You've learned how a single current meter can determine voltage and resistance. You should have no trouble, therefore, imagining the necessary resistors, a multi-position switch, a battery, and a milliammeter or microammeter in a single box!

Commercially available VOMs can measure current, voltage, and resistance within reasonable limits, but not all the way from "zero to infinity." The maximum limit for voltage typically ranges from 1000 V to 2000 V. The measurement of larger voltages requires special insulated probes and cables, as well as other safety precautions, to prevent the user from receiving a lethal shock. A common VOM can measure currents up to around 10 A. The maximum measurable resistance is several tens of megohms, while the lower limit is a fraction of 1 ohm.

FET Voltmeters

A good voltmeter disturbs the circuit under test as little as possible, and this capability requires that the meter have high internal resistance. Besides the electrostatic scheme described earlier in this chapter, there's another way to get high internal resistance: sample a tiny current, far too small for any meter to directly indicate, and then amplify this current so that a conventional milliammeter or microammeter can display it. When we draw a vanishingly small amount of current from an electrical circuit, the equivalent meter has extremely high resistance.

A device called a *field-effect transistor* (FET) provides an effective way to amplify currents on the order of *picoamperes*, or trillionths of an ampere, where a trillionth represents 0.000000000001 or 10^{-12}. (Don't worry about how FET amplifiers work right now. You'll learn all about amplifiers later in this book.) A voltmeter that uses a FET amplifier to minimize the current drain from the circuit under test is called a *FET voltmeter* (FETVM). Besides exhibiting "practically infinite" input resistance, a well-designed FETVM lets us accurately measure currents smaller than any conventional VOM can even detect.

Wattmeters

We can measure the DC electrical power that a component, circuit, or system dissipates by simultaneously measuring the voltage across it and the current through it. Remember that in a DC circuit, the power (P) in watts equals the product of the voltage (E) in volts and the current (I) in amperes, as follows:

$$P = EI$$

When we calculate DC power as a product of voltage and current, we call the resulting quantity *volt-amperes* (VA) or *volt-ampere power* (VA power). The VA figure provides an excellent indication of the actual power (or *true power*) that a component, circuit, or system consumes, as long as the circuit isn't too complicated, and as long as it works with pure DC.

Figure 3-10 illustrates a method of measuring the DC VA power consumed by a common lantern bulb. We connect a DC voltmeter in parallel with the bulb, thereby getting a reading of the voltage across it. We hook up a DC ammeter in series to get a reading of the current through

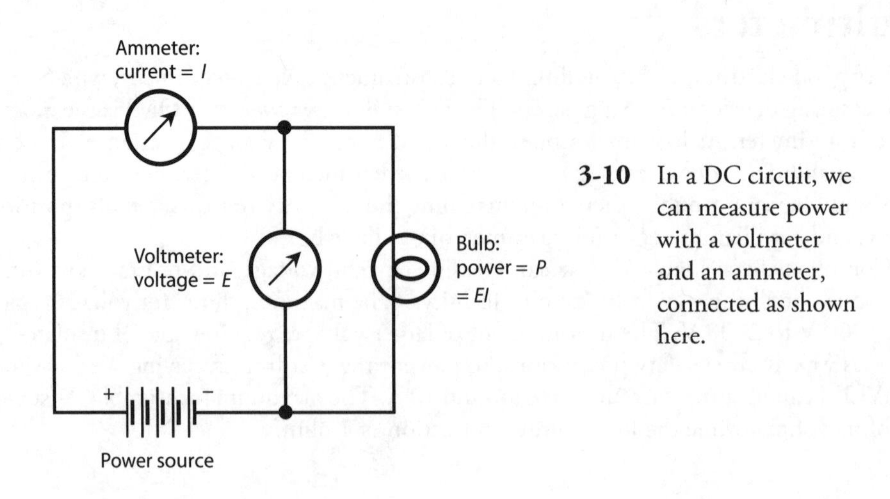

3-10 In a DC circuit, we can measure power with a voltmeter and an ammeter, connected as shown here.

the bulb. We take the voltage and current readings simultaneously, and then multiply volts times amperes to get the number of DC watts consumed by the bulb.

If we want to measure AC power, we must use specialized wattmeters having more sophisticated designs than the simple VA power meter just described. The same requirement obtains if we want to measure the peak audio power in a high-fidelity amplifier, or the average power produced or consumed over a period of time by any system.

Watt-Hour Meters

In small-scale, everyday systems, we can measure electrical energy in watt-hours (Wh) or kilowatt-hours (kWh). Not surprisingly, we call a metering device for these units a *watt-hour meter* or a *kilowatt-hour meter*.

A traditional (older) electrical energy meter employs a small motor, the speed of which depends on the current, and thereby, on the power at a constant voltage. The number of turns of the motor shaft, in a given length of time, varies in direct proportion to the number of watt-hours or kilowatt-hours consumed. We connect the motor at the point where the utility wires enter the building where the voltage is nominally 234 V AC. At this point the utility system splits into one group of circuits providing 234 V AC for heavy-duty appliances, such as the range, oven, washer, and dryer, and a second group of circuits providing 117 V AC for small appliances, such as lamps, clock radios, and television sets.

If you've observed this type of kilowatt-hour meter, you've seen a disk spinning, sometimes fast, and at other times slowly. Its speed depends on the power being used at any given time. The total number of times the disk rotates, hour after hour, day after day, during the course of each month determines the size of the bill that you get from the power company (which is also a function of the cost per kilowatt-hour, of course).

Kilowatt-hour meters count the number of disk turns by means of geared, rotary drums, or pointers. The drum-type meter gives a direct digital readout. The pointer type has several scales calibrated from 0 to 9 in circles, some going clockwise and others going counterclockwise. To read a pointer type device, you must "make your mind's eye go" in whatever direction (clockwise or counterclockwise) the scale goes for each individual meter. Figure 3-11 shows an example. You should read the meters from left to right. For each meter scale, take down the number that the pointer has most recently passed. Write down the rest as you go. This particular meter display shows a little more than 3875 kWh.

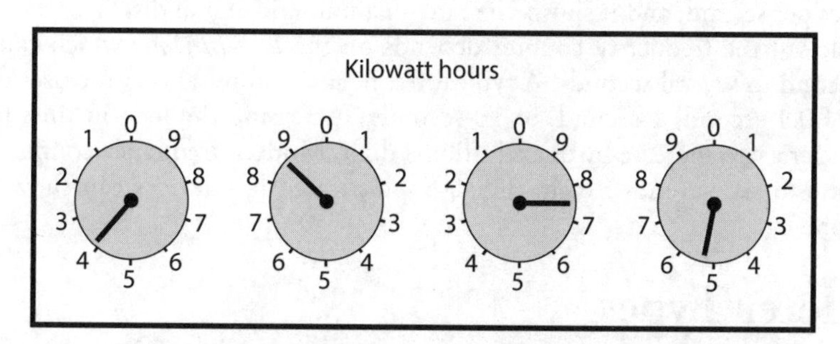

3-11 A utility meter with four rotary analog dials. In this example, the reading is a little more than 3875 kWh.

Digital Readout Meters

The display of a *numeric digital meter* tells you the size of a quantity in plain numerals. Anybody who can read numerals can read this type of meter without errors caused by guesswork, as can occur when people imprecisely read old-fashioned "needle-and-scale" *analog meters*. You'll find numeric digital readouts in utility power meters, clocks, and (in some situations) ammeters, voltmeters, and wattmeters. Numeric digital meters work well when the value of the measured quantity does not change often or fast.

In some applications, numeric digital meters can cause frustration. Consider, for example, the signal-strength indicator in a "shortwave" radio receiver. The intensity of a "shortwave" radio signal typically varies from moment to moment. In a situation like this, a numeric digital meter will display a constantly changing, meaningless jumble of numerals. Numeric digital meters require a certain length of time to "lock in" to the current or voltage under scrutiny. If this quantity never settles at any fixed value for a long enough time, the meter can never "lock in."

Analog meters, in contrast to numeric digital meters, work well in situations where the measured quantity fluctuates all the time. Analog meters give the user an intuitive sense of the "whereabouts" of the measured quantity relative to values slightly larger or smaller. Analog meters can follow along when a quantity changes from moment to moment, although not always with great accuracy. Some engineers and technicians, especially those who got their training in the 1970s or earlier, prefer analog meters even in situations where digital meters work as well.

With any metering device, you'd better know where the decimal point belongs! If you read the numerals in a digital meter correctly but you put the decimal point to the right or left by one digit, you'll get a reading that's off by a factor of 10. You must also be sure that you know the correct units. For example, a frequency indicator might read out in megahertz, but if you think it's telling you kilohertz, you'll be off by a factor of 1000.

Frequency Counters

Digital metering works well to measure the energy that a home or business uses over a period of time. Everyone finds a digital kilowatt-hour meter easier to read than the pointer-type meter. Digital metering also works well for measuring the frequencies of radio signals, but for a different reason.

A *frequency counter* measures the frequency of an AC wave by counting pulses or cycles, in a manner similar to the way a utility meter counts the number of turns of a motor. The frequency counter works electronically, without any moving parts. It can keep track of thousands, millions, or billions of pulses per second, and it shows the rate on a numeric digital display.

The accuracy of the frequency counter depends on the *lock-in time*, which can vary from a fraction of a second to several seconds. A typical frequency counter allows the user to select from lock-in times of 0.1 second, 1 second, or 10 seconds. Increasing the lock-in time by a factor of 10 causes the accuracy to increase by one additional digit. Modern frequency counters can provide readings accurate to six, seven, or eight digits. Sophisticated lab devices can show frequency to 10 digits or more.

Other Meter Types

Following are brief descriptions of some less common types of meters that you'll occasionally encounter in electricity and electronics.

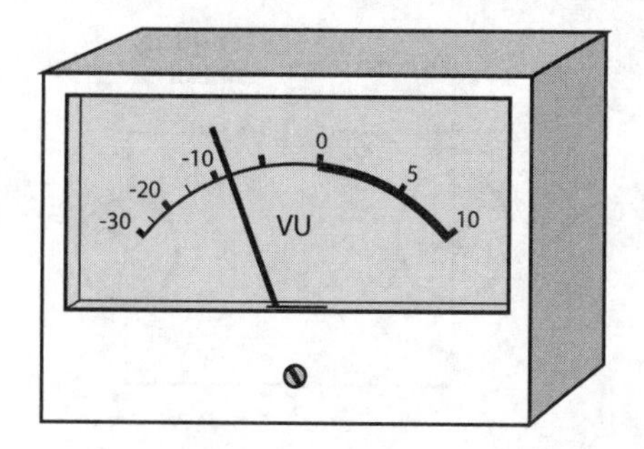

3-12 A VU (volume-unit) meter. The thick-line part of the scale (to the right of 0) indicates a high risk of audio distortion.

Loudness Meters

High-fidelity equipment, especially the more sophisticated amplifiers ("amps"), sometimes have *loudness meters* built-in. The meters express sound intensity in *decibels*, a unit that you will often have to use (and interpret) in reference to electronic signal levels. The decibel actually represents a specific difference in the intensity between two signals: an increase or decrease that you can just barely detect *when you expect it*. You'll learn more about decibels in Chap. 26.

Engineers sometimes measure audio loudness with a *volume-unit (VU) meter*. The typical VU meter scale has a zero marker with a heavy line or curve (red in some cases) to the right and a thinner, black line or curve to the left. The meter scale is calibrated in negative decibels (dB) below the zero marker and positive decibels above the zero marker. The zero marker itself corresponds to an audio power level of 2.51 mW. Figure 3-12 shows an example.

When music or a voice comes through a high-fidelity audio system, the VU meter needle kicks up. A competent audio engineer keeps the *gain* (that is, the volume control) set low enough so that the meter doesn't go past the zero mark and into the red range (shown in Fig. 3-12 as a thick black arc). If the meter does kick up into the red part of the scale, it means that the amplifier will likely produce distortion in the audio output, degrading the quality of the sound.

We can measure sound level in a more general way using a *sound-level meter*, calibrated in decibels and connected through a *semiconductor diode* to the output of a precision amplifier with a microphone of known sensitivity, as shown in Fig. 3-13. The diode "chops off" the negative-polarity part of every AC audio wave cycle, leaving only the positive part so that a DC meter can quantify it. Has anyone ever told you that a vacuum cleaner will produce "80 dB" of sound, and a large truck going by will subject your ears to "90 dB"? Acoustic engineers determine these figures using sound-level meters, and define the intensity levels relative to a faint sound at the *threshold of*

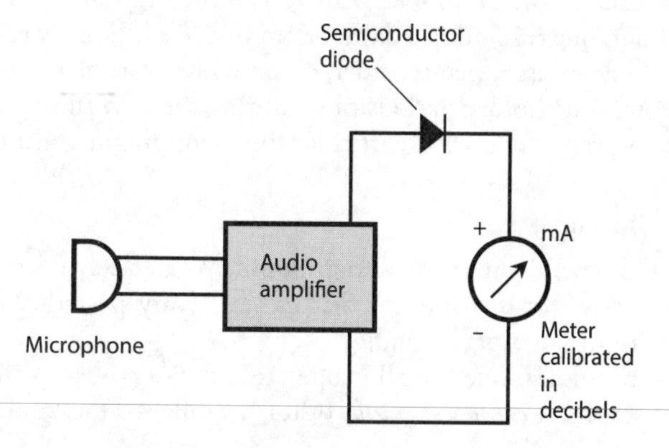

3-13 A simple device for measuring sound levels. The diode converts the AC audio signal to DC that the milliammeter can detect. We calibrate the meter scale in decibels relative to a predetermined reference level.

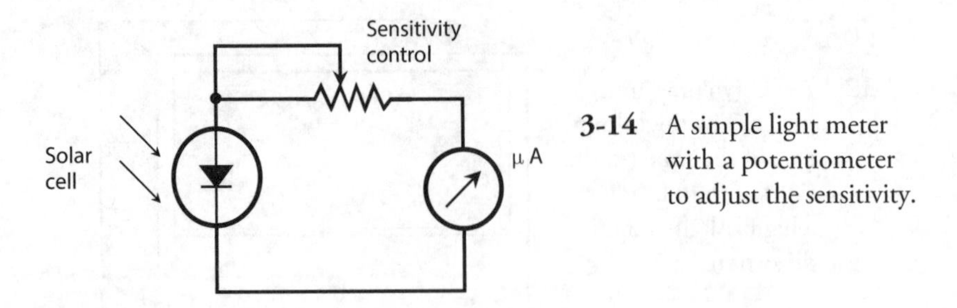

3-14 A simple light meter with a potentiometer to adjust the sensitivity.

hearing. That's the weakest sound that a person with good ears can hear in a room with excellent *acoustic insulation* to keep out stray noise.

Light Meters

Photographers commonly measure the intensity of visible light rays using a *light meter*, technically called an *illuminometer*. You can make an illuminometer by connecting a microammeter to a solar cell (technically called a *photovoltaic cell*), using a *potentiometer* (variable resistor) to control the meter sensitivity, as shown in Fig. 3-14. More sophisticated devices use DC amplifiers, similar to the type found in a FETVM, to enhance the sensitivity and to allow for several different ranges of readings.

Solar cells don't respond to light at exactly the same wavelengths as human eyes. An engineer would say that the *sensitivity-versus-wavelength function* of a typical solar cell differs from the sensitivity-versus-wavelength function of the human eye. We can overcome this problem by placing a specially designed color filter in front of the solar cell, so that the solar cell becomes sensitive to the same wavelengths, in the same proportions, as our eyes. Illuminometer manufacturers calibrate their products at the factory so that the meter displays visible-light intensity in standard illumination units, such as *lumens* or *candela*.

With appropriate modification, we can use a meter such as the one diagrammed in Fig. 3-14 to roughly determine the intensity of *infrared* (IR) or *ultraviolet* (UV) rays. Various specialized photovoltaic cells exhibit their greatest sensitivity at non-visible wavelengths, including IR and UV.

Pen Recorders

A D'Arsonval meter, equipped with a marking device at the needle's tip to keep a graphic record of the level of some quantity with respect to time, gives us a *pen recorder*. A flat sheet of graph paper, with a calibrated scale, is taped around a cylindrical drum. The drum is connected to a slowly rotating motor-driven shaft. The drum speed can vary; it might rotate at one revolution per minute, one revolution per hour, one revolution a day, or even one revolution a week! Figure 3-15 illustrates the principle.

You can use a pen recorder, along with a wattmeter, to get a reading of the power consumed by your household at various times during the day. In this way, you can find out when you use the most power, and at what particular times you might consume too much.

Oscilloscopes

Another graphic metering device, popular with electronics engineers, is the *oscilloscope*, which measures and records quantities that *oscillate* (vary periodically) at rates of hundreds, thousands, or millions of times per second.

An old-fashioned oscilloscope created a "graph" by throwing a beam of electrons at a phosphor screen. A *cathode-ray tube* (CRT), similar to the kind in a television set, was employed. More

Rotating drum
Armature and pen
Meter mechanism
Input

3-15 Functional drawing of a pen recorder.

modern oscilloscopes have electronic conversion circuits that allow for the use of a solid-state *liquid-crystal display* (LCD).

Oscilloscopes are useful for observing and analyzing the shapes of signal waveforms, and also for measuring peak signal levels (rather than only the effective levels). We can use an oscilloscope to indirectly measure the frequency of a waveform. The horizontal scale of an oscilloscope shows time, and the vertical scale shows the instantaneous signal voltage. We can also use an oscilloscope to indirectly measure power or current, by placing a known value of resistance across the input terminals.

Technicians and engineers develop a sense of what a signal waveform "ought to look like." Then they can often ascertain, by observing the oscilloscope display, whether or not the circuit under test is behaving the way it should.

Bar-Graph Meters

A cheap, simple kind of meter comprises a string of *light-emitting diodes* (LEDs) or an LCD along with a digital scale to indicate approximate levels of current, voltage, or power. This type of meter, like a digital meter, has no moving parts. To some extent, it offers the relative-reading feeling you get with an analog meter. Figure 3-16 shows a bar-graph meter designed to indicate the power output, in kilowatts, for a radio transmitter. This meter can follow along fairly well with fluctuations in the reading. In this example, the meter indicates about 0.8 kW, or 800 W.

The chief drawback of bar-graph meters is the fact that most of them don't give precise readings; they can only approximate. For this reason, engineers and technicians rarely use bar-graph meters in a laboratory environment. In addition, the individual LEDs or LCDs in some bar-graph meters flicker intermittently on and off when the signal level "falls between" two values given by the bars. Viewers find this phenomenon distracting or irritating.

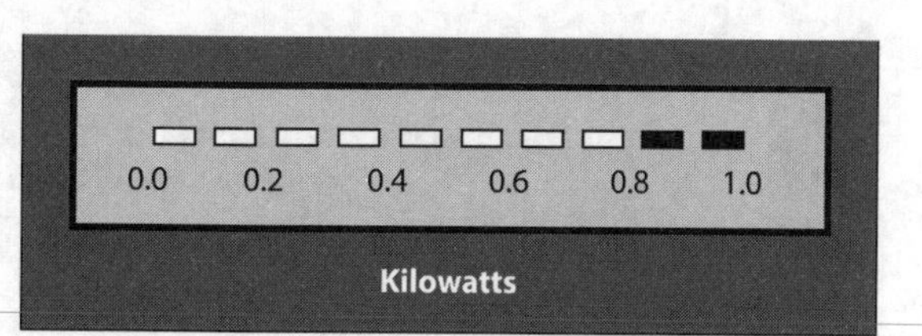

3-16 A bar-graph meter. In this case, the indication is about 80 percent of full-scale, representing 0.8 kW or 800 W.

Quiz

Refer to the text in this chapter if necessary. A good score is 18 out of 20 correct. Answers are in the back of the book.

1. You can use an oscilloscope to
   (a) see the shape of an AC wave.
   (b) detect an electrostatic charge.
   (c) measure an extremely high resistance.
   (d) measure electrical power.

2. An advantage of a meter that relies on electrostatic deflection rather than electromagnetic deflection is the fact that the electrostatic meter can measure
   (a) the frequency of an AC wave.
   (b) AC voltage as well as DC voltage.
   (c) magnetic field strength as well as electric field strength.
   (d) All of the above

3. An electronic thermometer works by measuring the DC output of
   (a) a solar cell.
   (b) an electroscope.
   (c) an illuminometer.
   (d) a thermocouple.

4. Which of the following voltages would produce the bar-graph-meter indication shown in Fig. 3-17?
   (a) 0.040 mV
   (b) 0.40 mV
   (c) 4.0 mV
   (d) 40 mV

5. Numeric digital meters work best when the measured quantity
   (a) is extremely large.
   (b) is current or voltage, but not power or resistance.
   (c) does not fluctuate rapidly.
   (d) constantly changes.

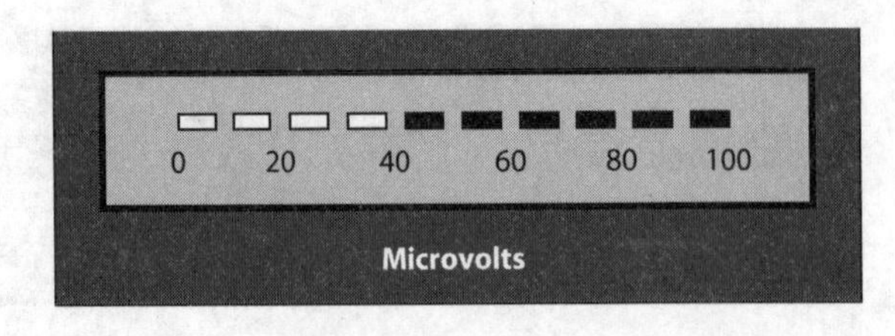

3-17 Illustration for Quiz Question 4.

6. Which of the following phenomena can you use an oscilloscope to measure or observe?
   (a) The waveform of an AC signal
   (b) The frequency of an AC signal
   (c) The peak-to-peak voltage of an AC signal
   (d) Any of the above

7. An electric utility meter measures, on a monthly-use basis,
   (a) energy.
   (b) voltage.
   (c) current.
   (d) power.

8. You place a 12-V battery in series with a resistor and a galvanometer. The resulting current causes the compass needle to deflect 20 degrees toward the west. How can you get the needle to deflect 30 degrees toward the west?
   (a) Maintain the battery polarity and decrease the resistance.
   (b) Maintain the battery polarity and increase the resistance.
   (c) Reverse the battery polarity and decrease the resistance.
   (d) Reverse the battery polarity and increase the resistance.

9. Electrostatic force can directly cause
   (a) two objects having *opposite* electric charges to *repel.*
   (b) two objects having *like* electric charges to *repel.*
   (c) electric current to stop flowing in a conductor if the voltage is too high.
   (d) a compass needle to veer to the right or left, depending on the polarity.

10. The VU meter shown in Fig. 3-18, assuming that we see the highest level to which the audio peaks ever get, tells us that
   (a) those peaks are about 6 dB too high to avoid distortion.
   (b) those peaks are about 6 dB too low to avoid distortion.
   (c) those peaks are about 6 dB below the point where distortion is likely.
   (d) nothing useful at all.

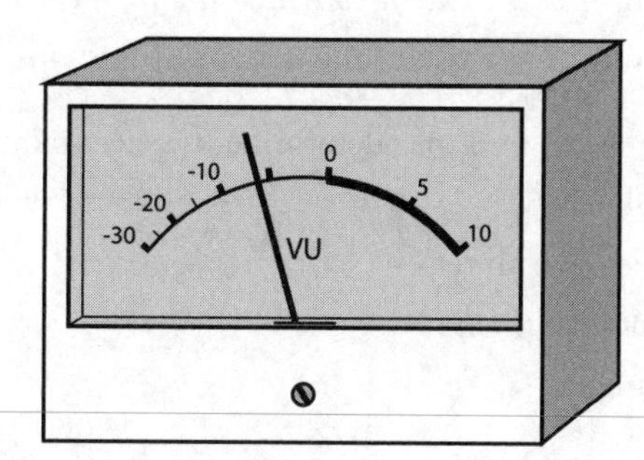

3-18 Illustration for Quiz Question 10.

11. You want to test a 330-ohm resistor to ensure that its actual resistance comes close to the specified value. You have an analog ohmmeter with a nonlinear scale that runs from "infinity" (at the far left) to 1 (at the far right) with 6 roughly in the middle (such as in Fig. 3-9), and that has range switches with six settings marked "x 1" to "x 100 k" in powers of 10. Which range switch will provide the most accurate reading? You can use Fig. 3-9 as a visual aid.
    (a) x 1
    (b) x 100
    (c) x 10 k
    (d) x 100 k

12. An *ideal ammeter* would have
    (a) infinite internal resistance.
    (b) moderate internal resistance.
    (c) low internal resistance.
    (d) zero internal resistance.

13. Where would you place a DC voltmeter if you wanted to directly measure the voltage of a battery connected to an electrical circuit?
    (a) Between either battery pole and electrical ground
    (b) Between the negative battery pole and the circuit input
    (c) Between the positive battery pole and the circuit input
    (d) Between the negative battery pole and the positive battery pole

14. An *ideal voltmeter* would have
    (a) infinite internal resistance.
    (b) moderate internal resistance.
    (c) low internal resistance.
    (d) zero internal resistance.

15. Why should a voltmeter have a high internal resistance?
    (a) To maximize the current that the meter draws from the circuit under test
    (b) To minimize the risk of electric shock to technicians who use the meter
    (c) To minimize the extent to which the meter disturbs the circuit under test
    (d) To minimize the risk of the meter burning out

16. A factor called *lock-in time* determines the accuracy of
    (a) an oscilloscope.
    (b) a frequency counter.
    (c) an analog voltmeter.
    (d) a VU meter.

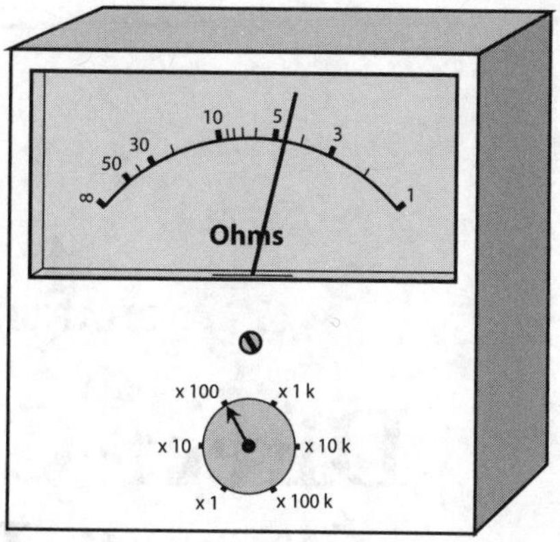

3-19 Illustration for Quiz Question 20.

17. In a general sense, bar-graph meters lack
    (a) useful range.
    (b) sensitivity.
    (c) precision.
    (d) physical ruggedness.
18. An analog ohmmeter has
    (a) a nonlinear scale.
    (b) a high current requirement.
    (c) a bar-graph display.
    (d) an AC power source.
19. You might find a D'Arsonval movement in an analog
    (a) voltmeter.
    (b) ammeter.
    (c) ohmmeter.
    (d) Any of the above
20. The ohmmeter shown in Fig. 3-19 displays
    (a) 4.7 ohms.
    (b) 47 ohms.
    (c) 470 ohms.
    (d) 4700 ohms.

4
CHAPTER

Direct-Current Circuit Basics

IN THIS CHAPTER, YOU'LL LEARN MORE ABOUT CIRCUIT DIAGRAMS, WHICH ENGINEERS AND TECHnicians call *schematic diagrams,* or simply *schematics,* because they detail the *schemes* for designing and assembling circuits. You'll also learn more about how current, voltage, resistance, and power interact in simple DC circuits.

Schematic Symbols

When we want to denote an *electrical conductor* such as a wire, we draw a straight, solid line either horizontally across, or vertically up and down, the page. We can "turn a corner" when we draw a conductor line, but we should strive to minimize the total number of "corners" in a diagram. By following this convention, we can keep schematics neat and ensure that they're easy for others to read.

When two conductor lines cross, assume that they *do not* connect at the crossing point unless you see a heavy, solid dot where they meet. Whenever you draw a "connecting dot," make it clearly visible, no matter how many conductors meet at the junction.

We portray a *resistor* by drawing a "zig-zag," as shown in Fig. 4-1A. We portray a two-terminal variable resistor or *potentiometer* by drawing a "zig-zag" with an arrow through it (Fig. 4-1B). We portray a three-terminal potentiometer by drawing a "zig-zag" with an arrow pointing sideways at it (Fig. 4-1C).

We symbolize an *electrochemical cell* by drawing two parallel lines, one longer than the other (Fig. 4-2A). The longer line represents the positive (+) terminal, while the shorter line represents the negative (−) terminal. We symbolize a *battery,* which is a combination of two or more cells in

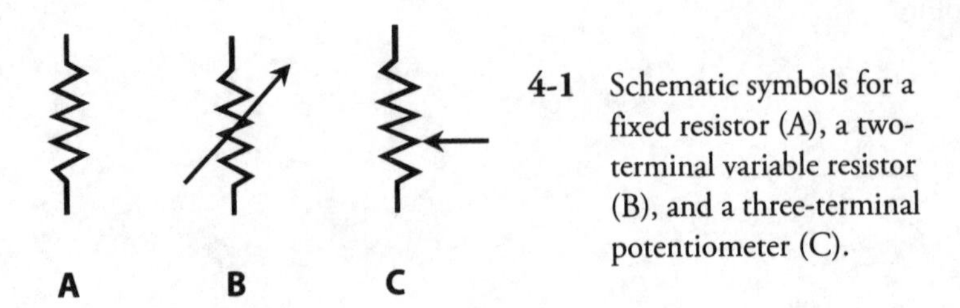

4-1 Schematic symbols for a fixed resistor (A), a two-terminal variable resistor (B), and a three-terminal potentiometer (C).

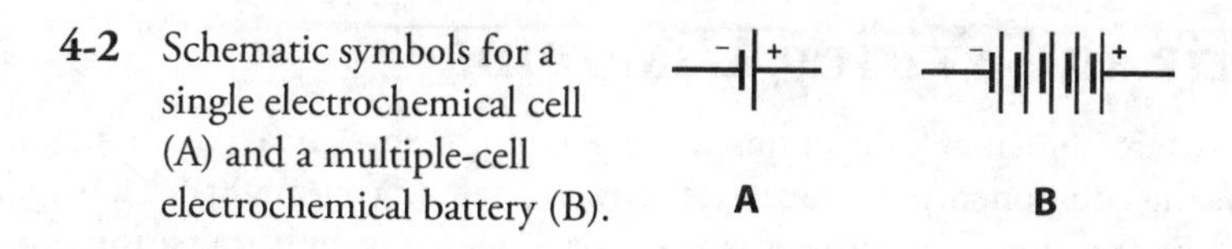

4-2 Schematic symbols for a single electrochemical cell (A) and a multiple-cell electrochemical battery (B).

series, by drawing several parallel lines, alternately long and short (Fig. 4-2B). As with the cell, the longer end line represents the positive terminal and the shorter end line represents the negative terminal.

We can portray meters as circles. Sometimes the circle has an arrow inside it, and the meter type, such as mA (milliammeter) or V (voltmeter) is written alongside the circle, as shown in Fig. 4-3A. Sometimes the meter type is denoted inside the circle, and no arrow appears (Fig. 4-3B). It doesn't matter which way you draw meters in your schematics, as long as you keep the style consistent throughout your work.

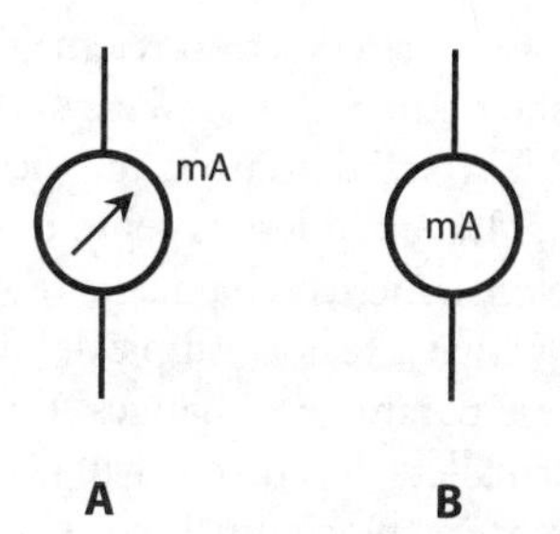

4-3 Meter symbols can have the designator either outside the circle (A) or inside (B). In this case, both symbols represent a milliammeter (mA).

Some other common symbols include the *incandescent lamp*, the *capacitor*, the *air-core coil*, the *iron-core coil*, the *chassis ground*, the *earth ground*, the *AC source*, the set of *terminals*, and the *black box* (specialized component or device), a rectangle with the designator written inside. These symbols appear in Fig. 4-4.

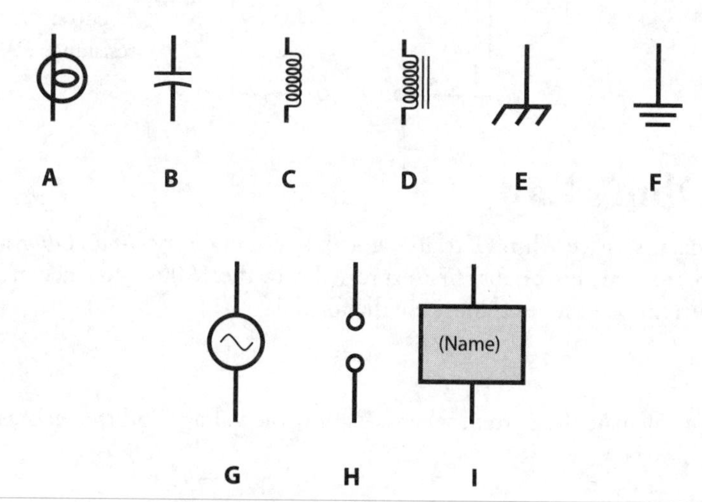

4-4 Schematic symbols for an incandescent lamp (A), a fixed capacitor (B), a fixed inductor with air core (C), fixed inductor with laminated-iron core (D), chassis ground (E), earth ground (F), a signal generator or source of alternating current (G), a pair of terminals (H), and a specialized component or device (I).

Schematic and Wiring Diagrams

Schematics illustrate the interconnections among components in a circuit or system, but the actual values of the components are not necessarily indicated. You might see a diagram of a radio-frequency (RF) power amplifier with resistors, capacitors, coils, and transistors, but without any data concerning the values or ratings of the components. That's a schematic diagram, but it's not a *wiring diagram*. The schematic tells you the *scheme* for the circuit, but you can't *wire up* the circuit and make it work because you don't have enough information.

Suppose you want to build a certain amplifier circuit. You go to an electronics store to get the parts. What values of resistors should you buy? How about capacitors? What type of transistor will work best? Must you wind the coils, or can you get them ready-made? Should you install *test points* for the benefit of technicians who might have to repair the amplifier someday? How many watts should the potentiometers be able to handle? A wiring diagram tells you all of these things.

Circuit Simplification

We can simplify most DC circuits to three major components: a voltage source, a set of conductors, and a resistance, as shown in Fig. 4-5. We call the source voltage E (or sometimes V), the current I, and the resistance R. The standard units for these components are the volt (V), the ampere (A), and the ohm, respectively. Italicized letters represent mathematical variables (voltage, current, and resistance in this case). Non-italicized characters represent abbreviations for physical units.

We already know that a relationship exists between the voltage, current, and resistance in a DC circuit. If one of these parameters changes, then one or both of the others will also change. If we make the resistance smaller, the current will get larger. If we increase the EMF, the current will also increase. If the current in the circuit increases, the voltage across the resistor will increase. Ohm's law comprises a simple set of formulas defining the relationship between these three quantities.

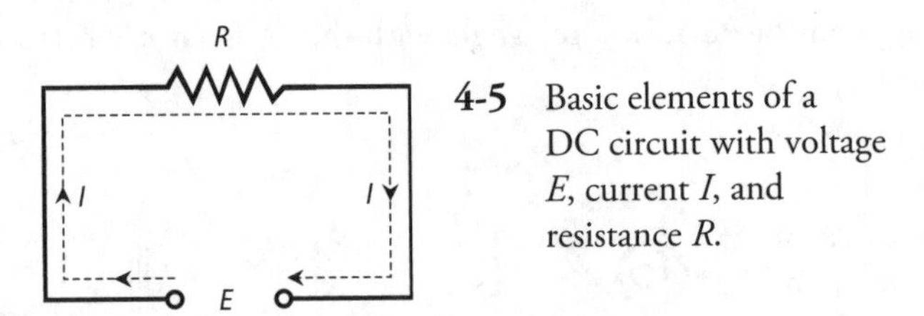

4-5 Basic elements of a DC circuit with voltage E, current I, and resistance R.

Ohm's Law

Scientists gave Ohm's Law its name in honor of *Georg Simon Ohm*, a German physicist who (according to some historians) first expressed it in the 1800s. To calculate the voltage when we know the current and the resistance, use the formula

$$E = IR$$

To calculate the current when we know the voltage and the resistance, use

$$I = E/R$$

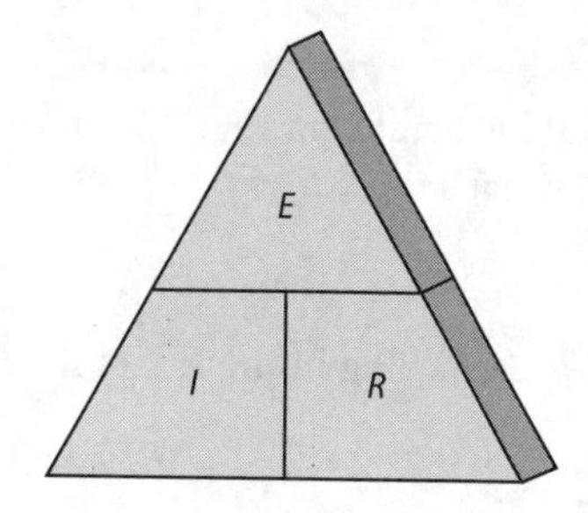

4-6 The Ohm's Law triangle showing voltage E, current I, and resistance R, expressed in volts, amperes, and ohms respectively.

To calculate the resistance when we know the voltage and the current, use

$$R = E/I$$

You need only remember one of these formulas to derive the other two. You can arrange the three variables geometrically into an *Ohm's Law triangle,* as shown in Fig. 4-6. When you want to find the formula for a particular parameter, cover up its symbol and read the positions of the others.

If you want Ohm's Law to produce the correct results, you must use the proper units. Under most circumstances, you'll want to use the *standard units* of volts, amperes, and ohms. If you use volts, milliamperes (mA), and ohms, or if you use kilovolts (kV), microamperes (μA), and megohms (M), you can't expect to get the right answers. If you see initial quantities in units other than volts, amperes, and ohms, you should convert to these standard units before you begin your calculations. After you've done all the arithmetic, you can convert the individual units to whatever you like. For example, if you get 13,500,000 ohms as a calculated resistance, you might prefer to call it 13.5 megohms. But in the calculation, you should use the number 13,500,000 (or 1.35×10^7) and stay with units of ohms.

Current Calculations

In order to determine the current in a circuit, we must know the voltage and the resistance, or be able to deduce them. Figure 4-7 illustrates a generic circuit with a variable DC generator, a voltmeter, some wire, an ammeter, and a potentiometer.

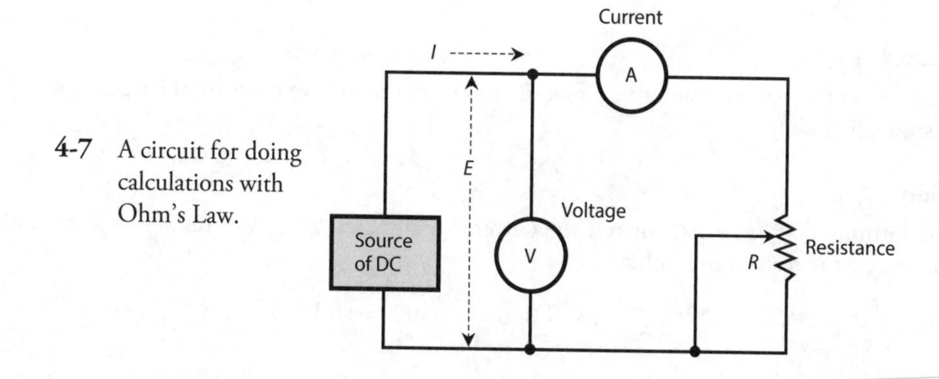

4-7 A circuit for doing calculations with Ohm's Law.

Problem 4-1

Suppose that the DC generator in Fig. 4-7 produces 36 V and we set the potentiometer to a resistance of 18 ohms. What's the current?

Solution

Use the formula $I = E/R$. Plug in the values for E and R in volts and ohms, getting

$$I = E/R = 36/18 = 2.0 \text{ A}$$

Problem 4-2

Imagine that the DC generator in Fig. 4-7 produces 72 V and the potentiometer is set to 12 k. What's the current?

Solution

First, convert the resistance to ohms, getting 12 k = 12,000 ohms. Then input the values in volts and ohms to get

$$I = E/R = 72/12{,}000 = 0.0060 \text{ A} = 6.0 \text{ mA}$$

Problem 4-3

Suppose that we adjust the DC generator in Fig. 4-7 so that it produces 26 kV, and we adjust the potentiometer so that it has a resistance of 13 M. What's the current?

Solution

First, change the resistance value from 13 M to 13,000,000 ohms. Then change the voltage value from 26 kV to 26,000 V. Finally, plug the voltage and resistance into the Ohm's Law formula, getting

$$I = E/R = 26{,}000/13{,}000{,}000 = 0.0020 \text{ A} = 2.0 \text{ mA}$$

Voltage Calculations

We can use Ohm's Law to calculate the DC voltage between two points when we know the current and the resistance.

Problem 4-4

Suppose we set the potentiometer in Fig. 4-7 to 500 ohms, and we measure the current as 20 mA. What's the DC voltage?

Solution

Use the formula $E = IR$. First, convert the current to amperes: 20 mA = 0.020 A. Then multiply the current by the resistance to obtain

$$E = IR = 0.\ 020 \times 500 = 10 \text{ V}$$

Problem 4-5

We set the potentiometer in Fig. 4-7 to 2.33 k, and we get 250 mA of current. What's the voltage?

Solution

Before doing any arithmetic, we convert the resistance and current to ohms and amperes. A resistance of 2.33 k equals 2,330 ohms, and a current of 250 mA equals 0.250 A. Now we can calculate the voltage as

$$E = IR = 0.250 \times 2{,}330 = 582.5 \text{ V}$$

We can round this result up to 583 V.

Problem 4-6

We set the potentiometer in Fig. 4-7 to get 1.25 A of current, and we measure the resistance as 203 ohms. What's the voltage?

Solution

These values are both in standard units. Input them directly to get

$$E = IR = 1.25 \times 203 = 253.75 \text{ V} = 254 \text{ V}$$

We can (and should) round off because we can't justify a result that claims more precision than the data that we start with.

The Rule of Significant Figures

Competent engineers and scientists go by the *rule of significant figures*, also called the *rule of significant digits*. After completing a calculation, we always round the answer off to the *least* number of digits given in the input data numbers.

If we follow this rule in Problem 4-6, we must round off the answer to three significant digits, getting 254 V. That's because the resistance (203 ohms) is specified only to that level of accuracy. If the resistance were given as 203.0 ohms, then we would again round off the answer to 254 V. If the resistance were given as 203.00 ohms, then we could still state the answer to only three significant digits, getting 254 V because we know the current only to three significant digits.

This rule takes some "getting-used-to" if you haven't known about it or practiced it before. But after a while, you'll use it "automatically" without giving it much thought.

Resistance Calculations

We can use Ohm's Law to calculate the resistance between two points when we know the voltage and the current.

Problem 4-7

If the voltmeter in Fig. 4-7 reads 12 V and the ammeter shows 2.0 A, what's the resistance of the potentiometer?

Solution

Use the formula $R = E/I$. We can plug in the values directly because they're expressed in volts and amperes. It works out as

$$R = E/I = 12/2.0 = 6.0 \text{ ohms}$$

Problem 4-8

What's the value of the resistance in Fig. 4-7 if the current equals 24 mA and the voltage equals 360 mV?

Solution

First, convert to amperes and volts, obtaining $I = 0.024$ A and $E = 0.360$ V. Then plug the numbers into the Ohm's Law equation to get

$$R = E/I = 0.360/0.024 = 15 \text{ ohms}$$

Problem 4-9

Suppose that the ammeter in Fig. 4-7 reads 175 μA and the voltmeter indicates 1.11 kV. What's the resistance?

Solution

Convert to amperes and volts, getting $I = 0.000175$ A and $E = 1110$ V. Then input these numbers, rounding off to get

$$R = E/I = 1110/0.000175 = 6{,}342{,}857 \text{ ohms} = 6.34 \text{ M}$$

Power Calculations

We can calculate the power P in a DC circuit, such as the one in Fig. 4-7, using the formula

$$P = EI$$

If we aren't given the voltage directly, we can calculate it if we know the current and the resistance. Recall the Ohm's Law formula for obtaining voltage:

$$E = IR$$

If we know I and R but we don't know E, we can get the power P as

$$P = EI = (IR)I = I^2R$$

If we know only E and R but don't know I, we can restate I as

$$I = E/R$$

Then we can substitute into the voltage-current power formula to obtain

$$P = EI = E(E/R) = E^2/R$$

Problem 4-10

Suppose that the voltmeter in Fig. 4-7 reads 15 V and the ammeter shows 70 mA. How much power does the potentiometer dissipate?

Solution

Use the formula $P = EI$. First, convert the current to amperes, getting $I = 0.070$ A. (The last 0 counts as a significant digit.) Then multiply by 15 V, getting

$$P = EI = 15 \times 0.070 = 1.05 \text{ W}$$

The input data only has two significant digits, while this answer, as it stands, has three. Rounding up gives 1.1 A. That's the number we should use.

Problem 4-11

If the resistance in the circuit of Fig. 4-7 equals 470 ohms and the voltage source delivers 6.30 V, what's the power dissipated by the potentiometer?

Solution

We don't have to do any unit conversions. Plug in the values directly and then do the arithmetic to get

$$P = E^2/R = 6.30 \times 6.30 \ / \ 470 = 0.0844 \text{ W} = 84.4 \text{ mW}$$

Problem 4-12

Suppose that the resistance in Fig. 4-7 is 33 k and the current is 756 mA. What's the power dissipated by the potentiometer?

Solution

We can use the formula $P = I^2R$ after converting to ohms and amperes: $R = 33{,}000$ and $I = 0.756$. Then calculate and round off to get

$$P = 0.756 \times 0.756 \times 33{,}000 = 18{,}861 \text{ W} = 18.9 \text{ kW}$$

Obviously, a common potentiometer can't dissipate that much power! Most potentiometers are rated at 1 W or so.

Problem 4-13

How much voltage would we need to drive 60.0 μA through 33.0 k?

Solution

These input numbers both have three significant figures because the zeros on the far right are important. (Without them, you'd only have two significant figures in your values.) Use Ohm's Law to find the voltage after converting to amperes and ohms, obtaining

$$E = IR = 0.0000600 \times 33{,}000 = 1.98 \text{ V}$$

Resistances in Series

When we connect two or more resistances in series, their ohmic values add up to get the total (or *net*) resistance.

Problem 4-14

We connect three resistors in series with individual resistances of 220 ohms, 330 ohms, and 470 ohms, as shown in Fig. 4-8. What's the net resistance of the combination?

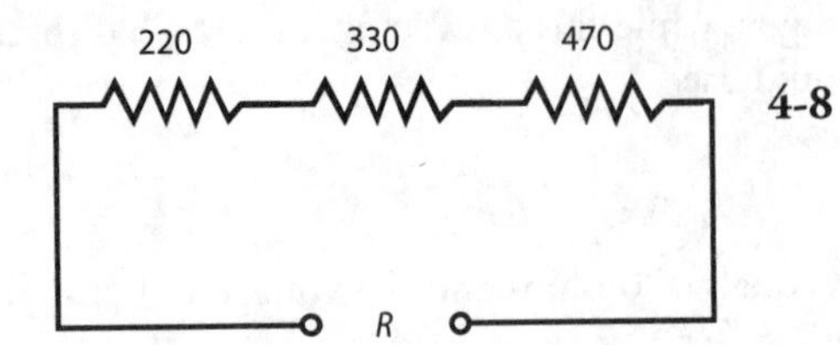

4-8 Three resistors in series. Illustration for Problem 4-14. All resistances are expressed in ohms.

Solution

Because we know all the values in ohms, we can add without doing any unit conversions to get

$$R = 220 + 330 + 470 = 1020 \text{ ohms} = 1.02 \text{ k}$$

That's the pure theory. But when we build a real-life circuit, the exact resistances depend on the component *tolerances*: how much we should expect the actual values to vary, as a result of manufacturing quirks, from the values specified by the vendor.

Resistances in Parallel

We can evaluate resistances in parallel by considering them as *conductances* instead. Engineers express conductance in units called *siemens*, symbolized S. (The word "siemens" serves both in the singular and the plural sense). Some older physics or engineering documents use the *mho* ("ohm" spelled backwards) as the fundamental unit of conductance; the mho and the siemens represent the same thing. When we connect conductances in parallel, their values add up, just as resistances add up in series. If we change all the ohmic values to siemens, we can add these figures up and convert the result back to ohms.

Engineers use the uppercase, italic letter G to symbolize conductance as a parameter or mathematical variable. The conductance in siemens equals the reciprocal of the resistance in ohms. We can express this fact using two formulas, assuming that neither R nor G ever equals zero:

$$G = 1/R$$

and

$$R = 1/G$$

Problem 4-15

Consider five resistors in parallel. Call the resistors R_1 through R_5, and call the total resistance R, as shown in Fig. 4-9. Suppose that the individual resistors have values of $R_1 = 10$ ohms, $R_2 = 20$ ohms,

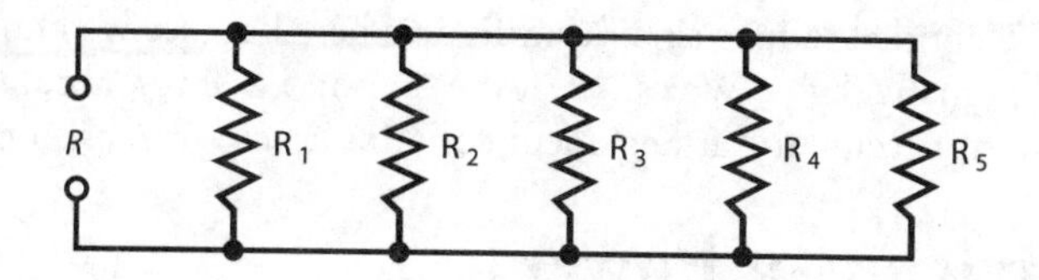

4-9 Five resistors R_1 through R_5, connected in parallel, produce a net resistance *R*. Illustration for Problems 4-15 and 4-16.

$R_3 = 40$ ohms, $R_4 = 50$ ohms, and $R_5 = 100$ ohms. What's the total resistance *R* of this parallel combination? (Note that we should not italicize R when it means "resistor" as a physical object, but we should italicize *R* when it means "resistance" as in a mathematical variable.)

Solution

To solve this problem, start by converting the resistances to conductances by taking their reciprocals. We'll get

$$G_1 = 1/R_1 = 1/10 = 0.10 \text{ S}$$
$$G_2 = 1/R_2 = 1/20 = 0.050 \text{ S}$$
$$G_3 = 1/R_3 = 1/40 = 0.025 \text{ S}$$
$$G_4 = 1/R_4 = 1/50 = 0.020 \text{ S}$$
$$G_5 = 1/R_5 = 1/100 = 0.0100 \text{ S}$$

When we add these numbers, we obtain

$$G = 0.10 + 0.050 + 0.025 + 0.020 + 0.0100 = 0.205 \text{ S}$$

The total resistance, rounded to two significant figures, turns out as

$$R = 1/G = 1/0.205 = 4.9 \text{ ohms}$$

We can calculate the net resistance of a parallel combination directly, but the arithmetic can get messy. Refer again to Fig. 4-9. The resistances combine according to the formula

$$R = 1/(1/R_1 + 1/R_2 + 1/R_3 + 1/R_4 + 1/R_5)$$

Once in a while, you'll encounter a situation where you have multiple resistances in parallel and their values are all equal. In a case of that sort, the total resistance equals the resistance of any one component divided by the number of components. For example, two 80-ohm resistors combine in parallel to yield a net resistance of $80/2 = 40$ ohms; four of the same resistors combine in parallel to produce $80/4 = 20$ ohms; five of them combine in parallel to give you $80/5 = 16$ ohms.

Problem 4-16

We have five resistors R_1 through R_5 connected in parallel, as shown in Fig. 4-9. Suppose that each one of the resistances is 1.800 k. What's the total resistance, *R*, of this combination?

Solution

Here, we can convert the resistances to 1800 ohms and then divide by 5 to get

$$R = 1800/5 = 360.0 \text{ ohms}$$

We're entitled to four significant figures here because we know the input value as stated, 1.800 K, to that many digits. We can treat the divisor 5 as *exact*, accurate to however many significant digits we want because the arrangement contains *exactly five* resistors.

Division of Power

When we connect sets of resistors to a source of voltage, each resistor draws some current. If we know the voltage, we can figure out how much current the entire set demands by calculating the net resistance of the combination, and then considering the combination as a single resistor.

If the resistors in the network all have the same ohmic value, the power from the source divides up equally among them, whether we connect the resistors in series or in parallel. For example, if we have eight identical resistors in series with a battery, the network consumes a certain amount of power, each resistor bearing 1/8 of the load. If we rearrange the circuit to connect the resistors in parallel with the same battery, the network *as a whole* dissipates more power than it does when the resistors are in series, but each *individual* resistor handles 1/8 of the total power, just as when they're in series.

If the resistances in a network do not all have identical ohmic values, then some resistors dissipate more power than others.

Resistances in Series-Parallel

We can connect sets of resistors, all having identical ohmic values, in parallel sets of series networks, or in series sets of parallel networks. In either case, we get a *series-parallel network* that can greatly increase the total power-handling capacity of the network over the power-handling capacity of a single resistor.

Sometimes, the total resistance of a series-parallel network equals the value of any one of the resistors. This happens if the components are all identical, and are arranged in a network called an *n-by-n* (or $n \times n$) *matrix*. That means when n is a whole number, we have n series-connected sets of n resistors with all the sets connected in parallel (Fig. 4-10A), or else n parallel-connected sets of n resistors, all connected in series (Fig. 4-10B). In practice, these arrangements yield identical results.

A series-parallel array of $n \times n$ resistors, all having identical ohmic values and identical power ratings, has n^2 times the power-handling capability of any resistor by itself. For example, a 3×3 series-parallel matrix of 2 W resistors can handle up to $3^2 \times 2 = 9 \times 2 = 18$ W. If we have a 10×10 array of 1/2 W resistors, then it can dissipate up to $10^2 \times 1/2 = 50$ W. Simply multiply the power-handling capacity of each individual resistor by the total number of resistors in the matrix.

The above-described scheme works if, *but only if*, all of the resistors have identical ohmic values and identical power-dissipation ratings. If the resistors have values and/or ratings that differ (even slightly), one of the components might draw more current than it can withstand, so it will burn out. Then the current distribution in the network will change, increasing the likelihood that a second resistor will fail. We can end up with a chain reaction of component destruction!

If we need a resistor that can handle 50 W and a certain series-parallel network will handle 75 W, that's fine. But we shouldn't "push our luck" and expect to get away with a network that will handle only 48 W in the same application. We should allow some extra tolerance, say 10 percent over the minimum rating. If we expect the network to dissipate 50 W, we should build it to handle 55 W, or a bit more. We don't have to engage in "overkill," however. We'll waste resources if we

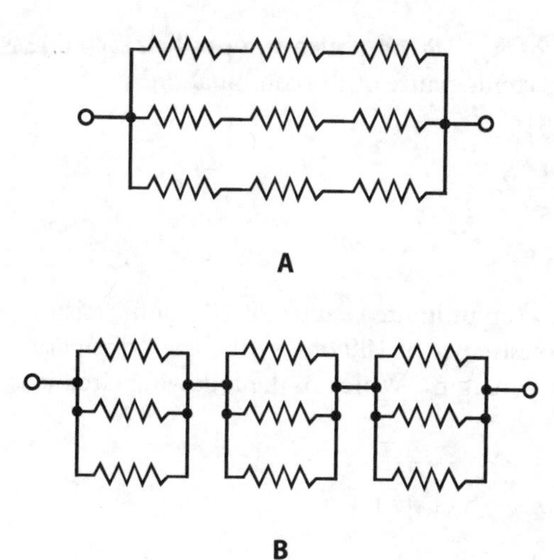

4-10 Two examples of series-parallel resistance matrices. At A, sets of series resistors join in parallel. At B, sets of parallel resistances join in series. These examples show symmetrical n-by-n matrices, where $n = 3$.

build a network that can handle 500 W when we only expect it to cope with 50 W—unless that's the only convenient combination we can cobble together with resistors we have on hand.

Non-symmetrical series-parallel networks, made up from identical resistors, can increase the power-handling capability over that of a single resistor. But in these cases, the total resistance differs from the value of any individual resistor. To obtain the overall power-handling capacity, we can always multiply the power-handling capacity of any individual resistor by the total number of resistors, whether the network is symmetrical or not—again, *if and only if*, all the resistors have identical ohmic values and identical power-dissipation ratings.

Quiz

Refer to the text in this chapter if necessary. A good score is at least 18 correct answers. The answers are in the back of the book.

1. We have an unlimited supply of 33-ohm resistors, each one capable of dissipating 0.50 W. We want a 33-ohm resistor that can dissipate 18 W (a figure that includes a 2-W safety margin). We can get that component by wiring up
   (a) a 6×6 series-parallel matrix of individual resistors.
   (b) a 9×4 series-parallel matrix of individual resistors.
   (c) a 3×12 series-parallel matrix of individual resistors.
   (d) Any of the above

2. We connect a 6.30-V lantern battery across a 330-ohm resistor. The resistor dissipates
   (a) 19.0 mW of power.
   (b) 8.31 mW of power.
   (c) 120 mW of power.
   (d) We need more information to calculate it.

3. If we connect 10 components in parallel, each one with a DC conductance of 0.15 S, what's the net DC conductance of the combination?
   (a) 0.015 S
   (b) 0.15 S
   (c) 1.5 S
   (d) 15 S

4. We have an unlimited supply of 100-ohm resistors, each one capable of dissipating 1.00 W. We want a resistance of 100 ohms capable of dissipating up to 12 W (a figure that includes a 2.5-W safety margin). Which of the following circuits is the smallest $n \times n$ matrix that will work here?
   (a) A 5×5 matrix
   (b) A 4×4 matrix
   (c) A 3×3 matrix
   (d) A 2×2 matrix

5. If we connect a 6.3-V battery across a 330-ohm resistor, the current is
   (a) 72 mA.
   (b) 36 mA.
   (c) 12 mA.
   (d) 19 mA.

6. The voltage across a resistor is 2.2 V. The resistor dissipates 400 mW. What's its resistance?
   (a) 12 ohms
   (b) 24 ohms
   (c) 48 ohms
   (d) 96 ohms

7. If we connect eight resistors in parallel, all identical and each with a value of 1.100 k, we get a component with a resistance of
   (a) 8800 ohms.
   (b) 4840 ohms.
   (c) 1100 ohms.
   (d) 137.5 ohms.

8. We wire up three resistors in parallel: 600 ohms, 300 ohms, and 200 ohms. Then we connect a 12-V battery across the combination. How much current does the 300-ohm resistor draw all by itself?
   (a) 80 mA
   (b) 40 mA
   (c) 33 mA
   (d) 11 mA

9. If we decrease the *conductance* of a resistor by a factor of 16 while leaving it connected to a source of constant DC voltage, then the power that the resistor dissipates will

    (a) decrease by a factor of 16.

    (b) decrease by a factor of 4.

    (c) increase by a factor of 4.

    (d) increase by a factor of 16.

10. If we double the DC voltage across a resistor and double its resistance as well, then the power that the resistor dissipates will

    (a) get cut in half.

    (b) stay the same.

    (c) double.

    (d) quadruple.

11. If we double the DC voltage across a resistor and double its resistance as well, then the current that the resistor draws will

    (a) get cut in half.

    (b) stay the same.

    (c) double.

    (d) quadruple.

12. If we know the current through a component (in amperes) and its resistance (in ohms), how can we calculate the energy (in joules) that the component consumes?

    (a) Square the current and then multiply by the resistance.

    (b) Multiply the current by the resistance.

    (c) Divide the resistance by the current.

    (d) We need more information to do it.

13. Suppose that 33.300 mA DC flows through a resistance of 3.333333 k. How can we *best* express the voltage across this resistance, taking significant figures into account?

    (a) 111 V

    (b) 111.0 V

    (c) 111.00 V

    (d) 110.999 V

14. If a potentiometer carries 18.5 mA DC and we set its resistance to 1.12 k, how much power does it dissipate?

    (a) 383 mW

    (b) 20.7 mW

    (c) 60.5 mW

    (d) 67.8 mW

15. We wire up seven 70.0-ohm resistors in parallel, and then connect a 12.6-V battery across the whole combination. How much current gets drawn from the battery?

(a) 25.7 mA

(b) 1.26 A

(c) 794 mA

(d) 180 mA

16. We remove three of the resistors from the circuit in Question 15. What will happen to the current drawn by any one of the remaining four resistors?

(a) It will go down to zero.

(b) It will become 4/7 of its previous value.

(c) It will stay the same.

(d) It will become 7/4 of its previous value.

17. We connect resistors with values of 180, 270, and 680 ohms in series with a 12.6-V battery. How much power does the set of resistors dissipate as a whole?

(a) 7.12 W

(b) 89.7 W

(c) 11.2 mW

(d) 140 mW

18. The three primary units that engineers use when working with DC systems are the

(a) ampere, volt, and ohm.

(b) watt, joule, and volt.

(c) siemens, ampere, and joule.

(d) erg, joule, and ohm.

19. A direct current of 3.00 A flows through a component whose conductance is 0.250 S. What's the voltage across the component?

(a) 0.750 V

(b) 12.0 V

(c) 36.0 V

(d) We need more information to calculate it.

20. A direct current of 3.00 A flows through a component whose conductance is 0.250 S. How much power does the component dissipate?

(a) 750 mW

(b) 2.25 W

(c) 36.0 W

(d) We need more information to calculate it.

5
CHAPTER

Direct-Current Circuit Analysis

IN THIS CHAPTER, YOU'LL LEARN MORE ABOUT DC CIRCUITS AND HOW THEY BEHAVE UNDER VARIOUS conditions. These principles apply to most AC utility circuits as well.

Current through Series Resistances

In a series circuit such as a string of light bulbs connected to a DC battery (Fig. 5-1), the current at any given point equals the current at any other point. In this situation, we connect an ammeter A between two of the bulbs. If we move the ammeter to any other point in the current path, it will indicate the same current. This uniformity of current holds true in any series DC circuit, no matter what the components are, and regardless of whether or not they all have the same resistance.

If the bulbs in Fig. 5-1 had different resistances, the current would still be the same at every point in the circuit, but some of the bulbs would consume more power than others. That scenario would likely present a problem; some of the lights would burn brightly and others would hardly

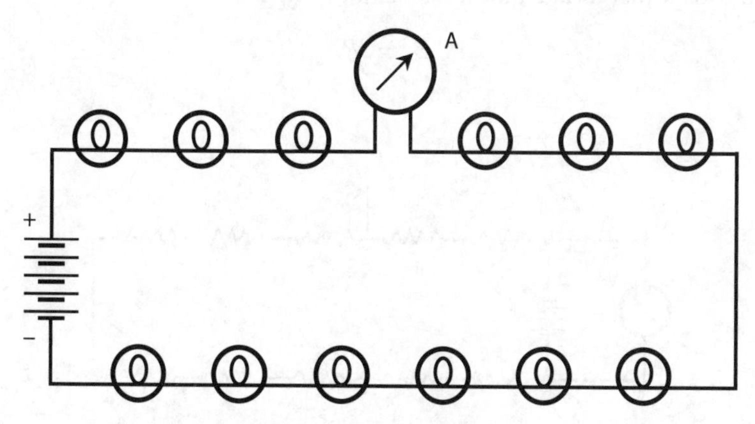

5-1 Light bulbs in series, with an ammeter (A) in the circuit.

glow at all. In a series circuit, even a slight discrepancy in the resistances of the individual components can cause major irregularity in the distribution of power.

Now suppose that one of the bulbs in Fig. 5-1 burns out. The entire string goes dark. We short out the faulty bulb's socket, hoping to get the lights working again. They do, but the current through the chain increases because its overall resistance goes down. Each remaining bulb carries a little more current than it should. Pretty soon, another bulb will burn out because of the excessive current. If we then replace it with a second short circuit, the current will rise still further. We should not be surprised if another bulb blows out almost right away thereafter.

Voltages across Series Resistances

Because the bulbs in the string of Fig. 5-1 all have identical resistances, the potential differences across them are all the same. If we have 12 identical series-connected bulbs in a 120 V circuit, each bulb gets 1/12 of the total, or 10 V. This even distribution of voltage will continue to hold true even if we replace all the bulbs with brighter or dimmer ones, as long as we make sure that all the bulbs in the string are identical.

Examine the schematic diagram of Fig. 5-2. Each resistor carries the same current, whether or not their resistances are all the same. Each resistance R_n has a potential difference E_n across it, equal to the product of the current and the resistance of that particular resistor. The voltages E_n appear in series, so they add up arithmetically. The voltages across all the resistors add up to the supply voltage E. If they did not, a "phantom EMF" would exist somewhere, adding or taking away the unaccounted-for voltage. But that cannot happen! Voltage can't come out of nowhere, nor can it vanish into "thin air."

Look at this situation another way. The voltmeter V in Fig. 5-2 shows the voltage E of the battery because we've connected the meter across the battery. The voltmeter V also shows the sum of the voltages E_n across the set of resistances because V is connected across the whole combination. The meter says the same thing whether we think of it as measuring the battery voltage E or as measuring the sum of the voltages E_n across the series combination of resistances. Therefore, E equals the sum of all the voltages E_n.

If you want to calculate the voltage E_n across any particular resistance R_n in a circuit like the one in Fig. 5-2, remember the Ohm's Law formula for finding voltage in terms of current and resistance. When you adapt that formula for this situation, you get

$$E_n = IR_n$$

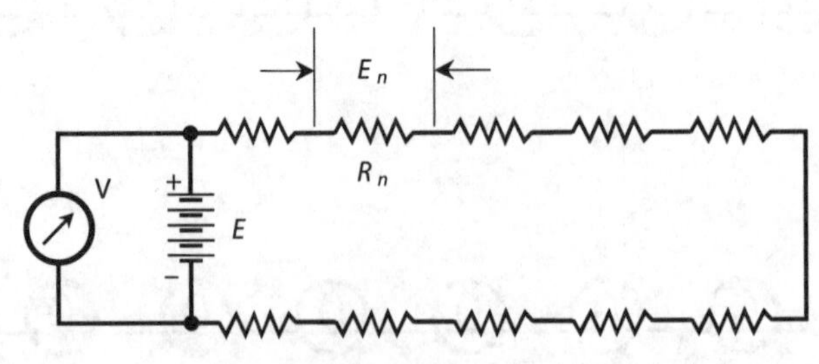

5-2 Analysis of voltages in a series circuit.

where E_n represents the potential difference in volts across the particular resistor, I represents the current in amperes through the whole circuit (and also, therefore, through the resistor of interest), and R_n represents the particular resistor's value in ohms. To determine the current I, you must know the total resistance R (the sum of all the resistances) and the supply voltage E. Then you can use the formula

$$I = E/R$$

If you're astute, you'll notice that you can substitute the value of I from the second of the foregoing formulas into the first one, getting

$$E_n = (E/R)\,R_n = E(R_n/R)$$

This new formula reveals an interesting fact. Each resistor in a series circuit receives voltage in direct proportion to the total supply voltage, and also in direct proportion to the ratio of its resistance to the total resistance. Engineers and technicians exploit these proportional relationships to build circuits called *voltage dividers*.

Problem 5-1

Figure 5-2 shows 10 resistors in series. Suppose that five of the resistors have values of 20 ohms, and the other five have values of 30 ohms. Further suppose that the battery provides 25 V DC. How much potential difference exists across any one of the 20-ohm resistors? How much potential difference exists across any one of the 30-ohm resistors?

Solution

Let's find the total resistance R of the entire series combination, so that we can calculate the current I on the basis of R and the battery voltage E. Once we know the current, we can find the voltage across any individual resistor. We have a total resistance of

$$R = (20 \times 5) + (30 \times 5) = 100 + 150 = 250 \text{ ohms}$$

The current at any point in the circuit is therefore

$$I = E/R = 25/250 = 0.10 \text{ A}$$

If we let $R_n = 20$ ohms, we have

$$E_n = IR_n = 0.10 \times 20 = 2.0 \text{ V}$$

and if we let $R_n = 30$ ohms, we have

$$E_n = IR_n = 0.10 \times 30 = 3.0 \text{ V}$$

Let's verify the fact that the voltages across all of the resistors add up to the supply voltage. We have five resistors with 2.0 V across each, for a total of 10 V; we have five resistors with 3.0 V across each, for a total of 15 V. The sum of the voltages across the resistors is therefore

$$E = 10 + 15 = 25 \text{ V}$$

Problem 5-2

In the circuit of Fig. 5-2, as described in Problem 5-1 and its solution, what will happen to the voltages across the resistances if we short-circuit three of the 20-ohm resistors and two of the 30-ohm resistors?

Solution

We've replaced three of the 20-ohm resistors with short circuits and two of the 30-ohm resistors with short circuits, leaving us, in effect, with two 20-ohm resistors and three 30-ohm resistors in series. Now we have

$$R = (20 \times 2) + (30 \times 3) = 40 + 90 = 130 \text{ ohms}$$

The current is therefore

$$I = E/R = 25/130 = 0.19 \text{ A}$$

The voltage E_n across any of the "unshorted" 20-ohm resistances R_n is

$$IR_n = 0.19 \times 20 = 3.8 \text{ V}$$

The voltage E_n across any of the "unshorted" 30-ohm resistances R_n is

$$IR_n = 0.19 \times 30 = 5.7 \text{ V}$$

Checking the total voltage, we add and round off to two significant figures, getting

$$E = (2 \times 3.8) + (3 \times 5.7) = 7.6 + 17.1 = 25 \text{ V}$$

Voltage across Parallel Resistances

Imagine a set of light bulbs connected in parallel with a DC battery (Fig. 5-3). If one bulb burns out, we'll have an easy time correcting the problem. In the parallel configuration, only the bad bulb goes dark, so we can identify it immediately. Parallel circuits have another advantage over series circuits. Variations in the resistance of one element have no effect on the power that the other elements receive.

In a parallel circuit, the voltage across each component equals the supply or battery voltage. The current drawn by any particular component depends only on the resistance of that component, and

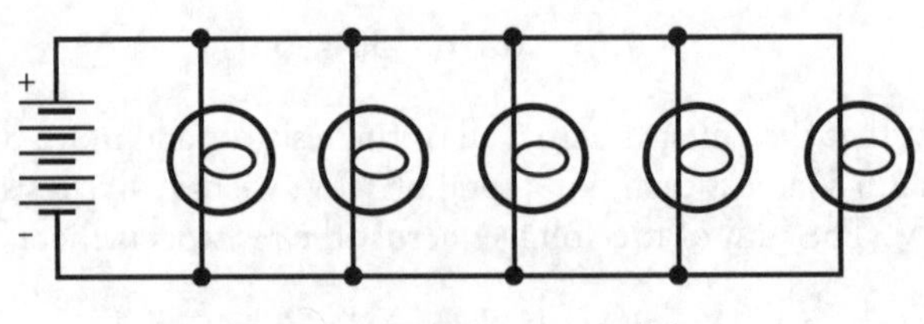

5-3 Light bulbs in parallel.

not on the resistances of any of the others. In this sense, the components in a parallel-wired circuit operate independently, as opposed to the series-wired circuit in which they interact.

If any one branch of a parallel circuit opens up, is disconnected, or is removed, the conditions in the other branches don't change. If we add new branches to a parallel circuit, assuming the power supply can handle the increased current demand, conditions in previously existing branches remain as they were. Parallel circuits exhibit better overall stability than series circuits. That's why engineers wire nearly all utility circuits in parallel.

Currents through Parallel Resistances

Figure 5-4 illustrates a generic circuit with resistors wired in parallel. Let's call the resistance values R_n, the total circuit resistance R, and the battery voltage E. We can measure the current I_n that flows through any particular branch n, containing resistance R_n, using an ammeter A.

In a parallel circuit such as the one in Fig. 5-4, the sum of all the currents I_n equals the total current I drawn from the power source. In a parallel circuit, the current divides among the individual components, in a manner similar to the way that voltage divides among the components in a series circuit.

Have you noticed that we portray the direction of current flow in Fig. 5-4 as going outward from the positive battery terminal? Don't let this confuse you. Electrons, which constitute most of the charge carriers in an ordinary wire, flow out of the negative terminal of a battery and toward the positive terminal. However, scientists consider *theoretical current*, more often called *conventional current*, as a flow of electricity from positive to negative.

Problem 5-3

Suppose that the battery in Fig. 5-4 delivers 24 V and we have 10 resistors, each with a value of 100 ohms, in the parallel circuit. What's the total current, I, drawn from the battery?

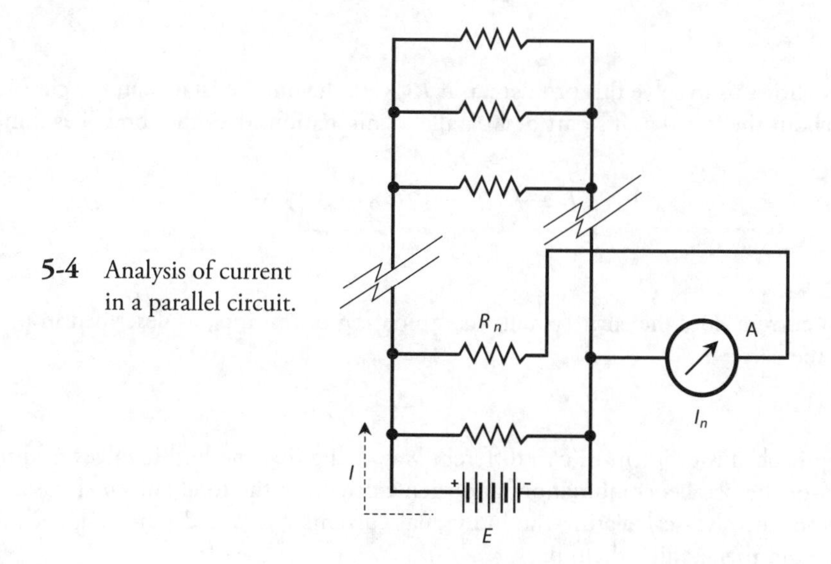

5-4 Analysis of current in a parallel circuit.

Solution

First, let's find the total resistance. Because all of the resistors have the same value, we can divide $R_n = 100$ by 10 to get $R = 10$ ohms. Then we can calculate the current as

$$I = E/R = 24/10 = 2.4 \text{ A}$$

Problem 5-4

In the circuit of Fig. 5-4 and described in Problem 5-3 and its solution, we have connected the ammeter A to indicate the current flowing through one particular resistor R_n. What does the meter show?

Solution

We must determine the current I_n that flows in any given branch of the circuit. The voltage equals 24 V across every branch, and every resistor has a value of $R_n = 100$ ohms. We can determine the ammeter reading using Ohm's Law, obtaining

$$I_n = E/R_n = 24/100 = 0.24 \text{ A}$$

Because we have a parallel circuit, all of the branch currents I_n should add up to the total current I. Each of the 10 identical branches carries 0.24 A, so the total current is

$$I = 0.24 \times 10 = 2.4 \text{ A}$$

This result agrees with the solution to Problem 5-3, as we would expect.

Problem 5-5

We connect three resistors in parallel across a battery that supplies $E = 12$ V. The resistance values are $R_1 = 24$ ohms, $R_2 = 48$ ohms, and $R_3 = 60$ ohms. These resistances carry currents I_1, I_2, and I_3 respectively. What is the current I_2 through R_2?

Solution

Ohm's Law allows us to solve this problem as if R_2 were the only resistance in the circuit. We need not worry about the fact that it's part of a parallel combination; the other branches don't affect I_2. Therefore

$$I_2 = E/R_2 = 12/48 = 0.25 \text{ A}$$

Problem 5-6

How much current does the entire parallel combination of resistors, as described in problem 5-5, draw from the battery?

Solution

We can approach this problem in two different ways. The first method involves finding the total resistance R of the parallel combination, and then calculating the total current I based on R. The second method involves calculating the individual currents I_1, I_2, and I_3 through R_1, R_2, and R_3, respectively, and then adding them up.

Using the first method, we begin by changing the resistances R_n into conductances G_n. This gives us

$$G_1 = 1/R_1 = 1/24 = 0.04167 \text{ S}$$
$$G_2 = 1/R_2 = 1/48 = 0.02083 \text{ S}$$
$$G_3 = 1/R_3 = 1/60 = 0.01667 \text{ S}$$

When we add these values together, we see that the net conductance of the parallel combination is $G = 0.07917$ S. The net resistance is the reciprocal of this value, or

$$R = 1/\text{G} = 1/0.07917 = 12.631 \text{ ohms}$$

We can use Ohm's Law to find

$$I = E/R = 12/12.631 = 0.95 \text{ A}$$

We kept some extra digits throughout the calculation, rounding off at the end of the process, in order to minimize *cumulative rounding error.*

Now let's try the second method. We calculate the currents through the individual resistors using Ohm's Law as follows:

$$I_1 = E/R_1 = 12/24 = 0.5000 \text{ A}$$
$$I_2 = E/R_2 = 12/48 = 0.2500 \text{ A}$$
$$I_3 = E/R_3 = 12/60 = 0.2000 \text{ A}$$

When we add these current values up, we get a total current of

$$I = I_1 + I_2 + I_3 = 0.\ 5000 + 0.2500 + 0.2000 = 0.95 \text{ A}$$

Again, we kept a few extra digits during the calculation, rounding off to two significant digits at the end.

Power Distribution in Series Circuits

When you want to calculate the power dissipated by resistors connected in series, you can calculate the total current I and then determine the power P_n dissipated by any one of the resistances R_n, using the formula

$$P_n = I^2 R_n$$

The total power (in watts) dissipated in a series combination of resistances equals the sum of the wattages dissipated in each resistance.

Problem 5-7

Imagine a DC series circuit with a power supply that provides 150 V and three resistances $R_1 = 200$ ohms, $R_2 = 400$ ohms, and $R_3 = 600$ ohms. How much power does R_2 dissipate?

Solution

First, let's find the current drawn from the battery by the whole circuit. The resistors are connected in series, so the total resistance is

$$R = 200 + 400 + 600 = 1200 \text{ ohms}$$

According to Ohm's Law, the current is

$$I = 150/1200 = 0.125 \text{ A}$$

The power dissipated by R_2 is

$$P_2 = I^2 R_2 = 0.125 \times 0.125 \times 400 = 6.25 \text{ W}$$

Problem 5-8

Calculate the total dissipated power P in the circuit of Problem 5-7 using two different methods.

Solution

First, let's figure out the power dissipated by each resistance separately, and then add the figures. We know the power P_2 from the solution to Problem 5-7. We calculated the current as 0.125 A. Now we can calculate using the power formula based on currents and resistances to get

$$P_1 = I^2 R_1 = 0.125 \times 0.125 \times 200 = 3.125 \text{ W}$$

and

$$P_3 = I^2 R_3 = 0.125 \times 0.125 \times 600 = 9.375 \text{ W}$$

Adding P_1, P_2, and P_3 yields a total power of $P = 3.125 + 6.25 + 9.375 = 18.75$ W, which we should round off to 18.8 W because we have input data accurate to only three significant figures.

The second method involves finding the total series resistance and then calculating the power from that value and the current value we calculated in Problem 5-7. The net resistance of the series combination is $R = 1200$ ohms, so

$$P = I^2 R = 0.125 \times 0.125 \times 1200 = 18.75 \text{ W}$$

We once again round this value to 18.8 W.

Power Distribution in Parallel Circuits

When we connect resistances in parallel, the currents in the individual resistors can, and often do, differ. The currents will turn out identical, if and only if, all of the resistances are identical. But the voltage across any given resistor always equals the voltage across any other resistor. We can find the power P_n dissipated by any particular resistance R_n using the formula

$$P_n = E^2 / R_n$$

where E represents the voltage of the supply. In a parallel DC circuit, just as in a series circuit, the total dissipated wattage equals the sum of the wattages dissipated by the individual resistances.

Problem 5-9

Suppose that a DC circuit contains three resistances $R_1 = 22$ ohms, $R_2 = 47$ ohms, and $R_3 = 68$ ohms connected in parallel across a battery that supplies a voltage of $E = 3.0$ V. Find the power dissipated by each resistance.

Solution

We can start by finding E^2, the square of the supply voltage. We'll need to use this figure several times:

$$E^2 = 3.0 \times 3.0 = 9.0$$

Resistance R_1 dissipates a power of

$$P_1 = 9.0/22 = 0.4091 \text{ W}$$

which we can round off to 0.41 W. Resistance R_2 dissipates a power of

$$P_2 = 9.0/47 = 0.1915 \text{ W}$$

which we can round off to 0.19 W. Resistance R_3 dissipates a power of

$$P_3 = 9.0/68 = 0.1324 \text{ W}$$

which we can round off to 0.13 W.

Problem 5-10

Find the total consumed power of the resistor circuit in Problem 5-9 using two different methods.

Solution

The first method involves adding the wattages P_1, P_2, and P_3 from the solution to Problem 5-9. If we use the four-significant-digit values, we get

$$P = 0.4091 + 0.1915 + 0.1324 = 0.7330 \text{ W}$$

which rounds off to 0.73 W. The second method involves finding the net resistance R of the parallel combination, and then calculating the power from the net resistance and the battery voltage. (As an exercise, calculate the net resistance yourself.) The net resistance works out to $R = 12.28$ ohms, accurate to four significant figures. Now we can calculate the total dissipated wattage as

$$P = E^2/R = 9.0/12.28 = 0.7329 \text{ W}$$

which we can round off to 0.73 W.

It's the Law!

In electricity and electronics, DC circuit analysis always follows certain axioms, or *laws*. The following rules merit your best efforts at memorization.

- In a series circuit, the current is the same at every point.
- In a parallel circuit, the voltage across any resistance equals the voltage across any other resistance, or across the whole set of resistances.

- In a series circuit, the voltages across all the resistances add up to the supply voltage.
- In a parallel circuit, currents through resistances always add up to the total current drawn from the power supply.
- In a series or parallel circuit, the total dissipated wattage equals the sum of the wattages dissipated in the individual resistances.

Now, let's get acquainted with two of the most famous laws that govern DC circuits. These rules have broad scope, allowing us to analyze complicated series-parallel DC networks.

Kirchhoff's First Law

The physicist, *Gustav Robert Kirchhoff* (1824–1887), conducted research and experiments before anyone understood much about how electric currents flow. Nevertheless, he used common sense to deduce two important properties of DC circuits.

Kirchhoff reasoned that in any DC circuit, the current going into any point ought to equal the current coming out of that point. This fact, Kirchhoff thought, must hold true no matter how many branches lead into or out of the point. Figure 5-5 shows two examples of this principle.

Examine Fig. 5-5A. At point X, the total current I going in equals $I_1 + I_2$, the total current coming out. At point Y, the total current $I_2 + I_1$ going in equals I, the total current coming out. Now examine Fig. 5-5B. At point Z, the total current $I_1 + I_2$ going in equals $I_3 + I_4 + I_5$, the total current coming out. We've just seen two examples of *Kirchhoff's First Law*. We can also call it *Kirchhoff's Current Law* or the *current-conservation principle*.

Problem 5-11

Refer to Fig. 5-5A. Suppose that all three resistors have values of 100 ohms, and that $I_1 = 2.0$ A, while $I_2 = 1.0$ A. What is the battery voltage?

Solution

First, let's find the total current I that the whole network of resistors demands from the battery. That's an easy task; we simply add the branch currents to get

$$I = I_1 + I_2 = 2.0 + 1.0 = 3.0 \text{ A}$$

Next, let's find the resistance of the entire network. The two 100 ohm resistances in series add to give us a net resistance of 200 ohms, which appears in parallel with a third resistance of 100 ohms. You can do the calculations and find that the net resistance R across the battery is 66.67 ohms. Then we can use Ohm's Law to calculate

$$E = IR = 3.0 \times 66.67 = 200 \text{ V}$$

Problem 5-12

In the circuit of Fig. 5-5B, imagine that both resistors below point Z have values of 100 ohms, and all three resistors above point Z have values of 10.0 ohms. Suppose that 500 mA flows through each 100-ohm resistor. How much current flows through any one of the 10.0-ohm resistors? How much potential difference appears across any one of the 10.0-ohm resistors?

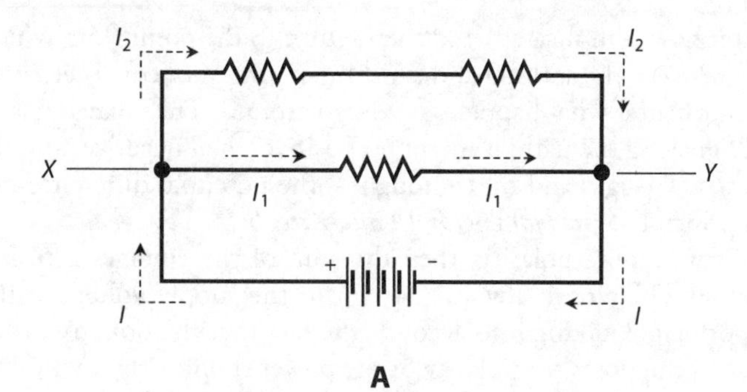

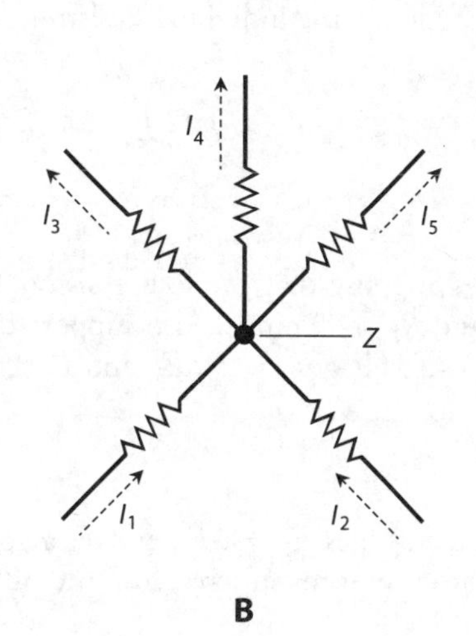

5-5 Kirchhoff's First Law. At A, the total current flowing into point *X* or point *Y* equals the total current flowing out of that point. That is, $I = I_1 + I_2$. At B, the total current flowing into point *Z* equals the current flowing out of point *Z*. That is, $I_1 + I_2 = I_3 + I_4 + I_5$.

Solution

The total current into point *Z* equals 500 mA + 500 mA = 1.00 A. According to Kirchhoff's First Law, this same current emerges from point *Z*, dividing equally three ways among the resistors because their values are all identical (10.0 ohms). The current through any one of those resistors is therefore 1.00/3 A = 0.333 A. We can now determine the potential difference *E* across any one of the 10.0 ohm resistances *R* using Ohm's Law, getting

$$E = IR = 0.333 \times 10.0 = 3.33 \text{ V}$$

Kirchhoff's Second Law

The sum of all the voltages, as you go around any DC circuit from some fixed point and return there from the opposite direction, and taking polarity into account, always equals zero. Does this notion seem counterintuitive? If so, let's try to "get into Kirchhoff's head" and figure out what he was thinking.

When Kirchhoff wrote his second law, he must have realized that potential differences can't appear out of nowhere, nor can they vanish into nothingness. If you go around a DC circuit (no matter how complicated) and then return to the point from which you started, the potential difference between the starting and finishing points *must* equal zero because those two points coincide. It doesn't matter what happens as you go around. The voltage relative to the starting point might rise and fall along the path; it might go positive, then negative, and then positive again. But in the end, all those voltages add up to nought—the potential difference between a point and itself. We call this principle *Kirchhoff's Second Law, Kirchhoff's Voltage Law,* or the *voltage-conservation principle.*

If we ignore polarity, then the sum of the voltages across all the individual resistances in a complex DC circuit always adds up to the supply voltage. Kirchhoff's Second Law expands on this principle, taking into account the fact that the polarity of the potential difference across each resistance opposes the polarity of the power supply. Figure 5-6 illustrates an example of Kirchhoff's Second Law. The polarity of the battery voltage E opposes the polarity of the sum of the voltages $E_1 + E_2 + E_3 + E_4$ across the individual resistors. Therefore, when we take polarity into account, we have

$$E + E_1 + E_2 + E_3 + E_4 = 0$$

Problem 5-13

Refer to Fig. 5-6. Suppose that the four resistors have values of $R_1 = 50$ ohms, $R_2 = 60$ ohms, $R_3 = 70$ ohms, and $R_4 = 80$ ohms. Also suppose that a current of $I = 500$ mA flows through the circuit. What are the voltages E_1, E_2, E_3, and E_4 across the individual resistors? What's the battery voltage E?

Solution

First, let's find the voltages E_1, E_2, E_3, and E_4 across each of the resistors, using Ohm's Law. We should convert the current to amperes, so $I = 0.500$ A. For R_1, we have

$$E_1 = IR_1 = 0.500 \times 50 = 25 \text{ V}$$

For R_2,

$$E_2 = IR_2 = 0.500 \times 60 = 30 \text{ V}$$

For R_3,

$$E_3 = IR_3 = 0.500 \times 70 = 35 \text{ V}$$

For R_4,

$$E_4 = IR_4 = 0.500 \times 80 = 40 \text{ V}$$

The battery voltage equals the sum

$$E = E_1 + E_2 + E_3 + E_4 = 25 + 30 + 35 + 40 = 130 \text{ V}$$

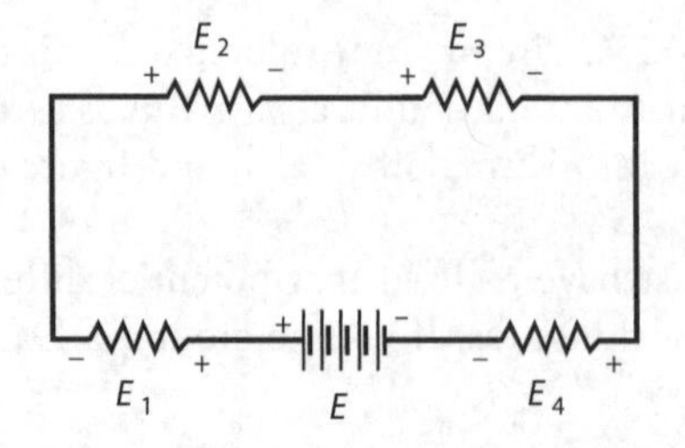

5-6 Kirchhoff's Second Law. The sum of the voltages across the resistances is equal to, but has opposite polarity from, the supply voltage. Therefore, $E + E_1 + E_2 + E_3 + E_4 = 0$.

Problem 5-14

In the situation shown by Fig. 5-6, suppose that the battery provides $E = 20$ V. Suppose the resistors labeled with voltages E_1, E_2, E_3, and E_4 have ohmic values in the ratio 1:2:3:4 respectively. What's the voltage E_3?

Solution

This problem doesn't provide any information about the current, nor does it tell us the exact resistances. But we need not know these things to solve for E_3. Regardless of the actual ohmic values, the ratio E_1:E_2:E_3:E_4 will always turn out the same as long as the resistances remain in the ratio 1:2:3:4. We can, therefore, "plug in" any ohmic values we want for the values of the resistors, as long as they exist in that ratio.

Let R_n be the resistance across which the voltage is E_n, where n can range from 1 to 4. Now suppose that the resistances are as follows:

- The resistance across which E_1 occurs is $R_1 = 1.0$ ohms
- The resistance across which E_2 occurs is $R_2 = 2.0$ ohms
- The resistance across which E_3 occurs is $R_3 = 3.0$ ohms
- The resistance across which E_4 occurs is $R_4 = 4.0$ ohms

These resistances follow the required ratio. The total resistance is

$$R = R_1 + R_2 + R_3 + R_4 = 1.0 + 2.0 + 3.0 + 4.0 = 10 \text{ ohms}$$

We can calculate the current through the entire series combination as

$$I = E/R = 20/10 = 2.0 \text{ A}$$

We can now calculate the potential difference E_3, which appears across the resistance R_3, with Ohm's Law, obtaining

$$E_3 = IR_3 = 2.0 \times 3.0 = 6.0 \text{ V}$$

Voltage Division

Resistances in series produce ratios of voltages, as we've already seen. We can tailor these ratios to meet certain needs by building *voltage-divider networks.*

When we want to design and build a voltage divider network, we should make the resistors' ohmic values as small as possible without imposing too much current demand on the battery or power supply. When we put the voltage divider to use with a real-world circuit, we don't want the *internal resistance* of that circuit to upset the operation of the divider. The voltage divider "fixes" the intermediate voltages most effectively when the resistance values are as small as the current-delivering capability of the power supply will allow.

Figure 5-7 illustrates the principle of voltage division. The individual resistances are R_1, R_2, R_3 and so on, all the way up to R_n. The total resistance is

$$R = R_1 + R_2 + R_3 + \ldots + R_n$$

The battery voltage equals E, so the current I in the circuit equals E/R. At the various points P_1, P_2, P_3, ..., P_n, let's call the potential differences relative to the negative battery terminal E_1, E_2, E_3 and so

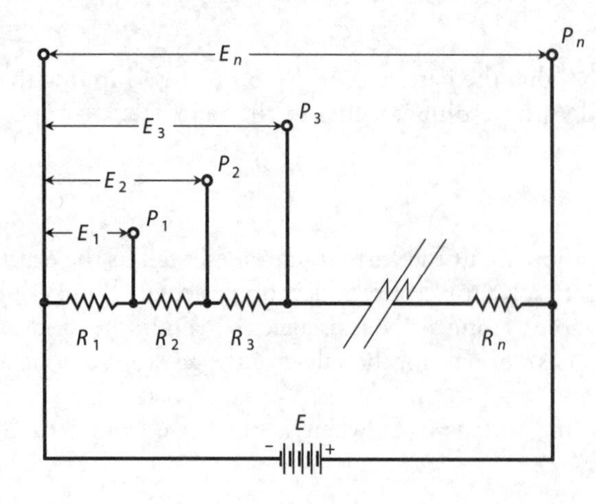

5-7 General arrangement for a voltage-divider circuit.

on, all the way up to E_n. The last voltage E_n equals the battery voltage, E, as we can see by looking at the diagram. All the other voltages are less than E, and ascend in succession. Mathematically, we write this fact as

$$E_1 < E_2 < E_3 < \ldots < E_n = E$$

where the mathematical symbol < means "is less than."

The voltages at the various points increase according to the sum total of the resistances up to each point, in proportion to the total resistance, multiplied by the supply voltage. Therefore, the following equations hold true:

$$E_1 = ER_1/R$$
$$E_2 = E(R_1 + R_2)/R$$
$$E_3 = E(R_1 + R_2 + R_3)/R$$

and so on. This process continues for each of the voltages at points all the way up to

$$E_n = E(R_1 + R_2 + R_3 + \ldots + R_n)/R = ER/R = E$$

Problem 5-15

Imagine that we've constructed an electronic device to do a specific task. Its battery supplies a voltage of $E = 9.0$ V. We connect its negative terminal to *common ground* (also called *chassis ground*). We want to build a voltage-divider network to obtain a terminal point where the DC voltage equals +2.5 V with respect to common ground. Provide an example of a pair of DC resistance values that we can connect in series so that +2.5 V appears at the point between them, as long as we don't connect any external circuit to the network.

Solution

Figure 5-8 illustrates the network. There exist infinitely many different combinations of resistances that will do our job. As long as we don't connect the voltage divider to any external circuit, we will always observe that, at the terminal point between the resistors,

$$R_1/(R_1 + R_2) = E_1/E$$

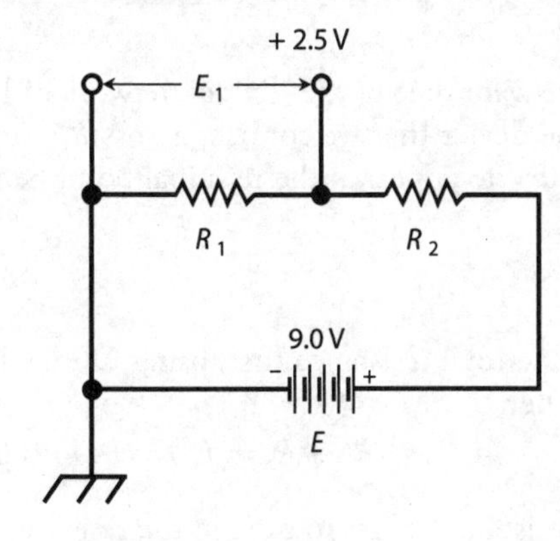

5-8 A voltage divider network that derives 2.5 V DC from a 9.0-V DC source.

Let's choose a total value of

$$R_1 + R_2 = 100 \text{ ohms}$$

In this case, we want to obtain a voltage $E_1 = +2.5$ V at the terminal point between the resistors. That means we want to have

$$E_1/E = 2.5/9.0 = 0.28$$

We must make the ratio of R_1 to the overall resistance equal to the voltage ratio, so that

$$R_1/(R_1 + R_2) = 0.28$$

We've chosen to construct our network such that the total resistance, $R_1 + R_2$, equals 100 ohms. Substituting 100 for $(R_1 + R_2)$ in the above formula, we get

$$R_1/100 = 0.28$$

which solves to

$$R_1 = 28 \text{ ohms}$$

so therefore

$$R_2 = 100 - 28 = 72 \text{ ohms}$$

In a practical situation, we'll want to choose the smallest possible value for R. This might be more or less than 100 ohms, depending on the nature of the circuit and the current-delivering capability of the battery. The *actual resistance values* don't determine the voltage at the terminal point; their *ratio* does.

Problem 5-16

How much current I, in milliamperes, flows through the series resistances in the situation described in Problem 5-15 and its solution?

Solution

Using the Ohm's Law formula for current in terms of resistance, we obtain

$$I = E/(R_1 + R_2) = 9.0/100 = 0.090 \text{ A} = 90 \text{ mA}$$

Problem 5-17

Imagine that we want the voltage-divider network of Fig. 5-8 to draw 600 mA to ensure the proper operation of the device that we connect across R_1. Suppose that the battery provides $E = 9.00$ V. We want +2.50 V to appear at the terminal point between the resistors. What values for R_1 and R_2 should we use?

Solution

Let's calculate the total resistance first, using Ohm's Law. Converting 600 mA to amperes, we get $I = 0.600$ A. Then

$$R_1 + R_2 = E/I = 9.00/0.600 = 15.0 \text{ ohms}$$

We want our resistance values to exist in the ratio

$$R_1/R_2 = 2.50/9.00 = 0.280$$

We should therefore choose

$$R_1 = 0.280 \times 15.0 = 4.20 \text{ ohms}$$

and

$$R_2 = 15.0 - 4.2\ 0 = 10.8 \text{ ohms}$$

Quiz

Refer to the text in this chapter if necessary. A good score is at least 18 correct answers. The answers are in the back of the book.

1. Consider a series-connected string of 10 light bulbs, all of which work properly, and all of which get their power from a single battery. Suddenly one of the bulbs burns out, leaving an open circuit in its place. What will happen?
   (a) All the other bulbs will go out.
   (b) The total current drawn from the battery will go up slightly.
   (c) The total current drawn from the battery will go down slightly.
   (d) The total current drawn from the battery will not change.

2. You connect four resistors in series with a 12.0-V battery: $R_1 = 47$ ohms, $R_2 = 22$ ohms, $R_3 = 33$ ohms, and $R_4 = 82$ ohms, as shown in Fig. 5-9. How much current flows through R_3?
   (a) 0.72 A
   (b) 0.36 A
   (c) 0.065 A
   (d) 0.015 A

3. In the circuit of Fig. 5-9, how much voltage appears across the series combination of resistances R_2 and R_3?
   (a) 7.5 V
   (b) 3.6 V
   (c) 8.8 V
   (d) 12 V

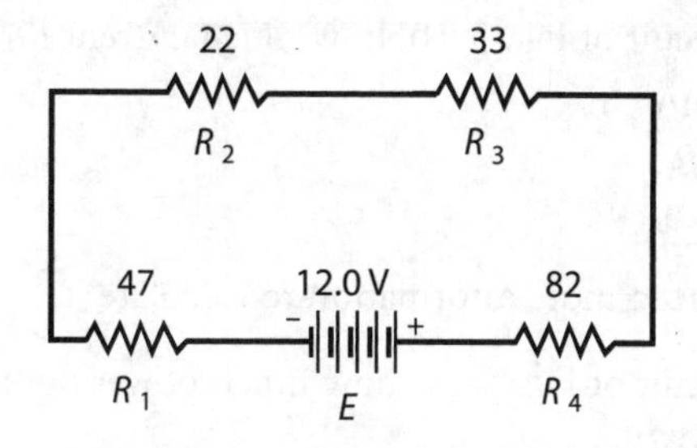

5-9 Illustration for Quiz Questions 2 through 5. Resistances are in ohms.

4. In the circuit of Fig. 5-9, how much power does the series combination of resistances R_2 and R_3 dissipate?
   (a) 3.6 W
   (b) 1.8 W
   (c) 0.46 W
   (d) 0.23 W

5. In the circuit of Fig. 5-9, how much power does R_3 dissipate?
   (a) 0.14 W
   (b) 0.28 W
   (c) 1.1 W
   (d) 2.2 W

6. Fill in the blanks in the following sentence to make it true: "In a parallel DC circuit containing a battery and two or more resistors, the ________ any resistor is the same as the ________ any other resistor."
   (a) potential difference across
   (b) current flowing through
   (c) power dissipated by
   (d) conductance of

7. Look back at Fig. 5-6. Suppose the battery supplies 12.0 V. We don't know any of the resistance values, but we do know that they're all the same. What's the voltage E_2?
   (a) 12.0 V
   (b) 6.00 V
   (c) 4.00 V
   (d) 3.00 V.

8. Three resistors are connected in parallel across a 4.5-V battery: $R_1 = 820$ ohms, $R_2 = 1.5$ k, and $R_3 = 2.2$ k, as shown in Fig. 5-10. How much voltage appears across R_2?
   (a) 3.0 mV
   (b) 1.5 V
   (c) 4.5 V
   (d) We need more information to calculate it.

9. In the circuit of Fig. 5-10, how much current flows through R_2?
   (a) 3.0 mA
   (b) 14 mA
   (c) 333 mA
   (d) We need more information to calculate it.

10. In the circuit of Fig. 5-10, how much power does R_2 draw from the battery?
   (a) 14 mW
   (b) 333 mW
   (c) 9.0 μW
   (d) We need more information to calculate it.

11. In the circuit of Fig. 5-10, how would we calculate the net conductance of the resistor network?
   (a) Add the conductances of the resistors.
   (b) Average the conductances of the resistors.
   (c) Take the reciprocal of the sum of the conductances of the resistors.
   (d) Take the reciprocal of the average of the conductances of the resistors.

12. In the circuit of Fig. 5-10, how much energy does the resistor network consume?
   (a) 5.5 joules
   (b) 9.2 joules
   (c) 47 joules
   (d) We need more information to figure it out.

13. In the circuit of Fig. 5-10, what will happen to the power dissipated by the network as a whole, if we change the value of R_1 from 820 to 8.2 ohms, but leave all the other values alone?
   (a) It will decrease a little.
   (b) It will decrease a lot.
   (c) It will increase a little.
   (d) It will increase a lot.

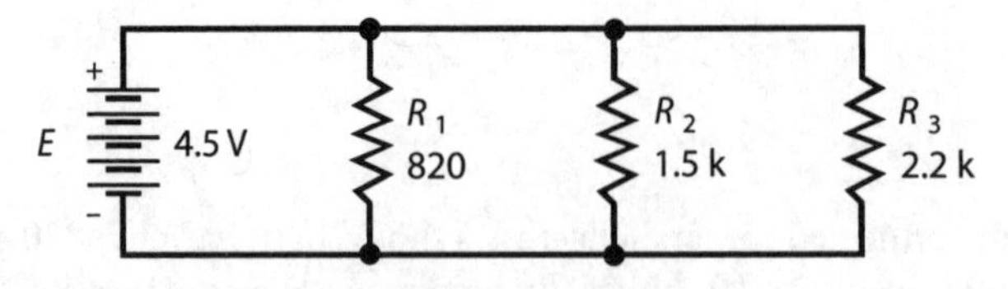

5-10 Illustration for Quiz Questions 8 through 13. Resistances are in ohms, where k indicates multiplication by 1000.

14. Refer back to Fig. 5-5B. Suppose that $I_3 + I_4 + I_5 = 250$ mA. If $I_1 = 100$ mA, what's the current I_2 through the resistor to the lower right of point Z?

    (a) 33 mA

    (b) 50 mA

    (c) 150 mA

    (d) 300 mA

15. Refer to Fig. 5-7. Suppose that the circuit has 10 resistors in total ($n = 10$), they all have values of 100 ohms, and the battery provides 6.3 V. If we double all the resistances to 200 ohms, what will happen to the voltage at point P_2?

    (a) It will double.

    (b) It will stay the same.

    (c) It will get cut in half.

    (d) We need more information to figure it out.

16. In the scenario of Fig. 5-7, suppose once more that $n = 10$; but instead of 100 ohms, the resistors all have values of 50 ohms. If we double the battery voltage to 12.6 V, in addition to changing the resistances, what will happen to the voltage at point P_2?

    (a) It will double.

    (b) It will stay the same.

    (c) It will get cut in half.

    (d) We need more information to figure it out.

17. Imagine four 100-ohm resistors in series, connected to a battery that supplies a voltage such that the entire network dissipates 4.00 W. How much power does each resistor consume?

    (a) 125 mW

    (b) 250 mW

    (c) 500 mW

    (d) 1.00 W

18. Imagine four 100-ohm resistors in parallel, connected to a battery that supplies a voltage such that the entire network dissipates 4.00 W. How much power does each resistor consume?

    (a) 125 mW

    (b) 250 mW

    (c) 500 mW

    (d) 1.00 W

19. Imagine four 100-ohm resistors in a 2 × 2 series-parallel matrix, connected to a battery that supplies a voltage such that the entire network dissipates 4.00 W. How much power does each resistor consume?

    (a) 125 mW
    (b) 250 mW
    (c) 500 mW
    (d) 1.00 W

20. When you design and build a voltage divider network, you should make the resistors' ohmic values as small as possible without imposing too much current demand on the power supply in order to

    (a) minimize the effect of external components on the network's behavior.
    (b) maximize the voltages at the various points in the network.
    (c) minimize the voltages at the various points in the network.
    (d) prevent overstressing external components connected to the network.

6
CHAPTER

Resistors

ALL ELECTRICAL COMPONENTS, DEVICES, AND SYSTEMS EXHIBIT RESISTANCE. IN PRACTICE, PERFECT conductors don't exist. You've seen some examples of circuits containing components designed to reduce or limit the current. We call these components *resistors.*

Purpose of the Resistor

Resistors play diverse roles in electrical and electronic equipment, despite the fact that their only direct action constitutes interference with the flow of current. Common applications include the following:

- Voltage division
- Biasing
- Current limiting
- Power dissipation
- Bleeding off charge
- Impedance matching

Voltage Division

You've learned how to build voltage dividers using resistors. The resistors dissipate some power in doing this job, but the resulting potential differences ensure that an external circuit or system operates properly. For example, a well-engineered voltage divider allows an amplifier to function efficiently, reliably, and with a minimum of distortion.

Biasing

In a *bipolar transistor*, a *field-effect transistor*, or an *electron tube*, the term *bias* means that we deliberately apply a certain DC voltage to one electrode relative to another, or relative to electrical ground. Networks of resistors can accomplish this function.

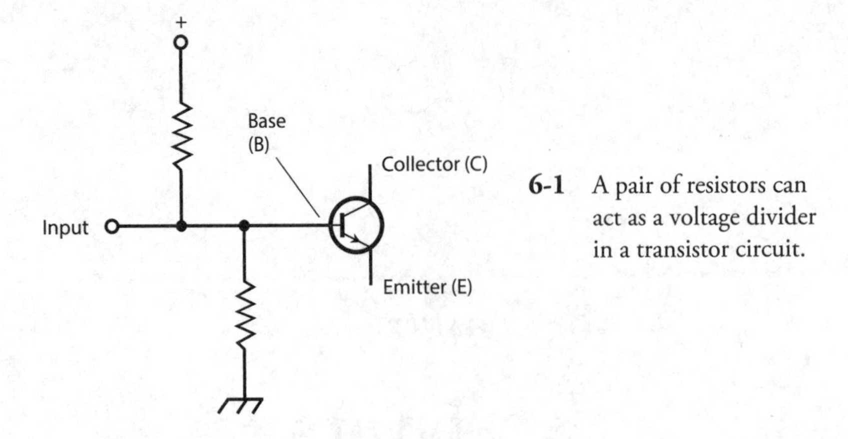

6-1 A pair of resistors can act as a voltage divider in a transistor circuit.

A radio transmitting amplifier works with a different bias than an oscillator or a low-level receiving amplifier. Sometimes we must build low-resistance voltage dividers to bias a tube or transistor; in some cases a single resistor will do.

Figure 6-1 shows a bipolar transistor that obtains its bias from a pair of resistors in a voltage-divider configuration. We'll learn about the transistor electrodes, called the *emitter* (E), the *base* (B), and the *collector* (C), later in this course.

Current Limiting

A sensitive amplifier designed for radio reception offers a good example of an application in which a *current-limiting resistor* in series with the power-supply or battery output keeps the transistor from dissipating too much power as heat. Without such a resistor to limit or control the current, the transistor would carry a lot of DC that wouldn't contribute to the signal amplification process, and might actually degrade it.

Figure 6-2 shows a current-limiting resistor connected between the emitter of a bipolar transistor and electrical ground, which also constitutes the negative power-supply connection (not shown). We can supply the signal input across that resistor (between E and ground) or at the base of the transistor (B). We would normally take the signal output from the collector (C).

Power Dissipation

In some applications, we want a resistor to dissipate power as heat. Such a resistor might constitute a "dummy" component, so that a circuit "sees" the resistor mimic the behavior of something more complicated.

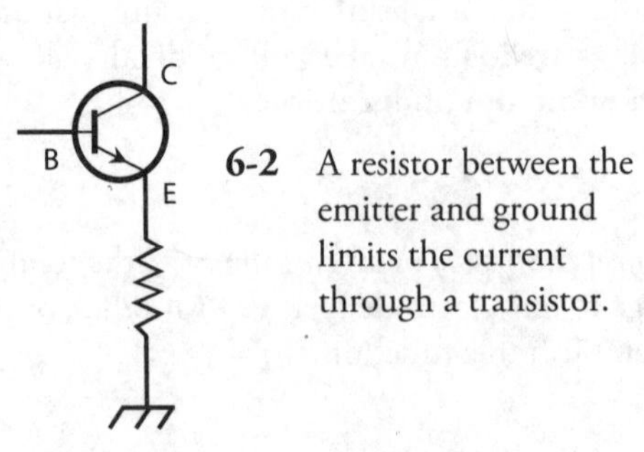

6-2 A resistor between the emitter and ground limits the current through a transistor.

When testing a radio transmitter, for example, we can install a massive resistor in place of the antenna (Fig. 6-3). This engineering trick allows us to test the transmitter for long periods at high power levels without interfering with on-the-air communications. The transmitter output heats the resistor without radiating any signal. However, the transmitter "sees" the resistor as if it were a real antenna—and a perfect one, too, if the resistor has the correct ohmic value.

We might take advantage of a resistor's power-dissipating ability at the input of a power amplifier, such as the sort used in hi-fi audio equipment. Sometimes the circuit *driving* the amplifier (supplying its input signal) produces too much power. A resistor, or network of resistors, can dissipate the excess power so that the amplifier doesn't receive too much input signal. In any type of amplifier, *overdrive* (an excessively strong input signal) can cause distortion, inefficiency, and other problems.

Bleeding Off Charge

A high-voltage, DC power supply employs capacitors (sometimes along with other components) to smooth out the current pulsations, known as *ripple*. These *filter capacitors* acquire an electric charge and store it for a while. In some power supplies, filter capacitors can hold the full output voltage of the supply, say something like 750 V, for a long time even after we power the whole system down. Anyone who attempts to repair or test such a power supply can receive a deadly shock from this voltage.

If we connect a *bleeder resistor* in parallel with each individual filter capacitor in a power supply, the resistors will drain the capacitors' stored charge, sparing personnel who service or test the

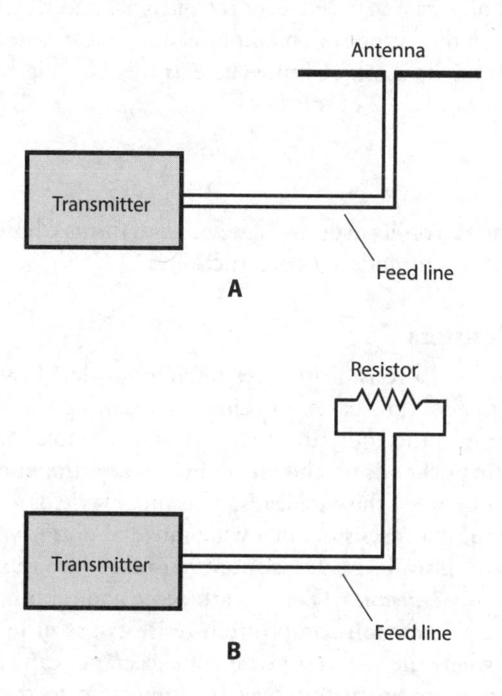

6-3 At A, a radio transmitter connected to a real antenna. At B, the same transmitter connected to a resistive "dummy" antenna.

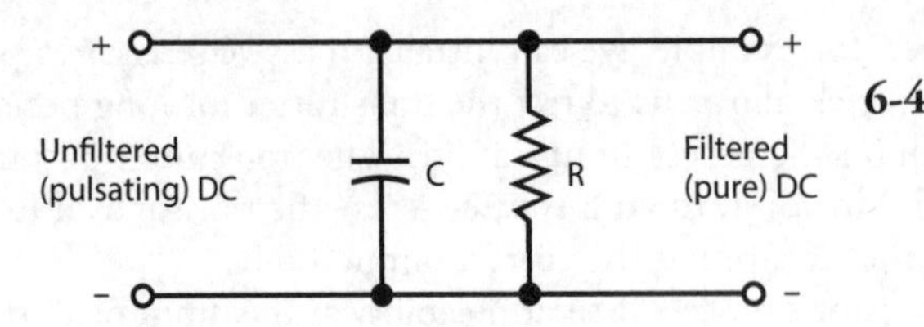

6-4 A bleeder resistor (R) connected across the filter capacitor (C) in a power supply.

power supply the risk of electrocution. In Fig. 6-4, the bleeder resistor R should have a value high enough so that it doesn't interfere with the operation of the power supply, but low enough so that it will discharge the capacitor C in a short time after power-down.

Even if a power supply has bleeder resistors installed, the wise engineer or technician, wearing heavy, insulated gloves, will short out all filter capacitors, using a screwdriver or other metal tool with an insulated handle, after power-down and before working on the circuit. Even if the supply has bleeder resistors, they can take awhile to get rid of the residual charge. Besides that, bleeder resistors sometimes fail!

Impedance Matching

We encounter a more sophisticated application for resistors in the *coupling* between two amplifiers, or in the input and output circuits of amplifiers. In order to produce the greatest possible amplification, the *impedances* between the output of a given amplifier and the input of the next must precisely agree. The same holds true between a source of signal and the input of an amplifier. The principle also applies between the output of an amplifier and a *load,* whether that load is a speaker, a headset, or whatever. We might think of impedance as the AC "big brother" of DC resistance. You'll learn about impedance in Part 2 of this book.

Fixed Resistors

In your engineering adventures, you'll encounter *fixed resistors* (units whose resistance never changes) in several different geometries and modes of construction.

Carbon-Composition Resistors

The cheapest method of making a resistor involves mixing powdered carbon (a fair electrical conductor) with a nonconductive solid or paste, pressing the resulting clay-like "goo" into a cylindrical shape, inserting wire leads in the ends, and then letting the whole mass harden (Fig. 6-5). The resistance of the final product depends on the ratio of carbon to the nonconducting material, and also on the physical distance between the wire leads. This process yields a *carbon-composition resistor.*

You'll find carbon-composition resistors in a wide range of ohmic values. This kind of resistor is *nonreactive,* meaning that it introduces almost pure resistance into the circuit, and essentially no *inductive reactance* or *capacitive reactance.* (You'll learn more about both types of reactance later in this book.) This property makes carbon-composition resistors useful in the construction of radio receivers and transmitters, where the slightest extraneous reactance can cause trouble.

Carbon-composition resistors dissipate power in proportion to their physical size and mass. Most of the carbon-composition resistors you see in electronics stores can handle ¼ W or ½ W. You can also find ⅛ W units for use in miniaturized, low-power circuitry, and 1 W or 2 W units

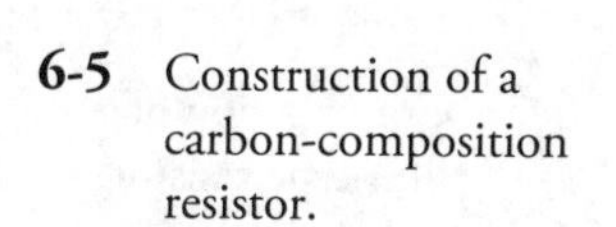
6-5 Construction of a carbon-composition resistor.

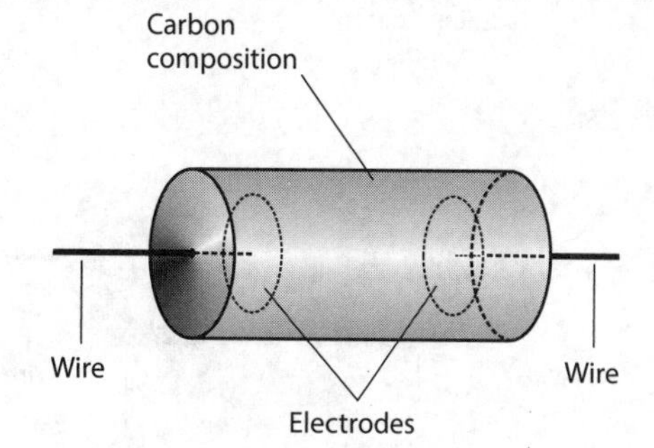

for circuits that require electrical ruggedness. Occasionally, you'll see a carbon-composition resistor with a power rating, such as 50 or 60 W, but not often.

Wirewound Resistors

We can also obtain resistance with a length of wire made from poorly conducting material. The wire can take the form of a coil wound around a cylindrical form, as shown in Fig. 6-6. The resistance depends on how well the wire conducts, on its diameter or *gauge*, and on its total stretched-out length. When we construct a component in this fashion, we have a *wirewound resistor*.

Wirewound resistors usually have low to moderate values of resistance. You'll find wirewound resistors whose ohmic values fall within a very narrow range, sometimes a fraction of 1 percent either way of the quoted value. We say that such components have *close tolerance* or *tight tolerance*. Some wirewound resistors can handle large amounts of power. That's their best asset. On the downside, wirewound resistors always exhibit some inductive reactance because of their coiled geometry, making them a poor choice for use in situations where high-frequency AC or radio-frequency (RF) current flows.

Film-Type Resistors

We can apply carbon paste, resistive wire, or some mixture of ceramic and metal to a cylindrical form as a film, or thin layer, to obtain a specific resistance. When we do this, we get a *carbon-film*

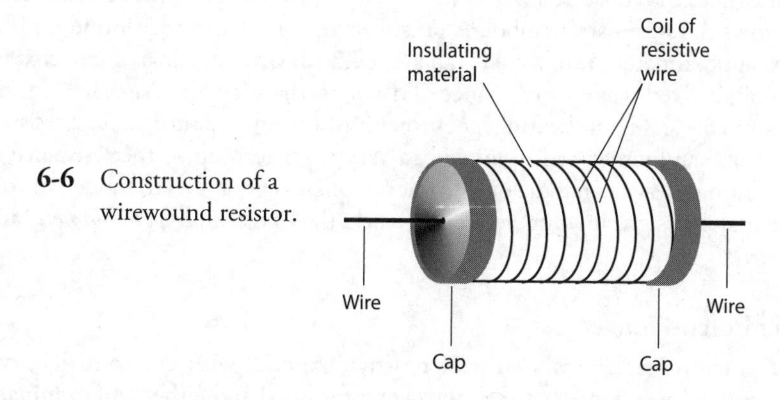

6-6 Construction of a wirewound resistor.

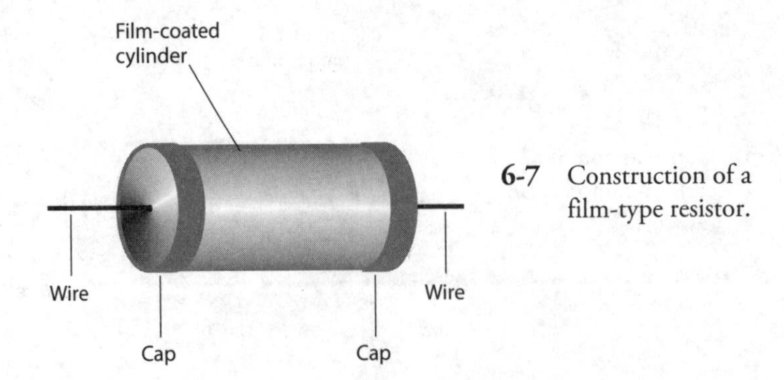

6-7 Construction of a film-type resistor.

resistor or *metal-film resistor*. Superficially, the component looks like a carbon-composition resistor, but the construction differs (Fig. 6-7).

The cylindrical form consists of an insulating substance, such as porcelain, glass, or thermoplastic. The film can be deposited on this form by various methods, and the value tailored as desired. Metal-film resistors can be manufactured to extremely close tolerances. Film-type resistors usually have low to medium-high resistance.

Film-type resistors, like carbon-composition resistors, have little or no inductive reactance—a big asset in high-frequency AC applications. However, film-type resistors generally can't handle as much power as carbon-composition or wirewound types of comparable physical size.

Integrated-Circuit (IC) Resistors

A semiconductor wafer known as an *integrated circuit* (IC), also called a *chip*, allows for the fabrication of resistors on its surface. The thickness of the resistive layer, and the types and concentrations of impurities added, determine the resistance of the component. Because of its microscopic size, a typical IC resistor can handle only a tiny amount of power. However, this limitation rarely poses a problem because the entire circuit on the chip functions at *nanopower levels* (on the order of nanowatts, units of 10^{-9} W) or *micropower levels* (on the order of microwatts, units of 10^{-6} W).

The Potentiometer

Figure 6-8A illustrates the construction geometry of a *potentiometer*, which acts as a variable resistor. Figure 6-8B shows the schematic symbol. A resistive strip, similar to that found on film-type fixed resistors, forms approximately ¾ of a circle (an arc of 270°), with terminals connected to either end. This strip exhibits a fixed value of resistance. To obtain the variable resistance, a sliding contact, attached to a rotatable shaft and bearing, goes to a third (middle) terminal. The resistance between the middle terminal and either end terminal can vary from zero up to the resistance of the whole strip. Most potentiometers can handle only low levels of current, at low to moderate voltages. You'll encounter two major designs in your electronics work: the *linear-taper potentiometer* and the *audio-taper potentiometer*.

Linear-Taper Potentiometer

A linear-taper potentiometer uses a strip of resistive material with constant density all the way around. As a result, the resistance between the center terminal and either end terminal changes at a

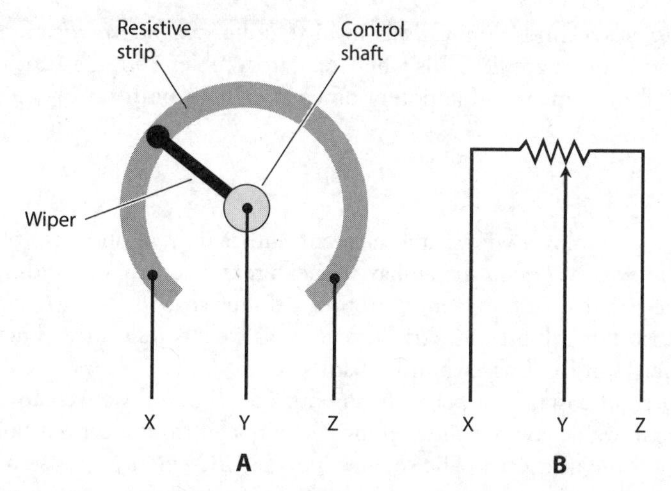

6-8 Simplified functional drawing of a rotary potentiometer (A), and the schematic symbol (B).

steady rate as the control shaft rotates. Engineers usually prefer linear-taper potentiometers in electronic test instruments. Linear-taper potentiometers also exist in some consumer electronic devices.

Suppose that a linear-taper potentiometer has a value of 0 to 270 ohms. In most units, the shaft rotates through a total angular displacement of about 270°. The resistance between the center and one end terminal increases along with the number of angular degrees that the shaft turns away from that end terminal. The resistance between the center and the other end terminal equals 270 minus the number of degrees that the control shaft subtends with respect to that terminal. The resistance between the middle terminal and either end terminal, therefore, constitutes a *linear function* of the angular shaft position.

Audio-Taper Potentiometer

In some applications, linear taper potentiometers don't work well. The volume control of a radio receiver or hi-fi audio amplifier provides a good example. Humans perceive sound intensity according to the *logarithm* of the actual sound power, not in direct proportion to the actual power. If you use a linear-taper potentiometer to control the volume (or *gain*) of a radio receiver or audio system, the sound volume (as you hear it) will change slowly in some parts of the control range, and rapidly in other parts of the control range. The device will work, but not in a user-friendly fashion.

An audio-taper potentiometer, if properly selected and correctly installed, can compensate for the way in which people perceive sound levels. The resistance between the center and either end terminal varies according to a *nonlinear function* of the angular shaft position. Some engineers call this type of device a *logarithmic-taper potentiometer* or *log-taper potentiometer* because the function of resistance versus angular displacement follows a logarithmic curve. As you turn the shaft, the sound intensity seems to increase in a linear manner, even though the actual power variation is logarithmic.

Slide Potentiometer

A potentiometer can employ a straight strip of resistive material rather than a circular strip so that the control moves up and down, or from side to side, in a straight line. This type of variable resistor, called a *slide potentiometer*, finds applications in hi-fi audio *graphic equalizers*, as gain controls in some amplifiers, and in other applications where operators prefer a straight-line control movement to a rotating control movement. Slide potentiometers exist in both linear-taper and audio-taper configurations.

Rheostat

A variable resistor can employ a wirewound element, rather than a solid strip of resistive material. We call this type of device a *rheostat*. It can have either a rotary control or a sliding control, depending on whether the resistive wire is wound around a donut-shaped form (*toroid*) or a cylindrical form (*solenoid*). Rheostats exhibit inductive reactance as well as resistance. They share the advantages and disadvantages of fixed wirewound resistors.

We can't adjust a rheostat in a perfectly smooth "continuum" as we can do with a potentiometer because the movable contact slides along the wire coil from a certain point on one turn to the adjacent point on the next turn. The smallest possible increment of resistance therefore equals the amount of resistance in one turn of the coil.

Rheostats find applications in heavy-duty systems, such as variable-voltage power supplies, designed for use with electron-tube amplifiers. This kind of supply contains a massive AC transformer that steps up the voltage from the 117 V utility mains. A rectifier circuit converts the AC to DC. The rheostat is usually placed between the utility outlet and the transformer, as shown in Fig. 6-9, providing adjustable DC voltage at the power-supply output.

The Decibel

As stated in the preceding paragraphs, perceived levels of sound change according to the logarithm of the actual sound power level. Engineers and technicians use *decibels* (symbolized dB) to express relative perceived sound-power levels. *Sound-power decibels* get along naturally with the use of audio-taper potentiometers for volume control.

Decibels in Terms of Power

One decibel (1 dB) represents the smallest *sudden* increase or decrease in sound power that you can detect, *if you anticipate an imminent change*. Positive decibel values represent power increases. Negative decibel values represent power decreases. If you *do not anticipate an imminent change* in the sound power and then it increases or decreases *suddenly*, you won't notice the fluctuation unless the difference amounts to at least +3 dB or −3 dB.

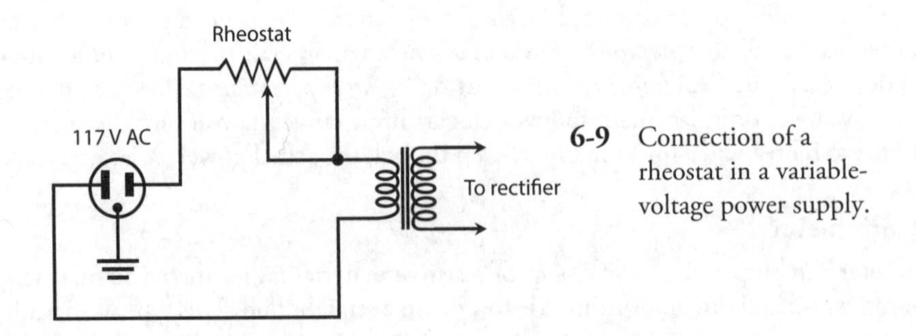

6-9 Connection of a rheostat in a variable-voltage power supply.

Changes in intensity, when expressed in decibels, are sometimes called *gain* and *loss*. Positive decibel changes represent gain, and negative decibel changes represent loss. You can omit the sign (plus or minus) when speaking of changes in terms of decibel gain or decibel loss because the words "gain" and "loss" imply the sign. If you say that a certain system causes 5 dB of sound-power loss, for example, you indicate that the circuit produces a sound-power change of −5 dB.

We calculate sound-power decibel values according to the logarithm of the ratio of change. Imagine that an acoustic disturbance produces P watts of power when it strikes your eardrums, and then the power increases or decreases to Q watts. To calculate the extent of the change in decibels, we use the formula

$$\text{dB} = 10 \log (Q/P)$$

where "log" stands for the base-10 logarithm.

As an example, suppose that a speaker emits 1 W of sound power, and then we turn up the gain so that the speaker emits 2 W of sound power. In this case, we have $P = 1$ and $Q = 2$, so

$$\text{dB} = 10 \log (2/1) = 10 \log 2 = 10 \times 0.3 = +3 \text{ dB}$$

If we turn the system gain back down so that the speaker once again emits only 1 W of sound power, then we get $P/Q = 1/2 = 0.5$, and we can calculate

$$\text{dB} = 10 \log (1/2) = 10 \times (-0.3) = -3 \text{ dB}$$

A gain or loss of 10 dB (a change of +10 dB or −10 dB, often shortened to ±10 dB) represents a 10-fold increase or decrease in sound power. A change of ±20 dB represents a 100-fold increase or decrease in power. You'll occasionally encounter sounds that vary in intensity over ranges of ±90 dB, which represents a thousand-million-fold increase or decrease in power—from a microwatt to a kilowatt, for example!

Power in Terms of Decibels

We can work the above-defined formula inside-out to determine the final sound power, given the initial power and the decibel change. To do this, we use the *inverse* of the base-10 logarithmic function, symbolized as $\log^{-1}$, or antilog. Any good scientific calculator can execute both the log and antilog functions.

> **Tip**
>
> If your calculator has a 10^x key, you can use it to find the antilog of any number x. Enter the number for which you want to find the antilog, and then hit 10^x.

Suppose that we have a sound of initial power P, and then it changes by x dB. We can find the final sound power Q using the formula

$$Q = P \text{ antilog } (x/10)$$

Consider an acoustic disturbance whose initial power equals 10 W, and then we perceive or measure a sudden level change of $x = -3$ dB. We can calculate the final sound power Q as

$$Q = 10 \text{ antilog } (-3/10) = 10 \text{ antilog } (-0.3) = 10 \times 0.5 = 5 \text{ W}$$

"Absolute" Decibels

We can specify "absolute" sound-power levels in decibels relative to the *threshold of hearing*, defined as the faintest possible acoustic disturbance that a person with normal hearing can detect in a quiet room. We assign a value of 0 dB to sounds at this threshold. We can then quantify the intensities of other sounds as figures, such as 30 dB or 75 dB.

If a certain noise has a loudness of 30 dB, that means it's 30 dB above the threshold of hearing, or 1000 times as loud (in terms of power) as the quietest detectable noise. A noise at 60 dB is 1,000,000 (or 10^6) times as powerful as a sound at the threshold of hearing. *Sound-level meters* can accurately determine and display the decibel levels of various noises and acoustic environments.

A typical conversation between two people a couple of meters apart occurs at a level of about 70 dB. This level represents 10,000,000 (or 10^7) times the threshold of hearing in terms of actual sound power. The roar of the crowd at a country-music concert might be 90 dB, or 1,000,000,000 (10^9) times the threshold of hearing. The 100-dB shriek of a tornado-warning siren a few meters away from your ears represents 10,000,000,000 (10^{10}) times the power contained in a whisper at the threshold of hearing.

Resistor Specifications

When we select a resistor for a particular application, we must obtain a unit that has the correct properties, or *specifications*. Here are some of the most important specifications to watch for.

Ohmic Value

In theory, a resistor can have any ohmic value from the lowest possible (such as a shaft of solid silver) to the highest (dry air). In practice, we'll rarely find resistors with values less than about 0.1 ohm or more than about 100 M.

Resistors are manufactured with ohmic values in power-of-10 multiples of numbers from the set

$$\{1.0, 1.2, 1.5, 1.8, 2.2, 2.7, 3.3, 3.9, 4.7, 5.6, 6.8, 8.2\}$$

We'll routinely see resistances such as 47 ohms, 180 ohms, 6.8 k, or 18 M, but we'll hardly ever find resistors with values such as 384 ohms, 4.54 k, or 7.297 M.

Additional basic resistances exist, intended especially for *tight-tolerance* (or *precision*) resistors: power-of-10 multiples of numbers from the set

$$\{1.1, 1.3, 1.6, 2.0, 2.4, 3.0, 3.6, 4.3, 5.1, 6.2, 7.5, 9.1\}$$

Tolerance

The first set of numbers above represents standard resistance values available in tolerances of plus or minus 10 percent ($\pm 10\%$). This means that the resistance might be as much as 10% more or less than the indicated amount. In the case of a 470-ohm resistor, for example, the value can be larger or smaller than the rated value by as much as 47 ohms, and still adhere to the rated tolerance. That's a range of 423 to 517 ohms.

Engineers calculate resistor tolerance figures on the basis of the *rated* resistance, not the measured resistance. For example, we might test a "470-ohm" resistor and find it to have an actual resistance of 427 ohms; this discrepancy would still put the component within $\pm 10\%$ of the specified value. But if we test it and find it to have a resistance of 420 ohms, its actual value falls outside the rated range, so it constitutes a "reject."

The second set of numbers listed above, along with the first set, represents all standard resistance values available in tolerances of plus or minus 5 percent (±5%). A 470-ohm, 5% resistor will have an actual value of 470 ohms plus or minus 24 ohms, or a range of 446 to 494 ohms.

For applications requiring exceptional precision, resistors exist that boast tolerances tighter than ±5%. We might need a resistor of such quality in a circuit or system where a small error can make a big difference. In most audio and RF oscillators and amplifiers, we'll usually do okay with resistors having ±10% or ±5% tolerances. In some applications, we can even get away with a ±20% tolerance.

Power Rating

A manufactured resistor always bears a specification that tells us how much power it can safely dissipate. The dissipation rating indicates *continuous duty*, which means that the component can dissipate a certain amount of power constantly and indefinitely.

We can calculate how much current a given resistor can handle using the formula for power P (in watts) in terms of current I (in amperes) and resistance R (in ohms), as follows:

$$P = I^2R$$

With algebra, we can change this formula to express the maximum allowable current in terms of the power dissipation rating and the resistance, as follows:

$$I = (P/R)^{1/2}$$

where the ½ power represents the positive square root.

We can effectively multiply the power rating for a given resistor by connecting identical units in series-parallel matrices of 2 × 2, 3 × 3, 4 × 4, or larger. If we need a 47-ohm, 45-W resistor but we have only a lot of 47-ohm, 1-W resistors available, we can connect seven sets of seven resistors in parallel (a 7 × 7 series-parallel matrix) and get a 47-ohm resistive component that can handle up to 7 × 7 W, or 49 W.

Resistor power dissipation ratings, like the ohmic values, are specified with a margin for error. A good engineer never tries to "push the rating" and use, say, a ¼-W resistor in a situation where it will need to draw 0.27 W. In fact, good engineers usually include their own safety margin, in addition to that offered by the vendor. Allowing a 10% safety margin, for example, we should never demand that a ¼-W resistor handle more than about 0.225 W, or expect a 1-W resistor to dissipate more than roughly 0.9 W.

Temperature Compensation

All resistors change value when the temperature rises or falls dramatically. Because resistors dissipate power by design, they get hot in operation. Sometimes the current that flows through a resistor does not rise high enough to appreciably heat the component. But in some cases it does, and the heat can cause the resistance to change. If this effect becomes great enough, a sensitive circuit will behave differently than it did when the resistor was still cool. In the worst-case scenario, an entire device or system can shut down because of a single "temperamental" resistor.

Resistor manufacturers do various things to prevent problems caused by resistors changing value when they get hot. In one scheme, resistors are specially manufactured so that they don't appreciably change value when they heat up. We call these components *temperature compensated.* As you might expect, a temperature-compensated resistor can cost several times as much as an ordinary resistor.

Rather than buy a single temperature-compensated resistor, we can employ a single resistor or a series-parallel matrix of resistors with a power rating several times higher than we ever expect the component to dissipate. This technique, called *over-engineering*, keeps the resistor or matrix from reaching temperatures high enough to significantly change the resistance. Alternatively, we might take several resistors, say five of them, each with five times the intended resistance, and connect them all in parallel. Or we can take several resistors, say four of them, each with about ¼ of the intended resistance, and connect them in series.

Whatever trick we employ to increase the power-handling capability of a component, we should never combine resistors with different ohmic values or power ratings into a single matrix. If we try that, then one of them might end up taking most of the load while the others "loaf," and the combination will perform no better than the single hot resistor we started with. Whenever we want to build a "resistor gang" to handle high current or keep cool under load, we should always procure a set of *identical* components.

Are You Astute?

You might ask, "If we want to build our own temperature-compensated resistor, can we use two resistors with half (or twice) the value we need, but with *opposite* resistance-versus-temperature characteristics, and connect them in series or parallel?" That's an excellent question. If we can find two such resistors, the component whose resistance decreases with increasing temperature (that is, the one that has a *negative temperature coefficient*) will partially or totally "undo" the thermal problem caused by the component whose resistance goes up with increasing temperature (the one that has a *positive temperature coefficient*). This scheme can sometimes work. Unfortunately, we'd likely spend more time trying to find two "ideally mismatched" resistors than we'd spend by resorting to "brute force" and building a series-parallel matrix.

The Color Code for Resistors

Some resistors have *color bands* that indicate their values and tolerances. You'll see three, four, or five bands around carbon-composition resistors and film resistors. Other resistors have enough physical bulk to allow for printed numbers that tell us the values and tolerances straightaway.

On resistors with *axial leads* (wires that come straight out of both ends), the first, second, third, fourth, and fifth bands are arranged as shown in Fig. 6-10A. On resistors with *radial leads* (wires that come off the ends at right angles to the axis of the component body), the colored regions are arranged as shown in Fig. 6-10B. The first two regions represent single digits 0 through 9, and the third region represents a multiplier of 10 to some power. (For the moment, don't worry about the fourth and fifth regions.) Table 6-1 indicates the numerals corresponding to various colors.

Suppose that you find a resistor with three bands: yellow, violet, and red, in that order. You can read as follows, from left to right, referring to Table 6-1:

- Yellow = 4
- Violet = 7
- Red = ×100

You conclude that the rated resistance equals 4700 ohms, or 4.7 k. As another example, suppose you find a resistor with bands of blue, gray, and orange. You refer to Table 6-1 and determine that

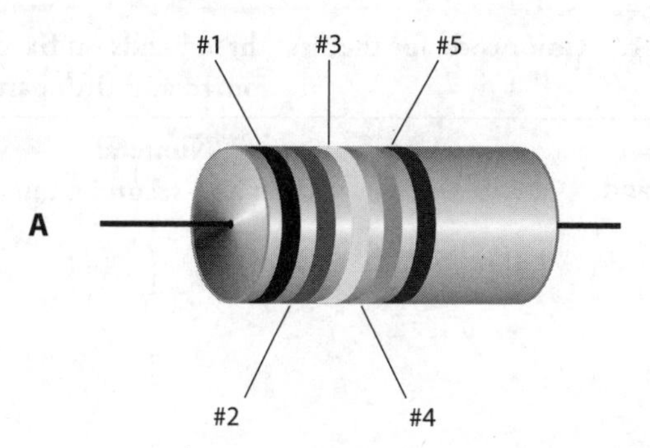

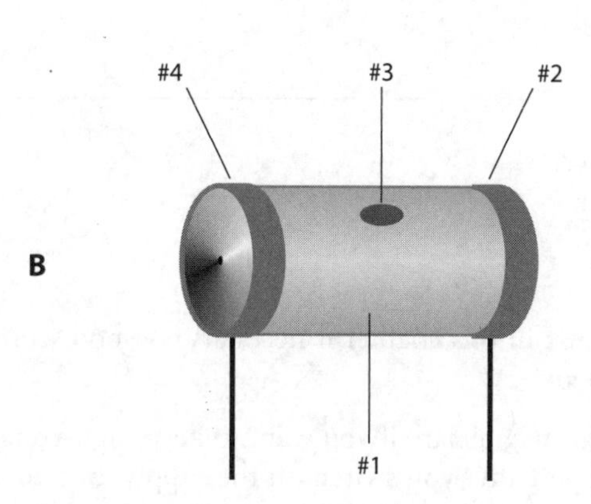

6-10 At A, locations of color-code bands on a resistor with axial leads. At B, locations of color code designators on a resistor with radial leads.

- Blue = 6
- Gray = 8
- Orange = ×1000

This sequence tells you that the resistor is rated at 68,000 ohms, or 68 k.

If a resistor has a fourth colored band on its surface (#4 as shown in Fig. 6-10 A or B), then that band tells you the tolerance. A silver band indicates ±10%. A gold band indicates ±5%. If no fourth band exists, then the tolerance is ±20%.

The fifth band, if any, indicates the maximum percentage by which you should expect the resistance to change after the first 1000 hours of use. A brown band indicates a maximum change of ±1% of the rated value. A red band indicates ±0.1%. An orange band indicates ±0.01%. A yellow band indicates ±0.001%. If the resistor lacks a fifth band, it tells you that the resistor might deviate by more than ±1% of the rated value after the first 1000 hours of use.

A savvy engineer or technician always tests a resistor with an ohmmeter before installing it in a circuit. If the component turns out defective or mislabeled, you can prevent potential future headaches by following this simple precaution. It takes only a few seconds to check a resistor's ohmic value. In contrast, once you've finished building a circuit and discover that it won't work because of some miscreant resistor, the troubleshooting process can take hours.

Table 6-1. Color code for the first three bands on fixed resistors. See text for discussion of the fourth and fifth bands.

Color of band	Numeral (first and second bands)	Multiplier (third band)
Black	0	1
Brown	1	10
Red	2	100
Orange	3	1000 (1 k)
Yellow	4	10^4 (10 k)
Green	5	10^5 (100 k)
Blue	6	10^6 (1 M)
Violet	7	10^7 (10 M)
Gray	8	10^8 (100 M)
White	9	10^9 (1000 M or 1 G)

Quiz

Refer to the text in this chapter if necessary. A good score is at least 18 correct. Answers are in the back of the book.

1. As a repair technician, if you want to keep high-voltage, power-supply filter capacitors from electrocuting you after you switch off the supply, but not let them interfere with the supply's performance when it's powered up, what should you do?
   (a) Wait 10 minutes after powering-down before you begin work.
   (b) Permanently short circuit all the filter capacitors.
   (c) Install inductors in series with all the capacitors.
   (d) None of the above

2. You have a package of resistors, all rated at 330 ohms ±10%. You test three of them with an ohmmeter, obtaining the readings in (a), (b), and (c) below. Which, if any, of these tested values lies outside the tolerance range, telling you that the component is a reject?
   (a) 299 ohms
   (b) 305 ohms
   (c) 362 ohms
   (d) They're all okay

3. Which of the following resistor types is a good choice for use in a circuit designed to operate at 14 MHz?
   (a) Carbon-composition
   (b) Carbon-film
   (c) Integrated-circuit
   (d) Any of the above

4. A musical note from your hi-fi system comes out at "20 dB." In terms of sound power, how does this volume level compare with the threshold of hearing?
   (a) It's twice as powerful.
   (b) It's 20 times as powerful.
   (c) It's 100 times as powerful.
   (d) None of the above! Decibels express frequency, not power.

5. The working part of a "dummy" antenna is
   (a) an inductor.
   (b) a resistor.
   (c) a capacitor.
   (d) a short circuit.

6. In the schematic diagram for a transistorized amplifier, you see a resistor between the base and ground, and another resistor between the base and the positive battery terminal. What purpose do these resistors serve if you choose their values correctly?
   (a) They maximize the current that flows through the transistor.
   (b) They bleed off any excess charge that might exist on the base.
   (c) They optimize the bias at the base.
   (d) They keep the transistor from shorting out.

7. A resistor has three colored bands, going from left to right in this order: green, red, brown. As rated by the manufacturer, its resistance is close to
   (a) 68 ohms.
   (b) 520 ohms.
   (c) 8.2 k.
   (d) 18 k.

8. The most common way to limit the current through a transistor is to connect a resistor between the
   (a) collector and emitter.
   (b) base and collector.
   (c) emitter and ground.
   (d) collector and ground.

9. In which of the following situations should we expect a 200-W wirewound resistor to work well?
   (a) In series with the emitter of a high-power RF amplifier transistor, for the purpose of current limiting
   (b) In series with the collector of a low-power RF amplifier transistor, for the purpose of voltage limiting
   (c) In series with the filter capacitor in a power supply, for the purpose of minimizing the output ripple
   (d) In any DC circuit needing a resistor that can dissipate far more power than a carbon-based resistor can do

10. A 470-ohm resistor carries 15 mA continuously. What resistor power rating is sufficient, but not needlessly high?

    (a) ¼ W
    (b) 1 W
    (c) 2 W
    (d) 5 W

11. How much voltage (rounded to two significant figures) appears across the resistor described in Question 10?

    (a) 0.15 V
    (b) 7.1 V
    (c) 10 V
    (d) 70 V

12. You find a resistor with a manufacturer's rated value of 470 ohms. You measure its resistance as 490 ohms. What's the percentage difference between the measured value and the manufacturer's rated value?

    (a) +4.08%
    (b) +4.26%
    (c) −4.08%
    (d) −4.26%

13. In an audio-taper potentiometer, the resistance varies in proportion to the

    (a) angular shaft displacement.
    (b) logarithm of the angular shaft displacement.
    (c) square of the angular shaft displacement.
    (d) square root of the angular shaft displacement.

14. Refer to Fig. 6-4. You connect this circuit in a DC power supply designed to provide 800 V DC. If you properly choose the resistor's ohmic value, it can

    (a) reduce the chance that a repair technician will get killed.
    (b) prevent capacitor C from shorting out or opening up.
    (c) maximize the efficiency of the power supply as a whole.
    (d) eliminate the output ripple whether the capacitor works or not.

15. A rheostat contains, among other things,

    (a) a coil of wire.
    (b) carbon paste.
    (c) carbon film.
    (d) Any of the above

16. A carbon-composition resistor has the following colored bands: red, red, red, silver. What's its resistance as rated by the manufacturer?

    (a) 22 ohms $\pm 10\%$
    (b) 220 ohms $\pm 10\%$
    (c) 2.2 k $\pm 10\%$
    (d) 22 k $\pm 10\%$

17. Which, if any, of the following characteristics describes an advantage of a potentiometer over a rheostat?

    (a) A potentiometer has reactance, but a rheostat lacks reactance and, therefore, won't work well in AC applications.
    (b) A potentiometer works well at DC, but a rheostat doesn't because it has too much reactance.
    (c) A potentiometer can work in high-voltage, high-current applications, but a rheostat can't.
    (d) None of the above

18. For the volume control in an audio amplifier, you would get the best results with a

    (a) log-taper potentiometer.
    (b) linear-taper potentiometer.
    (c) wirewound resistor.
    (d) rheostat.

19. If you want to build a graphic equalizer for your hi-fi system, you would probably want to use

    (a) rheostats.
    (b) voltage dividers.
    (c) slide potentiometers.
    (d) rotary potentiometers.

20. Which of the following resistor types has minimal reactance?

    (a) carbon-composition
    (b) rheostat
    (c) wirewound
    (d) capacitive

7
CHAPTER
Cells and Batteries

IN ELECTRICITY, WE CALL A UNIT SOURCE OF DC ENERGY A *CELL*. WHEN WE CONNECT TWO OR MORE cells in series, parallel, or series-parallel, we obtain a *battery*. Numerous types of cells and batteries exist, and inventors keep discovering more.

Electrochemical Energy

Early in the history of electricity science, physicists noticed that when metals came into contact with certain chemical solutions, a potential difference sometimes appeared between the pieces of metal. These experimenters had discovered the first *electrochemical cells*.

A piece of lead and a piece of lead dioxide immersed in an acid solution (Fig. 7-1) acquire a persistent potential difference. In the original experiments, scientists detected this voltage by connecting a galvanometer between the pieces of metal. A resistor in series with the galvanometer prevents excessive current from flowing, and keeps acid from "boiling out" of the cell. Nowadays, of course, we can use a laboratory voltmeter to measure the potential difference.

If we draw current from a cell, such as the one shown in Fig. 7-1, for a long time by connecting a resistor between its terminals, the current will gradually decrease, and the electrodes will become coated. Eventually, all the *chemical energy* in the acid will have turned into *electrical energy* and dissipated as *thermal energy* in the resistor and the cell's own chemical solution, escaping into the surrounding environment in the form of *kinetic energy*.

Primary and Secondary Cells and Batteries

Some electrical cells, once their chemical energy has been used up, must be thrown away. We call such a device a *primary cell*. Other kinds of cells, such as the lead-acid type, can get their chemical energy back again by means of *recharging*. Such a cell constitutes a *secondary cell*.

Primary cells include the ones you usually put in a flashlight, in a transistor radio, and in various other consumer devices. They use dry *electrolyte* (conductive chemical) pastes along with metal electrodes, and go by names, such as *dry cell*, *zinc-carbon cell*, or *alkaline cell*. When you encounter a shelf full of "batteries" in a department store, you'll see primary cells that go by names, such as

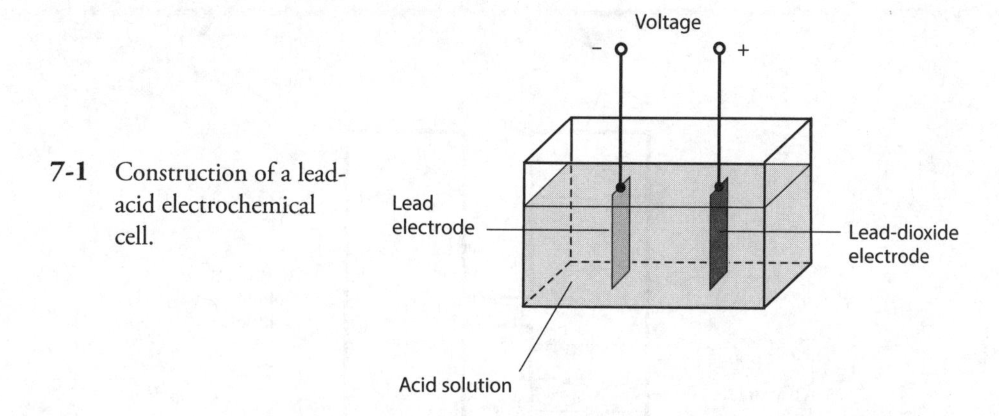

7-1 Construction of a lead-acid electrochemical cell.

AAA batteries, *D batteries*, *camera batteries*, and *watch batteries*. (These are actually cells, not true batteries.) You'll also see 9-V *transistor batteries* and large 6-V *lantern batteries*.

You can also find secondary cells in consumer stores. They cost several times as much as ordinary dry cells, and the requisite charging unit also costs a few dollars. But if you take care of rechargeable cells, you can use them hundreds of times, and they'll pay for themselves several times over.

The battery in your car or truck consists of several secondary cells connected in series. These cells recharge from the *alternator* (a form of generator) or from an external charging unit. A typical *automotive battery* has cells like the one in Fig. 7-1. You should never short-circuit the terminals of such a battery or connect a load to it that draws a large amount of current because the acid (sulfuric acid) can erupt out of the battery container. Serious skin and eye injuries can result. In fact, it's a bad idea to short-circuit any cell or battery because it can rupture and damage surrounding materials, wiring, and components. Some "shorted-out" cells and batteries can heat up enough to catch on fire.

Cells in Series and Parallel

When we want to make a battery from two or more electrochemical cells, we should always use cells having the same chemical composition and the same physical size and mass. In other words, all the cells in the set should be identical! Assuming we heed that principle, we can generalize as follows.

- When we connect cells in series, the *no-load output voltage* (when we don't make the cells deliver any current) multiplies by the number of cells, while the *maximum deliverable current* (when we make the cells produce as much current as they can) equals the maximum deliverable current from only one cell.
- When we connect cells in parallel, the no-load output voltage of the whole set equals the no-load output voltage of only one cell, while the maximum deliverable current from the set multiplies according to the number of cells.

The Weston Standard Cell

A *standard cell* produces a precise and predictable no-load output voltage for use in scientific laboratories. The *Weston standard cell* (or simply the *Weston cell*) generates 1.018 V DC at room

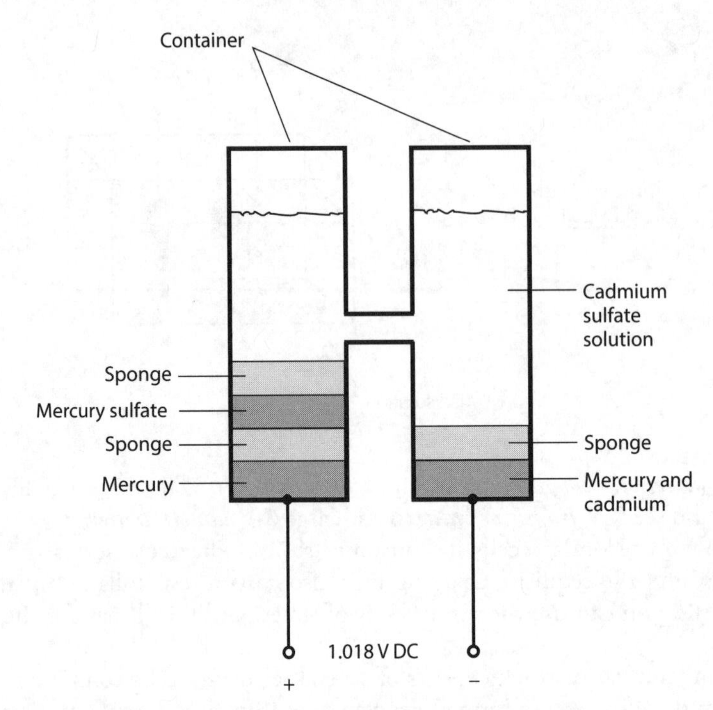

7-2 Construction of a Weston standard cell.

temperature. It employs a solution of cadmium sulfate, a positive electrode made from mercury sulfate, and a negative electrode made from mercury and cadmium. A two-chambered container holds the chemicals and electrodes (Fig. 7-2).

Most electrochemical cells intended for consumer use produce 1.2 V to 1.8 V DC. The exact output voltage from a new electrochemical cell when we test it under *no-load conditions* (zero current drain) depends on the chemicals used in its manufacture. Physical size or mass has nothing to do with a cell's no-load output voltage. Variables in the manufacturing process of a particular type of cell can affect its "new-off-the-shelf" output voltage slightly.

Storage Capacity

Engineers commonly work with two units of electrical energy: the watt-hour (Wh) and the kilowatt-hour (kWh), as we've learned. Any electrochemical cell or battery has a certain amount of electrical energy that we can "extract" before it "runs out of juice." We can quantify that energy in terms of watt-hours or kilowatt-hours. Some engineers express the capacity of a cell or battery of known voltage in units called *ampere-hours* (Ah).

As an example, a battery with a rating of 2 Ah can provide 2 A for 1 h, or 1 A for 2 h, or 100 mA for 20 h. Infinitely many possibilities exist, as long as the product of the current (in amperes) and the usage time (in hours) equals 2. Practical usage limitations are the *shelf life* at one extreme, and the *maximum deliverable current* at the other. We define shelf life as the length of time the battery will last if we never use it at all; this time period might be several years. We define the maximum deliverable current as the highest amount of current that the battery can provide at any moment without suffering a significant decrease in the output voltage because of its own *internal resistance*.

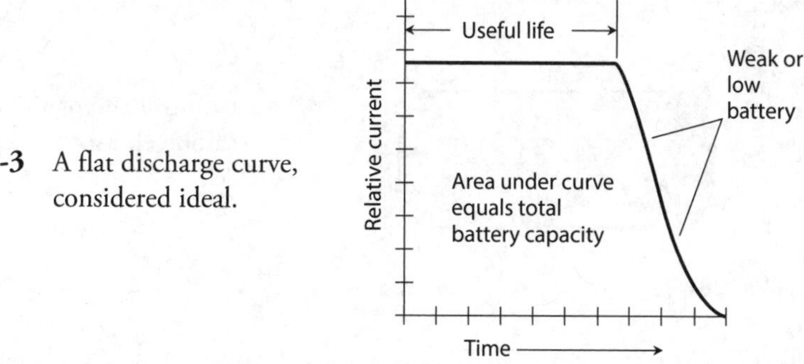

7-3 A flat discharge curve, considered ideal.

Small cells have storage capacity ratings of a few milliampere-hours (mAh) up to 100 or 200 mAh. Medium-sized cells can supply 500 mAh to 1 Ah. Large automotive batteries can provide upwards of 50 Ah. The energy capacity in watt-hours equals the ampere-hour capacity multiplied by the battery voltage. For a cell or battery having a particular chemical composition, the storage capacity varies directly in proportion to the physical volume of the device. A cell whose volume equals 20 cubic centimeters (cm^3), therefore, has twice the total energy storage capacity of a cell having the same chemical makeup, but that has a volume of only 10 cm^3.

An *ideal cell* or *ideal battery* (a theoretically perfect one) delivers a constant current for a while, and then the current drops fast (Fig. 7-3). Some types of cells and batteries approach this level of perfection, which we represent graphically as a *flat discharge curve*. Most cells and batteries are imperfect, and some are far from the ideal, delivering current that declines steadily from the start. When the deliverable current under constant load has tailed off to about half of its initial value, we say that the cell or battery has become "weak." At this time, we should replace it. If we allow such a cell or battery to run down until the current goes to zero, we call it "dead." The area under the curve in Fig. 7-3 represents the total capacity of the cell or battery in ampere-hours.

"Grocery Store" Cells and Batteries

The cells you see in retail stores provide approximately 1.5 V DC, and are available in sizes known as AAA (very small), AA (small), C (medium large), and D (large). You can also find batteries that deliver 6 or 9 V DC.

Zinc-Carbon Cells

Figure 7-4 is a "translucent" drawing of a *zinc-carbon cell*. The zinc forms the case, which serves as the negative electrode. A carbon rod constitutes the positive electrode. The electrolyte comprises a paste of manganese dioxide and carbon. Zinc-carbon cells don't cost very much. They work well at moderate temperatures, and in applications where the current drain is moderate to high. They don't perform well in extreme cold or extreme heat.

Alkaline Cells

The *alkaline cell* has granular zinc as the negative electrode, potassium hydroxide as the electrolyte, and an element called a *polarizer* as the positive electrode. The construction resembles that of the zinc-carbon cell. An alkaline cell can work at lower temperatures than a zinc-carbon cell can.

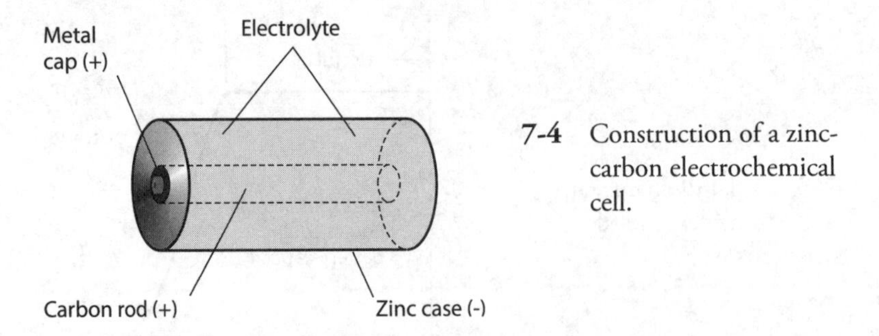

7-4 Construction of a zinc-carbon electrochemical cell.

The alkaline cell also lasts longer in most electronic devices. It's the cell of choice for use in transistor radios, calculators, and portable cassette players. The shelf life exceeds that of a zinc-carbon cell. As you might expect, it costs more than a zinc-carbon cell of comparable physical size.

Transistor Batteries

A *transistor battery* consists of six tiny zinc-carbon or alkaline cells connected in series and enclosed in a small box-shaped case. Each cell supplies 1.5 V, so the battery supplies 9 V. Even though these batteries have more voltage than individual cells, the energy capacity is less than that of a single size C or D cell. The electrical energy that we can get from a cell or battery varies in direct proportion to the amount of chemical energy stored in it—and that, in turn, is a direct function of the *volume* (physical size) of the cell or the *mass* (quantity of chemical matter) of the cell. Cells of size C or D have more volume and mass than a transistor battery does, and therefore, contain more stored energy for the same chemical composition. We can find transistor batteries in low-current electronic devices, such as remote-control garage-door openers, television (TV) and hi-fi remote-control units, and electronic calculators.

Lantern Batteries

A so-called *lantern battery* has much greater mass than a common dry cell or transistor battery, so it lasts much longer and can deliver more current. Lantern batteries are usually rated at 6 V. Two lantern batteries connected in series make a 12-V battery that can power a small Citizens Band (CB) or amateur ("ham") radio transceiver for a while. Lantern batteries work well in portable locations for medium-power needs.

Miniature Cells and Batteries

In recent years, cells and batteries have become available in many different sizes and shapes besides the old cylindrical cells, transistor batteries, and lantern batteries. Various interesting (and some strange-looking) cells and batteries operate wristwatches, small cameras, and other miniaturized electronic devices.

Silver-Oxide Cells and Batteries

A *silver-oxide cell* has a button-like shape, and can fit inside a small wristwatch. These types of cells come in various sizes, all with similar appearances. They supply 1.5 V, and offer excellent energy storage capacity considering their low mass. They also have a nearly flat discharge curve, like the one shown in the graph of Fig. 7-3. Zinc-carbon and alkaline cells and batteries, in contrast, have current output that declines more steadily with time, as shown in Fig. 7-5 (a so-called *declining discharge curve*). We can stack two or more silver-oxide cells to make a battery. Several of these miniature cells, one on top of the other, can provide 6, 9, or 12 V for a transistor radio or other light-duty electronic device.

Mercury Cells and Batteries

A *mercury cell,* also called a *mercuric-oxide cell,* has properties similar to those of silver-oxide cells. They're manufactured in the same general form. The main difference, often not of significance, is a somewhat lower voltage per cell: 1.35 V. If we stack up seven of these cells in series to make a battery, the resulting voltage will equal about 9.45 V, close to that of a standard 9-V transistor battery.

Mercury cells and batteries have fallen from favor in recent years because mercury acts as a toxin, even in trace amounts. The mercury concentration accumulates over time in animals and humans. When mercury cells and batteries run down, we must discard them, posing additional problems, because the mercury gradually leaks into the soil, and from there into our food and water supplies.

Lithium Cells and Batteries

Lithium cells gained popularity in the early 1980s. We can find several variations in the chemical makeup of these cells. They all contain lithium, a light, highly reactive metal. Lithium cells typically supply 1.5 V to 3.5 V, depending on the chemistry used in manufacture. These cells, like all other cells, can be stacked to make batteries.

Lithium batteries originally found application as memory backup power supplies for electronic microcomputers. Lithium cells and batteries have superior shelf life. They can last for years in very-low-current applications, such as memory backup or the powering of a digital *liquid-crystal-display* (LCD) watch or clock. These cells also provide high energy capacity per unit of volume or mass.

Lithium-Polymer (LiPo) Cells and Batteries

LiPo cells produce a voltage of 3.7 V rising to 4.35 V (fully charged) and are, therefore, often used as a single cell or a 7.4-V battery of two cells. Larger batteries are also found in laptops and other high-power devices.

The energy density of a LiPo cell is much higher than any of the other readily available battery technologies. For this reason, they have become the technology of choice for most applications in which a rechargeable battery is needed.

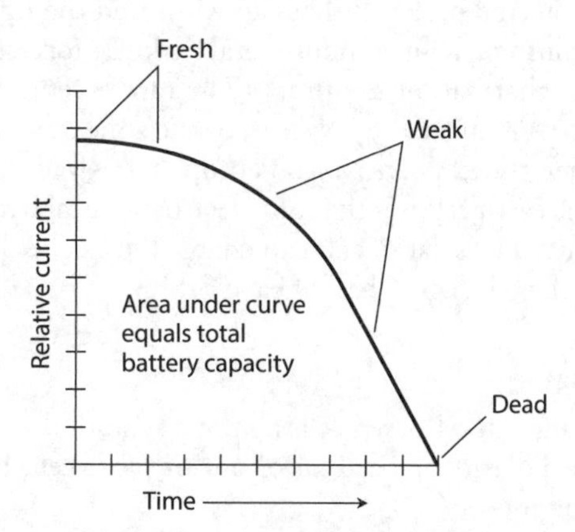

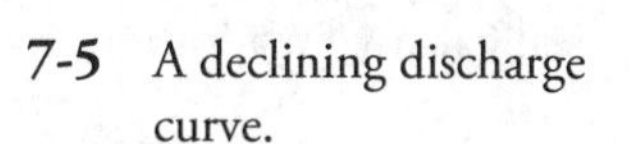
7-5 A declining discharge curve.

You must use care when charging LiPo cells, as they are prone to catching fire if overcharged. Overdischarging can easily destroy the battery. Some cells include a built-in IC that automatically prevents overcharging or undercharging. In a LiPo battery with two or more cells, each cell should be charged separately, using a special balanced charger.

Lead-Acid Batteries

You've seen the basic configuration for a lead-acid cell, which has a solution of sulfuric acid, along with a lead electrode (negative) and a lead-dioxide electrode (positive). These cells are rechargeable.

Automotive batteries comprise series-connected sets of lead-acid cells having a free-flowing liquid acid. You can't tip such a battery on its side, or turn it upside-down, without running the risk of having some of the acid electrolyte spill out. Some lead-acid batteries have semisolid electrolytes; they find applications in consumer electronic devices, notebook computers, and *uninterruptible power supplies* (UPSs) that can keep a desktop computer running for a few minutes if the utility power fails.

A large lead-acid battery, such as the one in your car or truck, can store several tens of ampere-hours. The smaller ones, like those in a UPS, have less capacity but more versatility. Their main attributes include the fact you can use and recharge them many times, they don't cost much money, and you don't have to worry about the irregular discharge characteristics that some rechargeable cells and batteries have.

Nickel-Based Cells and Batteries

Nickel-based cells include the *nickel-cadmium* (NICAD or NiCd) type and the *nickel-metal-hydride* (NiMH) type. *Nickel-based batteries* are available in packs of cells. You can sometimes plug these packs directly into consumer equipment. In other cases, the batteries actually form part of the device housing. All nickel-based cells are rechargeable. You can put them through hundreds or even thousands of *charge/discharge cycles* if you take good care of them.

Configurations and Applications

Nickel-based cells come in various sizes and shapes. *Cylindrical cells* look like ordinary dry cells. You'll find *button cells* in cameras, watches, memory backup applications, and other places where miniaturization matters. *Flooded cells* find application in heavy-duty electronic and electromechanical systems; some of these can store 1000 Ah or more. *Spacecraft cells* are manufactured in airtight, thermally protected packages that can withstand the rigors of a deep-space environment.

Most orbiting satellites endure total darkness for approximately half the time, and bask in direct sunlight the other half of the time. (The rare exception is the satellite with a carefully prescribed orbit that keeps it above the *gray line*, or the zone of surface sunrise or sunset. Such a satellite "sees" the sun all the time.) *Solar panels* can operate while the satellite receives sunlight, but during the times that the earth eclipses the sun, electrochemical batteries must power the electronic equipment on the satellite. The solar panels can charge the electrochemical battery, in addition to powering the satellite, for the "daylight" half of each orbit.

Precautions

Never discharge nickel-based cells all the way until they "totally die." If you make that mistake, you can cause the polarity of a cell, or of one or more cells in a battery, to permanently reverse, ruining the device for good.

Nickel-based cells and batteries, particularly the NICAD type, sometimes exhibit a bothersome characteristic called *memory* or *memory drain*. If you use such a device repeatedly, and you allow it to discharge to the same extent with every cycle, it seems to lose most of its capacity and "die too soon." You can sometimes "cure" a nickel-based cell or battery of this problem by letting it run down until it stops working properly, recharging it, running it down again, and repeating the cycle numerous times. In stubborn cases, you'll want to buy a new cell or battery instead of spending a lot of time trying to rejuvenate the old one.

Nickel-based cells and batteries work best if used with charging units that take several hours to fully replenish the charge. So-called *high-rate* or *quick* chargers are available, but some of these can force too much current through a cell or battery. It's best if the charger is made especially for the cell or battery type you use.

In recent years, concern has mounted about the toxic environmental effects of discarded heavy metals, including the cadmium in NICAD cells and batteries. For this reason, NiMH cells and batteries have largely supplanted NICAD types for consumer use. In most practical scenarios, you can directly replace a NICAD device with a NiMH device having the same output voltage and current-delivering capacity.

Photovoltaic Cells and Batteries

The *photovoltaic* (PV) *cell*, also called a *solar cell*, differs fundamentally from the electrochemical cell. A PV cell converts visible light, infrared (IR) rays, and/or ultraviolet (UV) rays directly into DC electricity.

Construction and Performance

Figure 7-6 shows the basic internal construction of a photovoltaic cell. A flat semiconductor *P-N junction* forms the active region within the device. It has a transparent housing so that radiant energy can directly strike the *P-type* silicon. The metal ribs, forming the positive electrode, are interconnected by means of tiny wires. The negative electrode consists of a metal backing, called the *substrate*, placed in contact with the *N-type* silicon.

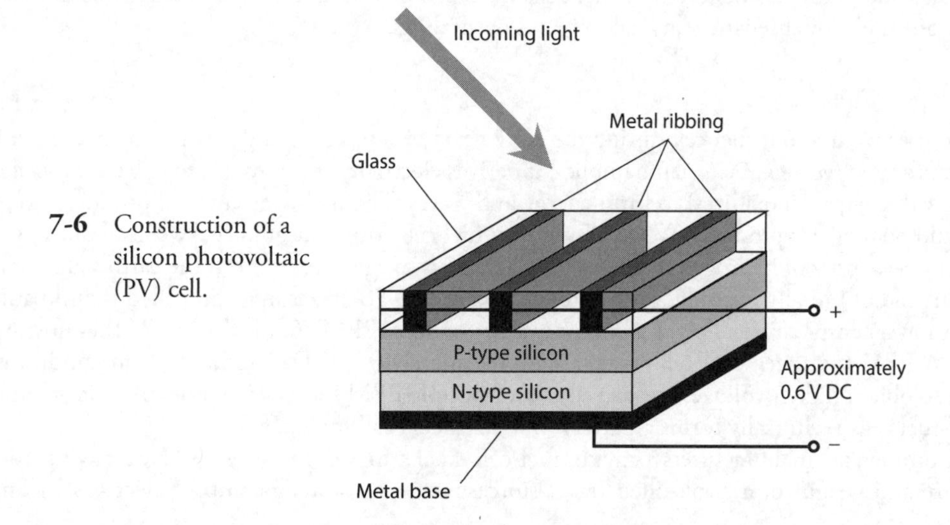

7-6 Construction of a silicon photovoltaic (PV) cell.

Most silicon-based solar cells provide about 0.6 V DC in direct sunlight. If the current demand is low, muted sunlight or artificial lamps can produce the full output voltage from a solar cell. As the current demand increases, the cell must receive more intense illumination to produce its full output voltage. A maximum limit exists to the current that a solar cell can deliver, no matter how bright the light. To obtain more current than that, we must connect multiple cells in parallel.

When we connect numerous photovoltaic cells in series-parallel, we obtain a *solar panel.* A large solar panel might consist of, say, 50 parallel sets of 20 series-connected cells. The series connection boosts the voltage, and the parallel connection increases the current-delivering ability. Sometimes, you'll find multiple solar panels connected in series or parallel to make vast arrays.

Practical Applications

Solar cells have become cheaper and more efficient in recent years, as researchers increasingly look to them as an alternative energy source. Solar panels are widely used in earth-orbiting satellites and interplanetary spacecraft. The famed Mars rovers could never have worked without them. Some alternative-energy enthusiasts have built systems that use solar panels in conjunction with rechargeable batteries, such as the lead-acid or nickel-cadmium types, to provide power independent of the commercial utilities.

A completely independent solar/battery power system is called a *stand-alone* system. It employs large solar panels, large-capacity lead-acid batteries, a *power inverter* to convert the DC into AC, and a sophisticated charging circuit. Obviously, these systems work best in environments where most days are sunny! The maximum deliverable power in full sunlight depends on the surface area of the panel.

Solar cells, either alone or supplemented with rechargeable batteries, can be connected into a home electric system in an *interactive* arrangement with the electric utilities. When the solar power system can't provide for the needs of the household all by itself, the utility company can make up for the shortage. Conversely, when the solar power system supplies more than enough for the needs of the home, the utility company can buy the excess energy from the consumer.

Fuel Cells

In the late 1900s, a new type of electrochemical power device, called the *fuel cell,* emerged. Many scientists and engineers believe that fuel cells hold promise as an alternative energy source to help offset our traditional reliance on coal, oil, and natural gas.

Hydrogen Fuel

The most talked-about fuel cell during the early years of research and development became known as the *hydrogen fuel cell.* As its name implies, it derives electricity from hydrogen. The hydrogen combines with oxygen (it oxidizes) to form energy and water. The hydrogen fuel cell produces no pollution and no toxic by-products. When a hydrogen fuel cell "runs out of juice," we need nothing more than a new supply of hydrogen to get it going again; its oxygen comes from the earth's atmosphere.

Instead of literally burning, the hydrogen in a fuel cell oxidizes in a controlled fashion and at a much lower temperature. The *proton exchange membrane* (PEM) *fuel cell* is one of the most widely used. A PEM hydrogen fuel cell generates approximately 0.7 V DC under no-load conditions. In order to obtain higher voltages, we can connect multiple PEM fuel cells in series. A series-connected set of fuel cells technically forms a battery, but engineers call it a *stack.*

Commercial manufacturers provide fuel-cell stacks in various sizes. A stack having roughly the size and weight of a book-filled travel suitcase can power a subcompact electric car. Smaller cells, called *micro fuel cells*, can provide DC to run devices that have historically operated from conventional cells and batteries. These include portable radios, lanterns, and notebook computers.

Other Fuels

Fuel cells can use energy sources other than hydrogen. Almost anything that will combine with oxygen to form energy can work. *Methanol*, a form of alcohol, is easier to transport and store than hydrogen because methanol exists as a liquid at room temperature. *Propane* powers some fuel cells. It can be stored in tanks for barbecue grills and some rural home heating systems. Still other fuel cells operate from *methane*, also known as natural gas. Theoretically, any combustible material will work: even oil or gasoline!

Some scientists object to the use of any energy source that employs so-called *fossil fuels* on which society has acquired a heavy dependence. To some extent we can dismiss these opinions as elitist, but in another sense, we must acknowledge a practical concern: Our planet has a finite supply of fossil fuels, the demand for which will grow for decades to come, especially in developing countries. Today's exotic energy alternative might become tomorrow's fuel of choice in the developed nations. The harder we try to make it so, the sooner it can happen.

A Promising Technology

As of this writing, fuel cells have not replaced conventional electrochemical cells and batteries in common applications, mainly because of the high cost. Hydrogen holds the honors as the most abundant and simplest chemical element in the universe, and it produces no toxic by-products when we liberate its stored energy. Hydrogen might, therefore, seem ideal as the choice for use in fuel cells. But storage and transport of hydrogen has proven difficult and expensive, especially for fuel cells and stacks intended for systems not affixed to permanent pipelines.

An interesting scenario, suggested by one of my physics teachers in the 1970s, involves piping hydrogen gas through lines already designed to carry methane. Some infrastructure modification would be necessary to safely handle hydrogen, which escapes through small cracks and openings more easily than methane. But hydrogen, if obtained at reasonable cost and in abundance, could power large fuel-cell stacks in households and businesses. Power inverters could convert the DC from such a stack to utility AC. A typical home power system of this sort would easily fit into a small room or a corner of the basement.

Quiz

Refer to the text in this chapter if necessary. A good score is 18 correct. Answers are in the back of the book.

1. Some interactive solar power systems for residential homes
   (a) operate from storage batteries during the day and recharge them at night.
   (b) can operate sophisticated systems such as computers, but not simple appliances such as lamps.
   (c) operate independently from the electric company.
   (d) allow the homeowner to sell energy to the electric company when the solar panels produce more power than the home needs.

2. Fill in the blank to make this statement true: "If you plot a battery discharge graph and see steady current for a while and then a rapid drop, then your battery has a ________ discharge characteristic."
   (a) uniform
   (b) flat
   (c) logarithmic
   (d) linear
3. A rechargeable battery, such as the one that starts your car, comprises
   (a) stand-alone cells.
   (b) primary cells.
   (c) secondary cells.
   (d) interactive cells.
4. If all other factors remain constant, then the total energy that an electrochemical battery can produce depends on
   (a) its voltage.
   (b) the number of cells that it has.
   (c) its size and mass.
   (d) the brightness of the light striking it.
5. The no-load voltage produced by several identical cells connected in series is
   (a) higher than the voltage produced by a single cell.
   (b) the same as the voltage produced by a single cell.
   (c) lower than the voltage produced by a single cell.
   (d) dependent on the current.
6. Under no-load conditions and bright sunlight, the output voltage from a PV cell
   (a) attains its maximum possible value.
   (b) declines with time.
   (c) increases with time.
   (d) equals zero.
7. What does a power inverter *not* do?
   (a) Allow household appliances to operate from batteries.
   (b) Convert AC to DC.
   (c) Convert DC to AC.
   (d) Work in solar power systems for home use.
8. Fill in the blank to make the following statement true: "Memory drain sometimes occurs in ________ cells and batteries."
   (a) primary
   (b) alkaline

(c) photovoltaic
(d) nickel-based

9. We connect five identical cells in parallel. Each individual cell produces 1.5 V under no-load conditions, and can deliver up to 12 A of current with a heavy load. Which of the following characteristics can we expect the whole battery to have?
   (a) A no-load voltage of 1.5 V and a maximum deliverable current of 12 A
   (b) A no-load voltage of 1.5 V and a maximum deliverable current of 60 A
   (c) A no-load voltage of 7.5 V and a maximum deliverable current of 12 A
   (d) A no-load voltage of 7.5 V and a maximum deliverable current of 60 A

10. Most automotive batteries contain, among other things,
   (a) sulfuric acid.
   (b) nickel.
   (c) cadmium.
   (d) P-type silicon.

11. You have a new 6.3-V lantern battery with an energy storage capacity of 5.2 Ah. If you connect a 63-ohm resistor to the battery but no other load, for how long should you expect current to flow through the resistor?
   (a) 31 minutes
   (b) 5 hours and 12 minutes
   (c) 2 days and 4 hours
   (d) 21 days and 16 hours

12. The maximum current that a battery can deliver depends on
   (a) its chemical composition.
   (b) the number of cells that it has.
   (c) its no-load output voltage.
   (d) its no-load output power.

13. In a portable lamp that consumes considerable power, you would probably find
   (a) a size-AA alkaline cell.
   (b) a lantern battery.
   (c) a fuel cell.
   (d) a PV panel.

14. The maximum power that a silicon PV panel can deliver depends on
   (a) its surface area.
   (b) the voltage of its cells.
   (c) its no-load output voltage.
   (d) its no-load output current.

15. Figure 7-3 represents a cell or battery with
    (a) nonlinear voltage output.
    (b) a nearly ideal discharge characteristic.
    (c) poor energy-handling capability.
    (d) nonlinear power storage capacity.

16. Fill in the blank to make the following sentence true: "A transistor battery has a voltage equal to that of ________ size AA 'grocery store' flashlight cells connected in series."
    (a) three
    (b) four
    (c) six
    (d) nine

17. You have two identical alkaline cells that each produce exactly 1.5 V as long as the current demand remains under 2.0 A. You connect them in series and then place a 1.5 k resistor across the combination. The resistor draws
    (a) 0.5 mA of current.
    (b) 1.0 mA of current.
    (c) 2.0 mA of current.
    (d) 4.0 mA of current.

18. You connect the same two cells (those described in the previous question) in parallel and then place a 1.5 k resistor across the combination. The resistor draws a current of
    (a) 0.5 mA.
    (b) 1.0 mA.
    (c) 2.0 mA.
    (d) 4.0 mA.

19. You have two identical alkaline cells that each produce 1.5 V as long as the current demand remains under 2.0 A. You connect them in series and then place a 1.5 k resistor across the combination. The resistor dissipates
    (a) 1.5 mW of power.
    (b) 3.0 mW of power.
    (c) 6.0 mW of power.
    (d) 12 mW of power.

20. You connect the same two cells (those described in the previous three questions) in parallel and then place a 1.5 k resistor across the combination. The resistor dissipates
    (a) 1.5 mW of power.
    (b) 3.0 mW of power.
    (c) 6.0 mW of power.
    (d) 12 mW of power.

8
CHAPTER

Magnetism

A *MAGNETIC FIELD* ARISES WHEN ELECTRIC CHARGE CARRIERS MOVE. CONVERSELY, WHEN AN ELECTRICAL conductor moves in a magnetic field, current flows in that conductor.

Geomagnetism

The earth has a core consisting largely of iron, heated to the extent that some of it liquefies. As the earth rotates on its axis, the iron in the core flows in convection patterns, generating the *geomagnetic field* that surrounds our planet and extends thousands of kilometers into space.

Earth's Magnetic Poles and Axis

The geomagnetic field has poles, just as an old-fashioned bar magnet does. On the earth's surface, these poles exist in the arctic and antarctic regions, but they are displaced considerably from the *geographic poles* (the points where the earth's axis intersects the surface). The *geomagnetic lines of flux* converge or diverge at the *geomagnetic poles*. The *geomagnetic axis* that connects the geomagnetic poles tilts somewhat with respect to the *geographic axis* on which the earth rotates.

Charged subatomic particles from the sun, constantly streaming outward through the solar system, distort the geomagnetic field. This so-called *solar wind* "blows" the geomagnetic field out of symmetry. On the side of the earth facing the sun, the lines of flux compress. On the side of the earth opposite the sun, the lines of flux dilate. Similar effects occur in other planets, notably Jupiter, that have magnetic fields. As the earth rotates, the geomagnetic field "dances" into space in the direction facing away from the sun.

The Magnetic Compass

Thousands of years ago, observant people noticed the presence of the geomagnetic field, even though they didn't know exactly what caused it. Certain rocks, called *lodestones*, when hung by strings, always orient themselves in a generally north-south direction. Long ago, seafarers and explorers correctly attributed this effect to the presence of a "force" in the air. The reasons for this phenomenon remained unknown for centuries, but adventurers put it to good use. Even today, a *magnetic compass* makes a valuable navigation aid. It can work when more sophisticated navigational devices, such as a *Global Positioning System* (GPS), fails.

The geomagnetic field interacts with the magnetic field around a compass needle, which comprises a small bar magnet. This interaction produces force on the compass needle, causing it to align itself parallel to the geomagnetic lines of flux in the vicinity. The force operates not only in a horizontal plane (parallel to the earth's surface), but also vertically in most locations. The vertical force component vanishes at the *geomagnetic equator*, a line running around the globe equidistant from both geomagnetic poles, so the force there is perfectly horizontal. But as the *geomagnetic latitude* increases, either towards the north or the south geomagnetic pole, the *magnetic force* pulls up and down on the compass needle more and more. We call the extent of the vertical force component at any particular place the *geomagnetic inclination*. Have you noticed this when using a magnetic compass? One end of the needle dips a little toward the compass face, while the other end tilts upward toward the glass.

Because the earth's geomagnetic axis and geographic axis don't coincide, the needle of a magnetic compass usually points somewhat to the east or west of true geographic north. The extent of the discrepancy depends on our surface location. We call the angular difference between geomagnetic north (north according to a compass) and geographic north (or *true north*) the *geomagnetic declination*.

Magnetic Force

As children, most of us discovered that magnets "stick" to some metals. Iron, nickel, a few other elements, and alloys or solid mixtures containing any of them constitute *ferromagnetic materials*. Magnets exert force on these metals. Magnets do not generally exert force on other metals unless those metals carry electric currents. Electrically insulating substances never "attract magnets" under normal conditions.

Cause and Strength

When we bring a *permanent magnet* near a sample of ferromagnetic material, the atoms in the material line up to a certain extent, temporarily magnetizing the sample. This atomic alignment produces a magnetic force between the atoms of the sample and the atoms in the magnet. Every single atom acts as a tiny magnet; when they act in concert with one another, the whole sample behaves as a magnet. Permanent magnets always attract samples of ferromagnetic material.

If we place two permanent magnets near each other, we observe a stronger magnetic force than we do when we bring either magnet near a sample of ferromagnetic material. The mutual force between two rod-shaped or bar-shaped permanent magnets is manifest as attraction if we bring two opposite poles close together (north-near-south or south-near-north) and repulsion if we bring two like poles into proximity (north-near-north or south-near-south). Either way, the force increases as the distance between the ends of the magnets decreases.

Some *electromagnets* produce fields so powerful that no human can pull them apart if they get "stuck" together, and no one can bring them all the way together against their mutual repulsive force. (We'll explore how electromagnets work later in this chapter.) Industrial workers can use huge electromagnets to carry heavy pieces of scrap iron or steel from place to place. Other electromagnets can provide sufficient repulsion to suspend one object above another, an effect known as *magnetic levitation*.

Electric Charge Carriers in Motion

Whenever the atoms in a sample of ferromagnetic material align to any extent rather than existing in a random orientation, a magnetic field surrounds the sample. A magnetic field can also result from the motion of electric *charge carriers*. In a wire, electrons move in incremental "hops" along the conductor from atom to atom. In a permanent magnet, the movement of orbiting electrons occurs in such a manner that an *effective current* arises.

Magnetic fields can arise from the motion of charged particles through space, as well as from the motion of charge carriers through a conductor. The sun constantly ejects protons and helium nuclei, both of which carry positive electric charges. These particles produce effective currents as they travel through space. The effective currents in turn generate magnetic fields. When these fields interact with the geomagnetic field, the subatomic particles change direction and accelerate toward the geomagnetic poles.

When an eruption on the sun, called a *solar flare,* occurs, the sun ejects far more charged subatomic particles than usual. As these particles approach the geomagnetic poles, their magnetic fields, working together, disrupt the geomagnetic field, spawning a *geomagnetic storm.* Such an event causes changes in the earth's upper atmosphere, affecting "shortwave radio" communications and producing the *aurora borealis* ("northern lights") and *aurora australis* ("southern lights"), well-known to people who dwell at high latitudes. If a geomagnetic storm reaches sufficient intensity, it can interfere with wire communications and electric power transmission at the surface.

Lines of Flux

Physicists consider magnetic fields to comprise *flux lines*, or *lines of flux*. The intensity of the field depends on the number of flux lines passing at right angles through a region having a certain cross-sectional area, such as a centimeter squared (cm^2) or a meter squared (m^2). The flux lines are not actual material fibers, but their presence can be shown by means of a simple experiment.

Have you seen the classical demonstration in which iron filings lie on a sheet of paper, and then the experimenter holds a permanent magnet underneath the sheet? The filings arrange themselves in a pattern that shows, roughly, the "shape" of the magnetic field in the vicinity of the magnet. A bar magnet has a field whose lines of flux exhibit a characteristic pattern (Fig. 8-1).

Another experiment involves passing a current-carrying wire through the paper at a right angle. The iron filings bunch up in circles centered at the point where the wire passes through the paper. This experiment shows that the lines of flux around a straight, current-carrying wire form concentric circles in any plane passing through the wire at a right angle. The center of every "flux circle" lies on the wire, which constitutes the path along which the charge carriers move (Fig. 8-2).

Polarity

A magnetic field has a specific orientation at any point in space near a current-carrying wire or a permanent magnet. The flux lines run parallel with the direction of the field. Scientists consider the magnetic field to begin, or originate, at a *north pole*, and to end, or terminate, at a *south pole*. These poles do not correspond to the geomagnetic poles; in fact, they're the opposite! The north geomagnetic pole is in reality a south pole because it attracts the north poles of magnetic compasses. Similarly, the south geomagnetic pole is really a north pole because it attracts the south poles of compasses. In the case of a permanent magnet, we can usually (but not always) tell where the magnetic poles are located. Around a current-carrying wire, the magnetic field revolves endlessly.

A charged electric particle (such as a proton) hovering in space forms an *electric monopole*, and the electric flux lines around it aren't closed. A positive charge does not have to mate with a negative

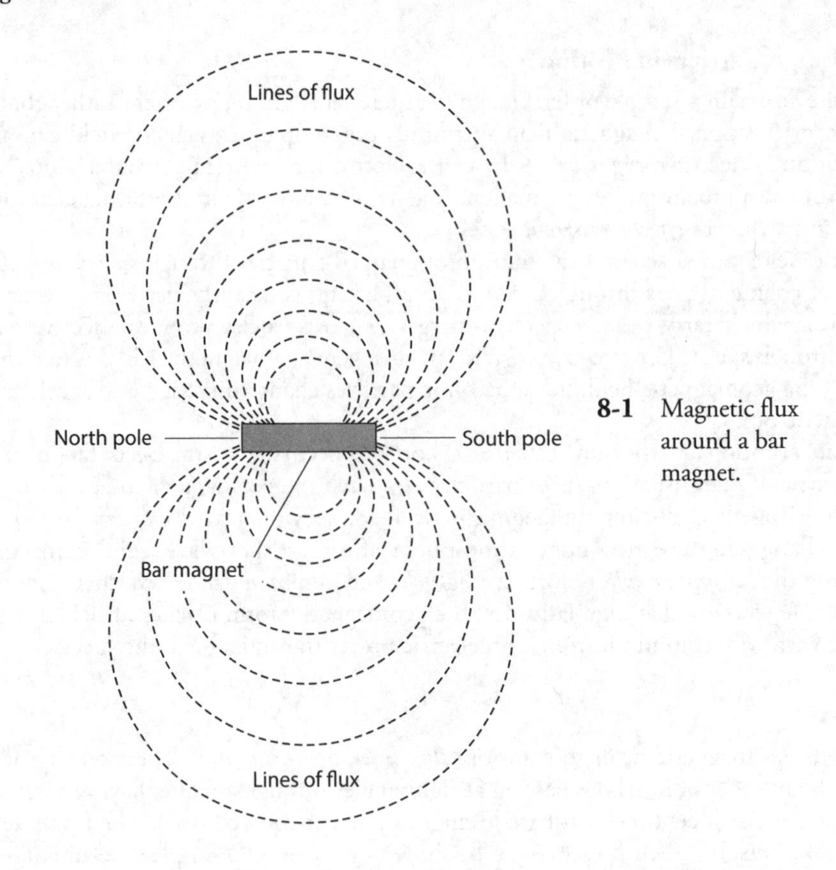

8-1 Magnetic flux around a bar magnet.

charge. The electric flux lines around any stationary, charged particle run outward in all directions for a theoretically infinite distance. But magnetic fields behave according to stricter laws. Under normal circumstances, all magnetic flux lines form closed loops. In the vicinity of a magnet, we can always find a starting point (the north pole) and an ending point (the south pole). Around a current-carrying wire, the loops form circles.

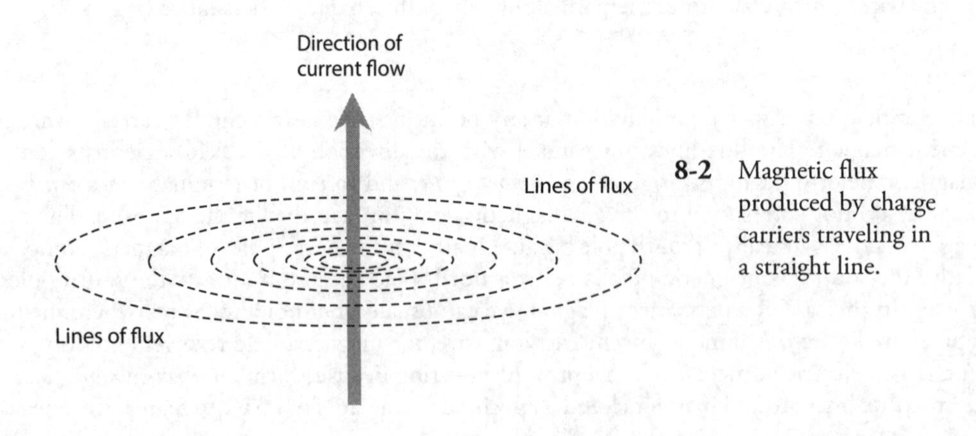

8-2 Magnetic flux produced by charge carriers traveling in a straight line.

Magnetic Dipoles

You might at first suppose that the magnetic field around a current-carrying wire arises from a monopole, or that no poles exist. The concentric flux circles don't seem to originate or terminate anywhere. But you can assign originating and terminating points to those circles, thereby defining a *magnetic dipole*—a pair of opposite magnetic poles in close proximity.

Imagine that you hold a flat piece of paper next to a current-carrying wire, so that the wire runs along one edge of the sheet. The magnetic circles of flux surrounding the wire pass through the sheet of paper, entering one side and emerging from the other side, so you have a "virtual magnet." Its north pole coincides with the face of the paper sheet from which the flux circles emerge. Its south pole coincides with the opposite face of the sheet, into which the flux circles plunge.

The flux lines in the vicinity of a magnetic dipole always connect the two poles. Some flux lines appear straight in a local sense, but in the larger sense they always form curves. The greatest magnetic field strength around a bar magnet occurs near the poles, where the flux lines converge or diverge. Around a current-carrying wire, the greatest field strength occurs near the wire.

Magnetic Field Strength

Physicists and engineers express the overall magnitude, or quantity, of a magnetic field in units called *webers*, symbolized Wb. We can employ a smaller unit, the *maxwell* (Mx), for weak fields. One weber equals 100,000,000 (10^8) maxwells. Therefore

$$1 \text{ Wb} = 10^8 \text{ Mx}$$

and

$$1 \text{ Mx} = 10^{-8} \text{ Wb}$$

The Tesla and the Gauss

If you have a permanent magnet or an electromagnet, you might see its "strength" expressed in terms of webers or maxwells. But more often, you'll hear or read about units called *teslas* (T) or *gauss* (G). These units define the concentration, or intensity, of the magnetic field as its flux lines pass at right angles through flat regions having specific cross-sectional areas.

The *flux density*, or number of "flux lines per unit of cross-sectional area," forms a more useful expression for magnetic effects than the overall quantity of magnetism. In equations, we denote flux density using the letter *B*. A flux density of one tesla equals one weber per meter squared (1 Wb/m^2). A flux density of one gauss equals one maxwell per centimeter squared (1 Mx/cm^2). As things work out, the gauss equals 0.0001 (10^{-4}) tesla, so we have the relations

$$1 \text{ G} = 10^{-4} \text{ T}$$

and

$$1 \text{ T} = 10^4 \text{ G}$$

If you want to convert from teslas to gauss (not gausses!), multiply by 10^4. If you want to convert from gauss to teslas, multiply by 10^{-4}.

Quantity versus Density

If the distinctions between webers and teslas, or between maxwells and gauss, confuse you, think of an ordinary light bulb. Suppose that a lamp emits 15 W of visible-light power. If you enclose the bulb completely, then 15 W of visible light strike the interior walls of the chamber, regardless of the size of the chamber. But this notion doesn't give you a useful notion of the brightness of the light. You know that a single bulb produces plenty of light if you want to illuminate a closet, but nowhere near enough light to illuminate a gymnasium. The important consideration is the number of watts *per unit of area*. When you say that a bulb gives off so-many watts of light *overall*, it's like saying that a magnet has a magnetic quantity of so-many webers or maxwells. When you say that the bulb produces so-many watts of light *per unit of area*, it's like saying that a magnetic field has a flux density of so-many teslas or gauss.

Magnetomotive Force

When we work with wire loops, *solenoidal* (helical) coils, and rod-shaped electromagnets, we can quantify a phenomenon called *magnetomotive force* with a unit called the *ampere-turn* (At). This unit describes itself well: the number of amperes flowing in a coil or loop, times the number of turns that the coil or loop contains.

If we bend a length of wire into a loop and drive 1 A of current through it, we get 1 At of magnetomotive force inside the loop. If we wind the same length of wire (or any other length) into a 50-turn coil and keep driving 1 A of current through it, the resulting magnetomotive force increases by a factor of 50, to 50 At. If we then reduce the current in the 50-turn loop to 1/50 A or 20 mA, the magnetomotive force goes back down to 1 At.

Sometimes, engineers employ a unit called the *gilbert* to express magnetomotive force. One gilbert (1 Gb) equals approximately 0.7958 At. The gilbert represents a slightly smaller unit than the At does. Therefore, if we want to determine the number of ampere-turns when we know the number of gilberts, we should multiply by 0.7958. To determine the number of gilberts when we know the number of ampere-turns, we should multiply by 1.257.

Magnetomotive force does not depend on core material or loop diameter. Even if we place a metal rod in a solenoidal coil, the magnetomotive force will not change if the current through the wire remains the same. A tiny 100-turn air-core coil carrying 1 A produces the same magnetomotive force as a huge 100-turn air-core coil carrying 1 A. Magnetomotive force depends *only* on the current and the number of turns.

Flux Density versus Current

In a straight wire carrying a steady, direct current and surrounded by air or a vacuum, we observe the greatest flux density near the wire, and diminishing flux density as we get farther away from the wire. We can use a simple formula to express magnetic flux density as a function of the current in a straight wire and the distance from the wire.

Imagine an infinitely thin, absolutely straight, infinitely long length of wire (that's the ideal case). Suppose that the wire carries a current of I amperes. Represent the flux density (in teslas) as B. Consider a point P at a distance r (in meters) from the wire, measured in a plane perpendicular to the wire, as shown in Fig. 8-3. We can find the flux density at point P using the formula

$$B = 2 \times 10^{-7} I / r$$

We can consider the value of the constant, 2×10^{-7}, mathematically exact to any desired number of significant figures.

Of course, we'll never encounter a wire with zero thickness or infinite length. But as long as the wire thickness constitutes a small fraction of r, and as long as the wire is reasonably straight near point P, this formula works quite well in most applications.

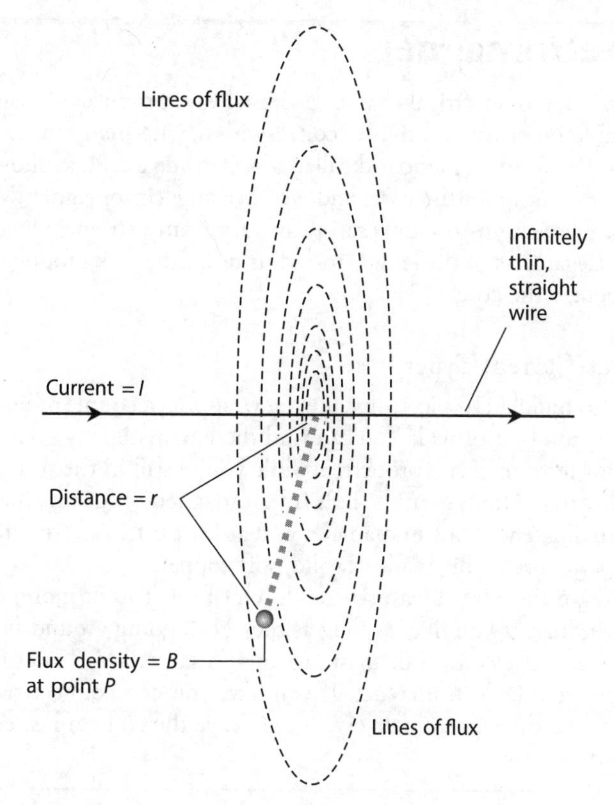

8-3 The magnetic flux density varies inversely with the distance from a wire carrying constant current.

Problem 8-1

What is the flux density B_t in teslas at a distance of 200 mm from a straight, thin wire carrying 400 mA of DC?

Solution

First, we must convert all quantities to units in the International System (SI). Thus, we have $r = 0.200$ m and $I = 0.400$ A. We can input these values directly into the formula for flux density to obtain

$$B_t = 2.00 \times 10^{-7} \times 0.400 / 0.200 = 4.00 \times 10^{-7}\ \mathrm{T}$$

Problem 8-2

In the above-described scenario, what is the flux density B_g (in gauss) at point P?

Solution

To figure this out, we must convert from teslas to gauss, multiplying the result in the solution to Problem 8-1 by 10^4 to get

$$B_g = 4.00 \times 10^{-7} \times 10^4 = 4.00 \times 10^{-3}\ \mathrm{G}$$

Electromagnets

The motion of electrical charge carriers always produces a magnetic field. This field can reach considerable intensity in a tightly coiled wire having many turns and carrying a large current. When we place a ferromagnetic rod called a *core* inside a coil, as shown in Fig. 8-4, the magnetic lines of flux concentrate in the core, and we have an electromagnet. Most electromagnets have cylindrical cores. The length-to-radius ratio can vary from extremely low (fat pellet) to extremely high (thin rod). Regardless of the length-to-radius ratio, the flux produced by current in the wire temporarily magnetizes the core.

Direct-Current Types

You can build a DC electromagnet by wrapping a couple of hundred turns of insulated wire around a large iron bolt or nail. You can find these items in any good hardware store. You should test the bolt for ferromagnetic properties while you're still in the store, if possible. (If a permanent magnet "sticks" to the bolt, then the bolt is ferromagnetic.) Ideally, the bolt should measure at least ⅜ inch (approximately 1 cm) in diameter and at least 6 inches (roughly 15 cm) long. You must use insulated wire, preferably made of solid, soft copper.

Wind the wire at least several dozen (if not 100 or more) times around the bolt. You can layer the windings if you like, as long as they keep going around in the same direction. Secure the wire in place with electrical or masking tape. A large 6-V "lantern battery" can provide plenty of DC to operate the electromagnet. If you like, you can connect two or more such batteries in parallel to increase the current delivery. Never leave the coil connected to the battery for more than a few seconds at a time.

Warning!

Do not use a lead-acid automotive battery for this experiment. The near short-circuit produced by an electromagnet can cause the acid from such a battery to boil out, resulting in serious injury.

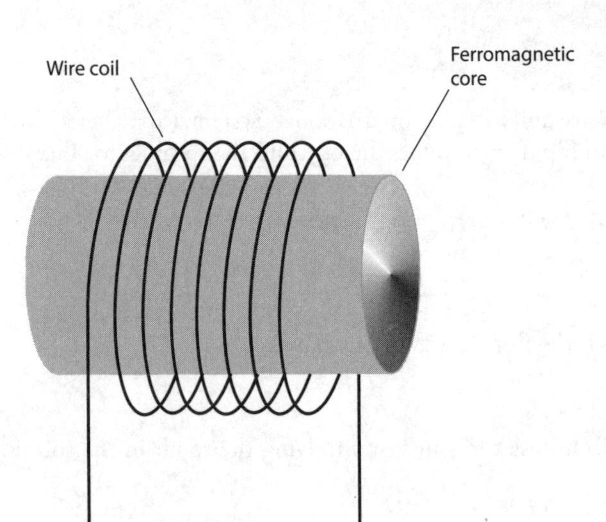

8-4 A simple electromagnet.

All DC electromagnets have defined north and south poles, just as permanent magnets have. However, an electromagnet can get much stronger than any permanent magnet. The magnetic field exists only as long as the coil carries current. When you remove the power source, the magnetic field nearly vanishes. A small amount of *residual magnetism* remains in the core after the current stops flowing in the coil, but this field has minimal intensity.

Alternating-Current Types

Do you suspect that you can make an electromagnet extremely powerful if, rather than using a lantern battery for the current source, you plug the ends of the coil directly into an AC utility outlet? In theory, you can do this. But don't! You'll expose yourself to the danger of electrocution, expose your house to the risk of electrical fire, and most likely cause a fuse or circuit breaker to open, killing power to the device anyway. Some buildings lack the proper fuses or circuit breakers to prevent excessive current from flowing through the utility wiring in case of an overload. If you want to build and test a safe AC electromagnet, my book, *Electricity Experiments You Can Do at Home* (McGraw-Hill, 2010), offers instructions for doing it.

Some commercially manufactured electromagnets operate from 60-Hz utility AC. These magnets "stick" to ferromagnetic objects. The polarity of the magnetic field reverses every time the direction of the current reverses, producing 120 fluctuations, or 60 complete north-to-south-to-north polarity changes, every second. In addition, the instantaneous intensity of the magnetic field varies along with the AC cycle, reaching alternating-polarity peaks at 1/120-second intervals and nulls of zero intensity at 1/120-second intervals. Any two adjacent peaks and nulls occur 1/4 cycle, or 1/240 second, apart.

If you bring a permanent magnet or DC electromagnet near either "pole" of an AC electromagnet, no net force results from the AC electromagnetism itself because equal and opposite attractive and repulsive forces occur between the alternating magnetic field and any steady external field. But the permanent magnet or the DC electromagnet attracts the core of the AC electromagnet, whether the AC device carries current or not.

Tech Note

With any electromagnet, a DC source, such as a battery, always produces a magnetic field that maintains the same polarity at all times. An AC source always produces a magnetic field that periodically reverses polarity.

Problem 8-3

Suppose that we apply 80-Hz AC to an electromagnet instead of the standard 60 Hz. What will happen to the interaction between the alternating magnetic field and a nearby permanent magnet or DC electromagnet?

Solution

Assuming that the behavior of the core material remains the same, the situation at 80 Hz will not change from the 60-Hz case. In theory, the AC frequency makes no difference in the behavior of an AC electromagnet. In practice, however, the magnetic field weakens at high AC frequencies because the AC electromagnet's *inductance* tends to impede the flow of current. This so-called *inductive reactance* depends on the number of coil turns, and also on the characteristics of the ferromagnetic core.

Magnetic Materials

Ferromagnetic substances cause magnetic lines of flux to bunch together more tightly than they exist in free space. A few materials cause the lines of flux to dilate compared with their free-space density. We call these substances *diamagnetic*. Examples of such materials include wax, dry wood, bismuth, and silver. No diamagnetic material reduces the strength of a magnetic field by anywhere near the factor that ferromagnetic substances can increase it. Usually, engineers use diamagnetic objects to keep magnets physically separated while minimizing the interaction between them.

Permeability

Permeability expresses the extent to which a ferromagnetic material concentrates magnetic lines of flux relative to the flux density in a vacuum. By convention, scientists assign a permeability value of 1 to a vacuum. If we have a coil of wire with an air core and we drive DC through the wire, then the flux inside the coil is about the same as it would be in a vacuum. Therefore, the permeability of air equals almost exactly 1. (Actually it's a tiny bit higher, but the difference rarely matters in practice.)

If we place a ferromagnetic core inside the coil, the flux density increases, sometimes by a large factor. By definition, the permeability equals that factor. For example, if a certain material causes the flux density inside a coil to increase by a factor of 60 compared with the flux density in air or a vacuum, that material has a permeability of 60. Diamagnetic materials have permeability values less than 1, but never very much less. Table 8-1 lists the permeability values for some common substances.

Retentivity

When we subject a substance, such as iron, to a magnetic field as intense as it can handle, say by enclosing it in a wire coil carrying high current, some residual magnetism always remains after

Table 8-1. Permeability Values for Some Common Materials

Substance	Permeability (approx.)
Air, dry, at sea level	1.0
Alloys, ferromagnetic	3000–1,000,000
Aluminum	Slightly more than 1
Bismuth	Slightly less than 1
Cobalt	60–70
Iron, powdered and pressed	100–3000
Iron, solid, refined	3000–8000
Iron, solid, unrefined	60–100
Nickel	50–60
Silver	Slightly less than 1
Steel	300–600
Vacuum	1.0 (exact, by definition)
Wax	Slightly less than 1
Wood, dry	Slightly less than 1

the current stops flowing in the coil. *Retentivity*, also known as *remanence*, quantifies the extent to which a substance "memorizes" a magnetic field imposed on it.

Imagine that we wind a wire coil around a sample of ferromagnetic material and then drive so much current through the coil that the magnetic flux inside the core reaches its maximum possible density. We call this condition *core saturation*. We measure the flux density in this situation, and get a figure of B_{max} (in teslas or gauss). Now suppose that we remove the current from the coil, and then we measure the flux density inside the core again, obtaining a figure of B_{rem} (in teslas or gauss, as before). We can express the retentivity B_r of the core material as a ratio according to the formula

$$B_r = B_{rem} / B_{max}$$

or as a percentage using the formula

$$B_{r\%} = 100\ B_{rem} / B_{max}$$

As an example, suppose that a metal rod can attain a flux density of 135 G when enclosed by a current-carrying coil. Imagine that 135 G represents the maximum possible flux density for that material. (For any substance, such a maximum always exists, unique to that substance; further increasing the coil current or number of turns will not magnetize it any further.) Now suppose that we remove the current from the coil, and 19 G remain in the rod. Then the retentivity B_r is

$$B_r = 19/135 = 0.14$$

As a percentage,

$$B_{r\%} = 100 \times 19/135 = 14\%$$

Certain ferromagnetic substances exhibit high retentivity, and therefore, make excellent permanent magnets. Other ferromagnetic materials have poor retentivity. They can sometimes work okay as the cores of electromagnets, but they don't make good permanent magnets.

If a ferromagnetic substance has low retentivity, it can function as the core for an AC electromagnet because the polarity of the magnetic field in the core follows along closely as the current in the coil alternates. If the material has high retentivity, the material acts "magnetically sluggish" and has trouble following the current reversals in the coil. Substances of this sort don't work well in AC electromagnets.

Problem 8-4

Suppose that we wind a coil of wire around a metal core to make an electromagnet. We find that by connecting a variable DC source to the coil, we can drive the magnetic flux density in the core up to 0.500 T but no higher. When we shut down the current source, the flux density inside the core drops to 500 G. What's the retentivity of this core material?

Solution

First, let's convert both flux density figures to the same units. We recall that 1 T = 10^4 G. Therefore, the flux density in gauss is $0.500 \times 10^4 = 5{,}000$ G when the current flows in the coil, and 500 G after we remove the current. "Plugging in" these numbers gives us the ratio

$$B_r = 500/5{,}000 = 0.100$$

or the percentage

$$B_{r\%} = 100 \times 500/5{,}000 = 100 \times 0.100 = 10.0\%$$

Permanent Magnets

Industrial engineers can make any suitably shaped sample of ferromagnetic material into a permanent magnet. The strength of the magnet depends on two factors:

- The retentivity of the material used to make it
- The amount of effort put into magnetizing it

The manufacture of powerful permanent magnets requires an alloy with high retentivity. The most "magnetizable" alloys derive from specially formulated mixtures of aluminum, nickel, and cobalt, occasionally including trace amounts of copper and titanium. Engineers place samples of the selected alloy inside heavy wire coils carrying high, continuous DC for an extended period of time.

You can magnetize any piece of iron or steel. Some technicians use magnetized tools when installing or removing screws from hard-to-reach places in computers, wireless transceivers, and other devices. If you want to magnetize a tool, stroke its metal shaft with the end of a powerful bar magnet several dozen times. But beware: Once you've imposed residual magnetism in a tool, it will remain magnetized to some extent forever!

Flux Density inside a Long Coil

Consider a long, helical coil of wire, commonly known as a *solenoid*, having n turns in a single layer. Suppose that it measures s meters in length, carries a steady direct current of I amperes, and has a ferromagnetic core of permeability μ. Assuming that the core has not reached a state of saturation, we can calculate the flux density B_t (in teslas) inside the material using the formula

$$B_t = 4\pi \times 10^{-7}\ \mu nI/s \approx 1.2566 \times 10^{-6}\ \mu nI/s$$

If we want to calculate the flux density B_g (in gauss), we can use the formula

$$B_g = 4\pi \times 10^{-3}\ \mu nI/s \approx 0.012566\ \mu nI/s$$

Problem 8-5

Imagine a DC electromagnet that carries a certain current. It measures 20 cm long, and has 100 turns of wire. The core, which has permeability $\mu = 100$, has not reached a state of saturation. We measure the flux density inside it as $B_g = 20$ G. How much current flows in the coil?

Solution

Let's start by ensuring that we use the proper units in our calculation. We're told that the electromagnet measures 20 cm in length, so we can set $s = 0.20$ m. The flux density equals 20 G. Using algebra, we can rearrange the second of the above formulas so that it solves for I. We start with

$$B_g = 0.012566\ \mu nI/s$$

Dividing through by I, we get

$$B_g/I = 0.012566\ \mu n/s$$

When we divide both sides by B_g, we obtain

$$I^{-1} = 0.012566\ \mu n/(sB_g)$$

Finally, we take the reciprocal of both sides to get

$$I = 79.580\ sB_g / (\mu n)$$

Now we can input the numbers from the statement of the problem. We calculate

$$I = 79.580\ sB_g / (\mu n) = 79.580 \times 0.20 \times 20 / (100 \times 100)$$
$$= 79.580 \times 4.0 \times 10^{-4} = 0.031832 \text{ A} = 31.832 \text{ mA}$$

We should round this result off to 32 mA.

Magnetic Machines

Electrical relays, bell ringers, electric "hammers," and other mechanical devices make use of the principle of the solenoid. Sophisticated electromagnets, sometimes in conjunction with permanent magnets, allow us to construct motors, meters, generators, and other electromechanical devices.

The Chime

Figure 8-5 illustrates a *bell ringer*, also known as a *chime*. Its solenoid comprises an electromagnet. The ferromagnetic core has a hollow region in the center along its axis, through which a steel rod, called the *hammer*, passes. The coil has many turns of wire, so the electromagnet produces a high flux density if a substantial current passes through the coil.

When no current flows in the coil, gravity holds the rod down so that it rests on the plastic base plate. When a pulse of current passes through the coil, the rod moves upward at high speed. The magnetic field "wants" the ends of the rod, which has the same length as the core, to align with the

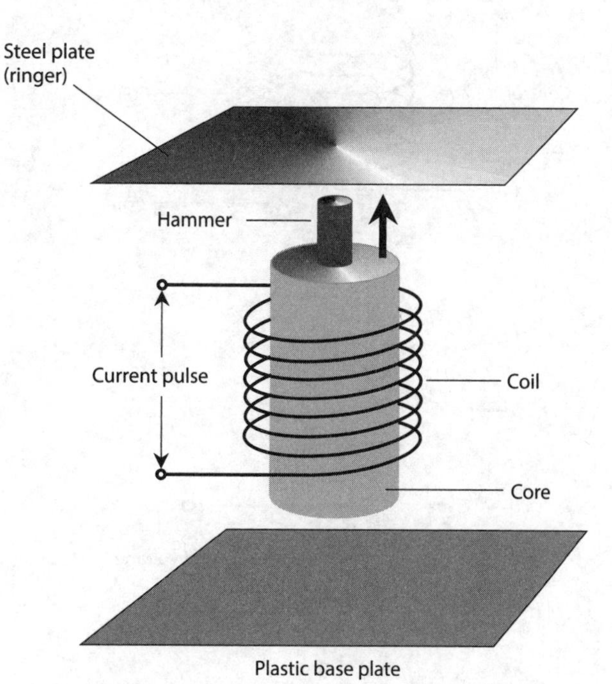

8-5 A bell ringer, also known as a chime.

ends of the core. But the rod's upward momentum causes it to pass through the core and strike the ringer. Then the steel rod falls back to its resting position, allowing the ringer to reverberate.

The Relay

We can't always locate switches near the devices they control. For example, suppose that you want to switch a communications system between two different antennas from a station control point 50 meters away. Wireless antenna systems carry high-frequency AC (the radio signals) that must remain within certain parts of the circuit. A *relay* makes use of a solenoid to allow remote-control switching.

Figure 8-6A illustrates a simple relay, and Fig. 8-6B shows the schematic diagram for the same device. A movable lever, called the *armature*, is held to one side (upward in this diagram) by a spring when no current flows through the coil. Under these conditions, terminal X contacts terminal Y, but X does not contact Z. When a sufficient current flows in the coil, the armature moves to the other side (downward in this illustration), disconnecting terminal X from terminal Y, and connecting X to Z.

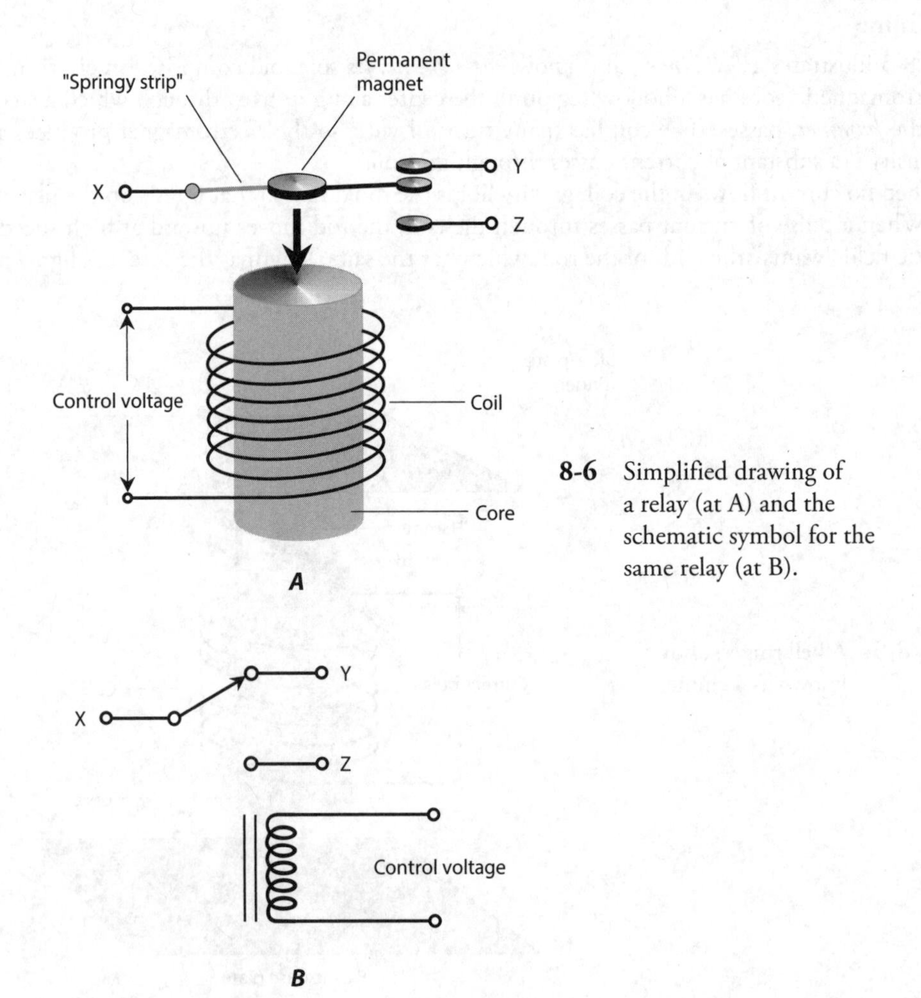

8-6 Simplified drawing of a relay (at A) and the schematic symbol for the same relay (at B).

A *normally closed relay* completes the circuit when no current flows in the coil, and breaks the circuit when coil current flows. ("Normal," in this sense, means the absence of coil current.) A *normally open relay* does the opposite, completing the circuit when coil current flows, and breaking the circuit when coil current does not flow. The relay shown in Fig. 8-6 can function either as a normally open relay or a normally closed relay, depending on which contacts we select. It can also switch a single line between two different circuits.

These days, engineers install relays primarily in circuits and systems that must handle large currents or voltages. In applications in which the currents and voltages remain low to moderate, electronic semiconductor switches, which have no moving parts, offer better performance and reliability than relays.

The DC Motor

Magnetic fields can produce considerable mechanical forces. We can harness these forces to perform useful work. A *DC motor* converts DC into rotating mechanical energy. In this sense, a DC motor constitutes a specialized *electromechanical transducer*. Such devices range in size from *nanoscale* (smaller than a bacterium) to *megascale* (larger than a house). Nanoscale motors can circulate in the human bloodstream or modify the behavior of internal body organs. Megascale motors can pull trains along tracks at hundreds of kilometers per hour.

In a DC motor, we connect a source of electricity to a set of coils, producing magnetic fields. The attraction of opposite poles, and the repulsion of like poles, is switched in such a way that a constant torque (rotational force) results. As the coil current increases, so does the torque that the motor can produce—and so does the energy it takes to operate the motor at a constant speed.

Figure 8-7 illustrates a DC motor in simplified form. The *armature coil* rotates along with the motor shaft. A set of two coils called the *field coil* remains stationary. Some motors use a pair of

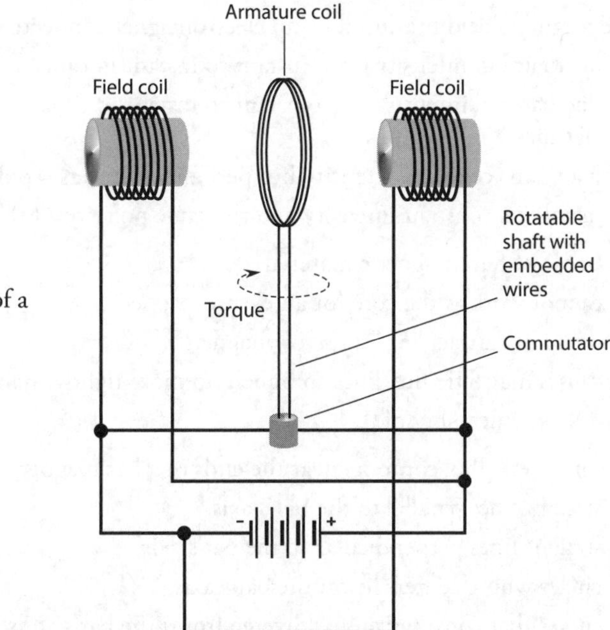

8-7 Simplified drawing of a DC motor.

permanent magnets instead of a field coil. Every time the shaft completes half a rotation, the *commutator* reverses the current direction in the armature coil, so the shaft torque continues in the same angular direction. The shaft's *angular* (rotational) *momentum* carries it around so that it doesn't "freeze up" at the points in time when the current reverses direction.

Electric Generator

The construction of an *electric generator* resembles that of an electric motor, although the two devices function in the opposite sense. We might call a generator a specialized *mechano-electrical transducer* (although I've never heard anybody use that term). Some generators can also operate as motors; we call such devices *motor-generators.*

A typical generator produces AC when a coil rotates in a strong magnetic field. We can drive the shaft with a gasoline-powered motor, a turbine, or some other source of mechanical energy. Some generators employ commutators to produce pulsating DC output, which we can *filter* to obtain pure DC for use with precision equipment.

Quiz

Refer to the text in this chapter if necessary. A good score is 18 correct. Answers are in the back of the book.

1. If a solenoidal wire coil has 50 turns and carries 500 mA, then it gives rise to a magnetomotive force of
   (a) 25 At.
   (b) 50 At.
   (c) 500 At.
   (d) None of the above

2. The magnetic field produced by an electromagnet connected to a lantern battery
   (a) fluctuates in intensity from instant to instant in time, and periodically reverses polarity.
   (b) fluctuates in intensity from instant to instant in time, but maintains the same polarity at all times.
   (c) maintains constant intensity but periodically reverses polarity.
   (d) maintains constant intensity and the same polarity all the time.

3. A sample of ferromagnetic material
   (a) cannot work as the core for an electromagnet.
   (b) does not "attract" or "stick to" magnets.
   (c) causes magnetic flux lines to bunch up more tightly than they do in free space.
   (d) has a permeability of 0.

4. The magnetic flux contours near the ends of a bar magnet take the form of
   (a) straight lines parallel to the bar's axis.
   (b) straight lines perpendicular to the bar's axis.
   (c) circles whose centers lie on the bar's axis.
   (d) curves that converge on (or diverge from) the bar's ends.

5. If you want to build an electromagnet that will produce an alternating magnetic field when you connect it to a lantern battery, you should choose a core material that has

(a) high permeability and high retentivity.
(b) high permeability and low retentivity.
(c) low permeability and high retentivity.
(d) no known characteristics; you can't build such an electromagnet.

6. A metal rod can support a flux density of up to 800 G when DC flows in a coil surrounding it. When you remove the rod leaving only air, the flux density in that air goes down to 20 G. What's the permeability of the rod?

(a) You need more information to figure it out.
(b) 40
(c) 0.025
(d) 410

7. What's the retentivity of the rod described in Question 6?

(a) You need more information to figure it out.
(b) 40
(c) 0.025
(d) 410

8. Which of the following units expresses magnetic flux density?

(a) Tesla
(b) Ampere
(c) Coulomb
(d) Siemens

9. What's the magnetic flux density at a point 2.00 m away from a straight, thin wire carrying 600 mA of DC?

(a) 6.00×10^{-7} T
(b) 6.00×10^{-8} T
(c) 3.00×10^{-7} T
(d) 3.00×10^{-8} T

10. You wind 70 turns of heavy copper insulated wire in a coil around a rod-shaped ferromagnetic core. You drive 22 A of DC through the coil. Then you double the current to 44 A. If the core reaches a state of saturation and if 3.3 A or more flows through the coil, the current increase described here causes the flux density inside the core to essentially

(a) stay the same.
(b) increase by a factor of the square root of 2.
(c) double.
(d) quadruple.

11. A *normally open* relay completes an external circuit
    (a) whether current flows in its coil or not.
    (b) only when current flows in its coil.
    (c) only when no current flows in its coil.
    (d) only when AC flows in its coil.

12. At a specific point on the earth's surface, the term *geomagnetic declination* refers to
    (a) the vertical tilt of the earth's magnetic flux lines.
    (b) the horizontal deflection of the earth's magnetic field.
    (c) the geomagnetic flux density through a horizontal plane.
    (d) the angular difference between geomagnetic north and true north.

13. If you want to make a powerful permanent magnet, you'll need an alloy that has
    (a) low permeability.
    (b) high density.
    (c) high retentivity.
    (d) low resistance per unit of length.

14. When you place a current-carrying wire coil in a vacuum, the magnetic flux density inside that coil
    (a) goes down to zero.
    (b) increases compared with the flux density when it has a ferromagnetic core.
    (c) decreases compared with the flux density when it has a ferromagnetic core.
    (d) remains the same as the flux density when it has a ferromagnetic core.

15. At the geomagnetic equator, the geomagnetic field's force on a compass needle is
    (a) horizontal.
    (b) vertical.
    (c) slanted.
    (d) nonexistent.

16. If a solenoidal wire coil has 1000 turns and carries 30.00 mA of DC, then its magnetomotive force is
    (a) dependent on the coil's length, diameter, and core material.
    (b) 37.71 Gb.
    (c) 1131 Gb.
    (d) 6.885 Gb.

17. If a solenoidal coil has 60 turns and you connect it to a 6.3-V lantern battery, how much magnetomotive force does that coil produce?
    (a) It depends on its diameter.
    (b) It depends on its DC conductance.
    (c) It depends on its core material.
    (d) All of the above

18. If you insert a ferromagnetic rod of permeability 16.0 inside the coil described in Question 17, the magnetomotive force will
    (a) increase by a factor of 4.00.
    (b) increase by a factor of 16.0.
    (c) increase by a factor of 256.
    (d) not change.

19. The magnetic force between the ends of two electromagnets depends on
    (a) the electromagnets' core material.
    (b) the distance between those ends.
    (c) the currents in the coils.
    (d) All of the above

20. Geomagnetic storms often accompany
    (a) "northern lights" (*aurora borealis*).
    (b) sudden changes in the sun's diameter.
    (c) polarity reversals in the earth's magnetic field.
    (d) thundershowers in their immediate vicinity.

Test: Part 1

DO NOT REFER TO THE TEXT WHEN TAKING THIS TEST. A GOOD SCORE IS AT LEAST 37 CORRECT. Answers are in the back of the book. It's best to have a friend check your score the first time, so you won't memorize the answers if you want to take the test again.

1. What can we call a solid medium with resistance so high that it's infinite for practical purposes?
   - (a) A static medium
   - (b) A separator
   - (c) An insulator
   - (d) A dynamic medium
   - (e) A diamagnet

2. Which of the following characteristics represents a problem with analog meters that have a needle or pointer that moves continuously along a scale on the meter face?
   - (a) They require the user to work with a conversion chart to figure out the reading.
   - (b) They have poor sensitivity at low frequencies.
   - (c) They don't display rapidly changing values.
   - (d) The user must sometimes "guesstimate" the readings between scale divisions.
   - (e) They can respond only to AC, not to DC.

3. We connect a 24-V battery at opposite ends of a set of six 4-ohm resistors in series. As a result, we see the same voltage across each individual resistor. If we replace the 4-ohm resistors with 8-ohm resistors, what will happen to the voltage across each resistor?
   - (a) It will go down to ¼ of its previous value.
   - (b) It will go down to half its previous value.
   - (c) It will stay the same.
   - (d) It will double.
   - (e) It will quadruple.

4. In an electrical conductor, true direct current
   (a) never varies in intensity.
   (b) periodically reverses direction.
   (c) periodically varies in intensity.
   (d) never reverses direction.
   (e) always comes from a battery.

5. In a set of resistors connected in series with a battery, the current through a particular resistor
   (a) equals the current through any other resistor.
   (b) equals the battery voltage divided by the resistance of the resistor.
   (c) equals the voltage across the resistor.
   (d) equals the battery current divided by the total number of resistors.
   (e) equals the battery current minus the combined current through all the other resistors.

6. If you live near the equator, you can expect to see the *aurora borealis* ("northern lights") at night
   (a) during a geomagnetic storm.
   (b) just before or after a tropical storm.
   (c) when a volcanic eruption occurs nearby.
   (d) during a lunar eclipse.
   (e) rarely, if ever.

7. A DC ohmmeter will give us a correct reading for a component in isolation (not connected to anything else) if that component
   (a) contains inductance and resistance in series.
   (b) contains capacitance and resistance in parallel.
   (c) contains pure resistance.
   (d) contains pure inductance.
   (e) Any of the above

8. In a DC circuit, the ratio of current to conductance in a length of wire expresses
   (a) voltage.
   (b) power.
   (c) energy.
   (d) resistance.
   (e) permeability.

9. In a DC circuit, 1000 μA flows through a resistor that dissipates 200 mW. What is the voltage drop (potential difference) across this resistor?
   (a) We need more information to calculate it.
   (b) 200 mV
   (c) 2 V
   (d) 20 V
   (e) 200 V

10. What purpose do the resistors serve in the arrangement shown by Fig. Test 1-1?
    (a) They short-circuit the input, protecting the point X from excessive current.
    (b) They bleed off excess electrical charge from point X.
    (c) They keep the battery charged from a DC source connected to point X.
    (d) They provide a specific DC voltage at point X between 0 V and + 12 V.
    (e) They turn the DC from the voltage source into AC for use with a common household appliance.

11. Zinc-carbon flashlight cells
    (a) can be discharged only once, and then you must discard them.
    (b) have higher voltages than other types of cells.
    (c) supply about 0.5 to 0.6 V DC.
    (d) can be recharged hundreds of times if you take good care of them.
    (e) contain liquid acid that can cause serious burns.

12. How long would a 50-W bulb have to remain continuously aglow to consume 6 kWh of energy?
    (a) 50 minutes
    (b) 1 hour and 12 minutes
    (c) 5 days
    (d) 12 days and 12 hours
    (e) We need more information to answer this question.

13. Suppose that a 24-V battery causes 200 mA of current to flow through a light bulb. What's the resistance of the bulb under these conditions?
    (a) 30 ohms
    (b) 60 ohms
    (c) 120 ohms

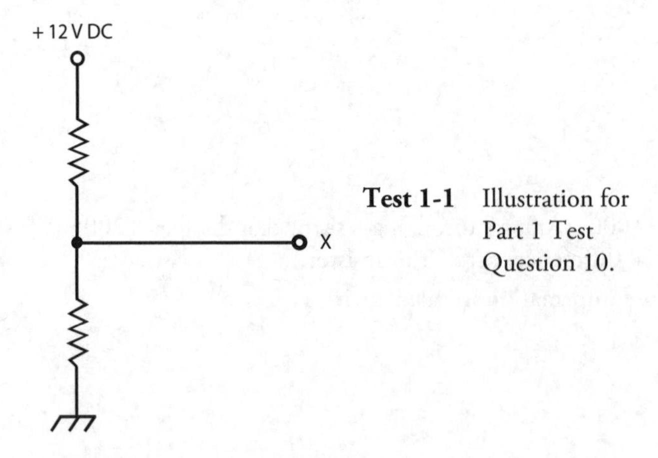

Test 1-1 Illustration for Part 1 Test Question 10.

(d) 240 ohms

(e) We need more information to figure it out.

14. We can use a series combination of resistors to

(a) build a voltage divider.

(b) limit the current through a device.

(c) get an ohmic value that none of our individual resistors have.

(d) bleed off the charge from a capacitor.

(e) All of the above

15. A phenomenon that shortens a cell's discharge cycle so that you must recharge it more often than normal, can occur in

(a) alkaline cells.

(b) zinc-carbon cells.

(c) lead-acid cells.

(d) mercury cells.

(e) None of the above

16. We connect a 6-V battery to a *series* combination of two 30-ohm resistors. How much power does either resistor dissipate all by itself?

(a) 150 mW

(b) 300 mW

(c) 1.2 W

(d) 4.8 W

(e) 9.6 W

17. We connect a 6-V battery to a *parallel* combination of two 30-ohm resistors. How much power does either resistor dissipate all by itself?

(a) 150 mW

(b) 300 mW

(c) 1.2 W

(d) 4.8 W

(e) 9.6 W

18. For a battery-powered DC device, let's say that E represents the battery voltage in volts, I represents the circuit current in amperes, and R represents the circuit resistance in ohms. Which of the following formulas correctly portrays Ohm's law?

(a) $I = R/E$

(b) $E = IR$

(c) $R = I/E$

(d) $R = EI$

(e) $I = ER$

19. The number of electrons in an atom's nucleus is
    (a) the atomic weight.
    (b) the atomic number.
    (c) the valence number.
    (d) the charge number.
    (e) zero.

20. We can express the rate at which moving charge carriers pass a fixed point in a wire as
    (a) amperes.
    (b) coulombs.
    (c) volt-seconds.
    (d) watts.
    (e) watt-seconds.

21. In the circuit of Fig. Test 1-2, how much potential difference exists across the 100-ohm resistor?
    (a) 3 V
    (b) 6 V
    (c) 9 V
    (d) 12 V
    (e) We need more information to calculate it.

22. If we connect 100 resistors, each rated at 47 ohms and ½ W, in a 10 × 10 series-parallel network, we'll end up with a 47-ohm resistor capable of dissipating up to
    (a) 2 W.
    (b) 5 W.
    (c) 20 W.
    (d) 50 W.
    (e) 100 W.

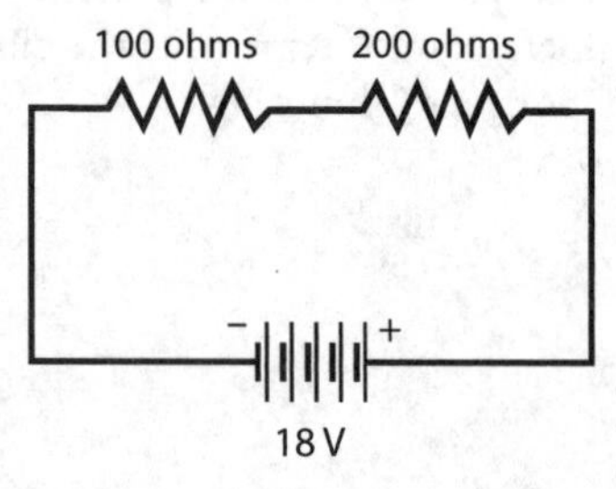

Test 1-2 Illustration for Part 1 Test Question 21.

23. The standard unit of DC potential difference is the theoretical equivalent of
    (a) an ohm-volt.
    (b) a volt-ampere.
    (c) an ohm-ampere.
    (d) a watt-ohm.
    (e) a watt-ampere.

24. Why should you never use a lead-acid automotive battery to power-up an electromagnet?
    (a) Oh, but you should! That type of battery is ideal for electromagnets.
    (b) The electromagnet will not have the correct polarity.
    (c) Too much residual magnetism will occur.
    (d) Dangerous acid might boil out of the battery.
    (e) The battery will not provide enough current.

25. We can wind a coil of wire around a hiker's compass to make a simple
    (a) voltmeter.
    (b) wattmeter.
    (c) galvanometer.
    (d) potentiometer.
    (e) electromagnet.

26. In order to get a readable, intuitive indication of a fluctuating current level, even if not especially precise, we would want
    (a) an electrometer.
    (b) an analog meter.
    (c) a bar-graph meter.
    (d) a digital meter.
    (e) a decibel meter.

27. We can use decibels to compare two different
    (a) magnetic fields.
    (b) AC frequencies.
    (c) DC resistances.
    (d) battery voltages.
    (e) sound levels.

28. The term *proton-exchange-membrane* refers to a type of
    (a) fuel cell.
    (b) dry cell.
    (c) standard cell.
    (d) wet cell.
    (e) solar cell.

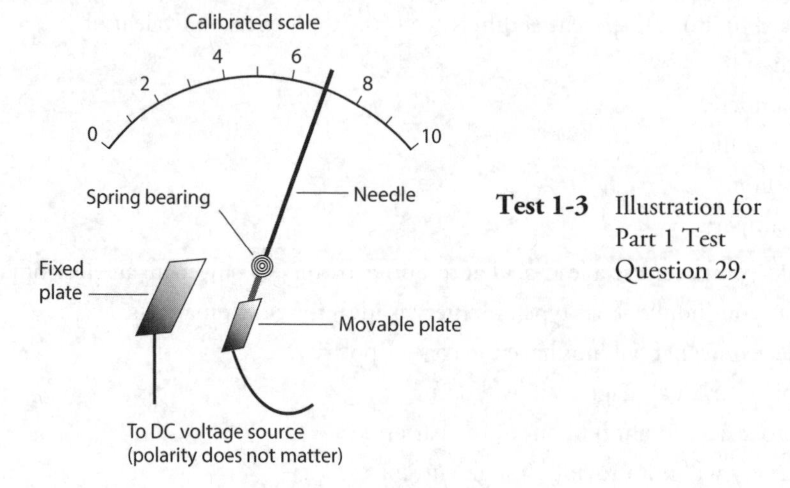

Test 1-3 Illustration for Part 1 Test Question 29.

29. The mechanical device shown in Fig. Test 1-3 forms the basis for
    (a) a D'Arsonval meter.
    (b) a digital meter.
    (c) a magnetic meter.
    (d) an electrostatic meter.
    (e) a decibel meter.

30. We might install a wirewound resistor
    (a) to dissipate DC power.
    (b) in a circuit in which inductance can't be tolerated.
    (c) across the primary winding of a step-down transformer.
    (d) across the secondary winding of a step-up transformer.
    (e) across a battery to increase its voltage.

31. A material with permeability greater than 1
    (a) concentrates magnetic flux relative to a vacuum.
    (b) allows a magnetic field to exist, whereas a vacuum does not.
    (c) exhibits lower retentivity than a vacuum.
    (d) dilates magnetic flux relative to a vacuum.
    (e) completely blocks magnetic fields.

32. The nucleus of an atom
    (a) contains at least one neutron.
    (b) contains at least one electron.
    (c) never contains any neutrons.
    (d) always has an electric charge.
    (e) always produces X rays.

33. Consider four resistors, each quoted at 220 ohms ±5% by the manufacturer. We measure their actual resistances with a precision ohmmeter and get the following readings. Which resistor fails to meet the manufacturer's claimed specification?
   (a) The first resistor, which tests at 208 ohms.
   (b) The second resistor, which tests at 233 ohms.
   (c) The third resistor, which tests at 234 ohms.
   (d) The fourth resistor, which tests at 207 ohms.
   (e) All four resistors fail to meet the manufacturers claimed specification.

34. We can theoretically represent an electrical charge quantity in terms of
   (a) volt-seconds.
   (b) watt-seconds.
   (c) ohm-seconds.
   (d) siemens-seconds.
   (e) ampere-seconds.

35. We connect a 24-V battery across a set of four resistors in parallel. The resistors have values of 100, 200, 300, and 400 ohms. How much current does the 300-ohm resistor draw?
   (a) 24 mA
   (b) 30 mA
   (c) 60 mA
   (d) 80 mA
   (e) We need more information to calculate it.

36. What purpose does capacitor C in Fig. Test 1-4 serve?
   (a) It smooths out the ripple from the input DC.
   (b) It increases the voltage across resistor R.
   (c) It limits the voltage across equipment connected to the output.
   (d) It protects resistor R against damage from current surges.
   (e) It enhances the ability of resistor R to limit the voltage.

37. We can theoretically express DC resistance in terms of
   (a) volts per ampere.
   (b) volt-amperes.
   (c) amperes per volt.
   (d) volt-siemens.
   (e) siemens per ampere.

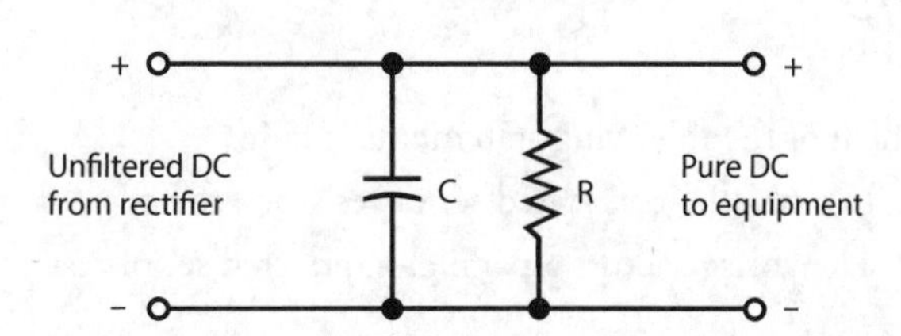

Test 1-4 Illustration for Part 1 Test Question 36.

38. Which, if any, of the following statements concerning potentiometers holds true?
   (a) All potentiometers function in high-power DC circuits.
   (b) All potentiometers exhibit inductive reactance.
   (c) Some potentiometers can measure extremely high voltages.
   (d) All potentiometers have wirewound elements.
   (e) None of the above

39. We connect the end terminals of a multiple-turn, circular wire loop to a resistive load, place a powerful bar magnet inside the loop, and then keep the loop fixed with respect to the magnet. We will observe
   (a) no current in the load.
   (b) a change in the load resistance.
   (c) an alternating magnetic field in the load.
   (d) a DC voltage across the load.
   (e) a steady magnetic field in the load.

40. A positively charged atom *always* has
   (a) more electrons than protons.
   (b) more protons than electrons.
   (c) the same number of protons and electrons.
   (d) more neutrons than protons.
   (e) more protons than neutrons.

41. In DC electrical circuits, the term *power* refers to the instantaneous rate at which
   (a) voltage changes.
   (b) current changes.
   (c) energy is expended.
   (d) charge quantity accumulates.
   (e) conductivity varies.

42. We connect a 36-V battery to a set of three resistors in series, having values of 3.0, 4.0, and 5.0 ohms. How much potential difference appears across the 4.0-ohm resistor?
   (a) 9.0 V
   (b) 12 V
   (c) 15 V
   (d) 18 V
   (e) 24 V

43. Which of the following statements is false?
   (a) In a parallel-connected set of resistors, the current is the same at every point.
   (b) The voltage across a parallel-connected set of resistors divides equally among them.

(c) The current in a series-connected set of resistors depends on which particular resistor we look at.

(d) The net ohmic value of a parallel-connected set of resistors exceeds the ohmic value of the largest individual resistor.

(e) All of the above statements are false.

44. What type of cell does Fig. Test 1-5 illustrate?

(a) A silicon-zinc cell

(b) A fuel cell

(c) A PV cell

(d) A barred metallic cell

(e) A Weston standard cell

45. Under ideal conditions and with no load, the cell shown in Fig. Test 1-5 will put out

(a) 0.6 V DC.

(b) 1.2 V DC.

(c) 1.5 V DC.

(d) 1.8 V DC.

(e) 2.1 V DC.

46. Suppose that an atom of carbon, whose nucleus usually harbors six protons and six neutrons, has eight neutrons instead. We can call it

(a) an isotope.

(b) a compound.

(c) a cation.

(d) an anion.

(e) an ion.

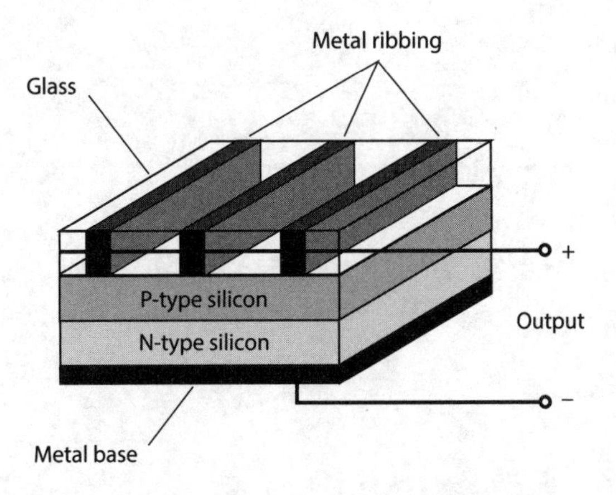

Test 1-5 Illustration for Part 1 Test Questions 44 and 45.

47. An advantage of a lead-acid battery over a zinc-carbon battery is the fact that the lead-acid battery
    (a) provides more voltage.
    (b) can be discharged and recharged many times.
    (c) can produce DC from visible light.
    (d) has greater mass per unit of energy output.
    (e) does not contain toxic carbon.

48. A DC electromagnet
    (a) has constant polarity.
    (b) requires an air core.
    (c) requires a high-resistance coil.
    (d) has polarity that periodically reverses.
    (e) behaves exactly like an AC electromagnet.

49. Which of the following properties is an asset of PV cells?
    (a) Their recharge capability makes them ideal for use in cell phones and tablets.
    (b) Their large energy-storage capacity makes them useful in large appliances.
    (c) Their lack of inductance or capacitance makes them useful in radio transmitters.
    (d) Their high voltage allows them to directly replace lantern and automotive batteries.
    (e) Their sensitivity to light makes them good for use in small solar-powered devices.

50. An analog voltmeter's mechanism *directly* measures
    (a) frequency.
    (b) current.
    (c) power.
    (d) energy.
    (e) charge.

2
PART
Alternating Current

9 CHAPTER

Alternating-Current Basics

WE CAN EXPRESS DC IN TERMS OF TWO VARIABLES: DIRECTION (POLARITY) AND INTENSITY (AMPLITUDE). If we want to gain a full understanding of alternating current (AC), we must work a little harder.

Definition of AC

Direct current or voltage has polarity that remains constant as time passes. Although the amplitude (the number of amperes, volts, or watts) can fluctuate from moment to moment, the charge carriers always flow in the same direction at any point in the circuit, and the charge poles always keep the same relative orientation.

In AC, the charge-carrier flow and the polarity reverse at regular intervals. The *instantaneous amplitude* (the amplitude at any given instant in time) of AC usually varies because of the repeated reversal of polarity. But in some cases, the amplitude remains constant, even though the polarity keeps reversing. The repeating change in polarity makes AC fundamentally different from DC.

Period and Frequency

In a *periodic AC wave*, the kind that we'll work with in this chapter (and throughout the rest of this book), the function of *instantaneous amplitude versus time* repeats, and the same pattern recurs indefinitely. The length of time between one complete iteration of the wave pattern, or one *cycle*, and the next iteration constitutes the *period* of the wave. Figure 9-1 shows two complete cycles, and therefore, two periods, of a simple AC wave. In theory, the period of a wave can range from a tiny fraction of a second to millions of years. When we want to express the period of an AC wave in seconds, we can denote it by writing T.

In the "olden days," scientists and engineers specified AC frequency in *cycles per second*, abbreviated *CPS*. They expressed medium and high frequencies in *kilocycles*, *megacycles*, or *gigacycles*, representing thousands, millions, or billions (thousand-millions) of cycles per second. Nowadays, we express frequency in *hertz*, abbreviated Hz. The hertz and the cycle per second refer to exactly the

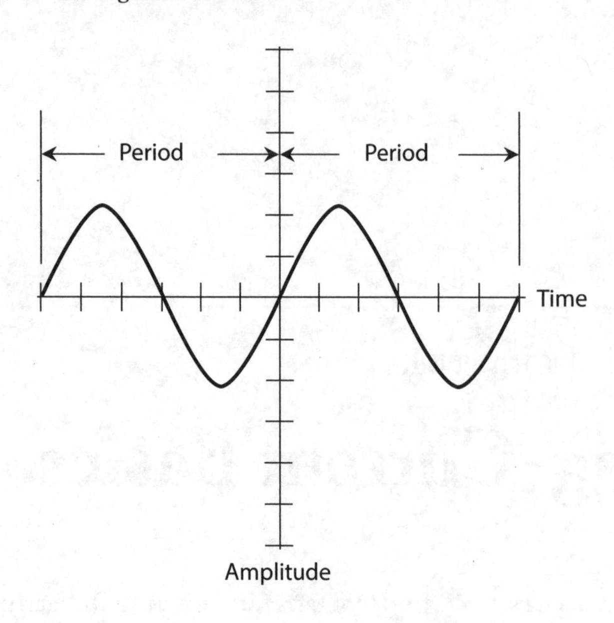

9-1 A sine wave. The period represents the time it takes for one cycle to complete itself.

same thing, so 1 Hz = 1 CPS, 10 Hz = 10 CPS, and so on. We can express medium and high frequencies in ***kilohertz*** (kHz), *megahertz* (MHz), or *gigahertz* (GHz), where

$$1 \text{ kHz} = 1000 \text{ Hz}$$
$$1 \text{ MHz} = 1000 \text{ kHz} = 1{,}000{,}000 \text{ Hz} = 10^6 \text{ Hz}$$
$$1 \text{ GHz} = 1000 \text{ MHz} = 1{,}000{,}000{,}000 \text{ Hz} = 10^9 \text{ Hz}$$

Sometimes we'll need an even bigger unit, the *terahertz* (THz), to specify AC frequency. A frequency of 1 THz constitutes a trillion (1,000,000,000,000 or 10^{12}) hertz. Electrical currents rarely attain such frequencies, although some forms of *electromagnetic radiation* do.

The frequency of an AC wave in hertz, denoted f, equals the reciprocal of the period T in seconds. Mathematically, we have

$$f = 1/T$$

and

$$T = 1/f$$

Some AC waves contain all their energy at a single frequency. We call such waves *pure*. But often, an AC wave contains energy components at multiples of the main, or *fundamental*, frequency. Components can sometimes also exist at frequencies that don't seem to bear any logical relation to the fundamental. Once in a while, we'll encounter a *complex AC wave* that contains energy at hundreds, thousands, or even infinitely many different component frequencies.

The Sine Wave

In its simplest form, AC has a *sine-wave*, or *sinusoidal*, nature. In a sine wave, the direction of the current reverses at regular intervals, and the current-versus-time curve follows the graph that we get when we plot the trigonometric *sine function* on a coordinate grid. Figure 9-1 shows the general shape of a sine wave—two complete cycles of it.

Whenever an AC wave contains all of its energy at a single frequency, it will exhibit a perfectly sinusoidal shape when we graph its amplitude as a function of time. Conversely, if an AC wave constitutes a perfect sinusoid, then all of its energy exists at a single frequency.

In practice, a wave might look like a sinusoid when displayed on an oscilloscope, even though it actually contains some imperfections too small to see. Nevertheless, a pure, single-frequency AC wave not only looks sinusoidal when we look at it on an oscilloscope screen, but it actually constitutes a *flawless* sinusoid.

Square Waves

As seen on an oscilloscope display, a *square wave* looks like a pair of parallel, dashed lines, one with positive polarity and the other with negative polarity (Fig. 9-2A). The scope shows a graph of voltage on the vertical scale and time on the horizontal scale. The transitions between negative and positive for a theoretically perfect square wave will not show up on the scope because they occur instantaneously (in no time at all). But in practice, we can sometimes see a square wave's transitions as vertical lines (Fig. 9-2B).

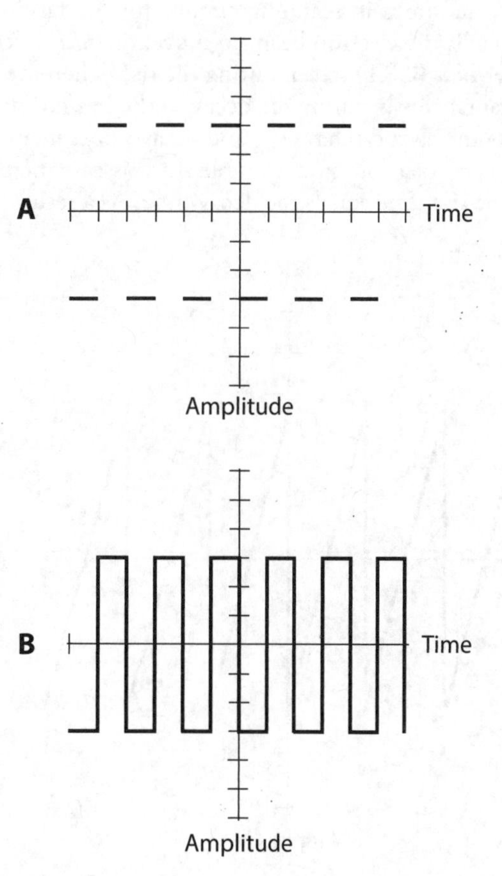

9-2 At A, the graphical rendition of a perfect square wave. The transitions occur instantaneously, so they do not appear visible. At B, the common rendition of a square wave, showing the transitions as vertical lines.

True square waves have equal negative and positive peaks, so the absolute amplitude of the wave never varies. Half of the time it's $+x$, and the other half of the time it's $-x$ (where x represents a certain fixed number of amperes or volts). Some squared-off waves appear "lopsided" because the negative and positive amplitudes differ. Still others remain at positive polarity longer than they remain at negative polarity (or vice-versa). They constitute examples of *asymmetrical square waves*, more properly called *rectangular waves*.

Sawtooth Waves

Some AC waves rise and/or fall in straight, sloping lines, as seen on an oscilloscope screen. The slope of the line indicates how fast the amplitude changes. We call them *sawtooth waves* because they resemble the "teeth" on a saw blade. Various electronic test devices and sound synthesizers can generate sawtooth waves with diverse frequencies and amplitudes.

Figure 9-3 shows a sawtooth wave in which the positive-going slope (called the *rise*) is essentially instantaneous as in a square or rectangular wave, but the negative-going slope (called the *decay*) is not so steep. The period of the wave equals the time between points at identical positions on two successive pulses.

Another form of sawtooth wave exhibits a defined, finite rise and an instantaneous decay. Engineers call it a *ramp* because each individual cycle looks like an incline going upwards (Fig. 9-4). Engineers use ramps in scanning circuits for old-fashioned television receivers and oscilloscopes. The ramp tells the electron beam to move, or *trace*, at constant speed from left to right across the *cathode-ray-tube* (CRT) screen during the rise. Then the ramp retraces, or brings the electron beam back, instantaneously during the decay so the beam can begin the next trace across the screen.

Sawtooth waves can have rise and decay slopes in an infinite number of different combinations. Figure 9-5 shows a common example. In this case, neither the rise nor the decay occurs instantaneously; the rise time equals the decay time. As a result, we get a so-called *triangular wave*.

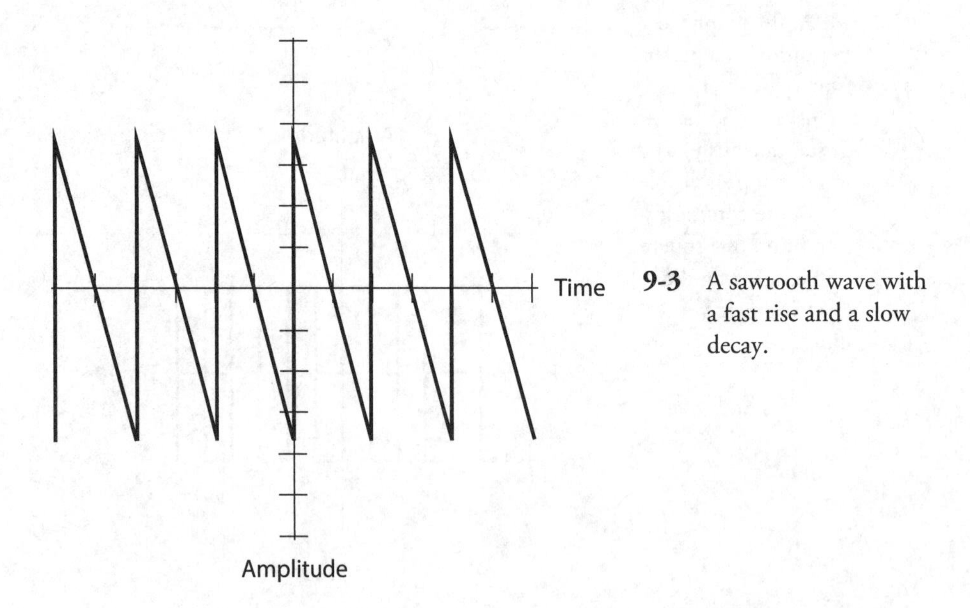

9-3 A sawtooth wave with a fast rise and a slow decay.

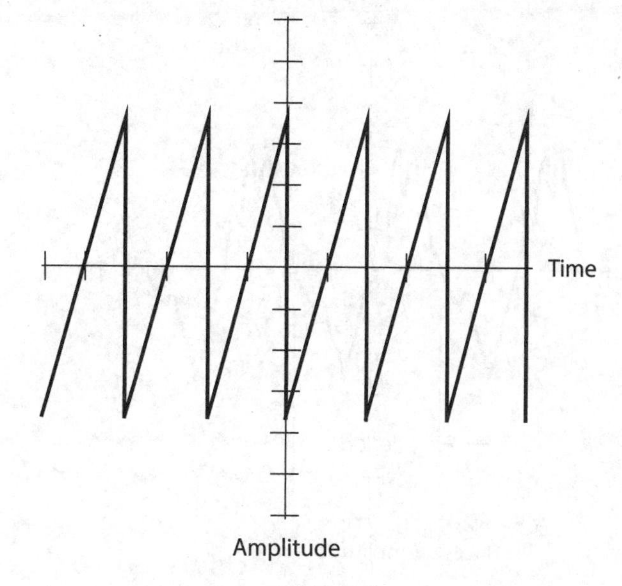

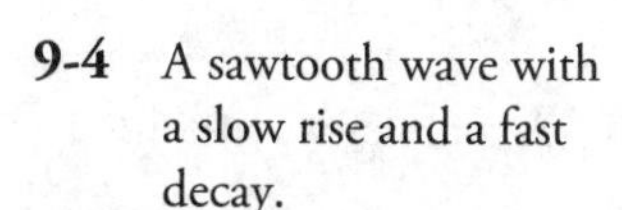

9-4 A sawtooth wave with a slow rise and a fast decay.

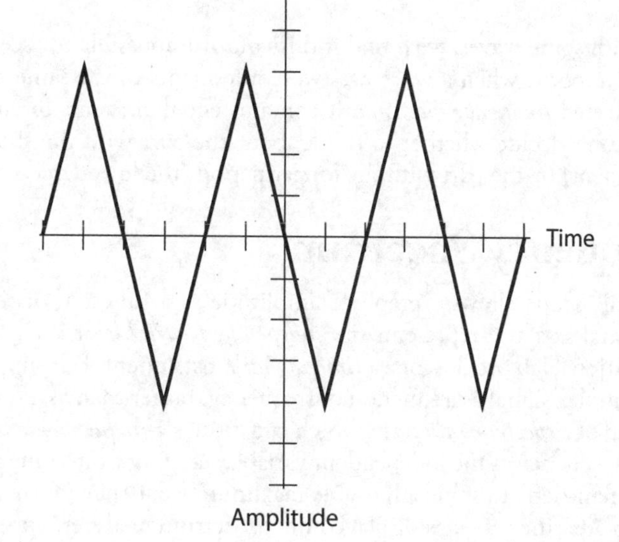

9-5 A triangular wave with equal rise and decay rates.

Complex Waveforms

As long as a wave has a definite period, and provided that the polarity keeps switching back and forth between positive and negative, it constitutes AC, no matter how complicated the actual shape of the waveform appears. Figure 9-6 shows an example of a complex AC wave. It has a definable period, and therefore, a definable frequency. The period equals the time between two points on succeeding wave repetitions.

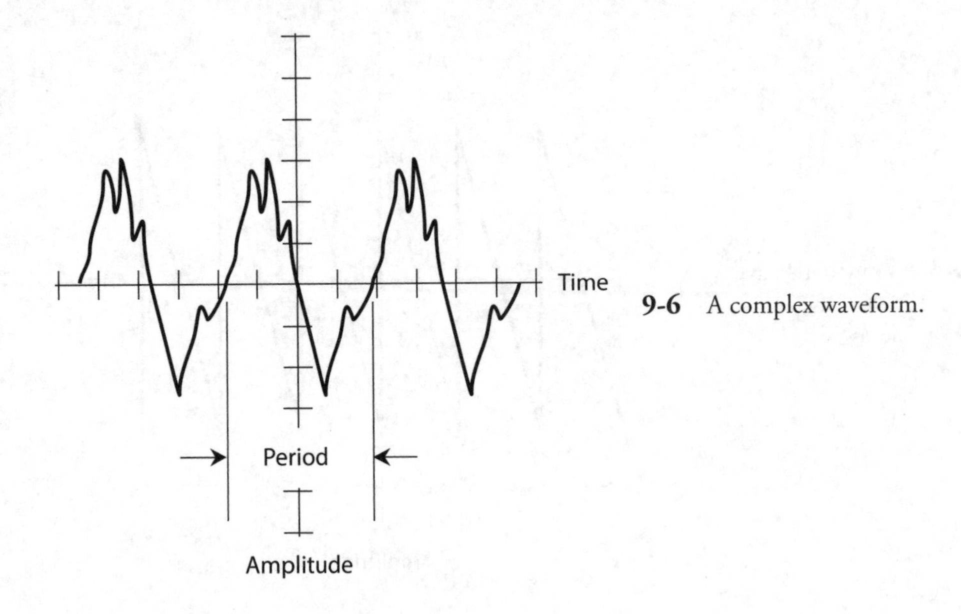

9-6 A complex waveform.

With some waves, we'll find it difficult or impossible to ascertain the period. This sort of situation can occur when a wave has two components of the same amplitude. Such a wave exhibits a multifaceted *frequency spectrum*; it contains equal amounts of energy at two different frequencies, so we can't decide whether to think about the part with the shorter period (the higher-frequency component) or the part with the longer period (the lower-frequency component).

Frequency Spectrum

An oscilloscope shows a graph of amplitude as a function of time. Because time appears on the horizontal axis and represents the *independent variable* or *domain* of the function, engineers call a conventional lab oscilloscope a *time-domain* instrument. But suppose you want to see the amplitude of a complex signal as a function of frequency, rather than as a function of time? You can do it with the help of a *spectrum analyzer*, which constitutes a *frequency-domain* instrument. Its horizontal axis shows frequency as the independent variable, ranging from some adjustable minimum frequency (at the extreme left) to some adjustable maximum frequency (at the extreme right).

An AC sine wave, as displayed on a spectrum analyzer, appears as a single *pip*, or vertical line, as shown in Fig. 9-7A. The wave concentrates all of its energy at a single frequency. But many, if not most, AC waves contain *harmonic* energy along with energy at the fundamental frequency. A harmonic is a secondary wave that occurs at a whole-number multiple of an AC wave's fundamental frequency. For example, if we have an AC wave whose fundamental frequency equals 60 Hz, then harmonics can exist at 120 Hz, 180 Hz, 240 Hz, and so on. The 120 Hz wave constitutes the *second harmonic*, the 180 Hz wave represents the *third harmonic*, the 240 Hz wave constitutes the *fourth harmonic*, and so on.

In general, if a wave has a frequency equal to *n* times the fundamental (where *n* equals some whole number), then we call that wave the *nth harmonic*. Figure 9-7B illustrates a wave's

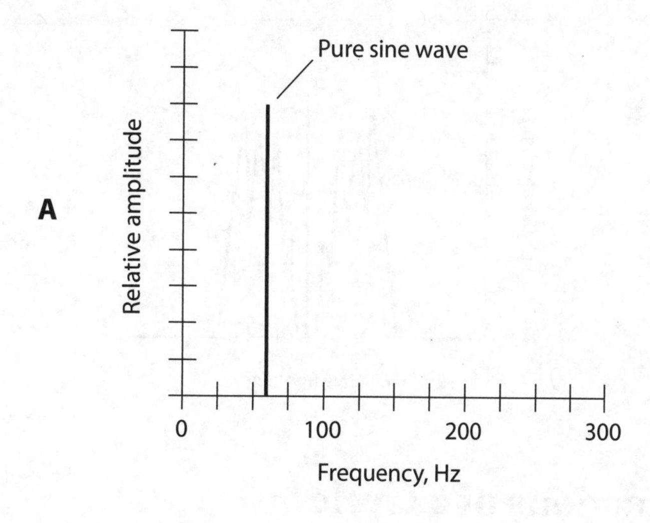

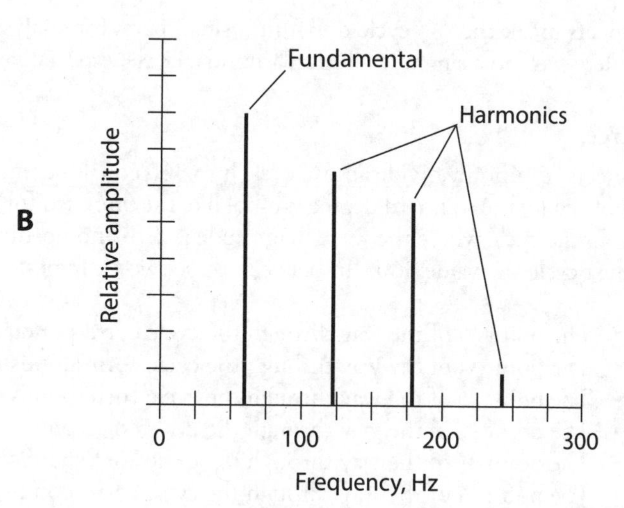

9-7 At A, a spectral diagram of a pure, 60-Hz sine wave. At B, a spectral diagram of a 60-Hz wave with three harmonics.

fundamental pip along with several harmonic pips as they would look on the display screen of a spectrum analyzer.

Square waves and sawtooth waves contain harmonic energy in addition to energy at the fundamental frequency. Other waves that contain a lot of harmonic energy can get more complicated. The exact shape of a wave depends on the amount of energy in the harmonics, and the way in which this energy is distributed among them.

Irregular waves can have any imaginable frequency distribution. Figure 9-8 shows a spectral (frequency-domain) display of an *amplitude-modulated* (AM) voice radio signal. Much of the energy concentrates at the frequency shown by the vertical line. That line portrays the signal's *carrier frequency*. We can also see evidence of energy near, but not exactly at, the carrier frequency. That part of the signal contains the voice information, technically called the signal's *intelligence*.

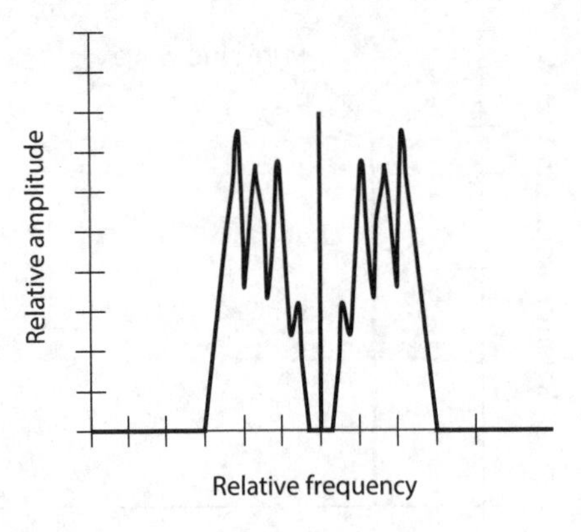

9-8 A spectral diagram of a modulated radio signal.

Fractions of a Cycle

Engineers break the AC cycle down into small parts for analysis and reference. We can compare a complete cycle to a single revolution around a circle, especially when we work with pure sine waves.

Degrees

Engineers commonly divide an AC cycle into 360 equal increments called *degrees* or *degrees of phase*, symbolized by the standard degree symbol like the one used for temperature (°). We assign 0° to the point in the cycle where the wave magnitude is zero and positive-going. We give the same point on the next cycle the value 360°. In between these two extremes, we have values such as the following:

- The point ⅛ of the way through the cycle corresponds to 45°.
- The point ¼ of the way through the cycle corresponds to 90°.
- The point ⅜ of the way through the cycle corresponds to 135°.
- The point ½ of the way through the cycle corresponds to 180°.
- The point ⅝ of the way through the cycle corresponds to 225°.
- The point ¾ of the way through the cycle corresponds to 270°.
- The point ⅞ of the way through the cycle corresponds to 315°.

You can doubtless imagine other points, or calculate them. You can multiply the fractional part of the cycle by 360° to get the number of degrees to which a particular point corresponds. Conversely, you can divide the number of degrees by 360 to get the fractional part of the cycle. Figure 9-9 illustrates a perfect sine wave cycle and the various degree points we encounter as we proceed through the cycle from beginning to end.

Radians

As an alternative to the degree scheme, we can break up a wave cycle into 2π equal parts, where π (pi) represents the circumference-to-diameter ratio of a circle. This constant equals approximately 3.1416. A *radian* (rad) of phase equals roughly 57.296°. Physicists often express the frequency of a wave in *radians per second* (rad/s) rather than in hertz. Because a complete 360° cycle comprises 2π radians, the *angular frequency* of a wave, in radians per second, equals 2π (approximately 6.2832) times the frequency in hertz.

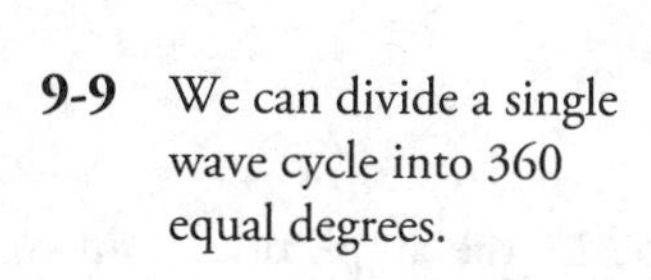
9-9 We can divide a single wave cycle into 360 equal degrees.

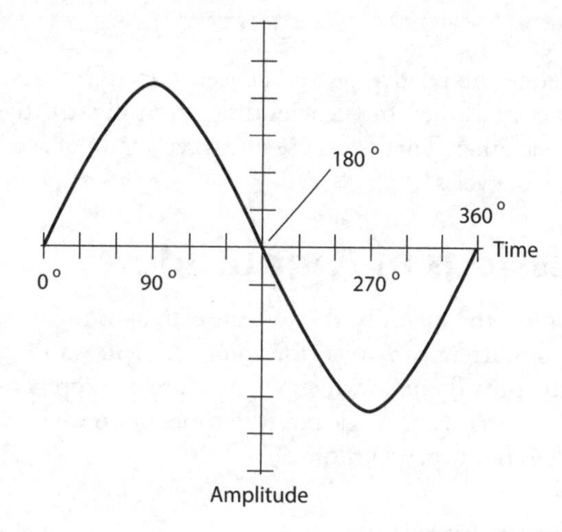

Phase Difference

Even if two AC waves have exactly the same frequency, they can produce different effects if they exist "out of sync" with each other. This phenomenon occurs in vivid fashion when AC waves combine to produce a third, or *composite*, wave. We can observe several "factoids" about combinations of pure sine waves, meaning AC waves that maintain constant frequency and that have equal (although opposite) positive and negative peak amplitudes.

- If two pure AC sine waves have the same frequency and the same intensity but differ in phase by 180° (½ cycle), then they precisely cancel each other out, and we don't observe any signal at all.
- If two pure AC sine waves have the same frequency and the same intensity, and if they coincide in phase (*phase coincidence*), then the composite wave has the same frequency and the same phase, but twice the intensity, of either wave alone.
- If two pure AC sine waves have the same frequency but different intensities, and if they differ in phase by 180°, then the composite signal has the same frequency as the originals, an intensity equal to the difference between the two, and a phase that coincides with the stronger of the two.
- If two pure AC sine waves have the same frequency and different intensities, and if they coincide in phase, then the composite wave has the same frequency and the same phase as the originals, and an intensity equal to the sum of the two.
- If two pure AC sine waves have the same frequency but differ in phase by some odd amount, such as 75° or 2.1 rad, then the resulting signal has the same frequency as the originals, but does not necessarily have the same wave shape, the same intensity, or the same phase as either original. As you can imagine, an infinite variety of such cases can occur.

In the United States, most household electricity from wall outlets consists of a 60-Hz sine wave with only one phase component. However, most electric utility companies send the energy over long distances as a combination of three separate 60-Hz waves, each differing by 120° of phase, which corresponds to ⅓ of a cycle. We call this mode *three-phase AC*. Each of the three waves carries ⅓ of the total power.

Caution!

We can define the relative phase between two AC waves if—*but only if*—they have precisely the same frequency. If their frequencies differ even slightly, then we can't define the phase of one wave relative to the other. That's because one wave's train of cycles keeps "catching and passing" the other wave's train of cycles.

Expressions of Amplitude

Depending on the quantity that we measure, we might specify the amplitude of an AC wave in amperes (for current), in volts (for voltage), or in watts (for power). In any case, we must remain aware of the time frame in which we measure or express the amplitude. Do we want to look at the amplitude of a wave at a single point in time, or do we want to express the amplitude as some quantity that does not depend on time?

Instantaneous Amplitude

The *instantaneous amplitude* of an AC wave equals the amplitude at some precise moment, or instant, in time. This value constantly changes as the wave goes through its cycle. The manner in which the instantaneous amplitude varies depends on the waveform. We can represent instantaneous amplitude values as individual points on the graphical display of a waveform.

Peak Amplitude

The *peak* (pk) *amplitude* of an AC wave is the maximum extent, either positive or negative, that the instantaneous amplitude attains. In many situations, the *positive peak amplitude* (pk+) and the *negative peak amplitude* (pk−) of an AC wave represent exact mirror images of each other. But sometimes things get more complicated. Figure 9-9 shows a wave in which the positive peak amplitude equals the negative peak amplitude, the only difference being in the polarity. Figure 9-10 shows a "lopsided" wave with different positive and negative peak amplitudes.

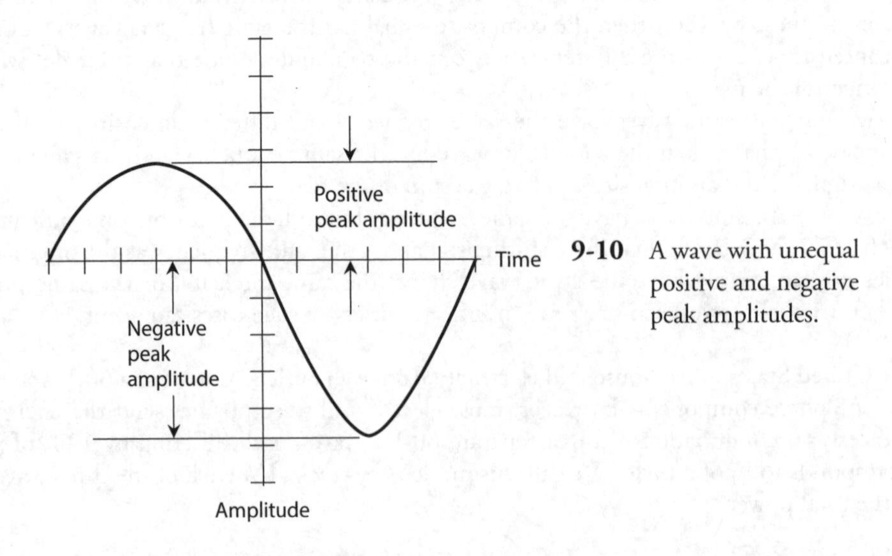

9-10 A wave with unequal positive and negative peak amplitudes.

In rigorous terms, we can define the positive peak amplitude of a waveform as the positive (upward) displacement from the horizontal axis to the *maximum* instantaneous voltage amplitude point on the graph (the *crest* of the waveform). Conversely, we can define the negative peak amplitude as the negative (downward) displacement from the horizontal axis to the *minimum* instantaneous amplitude point (the *trough* of the waveform).

Peak-to-Peak Amplitude

The *peak-to-peak* (pk-pk) *amplitude* of a wave equals the net difference between the positive peak amplitude and the negative peak amplitude, as shown in Fig. 9-11. When the positive and negative peak amplitudes of an AC wave have equal extent and opposite polarity (as they do in Fig. 9-9, for example), then the peak-to-peak amplitude equals exactly 2 times the positive peak amplitude, and exactly −2 times the negative peak amplitude. However, we can't make such simplistic statements for a "lopsided" wave (as we see in Fig. 9-10, for example).

Root-Mean-Square (RMS) Amplitude

Often, we'll want to quantify or express the *effective amplitude* of an AC wave. The effective amplitude of an AC wave constitutes the voltage, current, or power that a DC source would have to produce in order to cause the same general effect as the AC wave does. When someone tells you that a wall outlet provides 117 V, they usually mean 117 effective volts. Effective voltage usually differs from the peak voltage or the peak-to-peak voltage.

The most common expression for effective AC wave intensity is the *root-mean-square* (RMS) amplitude. This terminology reflects how we "operate on" the AC waveform. We start by squaring all the instantaneous amplitude values, point-by-point, to make everything positive. Then we average the resulting "all-positive" wave over one full cycle. Finally we take the square root of that average. The mathematics gets advanced, but the RMS scheme reflects real-world wave behavior. In practice, electronic metering devices do the "calculations" for us, providing direct RMS readings.

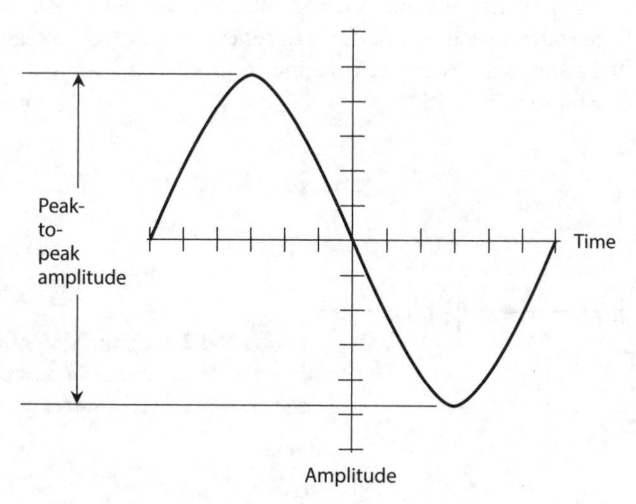

9-11 Peak-to-peak (pk-pk) amplitude of a sine wave.

Sine-Wave Value Basics

We can state five simple rules that apply to pure sine waves with equal and opposite positive and negative peak amplitudes:

1. The RMS value is approximately 0.707 times the positive peak value or −0.707 times the negative peak value.
2. The RMS value is approximately 0.354 times the peak-to-peak value.
3. The positive peak value is approximately 1.414 times the RMS value, and the negative peak value is approximately −1.414 times the RMS value.
4. The peak-to-peak value is approximately 2.828 times the RMS value.
5. The average value is always zero (the sum of the positive and negative peak values, taking the polarity signs into account).

We'll often specify RMS amplitude when talking about utility AC, radio-frequency (RF) AC, and audio-frequency (AF) AC signals.

Other RMS Values

Non-sine waves follow different rules for RMS amplitude. In the case of a perfect square wave, for example, the RMS value equals the positive or negative peak value, which in turn equals half the peak-to-peak value. For sawtooth and irregular waves, the relationship between the RMS value and the peak value depends on the exact shape of the wave.

> **Tech Tidbit**
>
> We can't work out the average voltage of an irregular wave unless we know the exact waveform. That calculation usually needs a computer and a sophisticated waveform analyzer.

Superimposed DC

Sometimes a wave has components of both AC and DC. We can get an AC/DC combination by connecting a DC voltage source, such as a battery, in series with an AC voltage source, such as the utility mains. Figure 9-12 shows an example. Imagine connecting a 12-V battery in series with the wall outlet. Imagine it—but don't do it!

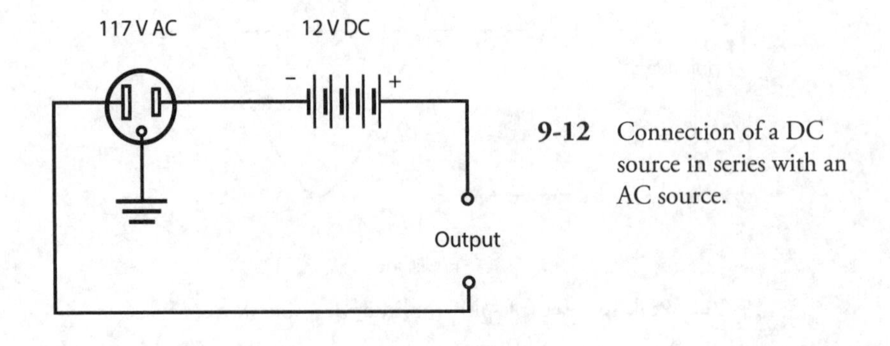

9-12 Connection of a DC source in series with an AC source.

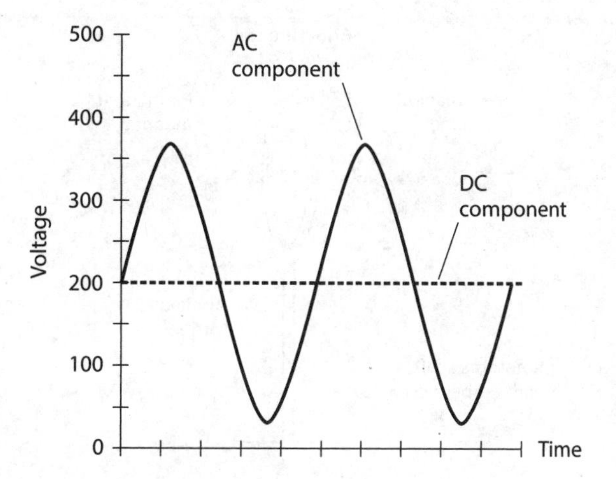

9-13 Waveform resulting from a 117-V AC sine-wave source connected in series with a +200-V DC source.

Warning!

Do not try this experiment.
The battery could rupture, and its chemical paste or fluid could injure you.

The AC wave is displaced either positively or negatively by 12 V, depending on the polarity of the battery. The voltage combination results in a sine wave at the output, but one peak exceeds the other peak by 24 V (twice the battery voltage).

Any AC wave can have a DC component superimposed. If the DC component exceeds the peak value of the AC wave, then fluctuating, or pulsating, DC will result. This would happen, for example, if a +200-V DC source were connected in series with the output of a common utility AC outlet, which has peak voltages of approximately ±165 V. Pulsating DC would appear, with an average value of +200 V but with instantaneous values much higher and lower (Fig. 9-13).

The Generator

We can generate AC by rotating a coil of wire in a magnetic field, as shown in Fig. 9-14. An AC voltage appears between the ends of the wire coil. The AC voltage that a generator can produce depends on the strength of the magnetic field, the number of turns in the wire coil, and the speed at which the coil rotates. The AC frequency depends only on the speed of rotation. Normally, for utility AC, this speed equals 3600 revolutions per minute (r/min), or 60 complete revolutions per second (RPS), so the AC output frequency equals 60 Hz.

When we connect a *load*, such as a light bulb or an electric space heater, to an AC generator, we will experience difficulty turning the generator shaft, as compared to *no-load conditions* (nothing connected to the output). As the amount of electrical power demanded from a generator increases, so does the mechanical power required to drive it. That's why we can't expect to connect a generator to a stationary bicycle and pedal an entire city into electrification. We can never get something for nothing. The electrical power that comes out of a generator can never exceed the mechanical power driving it. In fact, some energy always goes to waste, mainly as heat in the generator. Your

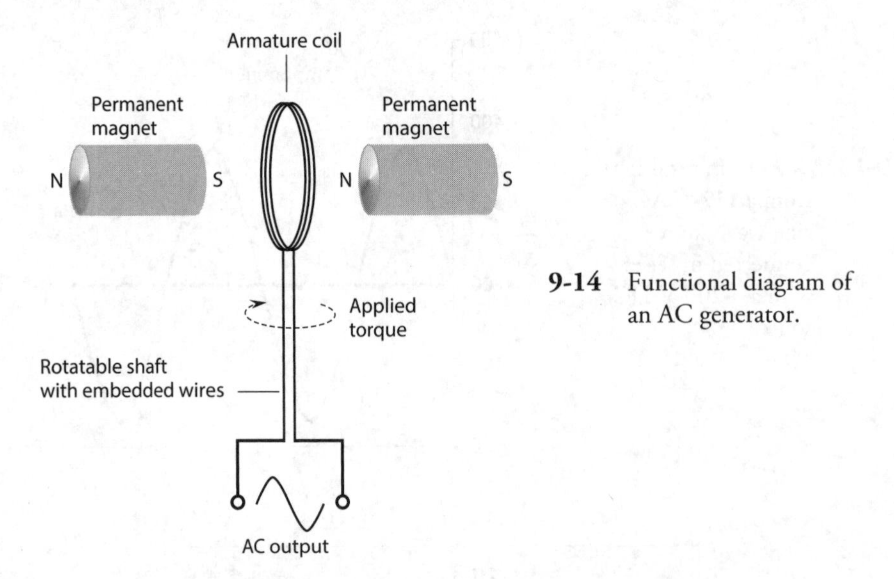

9-14 Functional diagram of an AC generator.

legs might generate enough power to run a small radio or television set, but nowhere near enough to provide electricity for a city.

The *efficiency* of a generator equals the ratio of the electrical output power to the mechanical driving power, both measured in the same units (such as watts or kilowatts), multiplied by 100 to get a percentage. No generator reaches 100% efficiency in the "real world," but a good one can come fairly close.

At power plants, massive turbines drive the electric generators. Heated steam, under pressure, forces the turbines to rotate. This steam derives from natural sources of energy such as fossil-fuel combustion, nuclear reactions, or heat from deep inside the earth. In some power plants, moving water directly drives the turbines. In still other facilities, wind drives them. Any of these energy sources, properly harnessed, can provide tremendous mechanical power, explaining why power plants can produce megawatts of electrical power.

Why AC and Not DC?

Do you wonder why the electric utility companies produce AC instead of DC? Well, AC may seem more complicated than DC in theory, but in practice AC has proven simpler to implement when we want to provide electricity to a large number of people. Electricity in the form of AC lends itself to voltage transformation, but in the form of DC it does not. Electrochemical cells produce DC directly, but they can't supply large populations. To serve millions of consumers, we need the immense power of falling or flowing water, the ocean tides, wind, fossil fuels, controlled nuclear reactions, or geothermal heat. All of these energy sources can drive turbines that turn AC generators.

Technology continues to advance in the realm of solar-electric energy. Someday a significant part of our electricity might come from *photovoltaic* power plants. These would generate DC. We could obtain high voltages by connecting giant arrays of solar panels in series. For now, however, photovoltaic energy works well only for individual consumers and small communities, mainly because of the expense of gigantic systems.

Thomas Edison favored DC over AC for electrical power transmission before the electric infrastructure had been designed and developed. His colleagues argued that AC would work better. But

Edison knew something that his contemporaries apparently preferred to ignore, if they knew it at all. At extremely high voltages, DC travels more efficiently over long distances than AC does. Long lengths of wire exhibit less *effective resistance* (also called *ohmic loss*) with DC than with AC, and less energy goes to waste in the form of magnetic fields surrounding the wire. Direct-current *high-tension* (high-voltage) transmission lines might, therefore, find favor someday. If engineers can find a way to bring the cost of such systems within reason, Edison might win a belated victory!

Quiz

Refer to the text in this chapter if necessary. A good score is at least 18 correct. Answers are in the back of the book.

1. In a perfect AC sine wave with no DC component, the peak-to-peak voltage equals
   (a) half the positive peak voltage.
   (b) the positive peak voltage.
   (c) twice the positive peak voltage.
   (d) four times the positive peak voltage.

2. In a perfect AC sine wave with no DC component, $\pi/2$ radians of phase represents
   (a) 1/6 of a full cycle.
   (b) 1/4 of a full cycle.
   (c) 1/3 of a full cycle.
   (d) 2/3 of a full cycle.

3. A rectangular wave has
   (a) an instantaneous rise time and a defined, finite decay time.
   (b) a defined, finite rise time and an instantaneous decay time.
   (c) a defined, finite rise time and an equal finite decay time.
   (d) an instantaneous rise time and an instantaneous decay time.

4. An engineer says that an AC sine wave has a frequency of 100 Hz. A physicist might say that the angular frequency is
   (a) 50π rad/s
   (b) 100π rad/s
   (c) 200π rad/s
   (d) $100\pi^2$ rad/s

5. Figure 9-15 shows an AC wave as it might appear on an oscilloscope screen. Each vertical division represents 100 mV. Each horizontal division represents 100 μs. What's the frequency of this wave?
   (a) 2.00 kHz
   (b) 5.00 kHz
   (c) 10.0 kHz
   (d) The graph lacks enough information to say.

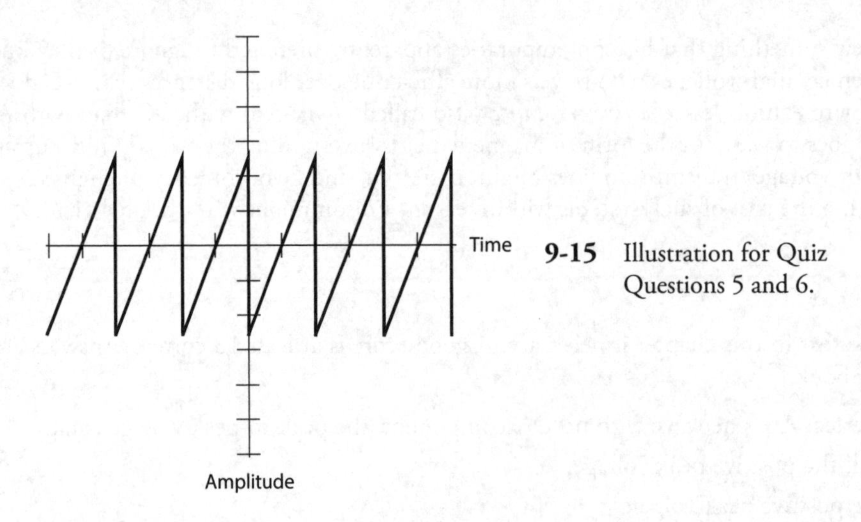

9-15 Illustration for Quiz Questions 5 and 6.

6. What's the negative peak voltage of the wave shown in Fig. 9-15, based on the appearance of the graph? Positive voltages go upward on the vertical scale, while negative voltages go downward.
   (a) −250 mV pk−
   (b) −500 mV pk−
   (c) −750 mV pk−
   (d) The graph lacks enough information to say.

7. Figure 9-16 shows another wave. Each vertical division represents 100 mV, and each horizontal division represents 100 μs. What's the period?
   (a) 100 μs
   (b) 200 μs
   (c) 400 μs
   (d) The graph lacks enough information to say.

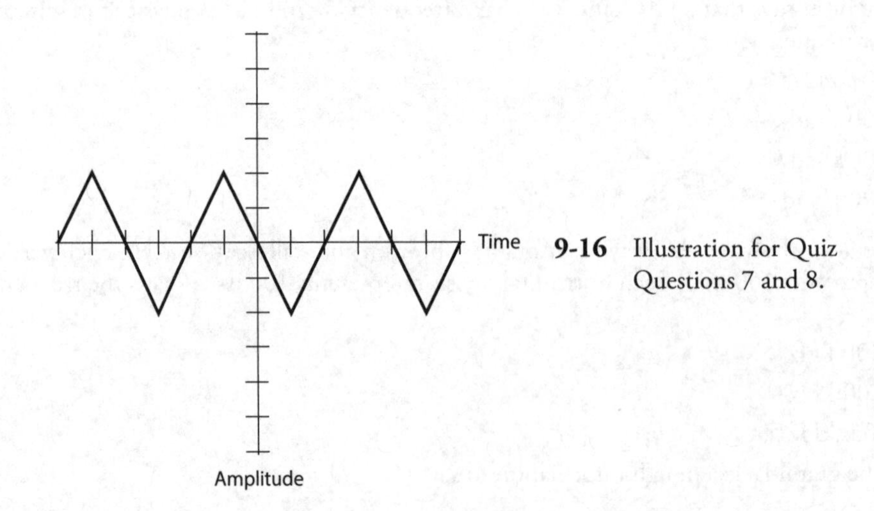

9-16 Illustration for Quiz Questions 7 and 8.

8. What's the peak-to-peak voltage of the wave shown in Fig. 9-16? Each vertical division represents 100 mV and each horizontal division represents 100 μs.
   (a) 100 mV pk-pk
   (b) 200 mV pk-pk
   (c) 400 mV pk-pk
   (d) The graph lacks enough information to say.

9. Figure 9-17 shows still another wave. Each vertical division represents 100 mV, and each horizontal division represents 100 μs. What's the period?
   (a) 100 μs
   (b) 200 μs
   (c) 400 μs
   (d) The graph lacks enough information to say.

10. What's the frequency of the wave shown in Fig. 9-17?
   (a) 1.25 kHz
   (b) 2.50 kHz
   (c) 500 kHz
   (d) The graph lacks enough information to say.

11. What's the positive peak voltage of the wave shown in Fig. 9-17? Positive voltages go upward on the vertical scale, while negative voltages go downward.
   (a) +240 mV pk+
   (b) +120 mV pk+
   (c) +480 mV pk+
   (d) The graph lacks enough information to say.

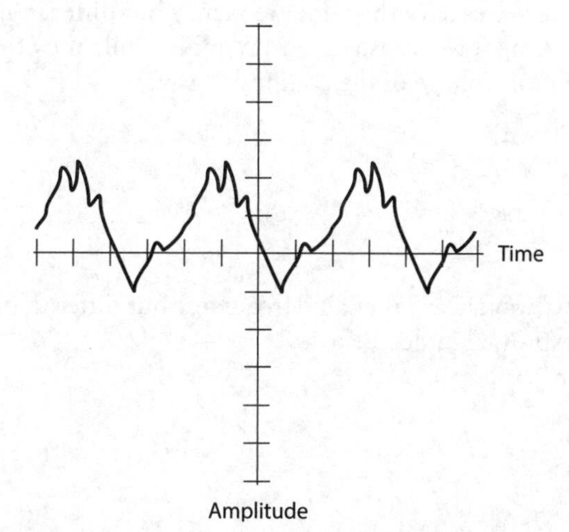

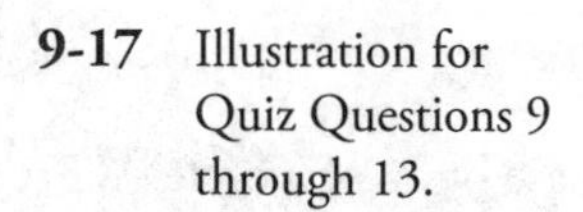
9-17 Illustration for Quiz Questions 9 through 13.

12. What's the negative peak voltage of the wave shown in Fig. 9-17?
    - (a) −100 mV pk−
    - (b) −200 mV pk−
    - (c) −500 mV pk−
    - (d) The graph lacks enough information to say.

13. What's the peak-to-peak voltage of the wave shown in Fig. 9-17?
    - (a) 170 mV pk-pk
    - (b) 141 mV pk-pk
    - (c) 340 mV pk-pk
    - (d) The graph lacks enough information to say.

14. Two sine waves have the same frequency along with DC components of unknown magnitude. The components are not the same for both waves, but the waves are in phase coincidence. One wave measures +21 V pk+ *not including* its DC component, while the other wave measures +17 V pk+ *not including* its DC component. What's the positive peak voltage of the composite wave?
    - (a) +38 V pk+
    - (b) +19 V pk+
    - (c) Zero
    - (d) We need more information to say.

15. Two sine waves have the same frequency, with DC components of +10 V each. The waves coincide in phase. One wave measures +21 V pk+ *including* its +10 V DC component, while the other wave measures +17 V pk+ *including* its +10 V DC component. What's the positive peak voltage of the composite wave?
    - (a) +58 V pk+
    - (b) +48 V pk+
    - (c) +38 V pk+
    - (d) We need more information to say.

16. Two sine waves have the same frequency but differ in phase by 180°. Neither wave has a DC component. One wave measures +21 V pk+, while the other wave measures +17 V pk+. What's the positive peak voltage of the composite wave?
    - (a) +38 V pk+
    - (b) +19 V pk+
    - (c) +8 V pk+
    - (d) +4 V pk+

17. Two AC waves have the same frequency but differ in phase by ⅙ of a cycle. What's this phase difference expressed in degrees?
    - (a) 120°
    - (b) 90°
    - (c) 60°
    - (d) 45°

18. The frequency of an AC wave in megahertz (MHz) equals
    (a) 1,000,000 times the reciprocal of its period in seconds.
    (b) 1000 times the reciprocal of its period in seconds.
    (c) 0.001 times the reciprocal of its period in seconds.
    (d) 0.000001 times the reciprocal of its period in seconds.

19. The 7th harmonic of an AC wave with a period of 10.0 microseconds (μs) has a frequency of
    (a) 7.00 MHz.
    (b) 3.50 MHz.
    (c) 700 kHz.
    (d) 350 kHz.

20. If you connect a 126-V DC battery in series with the output of a 120-V RMS (not positive peak or negative peak) sine-wave AC wall outlet, you'll get
    (a) pure DC.
    (b) AC with equal positive and negative peak voltages.
    (c) AC with different positive and negative peak voltages.
    (d) fluctuating DC.

10
CHAPTER

Inductance

IN THIS CHAPTER, YOU'LL LEARN ABOUT ELECTRICAL COMPONENTS THAT OPPOSE THE FLOW OF AC BY storing energy as magnetic fields. We call these devices *inductors*, and we call their action *inductance*. Inductors usually comprise wire coils, but even a length of wire or cable can form an inductor.

The Property of Inductance

Imagine a wire 1,000,000 miles (about 1,600,000 kilometers) long. Imagine that we make this wire into a huge loop, and then we connect its ends to the terminals of a battery, as shown in Fig. 10-1, driving current through the wire.

If we used a short wire for this experiment, the current would begin to flow immediately, and it would attain a level limited only by the resistance in the wire and the resistance in the battery. But because we have an extremely long wire, the electrons require some time to work their way from the negative battery terminal, around the loop, and back to the positive terminal. Therefore, it will take some time for the current to build up to its maximum level.

The magnetic field produced by the loop will start out small, during the first few moments when current flows in only part of the loop. The field will build up as the electrons get around the loop. Once the electrons reach the positive battery terminal so that a steady current flows around

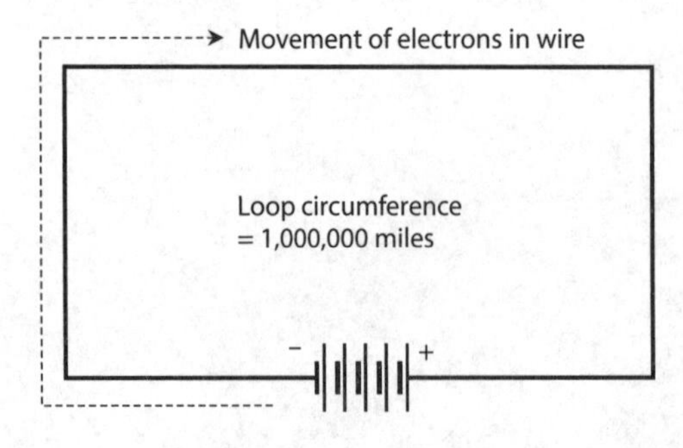

10-1 We can use a huge, imaginary loop of wire to illustrate the principle of inductance.

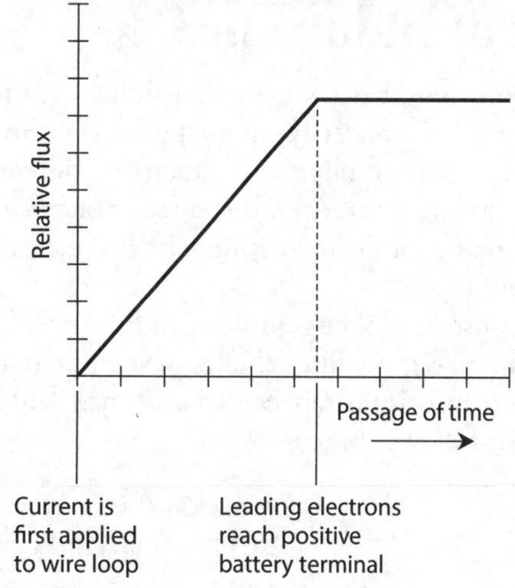

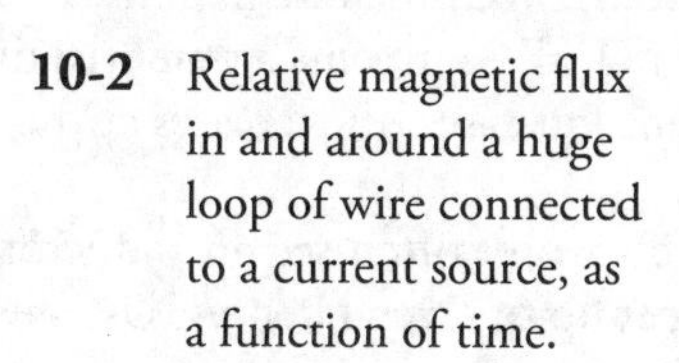
10-2 Relative magnetic flux in and around a huge loop of wire connected to a current source, as a function of time.

the entire loop, the magnetic field quantity will attain its maximum and level off, as shown in Fig. 10-2. At that time, we'll have a certain amount of energy stored in the magnetic field. The amount of stored energy will depend on the *inductance* of the loop, which depends on its overall size. We symbolize inductance, as a property or as a mathematical variable, by writing an italicized, uppercase letter *L*. Our loop constitutes an inductor. To abbreviate "inductor," we write an uppercase, non-italicized letter L.

Obviously, we can't make a wire loop measuring anywhere near 1,000,000 miles in circumference. But we can wind fairly long lengths of wire into compact coils. When we do that, the magnetic flux for a given length of wire increases compared with the flux produced by a single-turn loop, increasing the inductance. If we place a ferromagnetic rod called a *core* inside a coil of wire, we can increase the flux density and raise the inductance even more.

We can attain values of *L* many times greater with a ferromagnetic core than we can get with a similar-sized coil having an air core, a solid plastic core, or a solid dry wooden core. (Plastic and dry wood have permeability values that differ little from air or a vacuum; engineers occasionally use these materials as coil cores or "forms" in order to add structural rigidity to the windings without significantly changing the inductance.) The current that an inductor can handle depends on the diameter of the wire. But the value of *L* also depends on the number of turns in the coil, the diameter of the coil, and the overall shape of the coil.

If we hold all other factors constant, the inductance of a helical coil increases in direct proportion to the number of turns of wire. Inductance also increases in direct proportion to the diameter of the coil. If we "stretch out" a coil having a certain number of turns and a certain diameter while holding all other parameters constant, its inductance goes down. Conversely, if we "squash up" an elongated coil while holding all other factors constant, the inductance goes up.

Under normal circumstances, the inductance of a coil (or any other type of device designed to function as an inductor) remains constant regardless of the strength of the signal we apply. In this context, "abnormal circumstances" refer to an applied signal so strong that the inductor wire melts, or the core material heats up excessively. Good engineering sense demands that such conditions should never arise in a well-designed electrical or electronic system.

The Unit of Inductance

When we first connect a battery across an inductor, the current builds up at a rate that depends on the inductance. The greater the inductance, the slower the rate of current buildup for a given battery voltage. The unit of inductance quantifies the ratio between the rate of current buildup and the voltage across an inductor. An inductance of one *henry* (1 H) represents a potential difference of one volt (1 V) across an inductor within which the current increases or decreases at the rate of one ampere per second (1 A/s).

The henry constitutes a huge unit of inductance. You won't often see an inductor this large, although some power-supply filter chokes have inductances up to several henrys. Usually, engineers and technicians express inductances in *millihenrys* (mH), *microhenrys* (μH), or *nanohenrys* (nH). The units relate as follows:

$$1 \text{ mH} = 0.001 \text{ H} = 10^{-3} \text{ H}$$
$$1 \text{ μH} = 0.001 \text{ mH} = 10^{-6} \text{ H}$$
$$1 \text{ nH} = 0.001 \text{ μH} = 10^{-9} \text{ H}$$

Small coils with few turns of wire produce small inductances, in which the current changes quickly and the induced voltages are small. Large coils with ferromagnetic cores, and having many turns of wire, have high inductances in which the current changes slowly and the induced voltages are large. The current from a battery, building up or dying down through a high-inductance coil, can give rise to a large potential difference between the end terminals of the coil—many times the voltage of the battery itself. Spark coils, such as those used in internal combustion engines, take advantage of this principle. That's why large coils present a deadly danger to people ignorant of the wiles of inductance!

Inductors in Series

Imagine that we place two or more current-carrying inductors in close proximity and connect them in series. As long as the magnetic fields around those inductors don't interact, their inductances add exactly as resistances in series do. The total inductance, also called the *net inductance*, equals the sum of the individual inductances. We must use the same size unit for all the inductors if we want this simple rule to work.

Suppose that we have inductances $L_1, L_2, L_3, \ldots, L_n$ connected in series. As long as the magnetic fields of the inductors don't interact—that is, as long as no *mutual inductance* exists—we can calculate the total inductance L using the formula

$$L = L_1 + L_2 + L_3 + \ldots + L_n$$

Problem 10-1

Imagine three inductances L_1, L_2, and L_3 connected in series, as shown in Fig. 10-3. Suppose that no mutual inductance exists, and each inductance equals 40.0 mH. What's the total inductance L?

Solution

We simply add up the values to obtain

$$L = L_1 + L_2 + L_3 = 40.0 + 40.0 + 40.0 = 120 \text{ mH}$$

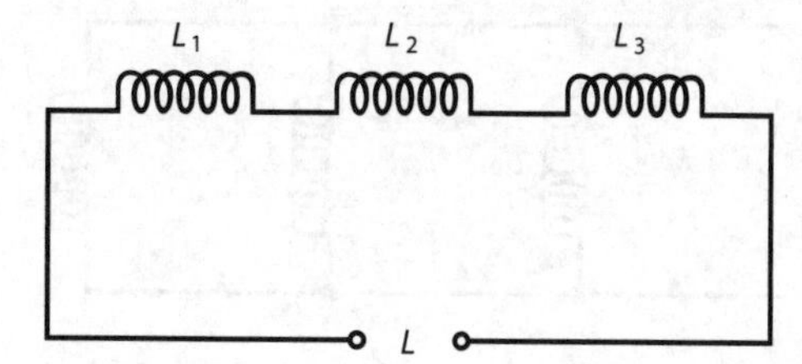

10-3 Inductances connected in series.

Problem 10-2

Consider three inductors having no mutual inductance, with values of $L_1 = 20.0$ mH, $L_2 = 55.0$ μH, and $L_3 = 400$ nH. What's the total inductance L, in millihenrys, of these components if we connect them in series, as shown in Fig. 10-3?

Solution

First, let's convert all the inductances to the same units. Microhenrys will do! In that case, we have

$$L_1 = 20.0 \text{ mH} = 20{,}000 \text{ μH}$$
$$L_2 = 55.0 \text{ μH}$$
$$L_3 = 400 \text{ nH} = 0.400 \text{ μH}$$

The total inductance equals the sum of these values, or

$$L = 20{,}000 + 55.0 + 0.400 = 20{,}055.4 \text{ μH}$$

After we convert to millihenrys, we get 20.0554 mH, which we can round off to 20.1 mH.

Inductors in Parallel

If no mutual inductance exists among two or more parallel-connected inductors, their values add like resistances in parallel. Consider several inductances $L_1, L_2, L_3, \ldots, L_n$ connected in parallel. We can calculate the net inductance L, with the formula

$$L = 1 / (1/L_1 + 1/L_2 + 1/L_3 + \ldots + 1/L_n)$$
$$= (1/L_1 + 1/L_2 + 1/L_3 + \ldots + 1/L_n)^{-1}$$

As with inductances in series, we must make sure that all the units agree.

Problem 10-3

Imagine three inductances L_1, L_2, and L_3 connected in parallel as shown in Fig. 10-4. Suppose that no mutual inductance exists, and each inductance equals 40 mH. What's the total inductance L?

Solution

According to the formula defined above, we have

$$L = 1 / (1/L_1 + 1/L_2 + 1/L_3) = 1 / (1/40 + 1/40 + 1/40)$$
$$= 1 / (3/40) = 40/3 = 13.333 \text{ mH}$$

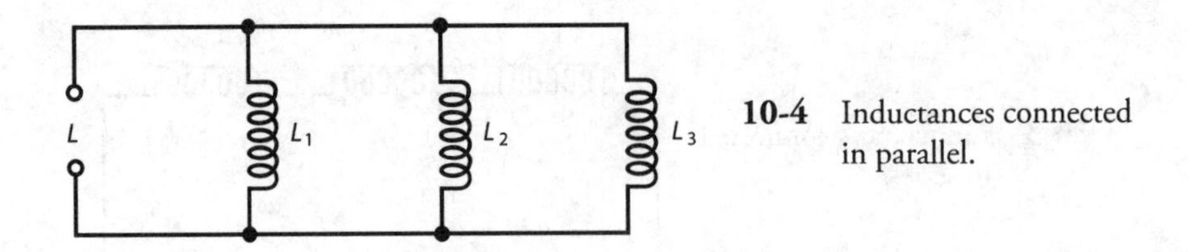

10-4 Inductances connected in parallel.

We should round this figure off to 13 mH because the original inductance values extend to only two significant digits.

Problem 10-4

Imagine four inductances $L_1 = 75.0$ mH, $L_2 = 40.0$ mH, $L_3 = 333$ μH, and $L_4 = 7.00$ H, all connected in parallel with no mutual inductance. What's the net inductance L?

Solution

Let's use henrys as the standard unit. Then we have

$$L_1 = 0.0750 \text{ H}$$
$$L_2 = 0.0400 \text{ H}$$
$$L_3 = 0.000333 \text{ H}$$
$$L_4 = 7.00 \text{ H}$$

When we plug these values into the parallel-inductance formula, we obtain

$$\begin{aligned} L &= 1/(1/0.0750 + 1/0.0400 + 1/0.000333 + 1/7.00) \\ &= 1/(13.33 + 25.0 + 3003 + 0.143) \\ &= 1/3041.473 = 0.00032879 \text{ H} = 328.79\ \mu\text{H} \end{aligned}$$

We should round this figure off to 329 μH. That's only a little less than the value of the 333-μH inductor all by itself!

Interaction among Inductors

In real-world circuits, we usually observe some mutual inductance between or among *solenoidal* (cylindrical or helical) coils. The magnetic fields extend significantly outside such coils, and mutual effects are difficult to avoid. The same holds true between nearby lengths of wire, especially at high AC frequencies. Sometimes, mutual inductance has no detrimental effect, but in some situations it does. We can minimize mutual inductance between coils by using *toroidal* (donut-shaped) windings instead of solenoidal windings. We can minimize mutual inductance between wires by *shielding* them—that is, by insulating them and then wrapping them with grounded sheets or braids of metal. The most common shielded wire takes a form known as *coaxial cable*.

Coefficient of Coupling

The *coefficient of coupling*, symbolized k, quantifies the extent to which two inductors interact, that is, whether their magnetic fields reinforce or oppose each other. We specify k as a number ranging from 0 (no interaction) to 1 (the maximum possible interaction).

Two coils separated by a huge distance have the minimum possible coefficient of coupling, which is zero ($k = 0$); two coils wound on the same form, one right over the other, exhibit the maximum possible coefficient of coupling ($k = 1$). As we bring two inductors closer and hold all other factors constant, k increases.

We can multiply k by 100 and add a percent-symbol (%) to express the coefficient of coupling as a percentage, defining the range $k_{\%} = 0\%$ to $k_{\%} = 100\%$.

Mutual Inductance

Engineers symbolize the *mutual inductance* between two inductors by writing an uppercase italic M. We can express this quantity in the same units as inductance: henrys, millihenrys, microhenrys, or nanohenrys. For any two particular inductors, M depends on the inductance values and the coefficient of coupling.

When we have two inductors with values of L_1 and L_2 (both expressed in the same size units) and with a coefficient of coupling equal to k, we can calculate the mutual inductance by multiplying the inductances, taking the square root of the result, and then multiplying by k. Mathematically, we have

$$M = k\,(L_1 L_2)^{1/2}$$

where the ½ power represents the positive square root.

Effects of Mutual Inductance

Mutual inductance can either increase or decrease the net inductance of a pair of series connected coils, compared with the condition of zero mutual inductance. The magnetic fields around the coils either reinforce or oppose each other, depending on the phase relationship of the AC applied to them. If the two AC waves (and thus the magnetic fields they produce) coincide in phase, the inductance increases as compared with the condition of zero mutual inductance. If the two waves oppose in phase, the net inductance decreases relative to the condition of zero mutual inductance.

When we have two inductors connected in series and we observe *reinforcing* mutual inductance, we can calculate the total inductance L with the formula

$$L = L_1 + L_2 + 2M$$

where L_1 and L_2 represent the inductances, and M represents the mutual inductance, all in the same size units. When we have two inductors connected in series and we observe *opposing* mutual inductance, we can calculate the total inductance L with the formula

$$L = L_1 + L_2 - 2M$$

where, again, L_1 and L_2 represent the values of the individual inductors, and M represents the mutual inductance, all in the same size units.

Problem 10-5

Imagine that we connect two coils, having inductances of 30 μH and 50 μH, in series so that their fields reinforce, as shown in Fig. 10-5. Suppose that the coefficient of coupling equals 0.500. What's the net inductance?

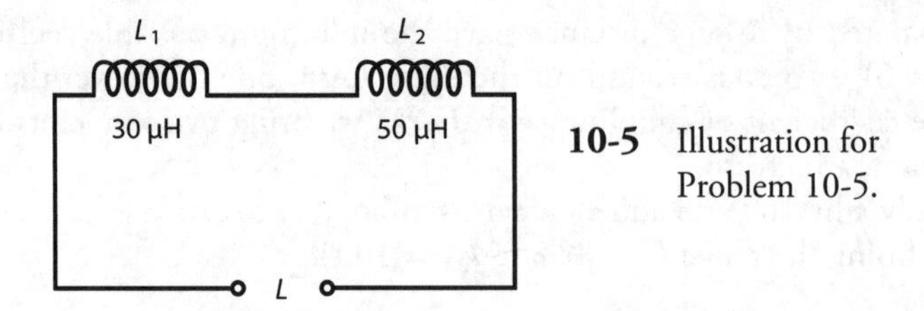

10-5 Illustration for Problem 10-5.

Solution

First, we must derive M from k. According to the formula for this purpose, we have

$$M = 0.500\ (50 \times 30)^{1/2} = 19.4\ \mu\text{H}$$

Now we can calculate the total inductance, getting

$$L = L_1 + L_2 + 2M = 30 + 50 + 38.8 = 118.8\ \mu\text{H}$$

which we should round off to 120 µH.

Problem 10-6

Imagine two coils with inductances of $L_1 = 835\ \mu\text{H}$ and $L_2 = 2.44$ mH. We connect them in series so that their magnetic fields oppose each other with a coefficient of coupling equal to 0.922, as shown in Fig. 10-6. What's the net inductance?

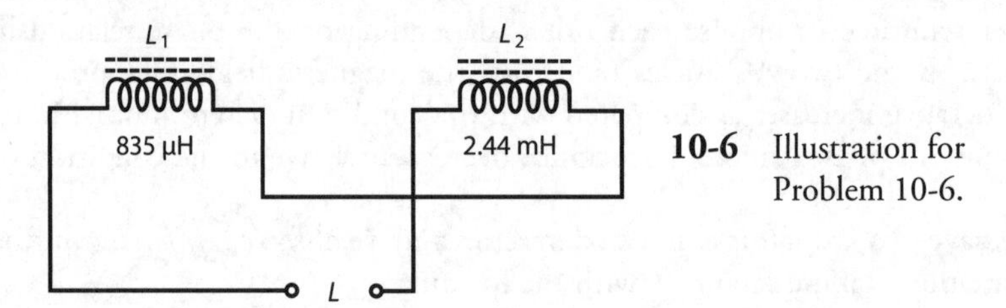

10-6 Illustration for Problem 10-6.

Solution

We know the coil inductances in different units. Let's use microhenrys for our calculations, so we have $L_1 = 835\ \mu\text{H}$ and $L_2 = 2440\ \mu\text{H}$. Now we can calculate M from k, obtaining

$$M = 0.922\ (835 \times 2440)^{1/2} = 1316\ \mu\text{H}$$

Finally, we calculate the total inductance as

$$L = L_1 + L_2 - 2M = 835 + 2440 - 2632 = 643\ \mu\text{H}$$

Air-Core Coils

The simplest inductors (besides plain, straight lengths of wire) are coils of insulated or enameled wire. You can wind a coil on a hollow cylinder made of plastic or other non-ferromagnetic material, forming an *air-core coil.* In practice, the attainable inductance for such coils can range from a few nanohenrys up to about 1 mH. The frequency of an applied AC signal does not affect the inductance of an air-core coil, but as the AC frequency increases, smaller and smaller values of inductance produce significant effects.

An air-core coil made of heavy-gauge wire, and having a large radius, can carry high current and can handle high voltages. Air dissipates almost no energy as heat, so air makes an efficient core material even though it has low permeability. For these reasons, air-core coil designs represent an excellent choice for the engineer who wants to build high-power RF transmitters, amplifiers, or tuning networks. However, air-core coils take up a lot of physical space in proportion to the available inductance, especially when designed to handle high currents and voltages.

Ferromagnetic Cores

Inductor manufacturers crush samples of ferromagnetic material into dust and then bind the powder into various shapes, providing cores that greatly increase the inductance of a coil having a given number of turns. Depending on the mixture used, the flux density can increase by a factor of up to about 1,000,000 (10^6). A physically small coil can thereby acquire a large inductance if we put a *powdered-iron core* inside it. Powdered-iron cores work well from the middle audio frequencies (AF) to well into the radio-frequency (RF) range.

Core Saturation

If a powdered-iron-core coil carries more than a certain amount of current, the core will *saturate.* When an inductor core operates in a state of *saturation*, the ferromagnetic material holds as much magnetic flux as it possibly can. Any further increase in the coil current will not produce an increase in the core's magnetic flux. In practical systems, this effect causes decreasing inductance with coil currents that exceed the critical value. In extreme cases, saturation can cause a coil to waste considerable power as heat, making the coil *lossy*.

Permeability Tuning

We can *tune* (vary the inductance of) a solenoidal coil without changing the number of turns if we slide a ferromagnetic core in and out of it. Because moving the core in and out of a coil changes the effective permeability within the coil, some engineers call this practice *permeability tuning*. We can precisely control the in/out core position by attaching the core to a screw shaft, as shown in Fig. 10-7. As we rotate the shaft clockwise, the core enters the coil, so the inductance increases. As we turn the shaft counterclockwise, the core moves out of the coil, so the inductance decreases.

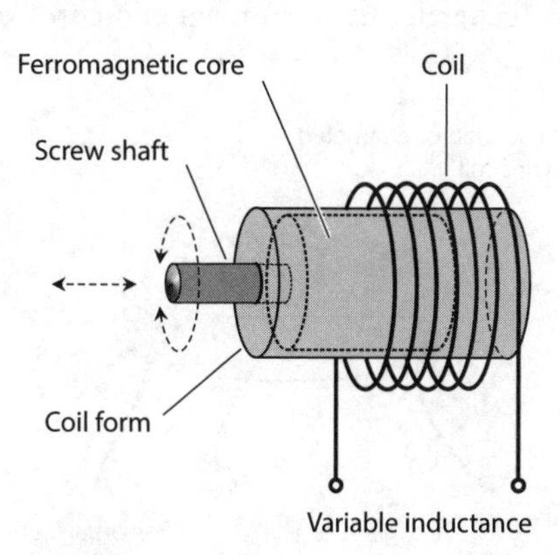

10-7 We can accomplish permeability tuning by moving a ferromagnetic core in and out of a solenoidal coil.

Toroids

When we want to construct a coil with a ferromagnetic core, we don't have to wind the wire on a rod-shaped core. We can use another core geometry called the *toroid*, whose shape resembles that of a donut. Figure 10-8 illustrates how we can wind a coil on a ferromagnetic toroid core.

Toroidal coils offer at least three advantages over solenoidal ones. First, it takes fewer turns of wire to get a certain inductance with a toroid, as compared with a solenoid. Second, we can make a toroid physically smaller for a given inductance and current-carrying capacity. Third, essentially all of the magnetic flux in a toroid remains within the core material. This phenomenon practically eliminates unwanted mutual inductance between a toroid and other components near it.

Toroidal coils have limitations. We can't permeability-tune a toroidal coil because the core constitutes a full circle, always "serving" the entire coil. Most people find toroidal coils harder to wind than solenoidal ones, especially when the winding must have a large number of turns. In some situations, we might actually *want* mutual inductance to exist between or among physically separate coils; if we use a toroid, we must wind both coils on the same core to make this happen.

Pot Cores

An alternative exists to the toroidal geometry for confining magnetic flux. We can surround a loop-shaped coil of wire with a ferromagnetic shell, as shown in Fig. 10-9, obtaining a *pot core*. A typical pot core has two halves, inside one of which we wind the coil. The wires emerge through small holes or slots.

Pot cores have advantages similar to those of toroids. The shell prevents the magnetic flux from extending outside the physical assembly, so mutual inductance between the coil and anything else in its vicinity is always zero. We can get far more inductance with a pot core than we can with a solenoidal coil of comparable physical size. In fact, pot cores work even better than toroids if we want to obtain a large inductance in a small space.

Pot-core coils can prove useful over the full AF range, even the lowest-frequency extreme (approximately 20 Hz). They don't function very well at frequencies above a few hundred kilohertz. Because of their geometry, we can't permeability-tune them. If we hold all other factors constant, then the inductance increases as the shell permeability increases. It doesn't matter, within reason, how strong or weak the signal is.

Inductors for RF Use

The RF spectrum ranges from a few kilohertz to well above 100 GHz. At the low end of this range, inductors generally use ferromagnetic cores. As the frequency increases, cores having low

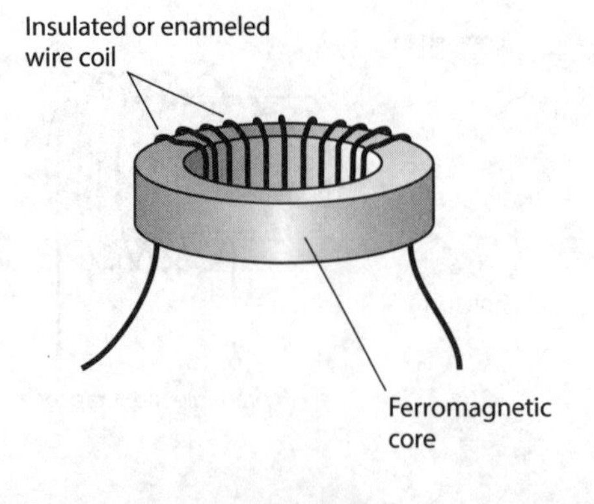

10-8 A toroidal coil surrounds a donut-shaped ferromagnetic core.

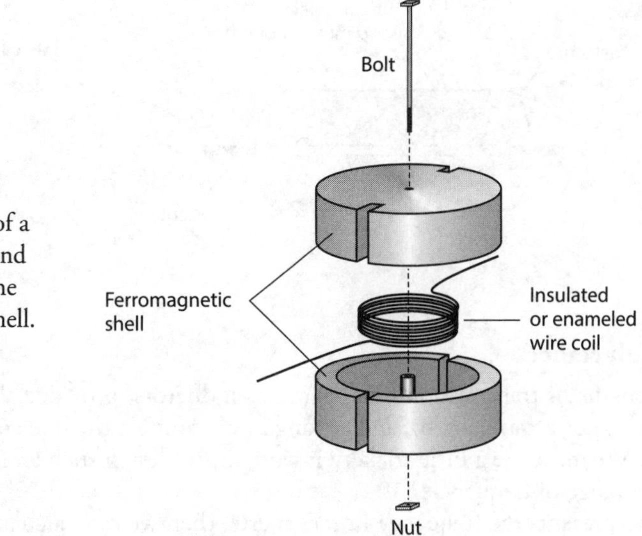

10-9 Exploded view of a pot core. We wind the coil inside the ferromagnetic shell.

permeability find favor. We'll commonly see toroids in RF systems designed for use at frequencies up through about 30 MHz. Above 30 MHz or so, we'll more often see air-core coils.

Transmission-Line Inductors

At frequencies about 100 MHz, we can make an inductor from a length of *transmission line*, rather than from a wire loop or coil. Most transmission lines exist in either of two geometries, the *parallel-wire* type or the *coaxial* type.

Parallel-Wire Line

A parallel-wire transmission line consists of two wires running alongside each other with constant spacing (Fig. 10-10). Polyethylene rods, molded at regular intervals to the wires, keep the spacing between the wires constant. A solid or "windowed" web of polyethylene can serve the same purpose. The substance separating the wires constitutes the *dielectric* of the transmission line.

Coaxial line

A coaxial transmission line contains a wire *center conductor* surrounded by a tubular braid or pipe called the *shield* (Fig. 10-11). Solid polyethylene *beads* (which resemble tiny toroids in shape), or a continuous hose-like length of foamed or solid polyethylene, separates the center conductor from the shield, maintains the spacing between them, and acts as a dielectric.

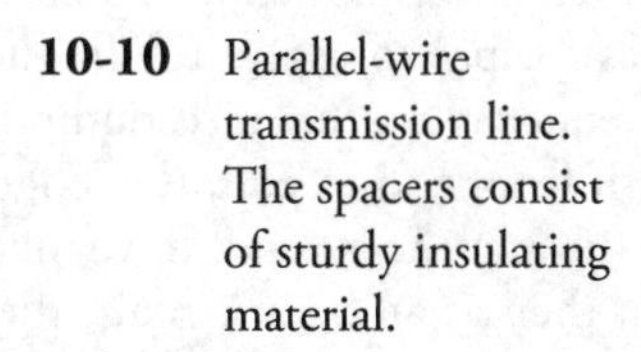

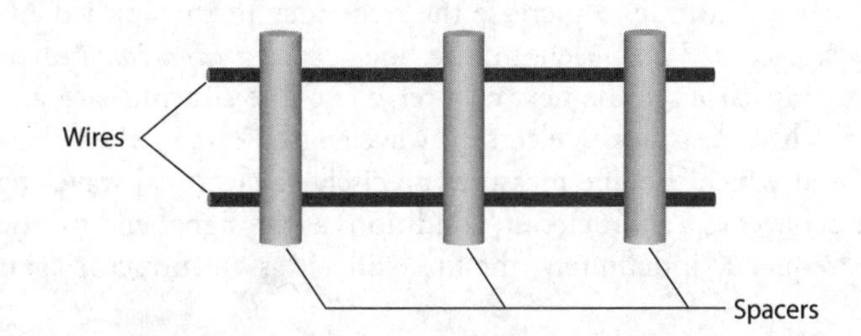

10-10 Parallel-wire transmission line. The spacers consist of sturdy insulating material.

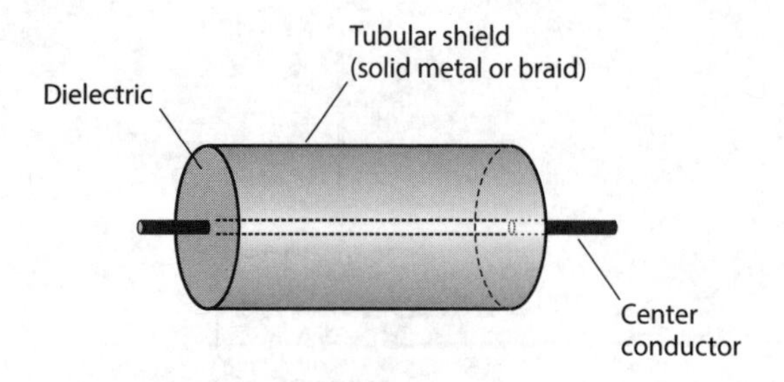

10-11 Coaxial transmission line. The dielectric material keeps the center conductor along the axis of the tubular shield.

Line Inductance

Short lengths of transmission line behave as inductors, provided that the line length remains less than 90° (¼ of a wavelength), and as long as we connect the line conductors *directly together* at the far end. We must use a little trickery if we want to design such an inductor. It will work only over a certain range of frequencies.

If f represents the frequency in megahertz, then we can calculate ¼ wavelength in *free space* (a vacuum), expressed in centimeters (s_{cm}), using the formula

$$s_{cm} = 7500/f$$

The length of a "real-world" quarter-wavelength transmission line is shortened from the free-space quarter wavelength because the dielectric reduces the speed at which the RF energy travels along the line. In practice, ¼ of an *electrical wavelength* along a transmission line can range anywhere from about 0.66 (or 66%) of the free-space quarter wavelength to 0.95 (or 95%) of the free-space quarter wavelength. Engineers call this "shortening factor" the *velocity factor* of the line because it represents the speed of RF waves in the line divided by the speed of RF waves in free space (the speed of light). If we let v represent the velocity factor of a particular transmission line, then the above formula for the length of a quarter-wave line, in centimeters, becomes

$$s_{cm} = 7500v/f$$

Very short lengths of line—a few electrical degrees—produce small values of inductance. As the length approaches ¼ wavelength, the inductance increases.

Transmission line inductors behave differently than coils in one important way. The inductance of a coil varies little, if at all, with changes in the frequency. But the value of a transmission-line inductor varies *drastically* as the frequency goes up or down. At first, the inductance increases as the frequency increases. As we approach a certain "critical" frequency, the inductance grows arbitrarily large, "approaching infinity." At the critical frequency, where the line measures precisely ¼ electrical wavelength from end to end, the line (as long as its conductors remain shorted out at the far end) acts like an open circuit at the signal-input end.

If we continue to increase the frequency of the applied AC signal so that the line's electrical length exceeds ¼ wavelength, the line acts as a *capacitor* rather than as an inductor. (You'll learn about capacitors in the next chapter.) The line continues to act as a capacitor up to the frequency at which it measures ½ electrical wavelength. Then, when the frequency reaches a second critical point at which the line measures precisely ½ electrical wavelength from end to end, the length of line behaves as a short circuit; conditions at the signal end mimic those at the far end. If we increase the frequency indefinitely, the line will act as an inductor again, then as an open circuit, then as a capacitor again, then as a short circuit, then as an inductor again, and so on. Each critical (or transition) frequency will exist when the electrical length of the line exactly equals a whole-number multiple of ¼ wavelength.

Stray Inductance

Any length of wire, no matter how short and no matter what the frequency, exhibits at least a little inductance. As with a transmission line, the inductance of any fixed-length wire increases as the frequency increases.

In wireless communications equipment, the inductance of, and among, wires can constitute a problem. A circuit might generate its own signal (oscillate) even if that's the last thing we want it to do. A receiver might respond to signals that we don't want it to intercept. A transmitter can send out signals on unauthorized frequencies. The frequency response of any circuit can change in unpredictable ways, degrading the performance of the equipment. Sometimes the effects of so-called *stray inductance* remain small or negligible; in other scenarios, stray inductance can cause serious malfunctions or even *catastrophic system failure.*

If we want to minimize stray inductance, we can use coaxial cables between and among sensitive circuits or components. We must connect the shield of each cable section to the *common ground* of the apparatus. In some systems, we might even have to enclose individual circuits in tight metal boxes to *electrically isolate* them.

Quiz

Refer to the text in this chapter if necessary. A good score is 18 correct. Answers are in the back of the book.

1. Which of the following statements represents an advantage of a toroidal coil over a solenoidal coil?
   (a) You can easily permeability-tune a toroid but not a solenoid.
   (b) A toroid can carry more current than a solenoid with the same gauge wire.
   (c) A toroid can function with a ferromagnetic core, but a solenoid cannot.
   (d) A toroid practically prevents unwanted mutual inductance, while solenoids allow it.

2. You connect four 44.0-mH toroidal inductors in parallel. They exhibit no mutual inductance. What's the net inductance of the combination?
   (a) 11.0 mH
   (b) 22.0 mH
   (c) 88.0 mH
   (d) 176 mH

3. You connect two 44.0-mH solenoidal inductors in series. They exhibit no mutual inductance. What's the net inductance of the combination?
   (a) 22.0 mH
   (b) 88.0 mH
   (c) 176 mH
   (d) 352 mH

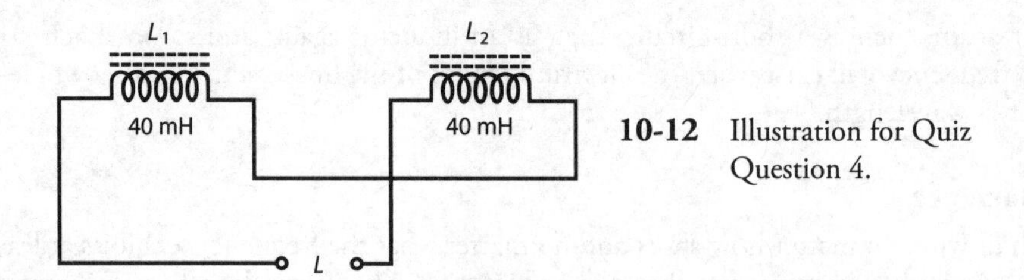

10-12 Illustration for Quiz Question 4.

4. In Fig. 10-12, both inductors have values of 40 mH. Their magnetic fields oppose each other. Some mutual inductance exists between them. The net inductance of the combination is
   (a) less than 80 mH.
   (b) exactly 80 mH.
   (c) more than 80 mH.
   (d) zero.

5. In Fig. 10-13, both inductors have values of 40 μH. Their magnetic fields reinforce each other. Some mutual inductance exists between them. The net inductance of the combination is
   (a) less than 80 μH.
   (b) exactly 80 μH.
   (c) more than 80 μH.
   (d) 160 μH.

6. To obtain the lowest possible inductance for a 100-turn coil, you should use
   (a) an air-core solenoid.
   (b) a powdered-iron-core solenoid.
   (c) a pot-core coil.
   (d) a powdered-iron-core toroid.

7. Which of the following factors affects the inductance of a pot-core coil, if all other factors stay the same?
   (a) The frequency of the applied signal
   (b) The amplitude of the applied signal
   (c) The wave shape of the applied signal
   (d) The permeability of the core (shell) material

8. As the number of turns increases in an air-core coil, then its inductance, assuming all other factors remain constant,
   (a) increases.
   (b) stays the same.

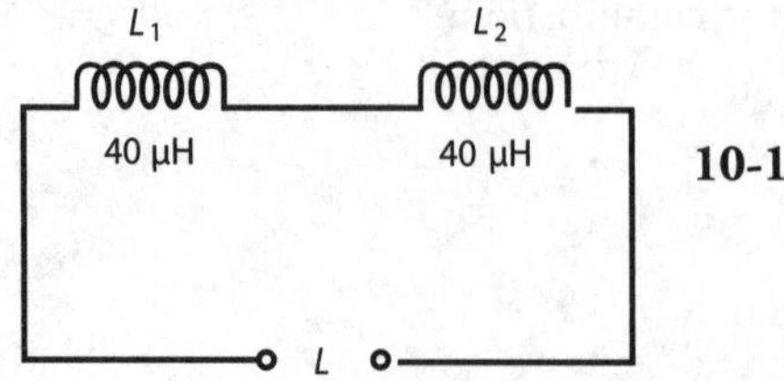

10-13 Illustration for Quiz Question 5.

(c) decreases.

(d) approaches zero.

9. As you increase the amplitude of the AC signal that you apply to a toroidal coil, leaving all other factors unchanged, the coil's inductance

(a) increases.

(b) stays the same.

(c) decreases.

(d) approaches zero.

10. To increase the inductance of a solenoidal coil without changing anything else, you can increase the

(a) signal frequency.

(b) core permeability.

(c) signal strength.

(d) wire diameter.

11. You connect a 500-μH inductor coil in series with a 900-μH inductor coil, winding them over each other so the coefficient of coupling is 1. Both coils are solenoids whose magnetic fields reinforce each other. What's the *mutual inductance* between the coils?

(a) 1.40 mH

(b) 700 μH

(c) 671 μH

(d) More information is needed to calculate it.

12. What's the *net (total) inductance* of the above combination?

(a) 2.74 mH

(b) 2.04 mH

(c) 2.01 mH

(d) More information is needed to calculate it.

13. At which of the following frequencies would you most likely use an air-core solenoidal coil to obtain useful inductance?

(a) 6 kHz

(b) 20 MHz

(c) 900 GHz

(d) Any of the above

14. You have two 50-turn, air-core, loop-like coils, each one measuring 2 centimeters in diameter. You align their axes, place them far from one another, and then gradually bring them closer together. What happens to the *coefficient of coupling* as you do this?

(a) It decreases.

(b) It stays the same.

(c) It increases.

(d) You'll need more information to answer this question.

15. What happens to the *mutual inductance* between the two coils as you carry out the exercise described in Question 14 above?
    (a) It decreases.
    (b) It stays the same.
    (c) It increases.
    (d) You'll need more information to answer this question.

16. You have two 50-turn loop-like coils, each one measuring 2 centimeters in diameter. You surround each coil with a pot-core shell, align their axes, place them far from one another, and then gradually bring them closer together. What happens to the *coefficient of coupling* as you do this?
    (a) It decreases.
    (b) It stays the same.
    (c) It increases.
    (d) You'll need more information to answer this question.

17. What happens to the *mutual inductance* between the two coils as you carry out the exercise described in Question 16 above?
    (a) It decreases.
    (b) It stays the same.
    (c) It increases.
    (d) You'll need more information to answer this question.

18. Consider a length of transmission line with its wires connected together at the far end. Suppose that the line's velocity factor is 0.750, and you apply a signal at 100 MHz to the open (near) end. To make this line measure ¼ electrical wavelength, you must cut it to a physical length of
    (a) 1.13 m.
    (b) 79.5 cm.
    (c) 56.3 cm.
    (d) 23.1 cm.

19. If you decrease the frequency to 90 MHz in the situation described in Question 18 above but don't change anything else, and if you keep the wires at the line's far end connected together, then the signal source at the open (near) end will "see"
    (a) a capacitance.
    (b) a short circuit.
    (c) an inductance.
    (d) an open circuit.

20. If you increase the frequency to 230 MHz in the situation described in Question 18 but don't change anything else, and if you keep the wires at the line's far end connected together, then the signal source at the open (near) end will "see"
    (a) a capacitance.
    (b) a short circuit.
    (c) an inductance.
    (d) an open circuit.

11
CHAPTER

Capacitance

ELECTRICAL RESISTANCE SLOWS THE FLOW OF AC OR DC CHARGE CARRIERS (USUALLY ELECTRONS) BY "brute force." Inductance impedes the flow of AC charge carriers by storing the energy as a magnetic field. *Capacitance* impedes the flow of AC charge carriers by storing the energy as an *electric field.*

The Property of Capacitance

Imagine two gigantic, flat, thin sheets of metal that conduct electricity well. Suppose that they have equal surface areas the size of the state of Nebraska. We place them one above the other, keep them parallel to each other, and separate them by a few centimeters of space. If we connect these two sheets of metal to the terminals of a battery, as shown in Fig. 11-1, the sheets will charge electrically, one with positive polarity and the other with negative polarity.

If the plates were small, they would both charge up almost instantly, attaining a relative voltage equal to the voltage of the battery. However, because the plates are so large and massive, it will take a little time for the negative one to "fill up" with extra electrons, and it will take an equal amount of time for the other one to have its excess electrons "drained out." Eventually, the voltage between the two plates will equal the battery voltage, and an electric field will exist in the space between them.

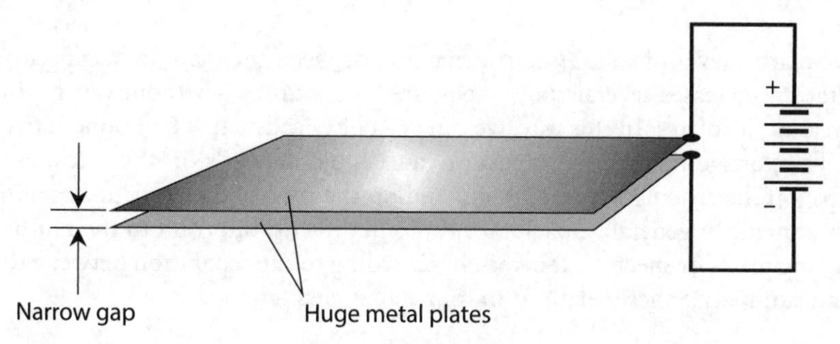

11-1 A hypothetical capacitor comprising two huge conducting plates.

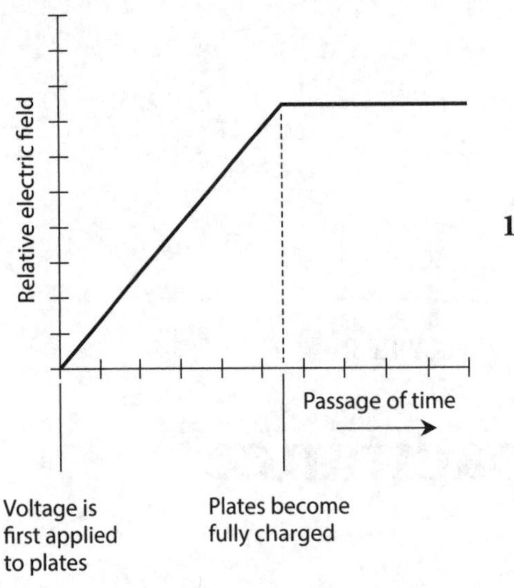

11-2 Relative electric field intensity, as a function of time, between two metal plates connected to a voltage source.

The electric field will start out small immediately after we connect the battery because the plates can't charge up right away. The charge will increase over a period of time; the rate of increase will depend on the plates' surface areas and on the spacing between them. Figure 11-2 portrays the intensity of this electric field as a function of time. We define *capacitance* as the ability of plates such as these, and of the space between them, to store electrical energy. As a quantity or variable, we denote capacitance by writing an uppercase, italic letter C.

Simple Capacitors

Obviously, we can't physically construct a capacitor of the dimensions described above! But we can place two sheets or strips of thin metal foil together, keeping them evenly separated by a thin layer of nonconducting material such as paper or plastic, and then roll up the assembly to obtain a large mutual surface area in a small physical volume. When we do this, the electric field quantity and intensity between the plates can get large enough so that the device exhibits considerable capacitance. Alternatively, we can take two separate sets of several plates and mesh them together with air between them.

When we place a layer of solid *dielectric* material between the plates of a capacitor, the electric flux concentration increases several times—perhaps many times—without our having to increase the surface areas of the plates. In this way, we can get a physically small component to exhibit a large capacitance. The voltage that such a capacitor can handle depends on the thickness of the metal sheets or strips, on the spacing between them, and on the type of dielectric material that we use to build the component. In general, capacitance varies in direct proportion to the mutual surface area of the conducting plates or sheets, but inversely according to the separation between the conducting sheets. We can summarize these relations in four statements:

1. If we maintain constant spacing between the sheets and increase their mutual area, the capacitance goes up.

2. If we maintain constant spacing between the sheets and decrease their mutual area, the capacitance goes down.
3. If we maintain constant mutual area and move the sheets closer together, the capacitance goes up.
4. If we maintain constant mutual area and move the sheets farther apart, the capacitance goes down.

The capacitance of a particular component also depends on the *dielectric constant* of the material between the metal sheets or plates. Dielectric constants are represented as numbers. By convention, scientists and engineers assign a dielectric constant of exactly 1 to a vacuum. If we take a capacitor in which a vacuum exists between the metal sheets or plates and then fill up all of the space with a material whose dielectric constant equals k, then the capacitance will increase by a factor of k. Dry air has a dielectric constant of almost exactly 1 (a little more, but rarely worth our worry). Table 11-1 lists the dielectric constants for several common substances.

The Unit of Capacitance

When we connect a battery between the plates of a capacitor, the potential difference between the plates builds up at a rate that depends on the capacitance. The greater the capacitance, the slower the rate of change of the voltage in the plates. The standard unit of capacitance quantifies the ratio between the current that flows and the rate of voltage change between the plates as the plates charge up. A capacitance of one *farad* (1 F) represents a current flow of 1 A while the voltage increases at the rate of 1 V/s. A capacitance of 1 F also results in 1 V of potential difference for an electric charge of 1 coulomb (1 C).

Table 11-1. Dielectric constants for several media that are used in capacitors. Except for air and a vacuum, these values are approximate.

Substance	Dielectric constant (approx.)
Air, dry, at sea level	1.0
Glass	4.8–8.0
Mica	4.0–6.0
Mylar	2.9–3.1
Paper	3.0–3.5
Plastic, hard, clear	3.0–4.0
Polyethylene	2.2–2.3
Polystyrene	2.4–2.8
Polyvinyl chloride	3.1–3.3
Porcelain	5.3–6.0
Quartz	3.6–4.0
Strontium titanate	300–320
Teflon	2.0–2.2
Titanium oxide	160–180
Vacuum	1.0

The farad represents a huge unit of capacitance. You'll almost never see a real-world capacitor with a value of 1 F. Engineers express capacitance values in terms of the *microfarad* (μF), the *nanofarad* (nF), and the *picofarad* (pF), where

$$1\ \mu\text{F} = 0.000001\ \text{F} = 10^{-6}\ \text{F}$$
$$1\ \text{nF} = 0.001\ \mu\text{F} = 10^{-9}\ \text{F}$$
$$1\ \text{pF} = 0.001\ \text{nF} = 10^{-12}\ \text{F}$$

The *millifarad*, which theoretically represents 0.001 F or 10^{-3} F, has never gained common usage.

Hardware manufacturers can produce physically small components that have fairly large capacitance values. Conversely, some capacitors with small values take up large physical volumes. The physical size of a capacitor, if all other factors are held constant, is proportional to the voltage that it can handle. The higher the rated voltage, the bigger the component.

Capacitors in Series

We rarely observe any mutual interaction among capacitors. We don't have to worry about *mutual capacitance* very often, the way we have to think about mutual inductance when working with wire coils.

When we connect two or more capacitors in series, their values combine just as resistances combine in parallel, assuming that no mutual capacitance exists among the components. If we connect two capacitors of the same value in series, the net capacitance equals half the capacitance of either component alone. In general, if we have several capacitors connected in series, we observe a net capacitance smaller than that of any of the individual components. As with resistances and inductances, we should always use the same size units when we calculate the net capacitance of any combination.

Consider several capacitors with values C_1, C_2, C_3, ..., C_n connected in series. We can find the net capacitance C using the formula

$$C = 1/(1/C_1 + 1/C_2 + 1/C_3 + \ldots + 1/C_n)$$

If we connect two or more capacitors in series, and if one of them has a value *many times* smaller than the values of all the others, then the net capacitance equals the *smallest* capacitance for most practical purposes.

Problem 11-1

Suppose that two capacitances, $C_1 = 0.10\ \mu\text{F}$ and $C_2 = 0.050\ \mu\text{F}$, appear in series, as shown in Fig. 11-3. What's the net capacitance?

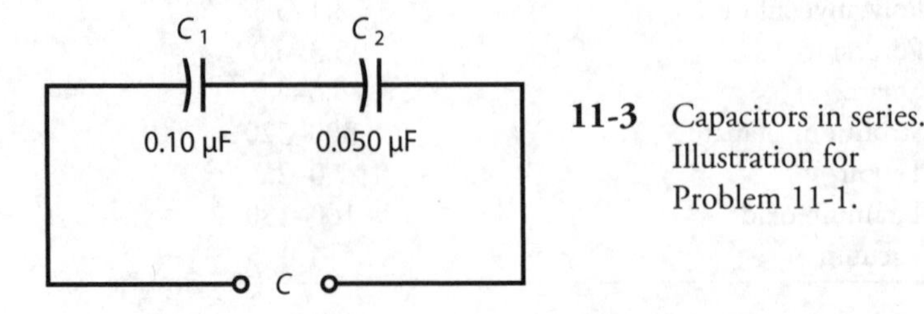

11-3 Capacitors in series. Illustration for Problem 11-1.

Solution

Let's use microfarads as the unit for our calculations. Using the above formula, we first find the reciprocals of the individual capacitances, getting

$$1/C_1 = 10$$

and

$$1/C_2 = 20$$

We add these numbers to obtain the reciprocal of the net series capacitance:

$$1/C = 10 + 20 = 30$$

Finally, we take the reciprocal of C^{-1} to obtain

$$C = 1/30 = 0.033\ \mu\text{F}$$

Problem 11-2

Imagine that we connect two capacitors with values of 0.0010 μF and 100 pF in series. What's the net capacitance?

Solution

First, let's convert both capacitances to microfarads. A value of 100 pF represents 0.000100 μF, so $C_1 = 0.0010\ \mu\text{F}$ and $C_2 = 0.000100\ \mu\text{F}$. The reciprocals are

$$1/C_1 = 1000$$

and

$$1/C_2 = 10{,}000$$

Now we can calculate the reciprocal of the series capacitance as

$$1/C = 1000 + 10{,}000 = 11{,}000$$

Therefore

$$C = 1/11{,}000 = 0.000091\ \mu\text{F}$$

We can also state this capacitance as 91 pF.

Problem 11-3

Suppose that we connect five 100-pF capacitors in series. What's the total capacitance?

Solution

If we have n capacitors in series, all of the same value, then the net capacitance C equals $1/n$ of the capacitance of any one of the components alone. In this case we have five 100-pF capacitors in series, so we get a net capacitance of

$$C = 100/5 = 20.0\ \text{pF}$$

Why the Curved Lines?

Do you wonder why the capacitor symbols in Fig. 11-3 (and everywhere else that they appear in this book) consist of one straight line and one curved line? Engineers commonly use this notation because, in many situations, one end of the capacitor either connects directly to, or faces toward, a *common ground* point (a neutral point, with a reference voltage of zero). In the circuit of Fig. 11-3, no common ground exists, so it doesn't matter which way the capacitors go. However, later in this book, we'll encounter some circuits in which the capacitor orientation does matter.

Capacitors in Parallel

Capacitances in parallel add like resistances in series. The net capacitance equals the sum of the individual component values, as long as we use the same units all the way through our calculations.

Suppose that we connect capacitors C_1, C_2, C_3, ..., C_n in parallel. As long as we observe no mutual capacitance between the components, we can calculate the net capacitance C with the formula

$$C = C_1 + C_2 + C_3 + \ldots + C_n$$

If we parallel-connect two or more capacitors, and if one of them has a value *many times* larger than the values of all the others, then for most practical purposes the net capacitance equals the *largest* capacitance.

Problem 11-4

Imagine three capacitors connected in parallel, having values of $C_1 = 0.100\ \mu F$, $C_2 = 0.0100\ \mu F$, and $C_3 = 0.00100\ \mu F$, as shown in Fig. 11-4. What's the net parallel capacitance?

Solution

We simply add the values up because all of the capacitances are expressed in the same size units (microfarads). We have

$$C = 0.100 + 0.0100 + 0.00100 = 0.11100\ \mu F$$

We can round off this result to $C = 0.111\ \mu F$.

Problem 11-5

Suppose that we connect two capacitors in parallel, one with a value of 100 μF and one with a value of 100 pF. What's the net capacitance?

Solution

In this case, we can say straightaway that the net capacitance is 100 μF for practical purposes. The 100-pF capacitor has a value that's only 1/1,000,000 of the capacitance of the 100-μF component. The smaller capacitance contributes essentially nothing to the net capacitance of this combination.

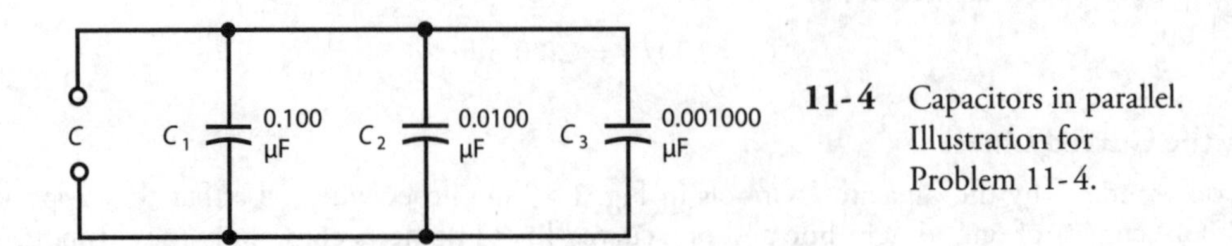

11-4 Capacitors in parallel. Illustration for Problem 11-4.

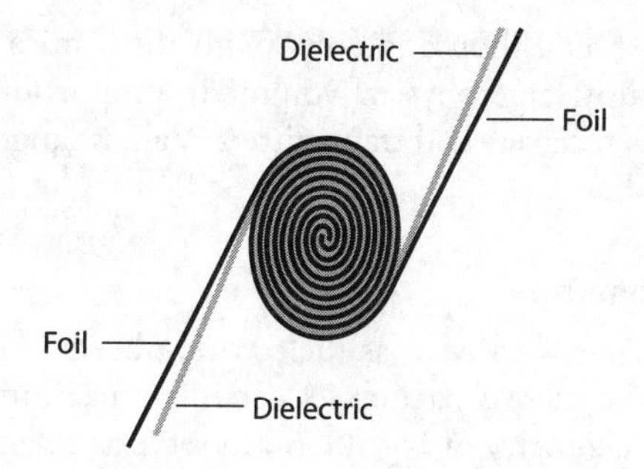

11-5 Cross-sectional drawing of a capacitor made of two foil sheets rolled up with dielectric material between them.

Fixed Capacitors

A *fixed capacitor* has a value that we can't adjust, and that (ideally) does not vary when environmental or circuit conditions change. Several common types of fixed capacitors have been used for decades.

Paper Capacitors

In the early days of electronics, capacitors were commonly made by placing paper, soaked with mineral oil, between two strips of foil, rolling the assembly up (Fig. 11-5), attaching wire leads to the two pieces of foil, and enclosing the rolled-up foil and paper in an airtight cylindrical case. *Paper capacitors* can still sometimes be found in older electronic equipment. They have values ranging from about 0.001 μF to 0.1 μF, and can handle potential differences of up to about 1000 V.

Mica Capacitors

Mica comprises a naturally occurring, solid, transparent mineral substance that flakes off in thin sheets. It makes an excellent dielectric for capacitors. *Mica capacitors* can be manufactured by alternately stacking metal sheets and layers of mica, or by applying silver ink to sheets of mica. The metal sheets are wired together into two meshed sets, forming the two terminals of the capacitor, as shown in Fig. 11-6.

Mica capacitors have low loss, so they exhibit high efficiency as long as we don't subject them to excessive voltage. Mica-capacitor voltage ratings can range from a few volts (with thin mica

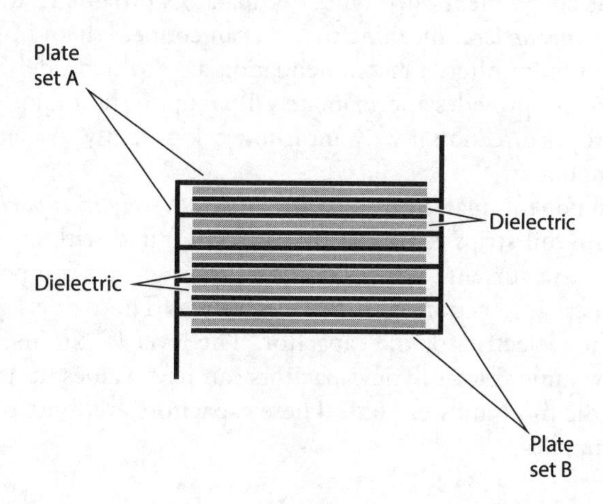

11-6 Cross-sectional drawing of a capacitor made of two meshed sets of several metal plates, separated by layers of dielectric material.

sheets) up to several thousand volts (with thick mica sheets and heavy-gauge metal plates). Mica capacitors occupy large physical volume in proportion to their capacitance. Mica capacitors work well in wireless receivers and transmitters. Values range from a few tens of picofarads up to approximately 0.05 μF.

Ceramic Capacitors

Ceramic materials work well as dielectrics. Sheets of metal are stacked alternately with wafers of ceramic to make these capacitors. A ceramic capacitor can have anywhere from one layer to dozens of layers. The geometry of Fig. 11-6 can serve as a general illustration. Ceramic, like mica, has low loss and allows for high efficiency.

For small values of capacitance, we need only one disk-shaped layer of ceramic material; we can glue two metal plates to the disk, one on each side, to obtain a *disk-ceramic capacitor*. To get larger capacitance values, we can stack layers of metal and ceramic, connecting alternate layers of metal together as the electrodes. The smallest single-layer ceramic capacitors exhibit only a few picofarads of capacitance. Large multilayer ceramic capacitors can have values ranging into the hundreds of microfarads.

Another geometry exists for the manufacture of ceramic capacitors. We can start with a cylinder (or "tube") of ceramic material and apply metal "ink" to its inside and outside surfaces to make a *tubular ceramic capacitor*. Ceramic capacitors, regardless of geometry, typically have values ranging from a few picofarads to about 0.5 μF. They have voltage ratings comparable to paper capacitors.

Plastic-Film Capacitors

Plastics make good dielectrics for the manufacture of capacitors. *Polyethylene* and *polystyrene* are commonly used. The method of manufacture resembles that for paper capacitors. Stacking methods work with plastic. The geometries can vary, so we'll find *plastic-film capacitors* in various shapes. Capacitance values range from about 50 pF to several tens of microfarads. Most often, we'll encounter values from approximately 0.001 μF to 10 μF. Plastic capacitors function well in electronic circuits at all frequencies, and at low to moderate voltages. They exhibit good efficiency, although not as high as that of mica-dielectric or air-dielectric capacitors.

Electrolytic Capacitors

All of the above-mentioned types of capacitors provide relatively small values of capacitance. They are also *nonpolarized*, meaning that we can connect them into a circuit in either direction (in some cases the vendor offers a recommendation as to which side should go to signal ground). An *electrolytic* capacitor provides approximately 1 μF up to thousands of microfarads, but we *must* connect it in the proper direction if we want it to work properly. An electrolytic capacitor constitutes a *polarized* component.

Component manufacturers assemble *electrolytic capacitors* by rolling up multiple layers of aluminum foil strips separated by paper saturated with an *electrolyte* liquid. The electrolyte conducts electric current. When DC flows through the component, the aluminum oxidizes because of chemical interaction with the electrolyte. The oxide layer does not conduct, and therefore, forms the dielectric for the capacitor. The layer is extremely thin, yielding high capacitance per unit of volume. Electrolytic capacitors can have values up to thousands of microfarads, and some can handle thousands of volts. These capacitors are most often seen in audio amplifiers and DC power supplies.

Tantalum Capacitors

Another type of electrolytic capacitor uses tantalum rather than aluminum. The tantalum can comprise foil like the aluminum in a conventional electrolytic capacitor. The tantalum can also take the form of a porous pellet, the irregular surface of which provides a large area in a small volume. An extremely thin oxide layer forms on the tantalum.

Tantalum capacitors have high reliability and excellent efficiency. We'll often see them used in military and aerospace environments—or anywhere where technicians find servicing inconvenient or impossible—because these devices almost never fail. Tantalum capacitors have values similar to those of aluminum electrolytics, and they work well in audio and digital circuits as replacements for aluminum types.

Transmission-Line Capacitors

In Chap. 10, we learned that sections of transmission line, cut to lengths shorter than ¼ electrical wavelength and shorted out at the far end, can function as inductors. Such line sections will act as capacitors if, instead of shorting the far end, we leave it open. The capacitance of such a transmission-line section increases with length until we get to ¼ electrical wavelength, when it behaves like a short circuit at the input end. Transmission-line capacitors are frequency-sensitive, just as transmission-line inductors are.

Semiconductor-Based Capacitors

Later in this book, you'll learn about *semiconductors*. These materials revolutionized electrical and electronic circuit design during the twentieth century. Today, nearly all electronic systems consists mainly of semiconductor-based components.

Manufacturers can employ semiconducting materials to build capacitors. A semiconductor *diode* conducts current in one direction, and refuses to conduct in the other direction. When a voltage appears across a diode so that it does not conduct, the diode acts as a capacitor. The capacitance varies depending on how much of this *reverse voltage* we apply to the diode. The greater the reverse voltage, the smaller the capacitance. This phenomenon makes the diode act as a *variable capacitor*. Some diodes are especially manufactured for this role. Their capacitances fluctuate rapidly along with pulsating DC. We call them *varactor diodes* or simply *varactors*.

Capacitors can be "etched" into the semiconductor materials of an *integrated circuit* (IC), also called a *chip*, as miniature varactors. A tiny capacitor can also be "etched" into an IC by sandwiching an oxide layer between two thin layers that conduct well. Most ICs look like little boxes with protruding metal prongs, which provide the electrical connections to external circuits and systems.

Semiconductor capacitors usually have small values of capacitance. They always have microscopic dimensions, and they can handle only low voltages. The advantages of semiconductor-based capacitors include miniaturization, and an ability (in the case of the varactor) to change in value at a rapid rate.

Variable Capacitors

We can vary the value of a capacitor at will by adjusting the mutual surface area between the plates, or by changing the spacing between the plates. Two main types of variable capacitors (besides varactors) exist: the *air variable capacitor* and the *trimmer capacitor*. We'll occasionally encounter a less common type known as a *coaxial capacitor*.

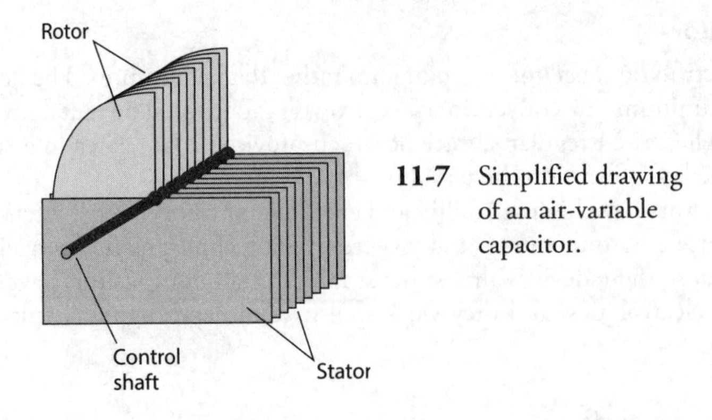

11-7 Simplified drawing of an air-variable capacitor.

Air-Variable Capacitors

We can assemble an air-variable capacitor by connecting two sets of metal plates so that they mesh, and by affixing one set to a rotatable shaft. The rotatable set of plates constitutes the *rotor*, and the fixed set constitutes the *stator*. We'll find this type of component in "vintage" radio receivers (particularly those that used *vacuum tubes* rather than semiconductor components), and in high-power wireless antenna tuning networks. Figure 11-7 illustrates an air-variable capacitor.

Air variables have maximum capacitance that depends on the number of plates in each set, and also on the spacing between the plates. Common maximum values range from 50 to 500 pF; occasionally we'll find an air variable that can go up to 1000 pF. Minimum values are generally on the order of a few picofarads. The voltage-handling capability depends on the spacing between the plates. Some air variables can handle several kilovolts at high AC frequencies. But the biggest advantage of air as a dielectric material is the fact that it has low loss—comparable to that of a vacuum.

We'll find air variables mainly in wireless equipment, designed to work at frequencies above approximately 500 kHz. These components offer high efficiency and excellent *thermal stability* (meaning that their values don't appreciably change with wild fluctuations in the ambient temperature). Although air variables technically lack polarization, we'll usually want to connect the rotor plates, along with the control shaft, to the metal chassis or circuit board perimeter, which constitutes the *common ground.*

Trimmer Capacitors

When we don't need to change the value of a capacitor often, we can use a *trimmer capacitor* in place of the more expensive, and bulkier, air-variable capacitor. A trimmer consists of two plates, mounted on a ceramic base and separated by a sheet of solid dielectric. We can vary the spacing between the plates with an adjusting screw as shown, in Fig. 11-8. Some trimmers contain two interleaved sets of multiple plates, alternating with dielectric layers to increase the capacitance.

We can connect a trimmer capacitor in parallel with an air-variable capacitor, facilitating exact adjustment of the latter component's minimum capacitance. Some air-variable capacitors have trimmers built in to serve this purpose. Typical maximum values for trimmers range from a few picofarads up to about 200 pF. They handle low to moderate voltages, exhibit excellent efficiency, and are free of polarizing characteristics.

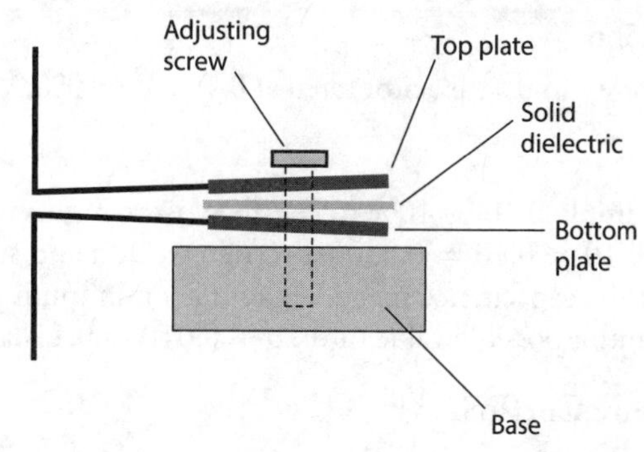

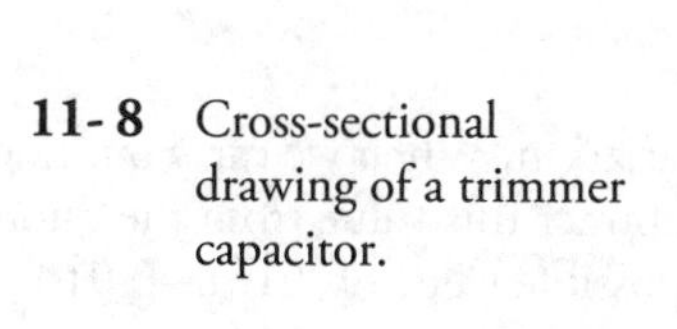
11-8 Cross-sectional drawing of a trimmer capacitor.

Coaxial Capacitors

We can use two telescoping sections of metal tubing to build a so-called *coaxial capacitor* (Fig. 11-9). The device works because of the variable effective surface area between the inner and outer tubing sections. A sleeve of plastic dielectric separates the sections of tubing, allowing us to adjust the capacitance by sliding the inner section in or out of the outer section. Coaxial capacitors work well at high-frequency AC applications, particularly in wireless antenna tuners. Their values range from a few picofarads up to approximately 100 pF.

Capacitor Specifications

When you seek a capacitor for a particular application, you should look for a component that has the proper specifications for the job you have in mind. Following are two especially significant capacitor specifications.

Tolerance

Component manufacturers rate capacitors according to how nearly we can expect their values to match the quoted capacitance. We call this rating the *tolerance*. The most common tolerance for fixed capacitors is ±10%; some capacitors are rated at ±5%, or even ±1%. As the tolerance figure decreases, we can expect the actual component value to more closely match the quoted value. For example, capacitor rated at 100 pF ±10 % can range from 90 pF to 110 pF. But if the tolerance equals ±1%, the manufacturer guarantees that the capacitance will not stray outside the range of 99 pF to 101 pF.

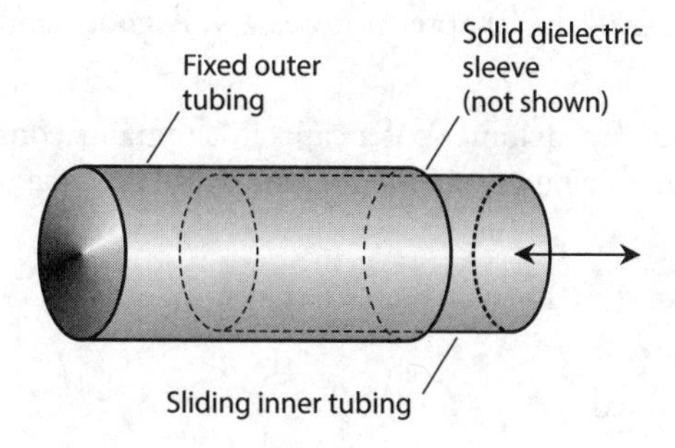

11-9 Simplified drawing of a coaxial variable capacitor.

Problem 11-6

Suppose that we find a capacitor rated at 0.10 μF ±10%. What's the guaranteed range of capacitance?

Solution

First, we multiply 0.10 by 10% to get the plus-or-minus variation. When we carry out that arithmetic, we get $0.10 \times 0.10 = 0.010$ μF. Then we add and subtract this value from the quoted capacitance to get the capacitance range, obtaining a minimum possible value of $0.10 - 0.010 = 0.09$ μF and a maximum possible value of $0.10 + 0.010 = 0.11$ μF.

Temperature Coefficient

Some capacitors increase in value as the temperature increases. These components have a *positive temperature coefficient.* Other capacitors decrease in value as the temperature rises; these exhibit a *negative temperature coefficient.* Some capacitors are specially manufactured (at considerable cost) so that their values remain constant over a certain temperature range. Within this span of temperatures, such capacitors have *zero temperature coefficient.*

Engineers commonly specify the temperature coefficient of a component in terms of *percent per degree Celsius* (%/°C). Sometimes, we can connect a capacitor with a negative temperature coefficient in series or parallel with a capacitor having a positive temperature coefficient, and the two opposite effects more or less nullify each other over a limited range of temperatures. In other instances, we can employ a capacitor with a positive or negative temperature coefficient to cancel out, or at least minimize, the effects of temperature on other components in a circuit, such as inductors and resistors.

Interelectrode Capacitance

Any two pieces of conducting material in close proximity can act as a capacitor. Often, such *interelectrode capacitance* is so small—a couple of picofarads or less—that we don't have to worry about it. In utility and audio-frequency (AF) circuits, interelectrode capacitance rarely poses any trouble, but it can cause problems in radio-frequency (RF) systems. The risk of trouble increases as the frequency increases.

The most common consequences of excessive interelectrode capacitance are *feedback* and unwanted changes in the characteristics of a circuit with variations in the operating frequency. We can minimize the interelectrode capacitance in an electronic device or system by keeping the interconnecting wires as short as possible within each individual circuit, by using shielded cables to connect circuits to each other, and by enclosing the most sensitive circuits in metal housings.

Quiz

Refer to the text in this chapter if necessary. A good score is 18 correct. Answers are in the back of the book.

1. If the value (capacitance) of a capacitor remains constant when the temperature changes within a reasonable range, then its temperature coefficient is

   (a) zero.

   (b) unity.

   (c) infinite.

   (d) undefined.

2. You can increase the value of a mica-dielectric capacitor, assuming all other factors remain constant, by replacing the mica with

   (a) strontium titanate.

   (b) polyethylene.

   (c) paper.

   (d) air.

3. Fill in the blank to make the following sentence true: "As you reduce the value of a capacitor, it ________ to completely charge after you connect a 6-V battery to its plates, assuming all other factors remain constant."

   (a) takes more time

   (b) becomes more difficult

   (c) takes less time

   (d) takes more energy

4. Which of the following values would you *most* likely find in an air-variable capacitor?

   (a) 1000 μF

   (b) 68 pF

   (c) 3.3 μF

   (d) 50,000 pF

5. Which of the following values would you *least* likely find in a single-layer disk-ceramic capacitor?

   (a) 150 pF

   (b) 680 pF

   (c) 0.01 μF

   (d) 330 μF

6. A capacitance of 0.01 nanofarad is the same as a capacitance of

   (a) 0.1 pF.

   (b) 1 pF.

   (c) 10 pF.

   (d) 100 pF.

7. If you connect four 100-pF capacitors in series, you get a net capacitance of

   (a) 25 pF.

   (b) 50 pF.

   (c) 100 pF.

   (d) 400 pF.

8. If you connect four 100-pF capacitors in parallel, you get a net capacitance of
   (a) 25 pF.
   (b) 50 pF.
   (c) 100 pF.
   (d) 400 pF.

9. If you connect four 100-pF capacitors in a 2 × 2 series-parallel matrix, you get a net capacitance of
   (a) 25 pF.
   (b) 50 pF.
   (c) 100 pF.
   (d) 400 pF.

10. If you connect nine 100-pF capacitors in a 3 × 3 series-parallel matrix, you get a net capacitance of
   (a) 25 pF.
   (b) 50 pF.
   (c) 100 pF.
   (d) 400 pF.

11. The main advantage of air as a dielectric material for capacitors is the fact that air
   (a) allows for a lot of capacitance in a small volume.
   (b) exhibits low loss.
   (c) works well at low voltages.
   (d) has a high dielectric constant.

12. A capacitance of 6800 nF is the same as
   (a) 0.06800 μF.
   (b) 0.6800 μF.
   (c) 6.800 μF.
   (d) 68.00 μF.

13. Which of the following properties is characteristic of tantalum capacitors?
   (a) High reliability
   (b) High capacitance per unit of volume
   (c) High efficiency
   (d) All of the above

14. Which of the following values would you most expect to find in an electrolytic capacitor?
   (a) 0.100 pF
   (b) 100 pF
   (c) 0.100 μF
   (d) 100 μF

15. You connect a 10-pF capacitor in parallel with a 20-pF capacitor. What's the net capacitance?
    (a) 30 pF
    (b) 15 pF
    (c) 6.7 pF
    (d) 5.5 pF

16. You connect a 10-pF capacitor in series with a 20-pF capacitor. What's the net capacitance?
    (a) 30 pF
    (b) 15 pF
    (c) 6.7 pF
    (d) 5.5 pF

17. You find a capacitor rated at 220 pF $\pm 10\%$. Which of the following capacitance values lies outside the acceptable range?
    (a) 180 pF
    (b) 195 pF
    (c) 246 pF
    (d) All of the above

18. A capacitor, rated at 330 pF, shows an actual value of 340 pF. By how much does its actual capacitance differ from its rated capacitance?
    (a) +2.94%
    (b) +3.03%
    (c) −2.94%
    (d) −3.03%

19. Which of the following values would you most expect to see in a paper capacitor?
    (a) 0.001 nF
    (b) 1.00 pF
    (c) 0.01 μF
    (d) 100 μF

20. If the number of plates in an air-variable capacitor decreases while all other factors remain constant,
    (a) the capacitance goes up.
    (b) the capacitance goes down.
    (c) the capacitance stays the same.
    (d) You'll need more information to answer this question.

12
CHAPTER

Phase

IN AN AC WAVE, EVERY FULL CYCLE REPLICATES EVERY OTHER FULL CYCLE; THE WAVE REPEATS INDEFInitely. In this chapter, you'll learn about the simplest possible shape (or waveform) for an AC disturbance. We call it a *sine wave* or *sinusoid.*

Instantaneous Values

When we graph its instantaneous amplitude as a function of time, an AC sine wave has the characteristic shape shown in Fig. 12-1. This illustration shows how the graph of the function $y = \sin x$ looks on an (x,y) coordinate plane. (The abbreviation *sin* stands for *sine* in trigonometry.) Imagine that the peak voltages equal $+1.0$ V and -1.0 V. Further imagine that the period equals exactly one second (1.0 s) so the wave has a frequency of 1.0 Hz. Let's say that the wave begins at time $t = 0.0$ s. In this scenario, each cycle begins every time the value of t "lands on" a whole number. At every such instant, the voltage is zero and *positive-going.*

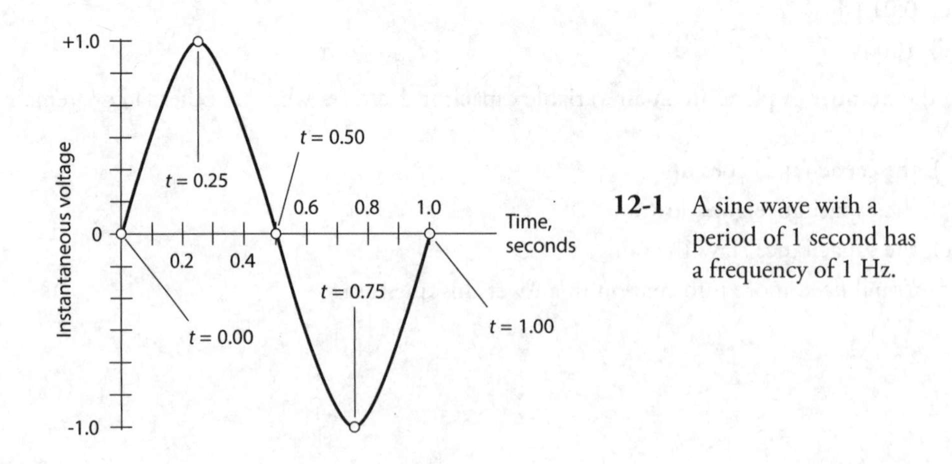

12-1 A sine wave with a period of 1 second has a frequency of 1 Hz.

If you freeze time at, say, $t = 446.00$ s and then measure the instantaneous voltage, you'll find that it equals 0.0 V. Looking at the diagram, you can see that the instantaneous voltage will also equal 0.0 V every so-many-and-a-*half* seconds, so it will equal 0.0 V at, say, $t = 446.50$ s. But instead of getting more positive at the "second-and-a-half" instants, the voltage trends negative. If you freeze time at so-many-and-a-*quarter* seconds, say, $t = 446.25$ s, the instantaneous voltage will equal +1.0 V. The wave will rest exactly at its positive peak. If you stop time at so-many-and-*three-quarter* seconds, say, $t = 446.75$ s, the instantaneous voltage will rest exactly at its negative peak, −1.0 V. At intermediate time points, such as so-many-and-*three-tenths* seconds, the voltage will have intermediate values.

Rate of Change

Figure 12-1 reveals the fact that the instantaneous voltage sometimes increases and sometimes decreases. *Increasing*, in this context, means "getting more positive," and *decreasing* means "getting more negative." In the situation shown by Fig. 12-1, the most rapid increases in voltage occur when $t = 0.00$ s and $t = 1.00$ s. The most rapid voltage decrease takes place when $t = 0.50$ s. When $t = 0.25$ s, and also when $t = 0.75$ s, the instantaneous voltage neither increases nor decreases. But these "unchanging voltages" exist only for vanishingly small instants in time.

Let n equal some positive whole number of seconds. No matter what whole number we choose for n, the situation at $t = n.25$ s appears the same as it does for $t = 0.25$ s. Also, for $t = n.75$ s, things appear the same as they are when $t = 0.75$ s. The single cycle shown in Fig. 12-1 represents every possible condition of an AC sine wave having a frequency of 1.0 Hz and peak values of +1.0 V and −1.0 V. The entire wave cycle repeats for as long as AC continues to flow in the circuit, assuming that we don't change the voltage or the frequency.

Now imagine that you want to observe the *instantaneous rate of change* in the voltage of the wave in Fig. 12-1, as a function of time. A graph of this function turns out as a sine wave, too—but it appears displaced to the left of the original wave by ¼ of a cycle. If you plot the instantaneous rate of change of a sine wave as a function of time (Fig. 12-2), you get the *derivative* of the waveform. The derivative of a sine wave turns out as a *cosine wave* because the mathematical derivative (in calculus) of the sine function equals the cosine function. The cosine wave has the same shape as the sine wave, but the *phase* differs by ¼ of a cycle.

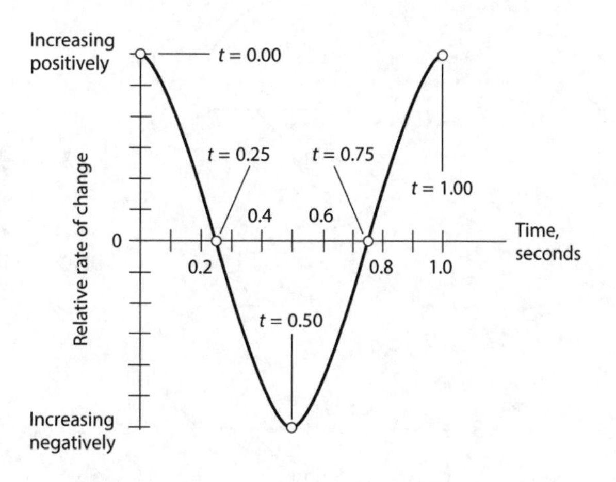

12-2 A sine wave representing the rate of change in the instantaneous voltage of the wave in Fig. 12-1.

Circles and Vectors

An AC sine wave represents the most efficient manner in which an electrical quantity can alternate. It has only one frequency component. All the wave energy concentrates into a single, smooth, swinging variation—a single frequency. When an AC wave has this characteristic, we can represent its fluctuations by comparing it to the motion of an object that follows a circular orbit at a constant speed around a fixed central point.

Circular Motion

Imagine that you revolve a ball around and around at the end of a string, at a rate of one revolution per second (1 r/s) so that the ball describes a horizontal circle in space, as shown in Fig. 12-3A. If a friend stands some distance away, with his or her eyes in the plane of the ball's path, she sees the ball oscillating back and forth, as shown in Fig. 12-3B, with a frequency of 1 Hz. That's one complete cycle per second because you cause the ball to "orbit" at 1 r/s.

If you graph the position of the ball, as seen by your friend, with respect to time, you'll get a sine wave, as shown in Fig. 12-4. This wave has the same fundamental shape as all sine waves. One sine wave might exhibit a greater distance between the peaks (peak-to-peak amplitude) than another, and one sine wave might appear "stretched out" lengthwise (wavelength) more than another. But every sine wave has the same general nature as every other. If we multiply or divide the peak-to-peak amplitude and/or the wavelength of any sine wave by the right numbers, we can make that sine wave fit exactly along the curve of any other sine wave. A *standard sine wave* has the following function in the (x,y) coordinate plane:

$$y = \sin x$$

In the situation of Fig. 12-3A, you might make the string longer or shorter. You might whirl the ball around faster or slower. These changes would alter the peak-to-peak amplitude and/or

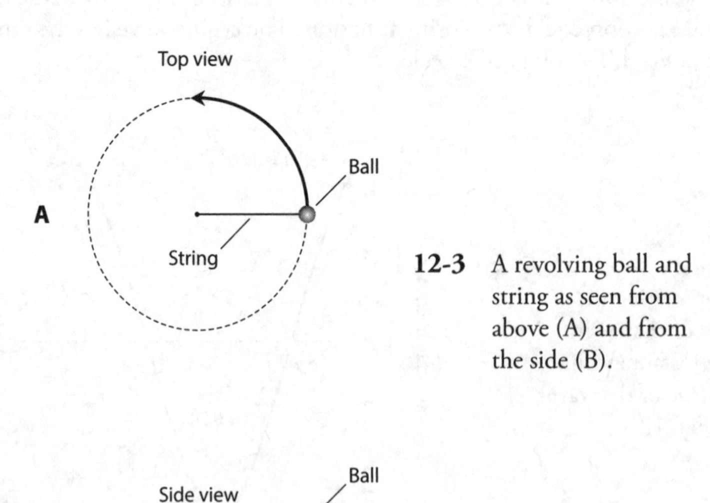

12-3 A revolving ball and string as seen from above (A) and from the side (B).

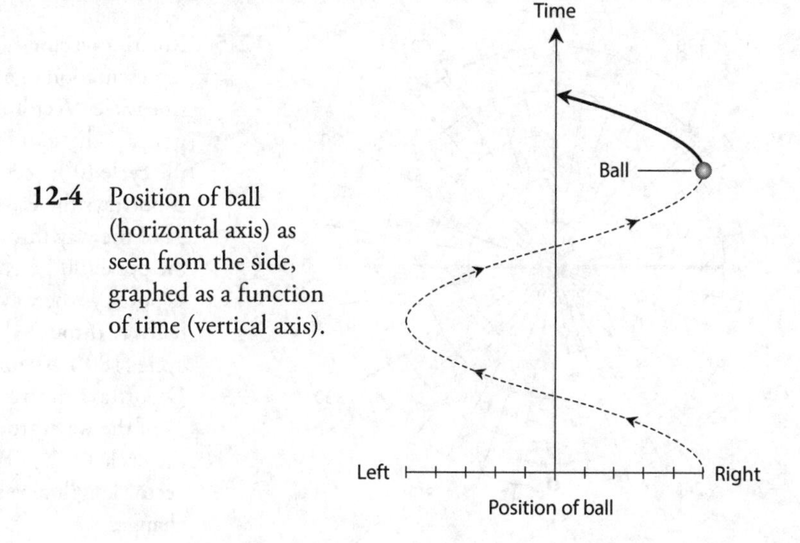

12-4 Position of ball (horizontal axis) as seen from the side, graphed as a function of time (vertical axis).

the frequency of the sine wave graphed in Fig. 12-4. But you can always portray a sine wave as the equivalent of constant circular revolution. Mathematicians and engineers call this trick the *circular-motion model* of a sine wave.

The Rotating Vector

In Chap. 9, you learned about *degrees of phase*. If you wondered why we discussed phase in terms of angles going around a circle, you should have a better grasp of the idea now! A circle contains 360 angular degrees (360°), as you know from your courses in basic geometry. Points along a sine wave intuitively correspond to angles, or positions, around a circle.

Figure 12-5 shows how we can use a *rotating vector* to represent a sine wave in a system of *polar coordinates*. In the polar coordinate system, we plot a point according to its distance (called the *radius*) from the *origin* (center of the graph) and its angle expressed counterclockwise from "due east" (called the *direction*). Compare this system with the more traditional system of *rectangular coordinates*, where we plot a point according to its horizontal and vertical displacement from the origin.

A *vector* constitutes a mathematical quantity with two independent properties, called *magnitude* (also called *length* or *amplitude*) and *direction* (or *angle*). Vectors lend themselves perfectly to polar coordinates. In the circular-motion model of Fig. 12-5, we can note the following specific situations:

- Vector **A** has a direction angle of 0°; it portrays the instant at which the wave amplitude equals zero and increases positively.
- Vector **B** points "north," representing the 90° phase angle, at which the wave has attained its maximum positive amplitude.
- Vector **C** points "west," representing a phase angle of 180°, the instant when the wave has gone back to zero amplitude while growing more negative.
- Vector **D** points "south," representing 270° of phase, the instant at which the wave has attained its maximum negative amplitude.
- When the vector has rotated counterclockwise through a full circle (360°), it once again becomes vector **A**, and the wave begins its next cycle.

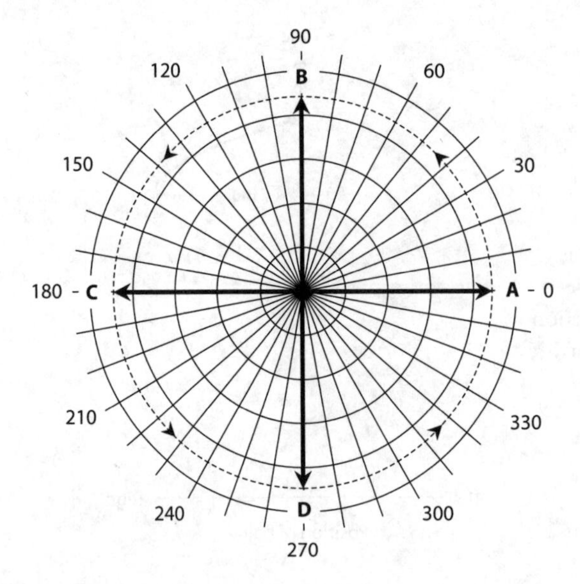

12-5 Rotating-vector representation of a sine wave. Vector **A** portrays the start of the cycle (0°); vector **B** portrays the wave ¼ of the way through the cycle (90°); vector **C** portrays the wave halfway through the cycle (180°); vector **D** portrays the wave ¾ of the way through the cycle (270°). The vector length never changes.

If you're astute, you'll notice that while the vector's direction constantly changes, its length always remains the same.

Vector "Snapshots"

Figure 12-6 shows the four points, on a sine wave, representing the *instantaneous vectors* **A, B, C,** and **D** from Fig. 12-5. Think of these four points as "snapshots" of the wave vector as it rotates counterclockwise at a constant *angular speed* that corresponds to one revolution per cycle of the wave. If the wave has a frequency of 1 Hz, the vector revolves at a rate of 1 r/s. We can increase or decrease the frequency and still use this model. If the wave has a frequency of 100 Hz, the speed of the vector will equal 100 r/s, or a revolution every 0.01 s. If the wave has a frequency of 1 MHz, then the speed of the vector will equal 1,000,000 r/s (10^6 r/s), and it will go once around the circle every 0.000001 s (10^{-6} s).

The peak amplitude (either positive or negative without the sign) of a pure AC sine wave corresponds to the length of its vector in the circular-motion model. Therefore, the peak-to-peak amplitude corresponds to twice the length of the vector. As the amplitude increases, the vector gets longer. In Fig. 12-5, we portray time during an individual cycle as an angle going counterclockwise from "due east."

In Fig. 12-5 and all other circular-model sine-wave vector diagrams, the vector length never changes, although it constantly rotates counterclockwise at a steady angular speed. The frequency of the wave corresponds to the speed at which the vector rotates. As the wave frequency increases, so does the vector's rotational speed. Just as the sine-wave's vector length remains independent of its rate of rotation, the amplitude of a sine wave is independent of its frequency.

Expressions of Phase Difference

The *phase difference*, also called the *phase angle*, between two sine waves can have meaning only when those two waves have the same frequency. If the frequencies differ, even by the slightest

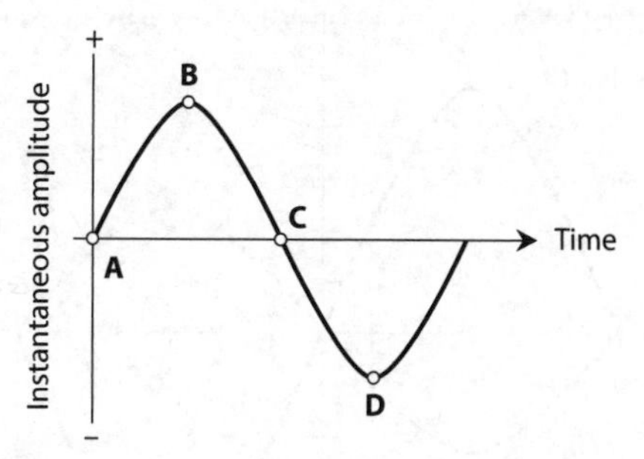

12-6 The four points for the vector model of Figure 12-5, shown in the standard amplitude-versus-time graphical manner for a sine wave.

amount, the relative phase constantly changes, and we can't specify a value for it. In the following discussions of phase angle, let's assume that the two waves always have identical frequencies.

Phase Coincidence

The term *phase coincidence* means that two waves begin at exactly the same moment. They "line up perfectly." Figure 12-7 illustrates a situation of this sort for two sine waves having different amplitudes. The phase difference in this case equals 0°. Alternatively, we could say that the phase difference equals some whole-number multiple of 360°, but engineers and technicians rarely speak of any phase angle of less than 0° or more than 360°.

If two sine waves exist in phase coincidence, and if neither wave has any DC superimposed on it, then the resultant wave constitutes a sine wave with positive peak (pk+) or negative peak (pk−) amplitudes equal to the sum of the positive and negative peak amplitudes of the composite waves. The phase of the resultant wave coincides with the phases of the two composite waves.

Waves ½ Cycle out of Phase

When two sine waves begin exactly ½ cycle (180°) apart, we get a situation like the one shown in Fig. 12-8. In this case, engineers sometimes say that the waves are *out of phase*, although this

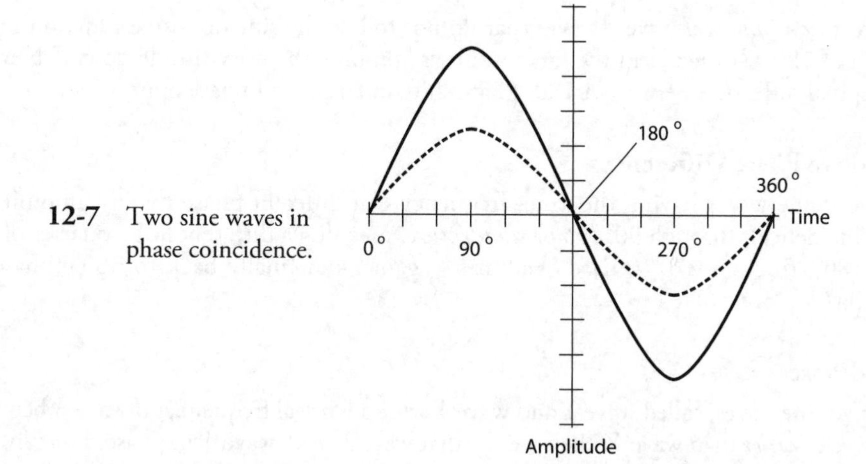

12-7 Two sine waves in phase coincidence.

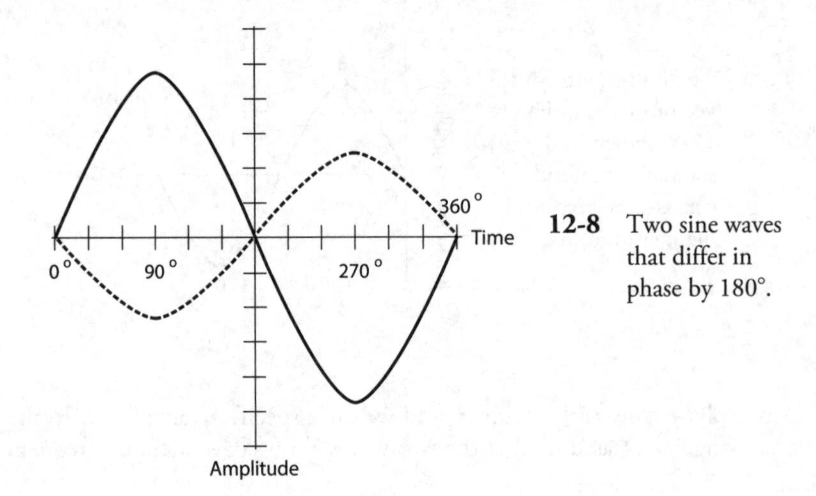

12-8 Two sine waves that differ in phase by 180°.

expression constitutes an imprecise statement because someone might take it to mean a phase difference other than 180°.

If two sine waves have the same amplitudes and exist 180° out of phase, and if neither wave has DC superimposed, they cancel each other out because the instantaneous amplitudes of the two waves are equal and opposite at every moment in time.

If two sine waves have different amplitudes and exist 180° out of phase, and if neither wave has DC superimposed, then the resultant wave is a sine wave with positive or negative peak amplitudes equal to the difference between the positive and negative peak amplitudes of the composite waves. The phase of the resultant wave coincides with the phase of the stronger composite.

Phase Opposition

Any perfect sine wave without superimposed DC has the property that, if we shift its phase by precisely 180°, we get a result identical to what we get by "flipping the original wave upside-down" (inverting it), a condition called *phase opposition*. Not all waveforms have this property. Perfect square waves do, but most rectangular and sawtooth waves don't, and irregular waveforms almost never do.

In most *nonsinusoidal* waves (waves that do not follow the sine or cosine function's graph), a phase shift of 180° *does not* yield the same result as "flipping the wave upside down." Never forget the conceptual difference between a 180° phase shift and a state of phase opposition!

Intermediate Phase Differences

Two perfect sine waves having the same frequency can differ in phase by any amount from 0° (phase coincidence), through 90° (*phase quadrature*, meaning a difference of a quarter of a cycle), through 180°, through 270° (phase quadrature again), and finally back to 360° (phase coincidence again).

Leading Phase

Imagine two sine waves, called wave *X* and wave *Y*, with identical frequency. If wave *X* begins a fraction of a cycle *earlier* than wave *Y*, then we say that wave *X leads* wave *Y* in phase. For this situation

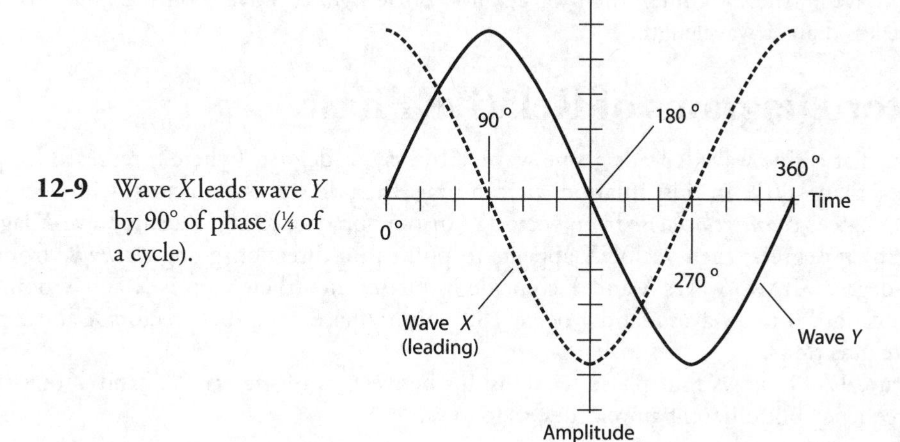

12-9 Wave *X* leads wave *Y* by 90° of phase (¼ of a cycle).

to hold true, wave *X* must begin its cycle less than 180° before wave *Y*. Figure 12-9 shows wave *X* leading wave *Y* by 90°.

When a particular wave *X* (the dashed line in Fig. 12-9) leads another wave *Y* (the solid line), then wave *X* lies to the *left* of wave *Y* on the time axis by some distance less than ½ wavelength. In a time-domain graph or display, displacement to the left represents earlier moments in time, and displacement to the right represents later moments in time; time "flows" from left to right.

Lagging Phase

Now imagine that some sine wave *X* begins its cycle more than 180° (½ cycle) but less than 360°(a full cycle) before wave *Y* starts. In this situation, we can imagine that wave *X* starts its cycle *later* than wave *Y* by some value between 0° and 180°. Then we say that wave *X lags* wave *Y*. Figure 12-10 shows wave *X* lagging wave *Y* by 90°. When a particular wave *X* (the dashed line in Fig. 12-10) lags

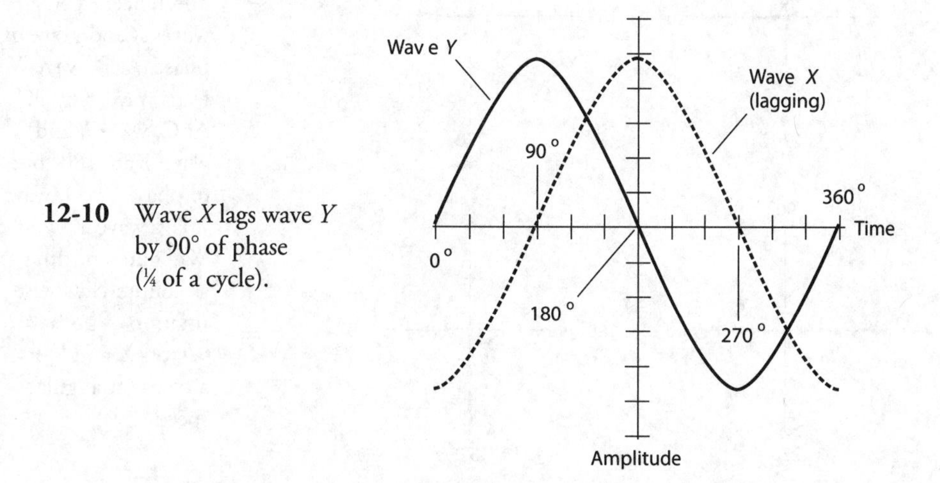

12-10 Wave *X* lags wave *Y* by 90° of phase (¼ of a cycle).

another wave *Y* (the solid line), then wave *X* lies to the *right* of wave *Y* on the time axis by some distance less than ½ wavelength.

Vector Diagrams of Relative Phase

Suppose that a sine wave *X* leads a sine wave *Y* by, say, *q* degrees (where *q* represents a positive angle less than 180°). In this situation, we can draw the two waves as vectors, with vector **X** oriented *q* degrees *counterclockwise* from vector **Y**. In the opposite sense, if a sine wave *X* lags a sine wave *Y* by *q* degrees, then vector **X** appears to point in a direction going *clockwise* from vector **Y** by *q* degrees. If two waves *X* and *Y* coincide in phase, then their vectors **X** and **Y** point in the same direction. If two waves *X* and *Y* occur 180° out of phase, then their vectors **X** and **Y** point in opposite directions.

Figure 12-11 shows four phase relationships between two sine waves *X* and *Y* that have the same frequency but different amplitudes, as follows:

1. At A, wave *X* exists in phase with wave *Y*, so vectors **X** and **Y** line up.
2. At B, wave *X* leads wave *Y* by 90°, so vector **X** points in a direction 90° counterclockwise from vector **Y**.
3. At C, waves *X* and *Y* exist 180° apart in phase, so vectors **X** and **Y** point in opposite directions.
4. At D, wave *X* lags wave *Y* by 90°, so vector **X** points in a direction 90° clockwise from vector **Y**.

In all of these examples, we should imagine that the vectors both rotate *counterclockwise* at a continuous, steady rate as time passes, always maintaining the same angle *with respect to each other*, and always staying at the same lengths. If the frequency in hertz equals *f*, then the pair of vectors rotates together, counterclockwise, at an angular speed of *f*, expressed in complete, full-circle rotations per second (r/s).

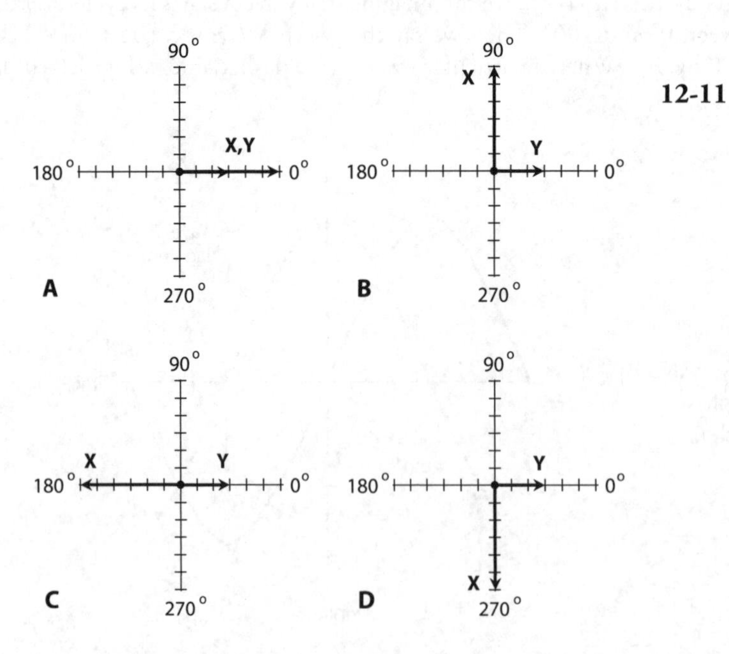

12-11 Vector representations of phase difference. At A, waves *X* and *Y* are in phase. At B, wave *X* leads wave *Y* by 90°. At C, wave *X* and wave *Y* are 180° out of phase. At D, wave *X* lags wave *Y* by 90°. We represent time as counterclockwise rotation of both vectors **X** and **Y** at a constant angular speed.

Quiz

Refer to the text in this chapter if necessary. A good score is 18 correct. Answers are in the back of the book.

1. A sine wave has a frequency of 50 kHz. Therefore, a complete cycle takes
   (a) 0.20 microseconds (μs). (Note: 1 μs = 0.000001 second.)
   (b) 2.0 μs.
   (c) 20 μs.
   (d) 200 μs.

2. Which of the following statements *is not* characteristic of a pure sine wave with a well-defined and constant period?
   (a) The electrical energy is distributed over a wide range (band) of frequencies.
   (b) The wave can be represented as a vector that rotates at a constant angular speed.
   (c) The wave has a well-defined, constant wavelength as long as the medium that carries it doesn't change.
   (d) The wave has a well-defined, constant frequency.

3. If someone says that two sine waves differ in phase by an amount that constantly changes, then we know that the waves have
   (a) different periods.
   (b) different wavelengths.
   (c) different frequencies.
   (d) All of the above

4. If wave X leads wave Y by ⅓ of a cycle, then
   (a) Y lags 120° behind X.
   (b) Y lags 90° behind X.
   (c) Y lags 60° behind X.
   (d) Y lags 30° behind X.

5. Fill in the blank to make the following statement true: "Suppose that two pure sine waves having identical frequencies and no DC components, coincide in phase. If you change the phase of one wave by ________, you'll get two waves in phase opposition."
   (a) 90°
   (b) 180°
   (c) 270°
   (d) 360°

6. We can change the phase of a pure sine wave having a constant frequency and no DC component by one of the following four phase angles, and end up with, in effect, the same wave. Which angle?
   (a) 45°
   (b) 90°
   (c) 180°
   (d) 360°

7. A phase difference of 22.5° in the circular-motion model of a sine wave represents
   (a) 1/16 of a revolution.
   (b) 1/8 of a revolution.
   (c) 1/4 of a revolution.
   (d) 1/2 of a revolution.

8. Two perfect sine waves exist in *phase opposition*. One wave has voltage peaks of +7 V pk+ and −7 V pk−, and the other wave has voltage peaks of +3 V pk+ and −3 V pk−. What's the *peak-to-peak* voltage of the composite wave?
   (a) 4 V pk-pk
   (b) 6 V pk-pk
   (c) 8 V pk-pk
   (d) 12 V pk-pk

9. A sine wave has a frequency of 60 Hz. How long does it take for 90° of phase to occur? (Note: 1 ms = 0.001 second.)
   (a) 2.1 ms
   (b) 4.2 ms
   (c) 8.3 ms
   (d) We need more information to calculate it.

10. In effect, a cosine wave is a sine wave shifted by
   (a) 60°.
   (b) 90°.
   (c) 120°.
   (d) 180°.

11. Two sine waves have the same frequency but differ in phase by 60°. Neither wave has a DC component. The two waves are offset by
   (a) 1/6 of a cycle.
   (b) 1/4 of a cycle.
   (c) 1/3 of a cycle.
   (d) 1/2 of a cycle.

12. Technically, the term *phase opposition* refers to two waves (whether sine or not) having the same frequency and
   (a) inverted with respect to each other.
   (b) displaced in phase by 90°.
   (c) displaced in phase by 180°.
   (d) displaced in phase by 270°.

13. In a polar-coordinate vector diagram in which the radius represents voltage, the length of the rotating vector for a pure sine wave containing no DC component represents
   (a) half the peak voltage (positive or negative).
   (b) the peak voltage (positive or negative).

(c) twice the peak voltage (positive or negative).
(d) the peak-to-peak voltage.

14. A sine wave *X* lags another sine wave *Y* by 45° of phase, so *Y* is
(a) 1⁄12 of a cycle ahead of *X*.
(b) 1⁄10 of a cycle ahead of *X*.
(c) 1⁄8 of a cycle ahead of *X*.
(d) 1⁄6 of a cycle ahead of *X*.

15. Figure 12-12 shows two sine waves *X* and *Y* that have the same frequency as a pair of polar vectors **X** and **Y**. Neither wave has a DC component. Which of the following statements holds true on the basis of this graph?
(a) wave *X* lags wave *Y* by 1⁄6 of a cycle.
(b) wave *X* lags wave *Y* by 1⁄3 of a cycle.
(c) wave *X* leads wave *Y* by 1⁄6 of a cycle.
(d) wave *X* leads wave *Y* by 1⁄3 of a cycle.

16. Which, if any, of the following conclusions can we make about the relative *peak-to-peak* amplitudes of the two waves shown in Fig. 12-12?
(a) The two waves have the same peak-to-peak amplitude.
(b) The peak-to-peak amplitude of wave *X* exceeds the peak-to-peak amplitude of wave *Y*.
(c) The peak-to-peak amplitude of wave *Y* exceeds the peak-to-peak amplitude of wave *X*.
(d) We can't say anything definitive about the relative peak-to-peak amplitudes of the two waves.

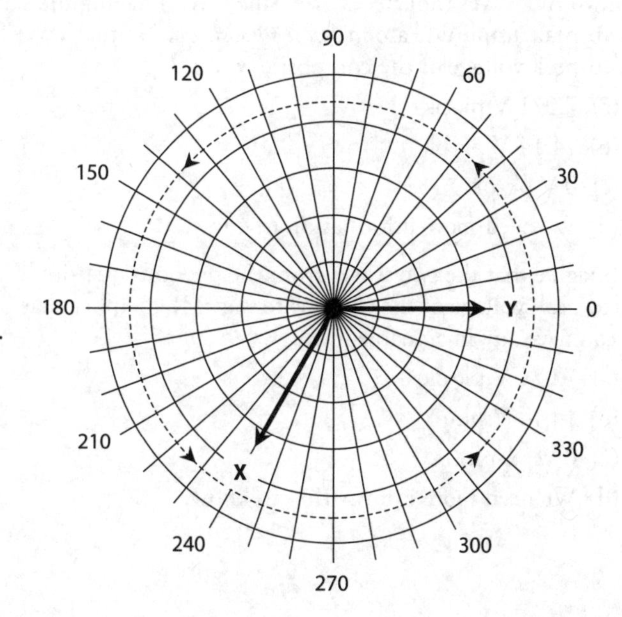

12-12 Illustration for Quiz Questions 15 and 16.

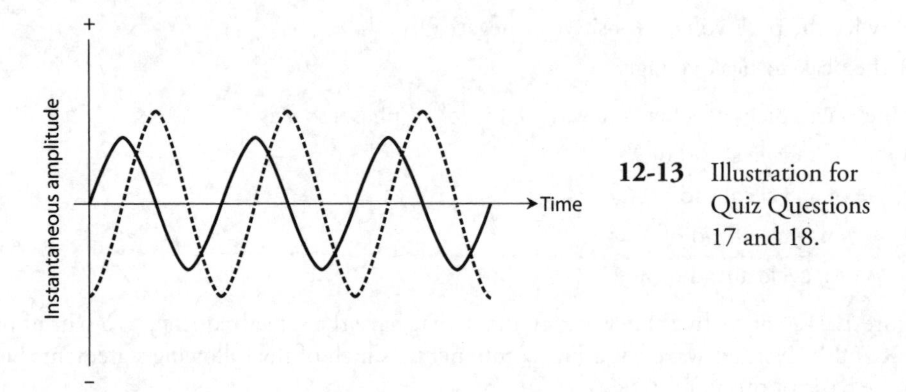

12-13 Illustration for Quiz Questions 17 and 18.

17. Figure 12-13 illustrates two sine waves, neither of which has a DC component, that are in phase
   (a) coincidence.
   (b) opposition.
   (c) quadrature.
   (d) reinforcement.

18. If we invert one of the waves in Fig. 12-13, then the two waves will be in phase
   (a) coincidence.
   (b) opposition.
   (c) quadrature.
   (d) reinforcement.

19. Two AC waves (not necessarily sine waves) having the same frequency and both with 10-V peak-to-peak amplitude are in *phase opposition*. Neither wave has a DC component. What's the peak-to-peak voltage of the composite wave?
   (a) 7.071 V pk-pk
   (b) 14.14 V pk-pk
   (c) 0 V pk-pk
   (d) We need more information to know.

20. Imagine that the two waves from the previous question differ in phase by 180°. What's the peak-to-peak voltage of the composite wave? Remember that the waves don't have to be sinusoids; they can have any imaginable form.
   (a) 7.071 V pk-pk
   (b) 14.14 V pk-pk
   (c) 0 V pk-pk
   (d) We need more information to know.

13
CHAPTER

Inductive Reactance

IN DC ELECTRICAL CIRCUITS, THE CURRENT, VOLTAGE, RESISTANCE, AND POWER RELATE ACCORDING to simple equations. The same equations work for AC circuits, provided that the components merely dissipate energy, and never store or release it. If a component stores or releases energy in an AC system, we say that the component has *reactance*. When we mathematically combine a component's reactance and resistance, we get an expression of the component's *impedance*, which fully quantifies how that component opposes, or *impedes*, the flow of AC.

Inductors and Direct Current

We can express DC resistance (in ohms, kilohms, megohms, or whatever other unit we want) as a number ranging from 0 (representing a perfect conductor) to extremely large values (representing poor conductors). Scientists call resistance a *scalar* quantity because we can portray its values as points on a *half-line,* or *ray,* having a one-dimensional *scale,* as shown in Fig. 13-1. However, when we add inductance to a circuit that already contains resistance and then drive AC through that circuit, things get more complicated.

Imagine that you have a supply wire that conducts electricity well. If you wind a length of the wire into a coil to make an inductor, and then you connect the coil to a battery or other source of DC (Fig. 13-2), the wire draws a small amount of current at first. But the current quickly becomes large, no matter how you configure the wire. You might wind it into a single-turn loop, let it lie in a mess on the floor, or wrap it around a wooden stick. In any case, you'll get a current equal to $I = E/R$, where I represents the current (in amperes), E represents the DC source voltage (in volts), and R represents the DC resistance of the wire (in ohms).

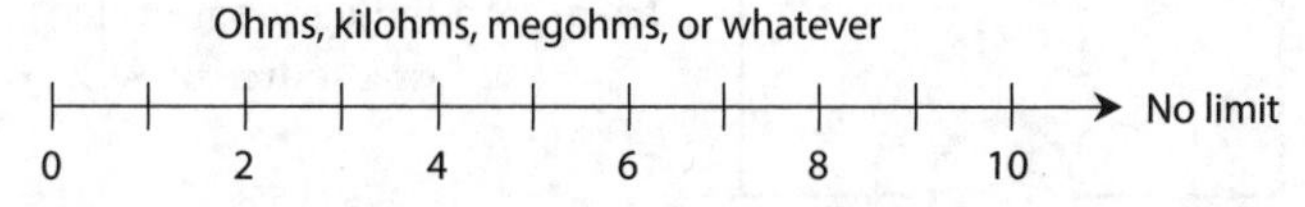

13-1 We can represent resistance values along a half-line or ray.

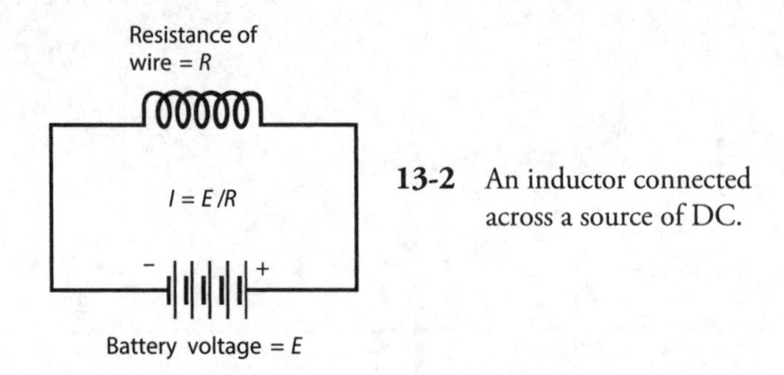

13-2 An inductor connected across a source of DC.

You can make an *electromagnet* by passing DC through a coil wound around an iron rod. You'll still observe a large, constant current in the coil, just as you would if the coil had no iron core. In a practical electromagnet, the coil heats up as some of the electrical energy dissipates in the wire; not all of the electrical energy contributes to the magnetic field. If you increase the DC source voltage and also increase the ability of the source to produce large currents, the wire in the coil will heat up more. Ultimately, if you increase the source voltage enough, and if it can deliver unlimited current, the wire will heat to the melting point.

Inductors and Alternating Current

Now suppose that you change the voltage source across the coil from DC to *pure AC* (that is, AC having no DC component), as shown in Fig. 13-3. Imagine that you can vary the frequency of the AC from a few hertz to hundreds of hertz, then kilohertz, then megahertz. At low frequencies, you'll see a large current in the coil, just as you did with the source of DC. But the coil exhibits a certain amount of inductance, and it takes a little time for current to establish itself in the coil. Depending on how many turns the coil has, and on whether the core consists of air or a ferromagnetic material, you'll reach a point, as you steadily increase the frequency, when the coil starts to get "sluggish." The current won't have time to fully establish itself in the coil before the AC polarity reverses.

At sufficiently high AC frequencies, the current through a coil will have trouble following the changes in the instantaneous voltage across it. Just as the coil starts to "think" that it can fully conduct, the AC voltage wave will pass its peak, go back to zero, and then "try" to pull the current the other way. In effect, this "sluggishness" will cause the coil to oppose the current in much the same way as a plain resistor would. As you raise the AC frequency, the coil's opposition to current will increase. Eventually, if you keep on increasing the frequency, the coil will fail to acquire

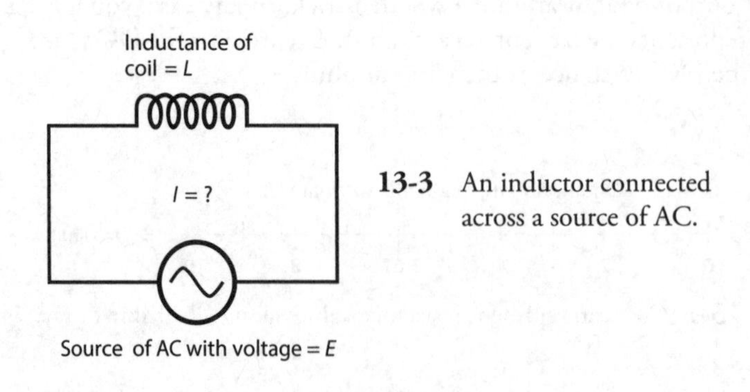

13-3 An inductor connected across a source of AC.

a significant current flow before the voltage polarity reverses. The coil will then act like a resistor with a high ohmic value.

With respect to AC, an inductor functions as a frequency-dependent resistor. We use the term *inductive reactance* to describe the opposition that the coil offers to AC. We express, or measure, inductive reactance in ohms. Inductive reactance can vary as resistance does, from almost nothing (a short piece of wire) to a few ohms (a small coil) to kilohms or megohms (coils having many turns, or coils with ferromagnetic cores operating at high AC frequencies). We can portray inductive reactance values along a half-line, just as we do with resistance. The numerical values on the half-line start at zero and increase without limit.

Reactance and Frequency

Inductive reactance constitutes one of two forms of reactance. (We'll examine the other form in the next chapter.) In mathematical expressions, we symbolize reactance in general as X, and we symbolize inductive reactance as X_L.

If the frequency of an AC source equals f (in hertz) and the inductance of a coil equals L (in henrys), then we can calculate the inductive reactance X_L (in ohms) using the formula

$$X_L = 2\pi f L \approx 6.2832\, fL$$

This same formula applies if we specify the frequency f in kilohertz and the inductance L in millihenrys. It also applies if we express f in megahertz and L in microhenrys. If we quantify frequency in thousands, we must quantify inductance in thousandths; if we quantify frequency in millions, we must quantify inductance in millionths.

Inductive reactance increases in a linear manner with (that is, in direct proportion to) increasing AC frequency, so the function of X_L versus f shows up as a straight line when we plot its graph on a rectangular coordinate plane. Inductive reactance also increases linearly with inductance, so the function of X_L versus L also appears as a straight line on a rectangular graph. Summarizing:

- If we hold L constant, then X_L varies in direct proportion to f.
- If we hold f constant, then X_L varies in direct proportion to L.

Figure 13-4 illustrates these relations on a generic rectangular coordinate grid.

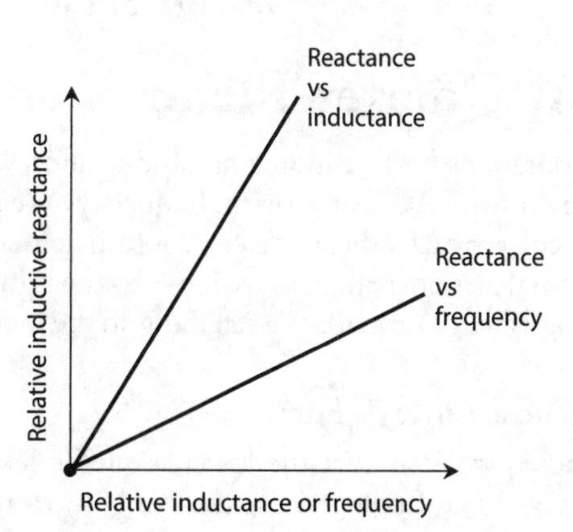

13-4 Inductive reactance varies in direct proportion to the inductance at a fixed frequency, and in direct proportion to the frequency for a fixed value of inductance.

Problem 13-1

Suppose that a coil has an inductance of 0.400 H, and the frequency of the AC passing through it equals 60.0 Hz. What's the inductive reactance?

Solution

Using the above formula, we can calculate and round off to three significant figures, getting

$$X_L = 6.2832 \times 60.0 \times 0.400 = 151 \text{ ohms}$$

Problem 13-2

How much inductive reactance will the above-described coil have if the power supply comprises a battery that supplies pure DC?

Solution

Because DC has a frequency of zero, we'll observe no inductive reactance at all. We can verify this by calculating

$$X_L = 6.2832 \times 0 \times 0.400 = 0 \text{ ohms}$$

Inductance has no practical effect with pure DC. The coil will exhibit a little bit of DC resistance because no wire constitutes a perfect electrical conductor, but that's not the same thing as AC reactance!

Problem 13-3

If a coil has an inductive reactance of 100 ohms at a frequency of 5.00 MHz, what's its inductance?

Solution

In this case, we must plug numbers into the formula and solve for the unknown L. Let's start out with the equation

$$100 = 6.2832 \times 5.00 \times L = 31.416\, L$$

Because we know the frequency in megahertz, the inductance will come out in microhenrys (μH). We can divide both sides of the above equation by 31.416 and then round off to three significant figures, getting

$$L = 100 / 31.416 = 3.18\ \mu\text{H}$$

The RX_L Quarter-Plane

In a circuit containing both resistance and inductance, we can't use a straight-line scale to portray the circuit's behavior with AC that varies in frequency. We must orient separate resistance and reactance rays perpendicular to each other to make a coordinate system, as shown in Fig. 13-5. Resistance appears on the horizontal axis, increasing as we move to the right. Inductive reactance appears on the vertical axis, increasing as we go upward. We call this grid the *resistance-inductive-reactance* (RX_L) *quarter-plane.*

What Are Those Little *j*'s For?

You're bound to wonder what the lowercase italic letters j represent in front of all the reactance numbers in Fig. 13-5. Engineers use the symbol j to represent a mathematical quantity called the

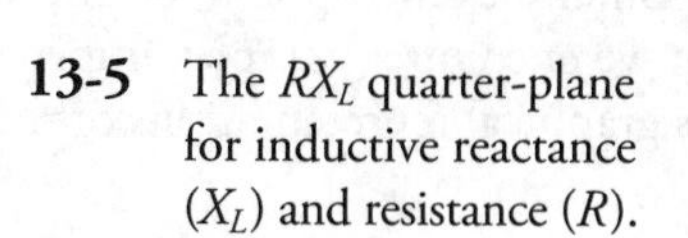

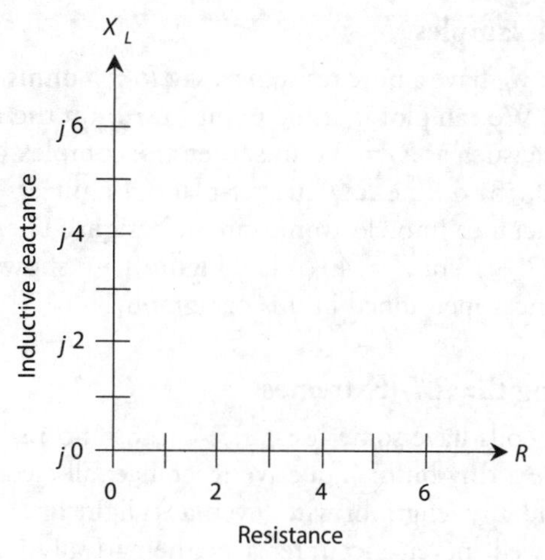

unit imaginary number. It's the positive square root of −1. (If you didn't already know that negative numbers can have square roots, you do now!) Electrical engineers call the positive square root of −1 the *j operator*. When we multiply *j* by itself over and over, we get the following four-way repeating sequence of quantities:

$$j \times j = -1$$
$$j \times j \times j = -j$$
$$j \times j \times j \times j = 1$$
$$j \times j \times j \times j \times j = j$$
$$j \times j \times j \times j \times j \times j = -1$$
$$j \times j \times j \times j \times j \times j \times j = -j$$
$$j \times j \times j \times j \times j \times j \times j \times j = 1$$
$$j \times j \times j \times j \times j \times j \times j \times j \times j = j$$
$$j \times j \times j \times j \times j \times j \times j \times j \times j \times j = -1$$
$$\downarrow$$
and so on, forever

When we multiply *j* by an ordinary number (that is, a *real number*), such as 2 or 5/2 or 7.764958, we get an *imaginary number*. All of the points on the vertical scale in Fig. 13-5 represent imaginary numbers. When we add an imaginary number to a real number, we get a *complex number*. All of the points in the entire quarter-plane of Fig. 13-5 represent complex numbers.

Complex Impedance

Each point on the RX_L quarter-plane corresponds to a unique *complex-number impedance* (or simply *complex impedance*). Conversely, each complex impedance value corresponds to a unique point on the quarter-plane. We express a complete impedance value Z, containing resistance and inductive reactance, on the RX_L quarter-plane in the form

$$Z = R + jX_L$$

where R represents the resistance (in ohms) and X_L represents the inductive reactance (also in ohms).

Some RX_L Examples

Suppose that we have a pure resistance, say $R = 5$ ohms. In this case, the complex impedance equals $Z = 5 + j0$. We can plot it at the point $(5, j0)$ on the RX_L quarter-plane. If we have a pure inductive reactance, such as $X_L = 3$ ohms, then the complex impedance equals $Z = 0 + j3$, and its point belongs at $(0, j3)$ on the RX_L quarter-plane. Engineers sometimes incorporate both resistance and inductive reactance into electronic circuit designs. Then we encounter complex impedance values, such as $Z = 2 + j3$ or $Z = 4 + j1.5$. Figure 13-6 shows graphical representations of the four complex impedances mentioned in this paragraph.

Approaching the RX_L Extremes

All practical coils have some resistance because no real-world wire conducts current perfectly. All resistors have a tiny bit of inductive reactance; all electrical components have wires called *leads* at each end, and any length of wire (even a straight one) exhibits some inductance. Therefore, in an AC circuit, we'll never encounter a mathematically perfect pure resistance, such as $5 + j0$, or a mathematically perfect pure reactance, such as $0 + j3$. We can approach these ideals, but we can never actually attain them (except in quiz and test problems).

How RX_L Points Move

Always remember that the values for X_L represent *reactances* (expressed in ohms), and not *inductances* (expressed in henrys). In an RX_L circuit, the reactance varies with the AC frequency, even if the inductance value never changes at all. Changing the frequency produces the graphical effect of making the points move in the RX_L quarter-plane. The points go vertically upward as the AC frequency increases, and downward as the AC frequency decreases. If the AC frequency goes all the way down to zero, thereby resulting in DC, the inductive reactance vanishes, and we're left with only a little bit of resistance, representing the DC ohmic loss in the inductor.

Some RX_L Impedance Vectors

Engineers sometimes represent points in the RX_L quarter-plane as vectors. Expressing a point in the RX_L quarter-plane as a vector gives that point a unique magnitude and a unique direction. Figure 13-6 shows four different points, each one represented by a certain distance to the right of the *origin*

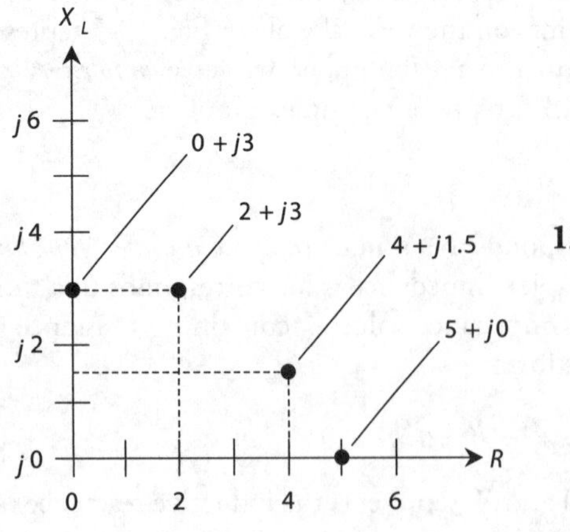

13-6 Four points in the RX_L quarter-plane.

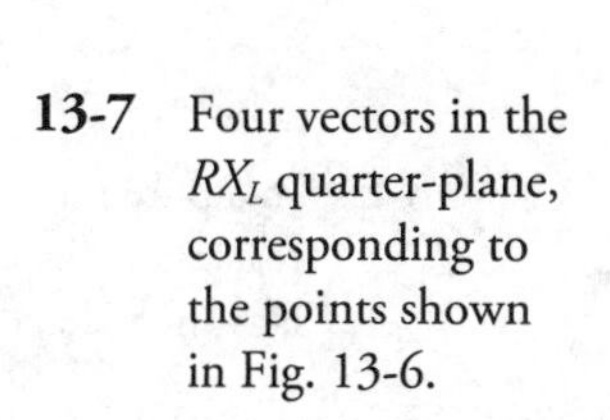

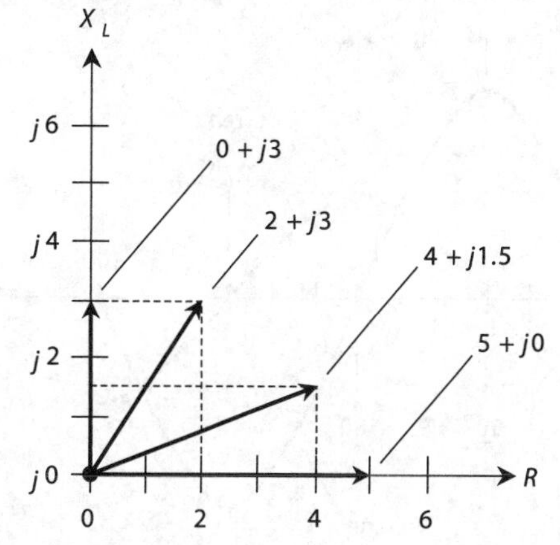

13-7 Four vectors in the RX_L quarter-plane, corresponding to the points shown in Fig. 13-6.

point (0, j0) that corresponds to the complex number $0 + j0$, and a certain distance upward from the origin. The first number in each complex sum represents the resistance R, and the second number represents the inductive reactance X_L. The RX_L combination constitutes a two-dimensional quantity. We can't define RX_L combinations as single numbers (scalar quantities) because any given RX_L combination possesses two quantities that can vary independently.

You can depict points, such as those shown in Fig. 13-6, by drawing straight rays from the origin out to those points. Then you can think of the rays instead of the points, with each ray having a certain length, or magnitude, and a certain direction, or angle counterclockwise from the resistance axis. These rays constitute *complex impedance vectors* (Fig. 13-7). When you think of complex impedances as vectors instead of mere points, you take advantage of a mathematical tool that can help you evaluate how AC circuits work under various conditions.

Current Lags Voltage

When we place an AC voltage source across an inductor and then power up the source so that the instantaneous voltage starts to increase (either positive or negative) from zero, it takes a fraction of a cycle for the current to follow. Later in the AC wave cycle, when the voltage starts decreasing from its maximum peak (either positive or negative), it again takes a fraction of a cycle for the current to follow. The instantaneous current can't quite keep up with the instantaneous voltage, as it does in a pure resistance, so we observe that in a circuit containing inductive reactance, the current *lags* (follows behind) the voltage. In some situations, this lag constitutes only a tiny fraction of an AC cycle, but it can range all the way up to ¼ of a cycle (90° of phase).

Pure Inductive Reactance

Suppose that we place an AC voltage source across a coil of wire made from an excellent conductor such as copper. Then we adjust the frequency of the AC source to a value high enough so that the inductive reactance X_L greatly exceeds the resistance R (by a factor of millions, say). In this situation, the coil acts as an essentially pure inductive reactance, and the current lags ¼ of a cycle (90°) behind the voltage for all intents and purposes, as shown in Fig. 13-8.

At low AC frequencies, we need a gigantic inductance if we want the current lag to approach 90°. As the AC frequency increases, we can get away with smaller inductances. If we could find some

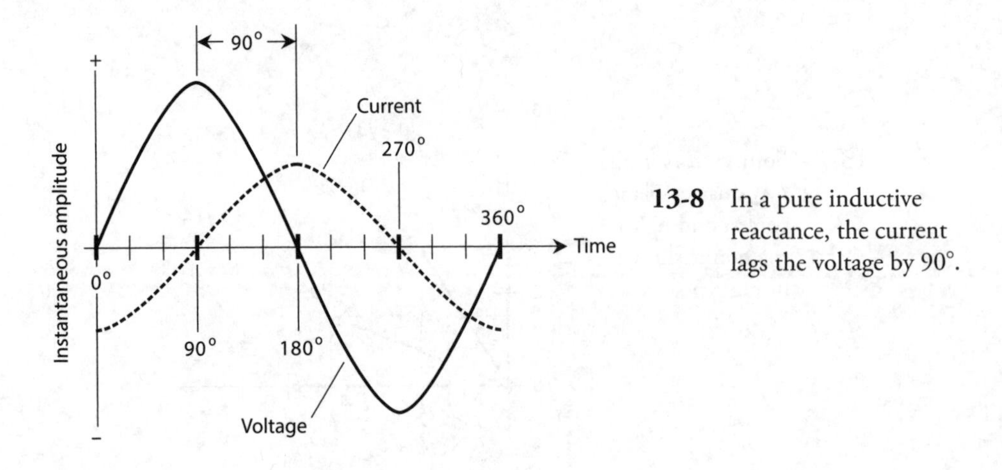

13-8 In a pure inductive reactance, the current lags the voltage by 90°.

wire that had no resistance whatsoever, and if we wound a coil with this wire, then the current would lag the voltage by exactly 90°, regardless of the AC frequency, and regardless of the coil size. In that case, we'd have an *ideal inductor* or a *pure inductive reactance*. No such thing exists in the "real world," but when the value of X_L greatly exceeds the value of R, the vector in the RX_L quarter-plane points almost exactly straight up along the X_L axis. The vector subtends an angle of just about 90° from the R axis.

Inductive Reactance with Resistance

When the resistance in a resistance-inductance (*RL*) circuit is significant compared with the inductive reactance, the current lags the voltage by something less than 90°, as shown in the example of Fig. 13-9. If R is small compared with X_L, the current lag equals almost 90°, but as R gets larger relative to X_L, the lag decreases.

The value of R in an *RL* circuit can increase relative to X_L if we deliberately place a pure resistance in series with the inductance. It can also happen because the AC frequency gets so low that X_L

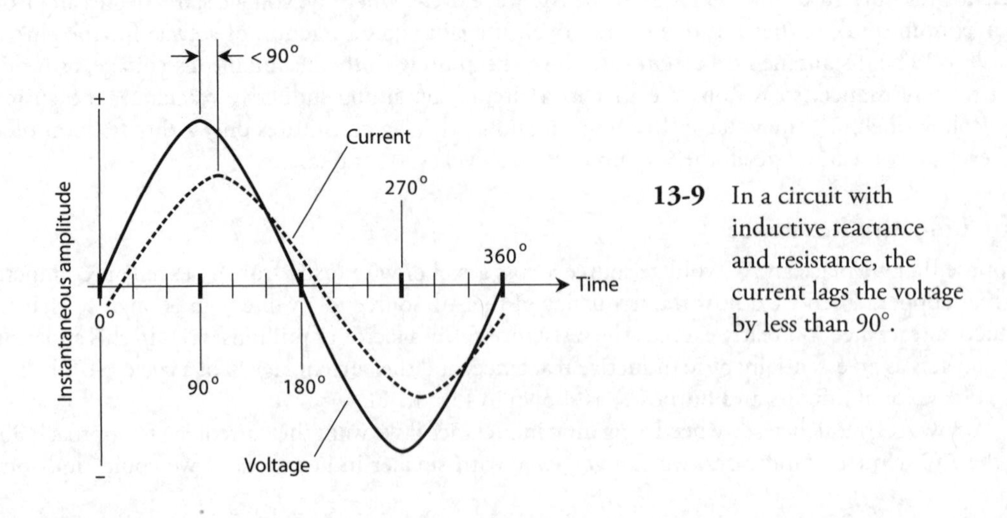

13-9 In a circuit with inductive reactance and resistance, the current lags the voltage by less than 90°.

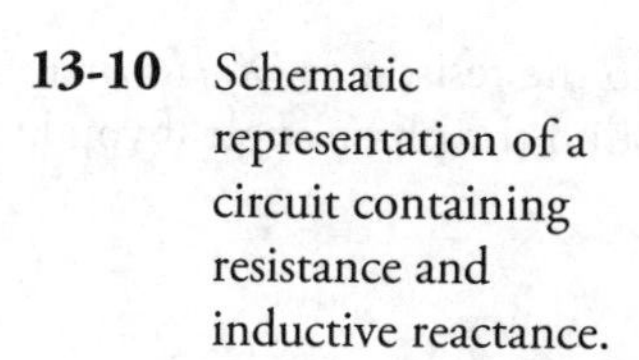

13-10 Schematic representation of a circuit containing resistance and inductive reactance.

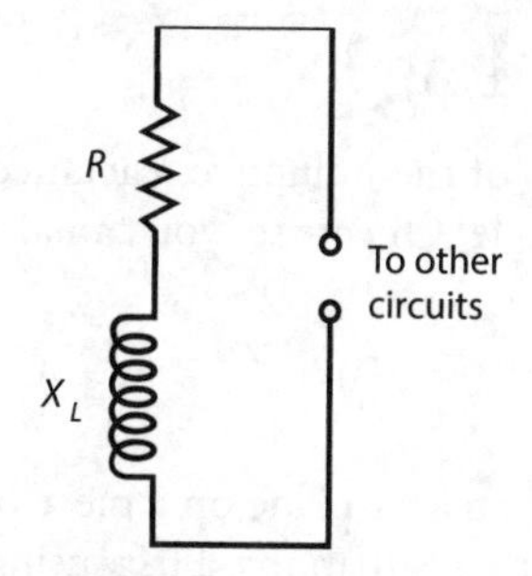

decreases until it reaches values comparable to the loss resistance R in the coil winding. In either case, we can schematically represent the circuit as an inductor in series with a resistor (Fig. 13-10).

If we know the values of X_L and R, we can find the *angle of lag*, also called the *RL phase angle* (or simply the *phase angle* if we know that we're dealing with resistance and inductance), by plotting the point $R + jX_L$ on the RX_L quarter-plane, drawing the vector from the origin out to that point, and then measuring the angle of the vector, counterclockwise from the resistance axis. We can use a protractor to measure this angle, or we can compute its value using trigonometry.

Actually, we don't have to know the actual values of X_L and R in order to find the angle of lag. All we need to know is their ratio. For example, if $X_L = 5$ ohms and $R = 3$ ohms, we get the same phase angle as we do if $X_L = 50$ ohms and $R = 30$ ohms, or if $X_L = 200$ ohms and $R = 120$ ohms. The angle of lag turns out the same for any values of X_L and R in the ratio 5:3.

Pure Resistance

As the resistance in an *RL* circuit becomes large with respect to the inductive reactance, the angle of lag gets small. The same thing happens if the inductive reactance gets small compared with the resistance. When R exceeds X_L by a large factor, the vector in the RX_L quarter-plane lies almost on the R axis, going "east," or to the right. The phase angle in this case is close to 0°. The current flows nearly in phase with the voltage fluctuations. In a pure resistance with no inductance whatsoever, the current would follow along exactly in phase with the voltage, as shown in Fig. 13-11. A pure resistance doesn't store and release energy as an inductive circuit does. It acquires and relinquishes all of its energy immediately, so no current lag occurs.

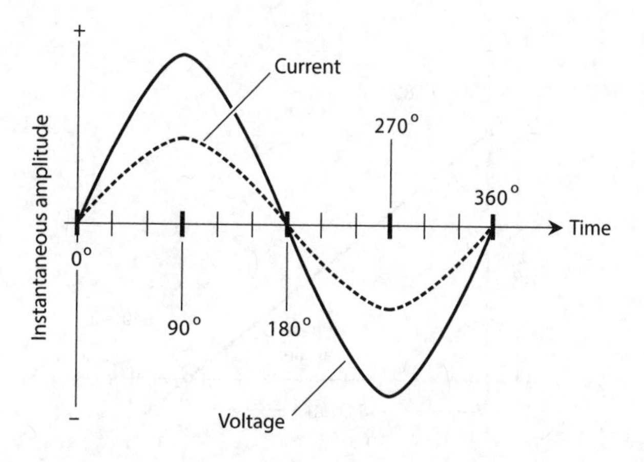

13-11 In a circuit with pure resistance (no reactance), the current tracks right along in phase with the voltage.

How Much Lag?

If you know the ratio of the inductive reactance to the resistance (X_L/R) in an *RL* circuit, then you can find the phase angle. Of course, you can also find the phase angle if you know the actual values of X_L and R.

Pictorial Method

You can draw an RX_L quarter-plane on a piece of paper and then use a ruler and a protractor to find a phase angle in most *RL* situations. First, using the ruler and a sharp pencil, draw a straight line a little more than 100 mm long, going from left to right. Then with the protractor, construct a line off the left end of this first line, going vertically upwards. Make this line at least 100 mm long. The horizontal line, or the one going to the right, constitutes the R axis of a coordinate system. The vertical line, or the one going upwards, forms the X_L axis.

If you know the values of X_L and R, divide them down or multiply them up so they're both between 0 and 100. For example, if $X_L = 680$ ohms and $R = 840$ ohms, you can divide them both by 10 to get $X_L = 68$ and $R = 84$. Plot these points lightly by making hash marks on the vertical and horizontal lines you've drawn. The R mark in this example should lie 84 mm to the right of the origin, and the X_L mark should lie 68 mm up from the origin.

Next, draw a line connecting the two hash marks, as shown in Fig. 13-12. This line will run at a slant, and will form a triangle along with the two axes. Your hash marks, and the origin of the coordinate system, form the three *vertices* (corner points) of a *right triangle*. We call the triangle "right" because one of its angles constitutes a *right angle* (90°). Measure the angle between the slanted line and the R axis. Extend one or both of the lines if necessary in order to get a good reading on the protractor. This angle will fall somewhere between 0° and 90°. It represents the phase angle in the *RL* circuit.

You can find the complex impedance vector, $R + jX_L$, by constructing a rectangle using the origin and your two hash marks as three of the rectangle's four vertices, and drawing new horizontal and vertical lines to complete the figure. The vector will show up as the diagonal of this rectangle (Fig. 13-13). The angle between this vector and the R axis will represent the phase angle. It should have the same measure as the angle of the slanted line relative to the R axis in Fig. 13-12.

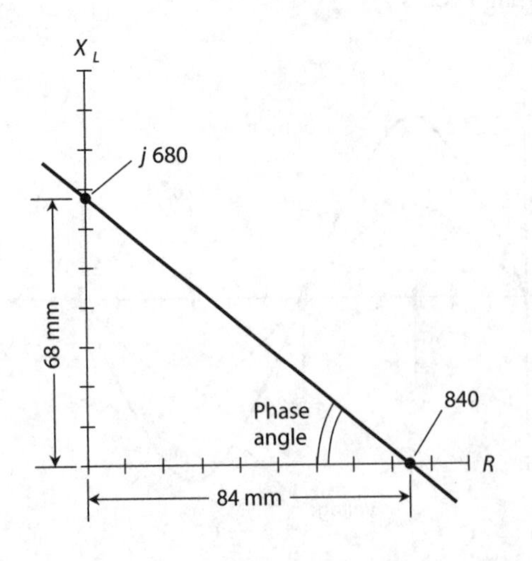

13-12 Pictorial method of finding phase angle in a circuit containing resistance and inductive reactance.

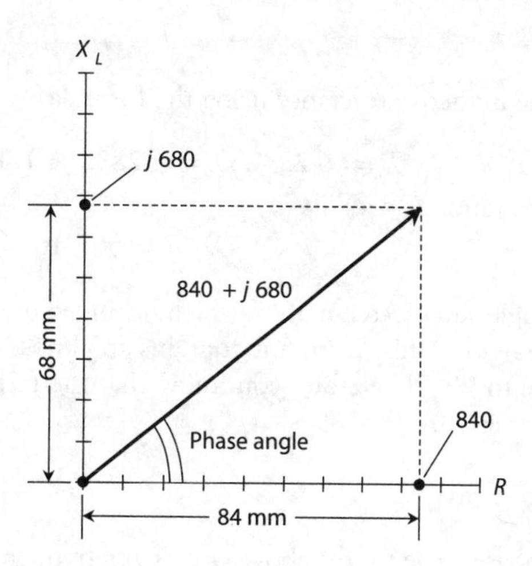

13-13 Another pictorial method of finding phase angle in a circuit containing resistance and inductive reactance. This method shows the impedance vector.

Trigonometric Method

If you have a scientific calculator that can find the *Arctangent* of a number (also called the *inverse tangent* and symbolized either as Arctan or $\tan^{-1}$), you can determine the phase angle more precisely than the pictorial method allows. Given the values of X_L and R, the phase angle equals the Arctangent of their ratio. We symbolize phase angle as a variable with the lowercase Greek letter phi (pronounced "fie" or "fee" and written ϕ). Expressed mathematically, the phase angle is

$$\phi = \tan^{-1} (X_L/R)$$

or

$$\phi = \text{Arctan } (X_L/R)$$

With most computers, you can find a number's Arctangent by setting the calculator program to work in the scientific mode, entering the number, hitting the key or checking the box marked "inv," and finally hitting the "tan" key.

Problem 13-4

Suppose that the inductive reactance in an *RL* circuit equals 680 ohms and the resistance equals 840 ohms. What's the phase angle?

Solution

The ratio X_L/R equals 680/840. A calculator will display this quotient as something like 0.8095 followed by some more digits. Find the Arctangent of this number. You should get 38.99 and some more digits. You can round this off to 39.0°.

Problem 13-5

Suppose that an *RL* circuit operates at a frequency of 1.0 MHz with a resistance of 10 ohms and an inductance of 90 μH. What's the phase angle? What does this result tell you about the nature of this *RL* circuit at this frequency?

Solution

First, find the inductive reactance using the formula

$$X_L = 6.2832\,fL = 6.2832 \times 1.0 \times 90 = 565 \text{ ohms}$$

Then find the ratio

$$X_L/R = 565/10 = 56.5$$

The phase angle equals Arctan 56.5, which, rounded to two significant figures, comes out to be 89°. Now you know that this *RL* circuit contains an almost pure inductive reactance because the phase angle is close to 90°. Therefore, you know that the resistance contributes little to the behavior of this *RL* circuit at 1.0 MHz.

Problem 13-6

What's the phase angle for the above circuit at a frequency of 10 kHz? With that information, what can you say about the behavior of the circuit at 10 kHz?

Solution

You must calculate X_L all over again for the new frequency. You can use megahertz as your unit of frequency because megahertz work in the formula with microhenrys. A frequency of 10 kHz equals 0.010 MHz. Calculating, you get

$$X_L = 6.2832\,fL = 6.2832 \times 0.010 \times 90 = 5.65 \text{ ohms}$$

Calculating the ratio of inductive reactance to resistance, you get

$$X_L/R = 5.65/10 = 0.565$$

The phase angle at the new frequency equals Arctan 0.565, which, rounded to two significant figures, turns out as 29°. This angle is not close to either 0° or 90°. Therefore, you know that at 10 kHz, the resistance and the inductive reactance both play significant roles in the behavior of the *RL* circuit.

Quiz

Refer to the text in this chapter if necessary. A good score is 18 correct. Answers are in the back of the book.

1. A coil has an inductive reactance of 120 ohms at 5.00 kHz. What's its inductance?
   (a) 19.1 mH
   (b) 1.91 mH
   (c) 38.2 mH
   (d) 3.82 mH

2. As a coil's inductance rises, its fixed-frequency reactance
   (a) alternately increases and decreases.
   (b) stays the same.

(c) increases.

(d) decreases.

3. An inductor has $X_L = 700$ ohms at $f = 2.50$ MHz. What is L?

(a) 223 μH

(b) 22.3 μH

(c) 446 μH

(d) 44.6 μH

4. In a coil having zero resistance and an AC signal applied, the *phase angle* is

(a) 0°.

(b) 45°.

(c) 90°.

(d) some value that depends on the signal frequency.

5. If the inductive reactance in ohms equals the resistance in ohms in an *RL* circuit with an AC signal applied, then the phase angle is

(a) between 0° and 45°.

(b) 45°.

(c) between 45° and 90°.

(d) some value that depends on the signal frequency.

6. In a pure resistance without inductance and with an AC signal applied, the phase angle is

(a) 0°.

(b) 45°.

(c) 90°.

(d) some value that depends on the signal frequency.

7. According to Fig. 13-14, X_L/R is

(a) 17.1.

(b) 8.57.

(c) 0.233.

(d) 0.117.

8. In Fig. 13-14, the R and X_L graph scale divisions differ in size, but we can determine the phase angle anyway. It's about

(a) 6.67°.

(b) 13.1°.

(c) 83.3°.

(d) 86.7°.

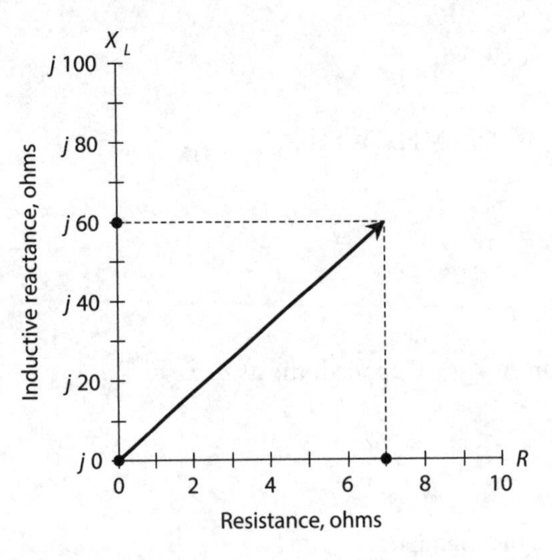

13-14 Illustration for Quiz Questions 7 and 8.

9. We apply an AC signal to a coil with an adjustable "roller tap" that lets us vary the number of coil turns through which the signal passes. (Engineers call this contraption a *roller inductor.*) When we set the tap so that the signal current must flow through the entire coil, we obtain a certain reactance that depends on the signal frequency. As we adjust the tap so the signal current passes through fewer and fewer turns, how must we change the signal frequency to maintain constant reactance? Assume that no resistance exists in the coil itself or in the components immediately external to it.
   (a) We must not change the frequency.
   (b) We must increase the frequency.
   (c) We must decrease the frequency.
   (d) We need more information to answer this question.

10. In the situation of Question 9, what happens to the phase angle as we adjust the coil in the manner described, assuming the coil is made of perfectly conducting wire?
   (a) It stays the same.
   (b) It gets larger.
   (c) It gets smaller.
   (d) We need more information to know.

11. The points along the vertical axis in the RX_L quarter-plane correspond one-to-one with values of
   (a) inductance.
   (b) inductive reactance.
   (c) resistance.
   (d) complex impedance.

12. A coil has an inductance of 50.0 mH. What's its reactance at 5.00 kHz?
    (a) 15.7 ohms
    (b) 31.4 ohms
    (c) 785 ohms
    (d) 1.57 k

13. A 1.0-mH inductor has a reactance of 3000 ohms. What's the frequency?
    (a) We need more information to calculate it.
    (b) 0.24 MHz
    (c) 0.48 MHz
    (d) 0.96 MHz

14. If we increase the resistance gradually from zero to unlimited values while keeping the inductive reactance constant in an *RL* circuit, the resulting points in the RX_L quarter-plane lie along
    (a) a straight ray pointing up from some point on the resistance axis.
    (b) a straight ray pointing to the right from some point on the reactance axis.
    (c) a straight ray ramping up and to the right from the origin.
    (d) a quarter-circle centered at the origin.

15. If we gradually increase both the resistance and the reactance in an *RL* circuit from zero to unlimited values at constant rates, the resulting points in the RX_L quarter-plane lie along
    (a) a straight ray pointing up from some point on the resistance axis.
    (b) a straight ray pointing to the right from some point on the reactance axis.
    (c) a straight ray ramping up and to the right from the origin.
    (d) a quarter-circle centered at the origin.

16. In a certain *RL* circuit, the ratio of the inductive reactance to the resistance starts out large and then decreases gradually to zero. The phase angle
    (a) increases and approaches 90°.
    (b) decreases and approaches 45°.
    (c) increases and approaches 45°.
    (d) decreases and approaches 0°.

17. In a certain *RL* circuit, the ratio of the inductive reactance to the resistance starts out at zero and gradually increases toward a limiting value of 1.732:1. The phase angle
    (a) increases and approaches 30°.
    (b) decreases and approaches 30°.
    (c) increases and approaches 60°.
    (d) decreases and approaches 60°.

18. A coil has an inductance of 100 nH at a frequency of 100 MHz. What's the inductive reactance?
    (a) We need more information to calculate it.
    (b) 126 ohms
    (c) 62.8 ohms
    (d) 31.4 ohms
19. An *RL* circuit comprises a 1.25-mH inductor and a 7.50-ohm resistor. The circuit's interconnecting wires conduct perfectly. What's the phase angle at 1.45 kHz?
    (a) 56.6°
    (b) 42.3°
    (c) 33.4°
    (d) 21.2°
20. What happens to the phase angle if we short out the resistor in the circuit described in the previous question?
    (a) It depends on the signal frequency.
    (b) It depends on the signal voltage.
    (c) It stays the same.
    (d) None of the above

14 CHAPTER

Capacitive Reactance

CAPACITIVE REACTANCE ACTS AS THE NATURAL COUNTERPART OF INDUCTIVE REACTANCE. WE CAN represent it graphically as a ray that goes in a negative direction. When we join the capacitive-reactance and inductive-reactance rays at their end points (both of which correspond to a reactance of zero), we get a complete number line, as shown in Fig. 14-1. This line depicts all possible values of reactance because any nonzero reactance must be either inductive or capacitive.

Capacitors and Direct Current

Imagine two huge, flat metal plates, both of which constitute excellent electrical conductors. If we connect them to a source of DC, as shown in Fig. 14-2, they draw a large amount of current while they become electrically charged. But as the plates reach equilibrium, the charging current goes down to zero.

If we increase the voltage of the battery or power supply, we eventually reach a point at which sparks jump between the plates of our capacitor. Ultimately, if the power supply can deliver the necessary voltage, this sparking, or *arcing*, becomes continuous. Under these conditions, the pair of plates no longer acts like a capacitor at all. When we place excessive voltage across a capacitor, the dielectric can't provide electrical separation between the plates. We call this undesirable condition *dielectric breakdown*.

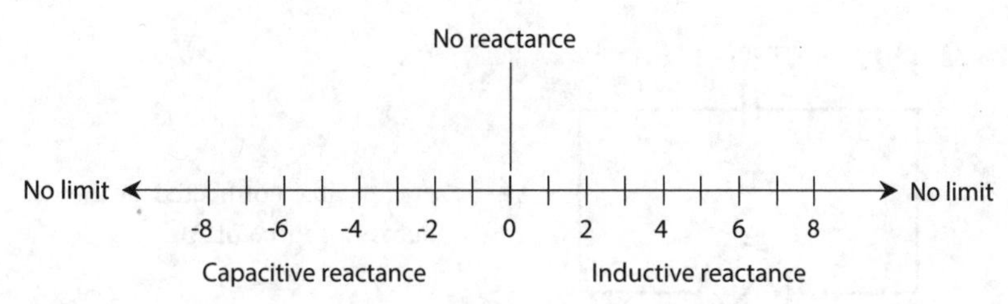

14-1 We can represent inductive and capacitive reactance values as points along a number line.

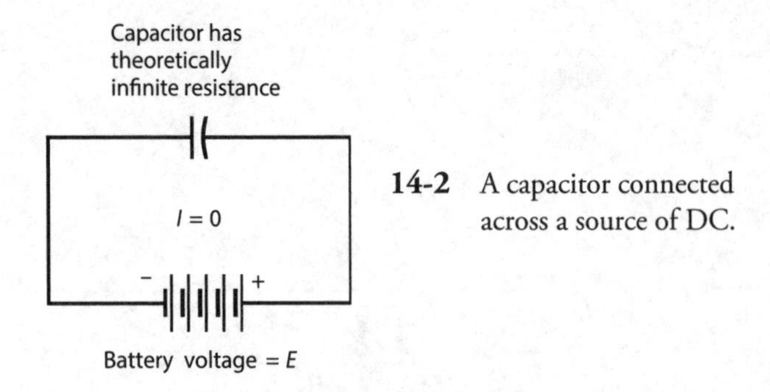

14-2 A capacitor connected across a source of DC.

In an air-dielectric or vacuum-dielectric capacitor, dielectric breakdown manifests itself as a temporary affair, rarely causing permanent damage to the component. The device operates normally after we reduce the voltage so that the arcing stops. However, in capacitors made with solid dielectric materials, such as mica, paper, polystyrene, or tantalum, dielectric breakdown can burn or crack the dielectric, causing the component to conduct current even after we reduce the voltage. If a capacitor suffers this sort of damage, we must remove it from the circuit, discard it, and replace it with a new one.

Capacitors and Alternating Current

Now suppose that we change the voltage source from DC to AC (Fig. 14-3). Imagine that we can adjust the frequency of this AC from a low initial value of a few hertz, up to hundreds of hertz, then to many kilohertz, megahertz, or gigahertz.

At first, the voltage between the plates follows the voltage of the power source as the AC polarity alternates. The plates can charge up quickly if they have small surface areas and/or if a lot of space exists between them, but they can't charge instantaneously. As we increase the frequency of the applied AC, we reach a point at which the plates can't charge up very much before the AC polarity reverses. Just as the plates begin to get a good charge, the AC passes its peak and starts to discharge them. As we raise the frequency still further, the set of plates acts increasingly like a short circuit. Eventually, if we keep raising the AC frequency, the period of the wave becomes much shorter than the charge/discharge time, and current flows in and out of the plates just as fast as it would if the plates were removed altogether and replaced with a plain piece of wire.

Capacitive reactance quantifies the opposition that a capacitor offers to AC. We express and measure capacitive reactance in ohms, just as we do with inductive reactance or pure resistance. But

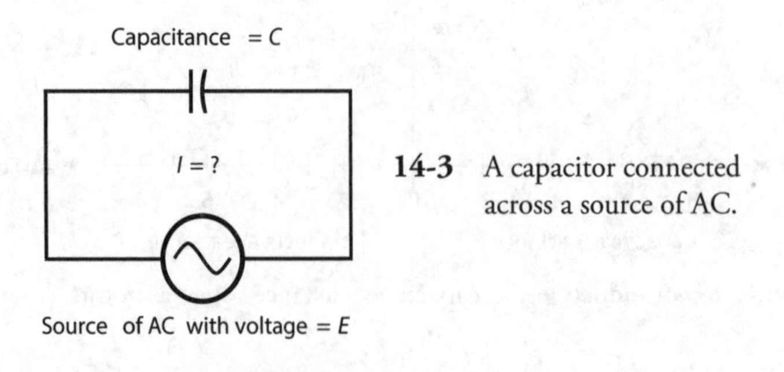

14-3 A capacitor connected across a source of AC.

capacitive reactance, by convention, has negative values rather than positive ones. Capacitive reactance, denoted X_C in mathematical formulas, can vary from near zero (when the plates are gigantic and close together, and/or the frequency is very high) to a few negative ohms, to many negative kilohms or megohms.

Capacitive reactance varies with frequency. It gets *larger negatively* as the AC frequency goes down, and *smaller negatively* as the applied AC frequency goes up. This behavior runs contrary to what happens with inductive reactance, which gets *larger positively* as the frequency goes up. Sometimes, non-technical people talk about capacitive reactance in terms of its *absolute value*, with the minus sign removed. Then we might say that X_C increases as the frequency goes down, or that X_C decreases as the frequency goes up. Nevertheless, we'll work with negative X_C values and stick with that convention. That way, the mathematics will give us the most accurate representation of how AC circuits behave when they contain capacitance.

Capacitive Reactance and Frequency

In a purely theoretical sense, capacitive reactance "mirrors" inductive reactance. In a geometrical or graphical sense, X_C constitutes a continuation of X_L into negative values, something like the extensions of the Celsius or Fahrenheit temperature scales to values "below zero."

If we specify the frequency of an AC source (in hertz) as f, and if we specify the capacitance of a component (in farads) as C, then we can calculate the capacitive reactance using the formula

$$X_C = -1/(2\pi f C) \approx -1/(6.2832 f C)$$

This formula also works if we input f in megahertz and C in microfarads (μF). It could even apply for values of f in kilohertz (kHz) and values of C in *millifarads* (mF)—but you'll almost never see capacitances expressed in millifarads.

For Quick Reference

1 microfarad = 1 μF = 10^{-6} F
1 nanofarad = 1 nF = 10^{-9} F
1 picofarad = 1 pF = 10^{-12} F

The function X_C versus f appears as a curve when we graph it in rectangular coordinates. The curve contains a *singularity* at $f = 0$; it "blows up negatively" as the frequency approaches zero. The function of X_C versus C also appears as a curve that attains a singularity at $C = 0$; it "blows up negatively" as the capacitance approaches zero. Summarizing:

- If we hold C constant, then X_C varies inversely with the negative of f.
- If we hold f constant, then X_C varies inversely with the negative of C.

Figure 14-4 illustrates these relations as graphs on a rectangular coordinate plane.

Read the Signs and Watch the Mess!

The arithmetic for dealing with capacitive reactance can give you trouble if you're not careful. You have to work with reciprocals, so the numbers can get awkward. Also, you have to watch those

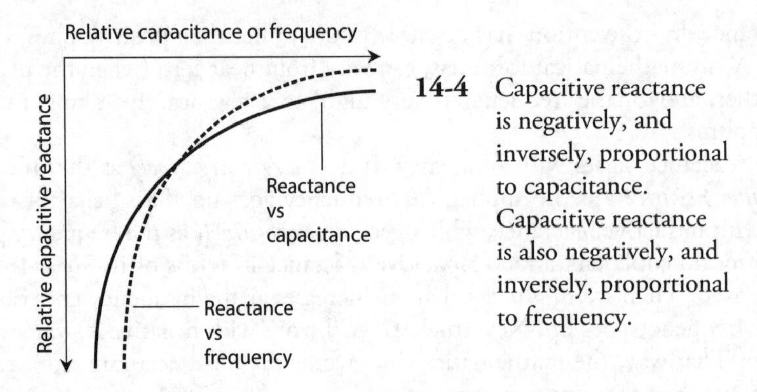

14-4 Capacitive reactance is negatively, and inversely, proportional to capacitance. Capacitive reactance is also negatively, and inversely, proportional to frequency.

negative signs. You can easily forget to include a minus sign when you should (I've done that more than once), and you might insert a negative sign when you shouldn't. The signs are critical when you want to draw graphs describing systems containing reactance. A minus sign tells you that you're working with capacitive reactance rather than inductive reactance.

Problem 14-1

Suppose that a capacitor has a value of 0.00100 μF at a frequency of 1.00 MHz. What's the capacitive reactance?

Solution

We can apply the foregoing formula directly because we know the input data in microfarads (millionths) and in megahertz (millions):

$$X_C = -1/(6.2832 \times 1.00 \times 0.00100) = -1/(0.0062832)$$
$$= -159 \text{ ohms}$$

Problem 14-2

What will happen to the capacitive reactance of the above-described capacitor if the frequency decreases to zero, so that the power source provides DC rather than AC?

Solution

In this case, we'll get an expression with 0 in the denominator if we plug the numbers into the capacitive-reactance formula, yielding a meaningless quantity. We might say, "The reactance of a capacitor at DC equals negative infinity," but a mathematician would wince at a statement like that. We'd do better to say, "We can't define the reactance of a capacitor at DC, but it doesn't matter because reactance applies only to AC circuits."

Problem 14-3

Suppose that a capacitor has a reactance of −100 ohms at a frequency of 10.0 MHz. What's its capacitance?

Solution

In this problem, we must put the numbers in the formula and then use algebra to solve for the unknown *C*. Let's start with the equation

$$-100 = -1/(6.2832 \times 10.0 \times C)$$

Dividing through by -100 and then multiplying through by *C*, we obtain

$$C = 1/(628.32 \times 10.0) = 1/6283.2 = 0.00015915$$

which rounds to $C = 0.000159$ μF. Because we input the frequency value in megahertz, this capacitance comes out in microfarads. We might also say that $C = 159$ pF, remembering that $1 \text{ pF} = 0.000001\ \mu\text{F} = 10^{-6}\ \mu\text{F}$.

The RX_C Quarter-Plane

In a circuit containing resistance along with capacitive reactance, the characteristics work in two dimensions, in a way that "mirrors" the situation with the RX_L quarter-plane. We can place resistance and capacitive-reactance half-lines end-to-end at right angles to construct an *RX_C quarter-plane*, as in shown Fig. 14-5. We plot resistance values horizontally, with increasing values toward the right. We plot capacitive reactance values vertically, with increasingly negative values as we move downward. We can denote complex impedances *Z* containing both resistance and capacitance in the form

$$Z = R + jX_C$$

keeping in mind that values of X_C never go into positive territory.

Some RX_C Examples

If we have a pure resistance, say $R = 3$ ohms, then the complex-number impedance equals $Z = 3 + j0$, which corresponds to the point $(3, j0)$ on the RX_C quarter-plane. If we have a pure capacitive

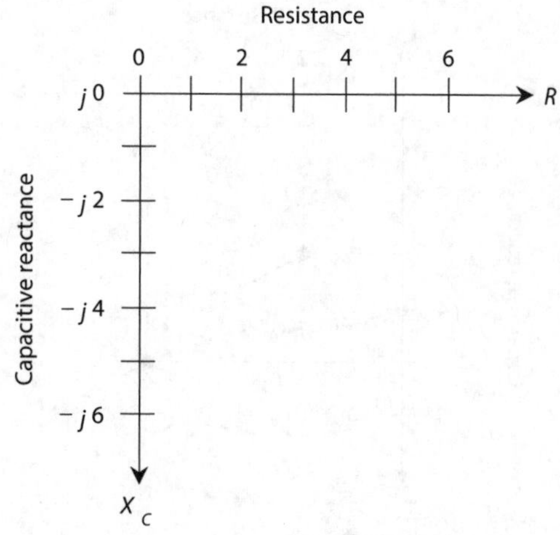

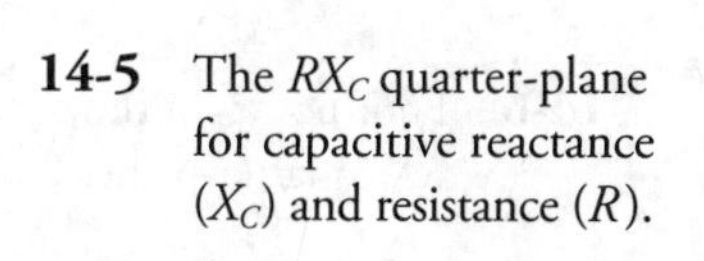
14-5 The RX_C quarter-plane for capacitive reactance (X_C) and resistance (*R*).

reactance, say $X_C = -4$ ohms, then the complex impedance equals $Z = 0 + j(-4)$, which we can write more simply as $Z = 0 - j4$ and plot at the point $(0,-j4)$ on the RX_C quarter-plane. The points representing $Z = 3 + j0$ (which we can also express as $Z = 3 - j0$ because the two values are identical) and $Z = 0 - j4$, along with two others, appear on the RX_C quarter-plane in Fig. 14-6.

Approaching the RX_C Extremes

In practical circuits, all capacitors exhibit some *leakage conductance*. If the frequency goes to zero—the source produces DC—a tiny current will inevitably flow because no real-world dielectric material constitutes a perfect electrical insulator (not even a vacuum). Some capacitors have almost no leakage conductance, but none are completely free of it. Conversely, all electrical conductors have a little capacitive reactance, simply because they occupy physical space. Therefore, we'll never see a mathematically pure conductor of AC, either. The impedances $Z = 3 - j0$ and $Z = 0 - j4$ both represent theoretical idealizations.

How RX_C Points Move

Remember that the values for X_C indicate reactance values, not capacitance values. Reactance varies with the frequency in an RX_C circuit. If we raise or lower the AC frequency that we apply to a particular capacitor, the value of X_C changes. Increasing the AC frequency causes X_C to get *smaller negatively* (closer to zero). Reducing the AC frequency causes X_C to get *larger negatively* (farther from zero, or lower down on the RX_C quarter-plane). If the frequency drops all the way to zero, the capacitive reactance drops off the bottom of the quarter-plane and loses meaning. Then we have two plates or sets of plates holding opposite electrical charges, but no "action" unless or until we discharge the component.

Some RX_C Impedance Vectors

We can represent points in the RX_C quarter-plane as vectors, just as we do in the RX_L quarter-plane. Figure 14-6 shows four different points, each one represented by a certain distance to the right of, and/or below, the origin (corresponding to the complex impedance $0 - j0$). The first number in each value represents the resistance R and the second number represents the capacitive reactance X_C. The

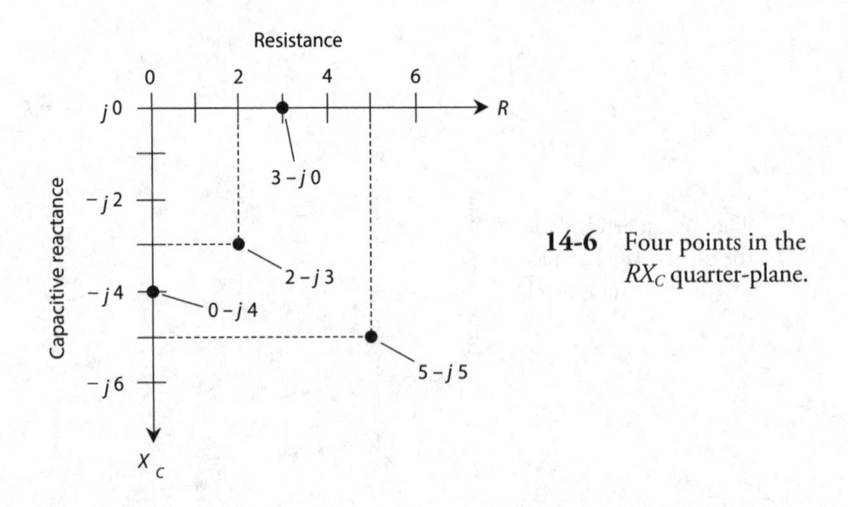

14-6 Four points in the RX_C quarter-plane.

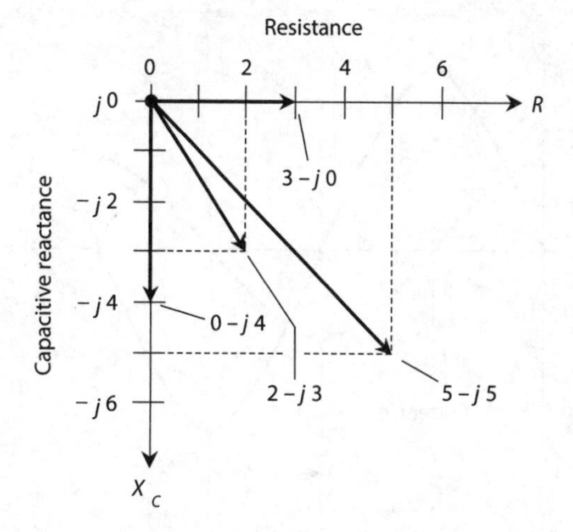

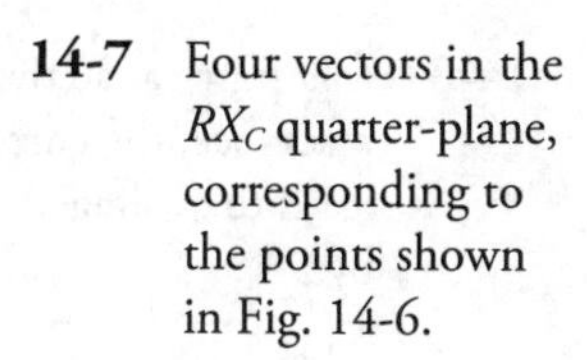
14-7 Four vectors in the RX_C quarter-plane, corresponding to the points shown in Fig. 14-6.

RX_C combination constitutes a two-dimensional quantity. We can depict points, such as those shown in Fig. 14-6, by drawing vectors from the origin out to those points, as shown in Fig. 14-7.

Capacitors and DC Revisited

If the plates of a practical capacitor have large surface areas, are placed close together, and are separated by a good solid dielectric, we will experience a sudden, dramatic bit of "action" when we discharge the component. A massive capacitor can hold enough charge to electrocute an unsuspecting person who comes into contact with its terminals. The well-known scientist and American statesman *Benjamin Franklin* wrote about an experience of this sort with a "home-brewed" capacitor called a *Leyden jar*, which he constructed by placing metal foil sheets inside and outside a glass bottle and then connecting a high-voltage battery to them for a short while. After removing the battery, Franklin came into contact with both foil sheets at the same time and described the consequent shock as a "blow" that knocked him to the floor. Luckily for himself and the world, he survived. If you ever encounter a Leyden jar, treat it with the respect that it deserves, however "innocent" it might look. If you get careless, it can kill you!

Current Leads Voltage

When we drive AC through a capacitor and the instantaneous current starts to increase (in either direction), it takes a fraction of a cycle for the voltage between the plates to follow. Once the current starts decreasing from its maximum peak (in either direction) in the cycle, it again takes a fraction of a cycle for the voltage to follow. The instantaneous voltage can't keep up with the instantaneous current, as it does in a pure resistance. Therefore, in a circuit containing capacitive reactance as well as resistance, the voltage lags the current in phase. A more often-used expression for this phenomenon says that the current *leads* the voltage.

Pure Capacitive Reactance

Suppose that we connect an AC voltage source across a capacitor. Imagine that the frequency is low enough, and/or the capacitance is small enough, so the absolute value of the capacitive reactance X_C greatly exceeds the resistance R (by a factor of millions, say). In this situation, the current leads

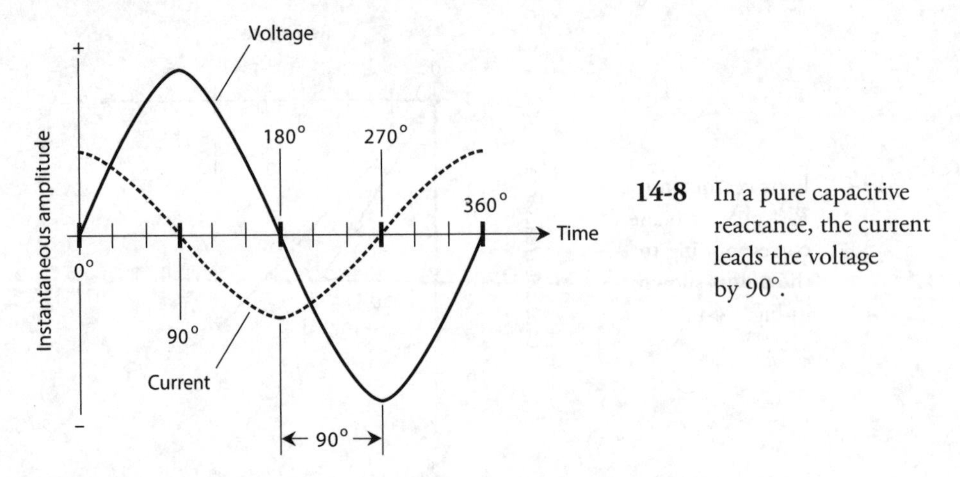

14-8 In a pure capacitive reactance, the current leads the voltage by 90°.

the voltage by just about 90° (Fig. 14-8). We have an essentially pure capacitive reactance, so the vector in the RX_C plane points almost exactly straight down, at an angle of almost exactly −90° with respect to the R axis.

Capacitive Reactance and Resistance

When the resistance in a resistance-capacitance circuit compares favorably with the absolute value of the capacitive reactance, the current leads the voltage by an angle of less than 90° (Fig. 14-9). If R is small compared with the absolute value of X_C, the difference equals almost 90°. As R gets larger, or as the absolute value of X_C becomes smaller, the phase angle decreases. If R becomes much larger than the absolute value of X_C, the phase angle approaches 0°. We call a circuit containing resistance and capacitance an *RC circuit.*

The value of R in an RC circuit might increase relative to the absolute value of X_C because we add resistance deliberately into a circuit. Or, it might happen because the frequency becomes so high that the absolute value of X_C drops to a value comparable to the loss resistance in the circuit conductors. In either case, we can represent the circuit as a resistance R in series with a capacitive reactance X_C (Fig. 14-10).

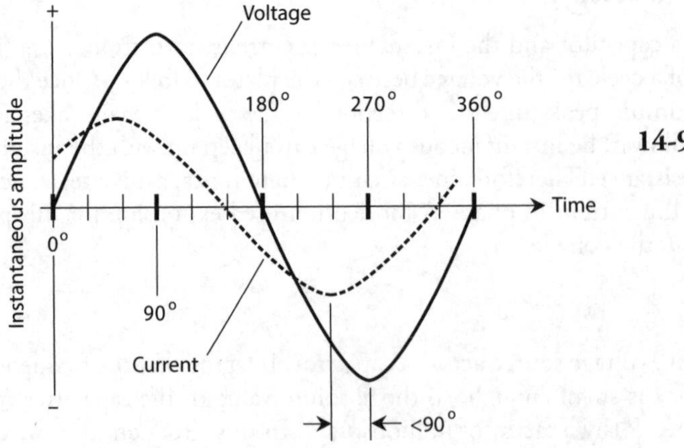

14-9 In a circuit with capacitive reactance and resistance, the current leads the voltage by less than 90°.

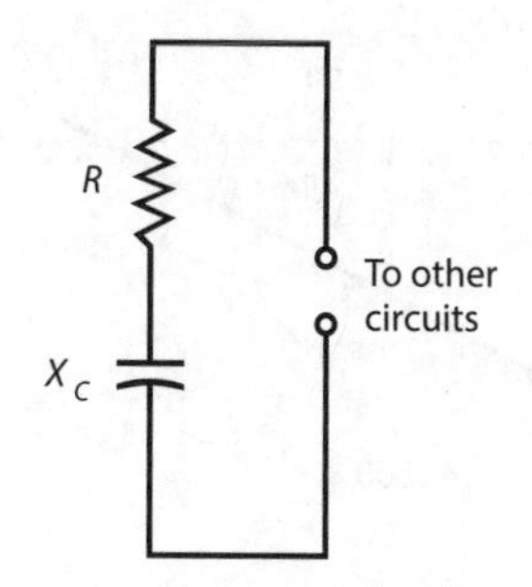

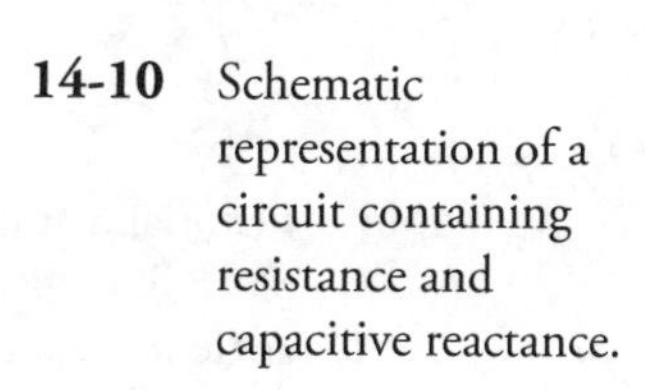
14-10 Schematic representation of a circuit containing resistance and capacitive reactance.

If we know the values of X_C and R, we can find the *angle of lead*, also called the *RC phase angle* (or simply the *phase angle* if we know that we're dealing with resistance and capacitance), by plotting the point for $R - jX_C$ on the RX_C plane, drawing the vector from the origin out to that point, and then measuring the angle of the vector clockwise from the R axis. We can use a protractor to measure this angle, as we did in the previous chapter for RL phase angles, or we can use trigonometry to calculate the angle.

As with RL circuits, we need to know only the ratio of X_C to R to determine the phase angle. For example, if $X_C = -4$ ohms and $R = 7$ ohms, you'll get the same angle as with $X_C = -400$ ohms and $R = 700$ ohms, or with $X_C = -16$ ohms and $R = 28$ ohms. The phase angle is the same whenever the ratio of X_C to R equals $-4{:}7$.

Pure Resistance

As the resistance in an RC circuit grows large compared with the absolute value of the capacitive reactance, the angle of lead grows smaller. The same thing happens if the absolute value of X_C gets small compared with the value of R. When R greatly exceeds the absolute value of X_C (regardless of their actual values), the vector in the RC plane points almost along the R axis. Then the RC phase angle is close to 0°. The voltage comes nearly into phase with the current. The plates of the capacitor do not come anywhere near getting fully charged with each cycle. The capacitor "passes the AC" with very little loss, as if it were shorted out. Nevertheless, it still has an extremely high value of X_C for any AC signals at much lower frequencies that might happen to exist across it at the same time. (Engineers put this property of capacitors to use in electronic circuits when they want to let high-frequency AC signals pass through a particular point while blocking signals at DC and at low AC frequencies.)

How Much Lead?

If you know the ratio of the capacitive reactance to the resistance X_C/R in an RC circuit, you can find the phase angle. Of course, you can find this angle if you know the precise values, too.

Pictorial Method

You can use a protractor and a ruler to find phase angles for RC circuits, just as you did with RL circuits in the previous chapter, as long as the angles aren't too close to 0° or 90°. First, draw a line somewhat longer than 100 mm, going from left to right on the paper. Then, use the protractor to construct a line going somewhat more than 100 mm vertically downward, starting at the left end of the horizontal line. The horizontal line forms the R axis of an RX_C quarter-plane. The line going down constitutes the X_C axis.

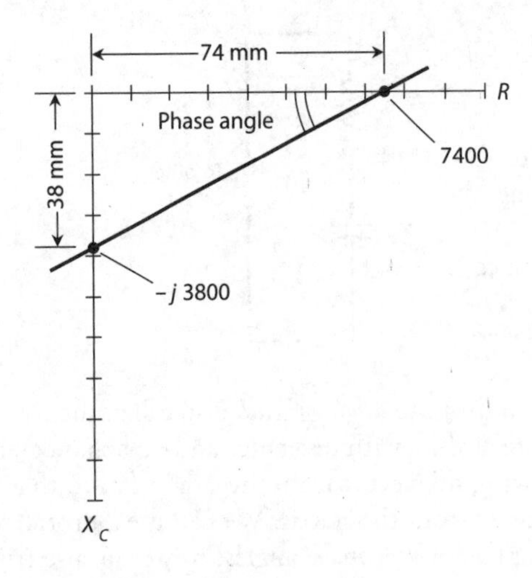

14-11 Pictorial method of finding phase angle in a circuit containing resistance and capacitive reactance.

If you know the actual values of X_C and R, divide or multiply them by a constant, chosen to make both values fall between −100 and 100. For example, if $X_C = -3800$ ohms and $R = 7400$ ohms, divide them both by 100, getting −38 and 74. Plot these points on the lines. The X_C point should lie 38 mm below the intersection point between your two axes. The R point should lie 74 mm to the right of the intersection point. Next, draw a line connecting the two points, as shown in Fig. 14-11. This line will lie at a slant, and will form a triangle along with the two axes. Therefore, you'll get a right triangle, with the right angle at the origin of the quarter-plane. Using a protractor, measure the angle between the slanted line and the R axis. Extend the lines, if necessary, to get a good reading on the protractor. This angle will fall somewhere between 0° and 90°. Multiply this reading by −1 to get the RC phase angle. That is, if the protractor shows 27°, the RC phase angle equals −27°.

You can draw the actual vector by constructing a rectangle using the origin and your two points, making new perpendicular lines to complete the figure. The diagonal of this rectangle represents the vector, which runs out from the origin (Fig. 14-12). The angle between the R axis and

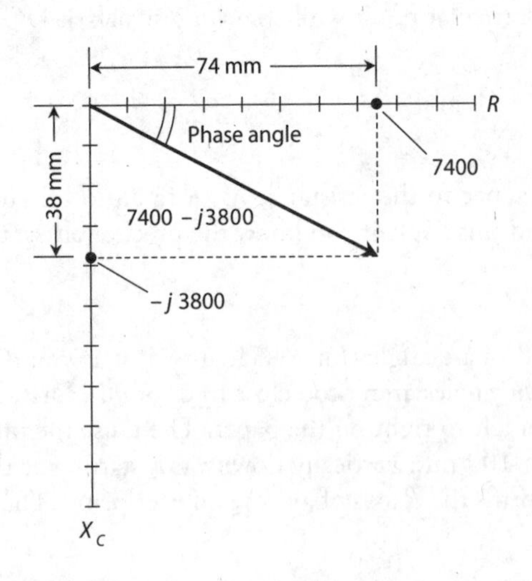

14-12 Another pictorial method of finding phase angle in a circuit containing resistance and capacitive reactance. This method shows the impedance vector.

this vector, multiplied by -1, gives you the phase angle. It has the same measure as the angle of the slanted line that you constructed in the process portrayed in Fig. 14-11.

Trigonometric Method

Using trigonometry, you can determine the *RC* phase angle more precisely than the pictorial method allows. Given the values of X_C and R, the *RC* phase angle equals the Arctangent of their ratio. We symbolize the phase angle in *RC* circuits by writing the lowercase Greek letter ϕ, just as we do in *RL* circuits. The formula is

$$\phi = \text{Arctan}\ (X_C / R)$$

When doing problems of this kind, remember to use the *capacitive reactance* values for X_C, and not the capacitance values. Also, use the actual value for X_C (a negative number) and not the absolute value of X_C (a positive number). If you know the capacitance but not the reactance, you must use the formula for X_C in terms of capacitance and frequency, and then calculate the phase angle. You should get angles that come out smaller than 0° but larger than $-90°$.

Don't Get Confused about the Angle!

By convention, phase angles in *RC* circuits always range from 0° down to $-90°$. This contrasts *RC* phase angles to *RL* phase angles, which always range from 0° up to 90°. You can avoid confusion about phase angles by remembering a simple rule: The phase angle always has the same sign (positive or negative) as the reactance.

Problem 14-4

Suppose that the capacitive reactance in an *RC* circuit equals –3800 ohms and the resistance equals 7400 ohms. What's the phase angle?

Solution

You can determine the ratio of the capacitive reactance to the resistance, getting

$$X_C / R = -3800/7400$$

The calculator display should show you something like –0.513513513. Find the Arctangent of this number, getting a phase angle of –27.18111109° on the calculator display. Round this result off to –27.18°.

Problem 14-5

Suppose that we operate an *RC* circuit at a frequency of 3.50 MHz. It has a resistance of 130 ohms and a capacitance of 150 pF. What's the phase angle to the nearest degree?

Solution

First, find the capacitive reactance for a capacitor of 150 pF at 3.50 MHz. Convert the capacitance to microfarads, getting $C = 0.000150$ μF. Remember that *micro*farads go with *mega*hertz. You'll get

$$X_C = -1/(6.2832 \times 3.50 \times 0.000150) = -1/0.00329868$$
$$= -303 \text{ ohms}$$

Now you can find the ratio

$$X_C/R = -303/130 = -2.33$$

The phase angle equals the Arctangent of –2.33, which works out to be – 67° to the nearest degree.

Problem 14-6

What's the phase angle in the above-described circuit if you increase the frequency to 8.10 MHz?

Solution

You need to find the new value for X_C because it will change as a result of the frequency change. Calculating, you get

$$X_C = -1/(6.2832 \times 8.10 \times 0.000150) = -1/0.007634$$
$$= -131 \text{ ohms}$$

The ratio X_C/R in this case equals –131/130, or –1.008. The phase angle equals the Arctangent of –1.008, which rounds off to –45°.

Quiz

Refer to the text in this chapter if necessary. A good score is at least 18 correct. Answers are in the back of the book.

1. In a circuit containing pure capacitive reactance and no resistance, the phase angle is always
   (a) +45°.
   (b) 0°.
   (c) −45°.
   (d) −90°.

2. In a circuit in which the resistance and the capacitive reactance are equal and opposite (the resistance positive, the reactance negative), the phase angle is always
   (a) +45°.
   (b) 0°.
   (c) −45°.
   (d) −90°.

3. In a circuit containing pure resistance and no reactance, the phase angle is always
   (a) +45°.
   (b) 0°.
   (c) −45°.
   (d) −90°.

4. A capacitor has a value of $C = 200$ pF. We apply a signal at $f = 4.00$ MHz. What's X_C?
   (a) −498 ohms
   (b) −995 ohms

(c) -199 ohms

(d) -3.98 k

5. In an *RC* circuit containing a finite nonzero resistance, as the ratio X_C/R approaches zero (from the negative side), the phase angle approaches

(a) $-90°$.

(b) $-45°$.

(c) $0°$.

(d) negative infinity.

6. A capacitor has a value of 0.0330 μF and a reactance of -123 ohms at a certain frequency. What frequency?

(a) 39.2 kHz

(b) 19.6 kHz

(c) 78.4 kHz

(d) We need more information to calculate it.

7. What happens to the value of a capacitor (in microfarads) as we decrease the spacing between the plates without changing anything else?

(a) It does not change.

(b) It increases.

(c) It decreases.

(d) We need more information to say.

8. What's the reactance of a 470-pF capacitor at 12.5 MHz?

(a) -2.71 k

(b) -271 ohms

(c) -27.1 ohms

(d) -2.71 ohms

9. What happens to the reactance of the capacitor described in Question 8 if we reduce the frequency by a factor of 10?

(a) It becomes 100 times what it was (negatively).

(b) It becomes 10 times what it was (negatively).

(c) It becomes 1/10 of what it was (negatively).

(d) It becomes 1/100 of what it was (negatively).

10. A capacitor has a reactance of -100 ohms at 200 kHz. What's its capacitance?

(a) 7.96 nF

(b) 79.6 nF

(c) 796 nF

(d) 7.96 μF

11. A series *RC* circuit comprises a capacitor whose reactance is −75 ohms at the frequency of operation, connected to a 50-ohm resistor. What's the phase angle?
    - (a) −34°
    - (b) −56°
    - (c) −85°
    - (d) −90°

12. A series *RC* circuit comprises a capacitor whose reactance is −50 ohms at the frequency of operation, connected to a 75-ohm resistor. What's the phase angle?
    - (a) −34°
    - (b) −56°
    - (c) −85°
    - (d) −90°

13. A series *RC* circuit comprises a capacitance of 0.01 μF along with a 4.7-ohm resistor. What's the phase angle for a signal with a constant frequency?
    - (a) −60°
    - (b) −45°
    - (c) −30°
    - (d) We need more information to answer this question.

14. What will happen to the phase angle in the circuit of Question 13 (whether or not we know its actual value) if we short out the resistor but leave the capacitor alone?
    - (a) It will become −90°.
    - (b) It will become −45°.
    - (c) It will become 0°.
    - (d) Nothing.

15. What will happen to the phase angle in the circuit of Question 14 (not 13!) if we short out the resistor and double the capacitance?
    - (a) It will become −60°.
    - (b) It will become −45°.
    - (c) It will become −30°.
    - (d) Nothing.

16. What will happen to the phase angle in the circuit of Question 15 (not 13 or 14!) if we triple the frequency while leaving all other factors constant?
    - (a) We need more information to answer this question.
    - (b) It will increase negatively (get closer to −90°).
    - (c) It will decrease negatively (get closer to 0°).
    - (d) Nothing.

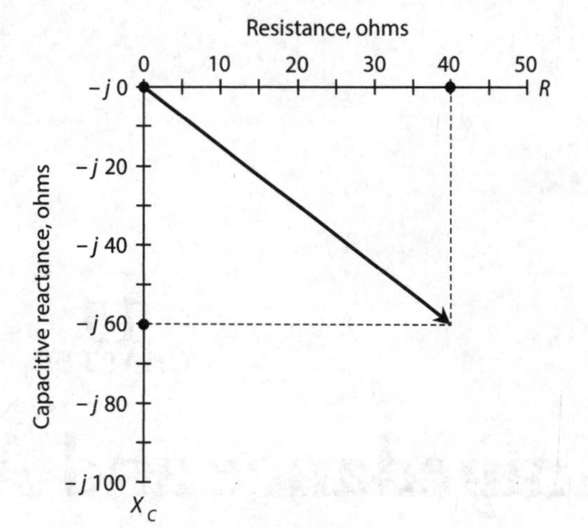

14-13 Illustration for Quiz Questions 19 and 20.

17. A 470-pF capacitor has a reactance of −800 ohms at a certain frequency. What frequency?
    (a) 423 kHz
    (b) 846 kHz
    (c) 212 kHz
    (d) We need more information to answer this question.

18. In the situation of Question 17, what happens to X_C if we cut the frequency in half?
    (a) It increases negatively by a factor of the square root of 2.
    (b) It increases negatively by a factor of 2.
    (c) It increases negatively by a factor of 4.
    (d) Nothing.

19. In the scenario portrayed by Fig. 14-13, the X_C/R ratio is roughly
    (a) −0.66.
    (b) −0.75.
    (c) −1.5.
    (d) −3.0.

20. In Fig. 14-13, the R and X_C scale divisions differ in size. We can nevertheless calculate the phase angle as roughly
    (a) −19°.
    (b) −56°.
    (c) −37°.
    (d) −33°.

15
CHAPTER

Impedance and Admittance

IN THIS CHAPTER, WE'LL DEVELOP A "RIGOROUS" WORKING MATHEMATICAL MODEL FOR COMPLEX impedance. We'll also learn about *admittance*, which quantifies how well AC circuits allow (or admit) the flow of current, rather than restraining (or impeding) it.

Imaginary Numbers Revisited

As we learned in Chap. 13, the engineering symbol j represents the unit imaginary number, technically defined as the positive square root of -1. Let's review this concept, because some people find it difficult to believe that negative numbers can have square roots. When we multiply j by itself, we get -1.

The term "imaginary" comes from the notion that j is somehow "less real" than the so-called *real numbers*. That's not true! All numbers are "unreal" in the sense that they're all abstract, however we classify them. You know that, if you've ever taken a course in number theory.

Actually, j isn't the "only" square root of -1. A negative square root of -1 also exists; it equals $-j$. When we multiply either j or $-j$ by itself, we get -1. (Pure mathematicians often denote these same numbers as i or $-i$.) The *set of imaginary numbers* comprises all possible real-number multiples of j. Examples include:

- $j \times 4$, which we write as $j4$
- $j \times 35.79$, which we write as $j35.79$
- $j \times (-25.76)$, which we write as $-j25.76$
- $j \times (-25{,}000)$, which we write as $-j25{,}000$

We can multiply j by any real number and portray it as a point on a line. If we do that for all the real numbers, we get an *imaginary-number line* (Fig. 15-1). We orient the imaginary number line vertically, at a right angle to the horizontal real number line, when we want to graphically render real and imaginary numbers at the same time. In electronics, real numbers represent resistances. Imaginary numbers represent reactances.

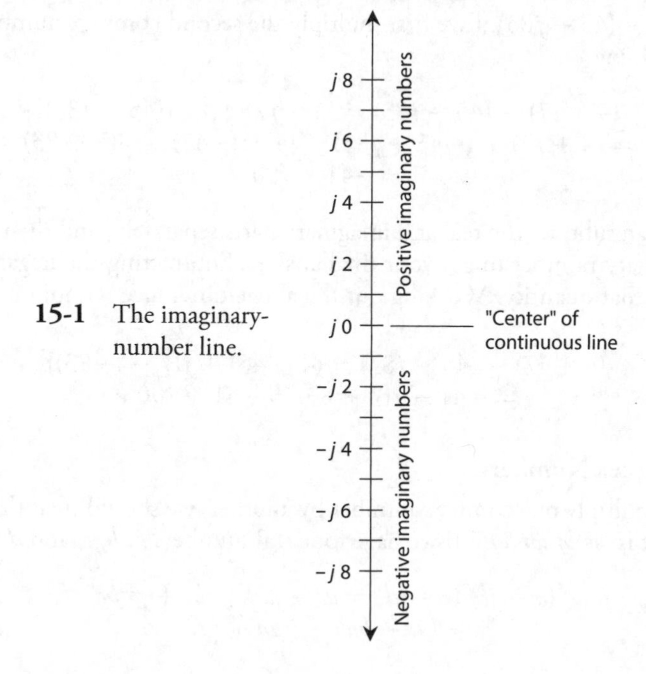

15-1 The imaginary-number line.

Complex Numbers Revisited (in Detail)

When we add a real number and an imaginary number, we get a *complex number*. In this context, the term "complex" does not mean "complicated"; a better word might be "composite." Examples include:

- The sum of 4 and $j5$, which equals $4 + j5$
- The sum of 8 and $-j7$, which equals $8 - j7$
- The sum of -7 and $j13$, which equals $-7 + j13$
- The sum of -6 and $-j87$, which equals $-6 - j87$

To completely portray the set of complex numbers in graphical form, we need a two-dimensional *coordinate plane.*

Adding and Subtracting Complex Numbers

When we want to add one complex number to another, we add the real parts and the complex parts separately and then sum up the total. For example, the sum of $4 + j7$ and $45 - j83$ works out as

$$(4 + 45) + j(7 - 83) = 49 + j(-76) = 49 - j76$$

Subtracting complex numbers involves a little trickery because we can easily confuse our signs. We can avoid this confusion by converting the difference to a sum. For example, we can find the difference $(4 + j7) - (45 - j83)$ if we first multiply the second complex number by -1 and then add the result, obtaining

$$\begin{aligned}(4 + j7) - (45 - j83) &= (4 + j7) + [-1(45 - j83)]\\ &= (4 + j7) + (-45 + j83) = [4 + (-45)] + j(7 + 83)\\ &= -41 + j90\end{aligned}$$

Alternatively, you can subtract the real and imaginary parts separately, and then combine the result back into an imaginary number to get your final answer. Subtracting the negative of a quantity is the same as adding that quantity. Working out the above difference without converting to a sum, you'll get

$$\begin{aligned}(4 + j7) - (45 - j83) &= (4 - 45) + j[7 - (-83)]\\ &= -41 + j(7 + 83) = -41 + j90\end{aligned}$$

Multiplying Complex Numbers

When we want to multiply one complex number by another, we should treat them both as sums of number pairs—that is, as *binomials*. If we have four real numbers a, b, c, and d, then

$$\begin{aligned}(a + jb)(c + jd) &= ac + jad + jbc + j^2 bd\\ &= (ac - bd) + j(ad + bc)\end{aligned}$$

The Complex Number Plane

We can construct a complete *complex-number plane* by taking the real and imaginary number lines and placing them together, at right angles, so that they intersect at the zero points, 0 and $j0$. Figure 15-2

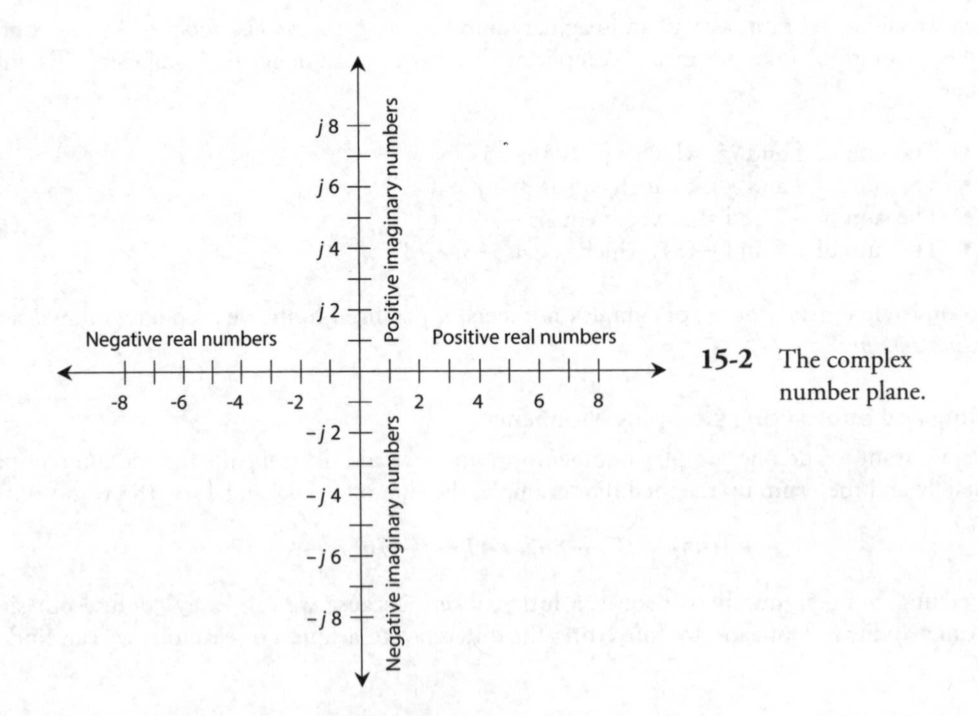

15-2 The complex number plane.

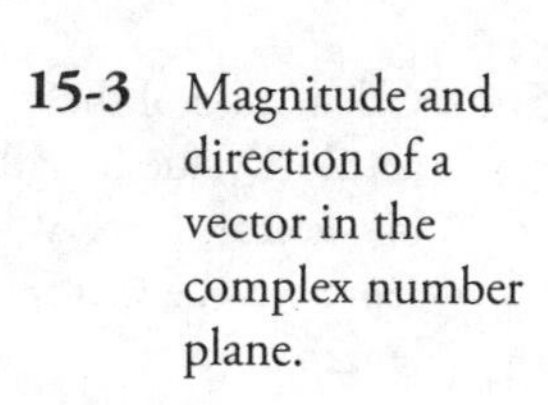

15-3 Magnitude and direction of a vector in the complex number plane.

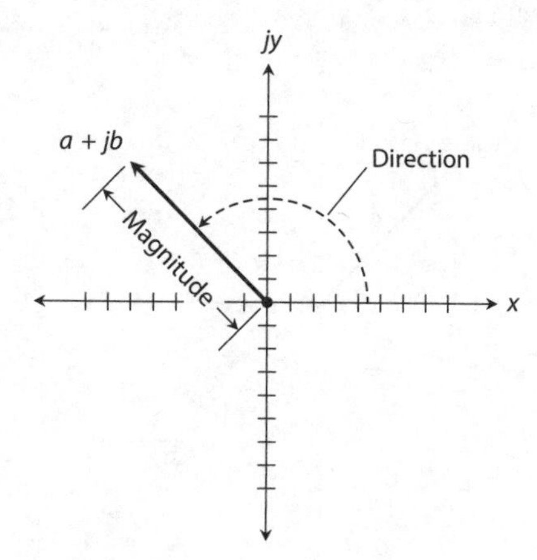

illustrates the arrangement, which gives us a *rectangular coordinate plane,* just like the ones that people use to graph everyday relations, such as temperature versus time.

Complex Number Vectors

Engineers sometimes represent complex numbers as vectors in the coordinate plane. This gives each complex number a unique *magnitude* and a unique *direction.* The magnitude of the vector for a complex number $a + jb$ equals the distance of the point (a, jb) from the origin $(0, j0)$. We represent the vector direction as the angle, expressed going counterclockwise from the positive real-number axis. Figure 15-3 illustrates how this scheme works.

Absolute Value

The *absolute value* of a complex number $a + jb$ equals the length, or magnitude, of its vector in the complex plane, measured from the origin $(0, j0)$ to the point (a, jb). Let's break this scenario down into three cases.

1. For a *pure real number* $a + j0$, the absolute value equals a, if a is positive. If a is negative, the absolute value of $a + j0$ equals $-a$.
2. For a *pure imaginary number* $0 + jb$, the absolute value equals b, if b (a real number) is positive. If b is negative, the absolute value of $0 + jb$ equals $-b$.
3. If the number $a + jb$ is neither a pure real nor a pure imaginary number, we must use a formula to find its absolute value. First, we square both a and b. Then we add those two squares. Finally, we take the square root of the sum of the squares to get the length c of the vector representing $a + jb$. Figure 15-4 shows the geometry of this method.

Problem 15-1

Find the absolute value of the complex number $-22 - j0$.

Solution

We have a pure real number in this case. Actually, $-22 - j0$ is the very same complex number as $-22 + j0$, because $-j0 = j0$. The absolute value equals $-(-22)$, or 22.

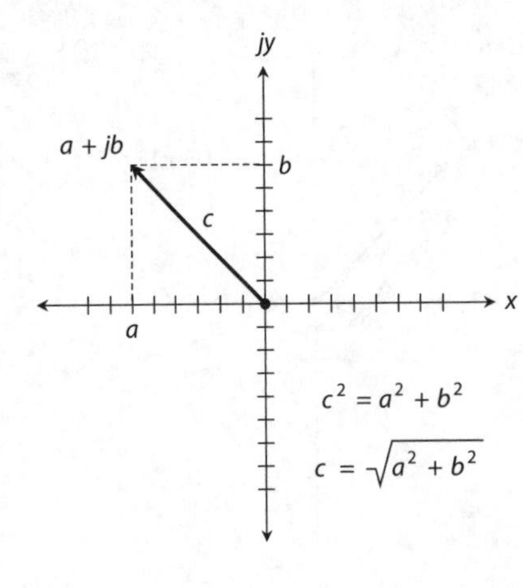

15-4 Calculation of the absolute value (length) of a vector. Here, we represent the vector length as *c*.

Problem 15-2

Find the absolute value of $0 - j34$.

Solution

This quantity is a pure imaginary number where $b = -34$ because $0 - j34 = 0 + j(-34)$. The absolute value equals $-(-34)$, or 34.

Problem 15-3

Find the absolute value of $3 - j4$.

Solution

In this case, we have $a = 3$ and $b = -4$. Using the formula described above and shown in Fig. 15-4, we get

$$[3^2 + (-4)^2]^{1/2} = (9 + 16)^{1/2} = 25^{1/2} = 5$$

The *RX* Half-Plane

Recall the quarter-plane for resistance R and inductive reactance X_L from Chap. 13. This region corresponds to the upper-right quadrant of the complex number plane shown in Fig. 15-2. Similarly, the quarter-plane for resistance R and capacitive reactance X_C corresponds to the lower-right quadrant of the complex number plane of Fig. 15-2. We represent resistances as nonnegative real numbers. We represent reactances as imaginary numbers.

No Negative Resistance

Strictly speaking, negative resistance can't exist because we can't have anything better than a perfect conductor. In some cases, we might treat a source of DC, such as a battery, as if it constitutes a "negative resistance." Once in a while, we might encounter a device in which the current drops as the applied voltage increases, producing a "reversed" resistance-behavior phenomenon that some engineers call "negative resistance." But for most practical applications, the resistance can never go

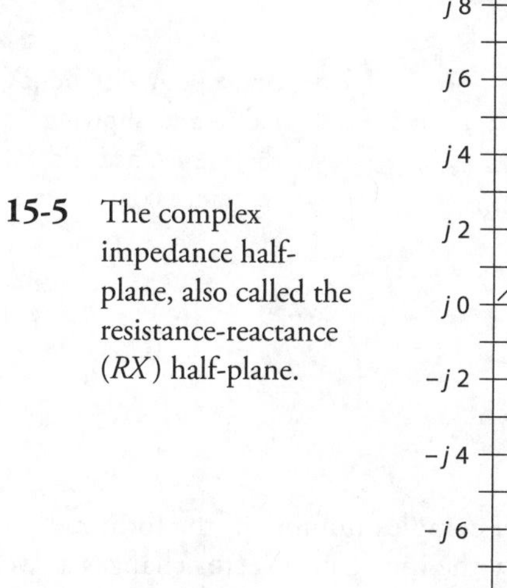

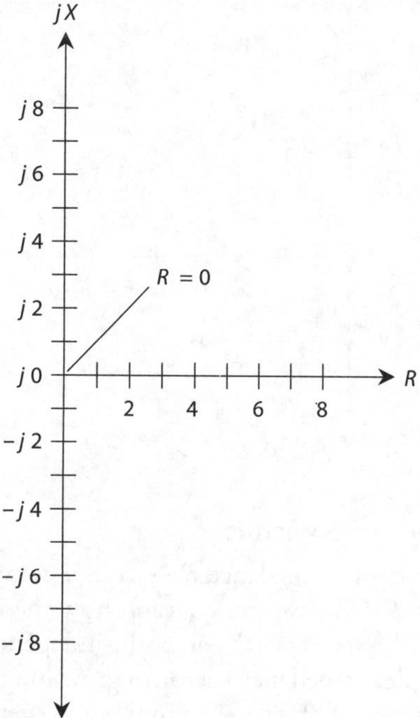

15-5 The complex impedance half-plane, also called the resistance-reactance (*RX*) half-plane.

"below zero." We can, therefore, remove the negative axis, along with the upper-left and lower-left quadrants, of the complex-number plane, obtaining an *RX half-plane,* as shown in Fig. 15-5. This system provides a complete set of coordinates for depicting complex impedance.

"Negative Inductors" and "Negative Capacitors"

Capacitive reactance X_C is effectively an extension of inductive reactance X_L into the realm of negatives. Capacitors act like "negative inductors." We can also say that inductors act like "negative capacitors" because the negative of a negative number equals a positive number. Reactance, in general, can vary from extremely large negative values, through zero, to extremely large positive values.

Complex Impedance Points

Imagine the point representing $R + jX$ moving around in the *RX* half-plane, and imagine where the corresponding points on the axes lie. We can locate these points by drawing dashed lines from the point $R + jX$ to the R and X axes, so that the dashed lines intersect the axes at right angles. Figure 15-6 shows several examples.

Now think of the points for R and X moving toward the right and left, or up and down, on their axes. Imagine what happens to the point representing $R + jX$ in various scenarios. This exercise will give you an idea of how impedance changes as the resistance and reactance vary in an AC electrical circuit.

Resistance constitutes a one-dimensional phenomenon. Reactance also manifests itself as one-dimensional. To completely define complex impedance, however, we must use two dimensions. The *RX* half-plane meets this requirement. Remember that the resistance and the reactance can vary independently of one another!

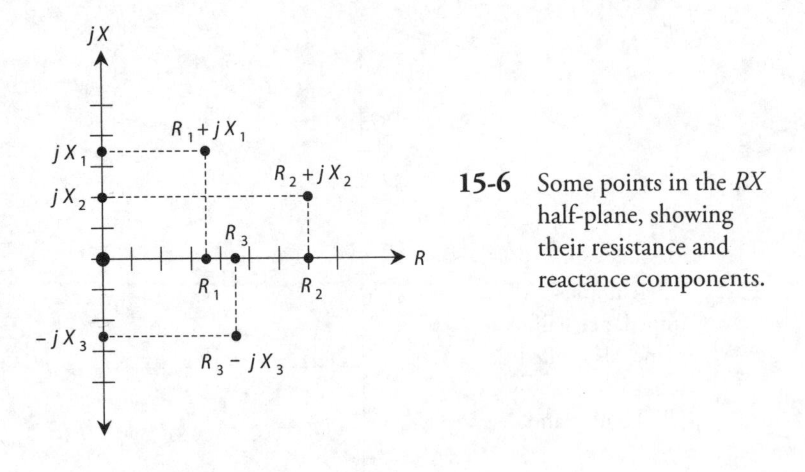

15-6 Some points in the *RX* half-plane, showing their resistance and reactance components.

Complex Impedance Vectors

We can represent any impedance $R + jX$ as a complex number of the form $a + jb$. We simply let $R = a$ and $X = b$. Now we can envision how the impedance vector changes as we vary either the resistance R or the reactance X, or both, independently. If X remains constant, an increase in R causes the complex impedance vector to grow longer. If R remains constant and X_L gets larger, the vector grows longer. If R stays the same but X_C gets larger negatively, the vector once again grows longer. Figure 15-7 illustrates the vectors corresponding to the points from Fig. 15-6.

Absolute-Value Impedance

You'll occasionally read or hear that the "impedance" of some device or component equals a certain number of "ohms." For example, in audio electronics, you'll encounter things like "8-ohm" speakers and "600-ohm" amplifier inputs. How, you ask, can manufacturers quote a single number for a quantity that needs two dimensions for its complete expression? That's a good question. Two answers exist.

First, specifications, such as "8 ohms" for a speaker or "600 ohms" for an amplifier input, refer to *purely resistive impedances,* also known as *non-reactive impedances.* Thus, the "8-ohm" speaker

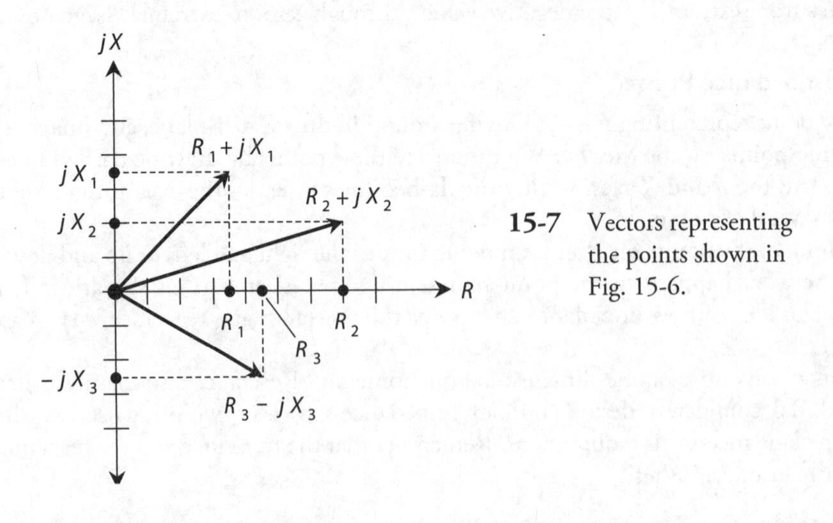

15-7 Vectors representing the points shown in Fig. 15-6.

really has a complex impedance of $8 + j0$, and the "600-ohm" input circuit is designed to operate with a complex impedance at, or near, $600 + j0$.

Second, you can talk about the length of the impedance vector (the absolute value of the complex impedance), calling this length a certain number of ohms. If you try to define impedance this way, however, you risk ambiguity and confusion because you can find infinitely many different vectors of a given length in the *RX* half-plane.

Sometimes, engineers and technicians write the uppercase, italic letter Z in place of the word "impedance," so you'll read expressions such as "$Z = 50$ ohms" or "$Z = 300$ ohms nonreactive." In this context, if no specific impedance is given, "$Z = 8$ ohms" can theoretically refer to $8 + j0$, $0 + j8$, $0 - j8$, or any other complex impedance whose point lies on a half-circle centered at the coordinate origin and having a radius of 8 units, as shown in Fig. 15-8.

Problem 15-4

Name seven different complex impedances that the expression "$Z = 10$ ohms" might mean.

Solution

We can easily name three such impedance values, each consisting of a pure reactance or a pure resistance, as follows:

$$Z_1 = 0 + j10$$
$$Z_2 = 10 + j0$$
$$Z_3 = 0 - j10$$

These impedances represent pure inductance, pure resistance, and pure capacitance, respectively.

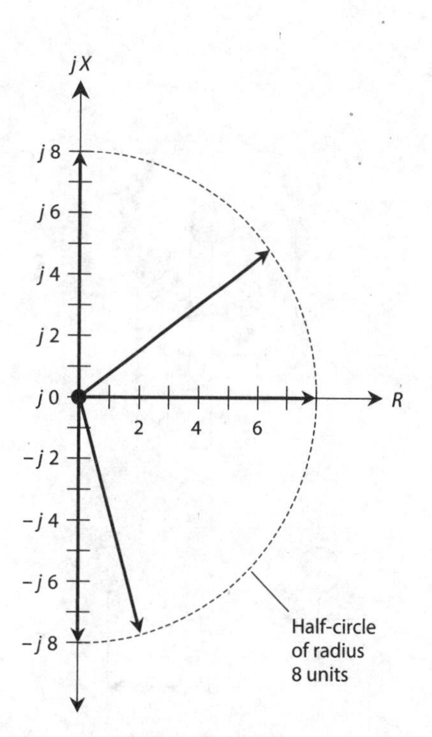

15-8 Some vectors representing an absolute-value impedance of 8 ohms. Infinitely many such vectors exist in theory, all of which terminate on the dashed circle.

A right triangle can exist having sides in a ratio of 6:8:10 units. We know this fact from basic coordinate geometry, because $6^2 + 8^2 = 10^2$. Therefore, we can have the following impedances, all of which have an absolute value of "10 ohms":

$$Z_4 = 6 + j8$$
$$Z_5 = 6 - j8$$
$$Z_6 = 8 + j6$$
$$Z_7 = 8 - j6$$

Characteristic Impedance

You'll sometimes encounter a property of certain electrical components known as *characteristic impedance* or *surge impedance*, symbolized Z_o. Characteristic impedance constitutes a fundamental property of *transmission lines.*

Transmission-Line Types

Engineers and technicians use transmission lines to get energy or signals from one place to another. We can always express transmission-line Z_o values in ohms as positive real numbers. We don't need complex numbers to define it.

Transmission lines usually take either of two forms, *coaxial* or *two-wire* (also called *parallel-wire*). Figure 15-9 illustrates cross-sectional renditions of both types. Examples of transmission lines include the "ribbon" that goes from an old-fashioned television (TV) antenna to the receiver set, the cable running from a hi-fi amplifier to the speakers, and the set of wires that carries electricity across the countryside.

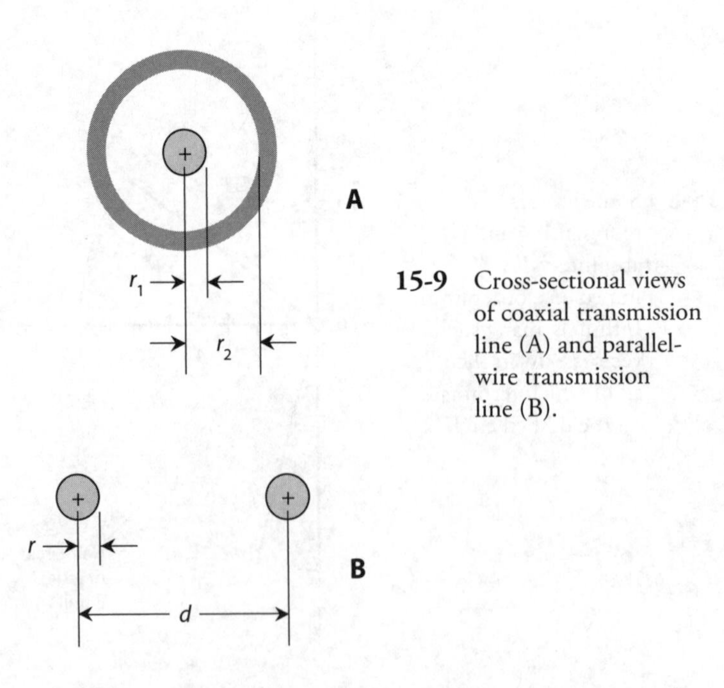

15-9 Cross-sectional views of coaxial transmission line (A) and parallel-wire transmission line (B).

Figure 15-9A shows a cross-section of a coaxial transmission line. The line has a wire *center conductor* that runs along a defined *axis*, and an *outer conductor* or *shield* in the shape of a conducting cylinder concentric with that same axis. The value of Z_o depends on the radius r_1 of the center conductor, on the inside radius r_2 of the shield, and on the type of *dielectric material* separating the center conductor from the shield. If we make the center conductor wire thicker (we increase r_1) while leaving all other factors unchanged, then Z_o decreases. If we enlarge the cylindrical shield (we increase r_2) while leaving all other factors unchanged, then Z_o increases.

Figure 15-9B shows a cross-sectional view of a parallel-wire transmission line. The value of Z_o depends on the radii r of the wires, on the spacing d between the centers of the wires, and on the nature of the dielectric material separating the wires. In general, the value of Z_o increases as the wire radii r get smaller, and decreases as the wire radii get larger, if we hold all other factors constant. (We assume that both wires have the same radius r.) If we leave the wire radii r constant but increase the spacing d between them, then Z_o increases. If we bring the wires closer together (decrease d) while leaving their radii r constant, then Z_o decreases.

Solid dielectric materials, such as polyethylene, when placed between the conductors, reduce the characteristic impedance of a transmission line, as compared with air or a vacuum. The extent of the reduction depends on the dielectric constant of the material. As we make the dielectric constant larger, while all other parameters remain constant in a transmission line, the extent of the reduction in Z_o (compared with air or a vacuum between the conductors) becomes greater.

Z_o in Practice

The ideal value of Z_o for a transmission line depends on the nature of the *load* into which the line delivers energy. If the load has a purely resistive impedance of R ohms, the value of the best line Z_o equals R ohms. If the load impedance is not a pure resistance, or if it's a pure resistance that differs considerably from the characteristic impedance of the transmission line, some energy goes to waste in heating up the transmission line. As the so-called *impedance mismatch* grows worse, the proportion of the energy wasted as heat goes up, and the *transmission-line efficiency* suffers.

Imagine a so-called "300-ohm" frequency-modulation (FM) receiving antenna, such as the folded-dipole type that you can install indoors. Suppose that you want to obtain the best possible reception. Of course, you should choose a good location for the antenna. You should make sure that the transmission line between your radio and the antenna remains as short as possible. But you should also ensure that you purchase "300-ohm" TV ribbon. The manufacturer of that ribbon has optimized its Z_o value for use with antennas whose impedances are close to $300 + j0$, representing a pure resistance of 300 ohms without any reactance.

Conductance

In an AC circuit, electrical *conductance* behaves exactly as it does in a DC circuit. We symbolize conductance (as a variable in equations) by writing an uppercase italic letter G. We express the relationship between conductance and resistance as the two formulas

$$G = 1/R$$

and

$$R = 1/G$$

The standard unit of conductance is the *siemens*, abbreviated as the uppercase non-italic letter S. In the above formulas, we get G in siemens if we input R in ohms, and vice-versa. As the conductance increases, the resistance decreases, and more current flows for a fixed applied voltage. Conversely, as G decreases, R goes up, and less current flows when we apply a fixed voltage.

Susceptance

Sometimes we'll come across the term *susceptance* in reference to AC circuits. We symbolize this quantity (as a variable in equations) by writing an uppercase italic letter B. Susceptance is the reciprocal of reactance, and it can occur in either the capacitive form or the inductive form. If we symbolize *capacitive susceptance* as B_C and *inductive susceptance* as B_L, then

$$B_C = 1/X_C$$

and

$$B_L = 1/X_L$$

The Reciprocal of *j*

All values of B theoretically contain the j operator, just as do all values of X. But when it comes to finding reciprocals of quantities containing j, things get tricky. The reciprocal of j actually equals its negative! That is,

$$1/j = -j$$

and

$$1/(-j) = j$$

As a result of these properties of j, the sign reverses whenever you find a susceptance value in terms of a reactance value. When expressed in terms of j, inductive susceptance is negative imaginary, and capacitive susceptance is positive imaginary—just the opposite situation from inductive reactance and capacitive reactance.

Imagine an inductive reactance of 2 ohms. We express this in imaginary terms as $j2$. To find the inductive susceptance, we must find $1/(j2)$. Mathematically, we convert this expression to a real-number multiple of j by breaking it down in steps as follows:

$$1/(j2) = (1/j)(1/2) = (1/j)0.5 = -j0.5$$

Now imagine a capacitive reactance of 10 ohms. We express this quantity in imaginary terms as $-j10$. To find the capacitive susceptance, we must find $1/(-j10)$. Here's how we can convert it to the product of j and a real number:

$$1/(-j10) = (1/-j)(1/10) = (1/-j)0.1 = j0.1$$

To find an imaginary value of susceptance in terms of an imaginary value of reactance, we must first take the reciprocal of the real-number part of the expression, and then multiply the result by −1.

Problem 15-5

Suppose that we have a capacitor of 100 pF at a frequency of 3.10 MHz. What's the capacitive susceptance B_C?

Solution

First, let's find X_C by using the formula for capacitive reactance. We have

$$X_C = -1/(6.2832\,fC)$$

Note that 100 pF = 0. 000100 μF. Therefore

$$X_C = -1/(6.2832 \times 3.10 \times 0.000100) = -1/0.00195$$
$$= -513 \text{ ohms}$$

The imaginary value of X_C equals $-j513$. The susceptance B_C equals $1/X_C$, so we have

$$B_C = 1/(-j513) = j0.00195 \text{ S}$$

The siemens quantifies susceptance, just as it defines conductance. Therefore, we can state the foregoing result as 0.00195 S of capacitive susceptance.

General Formula for B_C

We can now see that the general formula for capacitive susceptance in siemens, in terms of frequency in hertz and capacitance in farads, is

$$B_C = 6.2832\,fC$$

This formula also works for frequencies in megahertz and capacitance values in microfarads.

Problem 15-6

Suppose an inductor has $L = 163$ μH at a frequency of 887 kHz. What's the inductive susceptance B_L?

Solution

Note that 887 kHz = 0.887 MHz. We can calculate X_L from the formula for inductive reactance as follows:

$$X_L = 6.2832\,fL = 6.2832 \times 0.887 \times 163 = 908 \text{ ohms}$$

The imaginary value of X_L equals $j908$. The susceptance B_L equals $1/X_L$. It follows that

$$B_L = -1/(j908) = -j0.00110 \text{ S}$$

We can state this result as −0.00110 S of inductive susceptance.

General Formula for B_L

The general formula for inductive susceptance in siemens, in terms of frequency in hertz and inductance in henrys, is

$$B_L = -1/(6.2832\,fL)$$

This formula also works for frequencies in kilohertz and inductance values in millihenrys, and for frequencies in megahertz and inductance values in microhenrys.

Admittance

Real-number conductance and imaginary-number susceptance combine to form *complex admittance*, symbolized (as a variable in equations) as the uppercase italic letter Y. Admittance provides a complete expression of the extent to which a circuit allows AC to flow. As the absolute value of complex impedance gets larger, the absolute value of complex admittance becomes smaller, in general. Huge impedances correspond to tiny admittances, and vice-versa.

Complex Admittance

We can express admittance in complex form, just as we can do with impedance. However, we'd better keep careful track of which quantity we're talking about! We can avoid confusion if we take care to employ the correct symbol. We can get a complete expression of admittance by taking the complex composite of conductance and susceptance. Engineers usually write complex admittance in the form

$$Y = G + jB$$

when the susceptance is positive (capacitive), and in the form

$$Y = G - jB$$

when the susceptance is negative (inductive).

The "Parallel Advantage"

In Chaps. 13 and 14, we worked with series RL and RC circuits. Did you wonder, at that time, why we ignored parallel circuits in those discussions? We had a good reason: Admittance works far better than impedance when we want to mathematically analyze parallel AC circuits. In parallel AC circuits, resistance and reactance combine to make a mathematical mess. But conductance and susceptance add directly together in parallel circuits, yielding admittance. We'll analyze parallel RL and RC circuits in the next chapter.

The *GB* Half-Plane

We can portray complex admittance on a coordinate grid similar to the complex-impedance (RX) half-plane. We get a half-plane, not a complete plane because no such thing as negative conductance exists in the "real world." (We can't have a component that conducts worse than not at all!) We plot conductance values along a horizontal G axis. We plot susceptance along a vertical B axis. Figure 15-10 shows several points on the *GB half-plane*.

It's Inside-Out

Superficially, the GB half-plane looks identical to the RX half-plane. But mathematically, the two couldn't differ more! The GB half-plane is "inside-out" with respect to the RX half-plane. The center, or origin, of the GB half-plane represents the point at which no conductance exists for DC or for AC. It represents the *zero-admittance point* rather than the *zero-impedance point*. In the GB half-plane, the origin corresponds to a perfect open circuit. In the RX half-plane, the origin represents a perfect short circuit.

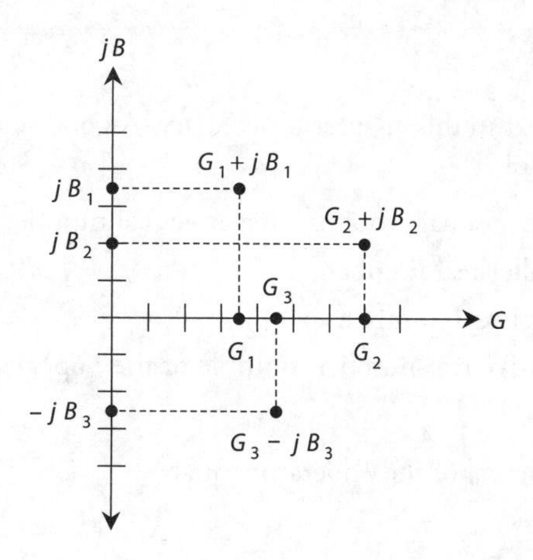

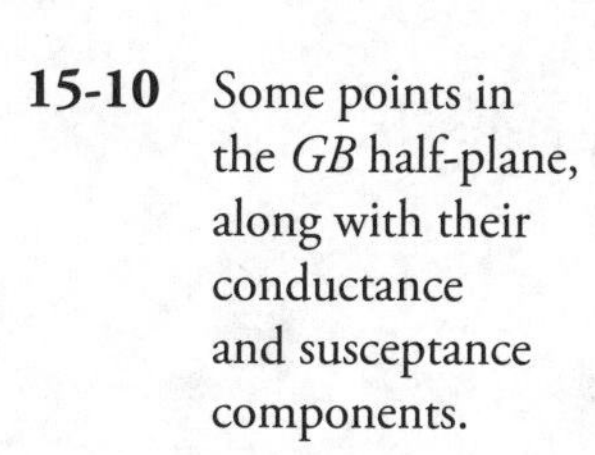
15-10 Some points in the *GB* half-plane, along with their conductance and susceptance components.

As you move out toward the right ("east") along the *G*, or conductance, axis of the *GB* half-plane, the conductance improves, and the current gets greater. When you move upward ("north") along the *jB* axis from the origin, you have ever-increasing positive (capacitive) susceptance. When you go downward ("south") along the *jB* axis from the origin, you encounter increasingly negative (inductive) susceptance.

Vector Representation of Admittance

We can denote specific complex admittance values as vectors, just as we can do with complex impedance values. Figure 15-11 shows the points from Fig. 15-10 as complex admittance vectors. Given a fixed applied AC voltage, long vectors in the *GB* half-plane generally indicate large currents, and short vectors indicate small currents.

Imagine a point moving around on the *GB* half-plane. Think of the vector getting longer and shorter, and changing direction as well. Vectors pointing generally "northeast," or upward and to the right, correspond to conductances and capacitances in parallel. Vectors pointing in a more or less "southeasterly" direction, or downward and to the right, portray conductances and inductances in parallel.

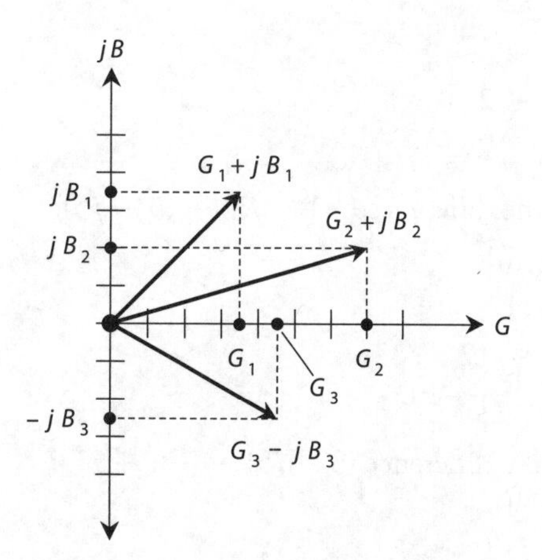

15-11 Vectors representing the points shown in Fig. 15-10.

Quiz

Refer to the text in this chapter if necessary. A good score is 18 or more correct. Answers are in the back of the book.

1. The positive square root of a negative real number equals
   (a) a smaller real number.
   (b) a larger real number.
   (c) a positive real-number multiple of the *j* operator.
   (d) 0.

2. The reciprocal of the *j* operator equals
   (a) itself.
   (b) its negative.
   (c) a real number.
   (d) 0.

3. If we add a real number to an imaginary number, we get
   (a) a real number.
   (b) an imaginary number.
   (c) a complex number.
   (d) −1.

4. What's the sum $(-1 + j7) + (3 - j5)$?
   (a) $2 + j2$
   (b) $2 - j2$
   (c) $-2 + j2$
   (d) $-2 - j2$

5. What's the sum $(3 - j5) + (-1 + j7)$?
   (a) $2 + j2$
   (b) $2 - j2$
   (c) $-2 + j2$
   (d) $-2 - j2$

6. What's the difference $(-1 + j7) - (3 - j5)$?
   (a) $4 + j12$
   (b) $4 - j12$
   (c) $-4 + j12$
   (d) $-4 - j12$

7. What's the difference $(3 - j5) - (-1 + j7)$?
   (a) $4 + j12$
   (b) $4 - j12$

(c) $-4 + j12$
(d) $-4 - j12$

8. If a specification paper tells you that a certain device has a nominal output impedance of "50 ohms," the manufacturer means that the load should ideally exhibit a complex impedance of
(a) $50 + j50$.
(b) $50 + j50$ or $50 - j50$.
(c) $0 + j50$ or $0 - j50$.
(d) None of the above

9. The complex impedance value $15 + j15$ could represent
(a) a pure resistance.
(b) a pure reactance.
(c) a resistor in series with an inductor.
(d) a resistor in series with a capacitor.

10. Which, if any, of the following complex numbers has an absolute value of 25?
(a) $15 - j20$
(b) $12.5 - j12.5$
(c) $5 - j5$
(d) None of the above

11. What's the absolute-value impedance of $4.50 + j5.50$?
(a) 4.50 ohms
(b) 5.50 ohms
(c) 7.11 ohms
(d) 50.5 ohms

12. What's the absolute-value impedance of $0.0 - j36$?
(a) 0.0 ohms
(b) 6.0 ohms
(c) 18 ohms
(d) 36 ohms

13. What's the magnitude of the vector whose end point lies at $(1000, -j1000)$ on the complex-number plane?
(a) 1000
(b) 1414
(c) 2000
(d) 2828

14. What's the magnitude of the vector whose end point lies at $(-1000, -j1000)$ on the complex-number plane?
(a) 1000
(b) 1414
(c) 2000
(d) 2828

15. If we enlarge the inside radius of the shield of a coaxial cable but don't change anything else about the cable, what happens to its characteristic impedance?
    (a) It increases.
    (b) It does not change.
    (c) It decreases.
    (d) We need more information to answer this question.

16. If we increase the radii of both wires in a two-wire transmission line but don't change anything else about the line, what happens to its characteristic impedance?
    (a) We need more information to answer this question.
    (b) It increases.
    (c) It does not change.
    (d) It decreases.

17. Suppose that a capacitor has a value of 0.010 μF at 1.2 MHz. What's the capacitive susceptance, stated as an imaginary number?
    (a) $B_C = j0.075$
    (b) $B_C = -j0.075$
    (c) $B_C = j13$
    (d) $B_C = -j13$

18. Absolute-value impedance equals the square root of
    (a) the real-number coefficient of the reactance plus the imaginary-number part of the admittance.
    (b) the real-number resistance plus the real-number coefficient of the reactance.
    (c) the real-number conductance plus the real-number coefficient of the susceptance.
    (d) None of the above

19. Suppose that an inductor has a value of 10.0 mH at 15.91 kHz. What's the inductive susceptance, stated as an imaginary number?
    (a) $B_L = -j1000$
    (b) $B_L = j1000$
    (c) $B_L = -j0.00100$
    (d) $B_L = j0.00100$

20. When we add the reciprocal of real-number resistance to the reciprocal of imaginary-number reactance, we get complex-number
    (a) impedance.
    (b) conductance.
    (c) susceptance.
    (d) admittance.

16
CHAPTER

Alternating-Current Circuit Analysis

WHEN YOU SEE AN AC CIRCUIT THAT CONTAINS COILS AND/OR CAPACITORS, YOU CAN ENVISION A complex-number half-plane, either *RX* (resistance-reactance) or *GB* (conductance-susceptance). The *RX* half-plane applies to series circuit analysis. The *GB* half-plane applies to parallel circuit analysis.

Complex Impedances in Series

In any situation where resistors, coils, and capacitors are connected in series, each component has an impedance that we can represent as a vector in the *RX* half-plane. The vectors for resistors remain constant regardless of the frequency. But the vectors for coils and capacitors vary as the frequency increases or decreases.

Pure Reactances

Pure inductive reactance (X_L) and capacitive reactance (X_C) simply add together when we connect coils and capacitors in series. That is,

$$X = X_L + X_C$$

In the *RX* half-plane, their vectors add, but because these vectors point in exactly opposite directions—inductive reactance upward and capacitive reactance downward, as shown in Fig. 16-1—the resultant sum vector inevitably points either straight up or straight down, unless the reactances are equal and opposite, in which case their sum equals the zero vector.

Problem 16-1

Suppose that we connect a coil and a capacitor in series with $jX_L = j200$ and $jX_C = -j150$. What's the net reactance?

Solution

We add the values to get

$$jX = j200 + (-j150) = j(200 - 150) = j50$$

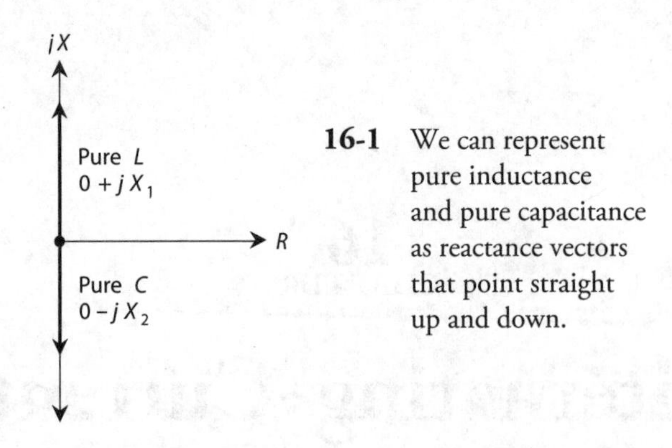

16-1 We can represent pure inductance and pure capacitance as reactance vectors that point straight up and down.

Problem 16-2

Suppose that we connect a coil and capacitor in series with $jX_L = j30$ and $jX_C = -j110$. What's the net reactance?

Solution

Again, we add the values, obtaining

$$jX = j30 + (-j110) = j(30 - 110) = -j80$$

Problem 16-3

Suppose that we connect a coil of $L = 5.00\ \mu\text{H}$ and a capacitor of $C = 200$ pF in series, and then drive an AC signal through the combination at $f = 4.00$ MHz. What's the net reactance?

Solution

First, we calculate the reactance of the inductor at 4.00 MHz to get

$$\begin{aligned} jX_L &= j6.2832fL \\ &= j(6.2832 \times 4.00 \times 5.00) = j125.664 \end{aligned}$$

Next, we calculate the reactance of the capacitor at 4.00 MHz (noting that 200 pF = 0.000200 μF) to get

$$\begin{aligned} jX_C &= -j[1/(6.2832fC)] \\ &= -j[1/(6.2832 \times 4.00 \times 0.000200)] = -j198.943 \end{aligned}$$

Finally, we add the reactances and round off to three significant figures, obtaining

$$jX = j125.664 + (-j198.943) = -j73.3$$

Beware of Cumulative Rounding Errors!

Do you wonder why we carried through some extra digits during the course of the preceding calculations, rounding off only at the very end of the process? The extra digits in the intermediate steps reduce our risk of ending up with a so-called *cumulative rounding error*. In situations where repeated rounding introduces many small errors, the final calculation can end up a full digit or two off, even after rounding to the appropriate number of significant figures. Let's use this precaution, when necessary, in all of our future calculations.

Problem 16-4

What's the net reactance of the above-described combination at $f = 10.0$ MHz?

Solution

First, we calculate the reactance of the inductor at 10.0 MHz to get

$$jX_L = j6.2832fL$$
$$= j(6.2832 \times 10.0 \times 5.00) = j314.16$$

Next, we calculate the reactance of the capacitor at 10.00 MHz to get

$$jX_C = -j[1/(6.2832fC)]$$
$$= -j[1/(6.2832 \times 10.0 \times 0.000200)] = -j79.58$$

Finally, we add the reactances and round off, obtaining

$$jX = j314.16 + (-j79.58) = j235$$

Adding Impedance Vectors

Whenever the resistance in a series circuit reaches values that are significant compared with the reactance, the impedance vectors no longer point straight up and straight down. Instead, they run off towards the "northeast" (for the inductive part of the circuit) and "southeast" (for the capacitive part). Figure 16-2 shows an example of this condition.

When two impedance vectors don't lie along a single line, we must use *vector addition* to get the correct net impedance vector. Figure 16-3 shows the geometry of vector addition. We construct a *parallelogram*, using the two vectors $Z_1 = R_1 + jX_1$ and $Z_2 = R_2 + jX_2$ as two adjacent sides of the figure. The diagonal of the parallelogram constitutes the vector representing the net complex impedance. In a parallelogram, pairs of opposite angles always have equal measures. These equalities are indicated by the pairs of single and double arcs in Fig. 16-3.

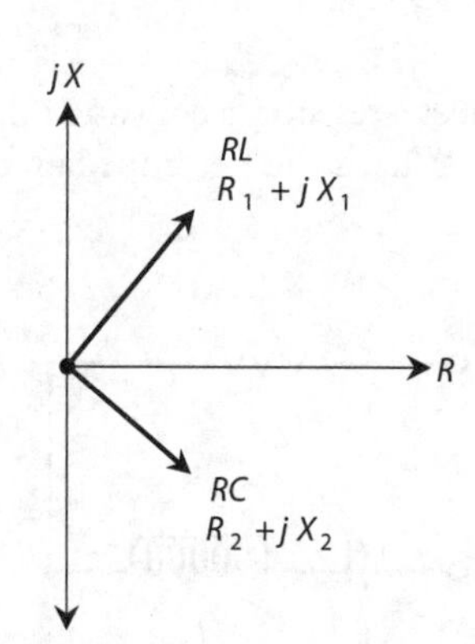

16-2 When we have resistance along with reactance, the impedance vectors are neither vertical nor horizontal.

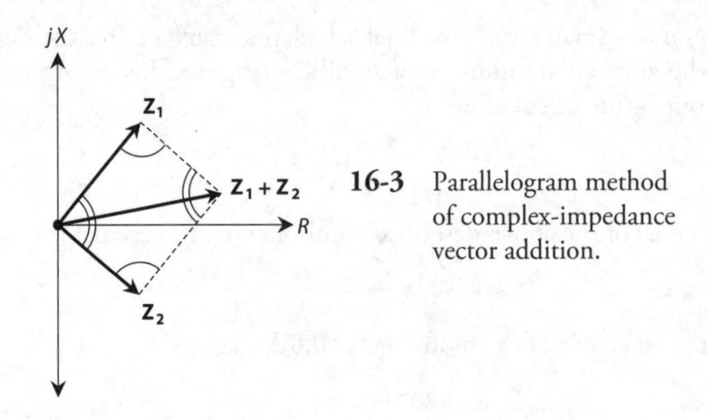

16-3 Parallelogram method of complex-impedance vector addition.

Formula for Complex Impedances in Series

Consider two complex impedances, $Z_1 = R_1 + jX_1$ and $Z_2 = R_2 + jX_2$. If we connect these impedances in series, we can represent the net impedance Z as the vector sum

$$Z = (R_1 + jX_1) + (R_2 + jX_2) = (R_1 + R_2) + j(X_1 + X_2)$$

The resistance and reactance components add separately. Remember that inductive reactances are positive imaginary while capacitive reactances are negative imaginary!

Series *RLC* Circuits

With an inductance, capacitance, and resistance in series (Fig. 16-4), you can imagine the resistance R as belonging entirely to the coil, if you want to take advantage of the above formulas. Then you have only two vectors to add (instead of three), when you calculate the impedance of the series *RLC* circuit. Mathematically, the situation works out as follows:

$$Z = (R + jX_L) + (0 + jX_C) = R + j(X_L + X_C)$$

Again, remember that X_C is never positive! So, although these general formulas contain addition symbols exclusively, you must add in a negative value (the equivalent of subtraction) when you include a capacitive reactance.

Problem 16-5

Suppose that we connect a resistor, a coil, and a capacitor in series with $R = 50$ ohms, $X_L = 22$ ohms, and $X_C = -33$ ohms. What's the net impedance Z?

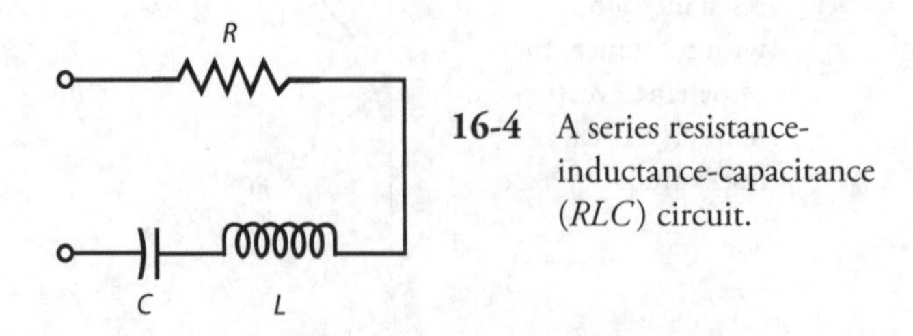

16-4 A series resistance-inductance-capacitance (*RLC*) circuit.

Solution

We can consider the resistor as part of the coil, obtaining $50 + j22$ and $0 - j33$. Adding these gives the resistance component of $50 + 0 = 50$, and the reactive component of $j22 - j33 = -j11$. Therefore, $Z = 50 - j11$.

Problem 16-6

Consider a resistor, a coil, and a capacitor in series with $R = 600$ ohms, $X_L = 444$ ohms, and $X_C = -444$ ohms. What's the net impedance, Z?

Solution

Again, we can imagine the resistor as part of the inductor. Then the complex impedance vectors are $600 + j444$ and $0 - j444$. Adding these, the resistance component equals $600 + 0 = 600$, and the reactive component equals $j444 - j444 = j0$. We have a net impedance $Z = 600 + j0$, a purely resistive impedance.

Series Resonance

When a series-connected *RLC* circuit has zero net reactance at a certain frequency, we say that the circuit exhibits *series resonance* at that frequency.

Problem 16-7

Consider a resistor, a coil, and a capacitor connected in series. The resistor has a value of 330 ohms, the capacitance equals 220 pF, and the inductance equals 100 μH. We operate the circuit at 7.15 MHz. What's the complex impedance?

Solution

First, we calculate the inductive reactance. Remember that

$$X_L = 6.2832 fL$$

and that megahertz and microhenrys go together in the formula. We multiply to obtain

$$jX_L = j(6.2832 \times 7.15 \times 100) = j4492$$

Next, we calculate the capacitive reactance using the formula

$$X_C = -1/(6.2832 fC)$$

We can convert 220 pF to microfarads, obtaining $C = 0.000220$ μF. Then we have

$$jX_C = -j[1/(6.2832 \times 7.15 \times 0.000220)] = -j101$$

Now, we can lump the resistance and the inductive reactance together, so one of the impedances becomes $330 + j4492$. The other impedance equals $0 - j101$. Adding these gives us

$$Z = 330 + j4492 - j101 = 330 + j4391$$

We can justify only three significant digits of accuracy here, so we might want to state this result as $Z = 330 + j4.39$k (remembering that "k" stands for "kilohms").

Problem 16-8

Suppose that we connect a resistor, a coil, and a capacitor in series. The resistance equals 50.0 ohms, the inductance equals 10.0 μH, and the capacitance equals 1000 pF. We operate the circuit at 1592 kHz. What's the complex impedance?

Solution

First, let's calculate the inductive reactance. Note that 1592 kHz = 1.592 MHz. Plugging in the numbers, we obtain

$$jX_L = j(6.2832 \times 1.592 \times 10.0) = j100$$

Next, we calculate the capacitive reactance. Let's convert picofarads to microfarads, and use megahertz for the frequency. Then we have

$$\begin{aligned} jX_C &= -j[1/(6.2832 \times 1.592 \times 0.001000)] \\ &= -j100 \end{aligned}$$

When we put the resistance and the inductive reactance together into a single complex number, we get $50.0 + j100$. We know that the capacitor's impedance is $0 - j100$. Adding the two complex numbers, we get

$$Z = 50.0 + j100 - j100 = 50.0 + j0$$

Our circuit exhibits a pure resistance of 50.0 ohms at 1592 kHz.

Complex Admittances in Parallel

When you see resistors, coils, and capacitors in parallel, remember that each component, whether a resistor, an inductor, or a capacitor, has an admittance that you can represent as a vector in the *GB* half-plane. The vectors for pure conductances remain constant, even as the frequency changes. But the vectors for the coils and capacitors vary with frequency.

Pure Susceptances

Pure inductive susceptance (B_L) and capacitive susceptance (B_C) add together when coils and capacitors appear in parallel. That is,

$$B = B_L + B_C$$

Remember that B_L is never positive, and B_C is never negative; we must invert the "sign scenario" with susceptance values as compared with reactance values.

In the *GB* half-plane, pure jB_L and jB_C vectors add. Because such vectors always point in exactly opposite directions—capacitive susceptance upward and inductive susceptance downward—the sum, jB, points either straight up or straight down, as shown in Fig. 16-5, unless the susceptances happen to be equal and opposite, in which case they cancel and the result equals the zero vector.

Problem 16-9

Consider a coil and capacitor connected in parallel with $jB_L = -j0.05$ and $jB_C = j0.08$. What's the net susceptance?

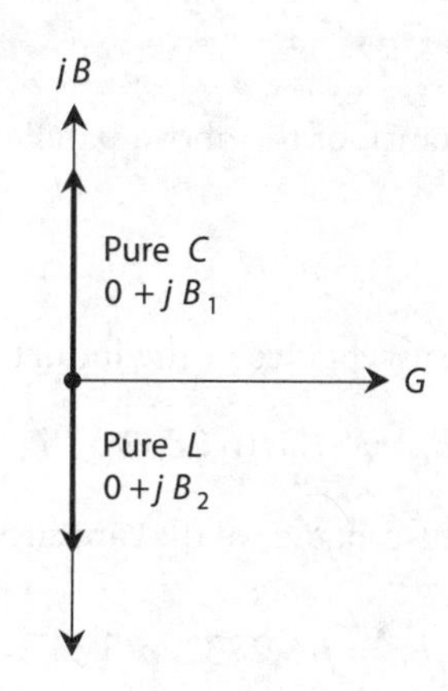

16-5 We can represent pure capacitance and pure inductance as susceptance vectors that point straight up and down.

Solution

We add the values to get

$$jB = jB_L + jB_C = -j0.05 + j0.08 = j0.03$$

Problem 16-10

Suppose that we connect a coil and capacitor in parallel with $jB_L = -j0.60$ and $jB_C = j0.25$. What's the net susceptance?

Solution

Again, we add the values to obtain

$$jB = -j0.60 + j0.25 = -j0.35$$

Problem 16-11

Suppose that we connect a coil of $L = 6.00$ μH and a capacitor of $C = 150$ pF in parallel and drive a signal through them at $f = 4.00$ MHz. What's the net susceptance?

Solution

First, we calculate the susceptance of the inductor at 4.00 MHz, obtaining

$$\begin{aligned} jB_L &= -j[1/(6.2832fL)] \\ &= -j[1/(6.2832 \times 4.00 \times 6.00)] = -j0.00663144 \end{aligned}$$

Next, we calculate the susceptance of the capacitor (converting its value to microfarads) at 4.00 MHz, getting

$$\begin{aligned} jB_C &= j(6.2832fC) \\ &= j(6.2832 \times 4.00 \times 0.000150) = j0.00376992 \end{aligned}$$

Finally, we add the inductive and capacitive susceptances and round off to three significant figures, ending up with

$$jB = -j0.00663144 + j0.00376992 = -j0.00286$$

Problem 16-12

What's the net susceptance of the above parallel-connected inductor and capacitor at a frequency of $f = 5.31$ MHz?

Solution

First, we calculate the susceptance of the inductor at 5.31 MHz to get

$$jB_L = -j[1/(6.2832 \times 5.31 \times 6.00)] = -j0.00499544$$

Next, we calculate the susceptance of the capacitor (converting its value to microfarads) at 5.31 MHz, getting

$$jB_C = j(6.2832 \times 5.31 \times 0.000150) = j0.00500457$$

Finally, we add the inductive and capacitive susceptances and round off to three significant figures, obtaining

$$jB = -j0.00499544 + j0.00500 = j0.00$$

Adding Admittance Vectors

When the conductance is significant in a parallel circuit containing inductance and capacitance, the admittance vectors don't point straight up and down. Instead, they run off towards the "north-east" (for the capacitive part of the circuit) and "southeast" (for the inductive part), as shown in Fig. 16-6. You've seen how vectors add in the *RX* half-plane. In the *GB* half-plane, things work in the same way. The net admittance vector equals the sum of the component admittance vectors.

Formula for Complex Admittances in Parallel

Imagine that we connect two admittances $Y_1 = G_1 + jB_1$ and $Y_2 = G_2 + jB_2$ in parallel. We can find the net admittance Y as the complex-number sum

$$Y = (G_1 + jB_1) + (G_2 + jB_2) = (G_1 + G_2) + j(B_1 + B_2)$$

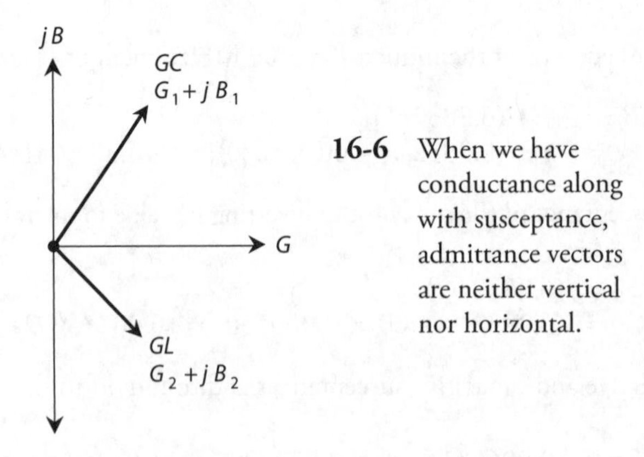

16-6 When we have conductance along with susceptance, admittance vectors are neither vertical nor horizontal.

Parallel *RLC* Circuits

When we connect a coil, capacitor, and resistor in parallel (Fig. 16-7), we can think of the resistance as a *conductance* in siemens (symbolized S), which equals the reciprocal of the value in ohms. If we consider the conductance as part of the inductor, we have only two complex numbers to add, rather than three, when finding the admittance of a parallel *RLC* circuit. We can use the formula

$$Y = (G + jB_L) + (0 + jB_C) = G + j(B_L + B_C)$$

Again, we must remember that B_L is never positive! So, although the formulas here have addition symbols in them, we actually subtract a value for the inductive susceptance by adding in a negative number.

Problem 16-13

Suppose that we connect a resistor, a coil, and a capacitor in parallel. The resistor has a conductance of $G = 0.10$ S. The susceptances are $jB_L = -j0.010$ and $jB_C = j0.020$. What's the complex admittance of this combination?

Solution

Let's consider the resistor as part of the coil, so we have two complex admittances in parallel: $0.10 - j0.010$ and $0.00 + j0.020$. Adding these values "part-by-part" gives us a conductance component of $0.10 + 0.00 = 0.10$ and a susceptance component of $-j0.010 + j0.020 = j0.010$. Therefore, the complex admittance equals $0.10 + j0.010$.

Problem 16-14

Consider a resistor, a coil, and a capacitor in parallel. The resistor has a conductance of $G = 0.0010$ S. The susceptances are $jB_L = -j0.0022$ and $jB_C = j0.0022$. What is the complex admittance of this combination?

Solution

Again, let's consider the resistor as part of the coil. Then the complex admittances are $0.0010 - j0.0022$ and $0.0000 + j0.0022$. Adding these, we get a conductance component of $0.0010 + 0.0000 = 0.0010$ and a susceptance component of $-j0.0022 + j0.0022 = j0.0000$. Thus, the admittance equals $0.0010 + j0.0000$, a pure conductance.

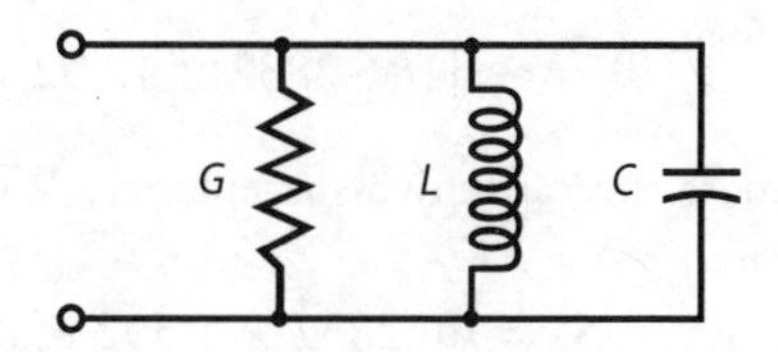

16-7 A parallel resistance-inductance-capacitance (*RLC*) circuit. Here, *G* represents conductance (the reciprocal of resistance), so we can just as well call this a *conductance-inductance-capacitance* (*GLC*) circuit.

Parallel Resonance

When a parallel *RLC* circuit lacks net susceptance at a certain frequency, we have a condition called *parallel resonance* at that frequency.

Problem 16-15

Suppose that we connect a resistor, a coil, and a capacitor in parallel. The resistor has a value of 100 ohms, the capacitance is 200 pF, and the inductance is 100 μH. We operate the circuit at a frequency of 1.00 MHz. What's the net complex admittance?

Solution

First, let's calculate the inductive susceptance. We recall the formula and plug in the numbers. Megahertz and microhenrys go together, so we have

$$\begin{aligned} jB_L &= -j[1/(6.2832fL)] \\ &= -j[1/(6.2832 \times 1.00 \times 100)] = -j0.00159155 \end{aligned}$$

Next, we calculate the capacitive susceptance. We can convert 200 pF to 0.000200 μF, so we have

$$\begin{aligned} jB_C &= j(6.2832fC) \\ &= j(6.2832 \times 1.00 \times 0.000200) = j0.00125664 \end{aligned}$$

We can consider the conductance, which equals $1/100 = 0.0100$ S, and the inductive susceptance together so that one of the parallel-connected admittances equals $0.0100 - j0.00159155$. The other admittance is $0 + j0.00125664$. When we add these complex numbers and then round off the susceptance coefficient to three significant figures, we get

$$\begin{aligned} Y &= 0.0100 - j0.00159155 + j0.00125664 \\ &= 0.0100 - j0.000335 \end{aligned}$$

Problem 16-16

Consider a resistor, a coil, and a capacitor in parallel. The resistance is 10.00 ohms, the inductance is 10.00 μH, and the capacitance is 1000 pF. The frequency is 1592 kHz. What's the complex admittance?

Solution

First, let's calculate the inductive susceptance. We can convert the frequency to megahertz; 1592 kHz = 1.592 MHz. Now we plug in the numbers to get

$$jB_L = -j[1/(6.2832 \times 1.592 \times 10.00)] = -j0.00999715$$

Next, we calculate the capacitive susceptance. We can convert 1000 pF to 0.001000 μF, so we obtain

$$jB_C = j(6.2832 \times 1.592 \times 0.001000) = j0.01000285$$

Finally, we consider the conductance, which equals $1/10.00 = 0.1000$ S, and the inductive susceptance in a single component, so that one of the parallel-connected admittances is $0.1000 - j0.00999715$.

The other is $0 + j0.01000285$. Adding these complex numbers and then rounding off the susceptance coefficient to four significant figures, we obtain

$$Y = 0.1000 - j0.00999715 + j0.01000285$$
$$= 0.1000 - j0.000$$

Converting Complex Admittance to Complex Impedance

As we've seen, the *GB* half-plane looks like the *RX* half-plane, although mathematically they differ. We can convert a quantity from a complex admittance $G + jB$ to a complex impedance $R + jX$ using the formulas

$$R = G/(G^2 + B^2)$$

and

$$X = -B/(G^2 + B^2)$$

If we know the complex admittance, we should find the resistance and reactance components individually, using the above formulas. Then we can assemble the two components into the complex impedance $R + jX$.

Problem 16-17

Suppose that a circuit has an admittance $Y = 0.010 - j0.0050$. What's the complex impedance, assuming the frequency never varies?

Solution

In this case, $G = 0.010$ S and $B = -0.0050$ S. We determine $G^2 + B^2$ as follows:

$$G^2 + B^2 = 0.010^2 + (-0.0050)^2$$
$$= 0.000100 + 0.000025 = 0.000125$$

Knowing this common denominator, we can calculate *R* and *X* as

$$R = G/0.000125 = 0.010/0.000125 = 80 \text{ ohms}$$

and

$$X = -B/0.000125 = 0.0050/0.000125 = 40 \text{ ohms}$$

Our circuit has a complex impedance of $Z = 80 + j40$.

Putting It All Together

When we encounter a parallel circuit containing resistance, inductance, and capacitance, and we want to determine the complex impedance of the combination, we should go through the following steps in order:

- Calculate the conductance *G* of the resistor.
- Calculate the susceptance B_L of the inductor.
- Calculate the susceptance B_C of the capacitor.
- Determine the net susceptance $B = B_L + B_C$.

- Determine the quantity $G^2 + B^2$.
- Compute R in terms of G and B using the appropriate formula.
- Compute X in terms of G and B using the appropriate formula.
- Write down the complex impedance as the sum $R + jX$.

Problem 16-18

Suppose that we connect a resistor of 10.0 ohms, a capacitor of 820 pF, and a coil of 10.0 μH in parallel. We operate the circuit at 1.00 MHz. What's the complex impedance?

Solution

Let's proceed according to the steps described above, leaving in some extra digits in the susceptance figures and then rounding off to three significant figures at the end of the process, as follows:

- Calculate $G = 1/R = 1/10.0 = 0.100$.
- Calculate $B_L = -1/(6.2832fL) = -1/(6.2832 \times 1.00 \times 10.0) = -0.0159155$.
- Calculate $B_C = 6.2832fC = 6.2832 \times 1.00 \times 0.000820 = 0.00515222$. (We must remember to convert the capacitance from picofarads to microfarads.)
- Calculate $B = B_L + B_C = -0.0159155 + 0.00515222 = -0.0107633$.
- Define $G^2 + B^2 = 0.100^2 + (-0.0107633)^2 = 0.0101158$.
- Calculate $R = G/0.0101158 = 0.100/0.0101158 = 9.89$.
- Calculate $X = -B/0.0101158 = 0.0107633/0.0101158 = 1.06$.
- The complex impedance equals $R + jX = 9.89 + j1.06$.

Problem 16-19

Suppose that we connect a resistor of 47.0 ohms, a capacitor of 500 pF, and a coil of 10.0 μH in parallel. What's their complex impedance at 2.25 MHz?

Solution

We proceed in the same fashion as we did when solving Problem 16-18, leaving in some extra digits in the conductance and susceptance figures until the end, as follows:

- Calculate $G = 1/R = 1/47.0 = 0.0212766$.
- Calculate $B_L = -1/(6.2832fL) = -1/(6.2832 \times 2.25 \times 10.0) = -0.00707354$.
- Calculate $B_C = 6.2832fC = 6.2832 \times 2.25 \times 0.000500 = 0.0070686$. (We convert the capacitance to microfarads.)
- Calculate $B = B_L + B_C = -0.00707354 + 0.0070686 = 0.00000$.
- Define $G^2 + B^2 = 0.0212766^2 + 0.00000^2 = 0.00045269$.
- Calculate $R = G/0.00045269 = 0.0212766/0.00045269 = 47.000$.
- Calculate $X = -B/0.00045269 = 0.00000/0.00045269 = 0.00000$.
- The complex impedance is $R + jX = 47.000 + j0.00000$. When we round both values off to three significant figures, we get $47.0 + j0.00$. This complex-number quantity represents a pure resistance equal to the value of the resistor in the circuit.

Reducing Complicated *RLC* Circuits

Sometimes we'll see circuits with several resistors, capacitors, and/or coils in series and parallel combinations. We can always reduce such a circuit to an equivalent series or parallel *RLC* circuit that contains one resistance, one capacitance, and one inductance.

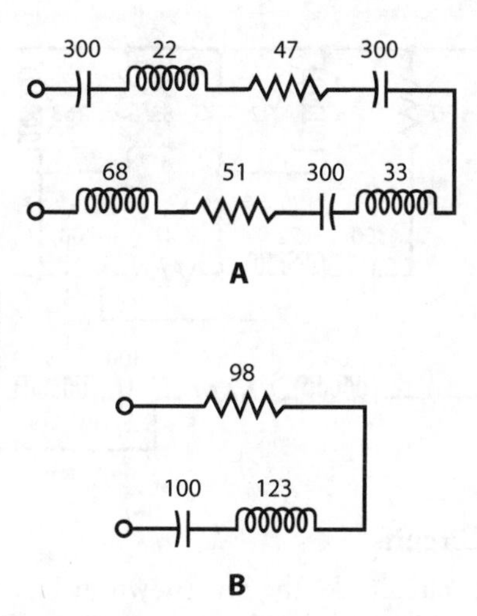

16-8 At A, a "complicated" series circuit containing multiple resistances and reactances. At B, the same circuit simplified. Resistances are in ohms; inductances are in microhenrys (μH); capacitances are in picofarads (pF).

Series Combinations

Resistances in series simply add. Inductances in series also add. Capacitances in series combine in a more complicated way, which you learned earlier. If you don't remember the formula, it is

$$C = 1/(1/C_1 + 1/C_2 + \ldots + 1/C_n)$$

where C_1, C_2, …, and C_n represent the individual capacitances, and C represents the net capacitance of the series combination. Figure 16-8A shows an example of a "complicated" series *RLC* circuit. Figure 16-8B shows the equivalent circuit with one resistance, one capacitance, and one inductance.

Parallel Combinations

Resistances and inductances combine in parallel just as capacitances combine in series. Capacitances in parallel simply add up. Figure 16-9A shows an example of a "complicated" parallel *RLC* circuit. Figure 16-9B shows the equivalent circuit with one resistance, one capacitance, and one inductance.

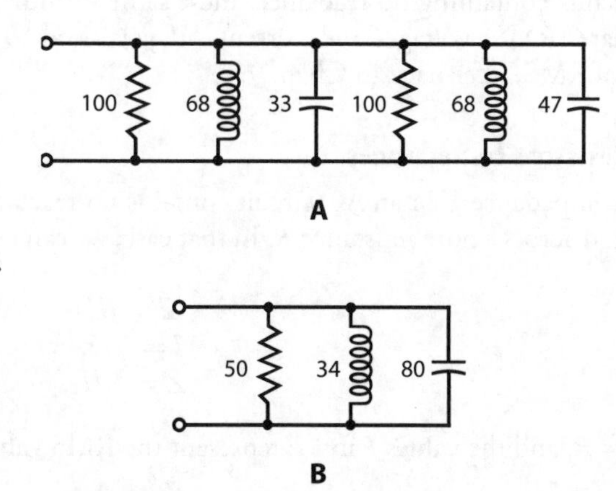

16-9 At A, a "complicated" parallel circuit containing multiple resistances and reactances. At B, the same circuit simplified. Resistances are in ohms; inductances are in microhenrys (μH); capacitances are in picofarads (pF).

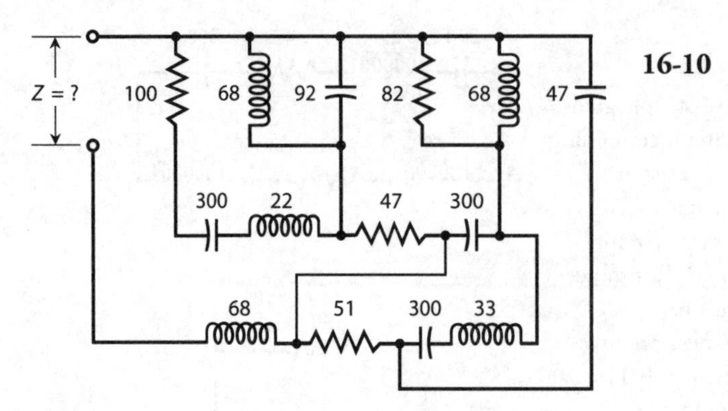

16-10 A series-parallel "nightmare circuit" containing multiple resistances and reactances. Resistances are in ohms; inductances are in microhenrys (μH); capacitances are in picofarads (pF).

"Nightmare" Circuits

Imagine an *RLC* circuit like the one shown in Fig. 16-10. How would you find the net complex impedance at, say, 8.54 MHz? You'll rarely encounter circuits such as this one in practical applications—or if you do, no one will likely ask you to *calculate* the net impedance at any particular frequency. But you can rest assured that, given a frequency, a complex impedance does exist, no matter how complicated the circuit might be.

A true electronics "geek" could use a computer to work out the theoretical complex impedance of a circuit, such as the one in Fig. 16-10, at a specific frequency. In practice, however, an engineer might take an *experimental* approach by building the circuit, connecting a signal generator to its input terminals, and measuring the resistance R and the reactance X at the frequency of interest using a lab instrument called an *impedance bridge.*

Ohm's Law for Alternating Current

We can state Ohm's Law for DC circuits as a simple relationship between three variables: the current I (in amperes), the voltage E (in volts), and the resistance R (in ohms). Here are the formulas, in case you don't recall them:

$$E = IR$$
$$I = E/R$$
$$R = E/I$$

In AC circuits containing no reactance, these same formulas apply, as long as we work with root-mean-square (RMS) voltages and currents. If you need to refresh your memory concerning the meaning of RMS, refer back to Chap. 9.

Purely Resistive Impedances

When the impedance Z in an AC circuit contains no reactance, all of the current and voltage exist through and across a pure resistance R. In that case, we can express Ohm's Law in terms of the three formulas

$$E = IZ$$
$$I = E/Z$$
$$Z = E/I$$

where $Z = R$, and the values I and E represent the RMS values for the current and voltage.

Complex Impedances

When you want to determine the relationship between current, voltage, and resistance in an AC circuit that contains reactance along with the resistance, things get quite interesting. Recall the formula for the square of the absolute-value impedance in a series *RLC* circuit:

$$Z^2 = R^2 + X^2$$

This equation tells us that

$$Z = (R^2 + X^2)^{1/2}$$

so Z equals the length of the vector $R + jX$ in the complex impedance plane. This formula applies only for series *RLC* circuits.

The square of the absolute-value impedance for a parallel *RLC* circuit, in which the resistance equals R and the reactance equals X, is defined as

$$Z^2 = R^2X^2 / (R^2 + X^2)$$

We can calculate Z, the absolute-value impedance, directly as

$$Z = [R^2X^2 / (R^2 + X^2)]^{1/2}$$

The ½ power of a quantity represents the positive square root of that quantity.

Problem 16-20

Imagine that a series *RX* circuit (shown by the generic block diagram of Fig. 16-11) has a resistance of $R = 50.0$ ohms and a capacitive reactance of $X = -50.0$ ohms. If we apply 100 V RMS AC to this circuit, what's the RMS current?

Solution

First, let's calculate the complex impedance using the above formula for series circuits, going to a few extra digits to prevent cumulative rounding errors in later calculations. We have

$$Z = (R^2 + X^2)^{1/2} = [50.0^2 + (-50.0)^2]^{1/2}$$
$$= (2500 + 2500)^{1/2} = 5000^{1/2} = 70.71068 \text{ ohms}$$

which we can round off to 70.7 ohms. Now we can calculate the current, using the original "unrounded" figure for Z, as

$$I = E/Z = 100 / 70.71068 = 1.414214 \text{ A RMS}$$

which we can round off to 1.41 A RMS.

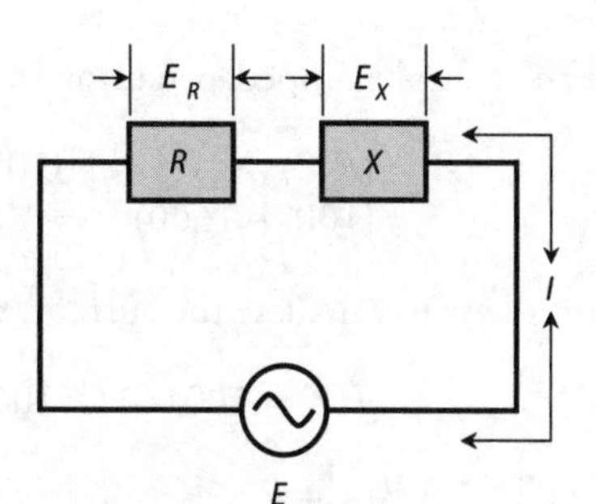

16-11 A series circuit containing resistance and reactance. Illustration for Problems 16-20 through 16-23.

Problem 16-21

What are the RMS AC voltages across the resistance and the reactance, respectively, in the circuit described in Problem 16-20?

Solution

The Ohm's Law formulas for DC will work here. We determined the current, going to several extra digits, as $I = 1.414214$ A RMS, so the voltage drop E_R across the resistance is

$$E_R = IR = 1.414214 \times 50.0 = 70.7107 \text{ V RMS}$$

which rounds off to 70.7 V RMS. The voltage drop E_X across the reactance is

$$E_X = IX = 1.41 \times (-50.0) = -70.7107 \text{ V RMS}$$

which rounds off to −70.7 V RMS. Do you see why we included all the extra digits in our intermediate calculations? If we'd used the rounded figure for the current (1.41 A RMS) in the foregoing calculations, we'd have gotten 70.5 and −70.5 V RMS for the answers we just found. That provides a great example of how cumulative rounding errors can mislead us if we're not careful!

Signs and Phase with RMS Values

Here's an important note in regards to the signs (plus or minus) in RMS figures. When we deal with RMS values, minus signs have no meaning. An RMS value can never be negative, so the minus sign in this result constitutes a mathematical artifact. We can consider the voltage across the reactance as 70.7 V RMS, equal in magnitude to the voltage across the resistance. *But the phase is different*, and that's what the minus sign tells us!

Note that the voltages across the resistance and the reactance (a capacitive reactance in the above-described case because it's negative) don't add up to 100 V RMS, which appears across the whole circuit. This phenomenon occurs because, in an AC circuit containing resistance and reactance, we always observe a difference in phase between the voltage across the resistance and the voltage across the reactance. The voltages across the components add up to the applied voltage *vectorially*, but not always *arithmetically*.

Problem 16-22

Suppose that a series *RX* circuit (Fig. 16-11) has $R = 10.0$ ohms and $X = 40.0$ ohms. The applied voltage is 100 V RMS AC. What's the current?

Solution

First, we calculate the complex impedance using the above formula for series circuits, obtaining

$$Z = (R^2 + X^2)^{1/2} = [10.0^2 + (40.0)^2]^{1/2}$$
$$= (100 + 1600)^{1/2} = 1700^{1/2} = 41.23106 \text{ ohms}$$

When we use Ohm's Law to calculate the current, we get

$$I = E/Z = 100/41.23106 = 2.425356 \text{ A RMS}$$

which rounds off to 2.43 A RMS.

Problem 16-23

What are the RMS AC voltages across the resistance and the reactance, respectively, in the circuit described in Problem 16-22?

Solution

Knowing the current (and using our "un-rounded" result), we can calculate the voltage across the resistance as

$$E_R = IR = 2.425356 \times 10.0 = 24.25356 \text{ V RMS}$$

which rounds off to 24.3 V RMS. The voltage across the reactance is

$$E_X = IX = 2.425356 \times 40.0 = 97.01424 \text{ V RMS}$$

which rounds off to 97.0 V RMS. If we take the arithmetic sum $E_R + E_X$, we get 24.25356 + 97.01424 = 121.2678 V RMS, which rounds off to 121.3 V RMS, as the total voltage across R and X. Again, this value comes out different from the actual applied voltage. The simple DC rule does not work here, for the same reason it didn't work in the scenario of Problem 16-21. The AC voltage across the resistance doesn't add up arithmetically with the AC voltage across the reactance because the two AC waves differ in phase.

Problem 16-24

Suppose that a parallel *RX* circuit (shown by the generic block diagram of Fig. 16-12) has $R = 30.0$ ohms and $X = -20.0$ ohms. We supply the circuit with $E = 50.0$ V RMS. What's the total current drawn from the AC supply?

Solution

First, we find the absolute-value impedance, remembering the formula for parallel circuits and going to a few extra digits to avoid cumulative rounding errors. We get

$$\begin{aligned} Z &= [R^2 X^2 / (R^2 + X^2)]^{1/2} \\ &= \{30.0^2 \times (-20.0)^2 / [30.0^2 + (-20.0)^2]\}^{1/2} \\ &= [900 \times 400 / (900 + 400)]^{1/2} = (360{,}000/1300)^{1/2} \\ &= 277^{1/2} = 16.64332 \text{ ohms} \end{aligned}$$

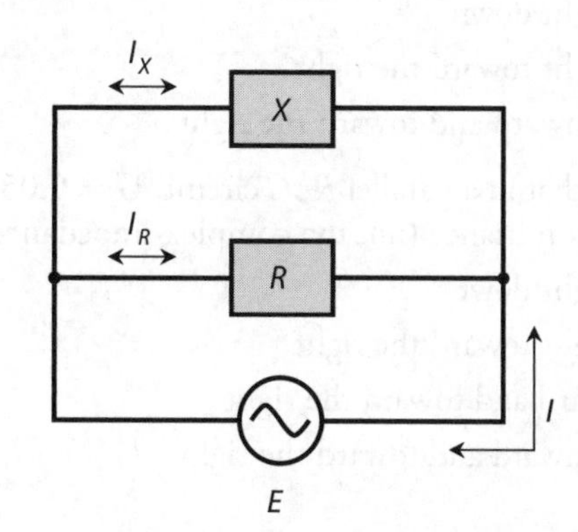

16-12 A parallel circuit containing resistance and reactance. Illustration for Problems 16-24 and 16-25.

The total current is therefore

$$I = E/Z = 50/16.64332 = 3.004208 \text{ A RMS}$$

which rounds off to 3.00 A RMS.

Problem 16-25

What are the RMS currents through the resistance and the reactance, respectively, in the circuit described in Problem 16-24?

Solution

The Ohm's Law formulas for DC will work here. We can calculate the current through the resistance as

$$I_R = E/R = 50.0/30.0 = 1.67 \text{ A RMS}$$

For the current through the reactance, we calculate

$$I_X = E/X = 50.0/(-20.0) = -2.5 \text{ A RMS}$$

As before, we can neglect the minus sign when we think in terms of RMS, so we can call this current 2.5 A RMS. Note that if we directly add the current across the resistance to the current across the reactance, we don't get 3.00 A, the actual total current. This effect takes place for the same reason that AC voltages don't add arithmetically in AC circuits that contain resistance and reactance. The constituent currents, I_R and I_X, differ in phase. Vectorially, they add up to 3.00 A RMS, but arithmetically, they don't.

Quiz

Refer to the text in this chapter if necessary. A good score is 18 correct. Answers are in the back of the book.

1. Suppose that in a series *RLC* circuit, $R = 50$ ohms and no net reactance exists. In which direction does the complex-impedance vector point?
   (a) Straight up
   (b) Straight down
   (c) Straight toward the right
   (d) Downward and toward the right

2. Suppose that in a parallel *RLC* circuit, $G = 0.05$ S and $B = -0.05$ S. In which direction does the complex-admittance (not the complex-impedance) vector point?
   (a) Straight down
   (b) Straight toward the right
   (c) Upward and toward the right
   (d) Downward and toward the right

3. Suppose that in a parallel *RLC* circuit, $R = 10$ ohms and $jX_C = -j10$. In which direction does the complex-admittance (not the complex-impedance) vector point?
   (a) Straight up
   (b) Straight toward the right
   (c) Upward and toward the right
   (d) Downward and toward the right

4. A vector pointing upward and toward the right in the *GB* half-plane would indicate
   (a) pure conductance.
   (b) conductance and inductive susceptance.
   (c) conductance and capacitive susceptance.
   (d) None of the above

5. A vector pointing upward and toward the left in the *RX* half-plane would indicate
   (a) pure resistance.
   (b) resistance and inductive reactance.
   (c) resistance and capacitive reactance.
   (d) None of the above

6. Suppose that a coil has a reactance of $j20$ ohms. What's the susceptance, assuming that the circuit contains nothing else?
   (a) $j0.050$ S
   (b) $-j0.050$ S
   (c) $j20$ S
   (d) $-j20$ S

7. Suppose that a capacitor has a susceptance of $j0.040$ S. What's the reactance, assuming that the circuit contains nothing else?
   (a) $j0.040$ ohms
   (b) $-j0.040$ ohms
   (c) $j25$ ohms
   (d) $-j25$ ohms

8. Suppose that we connect a coil and capacitor in series with $jX_L = j50$ and $jX_C = -j100$. What's the net reactance?
   (a) $j50$
   (b) $j150$
   (c) $-j50$
   (d) $-j150$

9. Suppose that we connect a coil of $L = 3.00\ \mu H$ and a capacitor of $C = 100$ pF in series, and then drive an AC signal through the combination at a frequency of $f = 6.00$ MHz. What's the net reactance?

   (a) $-j152$
   (b) $-j378$
   (c) $j152$
   (d) $j378$

10. Consider a resistor, a coil, and a capacitor in series with $R = 10$ ohms, $X_L = 72$ ohms, and $X_C = -83$ ohms. What's the net impedance Z?

   (a) $10 + j11$
   (b) $10 - j11$
   (c) $82 - j11$
   (d) $-73 - j11$

11. Consider a resistor, a coil, and a capacitor connected in series. The resistor has a value of 220.0 ohms, the capacitance equals 500.00 pF, and the inductance equals 44.00 μH. We operate the circuit at a frequency of 5.650 MHz. What's the complex impedance?

   (a) $220.0 + j1506$
   (b) $220.0 - j1506$
   (c) $0.000 + j1506$
   (d) $220.0 + j0$

12. Suppose that we connect a resistor, a coil, and a capacitor in series. The resistance equals 75.3 ohms, the inductance equals 8.88 μH, and the capacitance equals 980 pF. We operate the circuit at a frequency of 1340 kHz. What's the complex impedance?

   (a) $75.3 + j0.00$
   (b) $75.3 + j46.4$
   (c) $75.3 - j46.4$
   (d) $0.00 - j75.3$

13. Consider a coil and capacitor connected in parallel with $jB_L = -j0.32$ and $jB_C = j0.20$. What's the net susceptance?

   (a) $j0.52$
   (b) $-j0.52$
   (c) $j0.12$
   (d) $-j0.12$

14. Suppose that we connect a coil of 8.5 μH and a capacitor of 100 pF in parallel and drive a signal through them at 7.10 MHz. What's the net susceptance?

   (a) $-j0.0045$
   (b) $j0.0018$
   (c) $-j0.0026$
   (d) None of the above

15. What's the net susceptance of the parallel-connected inductor and capacitor described in Question 14 if we double the frequency to 14.2 MHz?

(a) $-j0.0090$

(b) $j0.0036$

(c) $-j0.0013$

(d) None of the above

16. Consider a resistor, a coil, and a capacitor in parallel. The resistance is 7.50 ohms, the inductance is 22.0 μH, and the capacitance is 100 pF. The frequency is 5.33 MHz. What's the complex admittance?

(a) $0.133 + j0.00199$

(b) $0.133 - j0.00199$

(c) $7.50 + j503$

(d) $7.50 - j503$

17. Suppose that a circuit has an admittance of $Y = 0.333 + j0.667$. What's the complex impedance, assuming the frequency does not change?

(a) $1.80 - j0.833$

(b) $1.80 + j0.833$

(c) $0.599 - j1.20$

(d) $0.599 + j1.20$

18. Suppose that we connect a resistor of 25 ohms, a capacitor of 0.0020 μF, and a coil of 7.7 μH in parallel (not in series!). We operate the circuit at 2.0 MHz. What's the complex impedance?

(a) $8.1 + j22$

(b) $8.1 - j22$

(c) $22 + j8.1$

(d) $22 - j8.1$

19. Suppose that a series *RX* circuit has a resistance of $R = 20$ ohms and a capacitive reactance of $X = -20$ ohms. Suppose that we apply 42 V RMS AC to this circuit. How much current flows?

(a) 0.67 A RMS

(b) 1.5 A RMS

(c) 2.3 A RMS

(d) 3.0 A RMS

20. Suppose that a parallel *RX* circuit has $R = 50$ ohms and $X = 40$ ohms. We supply the circuit with $E = 155$ V RMS. How much current does the entire circuit draw from the AC source?

(a) 5.0 A RMS

(b) 2.5 A RMS

(c) 400 mA RMS

(d) 200 mA RMS

17
CHAPTER

Alternating-Current Power and Resonance

WHEN WE WANT TO OPTIMIZE HOW POWER "TRAVELS," OR CHANGES FORM, WE FACE A CHALLENGE. A phenomenon called *resonance* can play an important role in efficient power transfer and conversion, especially at high AC frequencies.

Forms of Power

Scientists and engineers define power as the rate at which energy is expended, radiated, or dissipated. This definition applies to mechanical motion, chemical effects, electricity, sound waves, radio waves, heat, infrared (IR), visible light, ultraviolet (UV), X rays, gamma rays, and high-speed subatomic particles.

Units of Power

The standard unit of power is the *watt*, abbreviated W and equivalent to a *joule per second* (J/s). Sometimes, engineers specify power in *kilowatts* (kW or thousands of watts), *megawatts* (MW or millions of watts), *gigawatts* (GW or billions of watts), or *terawatts* (TW or trillions of watts). In numerical terms,

- 1 kW = 1000 W
- 1 MW = 1,000,000 W
- 1 GW = 1,000,000,000 W
- 1 TW = 1,000,000,000,000 W

We can also express power in *milliwatts* (mW or thousandths of watts), *microwatts* (μW or millionths of watts), *nanowatts* (nW or billionths of watts), or *picowatts* (pW or trillionths of watts). In numerical terms,

- 1 mW = 0.001 W
- 1 μW = 0.000001 W
- 1 nW = 0.000000001 W
- 1 pW = 0.000000000001 W

Volt-Amperes

In DC circuits, and also in AC circuits having no reactance, we can define power as the product of the voltage E across a device and the current I through that device; that is,

$$P = EI$$

If we express E in volts and I in amperes, then P comes out in *volt-amperes* (VA). Volt-amperes translate directly into watts when no reactance exists in a circuit (Fig. 17-1).

Volt-amperes, also called *VA power* or *apparent power*, can take various forms. A resistor converts electrical energy into heat energy at a rate that depends on the value of the resistance and the current through it. A light bulb converts electricity into light and heat. A radio antenna converts high-frequency AC into radio waves. A loudspeaker or headset converts low-frequency AC into sound waves. A microphone converts sound waves into low-frequency AC. Power figures in these forms give us measures of the intensity of the heat, light, radio waves, sound waves, or AC electricity.

Instantaneous Power

Engineers usually think of electrical power in RMS terms. But for VA power, peak values are sometimes used instead. If the AC constitutes a perfect sine wave with no DC component, the peak current (in either direction) equals 1.414 times the RMS current, and the peak voltage (of either polarity) equals 1.414 times the RMS voltage. If the current and the voltage follow along exactly in phase, then the product of their peak values equals twice the product of their RMS values.

In a reactance-free, sine-wave AC circuit, we see instants in time when the VA power equals twice the effective power. At other points in time, the VA power equals zero; at still other moments, the VA power falls somewhere between zero and twice the effective power level (Fig. 17-2). We call this constantly changing power level, measured or expressed at any particular point in time, the *instantaneous power*. In some situations, such as with an amplitude-modulated (AM) radio signal, the instantaneous power varies in a complicated fashion.

Reactive or Imaginary Power

In a pure resistance, the rate of energy expenditure per unit of time (or *true power*) equals the VA power (also known as *apparent power*). But when reactance exists, the VA power exceeds the power manifested as heat, light, radio waves, or whatever. The apparent power is greater than the true power. We call the difference *reactive power* or *imaginary power* because it exists only within the reactive part of the circuit, represented by the imaginary-number part of the complex-number impedance.

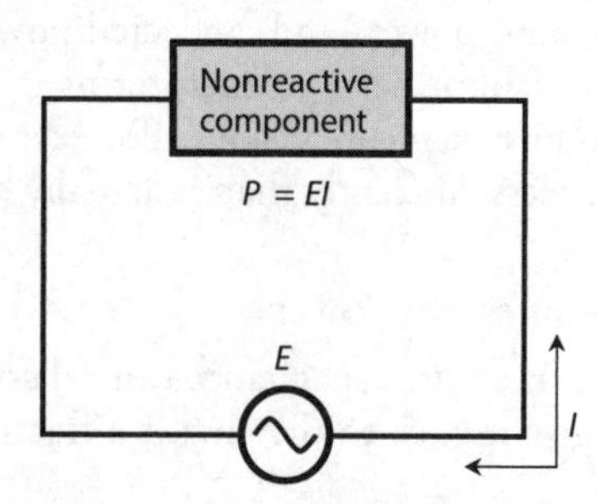

17-1 When an AC component contains no reactance, the power P is the product of the voltage E across the component and the current I through the component.

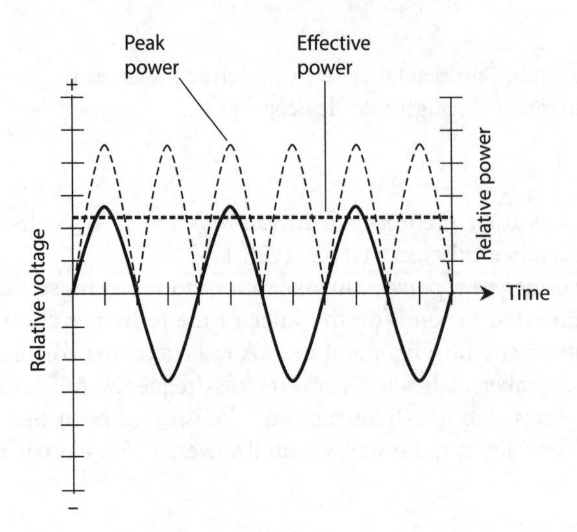

17-2 Peak versus effective power for a sine wave. The left-hand vertical scale shows relative voltage. The right-hand vertical scale shows relative power. The solid curve represents the voltage as a function of time. The light and heavy dashed waves show peak and effective power, respectively, as functions of time.

Inductors and capacitors store energy and release it a fraction of a cycle later, over and over, for as long as AC flows. This phenomenon, like true power, manifests itself as the rate at which energy changes from one form to another. But rather than existing in a form that we can employ in some practical way, imaginary power can only go in and out of "storage." We can't use it for anything. The storage/release cycle repeats along with the actual AC cycle.

True Power Doesn't Travel

If you connect a radio transmitter to a cable that runs to an antenna, you might say you're "feeding power" through the cable to the antenna. Even experienced radio-frequency (RF) engineers and technicians sometimes say that. However, true power always involves a *change in form,* such as from electrical current and voltage into radio waves, or from sound waves into heat. Power does not actually *travel* from place to place. It simply *happens* in one spot or another.

In a radio antenna system, some true power dissipates in a wasteful manner, ending up as heat in the transmitter amplifiers and in the transmission (feed) line, as shown in Fig. 17-3. Obviously, a competent RF engineer seeks to minimize the extent of this waste. The useful dissipation of true power occurs when the imaginary power, in the form of electric and magnetic fields, gets to the antenna, where it changes form and emerges from the antenna as *electromagnetic (EM) waves.*

If you do much work with wireless transmitting antenna systems, you'll hear or read expressions, such as "forward power" and "reflected power," or "power goes from this amplifier to these speakers." You can talk or write in such terms if you like, but keep in mind that the notion can sometimes lead to mistaken conclusions. For example, you might get the idea that an antenna system works more or less efficiently than it actually does.

Reactance Consumes No Power

A pure inductance or a pure capacitance can't dissipate any power. A pure reactance can only store energy and then give it back to the circuit a fraction of a cycle later. In real life, the dielectrics or

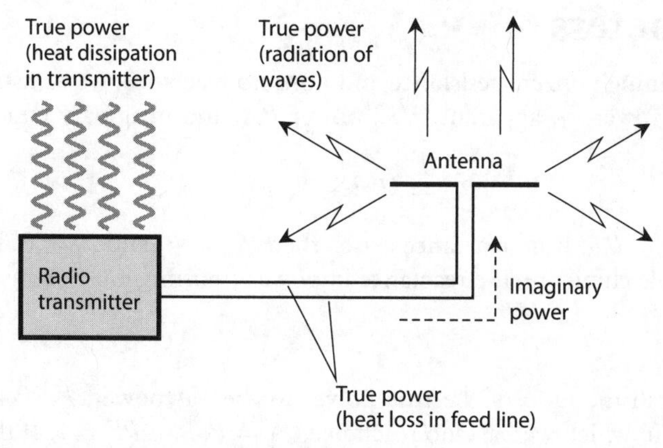

17-3 True power and imaginary power in a radio transmitter and antenna system.

wires in coils and capacitors dissipate some power as heat, but ideal components wouldn't do that. A capacitor, as we've learned, stores energy as an electric field. An inductor stores energy as a magnetic field.

A reactive component causes AC to shift in phase, so that the current does not follow in step with the voltage, as it would in a reactance-free circuit. In a circuit containing inductive reactance, the current lags the voltage by up to 90°, or one-quarter of a cycle. In a circuit with capacitive reactance, the current leads the voltage by up to 90°.

In a purely resistive circuit, the voltage and current precisely follow each other so that they combine in the most efficient possible way (Fig. 17-4A). But in a circuit containing reactance, the voltage and current do not follow exactly along with each other (Fig. 17-4B) because of the phase difference. In that case, the product of the voltage and current (the VA or apparent power) exceeds the actual energy expenditure (the true power).

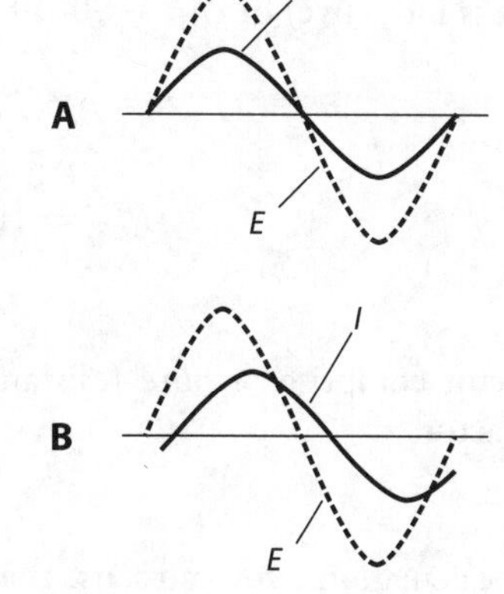

17-4 At A, the current *I* and voltage *E* are in phase in a nonreactive AC circuit. At B, *I* and *E* are not in phase when reactance exists.

Power Parameters

In an AC circuit containing nonzero resistance and nonzero reactance, we can summarize the relationship between true power P_T, apparent (VA) power P_{VA}, and imaginary (reactive) power P_X in terms of the formula

$$P_{VA} = (P_T^2 + P_X^2)^{1/2}$$

where $P_T < P_{VA}$ and $P_X < P_{VA}$. If no reactance exists, then $P_{VA} = P_T$ and $P_X = 0$. Engineers strive to minimize, and if possible eliminate, the reactance in power-transmission systems.

Power Factor

In an AC circuit, we call the ratio of the true power to the VA power, P_T / P_{VA}, the *power factor* (*PF*). In the ideal case in which we have no reactance, $P_T = P_{VA}$, so $PF = 1$. If the circuit contains reactance but lacks resistance or conductance (zero or infinite resistance), then $P_T = 0$, so $PF = 0$.

When a *load*, or a circuit in which we want power to dissipate or change form, contains both resistance and reactance, then *PF* falls between 0 and 1. We can also express the power factor as a percentage $PF_\%$ between 0 and 100. If we know P_T and P_{VA}, then we can calculate *PF* as

$$PF = P_T / P_{VA}$$

and $PF_\%$ as

$$PF_\% = 100 P_T / P_{VA}$$

When a load has nonzero, finite resistance and nonzero, finite reactance, then some of the power dissipates as true power, and some of the power gets "rejected" by the load as imaginary power.

We can determine the power factor in an AC circuit that contains reactance and resistance in two ways:

1. Find the cosine of the phase angle
2. Calculate the ratio of the resistance to the absolute-value impedance

Cosine of Phase Angle

Recall that in a circuit having reactance and resistance, the current and the voltage do not follow along exactly in phase. The phase angle (ϕ) constitutes the extent, expressed in degrees, to which the current and the voltage differ in phase. In a pure resistance, $\phi = 0°$. In a pure reactance, $\phi = +90°$ (if the net reactance is inductive) or $\phi = -90°$ (if the net reactance is capacitive). We can calculate the power factor as

$$PF = \cos \phi$$

and

$$PF_\% = 100 \cos \phi$$

Problem 17-1

Suppose that a circuit comprises a pure resistance of 600 ohms with no reactance whatsoever. What's the power factor?

Solution

Without doing any calculations, you can sense that $PF = 1$ because $P_{VA} = P_T$ in a pure resistance. It follows that $P_T / P_{VA} = 1$. But you can also look at this situation by noting that the phase angle equals 0°

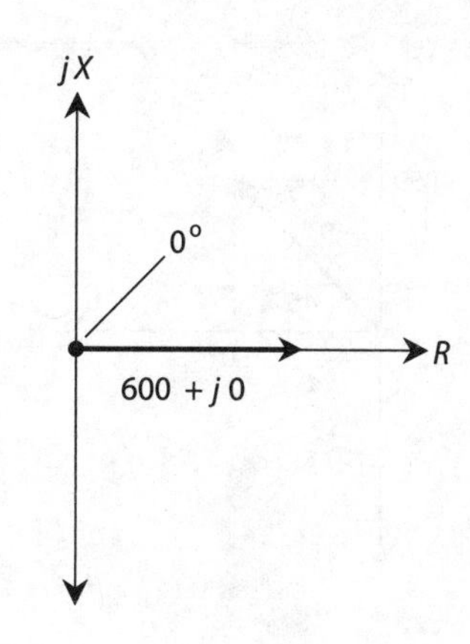

17-5 Vector diagram showing the phase angle for a purely resistive impedance of $600 + j0$. The R and jX scales are relative.

because the current follows in phase with the voltage. Using a calculator, you'll find that $\cos 0° = 1$. Therefore, $PF = 1 = 100\%$. Figure 17-5 illustrates the *RX* half-plane vector for this situation. Remember that you should express the phase angle with respect to the R axis.

Problem 17-2

Suppose that a circuit contains a pure capacitive reactance of -40 ohms, but no resistance. What's the power factor?

Solution

Here, the phase angle equals $-90°$, as shown in the *RX* half-plane vector diagram of Fig. 17-6. A calculator will tell you that $\cos -90° = 0$. Therefore, $PF = 0$, and $P_T/P_{VA} = 0 = 0\%$. None of the power is true; it's all reactive.

Problem 17-3

Consider a circuit that contains a resistance of 50 ohms and an inductive reactance of 50 ohms, connected in series. What's the power factor?

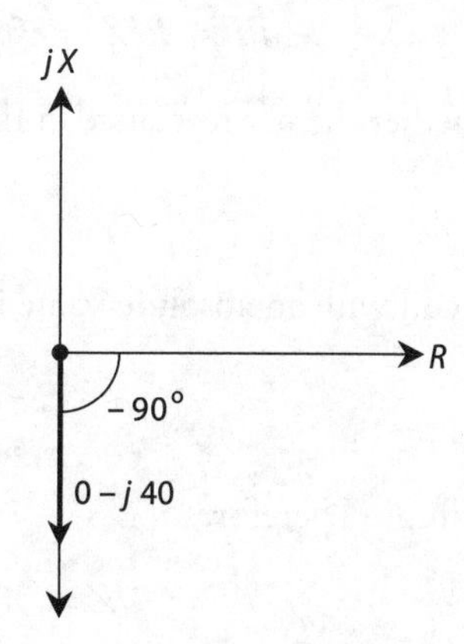

17-6 Vector diagram showing the phase angle for a purely capacitive impedance of $0 - j40$. The R and jX scales are relative.

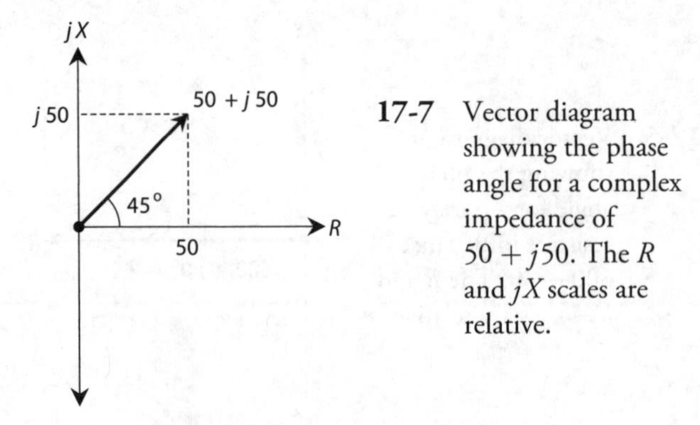

17-7 Vector diagram showing the phase angle for a complex impedance of $50 + j50$. The R and jX scales are relative.

Solution

The phase angle equals 45° (Fig. 17-7). The resistance and the reactance vectors have equal lengths, forming two sides of a right triangle. The complex impedance vector constitutes the *hypotenuse* (longest side) of the right triangle. To determine the power factor, you can use a calculator to find $\cos 45° = 0.707$, so you know that $P_T/P_{VA} = 0.707 = 70.7\%$.

The Ratio *R/Z*

We can calculate the power factor in an *RX* circuit by finding the ratio of the resistance R to the absolute-value impedance Z. Figure 17-7 provides an example. A right triangle is formed by the resistance vector R (the base), the reactance vector jX (the height), and the absolute-value impedance Z (the hypotenuse). The cosine of the phase angle equals the ratio of the base length to the hypotenuse length, or R/Z.

Problem 17-4

Suppose that a series circuit has an absolute-value impedance Z of 100 ohms, with a resistance R of 80 ohms. What's the power factor?

Solution

You can set up and calculate the ratio

$$PF = R/Z = 80/100 = 0.8 = 80\%$$

It doesn't matter whether the net reactance in this circuit happens to be capacitive or inductive.

Problem 17-5

Consider a series circuit with an absolute-value impedance of 50 ohms, purely resistive. What's the power factor?

Solution

Here, $R = Z = 50$ ohms. Therefore

$$PF = R/Z = 50/50 = 1 = 100\ \%$$

Problem 17-6

Consider a circuit with a resistance of 50 ohms and a capacitive reactance of −30 ohms. What's the power factor? Use the cosine method.

Solution

Remember the formula for phase angle in terms of reactance and resistance:

$$\phi = \text{Arctan } (X/R)$$

where X represents the reactance and R represents the resistance. Therefore

$$\phi = \text{Arctan } (-30/50) = \text{Arctan } (-0.60) = -31°$$

The power factor equals the cosine of this angle, so

$$PF = \cos\ (-31°) = 0.86 = 86\%$$

Problem 17-7

Consider a series circuit with a resistance of 30 ohms and an inductive reactance of 40 ohms. What's the power factor? Use the R/Z method.

Solution

First, find the absolute-value impedance using the formula for series circuits:

$$Z = (R^2 + X^2)^{1/2}$$

where R represents the resistance and X represents the net reactance. Plugging in the numbers, you get

$$Z = (30^2 + 40^2)^{1/2} = (900 + 1600)^{1/2} = 2500^{1/2} = 50 \text{ ohms}$$

Now you can calculate the power factor as

$$PF = R/Z = 30/50 = 0.60 = 60\%$$

You can graph this situation as a 30:40:50 right triangle (Fig. 17-8).

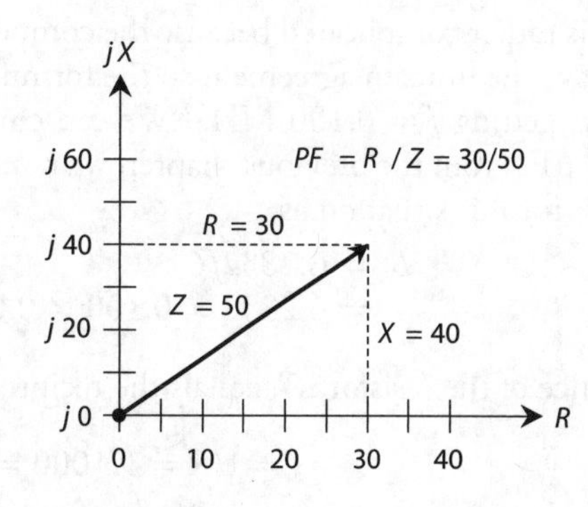

17-8 Illustration for Problem 17-7. The vertical and horizontal scale increments differ; this is a common practice in graphs.

How Much of the Power Is True?

The above formulas allow you to figure out, given the resistance, reactance, and VA power, how many watts constitute true, or real power, and how many watts constitute reactive, or imaginary power. Engineers must consider this situation when working with RF equipment because some RF wattmeters display VA power rather than true power. When reactance exists along with the resistance in a circuit or system, such wattmeters give artificially high readings.

Problem 17-8

Suppose that a circuit has 50 ohms of resistance and 30 ohms of inductive reactance in series. A wattmeter shows 100 W, representing the VA power. What's the true power?

Solution

We must determine the power factor to figure out the answer to this question. First, we calculate the phase angle as

$$\phi = \text{Arctan}\ (X/R) = \text{Arctan}\ (30/50) = 31°$$

The power factor equals the cosine of the phase angle, so

$$PF = \cos 31° = 0.86$$

The formula for the power factor in terms of true and VA power is

$$PF = P_{\text{T}}/P_{\text{VA}}$$

We can rearrange this formula to solve for true power, obtaining

$$P_{\text{T}} = PF \times P_{\text{VA}}$$

When we plug in $PF = 0.86$ and $P_{\text{VA}} = 100$, we get

$$P_{\text{T}} = 0.86 \times 100 = 86\ \text{W}$$

Problem 17-9

Suppose that a circuit has a resistance of 1000 ohms in *parallel* with a capacitance of 1000 pF. We operate the circuit at a frequency of 100 kHz. If a wattmeter designed to read VA power shows 88.0 W, what's the true power?

Solution

This problem is rather complicated because the components appear in parallel. To begin, let's make sure that we have the units in agreement so the formulas work right. We can convert the frequency f to megahertz, getting $f = 0.100$ MHz. We can convert the capacitance to microfarads, getting $C = 0.001000$ μF. From the previous chapter, we remember the formula for capacitive susceptance, and calculate it for this situation as

$$\begin{aligned} B_C &= 6.2832fC \\ &= 6.2832 \times 0.100 \times 0.001000 = 0.00062832\ \text{S} \end{aligned}$$

The conductance of the resistor, G, equals the reciprocal of the resistance, R, so

$$G = 1/R = 1/1000 = 0.001000\ \text{S}$$

Now let's use the formulas for calculating resistance and reactance in terms of conductance and susceptance in parallel circuits. First, we find the resistance as

$$R = G/(G^2 + B^2)$$
$$= 0.001000/(0.001000^2 + 0.00062832^2)$$
$$= 0.001000/0.0000013948 = 716.95 \text{ ohms}$$

Next, we find the reactance as

$$X = -B/(G^2 + B^2)$$
$$= -0.00062832/0.0000013948 = -450.47 \text{ ohms}$$

again rounded to four significant figures. Now we can calculate the phase angle as

$$\phi = \text{Arctan}\ (X/R)$$
$$= \text{Arctan}\ (-450.47/716.95) = \text{Arctan}\ (-0.62831) = -32.142°$$

We calculate the power factor as

$$PF = \cos \phi = \cos (-32.142°) = 0.84673$$

The VA power P_{VA} is given as 88.0 W. Therefore, we can calculate the true power, rounding off to three significant figures (because that's the extent of the accuracy of our input data), getting

$$P_T = PF \times P_{VA} = 0.84673 \times 88.0$$
$$= 74.5 \text{ W}$$

Power Transmission

Consider a radio broadcast or communications station. The transmitter produces high-frequency AC. We want to get the signal efficiently to an antenna located some distance from the transmitter. This process involves the use of an *RF transmission line*, also known as a *feed line*. The most common type is coaxial cable. Alternatively, we can use two-wire line, also called parallel-wire line, in some antenna systems. At ultra-high and microwave frequencies, another kind of transmission line, known as a *waveguide*, is often employed.

Loss: the Less, the Better!

The overriding challenge in the design and construction of any *power transmission* system lies in minimizing the loss. Power wastage occurs almost entirely as heat in the transmission line conductors and dielectric, and in objects near the line. Some loss can take the form of unwanted electromagnetic radiation from the line. Loss also occurs in transformers. Power loss in an electrical system is analogous to the loss of usable work produced by friction in a mechanical system.

In an ideal power-transmission system, all of the power constitutes VA power; that is, all of the power occurs as AC in the conductors and an alternating voltage between them. We don't want power in a transmission line or transformer to exist in the form of true power because that situation translates into heat loss, radiation loss, or both. True power dissipation or radiation should always take place in the load, usually at the opposite end of the transmission line from the source.

Power Measurement in a Transmission Line

In an AC transmission line, we can measure power by placing an AC voltmeter between the conductors and an AC ammeter in series with one of the conductors, as shown in Fig. 17-9. We should place both meters at the same point on the line (even though they don't look that way in this diagram!). In that case the power P (in watts) equals the product of the RMS voltage E (in volts) and the RMS current I (in amperes). We can use this technique in any transmission line at any AC frequency, from the 60 Hz of a utility system to many gigahertz in some wireless communications systems. However, when we measure power this way, we don't necessarily get an accurate indication of the true power dissipated by the load at the end of the line.

Recall that any transmission line has an inherent *characteristic impedance*. The value of this parameter Z depends on the diameters of the line conductors, the spacing between the conductors, and he type of dielectric material that separates the conductors. If the load constitutes a pure resistance R containing no reactance, and if $R = Z_o$, then the power indicated by the voltmeter/ammeter scheme will equal the true power dissipated by the load—provided that we place the voltmeter and ammeter directly across the load, at the end of the line opposite the source.

If the load constitutes a pure resistance that differs from the characteristic impedance of the line (that is, $R \neq Z_o$), then the voltmeter and ammeter will not give an indication of the true power. Also, if the load contains any reactance along with the resistance, the voltmeter/ammeter method will fail to give us an accurate reading of the true power, no matter what the resistance.

Impedance Mismatch

If we want a power transmission system to perform at its best, then the load impedance must constitute a pure resistance equal to the characteristic impedance of the line. When we don't have this ideal state of affairs, the system has an *impedance mismatch*.

Small impedance mismatches can usually—but not always—be tolerated in power transmission systems. In very-high-frequency (VHF), ultra-high-frequency (UHF) and microwave wireless transmitting systems, even a small impedance mismatch between the load (antenna) and the line can cause excessive power losses in the line.

We can usually get rid of an impedance mismatch by installing a *matching transformer* between the transmission line and the load. We can also correct impedance mismatches in some situations by deliberately placing a reactive component (inductor or capacitor) in series or parallel with the load to cancel out any existing load reactance.

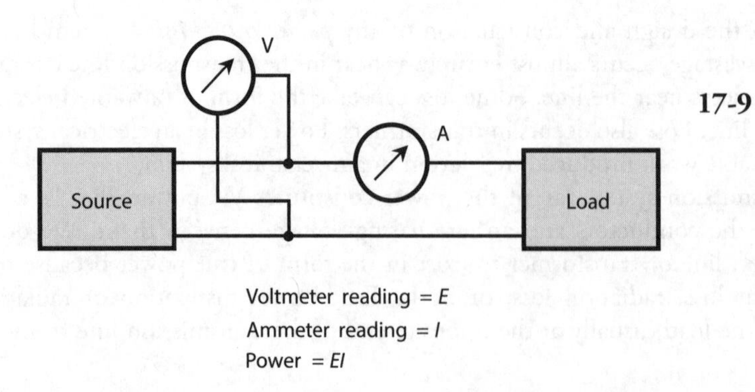

17-9 Power measurement in a transmission line. Ideally, we should measure the voltage and the current at the same physical point on the line.

Loss in a Mismatched Line

When we terminate a transmission line in a pure resistance R equal to the characteristic impedance Z_o of the line, then the RMS current I and the RMS voltage E remain constant all along the line, provided that the line has no ohmic loss and no dielectric loss. In such a situation, we have

$$R = Z_o = E/I$$

where we express R and Z_o in ohms, E in volts RMS, and I in amperes RMS. Of course, no transmission line is perfectly *lossless*. In a "real-world" transmission line, the current and voltage gradually decrease as a signal makes its way from the source to the load. Nevertheless, if the load constitutes a pure resistance equal to the characteristic impedance of the line, the current and voltage remain *in the same ratio* at all points along the line, as shown in Fig. 17-10.

Standing Waves

If a transmission line and its load aren't perfectly matched, then the current and voltage alternately rise and fall as they move along the line. We call the maxima and minima *loops* and *nodes*, respectively. At a maximum-current point (a *current loop*), the voltage reaches its minimum (a *voltage node*). Conversely, at a maximum-voltage point (a *voltage loop*), the current attains its minimum (a *current node*).

If we graph the current and voltage loops and nodes along a mismatched transmission line as functions of the position on the line, we see wavelike patterns that remain fixed over time. These orderly patterns of current and voltage don't move in either direction along the line; they simply "stand there." For this reason, engineers call them *standing waves*.

Losses Caused by Standing Waves

When a transmission line contains standing waves, we observe a corresponding pattern in the extent of the line loss, as follows:

- At current loops, the loss in the line conductors reaches a maximum
- At current nodes, the loss in the line conductors reaches a minimum
- At voltage loops, the loss in the line dielectric reaches a maximum
- At voltage nodes, the loss in the line dielectric reaches a minimum

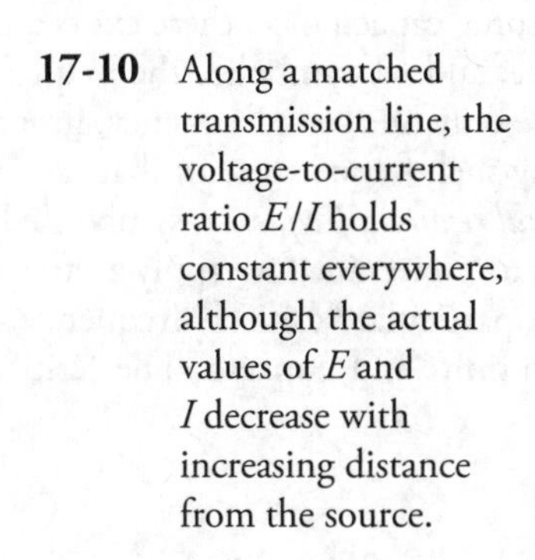

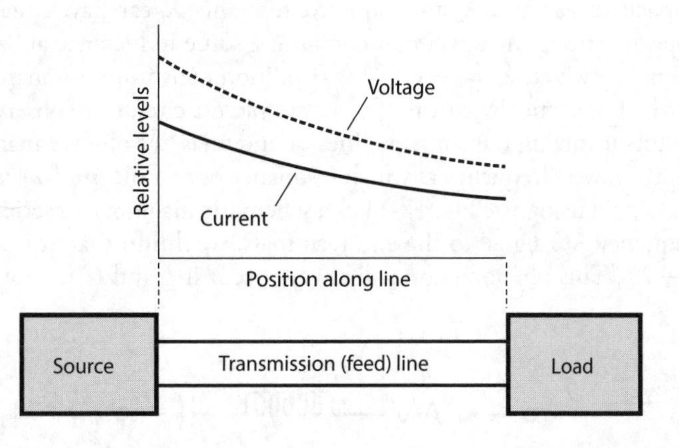

17-10 Along a matched transmission line, the voltage-to-current ratio E/I holds constant everywhere, although the actual values of E and I decrease with increasing distance from the source.

It's tempting to suppose that all of the loss variations ought to average out in a mismatched line, so that the excess loss in some places gets "paid back" in the form of reduced loss in other places. But things don't work that way. Overall, the losses in a mismatched line always exceed the losses in a perfectly matched line. The extra loss increases as the mismatch gets worse. We call it *transmission-line mismatch loss* or *standing-wave loss*; it occurs as heat dissipation, so it constitutes true power. Any true power that heats up a transmission line goes to waste because it never reaches the load. We want the true power to end up in the load—if not all of it, then as much of it as possible.

As the extent of the mismatch in a power-transmission system increases, so does the loss caused by the current and voltage loops in the standing waves. The more loss a line has when perfectly matched, the more loss a given amount of mismatch will cause. Standing-wave loss also increases as the frequency increases, if we hold all other factors constant.

Line Overheating

A severe mismatch between the load and the transmission line can cause another problem besides lost power: physical damage to, or destruction of, the line.

Suppose that a certain transmission line can effectively function with a radio transmitter that generates up to 1 kW of power, assuming that the line and the load are perfectly matched. If a severe mismatch exists and you try to feed 1 kW into the same line, the extra current at the current loops can heat the conductors to the point at which the dielectric material melts and the line shorts out. It's also possible for the voltage at the voltage loops to cause arcing between the line conductors. This arcing can perforate and/or burn the dielectric, ruining the line.

When we have no choice but to operate an RF transmission line with a significant impedance mismatch, we must refer to *derating functions* in order to determine how much power the line can safely handle. Manufacturers of prefabricated lines, such as coaxial cable, can (or should) provide this information.

Resonance

We observe resonance in an AC circuit when capacitive and inductive reactance both exist, but they have equal and opposite values so that they cancel each other out. We saw a couple of examples of this phenomenon in Chap. 16. Let's explore it in more detail.

Series Resonance

Capacitive reactance X_C and inductive reactance X_L can have equal magnitudes although they produce opposite effects. In any circuit containing some inductance and some capacitance, there exists a frequency at which $X_L = -X_C$. This condition constitutes resonance, and we symbolize the frequency at which it occurs by writing f_o. In a simple *LC* circuit, we observe only one such frequency; in some circuits involving transmission lines or antennas, we observe many such frequencies. In that case, we call the lowest frequency at which resonance occurs the *fundamental resonant frequency*, symbolized f_o.

You'll recognize Fig. 17-11 as a schematic diagram of a series *RLC* circuit. If we apply a variable-frequency AC signal to the end terminals, we'll find that at one particular "critical" frequency, $X_L = -X_C$. This phenomenon will always occur if *L* and *C* are both finite and nonzero. The "critical"

R
C
L

17-11 A series *RLC* circuit.

frequency represents f_o for the circuit. At f_o, the effects of capacitive reactance and inductive reactance cancel out so that the circuit appears as a pure resistance, with a value theoretically equal to R. We call this condition *series resonance.*

If $R = 0$ (a short circuit), then Fig. 17-11 represents a *series LC circuit*, and the impedance at resonance theoretically equals $0 + j0$. Under these conditions, the circuit offers no opposition to the flow of alternating current at f_o. Of course, "real-world" series *LC* circuits always contain at least a small amount of resistance so that a little loss occurs in the coil and capacitor; the real-number part of the complex impedance doesn't *exactly* equal 0. If the coil and capacitor have high quality and exhibit minimal loss, however, we can usually say that $R = 0$ "for all intents and purposes."

Parallel Resonance

Figure 17-12 shows a generic parallel *RLC* circuit. In this situation, we can think of the resistance R as a conductance G, with $G = 1/R$. For that reason, some people might say that we should call this arrangement a parallel *GLC* circuit. We can get away with either term as long as we include the word "parallel"!

At one particular frequency f_o, the inductive susceptance B_L will exactly cancel the capacitive susceptance B_C, so we have $B_L = -B_C$. This condition always occurs for some applied AC signal frequency f_o, as long as the circuit contains finite, nonzero inductance and finite, nonzero capacitance. At f_o, the susceptances cancel each other out, leaving theoretically zero susceptance. The admittance through the circuit theoretically equals the conductance G of the resistor. We call this condition *parallel resonance.*

If the circuit contains no resistor, but only a coil and capacitor, we have a *parallel LC circuit*, and the admittance at resonance theoretically equals $0 + j0$. The circuit will offer a lot of opposition to alternating current at f_o, and the complex impedance will, in an informal sense, equal "infinity" (∞). In a practical "real-world" *LC* circuit, the coil and capacitor always have a little bit of loss, so the real-number part of the complex impedance isn't really infinite. However, if we use low-loss components, we can get real-number coefficients of many megohms or even gigohms for parallel *LC* circuits at resonance, so that we can say that $R = \infty$ "for all intents and purposes."

Calculating the Resonant Frequency

We can calculate the resonant frequency f_o, of a series *RLC* or parallel *RLC* circuit in terms of the inductance L in henrys and the capacitance C in farads, using the formula

$$f_o = 1 / [2\pi(LC)^{1/2}]$$

Considering $\pi = 3.1416$, we can simplify this formula to

$$f_o = 0.15915 / (LC)^{1/2}$$

You might remember from your basic algebra or precalculus course that the ½ power of a quantity represents the positive square root of that quantity. The foregoing formula will also work if you want to find f_o in megahertz (MHz) when you know L in microhenrys (μH) and C in microfarads (μF).

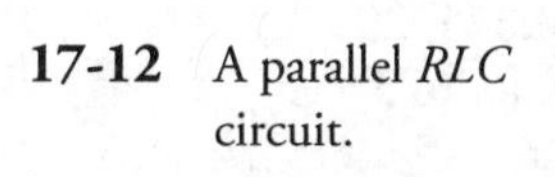

17-12 A parallel *RLC* circuit.

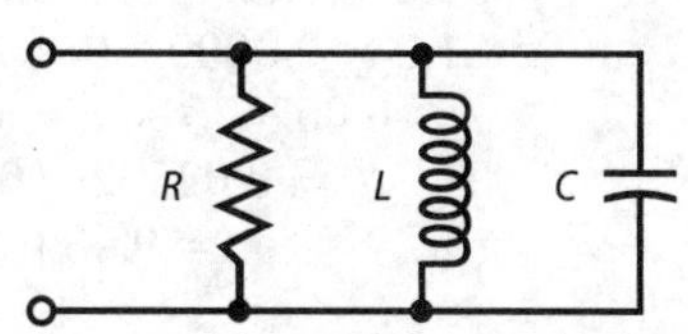

The Effects of R and G

Interestingly, the value of R or G does not affect the value of f_o in series or parallel *RLC* circuits. However, the presence of nonzero resistance in a series resonant circuit, or nonzero conductance in a parallel resonant circuit, makes f_o less well-defined than it would be if the resistance or conductance did not exist. If we short out R in the circuit of Fig. 17-11 or remove R from the circuit of Fig. 17-12, we have *LC* circuits that exhibit the most well-defined possible *resonant responses.*

In a series *RLC* circuit, the resonant frequency response becomes more "broad" as the resistance increases, and more "sharp" as the resistance decreases. In a parallel *RLC* circuit, the resonant frequency response becomes more "broad" as the *conductance* increases (R gets smaller), and more "sharp" as the conductance decreases (R gets larger). In theory, the "sharpest" possible responses occur when $R = 0$ in a series circuit, and when $G = 0$ (that is, $R = \infty$) in a parallel circuit.

Problem 17-10

Find the resonant frequency of a series circuit with a 100-μH inductor and a 100-pF capacitor.

Solution

We should convert the capacitance to 0.000100 μF. Then we can find the product $LC = 100 \times 0.000100 = 0.0100$. When we take the square root of this, we get 0.100. Finally, we can divide 0.15915 by 0.100, getting $f_o = 1.5915$ MHz. We should round this off to 1.59 MHz.

Problem 17-11

Find the resonant frequency of a parallel circuit consisting of a 33-μH coil and a 47-pF capacitor.

Solution

Let's convert the capacitance to 0.000047 μF. Then we find the product $LC = 33 \times 0.000047 = 0.001551$. Taking the square root of this, we get 0.0393827. Finally, we divide 0.15915 by 0.0393827 and round off, getting $f_o = 4.04$ MHz.

Problem 17-12

Suppose that we want to design a circuit so that it exhibits $f_o = 9.00$ MHz. We have a 33.0-pF fixed capacitor available. What size coil will we need to obtain the desired resonant frequency?

Solution

Let's use the formula for the resonant frequency and plug in the values. Then we can use arithmetic to solve for L. We convert the capacitance to 0.0000330 μF and calculate in steps as follows:

$$\begin{aligned}
f_o &= 0.15915 / (LC)^{1/2} \\
9.00 &= 0.15915 / (L \times 0.0000330)^{1/2} \\
9.00^2 &= 0.15915^2 / (0.0000330 \times L) \\
81.0 &= 0.025329 / (0.0000330 \times L) \\
81.0 \times 0.0000330 \times L &= 0.025329 \\
0.002673 \times L &= 0.025329 \\
L &= 0.025329 / 0.002673 \\
&= 9.48\ \mu H
\end{aligned}$$

Problem 17-13

Suppose that we want to design an *LC* circuit with $f_o = 455$ kHz. We have a 100-μH in our "junk box." What size capacitor do we need?

Solution

We should convert the frequency to 0.455 MHz. Then the calculation proceeds in the same way as with the preceding problem:

$$f_o = 0.15915 / (LC)^{1/2}$$
$$0.455 = 0.15915 / (100 \times C)^{1/2}$$
$$0.455^2 = 0.15915^2 / (100 \times C)$$
$$0.207025 = 0.025329 / (100 \times C)$$
$$0.207025 \times 100 \times C = 0.025329$$
$$20.7025 \times C = 0.025329$$
$$C = 0.025329 / 20.7025$$
$$= 0.00122\ \mu\text{F}$$

Adjusting the Resonant Frequency

In practical circuits, engineers often place variable inductors and/or variable capacitors in series or parallel *LC* circuits designed to function at resonance, thereby allowing for small errors in the actual resonant frequency (as opposed to the calculated value for f_o). We can design a circuit of this sort, called a *tuned circuit,* so that it exhibits a frequency slightly higher than f_o, and then install a *padding capacitance* (C_p) in parallel with the main capacitance *C*, as shown in Fig. 17-13. A padding capacitor is a small component with an adjustable value ranging from around 1 pF up to several picofarads, or several tens of picofarads. If the engineers want the circuit to offer a wider range of resonant frequencies, a variable capacitor, having a value ranging from a few picofarads up to several hundred picofarads, can serve the purpose.

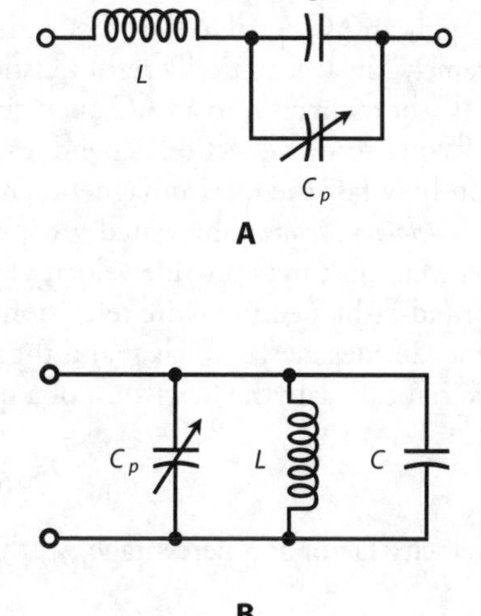

17-13 Padding capacitors (C_p) allow limited adjustment of the resonant frequency in a series *LC* circuit (as shown at A), or in a parallel *LC* circuit (as shown at B).

Resonant Devices

Resonant circuits often consist of coils and capacitors in series or parallel, but other kinds of hardware also exhibit resonance.

Piezoelectric Crystals

Pieces of the mineral *quartz*, when cut into thin wafers and subjected to voltages, will vibrate at high frequencies. Because of the physical dimensions of such a *piezoelectric crystal*, these vibrations occur at a precise frequency f_o, and also at whole-number multiples of f_o. We call these multiples—$2f_o$, $3f_o$, $4f_o$, and so on—*harmonic frequencies* or simply *harmonics*. The frequency f_o constitutes the *fundamental frequency* or simply the *fundamental*. The fundamental, f_o, is the lowest frequency at which resonance occurs. Quartz crystals can act like *LC* circuits in electronic devices. A crystal exhibits an impedance that varies with frequency. The reactance equals zero at f_o and the harmonic frequencies.

Cavities

Lengths of metal tubing, cut to specific dimensions, exhibit resonance at very-high, ultra-high, and microwave radio frequencies. They work in much the same way as musical instruments resonate with sound waves. However, the waves take the form of electromagnetic fields rather than acoustic disturbances. Such *cavities*, also called *cavity resonators*, have reasonable physical dimensions at frequencies above about 150 MHz. We can get a cavity to work below this frequency, but we'll find it difficult to construct because of its great length and clumsiness. Like crystals, cavities resonate at a specific fundamental frequency f_o, and also at all of the harmonic frequencies.

Sections of Transmission Line

When we cut a transmission line to any whole-number multiple of ¼ wavelength, it behaves as a resonant circuit. The most common length for a *transmission-line resonator* is ¼ wavelength, giving us a so-called *quarter-wave section*.

When we short-circuit a quarter-wave section at one end and apply an AC signal to the other end, the section acts like a parallel-resonant *LC* circuit, and it has an extremely high (theoretically infinite) resistive impedance at the fundamental resonant frequency f_o. When we leave one end of a quarter-wave section open and apply an AC signal to the other end, the section acts like a series-resonant *LC* circuit, and it has an extremely low (theoretically zero) resistive impedance at f_o. In effect, a quarter-wave section converts an AC short circuit into an AC open circuit and vice-versa—at a specific frequency f_o.

The length of a quarter-wave section depends on the desired fundamental resonant frequency f_o. It also depends on how fast the electromagnetic energy travels along the line. We can define this speed in terms of a *velocity factor*, abbreviated v, expressed as a fraction or percentage of the speed of light in free space. Manufacturers provide velocity factors for prefabricated transmission lines, such as coaxial cable or old-fashioned two-wire television "ribbon."

If the frequency in megahertz equals f_o and the velocity factor of a line equals v (expressed as a fraction), then we can calculate the length L_{ft} of a quarter-wave section of transmission line in feet using the formula

$$L_{ft} = 246v / f_o$$

If we know the velocity factor as a percentage $v_{\%}$, then the above formula becomes

$$L_{ft} = 2.46v_{\%} / f_o$$

If we know the velocity factor as a fraction v, then the length L_m of a quarter-wave section in meters is

$$L_m = 75.0v / f_o$$

For the velocity factor as a percentage $v_{\%}$, we have

$$L_m = 0.750v_{\%} / f_o$$

Note that we use L to stand for "length," not "inductance," in this context!

Antennas

Many types of antennas exhibit resonant properties. The simplest type of resonant antenna, and the only kind that we'll consider here, is the center-fed, half-wavelength *dipole antenna* (Fig. 17-14).

We can calculate the approximate length L_{ft}, in feet, for a dipole antenna at a frequency of f_o using the formula

$$L_{ft} = 467 / f_o$$

taking into account the fact that electromagnetic fields travel along a wire at about 95% of the speed of light. A straight, thin wire in free space, therefore, has a velocity factor of approximately $v = 0.95$. If we specify the approximate length of the half-wave dipole in meters as L_m, then

$$L_m = 143 / f_o$$

A half-wave dipole has a purely resistive impedance of about 73 ohms at its fundamental frequency f_o. But this type of antenna also resonates at harmonics of f_o. The dipole measures a full wavelength from end to end at $2f_o$; it measures 3/2 wavelength from end to end at $3f_o$; it measures two full wavelengths from end to end at $4f_o$, and so on.

Radiation Resistance

At f_o and all of the *odd-numbered harmonics,* a dipole antenna behaves like a series resonant *RLC* circuit with a fairly low resistance. At all *even-numbered* harmonics, the antenna acts like a parallel resonant *RLC* circuit with a high resistance.

Does the mention of resistance in a half-wave dipole antenna confuse you? Maybe it should! Figure 17-14 shows no resistor. Where, you ask, does the resistance in a half-wave dipole come from? The answer gets into some rather esoteric electromagnetic-wave theory, and it brings to light an interesting property that all antennas have: *radiation resistance.* This parameter constitutes a crucial factor in the design and construction of all RF antenna systems.

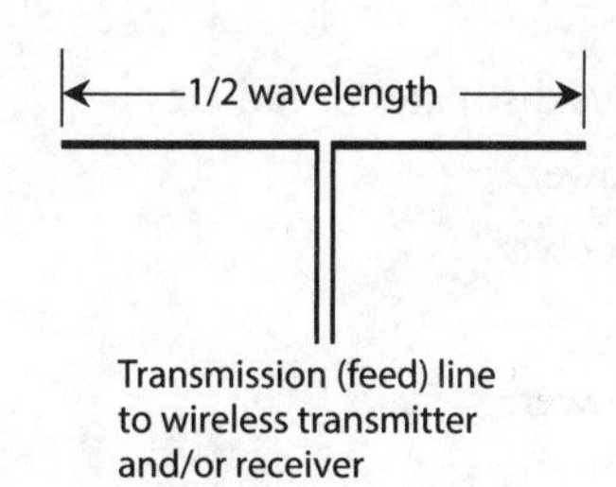

17-14 The half-wave, center-fed dipole constitutes a simple and efficient antenna.

When we connect a radio transmitter to an antenna and send out a signal, energy radiates into space in the form of radio waves from the antenna. Although no physical resistor exists anywhere in the antenna system, the radiation of radio waves acts just like power dissipation in a pure resistance. In fact, if we replace a half-wave dipole antenna with a 73-ohm non-reactive resistor that can safely dissipate enough power, we'll discover that a wireless transmitter connected to the opposite end of the line won't "know" the difference between that resistor and the dipole antenna.

Problem 17-14

How many feet long, to the nearest foot, is a quarter-wave section of transmission line at 7.1 MHz if the velocity factor equals 80%?

Solution

We can use the formula for the length L_{ft} of a quarter-wave section based on the velocity factor $v_{\%} = 80$, as follows:

$$\begin{aligned} L_{ft} &= 2.46\ v_{\%} / f_o \\ &= (2.46 \times 80) / 7.1 \\ &= 28 \text{ ft} \end{aligned}$$

Quiz

Refer to the text in this chapter if necessary. A good score is 18 or more correct. Answers are in the back of the book.

1. A transmission line operates at its best efficiency when
   (a) the load impedance constitutes a pure resistance equal to the characteristic impedance of the line.
   (b) the load impedance constitutes a pure inductive reactance equal to the characteristic impedance of the line.
   (c) the load impedance constitutes a pure capacitive reactance equal to the characteristic impedance of the line.
   (d) the absolute-value impedance of the load equals the characteristic impedance of the line.

2. The ninth harmonic of 900 kHz is
   (a) 100 kHz.
   (b) 300 kHz.
   (c) 1.20 MHz.
   (d) 8.10 MHz.

3. A pure resistance dissipates or radiates
   (a) complex power.
   (b) imaginary power.
   (c) true power.
   (d) apparent power.

4. Suppose that we want to build a ½-wave dipole antenna designed to have a fundamental resonant frequency of 14.3 MHz. How long should we make the antenna, as measured from end to end in meters?
   (a) 32.7 m
   (b) 10.0 m
   (c) 16.4 m
   (d) We need more information to answer this question.

5. When a transmission line exhibits standing waves, we find a voltage maximum
   (a) wherever we find a current maximum.
   (b) wherever we find a current minimum.
   (c) at the transmitter end of the line.
   (d) at the load end of the line.

6. Standing waves on a transmission line (as compared with a line operating without any impedance mismatch) increase the loss in the wire conductors at the
   (a) current maxima.
   (b) voltage maxima.
   (c) current minima and voltage maxima.
   (d) current minima and voltage minima.

7. When we take the cosine of the phase angle in an AC circuit or system that contains both resistance and reactance, we get the
   (a) true power.
   (b) imaginary power.
   (c) apparent power.
   (d) power factor.

8. Which of the following parameters is an example of true power in an AC circuit or system?
   (a) The AC that appears between the plates of a capacitor
   (b) The AC that passes through a wire inductor
   (c) The AC that dissipates as heat in a transmission line
   (d) The AC that travels along a transmission line

9. Suppose that the apparent power in a circuit equals 40 W and the true power equals 30 W. What's the power factor?
   (a) 60%
   (b) 75%
   (c) 80%
   (d) We need more information to calculate it.

10. Suppose that the true power in a circuit equals 40 W and the imaginary power equals 30 W. What's the power factor?
    (a) 60%
    (b) 75%
    (c) 80%
    (d) We need more information to calculate it.

11. Consider a series circuit with a resistance of 24 ohms and an inductive reactance of 10 ohms. What's the power factor?
    (a) 42%
    (b) 58%
    (c) 92%
    (d) 18%

12. Imagine that you encounter a series circuit with a resistance of 24 ohms and a capacitive reactance of −10 ohms. What's the power factor?
    (a) 42%
    (b) 58%
    (c) 92%
    (d) 18%

13. Suppose that a circuit has 24 ohms of resistance and 10 ohms of inductive reactance in series. A meter shows 100 W, representing the VA power. What's the true power?
    (a) 18 W
    (b) 34 W
    (c) 85 W
    (d) 92 W

14. Suppose that the true power equals 100 W in a circuit that consists of a resistance of 60.0 ohms in series with an inductive reactance of 80.0 ohms. What's the VA power?
    (a) 167 W
    (b) 129 W
    (c) 60.0 W
    (d) 36.0 W

15. Suppose that the true power equals 100 W in a circuit that consists of a resistance of 80.0 ohms in series with an inductive reactance of 60.0 ohms. (The resistance and reactance numbers here are transposed from the values in Question 14.) What's the VA power?
    (a) 64.0 W
    (b) 80.0 W
    (c) 125 W
    (d) 156 W

16. Suppose that we connect a coil and capacitor in series. The inductance is 36 μH and the capacitance is 0.0010 μF. What's the resonant frequency?
    (a) 36 kHz
    (b) 0.84 MHz
    (c) 2.4 MHz
    (d) 6.0 MHz

17. What will happen to the resonant frequency of the circuit described in Question 16 if we connect a 100-ohm resistor in series with the existing coil and capacitor?
    (a) It will increase.
    (b) It will stay the same.
    (c) It will decrease.
    (d) We need more information to answer this question.

18. Suppose that we connect a coil and capacitor in parallel, with $L = 75$ μH and $C = 150$ pF. What's f_o?
    (a) 1.5 MHz
    (b) 2.2 MHz
    (c) 880 kHz
    (d) 440 kHz

19. What will happen to the resonant frequency of the circuit described in Question 18 if we connect a 22-pf capacitor in parallel with the existing coil and capacitor?
    (a) It will increase.
    (b) It will stay the same.
    (c) It will decrease.
    (d) We need more information to answer this question.

20. We want to cut a ¼-wave section of transmission line for use at 18.1 MHz. The line has a velocity factor of 0.667. How long should we make the section?
    (a) 9.05 m
    (b) 3.62 m
    (c) 3.00 m
    (d) 2.76 m

18
CHAPTER

Transformers and Impedance Matching

WE CAN USE A TRANSFORMER TO OBTAIN THE OPTIMUM VOLTAGE FOR A CIRCUIT, DEVICE, OR SYSTEM. Transformers have various uses in electricity and electronics. For example, they can:

- Match the impedances between a circuit and a load
- Match the impedances between two different circuits or devices
- Provide DC isolation between circuits or devices while letting AC pass
- Make balanced and unbalanced circuits, feed systems, and loads compatible

Principle of the Transformer

When we place two wires near and parallel to each other and then drive a fluctuating current through one of them, a fluctuating current appears in the other, even though no direct physical connection exists between them. We call this effect *electromagnetic induction.* All AC transformers work according to this principle.

Induced Current and Coupling

If one wire carries sine-wave AC of a certain frequency, then the *induced current* shows up as sine-wave AC of the same frequency in the other wire. As we reduce the spacing between the two wires, keeping them straight and parallel at all times, the induced current increases for a given current in the first wire. If we wind the wires into coils (making certain that the wires are insulated or enameled so that they can't "short out" between the coil turns or to anything else nearby) and place the coils along a common axis, as shown in Fig. 18-1, we observe more induced current than we do if the same wires run straight and parallel. We say that the *coupling* improves when we coil the wires up and place them along a common axis. We can improve the coupling (efficiency of induced-current transfer) still more if we wind one coil directly over the other.

Primary and Secondary

A transformer comprises two coils of insulated or enameled wire, along with the *core*, or *form*, on which we wind them. We call the first coil, through which we deliberately drive current, the *primary*

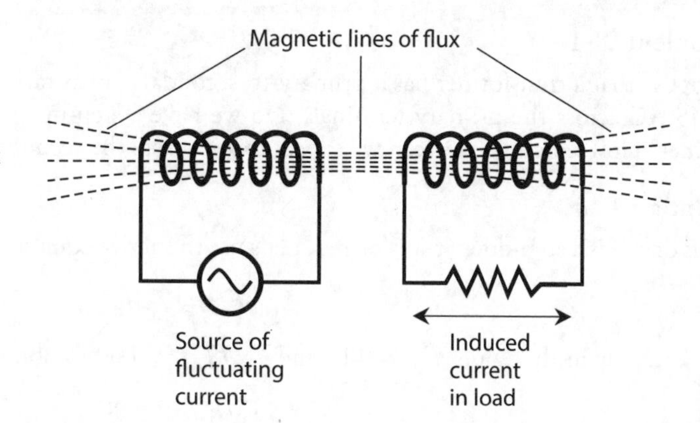

18-1 Magnetic lines of flux between two aligned coils, when one coil carries fluctuating or alternating current.

winding. We call the second coil, in which the induced current appears, the *secondary winding*. Engineers and technicians usually call them simply the *primary* and the *secondary*.

When we apply AC to a primary winding, the coil currents are attended by potential differences between the coil ends, constituting the *primary voltage* and *secondary voltage*. In a *step-down transformer*, the primary voltage exceeds the secondary voltage. In a *step-up transformer*, the secondary voltage exceeds the primary voltage. Let's abbreviate the primary voltage as E_{pri} and the secondary voltage as E_{sec}. Unless otherwise stated, we specify effective (RMS) voltages when talking about the AC in a transformer.

The windings of a transformer exhibit inductance because they're coils. The optimum inductance values for the primary and secondary depend on the frequency of operation, and also on the resistive components of the impedances of the circuits to which we connect the windings. As the frequency increases and the resistive component of the impedance remains constant, the optimum inductance decreases. If the resistive component of the impedance increases and the frequency remains constant, the optimum inductance increases.

Turns Ratio

We define the *primary-to-secondary turns ratio* in a transformer as the ratio of the number of turns in the primary T_{pri} to the number of turns in the secondary T_{sec}. We can denote this ratio as $T_{pri}:T_{sec}$ or T_{pri}/T_{sec}. In a transformer with optimum primary-to-secondary coupling (Fig. 18-2), we always find that

$$E_{pri}/E_{sec} = T_{pri}/T_{sec}$$

Stated in words, the primary-to-secondary voltage ratio equals the primary-to-secondary turns ratio.

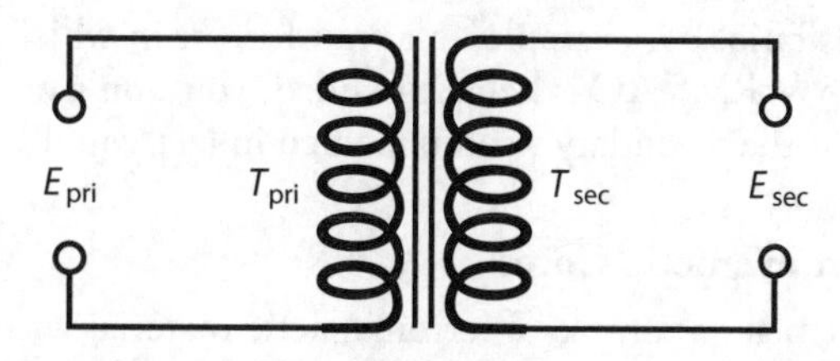

18-2 The primary voltage (E_{pri}) and secondary voltage (E_{sec}) in a transformer depend on the number of turns in the primary winding (T_{pri}) versus the number of turns in the secondary winding (T_{sec}).

Problem 18-1

Suppose that a transformer has a primary-to-secondary turns ratio of exactly 9:1. We apply 117 V RMS AC across the primary terminals. Do we have a step-up transformer or a step-down transformer? How much voltage can we expect to see across the secondary?

Solution

This device is a step-down transformer. Let's use the above equation and solve for E_{sec}. We start with

$$E_{pri} / E_{sec} = T_{pri} / T_{sec}$$

We can plug in the values $E_{pri} = 117$ and $T_{pri} / T_{sec} = 9.00$ to obtain

$$117 / E_{sec} = 9.00$$

Using a little bit of algebra, we can solve this equation to obtain

$$E_{sec} = 117 / 9.00 = 13.0 \text{ V RMS}$$

Problem 18-2

Consider a transformer with a primary-to-secondary turns ratio of exactly 1:9. The voltage across the primary equals 121.4 V RMS. Do we have a step-up transformer or a step-down transformer? What's the voltage across the secondary?

Solution

In this case, we have a step-up transformer. As before, we can plug in the numbers and solve for E_{sec}. We start with

$$E_{pri} / E_{sec} = T_{pri} / T_{sec}$$

Inputting $E_{pri} = 121.4$ and $T_{pri} / T_{sec} = 1/9.000$, we get

$$121.4 / E_{sec} = 1/9.000$$

which solves to

$$E_{sec} = 9.000 \times 121.4 = 1093 \text{ V RMS}$$

Which Ratio Is Which?

Sometimes, when you read the specifications for a transformer, the manufacturer will quote the *secondary-to-primary turns ratio* rather than the primary-to-secondary turns ratio. You can denote the secondary-to-primary turns ratio as T_{sec} / T_{pri}. In a step-down transformer, T_{sec} / T_{pri} is less than 1, but in a step-up unit, T_{sec} / T_{pri} is greater than 1.

When you hear someone say that a transformer has a certain "turns ratio" (say 10:1), you'd better make sure that you know which ratio they're talking about. Do they mean T_{pri} / T_{sec} or T_{sec} / T_{pri}? If you get it wrong, you'll miscalculate the secondary voltage by a factor equal to the *square* of the turns ratio! For example, in a transformer in which $T_{sec} / T_{pri} = 10:1$ and to which an input voltage of 25 V RMS AC is being provided, you don't want to think that you'll see only 2.5 V RMS AC across the secondary winding, when in fact you'll have to contend with 250 V RMS AC.

Ferromagnetic Cores

If we place a sample of ferromagnetic material within a pair of coils composing a transformer, the extent of coupling increases beyond what we get with an air core. However, some energy is invari-

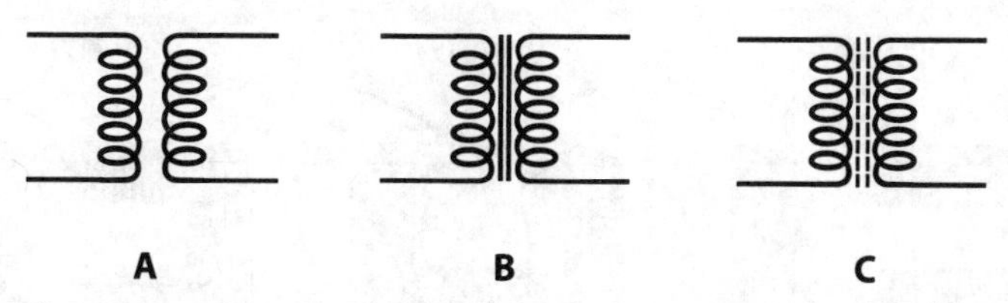

18-3 Schematic symbols for transformers. At A, air core. At B, laminated-iron core. At C, powdered-iron core.

ably lost as heat in a ferromagnetic transformer core. Also, ferromagnetic cores limit the maximum frequency at which a transformer will work efficiently.

The schematic symbol for an air-core transformer looks like two inductor symbols placed back-to-back, as shown in Fig. 18-3A. If the transformer has a *laminated-iron* (layered-iron) core, we add two parallel lines to the schematic symbol (Fig. 18-3B). If the core consists of *powdered iron*, we break up the two parallel lines (Fig. 18-3C).

In transformers intended for use with 60-Hz utility AC, and also for low audio-frequency (AF) applications, sheets of an alloy called *silicon steel*, glued together in layers, are often employed as transformer cores. The silicon steel goes by the nickname *transformer iron*. The layering breaks up the electrical currents that tend to circulate in solid-iron cores. These so-called *eddy currents* flow in loops, serving no useful purpose, but they heat up the core, thereby wasting energy that we could otherwise obtain from the secondary winding. We can "choke off" eddy currents by breaking up the core into numerous thin, flat layers with insulation between them.

Transformer Loss

A rather esoteric form of loss, called *hysteresis loss*, occurs in all ferromagnetic transformer cores, but especially in laminated iron. Hysteresis is the tendency for a core material to act "sluggish" in accepting a fluctuating magnetic field. Air cores essentially never exhibit this type of loss. In fact, air has the lowest overall loss of any known transformer core material. Laminated cores exhibit high hysteresis loss above the AF range, so they don't work well above a few kilohertz.

At frequencies up to several tens of megahertz, powdered iron can serve as an efficient RF transformer core material. It has high magnetic permeability and concentrates the flux considerably. High-permeability cores minimize the number of turns needed in the coils, thereby minimizing the ohmic (resistive) loss that can take place in the wires.

At the highest radio frequencies (more than a couple of hundred megahertz), air represents the best overall choice as a transformer core material because of its low loss and low permeability.

Transformer Geometry

The properties of a transformer depend on the shape of its core, and on the way in which the wires surround the core. In electricity and electronics practice, you'll encounter several different types of *transformer geometry*.

E Core

The *E core* gets its name from the fact that it has the shape of a capital letter E. A bar, placed at the open end of the E, completes the core assembly (Fig. 18-4A). We can wind a primary and secondary on an E core in either of two ways.

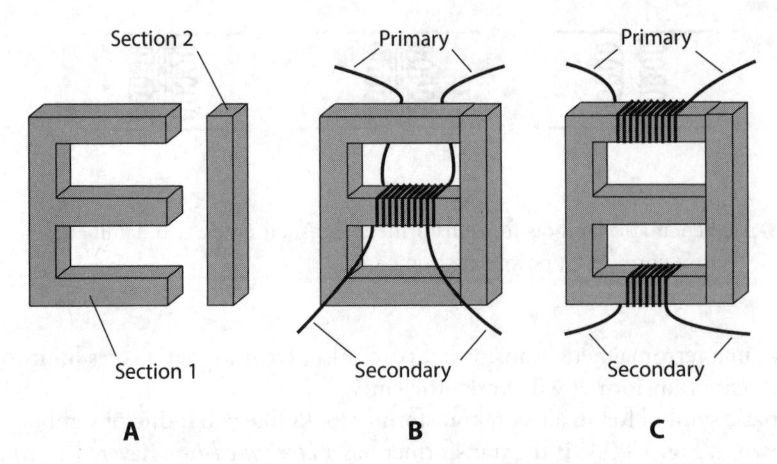

18-4 At A, a utility transformer E core, showing both sections. At B, the shell winding method. At C, the core winding method.

The simplest winding method involves winding both the primary coil and the secondary coil around the middle bar of the E, as shown in Fig. 18-4B. We call this scheme the *shell method* of transformer winding. It provides maximum coupling, but it also results in considerable capacitance between the primary and the secondary. Such *inter-winding capacitance* can sometimes be tolerated, but often it cannot. Another disadvantage of the shell geometry is the fact that, when we wind coils one on top of the other, the transformer can't handle very much voltage. High voltages cause *arcing* (sparking attended by unwanted current) between the windings, which can destroy the insulation on the wires, lead to permanent short circuits, and even set the transformer on fire.

If we don't want to use the shell method, we can employ the *core method* of transformer winding. In this scheme, we place one winding at the bottom of the E section, and the other winding at the top as shown in Fig. 18-4C. The coupling occurs by means of magnetic flux in the core. The core method results in lower inter-winding capacitance than we observe in a shell-wound transformer designed for the same voltage-transfer ratio because the windings are located physically farther apart. A core-wound transformer can handle higher voltages than a shell-wound transformer of the same physical size. Sometimes the center part of the E is left out of the core, in which case we have a transformer with an *O core* or a *D core.*

Shell-wound and core-wound transformers are commonly used at 60 Hz in electrical and electronic appliances and devices of all kinds. We can also find transformers of this sort in some older AF systems.

Solenoidal Core

A pair of cylindrical coils, wound around a rod-shaped piece of powdered iron, can operate as an RF transformer, usually as a *loopstick antenna* in portable radio receivers and *radio direction-finding* (RDF) equipment. We can wind one of the coils directly over the other, or we can separate them, as shown in Fig. 18-5, to reduce the *interwinding capacitance* between the primary and secondary.

In a loopstick antenna, the primary winding intercepts the electromagnetic waves that carry the wireless signals. The secondary winding provides an optimum impedance match to the first amplifier stage, or *front end,* of the radio receiver or direction finder. We'll explore the use of transformers for impedance matching later in this chapter.

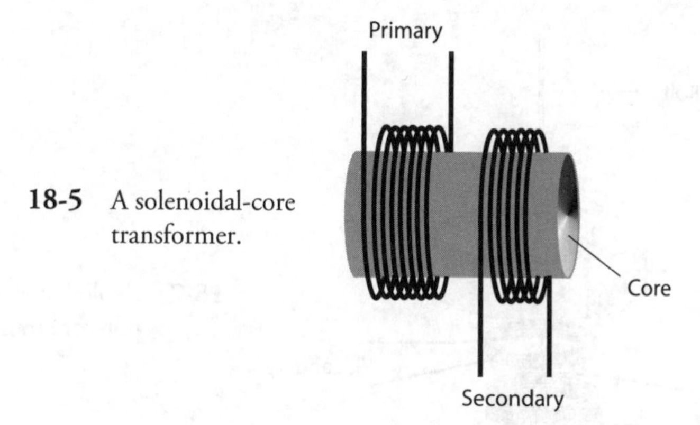

18-5 A solenoidal-core transformer.

Toroidal Core

In recent decades, a donut-shaped transformer core called a *toroidal core* (or *toroid*) has become common for winding RF transformers. When we want to construct a *toroidal transformer*, we can wind the primary and secondary directly over each other, or we can wind them over different parts of the core, as shown in Fig. 18-6. As with other transformers, we observe more interwinding capacitance if we place the coils directly over each other than we do when we keep them physically separated.

In a toroidal inductor or transformer, practically all of the magnetic flux remains within the core material; almost none ventures outside. This property allows circuit designers to place toroids near other components without worrying about unintended mutual inductance. Also, a toroidal coil or transformer can be mounted directly on a metal chassis, and the external metal has no effect on transformer operation.

A toroidal core provides considerably more inductance per turn, for the same kind of ferromagnetic material, than a solenoidal core offers. We'll often see toroidal coils or transformers that have inductance values up to several hundred millihenrys.

Pot Core

A *pot core* takes the form of a ferromagnetic shell that completely surrounds a loop-shaped wire coil. The core is manufactured in two halves (Fig. 18-7). You wind the coil inside one of the shell halves, and then bolt the two shell halves together. In the resulting device, all of the magnetic flux remains confined to the core material.

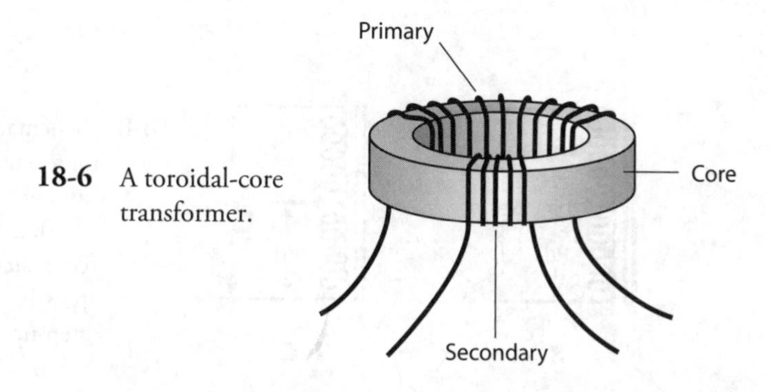

18-6 A toroidal-core transformer.

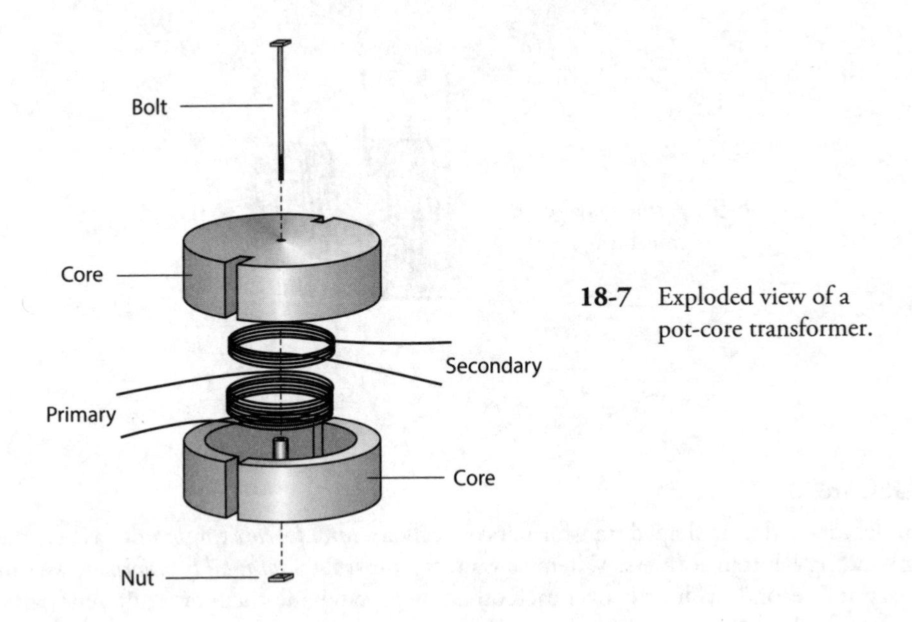

18-7 Exploded view of a pot-core transformer.

Like a toroid, a pot core is *self-shielding*. Essentially no magnetic coupling takes place between the windings and external components. You can use a pot core to wind a single, high-inductance coil. You can obtain inductance values of more than 1 H with a reasonable number of wire turns. However, you must wind the primary and secondary right next to each other; the geometry of the core prevents significant physical separation of the windings. Therefore, you'll always get a lot of interwinding capacitance.

Pot cores find diverse applications at AF, and also in the lowest-frequency part of the RF spectrum. You'll rarely, if ever, find these types of coils in high-frequency RF systems because other geometries can provide the necessary inductance values without the undesirable interwinding capacitance.

Autotransformer

In some situations, you might not need (or even want) DC isolation between the primary and secondary windings of a transformer. In a case of this sort, you can use an *autotransformer* that consists of a single, tapped winding. Figure 18-8 shows three autotransformer configurations:

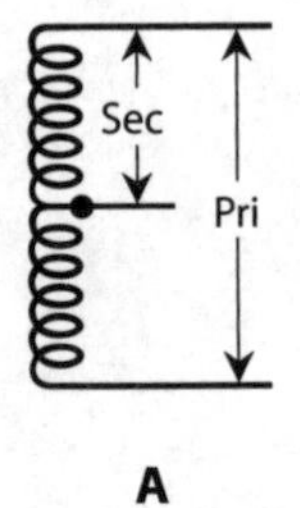

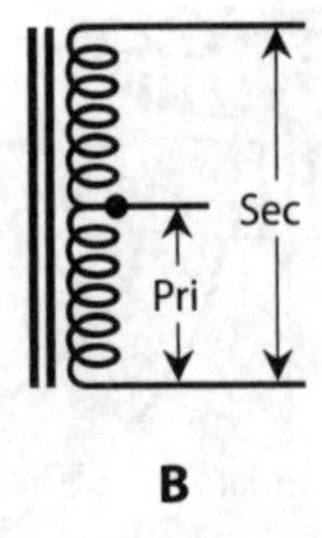

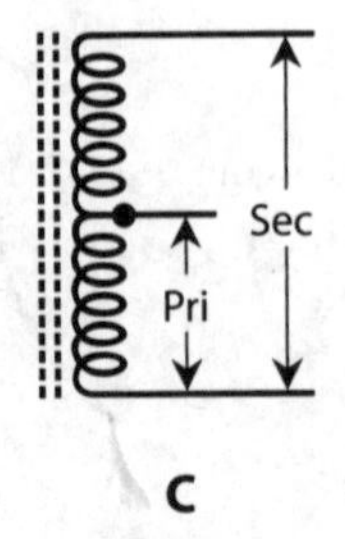

18-8 Schematic symbols for autotransformers. At A, air core, step-down. At B, laminated iron core, step-up. At C, powdered iron core, step-up.

- The unit at A has an air core, and operates as a step-down transformer.
- The unit at B has a laminated-iron core, and operates as a step-up transformer.
- The unit at C has a powdered-iron core, and operates as a step-up transformer.

You'll sometimes find autotransformers in older radio receivers and transmitters. Autotransformers work well in impedance-matching applications. They also work as solenoidal loopstick antennas. Autotransformers are occasionally, but not often, used in AF applications and in 60-Hz utility wiring. In utility circuits, autotransformers can step the voltage down by a large factor, but they can't efficiently step voltages up by more than a few percent.

Power Transformers

Any transformer used in the 60-Hz utility line, intended to provide a certain RMS AC voltage for the operation of electrical circuits, constitutes a *power transformer*. Power transformers exist in a vast range of physical sizes, from smaller than a grapefruit to bigger than your living room.

At the Generating Plant

We'll find the largest transformers at locations where electricity is generated. Not surprisingly, high-energy power plants have bigger transformers that develop higher voltages than low-energy, local power plants have. These transformers handle extreme voltages and currents simultaneously.

When we want to transmit electrical energy over long distances, we must use high voltages. That's because, for a given amount of power ultimately dissipated by the loads, the current goes down as the voltage goes up. Lower current translates into reduced ohmic loss in the transmission line. Recall the formula for power in nonreactive circuits in terms of the current and the voltage:

$$P = EI$$

where P represents the power (in watts), E represents the voltage (in volts), and I represents the current (in amperes). If we can make the voltage 10 times larger, for a given power level, then the current goes down to only 1/10 as much. The ohmic losses in the wires vary in proportion to the *square* of the current. To understand why, remember that

$$P = I^2 R$$

where P represents the power (in watts), I represents the current (in amperes), and R represents the resistance (in ohms). Engineers can't do very much about the wire resistance or the power consumed by the loads in a large electrical grid, but the engineers can adjust the voltage, and thereby control the current.

Suppose that we increase the voltage in a power transmission line by a factor of 10, while the load at the end of the line draws constant power. The increase in voltage reduces the current to 1/10 of its previous value. As a result, we cut the ohmic loss to $(1/10)^2$, or 1/100, of its previous amount. We, therefore, enjoy a big improvement in the efficiency of the transmission line (at least in terms of the ohmic loss in the wires).

Now we know why regional power plants have massive transformers capable of generating hundreds of thousands—or even millions—of volts! Up to a certain limit, we get better results for our money if we use high RMS voltage than we do if we use heavy-gauge wire for long-distance utility transmission lines.

Along the Line

Extreme voltages work well for *high-tension* power transmission, but a 200-kV RMS AC electrical outlet would not interest the average consumer! The wiring in a high-tension system requires considerable precautions to prevent arcing (sparking) and short circuits. Personnel must remain at least several meters away from the wires to avoid electrocution.

In a utility grid, medium-voltage power lines branch out from the major lines, and step-down transformers are used at the branch points. These lines fan out to lower-voltage lines, and step-down transformers are employed at these points, too. Each transformer must have windings heavy enough to withstand the product $P = EI$, the amount of VA power delivered to all the subscribers served by that transformer, at periods of peak demand.

Sometimes, such as during a heat wave, the demand for electricity rises above the normal peak level, "loading down" the circuit to the point that the voltage drops several percent. Then we have a *brownout.* If consumption rises further still, a dangerous current load appears at one or more intermediate power transformers. Circuit breakers in the transformers protect them from destruction by opening the circuit. Then we experience a *blackout.*

In most American homes, transformers step the voltage down to approximately 234 V RMS or 117 V RMS. Usually, 234-V RMS electricity appears in the form of three sine waves, called *phases*, each separated by 120°, and each appearing at one of the three slots in the outlet (Fig. 18-9A). We'll commonly see this system employed with heavy appliances, such as ovens, air conditioners, and washing machines. A 117-V RMS outlet, in contrast, supplies only one phase, appearing between two of the three slots in the outlet. The third opening should go directly to a substantial earth ground (Fig. 18-9B). This system is commonly used for basic household appliances, such as lamps, television sets, and computers.

In Electronic Equipment

Most consumer electronic systems have physically small power transformers. Most solid-state devices use low DC voltages, ranging from roughly 5 V to 50 V. For operation from 117-V RMS AC utility mains, therefore, such equipment requires step-down transformers in its power supplies.

Solid-state equipment usually (but not always) consumes relatively little power, so the transformers are rarely bulky. In high-powered AF or RF amplifiers, whose transistors can demand more than 1000 watts (1 kW) in some cases, the transformers require heavy-duty secondary windings, capable of delivering RMS currents of 90 A or more.

Older television receivers have *cathode-ray tube* (CRT) displays that need several hundred volts, derived from a step-up transformer in the power supply. Such transformers don't have to deliver a lot of current, so they have relatively little bulk or mass. Another type of device that needs high

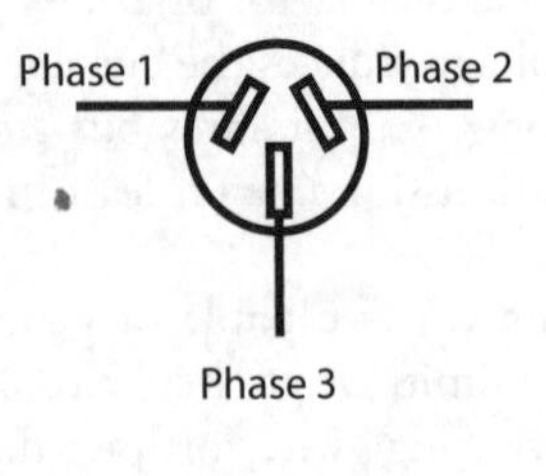

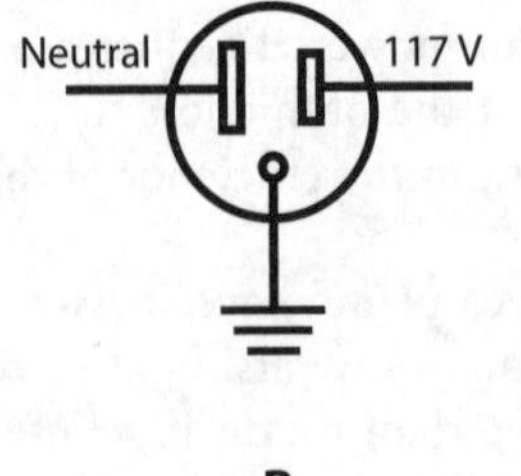

18-9 At A, an outlet for three-phase, 234-V RMS utility AC. At B, a conventional single-phase utility outlet for 117-V RMS utility AC.

voltage is a vacuum-tube RF amplifier such as the sort used by some amateur radio operators. Such an amplifier may demand 2 kV to 5 kV at around 500 mA.

> **Warning!**
>
> Treat any voltage higher than 12 V as dangerous. The voltage in a television set or some ham radios can present an electrocution hazard even after you power the system down. Do not try to service such equipment unless you have the necessary training.

At Audio Frequencies

Audio-frequency (AF) power transformers resemble those employed for 60-Hz electricity, except that the frequency is somewhat higher (up to 20 kHz), and audio signals exist over a range, or *band*, of frequencies (20 Hz to 20 kHz) rather than at only one frequency.

Most AF transformers are constructed like miniature utility transformers. They have laminated E cores with primary and secondary windings wound around the cross bars, as shown in Fig. 18-4. Audio transformers can function in either the step-up or step-down mode, and are designed to match impedances rather than to produce specific output voltages.

Audio engineers strive to minimize the system reactance so that the absolute-value impedance, Z, is close to the resistance R for both the input and the output. For that ideal condition to exist, the reactance X must be zero or nearly zero. In the following discussion of impedance-matching transformers (both for AF and RF applications), let's assume that the impedances always constitute pure resistances of the form $Z = R + j0$.

Isolation and Impedance Matching

Transformers can provide *isolation* between electronic circuits. While a transformer can and should provide *inductive coupling*, it should exhibit relatively little *capacitive coupling*. We can minimize the amount of capacitive coupling by using cores that minimize the number of wire turns needed in the windings, and by keeping the windings physically separated from each other (rather than overlapping).

Balanced and Unbalanced Loads and Lines

When we connect a device to a *balanced load*, we can reverse the terminals without significantly affecting the operating behavior. A plain resistor offers a good example of a balanced load. The two-wire antenna input in an old-fashioned analog television receiver provides another example. A *balanced transmission line* usually has two wires running alongside each other and separated by a constant physical distance, such as old-fashioned *TV ribbon*, also called *twinlead*.

An *unbalanced load* must be connected a certain way; we can't reverse the terminals. Switching the leads will result in improper circuit operation. In this sense, an unbalanced load resembles a polarized component, such as a battery, a diode, or an electrolytic capacitor. Many wireless antennas constitute unbalanced loads. Usually, unbalanced sources, transmission lines, and loads have one side connected to ground. The coaxial input of a television receiver is unbalanced; the shield (braid) of the cable connects to ground. An *unbalanced transmission line* usually comprises a coaxial cable of the sort that you see in a cable television system.

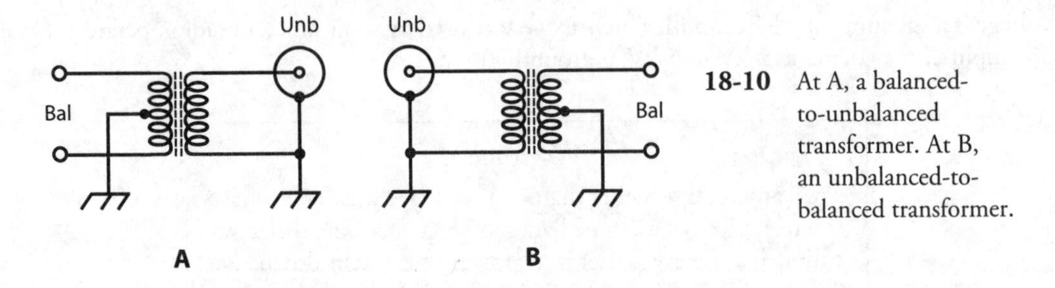

18-10 At A, a balanced-to-unbalanced transformer. At B, an unbalanced-to-balanced transformer.

Normally, you can't connect an unbalanced line to a balanced load, or a balanced line to an unbalanced load, and expect to obtain optimum performance. However, a transformer can provide compatibility between these two types of systems. Figure 18-10A illustrates a *balanced-to-unbalanced transformer*. The balanced (input) side of the transformer has a grounded center tap. Figure 18-10B shows an *unbalanced-to-balanced transformer*. The balanced (output) side has a grounded center tap.

The turns ratio of a balanced-to-unbalanced transformer (also called a *balun*) or an unbalanced-to-balanced transformer (also known as an *unbal*) can equal 1:1, but that's not a requirement. If the impedances of the balanced and unbalanced parts of the systems are the same, then a 1:1 turns ratio works well. But if the impedances differ, the turns ratio should match the impedances. Shortly, we'll see how we can adjust the turns ratio of a transformer to convert a purely resistive impedance into another purely resistive impedance.

Transformer Coupling

Engineers sometimes use transformers between amplifier stages in electronic equipment to obtain a large *amplification factor* from the system. Part of the challenge in making a radio receiver or transmitter perform well involves getting the amplifiers to operate together in a stable manner. If too much feedback occurs, a series of amplifiers will oscillate, degrading or ruining the performance of the radio. Transformers that minimize the capacitance between the amplifier stages, while still transferring the desired signals, can help to prevent such oscillation.

Impedance-Transfer Ratio

The *impedance-transfer ratio* of a transformer varies according to the square of the turns ratio, and also according to the square of the voltage-transfer ratio. If the primary (source) and secondary (load) impedances are purely resistive and are denoted Z_{pri} and Z_{sec}, then

$$Z_{pri} / Z_{sec} = (T_{pri} / T_{sec})^2$$

and

$$Z_{pri} / Z_{sec} = (E_{pri} / E_{sec})^2$$

The inverses of these formulas, in which we express the turns ratio or voltage-transfer ratio in terms of the impedance-transfer ratio, are

$$T_{pri} / T_{sec} = (Z_{pri} / Z_{sec})^{1/2}$$

and

$$E_{pri} / E_{sec} = (Z_{pri} / Z_{sec})^{1/2}$$

Problem 18-3

Consider a situation in which we need a transformer to match an input impedance of 50.0 ohms, purely resistive, to an output impedance of 200 ohms, also purely resistive. What's the required turns ratio T_{pri}/T_{sec}?

Solution

The transformer must have a step-up impedance ratio of

$$Z_{pri}/Z_{sec} = 50.0/200 = 1/4.00$$

From the above information, we can calculate

$$T_{pri}/T_{sec} = (1/4.00)^{1/2} = 0.250^{1/2} = 0.5 = 1/2$$

Problem 18-4

Suppose that a transformer has a primary-to-secondary turns ratio of 9.00:1. The load, connected to the transformer output, constitutes a pure resistance of 8.00 ohms. What's the impedance at the primary?

Solution

The impedance-transfer ratio equals the square of the turns ratio. Therefore

$$Z_{pri}/Z_{sec} = (T_{pri}/T_{sec})^2 = (9.00/1)^2 = 9.00^2 = 81.0$$

We know that the secondary impedance Z_{sec} equals 8.00 ohms, so

$$Z_{pri} = 81.0 \times Z_{sec} = 81.0 \times 8.00 = 648 \text{ ohms}$$

Radio-Frequency Transformers

Some RF transformers have primary and secondary windings, just like utility transformers. Others employ transmission-line sections. Let's look at these two types, which, taken together, account for most RF transformers in use today.

Wirewound Types

In the construction of wirewound RF transformers, we can use powdered-iron cores up to quite high frequencies. Toroidal cores work especially well because of their *self-shielding* characteristic (all of the magnetic flux stays within the core material). The optimum number of turns depends on the frequency, and also on the permeability of the core.

In high-power applications, air-core coils are often preferred. Although air has low permeability, it has negligible hysteresis loss, and will not heat up or fracture as powdered-iron cores sometimes do. However, some of the magnetic flux extends outside of an air-core coil, potentially degrading the performance of the transformer when it must function in close proximity to other components.

A major advantage of coil type transformers, especially when wound on toroidal cores, lies in the fact that we can get them to function efficiently over a wide band of frequencies, such as from 3.5 MHz to 30 MHz. A transformer designed to work well over a sizable frequency range is called a *broadband transformer*.

Transmission-Line Types

As you recall, any transmission line has a characteristic impedance, denoted as Z_o, that depends on the line construction. This property allows us to construct impedance transformers out of coaxial or parallel-wire line for operation at some radio frequencies.

Transmission-line transformers usually consist of quarter-wave sections. From the previous chapter, remember the formula for the length of a quarter-wave section:

$$L_{ft} = 246v / f_o$$

where L_{ft} represents the length of the section in feet, v represents the velocity factor expressed as a fraction or ratio, and f_o represents the operating frequency in megahertz. If we want to specify the length L_m in meters, then

$$L_m = 75v / f_o$$

Suppose that a quarter-wave section of line, with characteristic impedance Z_o, is terminated in a purely resistive impedance R_{out}. In this situation (Fig. 18-11), the impedance that appears at the input end of the line, R_{in}, also constitutes a pure resistance, and the following relations hold:

$$Z_o^2 = R_{in} R_{out}$$

and

$$Z_o = (R_{in} R_{out})^{1/2}$$

We can rearrange the first formula to solve for R_{in} in terms of R_{out}, and vice versa:

$$R_{in} = Z_o^2 / R_{out}$$

and

$$R_{out} = Z_o^2 / R_{in}$$

These equations hold true at the frequency f_o for which the line length measures ¼ wavelength. We can replace the word "wavelength" by the italic lowercase Greek letter lambda (λ), denoting the length of a quarter-wave section as (¼)λ or 0.25λ.

Neglecting line losses, the above relations hold at all *odd harmonics* of f_o, that is, at $3f_o$, $5f_o$, $7f_o$, and so on, at which we have sections measuring 0.75λ, 1.25λ, 1.75λ, and so on. At other frequencies, a length of transmission line fails to function as a simple transformer. Instead, it behaves in a complex manner, the mathematical details of which would take us beyond the scope of this discussion.

Quarter-wave transmission-line transformers work well in antenna systems, especially at the higher frequencies (above a few megahertz) at which their dimensions become practical. A quarter-wave matching section should be constructed from an unbalanced line if the load is unbalanced, and from a balanced line if the load is balanced.

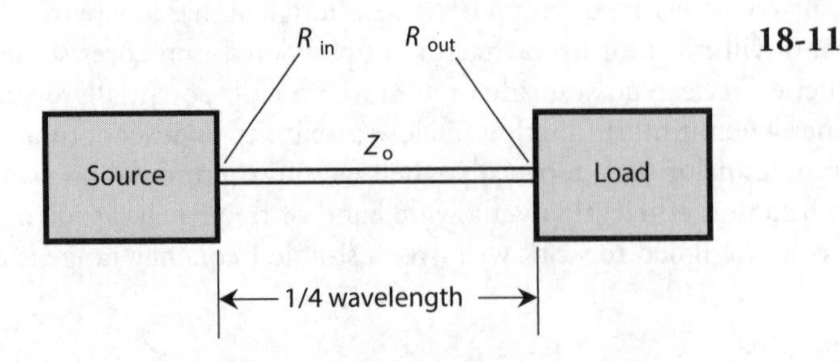

18-11 A quarter-wave matching section of transmission line. The input impedance equals R_{in}, the output impedance equals R_{out}, and the characteristic impedance equals Z_o.

A disadvantage of quarter-wave sections arises from the fact that they work only at specific frequencies, depending on their physical length and on the velocity factor. But this shortcoming is often offset by the ease with which we can construct them, if we intend to use a piece of radio equipment at only one frequency, or at odd-numbered harmonics of that frequency.

Problem 18-5

Suppose an antenna has a purely resistive impedance of 100 ohms. We connect it to a ¼-wave section of 75-ohm coaxial cable. What's the impedance at the input end of the section?

Solution

Use the formula from above to calculate

$$R_{in} = Z_o{}^2 / R_{out} = 75^2 / 100 = 5625 / 100 = 56.25 \text{ ohms}$$

We can round this result off to 56 ohms.

Problem 18-6

Consider an antenna known to have a purely resistive impedance of 300 ohms. You want to match it to the output of a radio transmitter designed to work into a 50.0-ohm pure resistance. What's the characteristic impedance needed for a quarter-wave matching section?

Solution

You can calculate it using the formula for that purpose, as follows:

$$Z_o = (R_{in}\ R_{out})^{1/2} = (300 \times 50.0)^{1/2} = 15{,}000^{1/2} = 122 \text{ ohms}$$

Few, if any, commercially manufactured transmission lines have this particular characteristic impedance. Prefabricated lines come in standard Z_o values, and a perfect match might not be obtainable. In that case, you can try the closest obtainable Z_o. In this case, it would probably be 92 ohms or 150 ohms. If you can't find anything near the characteristic impedance needed for a quarter-wave matching section, then you're better off using a coil-type transformer.

What about Reactance?

When no reactance exists in an AC circuit using transformers, things stay simple. But often, especially in RF antenna systems, pure resistance doesn't occur naturally. We have to "force it" by inserting inductors and/or capacitors to cancel the reactance out. The presence of reactance in a load makes a perfect match impossible with an impedance-matching transformer alone.

Inductive and capacitive reactances effectively oppose each other, and their magnitudes can vary. If a load presents a complex impedance $R + jX$, then we can cancel the reactance X by introducing an equal and opposite reactance $-X$ in the form of an inductor or capacitor connected in series with the load. This action gives us a pure resistance with a value equal to $(R + jX) - jX$, or simply R. The reactance-canceling component should always be placed at the point where the load connects to the line.

When we want to conduct wireless communications over a wide band of frequencies, we can place adjustable impedance-matching and reactance-canceling networks between the transmission line and the antenna. Such a circuit is called a *transmatch* or an *antenna tuner*. These devices not only match the resistive portions of the transmitter and load impedances, but they cancel reactances in the load. Transmatches are popular among amateur radio operators, who use equipment capable of operation from less than 2 MHz up to the highest known radio frequencies.

Optimum Location

Whenever we use a transformer or ¼-wave section to match the characteristic impedance of an RF feed line to the purely resistive impedance of a well-designed antenna, that transformer or section should be placed between the line and the antenna. We call this location the *feed point.* If we place the transformer or ¼-wave section anywhere else, it won't work at its best. In some cases, an improperly located transformer or ¼-wave section, even when tailored to the correct specifications, will make the overall impedance-mismatch situation worse!

Quiz

Refer to the text in this chapter if necessary. A good score is 18 or more correct. Answers are in the back of the book.

1. An autotransformer
   (a) can automatically match impedances over a wide range.
   (b) can effectively step down an AC voltage or impedance.
   (c) has an automatically variable center tap.
   (d) consists of a variable length of transmission line.

2. Which of the following core types works best if you need a coil winding inductance of 35 nH?
   (a) Air
   (b) Powdered-iron solenoid
   (c) Toroid
   (d) Pot core

3. Which of the following statements concerning air cores as compared with powdered-iron cores is true?
   (a) Air concentrates the magnetic lines of flux more than powdered iron, allowing for higher inductance for a given number of turns per winding.
   (b) Air-core transformers have greater loss than powdered-iron-core transformers, so air-core transformers work poorly for impedance matching.
   (c) Air-core transformers require toroidal windings, limiting the possible configurations, while powdered-iron-core transformers have no such constraints.
   (d) Air-core transformers operate best at the highest frequencies, while powdered-iron-core transformers are generally necessary at the lowest frequencies.

4. If a load contains no reactance whatsoever, then in theory we can use a transformer to match it perfectly to a transmission line
   (a) having any characteristic impedance within reason.
   (b) whose characteristic impedance is the same as the load impedance or lower, but not higher.
   (c) whose characteristic impedance is the same as the load impedance or higher, but not lower.
   (d) at only one frequency.

5. The primary-winding impedance always exceeds the secondary-winding impedance in
   (a) an autotransformer.
   (b) a step-up transformer.

(c) a step-down transformer.

(d) a balanced-to-unbalanced transformer.

6. We can minimize eddy currents in a 60-Hz AC utility transformer by

(a) using an air core.

(b) using a laminated-iron core.

(c) using the core winding method.

(d) winding the primary directly over the secondary.

7. If we wind the secondary directly over the primary in an air-core transformer, we should always expect to see

(a) a large impedance-transfer ratio.

(b) excellent performance at high frequencies.

(c) relatively high capacitance between the windings.

(d) significant hysteresis loss.

8. Suppose that a transformer has a primary-to-secondary turns ratio of exactly 4.00:1. We apply 20.0 V RMS AC across the primary terminals. How much AC RMS voltage can we expect to see across the secondary?

(a) 80.0 V RMS

(b) 40.0 V RMS

(c) 10.0 V RMS

(d) 5.00 V RMS

9. Suppose that a transformer has a primary-to-secondary turns ratio of exactly 1:4.00. The voltage at the primary equals 20.0 V RMS. What's the AC RMS voltage at the secondary?

(a) 80.0 V RMS

(b) 40.0 V RMS

(c) 10.0 V RMS

(d) 5.00 V RMS

10. Suppose that a transformer has a secondary-to-primary turns ratio of exactly 2.00:1. We apply 20.0 V RMS AC across the primary terminals. How much AC RMS voltage can we expect to see across the secondary?

(a) 80.0 V RMS

(b) 40.0 V RMS

(c) 10.0 V RMS

(d) 5.00 V RMS

11. Suppose that a transformer has a secondary-to-primary turns ratio of exactly 1:2.00. The voltage at the primary equals 20.0 V RMS. What's the AC RMS voltage at the secondary?

(a) 80.0 V RMS

(b) 40.0 V RMS

(c) 10.0 V RMS

(d) 5.00 V RMS

12. We want a transformer to match an input impedance of 300 ohms, purely resistive, to an output impedance of 50.0 ohms, also purely resistive. What's the required primary-to-secondary turns ratio?

    (a) 36.0:1
    (b) 6.00:1
    (c) 2.45:1
    (d) 2.00:1

13. Suppose that a transformer has a primary-to-secondary turns ratio of 4.00:1. The load, connected to the transformer output, constitutes a pure resistance of 50.0 ohms. What's the impedance at the primary?

    (a) 12.8 k
    (b) 800 ohms
    (c) 400 ohms
    (d) 200 ohms

14. A transformer has a primary-to-secondary impedance-transfer ratio of 4.00:1. We apply an AC signal of 200 V RMS to the primary. What's the RMS AC signal voltage at the secondary?

    (a) 800 V RMS
    (b) 400 V RMS
    (c) 141 V RMS
    (d) 100 V RMS

15. An antenna has a purely resistive impedance of 600 ohms. We connect it to a ¼-wave section of 92-ohm coaxial cable. What's the impedance at the input end of the section?

    (a) 14 ohms
    (b) 55 k
    (c) 6.5 ohms
    (d) 346 ohms

16. Suppose that we operate the system described in Question 15 at a frequency of 14 MHz. Our 92-ohm coaxial cable has a velocity factor of 0.75. How much cable will we need to construct a ¼-wave section?

    (a) 8.0 m
    (b) 7.5 m
    (c) 5.3 m
    (d) 4.0 m

17. A radio transmitter is designed to operate into a purely resistive impedance of 50 ohms. We have an antenna that exhibits a purely resistive impedance of 800 ohms. If we want to build an impedance-matching transformer to match these two impedances, what should its primary-to-secondary turns ratio be if we connect the transmitter to the primary and the antenna to the secondary?

    (a) 1:16
    (b) 1:8

(c) 1:4

(d) 1:2

18. Suppose that, in the situation described by Question 17, we want to use a ¼-wave section of transmission line to match the transmitter impedance to the antenna impedance. We need a line whose characteristic impedance equals

(a) 400 ohms

(b) 200 ohms

(c) 141 ohms

(d) 100 ohms

19. Imagine that the situation in the preceding two questions gets more complicated. The antenna has reactance in addition to the 800-ohm resistive component. That reactance results in a complex antenna impedance of $800 + j35$ at a frequency of 14 MHz. Our antenna constitutes an unbalanced system, so it's designed to work with a coaxial-cable transmission line or an unbalanced transformer secondary. How can we modify the antenna so that it will work properly at 14 MHz with either of the impedance-matching systems described in Questions 17 and 18?

(a) We can connect a capacitor in series with the antenna at the point where the transformer secondary or ¼-wave section output meets the antenna, such that the capacitive reactance equals -35 ohms at 14 MHz.

(b) We can connect an inductor in series with the antenna at the point where the transformer secondary or ¼-wave section output meets the antenna, such that the inductive reactance equals 35 ohms at 14 MHz.

(c) We can connect a capacitor in series with the system at the point where the transmitter output meets the transformer primary or ¼-wave section input, such that the capacitive reactance equals -35 ohms at 14 MHz.

(d) We can connect an inductor in series with the system at the point where the transmitter output meets the transformer primary or ¼-wave section input, such that the inductive reactance equals 35 ohms at 14 MHz.

20. Suppose that we want to use the antenna system described in Questions 17 through 19 over a continuous range of frequencies from 10 MHz to 20 MHz. In that case, what can we do to obtain a perfect impedance match at any frequency in that range?

(a) We can place a variable inductor in series with the antenna at the point where the transformer secondary or ¼-wave section output meets the antenna.

(b) We can place a variable capacitor in series with the antenna at the point where the transformer secondary or ¼-wave section output meets the antenna.

(c) We can place a well-engineered transmatch in series with the antenna at the point where the transformer secondary or ¼-wave section output meets the antenna.

(d) We can't do anything. We can never expect to get a perfect impedance match over a continuous range of frequencies in a situation of this sort.

Test: Part 2

DO NOT REFER TO THE TEXT WHEN TAKING THIS TEST. A GOOD SCORE IS AT LEAST 37 CORRECT. Answers are in the back of the book. It's best to have a friend check your score the first time, so you won't memorize the answers if you want to take the test again.

1. If two pure sine waves of the same frequency are 135° out of phase, it's equivalent to
   - (a) 1/16 of a cycle.
   - (b) 1/10 of a cycle.
   - (c) 1/6 of a cycle.
   - (d) 3/8 of a cycle.
   - (e) 3/4 of a cycle.

2. What should we do with the individual resistance values in a series *RLC* circuit if we want to determine the net resistance?
   - (a) Add them all up.
   - (b) Multiply them by each other.
   - (c) Convert them to conductances, add those values to each other, and then convert the result back to resistance.
   - (d) Convert them to susceptances, multiply them by each other, and then convert the result back to resistance.
   - (e) Multiply them by each other and then take the square root of the result.

3. When two identical capacitors are connected in series and no mutual capacitance exists, the net capacitance is
   - (a) 1/4 of the value of either individual capacitor.
   - (b) half the value of either individual capacitor.
   - (c) the same as the value of either individual capacitor.
   - (d) twice the value of either individual capacitor.
   - (e) four times the value of either individual capacitor.

4. At any particular frequency, a capacitor with a negative temperature coefficient
   (a) gets less reliable as the temperature rises.
   (b) gets more reliable as the temperature rises.
   (c) heats up more as its reactance gets farther from zero (increases negatively).
   (d) has reactance that gets farther from zero (increases negatively) as the temperature rises.
   (e) has reactance that gets closer to zero (decreases negatively) as the temperature rises.

5. Refer to Fig. Test 2-1. What's the phase relationship between the two pure sine waves *A* and *B*?
   (a) Wave *A* lags wave *B* by 90°.
   (b) Wave *A* lags wave *B* by 45°.
   (c) Wave *A* lags wave *B* by 30°.
   (d) Waves *A* and *B* coincide in phase.
   (e) Waves *A* and *B* oppose in phase.

6. Refer to Fig. Test 2-1 again. What can we say about the period of wave *A* with respect to the period of wave *B*?
   (a) The period of wave *A* equals the period of wave *B*.
   (b) The period of wave *A* equals half the period of wave *B*.
   (c) The period of wave *A* equals one and a half times the period of wave *B*.
   (d) The period of wave *A* equals twice the period of wave *B*.
   (e) Nothing, because we can't define the relative period.

7. If we connect a 12.6-V automotive battery directly to a 100-ohm resistor, then that resistor dissipates 1.59 W of
   (a) conductive power.
   (b) apparent power.
   (c) true power.
   (d) resistive power.
   (e) imaginary power.

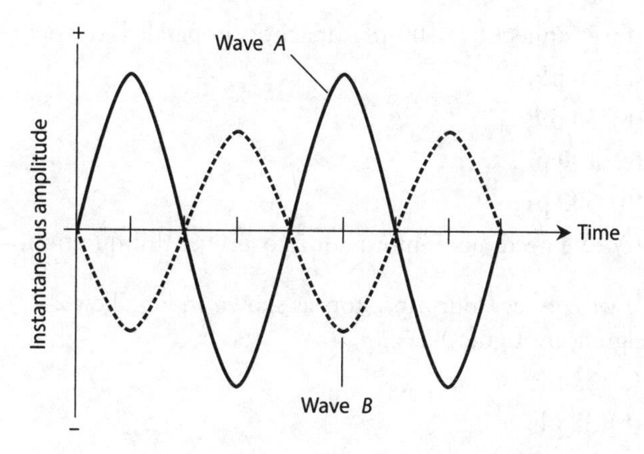

Test 2-1 Illustration for Part 2 Test Questions 5 and 6.

8. How long does it take for 36.00° of a cycle to pass in a 60.00-Hz sine wave?
   (a) 0.01667 ms
   (b) 0.1667 ms
   (c) 1.667 ms
   (d) 16.67 ms
   (e) 166.7 ms

9. What's the period of an AC wave whose frequency is 1.00 kHz?
   (a) 1.00 ms
   (b) 1.25 ms
   (c) 2.50 ms
   (d) 5.00 ms
   (e) 25.0 ms

10. Consider a series circuit comprising a pure resistance of 60 ohms and an inductive reactance of 100 ohms. What's the circuit's complex-number impedance?
   (a) $60 - j100$
   (b) $-100 - j60$
   (c) $60 + j100$
   (d) $-100 + j60$
   (e) 160 ohms

11. Which of the following capacitor types would most likely have a value of 0.01 μF?
   (a) Ceramic
   (b) Air variable
   (c) Photovoltaic
   (d) Electrolytic
   (e) Electrodynamic

12. If we connect five 100-pF capacitors in parallel, the net capacitance is
   (a) 20 pF.
   (b) 50 pF.
   (c) 100 pF.
   (d) 500 pF.
   (e) We need more information to answer this question.

13. If we connect four capacitors as shown in Fig. Test 2-2, the net capacitance C (rounded to two significant figures) is
   (a) 33 pF.
   (b) 38 pF.

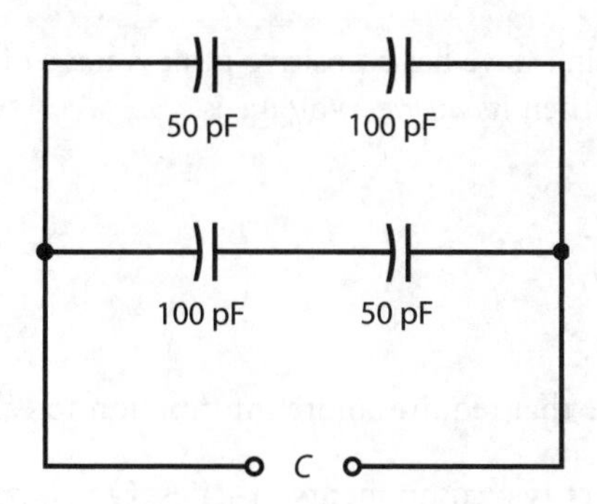

Test 2-2 Illustration for Part 2 Test Question 13.

(c) 50 pF.
(d) 67 pF.
(e) 88 pF.

14. When we want an antenna system to work at its best, which of the following properties should we make sure that it has?
(a) The characteristic impedance of the feed line should equal the resistive part of the antenna impedance, and the antenna itself should have no reactance.
(b) The characteristic impedance of the feed line should equal the reactive part of the antenna impedance, and the antenna itself should be a pure reactance.
(c) The characteristic impedance of the feed line should equal the sum of the resistive and reactive parts of the antenna impedance.
(d) The standing-wave ratio (SWR) on the feed line should be zero.
(e) The SWR on the feed line should be extremely high (ideally, infinite).

15. If a pure sine wave has a positive peak voltage of +170 V pk+ and a negative peak voltage of −170 V pk−, then its *peak-to-peak* voltage is
(a) zero.
(b) 240 V pk-pk.
(c) 340 V pk-pk.
(d) 120 V pk-pk.
(e) a value that requires more information to calculate.

16. If a pure sine wave has a positive peak voltage of +170 V pk+ and a negative peak voltage of −170 V pk−, then its *effective* or *RMS* voltage is
(a) zero.
(b) 240 V RMS.
(c) 340 V RMS.
(d) 120 V RMS.
(e) a value that requires more information to calculate.

17. If a pure sine wave has a positive peak voltage of +170 V pk+ and a negative peak voltage of −170 V pk−, then its *average* voltage is

(a) zero.
(b) 240 V.
(c) 340 V.
(d) 120 V.
(e) a value that requires more information to calculate.

18. We connect two components in series. One has a complex impedance of $10 + j20$, the other $40 - j20$. What's the net complex-number impedance of the combination?

(a) $50 + j40$
(b) $50 - j40$
(c) $50 + j0$
(d) $30 + j40$
(e) $30 - j40$

19. The situation described in Question 18 represents an example of

(a) reactance.
(b) quadrature.
(c) dissipation.
(d) dissonance.
(e) resonance.

20. We run a small electric generator with a moderate load. Then we disconnect that load to let the generator "run free." What happens?

(a) The generator motor speeds up.
(b) The generator motor slows down.
(c) The output voltage decreases.
(d) It takes more mechanical power to turn the generator's motor shaft.
(e) The output current increases.

21. Consider a series *RLC* circuit, where *R* represents the resistance and *X* represents the net reactance. We can find the absolute-value impedance according to one of the following formulas. Which one?

(a) $Z = R + X$
(b) $Z = (R^2 + X^2)^{1/2}$
(c) $Z = [RX / (R^2 + X^2)]^{1/2}$
(d) $Z = 1 / (R^2 + X^2)$
(e) $Z = R^2X^2 / (R + X)$

22. The resistance is less than the reactance in an *RL* circuit, but neither value is zero or "infinity." What's the phase angle?
    (a) 0°
    (b) Something between 0° and 45°
    (c) 45°
    (d) Something between 45° and 90°
    (e) 90°

23. The resistance equals the reactance in an *RL* circuit; both values are finite and nonzero. What's the phase angle?
    (a) 0°
    (b) Something between 0° and 45°
    (c) 45°
    (d) Something between 45° and 90°
    (e) 90°

24. The resistance is greater than the reactance in an *RL* circuit, but neither value is zero or "infinity." What's the phase angle?
    (a) 0°
    (b) Something between 0° and 45°
    (c) 45°
    (d) Something between 45° and 90°
    (e) 90°

25. In a series-resonant *RLC* circuit, which (if any) of the following formulas *always* holds true? (Remember that L and C stand for actual inductance and capacitance values, not reactance values.)
    (a) $R = 0$
    (b) $L = C$
    (c) $L = 0$
    (d) $C = 0$
    (e) None of the above

26. A perfectly conducting (loss-free) AC transmission line carries 200 mA RMS and 100 V RMS. The load contains no reactance, but its resistance equals the characteristic impedance of the line. How much *reactive power* does the load dissipate?
    (a) None
    (b) 7.07 W
    (c) 14.1 W
    (d) 20.0 W
    (e) 28.3 W

27. We connect a 400-pF capacitor in *parallel* with a 20-μH inductor. Then we increase the capacitance to 800 pF and decrease the inductance to 10 μH. What happens to the resonant frequency of the circuit?
   (a) It stays the same.
   (b) It increases by a factor of 2π.
   (c) It increases by a factor of $4\pi^2$.
   (d) It decreases by a factor of 2π.
   (e) It decreases by a factor of $4\pi^2$.

28. We connect a 400-pF capacitor in *series* with a 20-μH inductor. Then we increase the capacitance to 800 pF and decrease the inductance to 10 μH. What happens to the resonant frequency of the circuit?
   (a) It stays the same.
   (b) It increases by a factor of 2π.
   (c) It increases by a factor of $4\pi^2$.
   (d) It decreases by a factor of 2π.
   (e) It decreases by a factor of $4\pi^2$.

29. In Fig. Test 2-3, vector **A** represents
   (a) pure inductance.
   (b) pure resistance.
   (c) pure capacitance.
   (d) a combination of resistance and inductance.
   (e) a combination of resistance and capacitance.

30. In Fig. Test 2-3, vector **B** represents
   (a) pure inductance.
   (b) pure resistance.
   (c) pure capacitance.
   (d) a combination of resistance and inductance.
   (e) a combination of resistance and capacitance.

31. In Fig. Test 2-3, vector **C** represents
   (a) pure inductance.
   (b) pure resistance.
   (c) pure capacitance.
   (d) a combination of resistance and inductance.
   (e) a combination of resistance and capacitance.

32. In Fig. Test 2-3, vector **D** represents
   (a) pure inductance.
   (b) pure resistance.

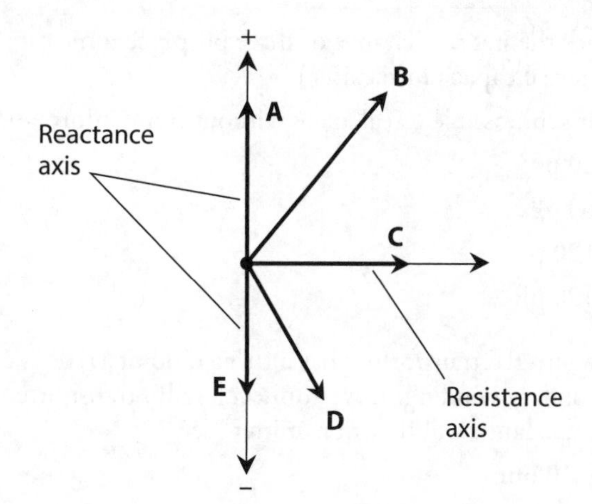

Test 2-3 Illustration for Part 2 Test Questions 29 through 33.

(c) pure capacitance.
(d) a combination of resistance and inductance.
(e) a combination of resistance and capacitance.

33. In Fig. Test 2-3, vector **E** represents
    (a) pure inductance.
    (b) pure resistance.
    (c) pure capacitance.
    (d) a combination of resistance and inductance.
    (e) a combination of resistance and capacitance.

34. An air-core coil
    (a) works well as a 60-Hz AC transformer.
    (b) has less inductance than a toroid-core coil with the same number of turns.
    (c) provides high inductance with relatively few turns.
    (d) works best at very-low frequencies (VLF) and low frequencies (LF).
    (e) makes an efficient loop antenna for VLF and LF transmitting.

35. What's the net inductance of three 60-mH toroid-core inductors connected in parallel? (Remember that mutual inductance is not a factor with toroidal coils because all the magnetic flux is confined to the core.)
    (a) It's impossible to calculate without more information.
    (b) 20 mH
    (c) 60 mH
    (d) 120 mH
    (e) 180 mH

36. What's the net capacitance of three 60-pF ceramic capacitors connected in parallel? (Assume that no mutual capacitance exists.)
   (a) It's impossible to calculate without more information.
   (b) 20 pF
   (c) 60 pF
   (d) 120 pF
   (e) 180 pF

37. A step-up RF transformer has a turns ratio of 1:10. We connect a purely resistive load of 10k to the secondary winding. If we connect a radio transmitter to the primary winding, what purely resistive impedance will that transmitter "see"?
   (a) 0.10 ohm
   (b) 1.0 ohms
   (c) 10 ohms
   (d) 0.10k
   (e) 1.0k

38. Consider again the transformer described in Question 37. If the transmitter is designed to operate into a purely resistive impedance of 50 ohms (as most ham radios are, for example), what purely resistive impedance should we connect to the secondary to get optimum performance from the transmitter, which remains connected to the primary?
   (a) 5.0k
   (b) 0.50k
   (c) 50 ohms
   (d) 5.0 ohms
   (e) 0.50 ohms

39. When we translate 100 nH into microhenrys, we get
   (a) 0.001 μH.
   (b) 0.010 μH.
   (c) 0.100 μH.
   (d) 1.00 μH.
   (e) 10.0 μH.

40. We can denote a specific pure inductive reactance on the RX_L half-plane as a point
   (a) on the positive real-number axis.
   (b) on the positive imaginary-number axis.
   (c) in the lower-right part of the half-plane, but not on either axis.
   (d) in the upper-right part of the half-plane, but not on either axis.
   (e) on the negative imaginary-number axis.

41. How many degrees of phase indicate the time-point in a cycle at which a pure sine wave, with no DC component, has a negative-going voltage of zero? (A cycle begins when the voltage is zero and positive-going.)

(a) 0°
(b) 45°
(c) 90°
(d) 180°
(e) 270°

42. How many degrees of phase indicate the time-point in a cycle at which a pure sine wave, with no DC component, reaches its maximum negative instantaneous voltage?

(a) 0°
(b) 45°
(c) 90°
(d) 180°
(e) 270°

43. You want to use a quarter-wave transmission-line section to match a feed line with a characteristic impedance of 50 ohms to an antenna with a purely resistive impedance of 113 ohms. What characteristic impedance should that matching section have?

(a) 63 ohms
(b) 69 ohms
(c) 75 ohms
(d) 82 ohms
(e) 91 ohms

44. What waveform does AC with a lot of harmonic energy have?

(a) Sine
(b) Triangular
(c) Ramp
(d) Square
(e) We need more information to say.

45. In a resonant *RLC* circuit, the complex-number reactance values are both nonzero, but when you add them, you get

(a) $j0$.
(b) a positive imaginary number.
(c) a negative imaginary number.
(d) a positive real number.
(e) a negative real number.

46. In an *RLC* circuit comprising a discrete resistor, inductor, and capacitor, resonance occurs at
    (a) a specific frequency and all its even-numbered harmonics.
    (b) a specific frequency and all its odd-numbered harmonics.
    (c) a specific frequency and all its harmonics.
    (d) a single specific frequency.
    (e) None of the above

47. Which of the following factors (a), (b), (c), or (d), if any, *does not* affect the characteristic impedance of coaxial cable?
    (a) The outside diameter of the center conductor
    (b) The inside diameter of the shield
    (c) The spacing between the center conductor and shield
    (d) The nature of the insulating material (dielectric) between the center conductor and shield
    (e) Any or all of the above factors can affect the characteristic impedance of coaxial cable.

48. In a series *RL* circuit in which $R = 910$ ohms and $X_L = 910$ ohms, the current lags the voltage by
    (a) 90°.
    (b) 60°.
    (c) 45°.
    (d) 30°.
    (e) 0°.

49. Two pure sine-wave signals both lack DC components, have the same frequency, and have the same peak-to-peak voltage; yet when we combine them we get no signal whatsoever. What's the phase difference between these two sine waves?
    (a) 0°
    (b) 180°
    (c) 360°
    (d) Any whole-number multiple of 180°
    (e) We need more information to say.

50. A series *RC* circuit has a resistance of 10k and a capacitive reactance of $-j13$ ohms at a particular frequency. The current leads the voltage by
    (a) 90°.
    (b) 60°.
    (c) 45°.
    (d) 30°.
    (e) None of the above

3
PART
Basic Electronics

19 CHAPTER

Introduction to Semiconductors

SINCE THE 1960S WHEN THE TRANSISTOR BECAME COMMON IN CONSUMER DEVICES, *SEMICONDUCTORS* have acquired a dominating role in electronics. The term *semiconductor* arises from the ability of certain materials to conduct some of the time, but not all of the time. We can control the conductivity to obtain amplification, rectification, oscillation, signal mixing, switching, and other effects.

The Semiconductor Revolution

Decades ago, *vacuum tubes*, also known as *electron tubes*, were the only amplifying devices available for use in electronic systems. A typical tube (called a *valve* in England) ranged from the size of your thumb to the size of your fist. You'll still find tubes in some power amplifiers, microwave oscillators, and video display units.

Tubes generally required high voltage in order to operate efficiently. Even in modest radio receivers, it took at least 50 V DC, and more often 100 V DC or more, to get a tube to work. Such voltages mandated bulky, massive power supplies and created an electrical shock hazard.

Nowadays, a transistor of microscopic dimensions can perform the functions of a tube in most low-power electronic circuits. The power supply can comprise a couple of AA "flashlight cells" or a 9-V "transistor battery." Even in high-power applications, transistors have smaller dimensions and weigh less than tubes having similar signal-output specifications (Fig. 19-1).

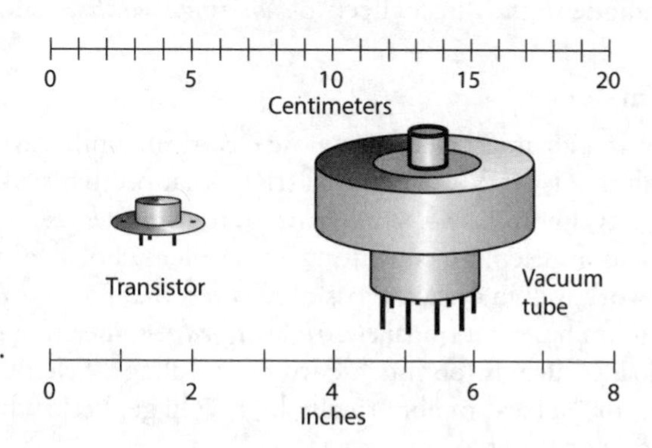

19-1 A power-amplifier transistor (at left) has much smaller volume and mass than a vacuum tube of comparable signal-output capacity (right).

Integrated circuits (ICs), hardly larger than individual transistors, can do the work of hundreds, thousands, or millions of vacuum tubes. You can find an excellent example of IC technology in personal computers and in the peripheral devices you use with them, such as displays, external disk drives, printers, and modems.

Vacuum tubes enjoy some advantages over semiconductor devices, even today. We can momentarily exceed the voltage, current, or power rating of a vacuum tube, and the device will usually "forgive" us, while a semiconductor device in a similar application might "die" immediately. A few audio enthusiasts, especially in popular music bands, insist that amplifiers made with vacuum tubes produce better quality sound than similar amplifiers made with semiconductor devices.

Semiconductor Materials

Various elements, compounds, and mixtures can function as semiconductors. Two common semiconductor materials are the element *silicon* and a compound of gallium and arsenic known as *gallium arsenide* (abbreviated GaAs). In the early years of semiconductor technology, *germanium* formed the basis for many semiconductors, but today we rarely see it. Other substances that work as semiconductors include *selenium*, *cadmium* compounds, *indium* compounds, and the oxides of certain metals.

Silicon

Silicon is an element with atomic number 14 and atomic weight 28. In its pure state, silicon appears as a light-weight metal similar to aluminum. Pure silicon conducts electric currents better than a dielectric material does, but not as well as most metallic conductors, such as silver, copper, or aluminum.

The earth's crust contains silicon in great abundance, and we can mine it from some crustal rocks and sand. In its natural state, silicon almost always exists "tied up" as compounds with other elements. Industrial vendors extract the pure element. Electronic-component manufacturers mix other substances, known as *impurities*, with silicon to give the semiconductor material specific properties. The resulting solids are cut into thin slices known as *chips* or *wafers*.

Gallium Arsenide

Another common semiconductor is the compound gallium arsenide. Engineers and technicians write its acronym-like chemical symbol, GaAs, and pronounce it aloud as "gas." If you hear about "gasfets" and "gas ICs," you're hearing about gallium-arsenide technology.

GaAs devices perform well at higher frequencies than silicon-based devices do because electric currents can travel faster through GaAs than through silicon compounds. GaAs devices are relatively immune to the direct effects of *ionizing radiation*, such as X rays and gamma rays.

Selenium

Selenium is a chemical element whose electrical conductivity varies depending on the intensity of visible, ultraviolet (UV), or infrared (IR) radiation that strikes it. All semiconductor materials have this property, known as *photoconductivity*, to some degree, but in selenium, the effect is pronounced. For this reason, selenium constitutes an excellent choice for the manufacture of *photocells*. Selenium can also work well in certain types of *rectifiers* that convert AC to pulsating DC.

Selenium has exceptional *electrical ruggedness*, meaning that it can withstand short-lived electrical overloads, such as too much current or voltage. Selenium-based components can survive brief *transients*, or "spikes" of abnormally high voltage, better than components made with most other semiconductor materials.

Germanium

Pure elemental germanium constitutes a poor electrical conductor. It becomes a semiconductor only when we add impurities. In the early years of semiconductor technology, engineers used germanium-based components far more often than they do now. In fact, we'll rarely encounter a germanium device in any electronic system today. High temperatures, such as soldering tools generate, can destroy a germanium-based diode or transistor.

Metal Oxides

Certain metal oxides have properties that make them useful in the manufacture of semiconductor devices. When you hear about MOS (pronounced "moss") or CMOS (pronounced "sea moss") technology, you're hearing about *metal-oxide semiconductor* and *complementary metal-oxide semiconductor* devices, respectively.

Certain types of transistors, and many kinds of ICs, make use of MOS technology. In integrated circuits, MOS and CMOS construction allows for a large number of discrete components, such as resistors, inductors, diodes, and transistors, on a single chip. Engineers say that MOS/CMOS has *high component density*.

Metal-oxide components need almost no current in order to function. When we use a battery to power a small MOS-based device, that battery will last almost as long as it would if we let it sit on the shelf and didn't use it for anything. Most MOS-based devices can work at extreme speeds, allowing for operation at high frequencies in radio-frequency (RF) equipment, and facilitating the rapid switching that's important in today's computers.

All MOS and CMOS components suffer from one outstanding limitation: A discharge of static electricity can destroy one of them in an instant. We must use care when handling components of this type. Any technician working with MOS and CMOS components must wear a metal wrist strap that connects one wrist to a good earth ground, so static-electric buildup cannot occur.

Doping and Charge Carriers

Impurities give a semiconductor material the properties that it needs to function as an electronic component. The impurities cause the material to conduct current in certain ways. When manufacturers add an impurity to a semiconductor element, they call the process *doping*. The impurity material itself is called a *dopant*.

Donor Impurities

When a dopant contains an inherent excess of electrons, we call it a *donor impurity*. Adding such a substance causes conduction mainly by means of electron flow, as in an ordinary metal such as copper. The excess electrons move from atom to atom when a potential difference exists between two different points in the material. Elements that serve as donor impurities include antimony, arsenic, bismuth, and phosphorus. A material with a donor impurity is called an *N type semiconductor*, because an electron carries a unit of negative (N) electric charge.

Acceptor Impurities

If an impurity has an inherent deficiency of electrons, we call it an *acceptor impurity*. When we add an element, such as aluminum, boron, gallium, or indium, to a semiconductor element, the resulting material conducts mainly by means of *hole flow*. A *hole* comprises an "assigned spot" in an atom where an electron would exist under normal conditions, but in fact doesn't show up there.

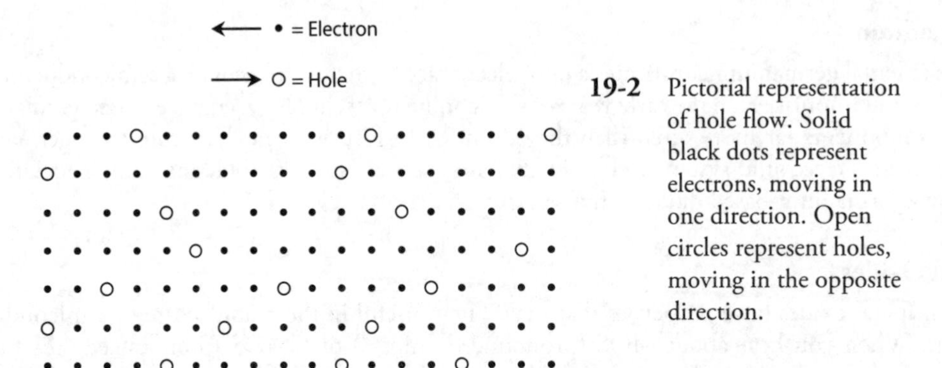

19-2 Pictorial representation of hole flow. Solid black dots represent electrons, moving in one direction. Open circles represent holes, moving in the opposite direction.

A semiconductor with an acceptor impurity is called a *P type semiconductor* because a hole has, in effect, a unit of positive (P) electric charge.

Majority and Minority Carriers

Charge carriers in semiconductor materials always constitute either electrons or holes. We never see "oddball" charge carriers, such as protons or helium nuclei, in electronic devices (although we do encounter them in high-energy physics). In any semiconductor substance, some of the current takes the form of electrons passed from atom to atom in a negative-to-positive direction, and some of the current occurs as holes that move from atom to atom in a positive-to-negative direction.

Sometimes electrons account for most of the current in a semiconductor. This situation exists if the material has donor impurities, that is, if it's of the N type. In other cases, holes account for most of the current. This phenomenon occurs when the material has acceptor impurities, making it P type. We call the more abundant, or dominating, charge carriers (either electrons or holes) the *majority carriers*. We call the less abundant ones the *minority carriers*. The ratio of majority to minority carriers can vary, depending on the exact chemical composition of the semiconductor material.

Figure 19-2 shows a simplified illustration of electron flow versus hole flow in a sample of N type semiconductor material, in which the majority carriers constitute electrons and the minority carriers constitute holes. Each point location in the grid represents an atom. The solid black dots represent electrons. Imagine them moving from right to left as they "jump" from atom to atom. Small open circles represent holes. Imagine them moving from left to right as they "jump" from atom to atom. In the example, the positive battery or power-supply terminal (the "source of holes") would lie somewhere out of the picture toward the left, and the negative battery or power-supply terminal (the "source of electrons") would lie out of the picture toward the right.

The P-N Junction

Connecting a piece of semiconducting material, either P or N type, to a source of current can provide us with phenomena for scientific observations and experiments. But when the two types of material come into direct contact, the boundary between the P type sample and the N type sample, called the *P-N junction*, behaves in ways that make semiconductor materials truly useful.

The Semiconductor Diode

Figure 19-3 shows the schematic symbol for a *semiconductor diode*, formed by joining a piece of P type material to a piece of N type material. The N type semiconductor is represented by the short,

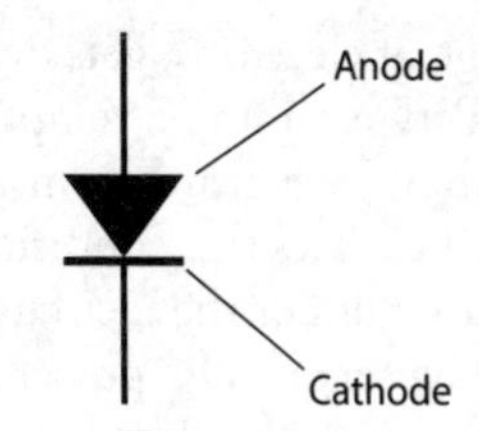

19-3 Schematic symbol for a semiconductor diode.

straight line in the symbol; we call it the *cathode*. The P type semiconductor is represented by the arrow; we call it the *anode*. Electrons can move easily in the direction opposite the arrow, and holes can move easily in the direction in which the arrow points. Electrons normally do not move with the arrow, and holes normally do not move against the arrow.

If you connect a battery and a resistor in series with the diode, you'll get a current to flow if you connect the negative battery terminal to the diode's cathode and the positive terminal to the anode, as shown in Fig. 19-4A. A series-connected resistor eliminates the risk of diode destruction by excessive current. No current will flow if you reverse the battery polarity, as shown in Fig. 19-4B.

It takes a specific, well-defined minimum applied voltage for conduction to occur through a semiconductor diode in the situation shown by Fig. 19-4A. We call this threshold potential difference the *forward breakover voltage*. Depending on the type of semiconductor material, the forward breakover voltage for a particular diode can vary from about 0.3 V to 1 V. If the voltage across the P-N junction falls short of the forward breaker voltage, the diode will fail to conduct current, even when we connect it as shown in Fig. 19-4A. This effect, known as the *forward breakover effect* or the *P-N junction threshold effect*, allows us to build circuits to limit the maximum positive and/or negative peak voltages that signals can attain. We can also take advantage of this effect to construct a device called a *threshold detector*, in which a signal's positive or negative peak amplitude must equal or exceed a certain minimum in order to pass through.

How the Junction Works

When the N type material has a negative voltage with respect to the P type (as in Fig. 19-4A) that exceeds the forward breakover voltage, electrons flow easily from N to P. The N type semiconductor, which already has an excess of electrons, receives more; the P type semiconductor, already

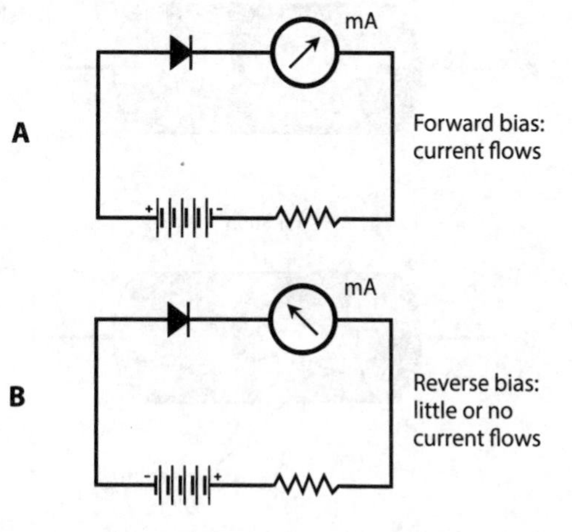

19-4 Series connection of a battery, a resistor, a current meter, and a diode. At A, forward bias results in a flow of current. At B, reverse bias results in no current.

"suffering" from a shortage of electrons, gets deprived of still more. The N type material constantly "feeds" electrons to the P type in an "attempt" to create an electron balance, and the battery or power supply keeps "robbing" electrons from the P type material in order to sustain the electron imbalance. Figure 19-5A illustrates this condition, known as *forward bias*. Current can flow through a forward-biased diode easily under these circumstances.

When we reverse the battery or DC power-supply polarity so that the N type material acquires a positive voltage with respect to the P type material, we have a condition called *reverse bias*. Electrons in the N type material migrate towards the positive charge pole, away from the P-N junction. In the P type material, holes drift toward the negative charge pole, also away from the P-N junction. The electrons constitute the majority carriers in the N type material, and the holes are the majority carriers in the P type material. The charge, therefore, disappears in the vicinity of the P-N junction, as shown in Fig. 19-5B. This "charge-free zone," where majority carriers are deficient, is called the *depletion region*. A shortage of majority carriers in any semiconductor substance means that the substance cannot conduct well, so a depletion region acts like an electrical insulator. This phenomenon reveals the reason why a semiconductor diode will not normally conduct when reverse-biased. A diode forms a "one-way current gate"—usually!

When the cathode and anode of a P-N junction have the same electrical potential as applied from an external source, we call the condition *zero bias*.

Junction Capacitance

Some P-N junctions can alternate between conduction (in forward bias) and non-conduction (in reverse bias) millions or billions of times per second. Other P-N junctions can't work so fast. The maximum switching speed depends on the capacitance at the P-N junction during conditions of reverse bias. As the *junction capacitance* of a diode increases, the highest frequency at which it can alternate between the conducting state and the non-conducting state decreases.

The junction capacitance of a diode depends on several factors, including the operating voltage, the type of semiconductor material, and the cross-sectional area of the P-N junction. If you examine Fig. 19-5B, you might get the idea that the depletion region, sandwiched between two semiconducting sections, can play a role similar to that of the dielectric in a capacitor. If so, you're right!

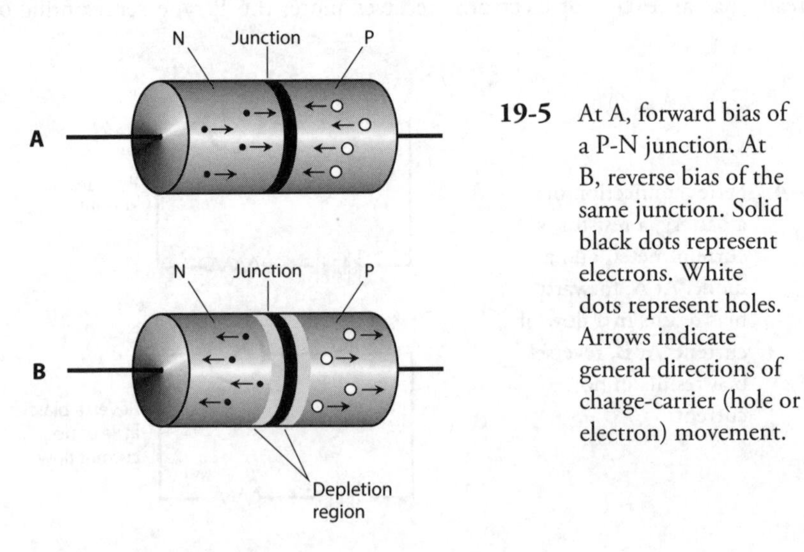

19-5 At A, forward bias of a P-N junction. At B, reverse bias of the same junction. Solid black dots represent electrons. White dots represent holes. Arrows indicate general directions of charge-carrier (hole or electron) movement.

A reverse-biased P-N junction forms a capacitor. Some semiconductor components, called *varactor diodes*, are manufactured with this property in mind.

We can vary the junction capacitance of a diode by changing the reverse-bias voltage because this voltage affects the width of the depletion region. As we increase the reverse voltage, the depletion region gets wider, and the capacitance goes down.

Avalanche Effect

Sometimes, a diode conducts even when reverse-biased. The greater the reverse bias voltage, the more like an electrical insulator a P-N junction gets—up to a point. But if the reverse bias reaches or exceeds a specific critical value, the voltage overcomes the ability of the junction to prevent the flow of current, and the junction conducts as if forward-biased. Engineers call this phenomenon the *avalanche effect* because conduction occurs in a sudden and massive way, something like an avalanche on a snowy mountain.

Avalanche effect does not damage a P-N junction (unless the applied reverse voltage is extreme). It's a temporary thing. The depletion region disappears while avalanche breakdown occurs. But when the voltage drops back below the critical value, the junction behaves normally again. If the bias remains reversed after the avalanche breakdown condition ends, the depletion region returns.

Some diodes are designed to take advantage of the avalanche effect. In other cases, avalanche effect limits the performance of a circuit. In a device called a *Zener diode*, designed specifically to regulate DC voltages, you'll hear or read about the *Zener voltage* specification. This value can range from a couple of volts to well over 100 V. Zener voltage is theoretically equivalent to avalanche voltage, but in the case of a Zener diode, the manufacturer tailors the semiconductor material so as to produce an exact, predictable avalanche voltage.

For *rectifier diodes* in power supplies, you'll hear or read about the *peak inverse voltage* (PIV) or *peak reverse voltage* (PRV) specification. That's the highest instantaneous reverse-bias voltage that we can expect the device to withstand without risking avalanche breakdown. In practical applications, rectifier diodes must have PIV ratings great enough so that the avalanche effect will never occur (or even come close to happening) during any part of the AC cycle.

Quiz

Refer to the text in this chapter if necessary. A good score is at least 18 correct. Answers are in the back of the book.

1. When we apply reverse bias to a P-N junction at less than the avalanche voltage, the junction
   (a) does not conduct.
   (b) conducts intermittently.
   (c) conducts fairly well.
   (d) conducts very well.
2. Which, if any, of the following statements is *false*?
   (a) Some audio enthusiasts think that power amplifiers made with vacuum tubes sound better than power amplifiers made with semiconductor devices.
   (b) Tubes are physically larger than transistors that do the same things.
   (c) Transistors need more voltage, in general, than vacuum tubes to operate properly.
   (d) All of the above statements are true.

3. When we dope a semiconductor with an acceptor impurity, we get
   (a) E type material.
   (b) N type material.
   (c) P type material.
   (d) H type material.

4. When we dope a semiconductor with an acceptor impurity, that material ends up with a surplus of
   (a) protons.
   (b) neutrons.
   (c) electrons.
   (d) holes.

5. Pure silicon is
   (a) a compound.
   (b) an element.
   (c) a mixture.
   (d) a liquid.

6. When we forward-bias a P-N junction, it fails to conduct
   (a) unless the junction has enough capacitance.
   (b) if the applied voltage is less than the forward breakover voltage.
   (c) if the applied voltage exceeds the avalanche voltage.
   (d) unless the voltage remains constant.

7. Donor impurities have an inherent excess of
   (a) neutrons.
   (b) protons.
   (c) electrons.
   (d) holes.

8. Imagine a "bucket" containing more electrons than holes. The net electrical charge of the "bucket's" contents is
   (a) positive.
   (b) negative.
   (c) zero.
   (d) impossible to determine.

9. We apply DC reverse bias to a P-N junction, but it's not enough to cause avalanche effect. Then we double the voltage, but it still doesn't cause an avalanche effect. What happens to the capacitance at the depletion region?
   (a) It stays the same.
   (b) It increases.
   (c) It decreases.
   (d) None of the above; a depletion region has inductance, not capacitance.

10. We apply DC reverse bias to a P-N junction, but it's not enough to cause avalanche effect. Then we increase the voltage past the point where avalanche effect occurs. What happens?
    (a) The junction conducts.
    (b) The depletion region expands to take up the whole diode.
    (c) The junction's reactance increases.
    (d) The diode burns out.

11. Gallium arsenide is
    (a) a compound.
    (b) a liquid.
    (c) an element.
    (d) a mixture.

12. Which of the following materials makes the best choice for a photocell?
    (a) Bismuth
    (b) Indium
    (c) Aluminum
    (d) Selenium

13. Semiconductor manufacturers can make pure germanium into an N type semiconductor material by
    (a) adding a donor impurity.
    (b) adding an acceptor impurity.
    (c) applying a negative charge.
    (d) None of the above; they can't!

14. Which of the following things can happen when we dope a semiconductor material?
    (a) We get a pure chemical element.
    (b) Current flows mostly as holes.
    (c) We get a conductor.
    (d) We get a dielectric.

15. In the circuit of Fig. 19-6, the milliammeter (mA) indicates no current. By examining the polarities of the diode and the DC source, we can see that the diode is
    (a) reverse-biased beyond the avalanche voltage.
    (b) reverse-biased below the avalanche voltage.
    (c) forward-biased beyond the forward breakover voltage.
    (d) forward-biased at less than the forward breakover voltage.

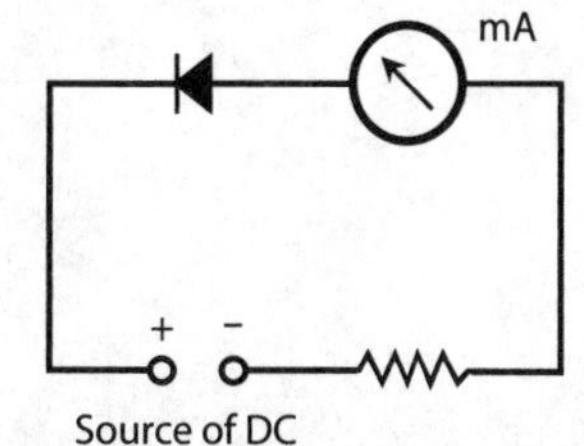

19-6 Illustration for Quiz Question 15.

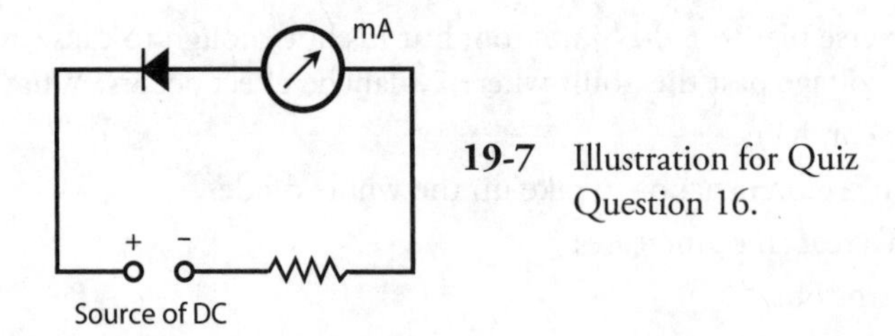

19-7 Illustration for Quiz Question 16.

16. In the circuit of Fig. 19-7, the milliammeter (mA) indicates significant current. By examining the polarities of the diode and the DC source, we can see that the diode is
    - (a) reverse-biased beyond the avalanche voltage.
    - (b) reverse-biased below the avalanche voltage.
    - (c) forward-biased beyond the forward breakover voltage.
    - (d) forward-biased below the forward breakover voltage.

17. Charge carriers move faster through some semiconductors than others. In general, as the charge-carrier speed increases, so does the maximum
    - (a) speed at which a device made with that substance can operate.
    - (b) voltage that the substance can withstand.
    - (c) current that the substance can handle.
    - (d) resistance of a component made with that substance.

18. In a P-N junction forward-biased beyond the forward breakover voltage, the junction
    - (a) is surrounded by a zone devoid of charge carriers.
    - (b) acts like a capacitor.
    - (c) does not conduct current.
    - (d) None of the above

19. Which of the following things commonly serves as a charge carrier in a semiconductor?
    - (a) An atomic nucleus
    - (b) A proton
    - (c) A neutron
    - (d) An electron

20. Fill in the blank to make the following statement true: "In a semiconductor material, ________ account(s) for most of the current."
    - (a) minority carriers
    - (b) majority carriers
    - (c) electrons
    - (d) holes

20
CHAPTER

Diode Applications

IN THE EARLY YEARS OF ELECTRONICS, MOST DIODES WERE VACUUM TUBES. TODAY, MOST DIODES are made from semiconductor materials. Contemporary diodes can do almost everything that the old vacuum-tube ones could, and a few things that people in the tube era never imagined.

Rectification

A *rectifier diode* passes current in only one direction, as long as we don't exceed its specifications. This property makes the device useful for changing AC to DC. Generally speaking:

- When the cathode has a more negative charge than the anode, current flows.
- When the cathode has a more positive charge than the anode, current does not flow.

The constraints on this behavior are, as we've learned, the forward breakover and avalanche voltages.

Examine the circuit shown in Fig. 20-1A. Suppose that we apply a 60-Hz AC sine wave to the input terminals. During half of the cycle, the diode conducts, and during the other half, it doesn't. This behavior cuts off half of every cycle. Depending on which way we connect the diode, we can

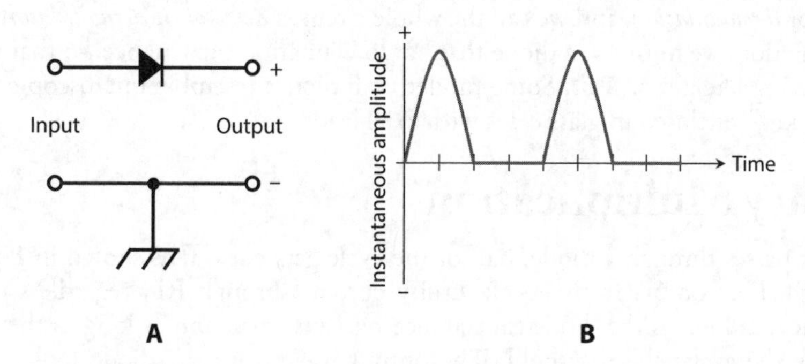

20-1 At A, a half-wave rectifier circuit. At B, the output of the circuit shown at A when we apply an AC sine wave to the input.

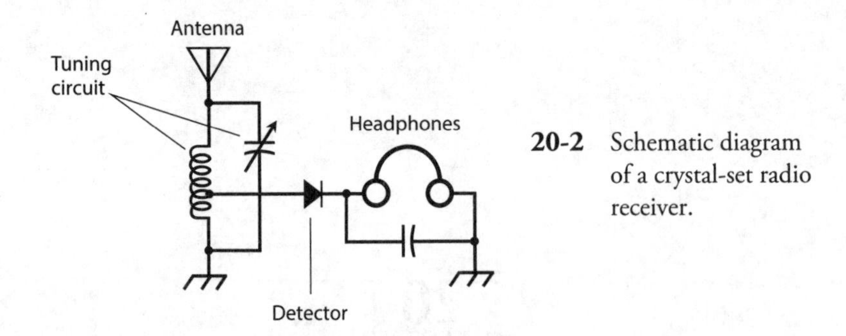

20-2 Schematic diagram of a crystal-set radio receiver.

make it cut off either the positive half of the AC cycle or the negative half. Figure 20-1B is a graph of the output of the circuit shown at A.

The circuit and wave diagrams of Fig. 20-1 involve a *half-wave rectifier* circuit, which is the simplest possible rectifier. Simplicity constitutes the chief advantage of the half-wave rectifier over other rectifier circuits. You'll learn about various types of rectifier diodes and circuits in the next chapter.

Detection

One of the earliest diodes, existing even before vacuum tubes, was made partly with semiconductor material. Known as a "cat's whisker," the device comprised a fine piece of wire in contact with a small fragment of the mineral *galena.* This contraption had the ability to act as a rectifier for extremely weak RF currents. When experimenters connected the "cat's whisker" in a configuration such as the circuit of Fig. 20-2, the resulting device could receive amplitude-modulated (AM) radio signals and produce audible output in the headset.

The galena fragment, sometimes called a "crystal," gave rise to the nickname *crystal set* for this primitive radio receiver. You can build a crystal set today using an RF diode, a coil, a tuning capacitor, a headset, and a long-wire antenna. The circuit needs no battery or other source of electrical power! If a broadcast station exists within a few miles of the antenna, the received signal alone produces enough audio to drive the headset. For ideal performance, the headset should be *shunted* with a capacitor whose value is large enough to "short out" residual RF current to ground, but not so large as to "short out" the audio signal. (When we say *shunted* in this context, we mean that the capacitor goes in parallel, or in *shunt,* with the headset.)

In the circuit of Fig. 20-2, the diode recovers the audio from the radio signal. We call this process *detection* or *demodulation*, and we call the whole circuit a *detector* or *demodulator*. If we want the detector to function, we must use a diode that has low junction capacitance, so that it can rectify at RF without acting like a capacitor. Some modern RF diodes resemble microscopic versions of the old "cat's whisker," enclosed in glass cases with axial leads.

Frequency Multiplication

When current passes through a diode, half of the cycle gets cut off, as shown in Fig. 20-1B. This "chopping-off" effect occurs from 60-Hz utility current through RF, regardless of the applied-signal frequency, as long as the diode capacitance remains small and as long as the reverse voltage remains below the avalanche threshold. The output wave from the diode looks much different than the input wave. We call this condition *nonlinearity*. Whenever a circuit exhibits nonlinearity, harmonics appear in the output. The harmonics show up as signals at integer multiples of the input frequency, as we learned in Chap. 9.

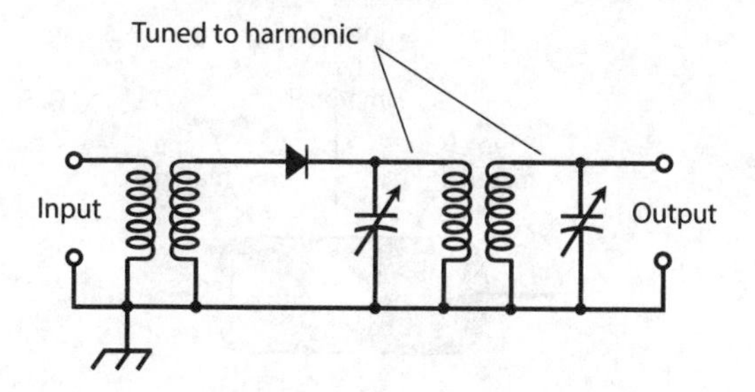

20-3 A frequency-multiplier circuit using a semiconductor diode.

In situations where nonlinearity represents an undesirable state of affairs, engineers strive to make electronic circuits *linear*, so the output waveform has exactly the same shape as the input waveform (even if the amplitudes differ). But in some applications, we want a circuit to act in a nonlinear fashion—for example, when we intend to generate harmonics. We can deliberately introduce nonlinearity in a circuit to obtain *frequency multiplication*. Diodes work well for this purpose. Figure 20-3 illustrates a simple *frequency-multiplier* circuit. We tune the output *LC* circuits to the desired *n*th harmonic frequency, nf_o, rather than to the input or fundamental frequency, f_o.

For a diode to work as a frequency multiplier in RF systems, it must be of a type that would also work well as a detector at the same frequencies. This means that the component should act like a rectifier, but not like a capacitor.

Signal Mixing

When we combine two waves having different frequencies in a nonlinear circuit, we get new waves at frequencies equal to the sum and difference of the frequencies of the input waves. Diodes can provide the nonlinearity that we need to make this happen.

Consider two AC signals with frequencies f_1 and f_2. Let's assign f_2 to the wave with the higher frequency, and f_1 to the wave with the lower frequency. If we combine these two signals in a nonlinear circuit, new waves result. One of the new waves has a frequency of $f_2 + f_1$, and the other has a frequency of $f_2 - f_1$. We call these sum and difference frequencies *beat frequencies*. We call the signals themselves *mixing products* or *heterodynes*. The heterodynes appear in the output along with the original signals at frequencies f_1 and f_2.

Figure 20-4 shows hypothetical input and output signals for a *mixer* circuit on a *frequency-domain* display. The amplitude (on the vertical scale or axis) constitutes a function of the frequency (on the horizontal scale or axis). Engineers see this sort of display when they look at the screen of a lab instrument known as a *spectrum analyzer*. In contrast, an ordinary oscilloscope displays amplitude (on the vertical scale or axis) as a function of time (on the horizontal scale or axis), so it provides a *time domain* display.

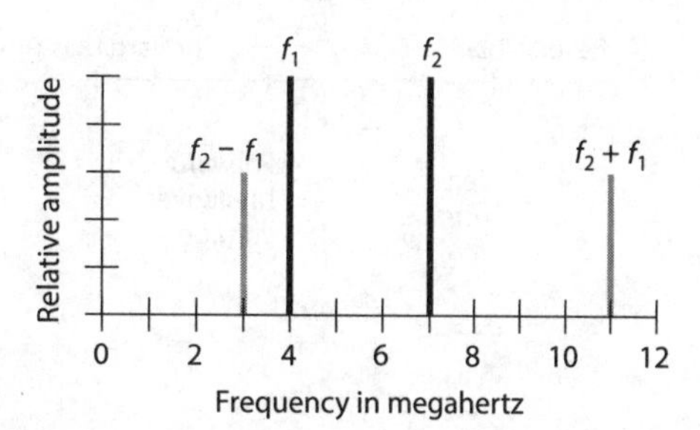

20-4 Spectral (frequency-domain) illustration of signal mixing.

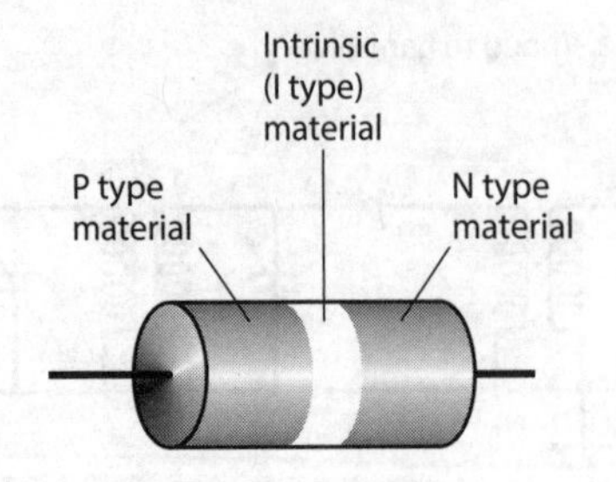

20-5 The PIN diode has a layer of intrinsic (I type) semiconductor material at the P-N junction.

Switching

The ability of diodes to conduct currents when forward-biased and block currents when reverse-biased makes them useful for switching in some applications. Diodes can perform switching operations much faster than any mechanical device—up to millions or even billions of on/off operations per second.

One type of diode, made for use as an RF switch, has a special semiconductor layer sandwiched in between the P type and N type material. The material in this layer is called an *intrinsic* (or *I type*) *semiconductor*. The *intrinsic layer* (or *I layer*) reduces the capacitance of the diode, allowing the device to function effectively at higher frequencies than an ordinary diode can. A diode with an I type semiconductor layer sandwiched in between the P and N type layers is called a *PIN diode* (Fig. 20-5).

Direct-current bias, applied to one or more PIN diodes, allows us to effectively channel RF currents to desired points without using relays and cables. A PIN diode also makes a good RF detector, especially at very high frequencies.

Voltage Regulation

Most diodes have an avalanche breakdown voltage much higher than the reverse-bias voltage ever gets. The value of the avalanche voltage depends on the internal construction of the diode, and on the characteristics of the semiconductor materials that compose it. *Zener diodes* are specially made to exhibit well-defined, constant avalanche voltages.

Suppose that a certain Zener diode has an avalanche voltage, also called the *Zener voltage*, of 50 V. If we apply a reverse bias to the P-N junction, the diode acts as an open circuit as long as the

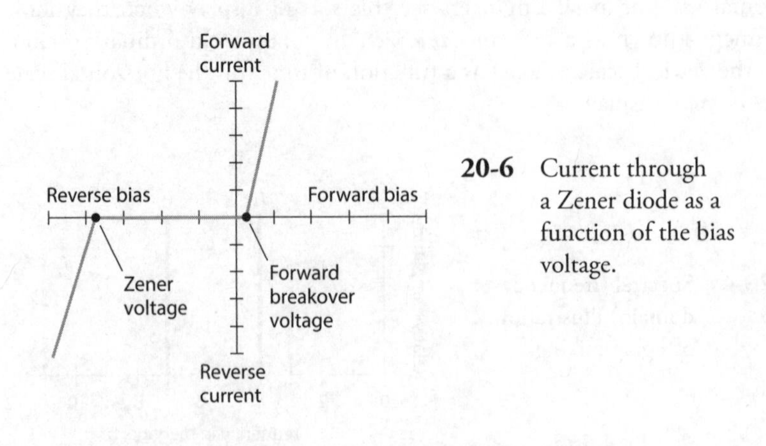

20-6 Current through a Zener diode as a function of the bias voltage.

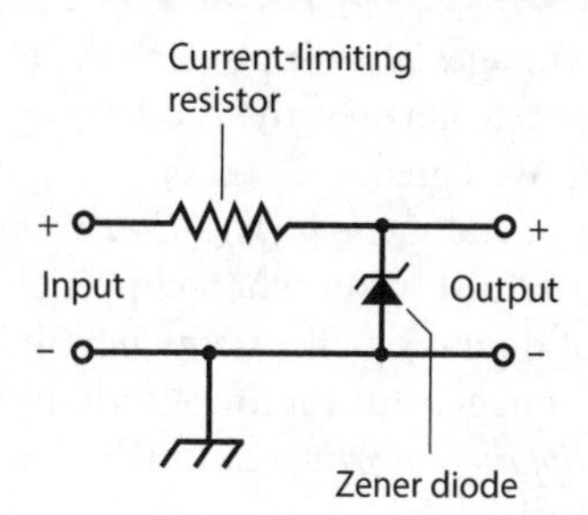

20-7 Connection of a Zener diode for voltage regulation. The series-connected resistor limits the current to prevent destruction of the diode.

potential difference between the P and N type materials remains less than 50 V. But if the reverse-bias voltage reaches 50 V, even for a moment, the diode conducts. This phenomenon prevents the instantaneous reverse-bias voltage from exceeding 50 V.

Figure 20-6 shows a graph of the current through a hypothetical Zener diode as a function of the voltage. The Zener voltage shows up as an abrupt rise in the reverse current as the reverse-bias voltage increases (that is, as we move toward the left along the horizontal axis).

Figure 20-7 shows a simple Zener-diode *voltage-regulator* circuit. Note the polarity of the diode: we connect the cathode to the positive pole and the anode to the negative pole, opposite from the way we use a diode in a rectifier circuit. The series-connected resistor limits the current that can flow through the Zener diode. Without that resistor, the diode would conduct excessive current and burn out.

Amplitude Limiting

In Chap. 19, we learned that a forward-biased diode will not conduct until the voltage reaches or exceeds the forward breakover voltage. We can state a corollary to this principle: A diode will always conduct when the forward-bias voltage reaches or exceeds the forward breakover voltage. In a diode, the potential difference between the P and N type wafers remains fairly constant—roughly equal to the forward breakover voltage—as long as current flows in the forward direction. In the case of silicon diodes, this potential difference or *voltage drop* is approximately 0.6 V. For germanium diodes, the voltage drop is roughly 0.3 V, and for selenium diodes it's around 1 V.

We can take advantage of the "constant-voltage-drop" property of semiconductor diodes when we want to build a circuit to limit the amplitude of a signal. Figure 20-8A shows how we can connect two identical diodes back-to-back in parallel with the signal path to *limit*, or *clip*, the positive

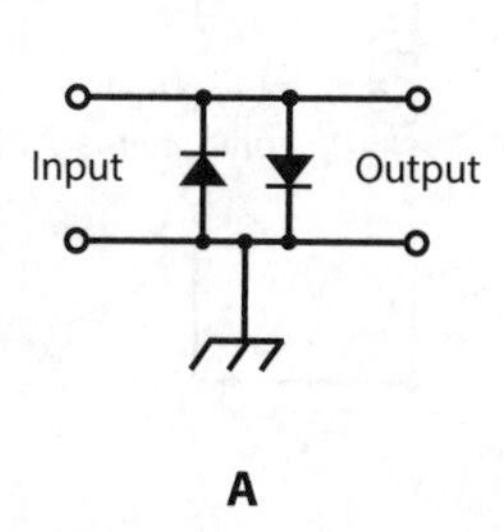

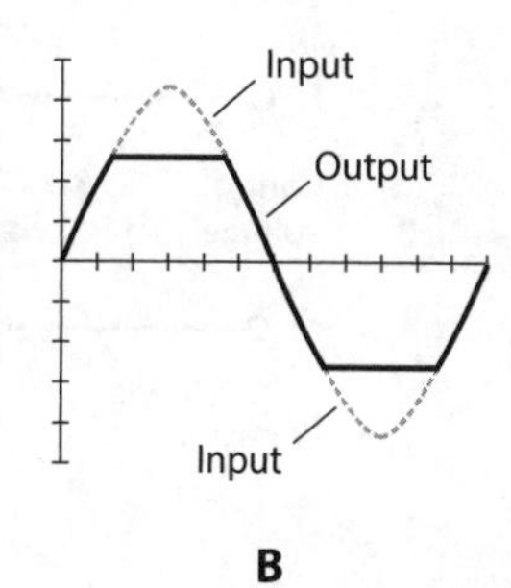

20-8 At A, connection of two diodes to act as an AC limiter. At B, illustration of sine-wave peaks cut off by the action of the diodes in an AC limiter.

and negative peak voltages of an input signal. In this configuration, the peak voltages are limited to the forward breakover voltage of the diodes. Figure 20-8B shows the input and output waveforms of a typical clipped AC signal.

The downside of the *diode voltage-limiter* circuit, such as the one shown in Fig. 20-8A, is the fact that it introduces distortion when clipping occurs. This distortion might not cause a problem for the reception of digital signals, frequency-modulated (FM) signals, or analog signals that rarely reach the limiting voltage. But for amplitude-modulated (AM) signals with peaks that rise past the limiting voltage, *clipping distortion* can make voices difficult to understand, and it utterly ruins the sound quality of music!

Frequency Control

When we reverse-bias a diode, we observe a region at the P-N junction with dielectric (insulating) properties. As you know from the last chapter, this zone is called the depletion region because it has a shortage (depleted supply) of majority charge carriers. The width of the depletion region depends on several parameters, including the reverse-bias voltage.

As long as the reverse bias remains lower than the avalanche voltage, varying the bias affects the width of the depletion region. The fluctuating width in turn varies the junction capacitance. This capacitance, which is always on the order of only a few picofarads, varies inversely with the square root of the reverse-bias voltage—again, as long as the reverse bias remains less than the avalanche voltage. For example, if we quadruple the reverse-bias voltage, the junction capacitance drops to half; if we decrease the reverse-bias voltage by a factor of 9, then the junction capacitance increases by a factor of 3.

Some diodes are manufactured especially for use as variable capacitors. Such a device is known as a varactor diode, as you learned in the last chapter. Varactors find their niche in a special type of circuit called a *voltage-controlled oscillator* (VCO). Figure 20-9 shows an example of a parallel-tuned *LC* circuit in a VCO, using a coil, a fixed capacitor, and a varactor. The fixed capacitor, whose value should greatly exceed the capacitance of the varactor, keeps the coil from short-circuiting the control voltage across the varactor. The schematic symbol for a varactor diode has two lines on the cathode side, as opposed to one line in the symbol for a conventional diode.

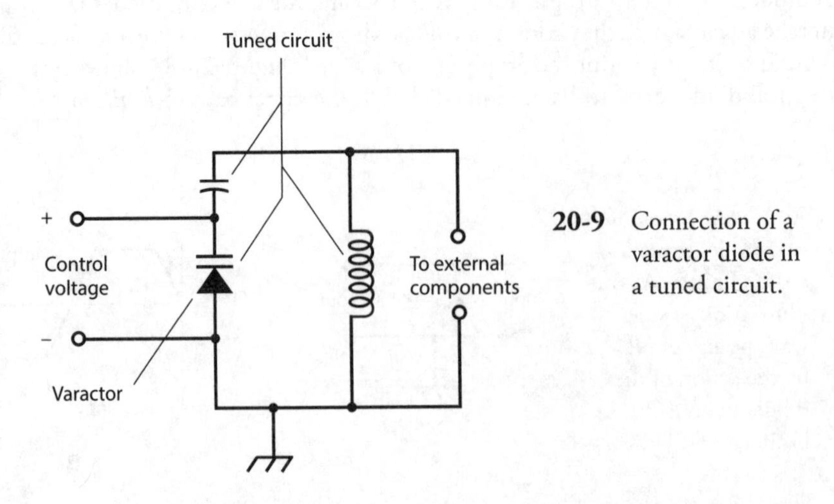

20-9 Connection of a varactor diode in a tuned circuit.

Oscillation and Amplification

Under certain conditions, diodes can generate or amplify *microwave* RF signals—that is, signals at extremely high AC frequencies. Devices commonly employed for these purposes include the *Gunn diode*, the *IMPATT diode*, and the *tunnel diode*.

Gunn Diodes

A Gunn diode can produce from 100 mW to 1 W of RF power output, and is manufactured with gallium arsenide (GaAs). A Gunn diode oscillates because of the *Gunn effect*, named after the engineer *J. Gunn* who first observed it in the 1960s while working for the International Business Machines (IBM) Corporation.

A Gunn diode doesn't work like a rectifier, detector, or mixer. Instead, the oscillation takes place as a result of a quirk called *negative resistance*, in which an increase in the instantaneous applied voltage causes a decrease in the instantaneous current flow under specific conditions.

Gunn-diode oscillators are often tuned using varactor diodes. A Gunn-diode oscillator, connected directly to a horn-shaped antenna, gives us a device known as a *Gunnplexer*. Amateur-radio experimenters use Gunnplexers for low-power wireless communication at frequencies of 10 GHz and above.

IMPATT Diodes

The acronym *IMPATT* comes from the words *imp*act *a*valanche *t*ransit *t*ime. This type of diode, like the Gunn diode, works because of the negative resistance phenomenon. An *IMPATT diode* constitutes a microwave oscillating device like a Gunn diode, except that it's manufactured from silicon rather than gallium arsenide. An IMPATT diode can operate as an amplifier for a microwave transmitter that employs a Gunn-diode oscillator. As an oscillator, an IMPATT diode produces about the same amount of output power, at comparable frequencies, as a Gunn diode does.

Tunnel Diodes

Another type of diode that will oscillate at microwave frequencies is the *tunnel diode*, also known as the *Esaki diode*. Made from GaAs semiconductor material, the tunnel diode produces only enough power to function as a local oscillator in a microwave radio receiver or transceiver. Tunnel diodes work well as weak-signal amplifiers in microwave receivers because they generate very little unwanted noise. The low-noise characteristic is typical of GaAs devices.

Energy Emission

Some semiconductor diodes emit radiant energy when current passes through the P-N junction in a forward direction. This phenomenon occurs as electrons "fall" from higher to lower energy states within atoms.

LEDs and IREDs

Depending on the exact mixture of the semiconductors used in manufacture, visible light of almost any color can be produced by forward-biased diodes. Infrared-emitting devices also exist. An *infrared-emitting diode* (IRED) produces energy at wavelengths slightly longer than those of visible red light.

The intensity of the radiant energy from an LED or IRED depends to some extent on the forward current. As the current rises, the brightness increases, but only up to a certain point. If the current continues to rise, no further increase in brilliance takes place, and we say that the LED or IRED is working in a state of *saturation.*

Digital Displays

Because LEDs can be made in various different shapes and sizes, they work well in digital displays. You've seen digital clock radios, hi-fi radios, calculators, and car radios that use LEDs. They make good indicators for "on/off," "a.m./p.m.," "battery low," and other conditions.

In recent years, LED displays have been largely replaced by *liquid-crystal displays* (LCDs). The LCD technology has advantages over LED technology, including lower power consumption and better visibility in direct sunlight. However, LCDs require *backlighting* in dim or dark environments.

Communications

Both LEDs and IREDs work well in communications systems because we can modulate their intensity to carry information. When the current through the device is sufficient to produce output, but not so great as to cause saturation, the LED or IRED output follows along with rapid current changes. This phenomenon allows engineers to build circuits for transmitting analog and digital signals over visible-light and IR energy beams. Some modern telephone systems make use of modulated light transmitted through clear glass or plastic cables, a technology called *fiberoptics.*

Special LEDs and IREDs, known as *laser diodes,* produce *coherent radiation.* The rays from these diodes aren't the intense, parallel beams that most people imagine when they think about lasers. A laser LED or IRED generates a cone-shaped beam of low intensity. However, we can use lenses to focus the emission into a parallel beam. The resulting rays have some of the same properties as the beams from large lasers, including the ability to travel long distances with minimal decrease in intensity.

Photosensitive Diodes

Most P-N junctions exhibit conductivity that varies with exposure to radiant energy, such as IR, visible light, and UV. Conventional diodes aren't normally affected by these rays because they're enclosed in opaque packages! Some *photosensitive diodes* have variable DC resistance that depends on the intensity of the visible, IR, or UV rays that strike their P-N junctions. Other types of diodes produce their own DC in the presence of radiant energy.

Silicon Photodiodes

A silicon diode, housed in a transparent case and constructed so that visible light can strike the barrier between the P and N type materials, forms a *silicon photodiode.* If we apply a reverse-bias voltage to the device at a certain level below the avalanche threshold, no current flows when the junction remains in darkness, but current flows when sufficient radiant energy strikes.

At constant reverse-bias voltage, the current varies in direct proportion to the intensity of the radiant energy, within certain limits. When radiant energy of variable intensity strikes the P-N junction of a reverse-biased silicon photodiode, the output current follows the light-intensity variations. This property makes silicon photodiodes useful for receiving modulated-light signals of the kind used in fiberoptic communications systems.

Silicon photodiodes exhibit greater sensitivity to radiant energy at some wavelengths than at others. The greatest sensitivity occurs in the *near infrared* part of the spectrum, at wavelengths slightly longer than the wavelength of visible red light.

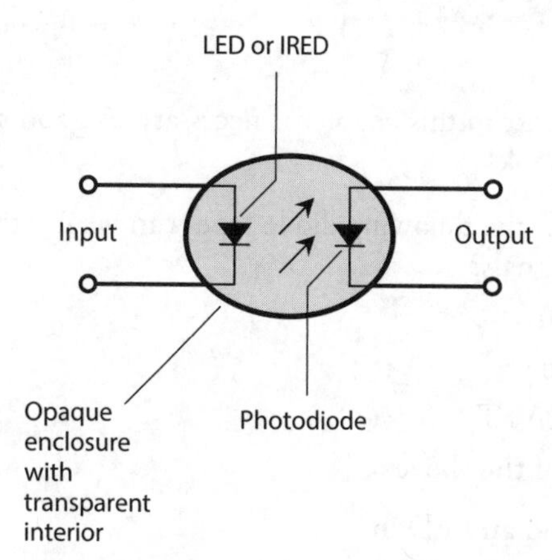

20-10 An optoisolator contains an LED or IRED at the input and a photodiode at the output with a transparent medium between them.

The Optoisolator

An LED or IRED and a photodiode can be combined in a single package to construct a component called an *optoisolator*. This device, shown schematically in Fig. 20-10, creates a modulated-light signal and sends it over a small, clear gap to a receptor. An LED or IRED converts the electrical input signal to visible light or IR. A photodiode changes the visible light or IR back into an electrical signal, which appears at the output. An opaque enclosure prevents external light or IR from reaching the photodiode, ensuring that the photodiode receives only the radiation from the internal source.

When we want to transfer, or *couple*, an AC signal from one circuit to another, the two stages interact if we make the transfer through a direct electrical connection. The input impedance of a given stage, such as an amplifier, can affect the behavior of the circuits that "feed power" to it, leading to problems. Optoisolators overcome this effect because the coupling occurs optically rather than electrically. If the input impedance of the second circuit changes, the first circuit "sees" no change in the output impedance, which comprises only the impedance of the LED or IRED. The circuits' impedances, and the electrical effects thereof, are literally isolated from each other.

Photovoltaic Cells

A silicon diode, with no bias voltage applied, can generate DC all by itself if enough IR, visible, or UV energy strikes its P-N junction. We call this phenomenon the *photovoltaic effect*. Solar cells work because of this effect.

Photovoltaic cells are specially manufactured to have the greatest possible P-N junction surface area, thereby maximizing the amount of radiant energy that strikes the junction. A single silicon photovoltaic cell can produce about 0.6 V of DC electricity in daylight. The amount of current that it can deliver, and therefore, the amount of power it can provide, depends on the surface area of the junction.

We can connect photovoltaic cells in series-parallel combinations to provide power for solid-state electronic devices, such as portable radios. The DC from these arrays can charge batteries, allowing for use of the electronic devices when radiant energy is not available (for example, at night or on dark days). A large assembly of solar cells, connected in series-parallel, is called a *solar panel*.

The power produced by a solar panel depends on the power from each individual cell, the number of cells in the panel, the intensity of the radiant energy that strikes the panel, and the angle at which the rays hit the surface of the panel. Some solar panels can produce several kilowatts of electrical power when the midday sun's unobstructed rays arrive perpendicular to the surfaces of all the cells.

Quiz

Refer to the text in this chapter if necessary. A good score is at least 18 correct. Answers are in the back of the book.

1. Which of the following diode types can we use to produce ultra-high frequency (UHF) or microwave signals?
   (a) Gunn
   (b) Tunnel
   (c) IMPATT
   (d) All of the above

2. We'll find an LED in
   (a) a microwave oscillator.
   (b) an optoisolator.
   (c) a rectifier.
   (d) a voltage regulator.

3. We can connect two diodes, as shown in Fig. 20-11, to obtain
   (a) oscillation.
   (b) demodulation.
   (c) amplification.
   (d) clipping.

4. In a crystal-set radio receiver, the diode should have the smallest possible
   (a) junction capacitance.
   (b) forward-breakover voltage.
   (c) avalanche voltage.
   (d) reverse bias.

5. Diodes can work as frequency multipliers because diodes
   (a) are nonlinear devices.
   (b) can demodulate signals.

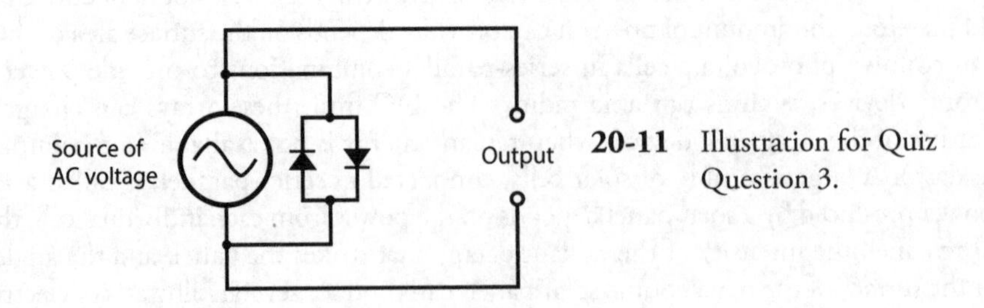

20-11 Illustration for Quiz Question 3.

(c) require no external power.

(d) All of the above

6. What do we call a condition in which the current through a component goes down as the voltage across it goes up?

(a) Nothing! It can't happen, so it has no name.

(b) Transconductance

(c) Negative resistance

(d) Current inversion

7. We input signals at 0.700 MHz and 1.300 MHz to a diode-based mixer. We get output at

(a) 0.500 MHz.

(b) 1.00 MHz.

(c) 0.600 MHz.

(d) All of the above

8. Which of the following diode types would we use in a circuit designed to measure the brilliance of a visible light source?

(a) A rectifier diode

(b) A photodiode

(c) A Zener diode

(d) An RF diode

9. Which of the following diode types would we use in a crystal set?

(a) A rectifier diode

(b) A photodiode

(c) A Zener diode

(d) An RF diode

10. When we apply negative voltage to a diode's anode and positive voltage to the cathode, we *sometimes* get

(a) avalanche breakdown.

(b) forward bias.

(c) junction depletion.

(d) forward breakover.

11. In the circuit of Fig. 20-12, the diode's nonlinearity allows it to function as a frequency multiplier. Nonlinearity also allows a diode to function as

(a) a microwave oscillator.

(b) a signal mixer.

(c) an optoisolator.

(d) a PV cell.

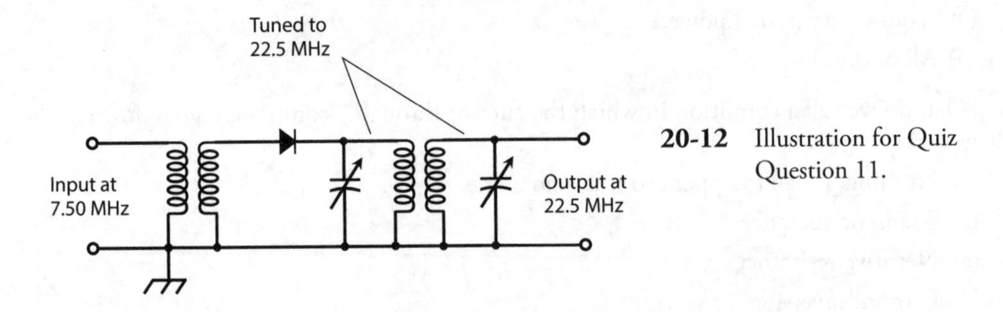

20-12 Illustration for Quiz Question 11.

12. Which of the following statements applies to the diode in a voltage-controlled oscillator (VCO)?
    (a) It requires high voltage.
    (b) It demodulates the signals.
    (c) It operates in a state of reverse bias.
    (d) It has low (almost no) resistance.

13. We'll usually find a varactor diode in
    (a) a voltage regulator.
    (b) a rectifier.
    (c) an optoisolator.
    (d) a VCO.

14. We input a pure sine-wave signal to a circuit with a diode that introduces nonlinearity. Which of the following statements holds true for the output?
    (a) It contains numerous signals at whole-number multiples (1, 2, 3, 4, etc.) of the input signal frequency.
    (b) It contains numerous signals at whole-number fractions of the input signal frequency (1, ½, ⅓, ¼, etc.).
    (c) It contains a signal at one and only one frequency: that of the input signal.
    (d) We need more information to answer this question.

15. The maximum voltage that a solar panel assembled with silicon PV cells can produce depends on
    (a) the number of cells in series, or in series-connected sets of parallel-connected cells.
    (b) the number of cells in parallel, or in parallel-connected sets of series-connected cells.
    (c) the surface area of the entire panel, regardless of how the cells are connected.
    (d) Any of the above

16. The visible light emitted by an LED occurs as
    (a) a result of avalanche effect.
    (b) the reverse-bias voltage decreases.
    (c) electrons lose energy in atoms.
    (d) the PN junction gets hot.

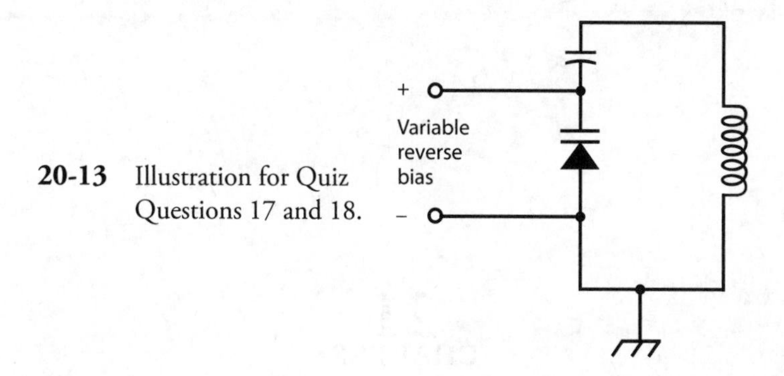

20-13 Illustration for Quiz Questions 17 and 18.

17. What happens to the *LC* resonant frequency as we increase the reverse-bias voltage across the varactor in the circuit of Fig. 20-13, assuming that avalanche breakdown never occurs?
   (a) It fluctuates.
   (b) It goes up.
   (c) It goes down.
   (d) Nothing.

18. What would happen if the fixed capacitor in the circuit of Fig. 20-13 shorted out?
   (a) The circuit would no longer amplify.
   (b) Excessive current would flow through the varactor.
   (c) The output voltage would become unstable.
   (d) The inductor would short out the source of reverse bias meant for the varactor.

19. Which of the following diode types is manufactured for minimum junction capacitance so it can work as an RF switch?
   (a) An IMPATT diode
   (b) A "cat's whisker"
   (c) A PIN diode
   (d) A Gunn diode

20. In which of the following devices does a photodiode serve an essential function?
   (a) A rectifier
   (b) An optoisolator
   (c) A frequency multiplier
   (d) A VCO

21
CHAPTER

Bipolar Transistors

THE WORD *TRANSISTOR* IS A CONTRACTION OF "CURRENT-*TRANS*FERRING RES*ISTOR*." A *BIPOLAR TRANSistor* has two P-N junctions. Two configurations exist for bipolar transistors: a P type layer between two N type layers (called *NPN*), or an N type layer between two P type layers (called *PNP*).

NPN versus PNP

Figure 21-1A is a simplified drawing of an *NPN bipolar transistor*, and Fig. 21-1B shows the symbol that engineers use for it in schematic diagrams. The P type, or center, layer constitutes the *base*. One of the N type semiconductor layers forms the *emitter*, and the other N type layer forms the *collector*. We label the base, the emitter, and the collector B, E, and C, respectively.

A *PNP bipolar transistor* has two P type layers, one on either side of a thin, N type layer as shown in Fig. 21-2A. The schematic symbol appears at Fig. 21-2B. The N type layer constitutes the base. One of the P type layers forms the emitter, and the other P type layer forms the collector. As with the NPN device, we label these electrodes B, E, and C.

We can tell from a schematic diagram whether the circuit designer means for a transistor to be NPN type or PNP type. Once we realize that the arrow always goes with the emitter, we can identify the three electrodes without having to label them. In an NPN transistor, the arrow at the emitter points outward. In a PNP transistor, the arrow at the emitter points inward.

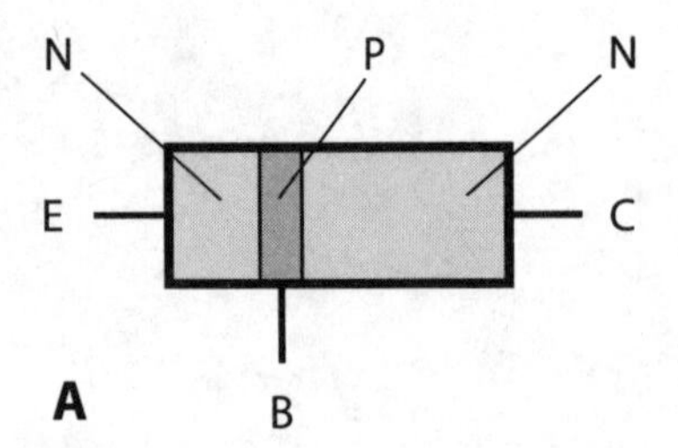

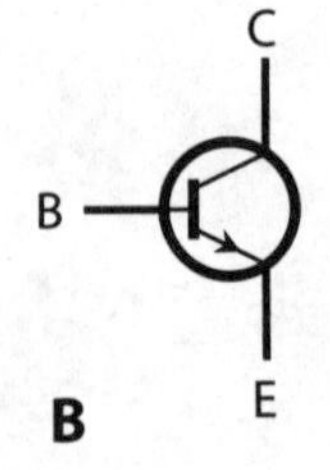

21-1 At A, a simplified structural drawing of an NPN transistor. At B, the schematic symbol. We identify the electrodes as E = emitter, B = base, and C = collector.

21-2 At A, a simplified structural drawing of a PNP transistor. At B, the schematic symbol. We identify the electrodes as E = emitter, B = base, and C = collector.

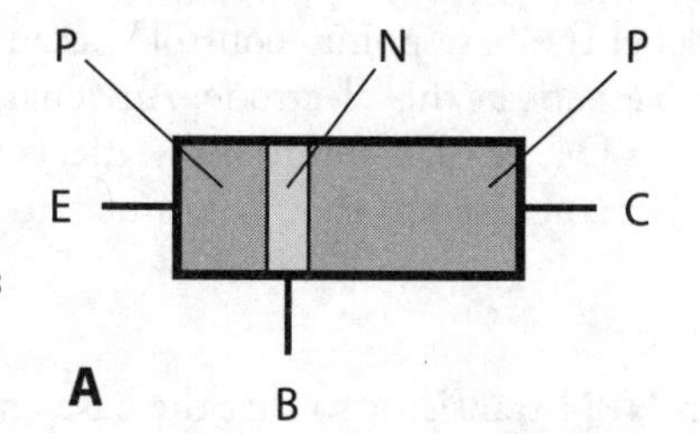

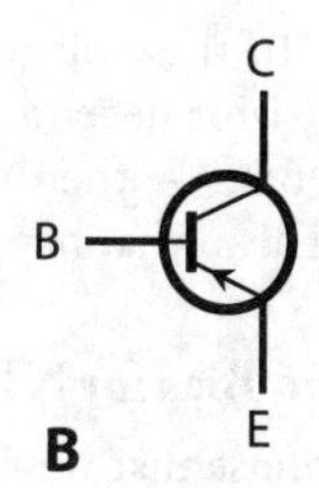

Generally, PNP and NPN transistors can perform the same electronic functions. However, they require different voltage polarities, and the currents flow in different directions. In many situations, we can replace an NPN device with a PNP device or vice versa, reverse the power-supply polarity, and expect the circuit to work with the replacement device if it has the appropriate specifications.

Biasing

For a while, let's imagine a bipolar transistor as two diodes connected in *reverse series* (that is, in series but in opposite directions). We can't normally connect two diodes together this way and get a working transistor, but the analogy works for *modeling* (technically describing) the behavior of bipolar transistors. Figure 21-3A shows a dual-diode NPN transistor model. The base is at the connection point between the two anodes. The cathode of one diode forms the device's emitter, while the cathode of the other diode forms the collector. Figure 21-3B shows the equivalent "real-world" NPN transistor circuit.

NPN Biasing

In an NPN transistor, we normally bias the device so that the collector voltage is positive with respect to the emitter. We illustrate this scheme by indicating the battery's polarity in Figs. 21-3A and 21-3B. Typical DC voltages for a transistor's power supply range between about 3 V and 50 V, meaning that the collector electrode and the emitter electrode differ in potential by 3 V to 50 V.

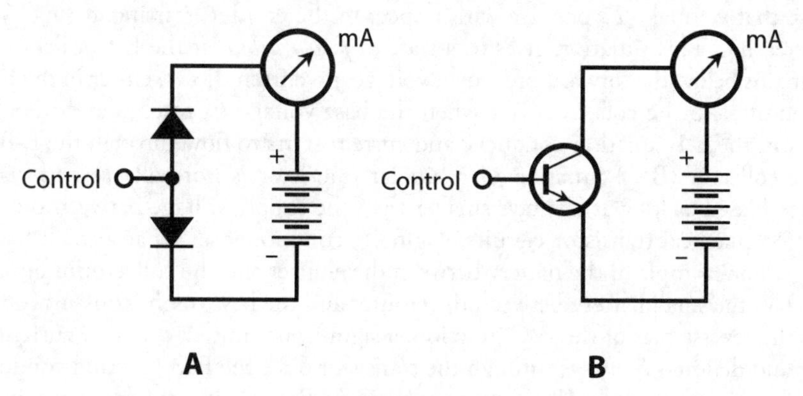

21-3 At A, the dual-diode model of a simple NPN circuit. At B, the actual transistor circuit.

In these diagrams, we label the base point "control" because the flow of current through the transistor depends on what happens at this electrode. Any change in the voltage that we apply to the base—either in the form of DC or AC—profoundly affects what happens inside the transistor, and also what happens in other components that we connect to it.

Zero Bias for NPN

Suppose that we connect an NPN transistor so that the base and the emitter are at the same voltage. We call this condition *zero bias* because the potential difference between the two electrodes equals 0 V. In this situation, the *emitter-base current*, often called simply the *base current* and denoted I_B, equals zero. The *emitter-base* (E-B) *junction*, which is a P-N junction, operates below its forward breakover voltage, preventing current from flowing between the emitter and the base. Zero bias also prevents any current from flowing between the emitter and the collector unless we inject an AC signal at the base to change things. Such a signal must, at least momentarily, attain a positive voltage equal to or greater than the forward breakover voltage of the E-B junction. When no current flows between the emitter and the collector in a bipolar transistor under no-signal conditions, we say that the device is operating in a state of *cutoff*.

Reverse Bias for NPN

Imagine that we connect an extra battery between the base and the emitter in the circuit of Fig. 21-3B, with the polarity set such as to force the base voltage E_B to become negative with respect to the emitter. The presence of the new battery causes the E-B junction to operate in a state of reverse bias. No current flows through the E-B junction in this situation (as long as the new battery voltage is not so great that avalanche breakdown occurs at the E-B junction). In that sense, the transistor behaves in the same way when reverse-biased as it does when zero-biased. If we inject an AC signal at the base with the intent to cause a flow of current during part of the cycle, the signal must attain, at least momentarily, a positive voltage high enough to overcome the sum of the reverse bias voltage (produced by the battery) and the forward breakover voltage of the E-B junction. We'll have a harder time getting a reverse-biased transistor to conduct than we'll have getting a zero-biased transistor to conduct.

Forward Bias for NPN

Now suppose that we make E_B positive with respect to the emitter, starting at small voltages and gradually increasing. This situation gives us a state of *forward bias* at the E-B junction. If the forward bias remains below the forward breakover voltage, no current flows, either in the E-B junction or from the emitter to the collector. But when the base voltage E_B reaches and then exceeds the breakover point, the E-B junction conducts, and current starts to flow through the E-B junction.

The base-collector (B-C) junction of a bipolar transistor is normally reverse-biased. It will remain reverse-biased as long as E_B stays smaller than the supply voltage between the emitter and the collector. In practical transistor circuits, engineers commonly set E_B at a small fraction of the supply voltage. For example, if the battery between the emitter and the collector in Figs. 21-3A and B provides 12 V, the small battery between the emitter and the base might consist of a single 1.5-V cell. Despite the reverse bias of the B-C junction, a significant emitter-collector current, called *collector current* and denoted I_C, flows through the transistor once the E-B junction conducts.

In a real transistor circuit, such as the one shown in Fig. 21-3B, the meter reading will jump when we apply DC base bias to reach and then exceed the forward breakover voltage of the E-B

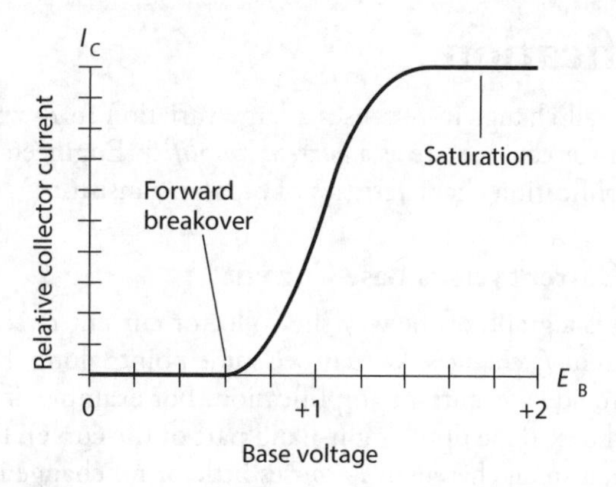

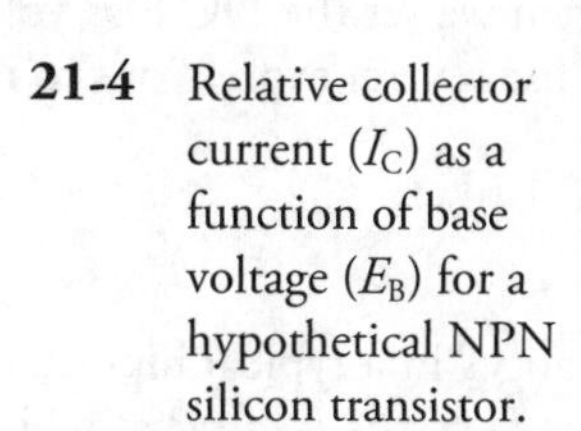

21-4 Relative collector current (I_C) as a function of base voltage (E_B) for a hypothetical NPN silicon transistor.

junction. If we continue to increase the forward bias at the E-B junction, even a small rise in E_B, attended by a rise in the base current I_B, will cause a large increase in the collector current I_C. Figure 21-4 portrays the situation as a graph. Once current starts to flow in the collector, increasing E_B a tiny bit will cause the I_C to go up a lot! However, if E_B continues to rise, we'll eventually arrive at a voltage where the curve for I_C versus E_B levels off. Then we say that the transistor has reached a state of *saturation*. It's "running wide open," conducting as much as it possibly can, given a fixed potential difference between the collector and the emitter.

PNP Biasing

We can describe the situation inside a PNP transistor, as we vary the voltage of the small battery or cell between the emitter and the base, as a "mirror image" of the case for an NPN device. The diodes are reversed, the arrow points inward rather than outward in the transistor symbol, and all the polarities are reversed. The dual-diode PNP model, along with the "real-world" transistor circuit, appear in Fig. 21-5. We can repeat the foregoing discussion (for the NPN case) almost verbatim, except that we must replace every occurrence of the word "positive" with the word "negative." Qualitatively, the same things happen in the PNP device as in the NPN case.

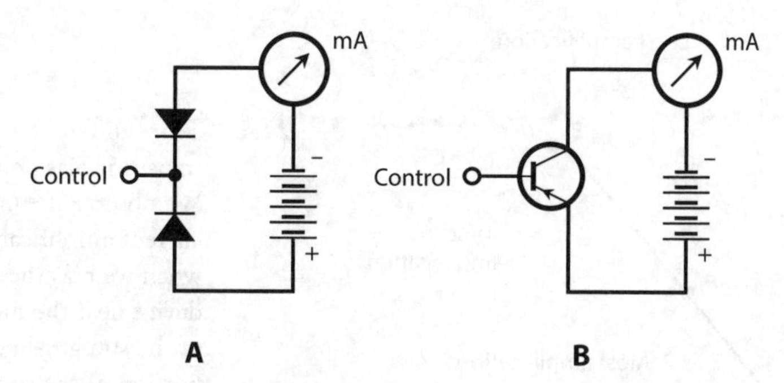

21-5 At A, the dual-diode model of a simple PNP circuit. At B, the actual transistor circuit.

Amplification

Because a small change in I_B causes a large variation in I_C when we set the DC bias voltages properly, a transistor can operate as a *current amplifier*. Engineers use several expressions to describe the current-amplification characteristics of bipolar transistors.

Collector Current versus Base Current

Figure 21-6 is a graph of the way the collector current I_C changes in a typical bipolar transistor as the base current I_B changes. We can see some points along this I_C versus I_B curve at which a transistor won't provide any current amplification. For example, if we operate the transistor in saturation (shown by the extreme upper right-hand part of the curve), the I_C versus I_B curve runs horizontally. In this zone, a small change in I_B causes little or no change in I_C. But if we bias the transistor near the middle of the "ramped-up" straight-line part of the curve in Fig. 21-6, the transistor will work as a current amplifier.

Whenever we want a bipolar transistor to amplify a signal, we must bias the device so that a small change in the current between the emitter and the base will result in a large change in the current between the emitter and the collector. The ideal voltages for E_B (the base bias) and E_C (the power-supply voltage) depend on the internal construction of the transistor, and also on the chemical composition of the semiconductor materials that make up its N type and P type sections.

Static Current Characteristics

We can describe the current-carrying characteristics of a bipolar transistor in simplistic terms as the *static forward current transfer ratio*. This parameter comes in two "flavors," one that describes the collector current versus the emitter current when we place the base at electrical ground (symbolized H_{FB}), and the other that describes the collector current versus the base current when we place the emitter at electrical ground (symbolized H_{FE}).

The quantity H_{FB} equals the ratio of the collector current to the emitter current at a given instant in time with the base grounded:

$$H_{FB} = I_C / I_E$$

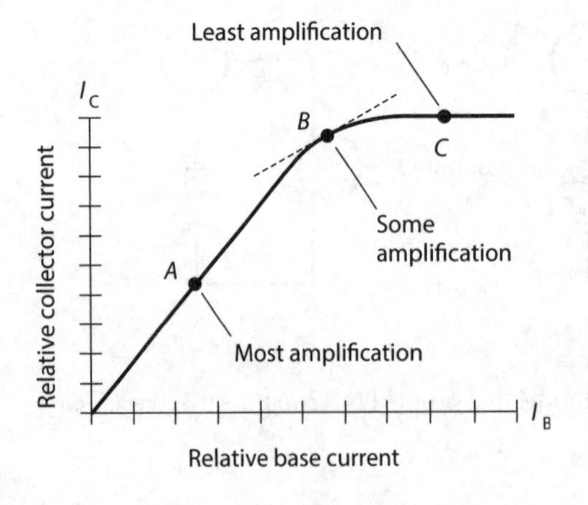

21-6 Three different transistor bias points. We observe the most current amplification when we bias the device near the middle of the straight-line portion of the curve.

For example, if an emitter current I_E of 100 mA results in a collector current I_C of 90 mA, then we can calculate

$$H_{FB} = 90/100 = 0.90$$

If $I_E = 100$ mA and $I_C = 95$ mA, then

$$H_{FB} = 95/100 = 0.95$$

The quantity H_{FE} equals the ratio of the collector current to the base current at a given instant in time with the emitter grounded:

$$H_{FE} = I_C / I_B$$

For example, if a base current I_B of 10 mA results in a collector current I_C of 90 mA, then we can calculate

$$H_{FE} = 90/10 = 9.0$$

If $I_B = 5.0$ mA and $I_C = 95$ mA, then

$$H_{FE} = 95/5.0 = 19$$

Alpha

We can describe the current variations in a bipolar transistor by dividing the *difference* in I_C by the *difference* in I_E that occurs when we apply a small signal to the emitter of a transistor with the base connected to electrical ground (or placed at the same potential difference as electrical ground). We call this ratio the *alpha*, symbolized as the lowercase Greek letter alpha (α). Let's abbreviate the words "the difference in" by writing *d*. Then mathematically, we can define

$$\alpha = dI_C / dI_E$$

We call this quantity the *dynamic current gain* of the transistor for the grounded-base situation. The alpha for any transistor is always less than 1, because whenever we apply a signal to the input, the base "bleeds off" at least a little current from the emitter before it shows up at the collector.

Beta

We get an excellent definition of current amplification for "real-world signals" when we divide the difference in I_C by the difference in I_B as we apply a small signal to the base of a transistor with the emitter at electrical ground. Then we get the dynamic current gain for the grounded-emitter case. We call this ratio the *beta*, symbolized as the lowercase Greek letter beta (β). Once again, let's abbreviate the words "the difference in" as *d*. Then we have

$$\beta = dI_C / dI_B$$

The beta for any transistor can exceed 1—and often does, greatly!—so this expression for "current gain" lives up to its name. However, under some conditions, we might observe a beta of less than 1. This condition can occur if we improperly bias a transistor, if we choose the wrong type of transistor for a particular application, or if we attempt to operate the transistor at a signal frequency that's far higher than the maximum frequency for which it is designed.

How Alpha and Beta Relate

Whenever base current flows in a bipolar transistor, we can calculate the beta in terms of the alpha with the formula

$$\beta = \alpha / (1 - \alpha)$$

and we can calculate the alpha in terms of the beta using the formula

$$\alpha = \beta / (1 + \beta)$$

With a little bit of algebra, we can derive these formulas from the fact that, at any instant in time, the collector current equals the emitter current minus the base current; that is,

$$I_C = I_E - I_B$$

"Real-World" Amplification

Let's look at Fig. 21-6 again. It's a graph of the collector current as a function of the base current (I_C versus I_B) for a hypothetical transistor. We can infer both H_{FE} and β from this graph. We can find H_{FE} at any particular point on the curve when we divide I_C by I_B at that point. Geometrically, the value of β at any given point on the curve equals the *slope* ("rise over run") of a *tangent line* at that point. On a two-dimensional coordinate grid, the tangent to a curve at a point constitutes the straight line that intersects the curve at that point without crossing the curve.

In Fig. 21-6, the tangent to the curve at point *B* appears as a dashed, straight line; the tangents to the curve at points *A* and *C* lie precisely along the curve (and, therefore, don't show up visually). As the slope of the line tangent to the curve increases, the value of β increases. Point *A* provides the highest value of β for this particular transistor, as long as we don't let the input signal get too strong. Points in the immediate vicinity of A provide good β values as well.

For small-signal amplification, point *A* in Fig. 21-6 represents a good bias level. Engineers would say that it's a favorable *operating point*. The β figure at point *B* is smaller (the curve slopes less steeply upward as we move toward the right) than the β figure at point *A*, so point *B* represents a less favorable operating point for small-signal amplification than point *A* does. At point *C*, we can surmise that $\beta = 0$ because the slope of the curve equals zero in that vicinity (it doesn't "rise" at all as we "run" toward the right). The transistor won't amplify weak signals when we bias it at point *C* or beyond.

Overdrive

Even when we bias a transistor so that it can produce the greatest possible current amplification (at or near point *A* in Fig. 21-6), we can encounter problems if we inject an AC input signal that's too strong. If the input-signal amplitude gets large enough, the transistor's operating point might move to or beyond point *B*, off the coordinate grid to the left, or both, during part of the signal cycle. In that case, the effective value of β will decrease. Figure 21-7 shows why this effect occurs. Points *X* and *Y* represent the instantaneous current extremes during the signal cycle in this particular case. Note that the slope of the line connecting points *X* and *Y* is less than the slope of the straight-line part of the curve at and near point *A*.

When our AC input signal is so strong that it drives the transistor to the extreme points *X* and *Y*, as shown in Fig. 21-7, a transistor amplifier introduces *distortion* into the signal, meaning that the output wave does not have the same shape as the input wave. We call this phenomenon *nonlinearity*. We can sometimes tolerate this condition, but often it's undesirable. Under most circumstances, we'll want our amplifier to remain *linear* (or to exhibit excellent *linearity*), meaning that the output wave has the same shape as (although probably stronger than) the input wave.

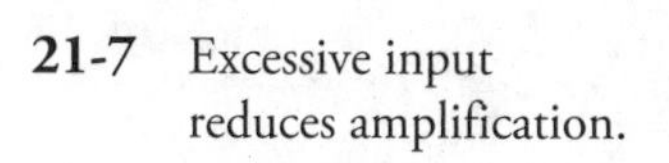
21-7 Excessive input reduces amplification.

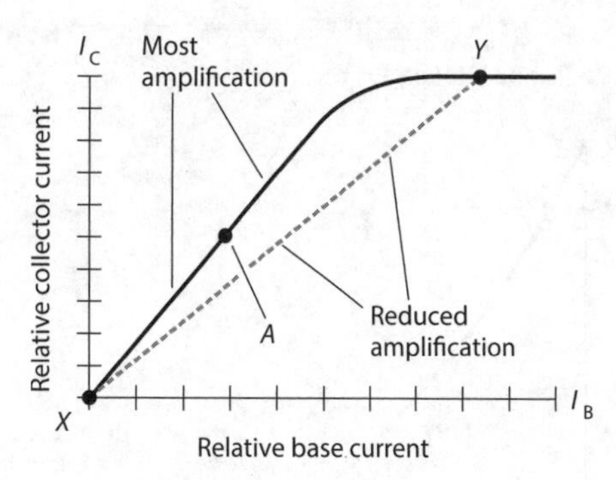

When the input signal to a transistor amplifier exceeds a certain critical maximum, we get a condition called *overdrive*. An overdriven transistor operates in or near saturation during part of the input signal cycle. Overdrive reduces the overall circuit efficiency, causes excessive collector current to flow, and can overheat the base-collector (B-C) junction. Sometimes overdrive can physically destroy a transistor.

Gain versus Frequency

A bipolar transistor exhibits an amplification factor (gain) that decreases as the signal frequency increases. Some bipolar transistors can amplify effectively at frequencies up to only a few megahertz. Other devices can work into the gigahertz range. The maximum operating frequency for a particular bipolar transistor depends on the capacitances of the P-N junctions inside the device. A low *junction capacitance* value translates into a high *maximum usable frequency*.

Expressions of Gain

You've learned about *current gain* expressed as a ratio. You'll also hear or read about *voltage gain* or *power gain* in amplifier circuits. You can express any gain figure as a ratio. For example, if you read that a circuit has a voltage gain of 15, then you know that the output signal voltage equals 15 times the input signal voltage. If someone tells you that the power gain of a circuit is 25, then you know that the output signal power equals 25 times the input signal power.

Alpha Cutoff

Suppose that we operate a bipolar transistor as a current amplifier, and we deliver an input signal to it at 1 kHz. Then we steadily increase the input-signal frequency so that the value of α declines. We define the *alpha cutoff frequency* of a bipolar transistor, symbolized f_α, as the frequency at which α decreases to 0.707 times its value at 1 kHz. (Don't confuse this use of the term "cutoff" with the state of "cutoff" that we get when we zero-bias or reverse-bias a transistor under no-signal conditions!) A transistor can have considerable gain at its alpha cutoff frequency. By looking at this specification for a particular transistor, we can get an idea of how rapidly it loses its ability to amplify as the frequency goes up.

Beta Cutoff Frequency

Imagine that we repeat the above-described variable-frequency experiment while watching β instead of α. We discover that β decreases as the frequency increases. We define the *beta cutoff frequency* (also

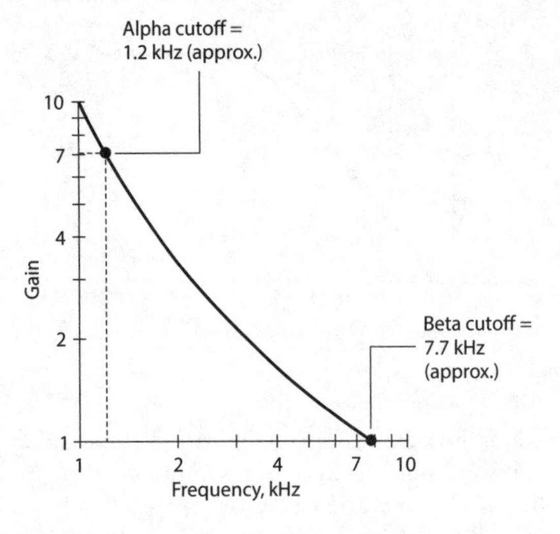

21-8 Alpha cutoff and beta cutoff frequencies for a hypothetical transistor.

called the *gain bandwidth product*) for a bipolar transistor, symbolized f_β or f_T, as the frequency at which β gets down to 1. If we try to make a transistor amplify above its beta cutoff frequency, we'll fail!

Figure 21-8 shows the alpha cutoff and beta cutoff frequencies for a hypothetical transistor on a graph of gain versus signal frequency. Note that the scales of this graph are not linear and the divisions are unevenly spaced. We call this type of plot a *log-log* graph because both scales are *logarithmic* rather than linear. The value on either scale increases in proportion to the *base-10 logarithm* of the distance from the origin, rather than varying in direct proportion to that distance.

Common-Emitter Configuration

A bipolar transistor can be "wired up" in three general ways. We can ground the emitter for the signal, we can ground the base for the signal, or we can ground the collector for the signal. An often-used arrangement is the *common-emitter circuit.* "Common" means "grounded for the signal." Figure 21-9 shows the basic configuration.

Even if a circuit point remains at ground potential for signals, it can have a significant DC voltage with respect to electrical ground. In the circuit shown, capacitor C_1 appears as a short circuit to the AC signal, so the emitter remains at *signal ground.* But resistor R_1 causes the emitter to attain and hold a certain positive DC voltage with respect to electrical ground (or a negative voltage, if we replace the NPN transistor with a PNP device). The exact DC voltage at the emitter depends on the resistance of R_1, and on the bias at the base. We set the DC base bias by adjusting the ratio of the values of resistors R_2 and R_3. The DC base bias can range from 0 V, or ground potential, to +12 V, which equals the power-supply voltage. Normally it's a couple of volts.

Capacitors C_2 and C_3 block DC to or from the external input and output circuits, while letting the AC signal pass. Resistor R_4 keeps the output signal from "shorting out" through the power supply. A signal enters the common-emitter circuit through C_2, so the signal causes the base current I_B to vary. The small fluctuations in I_B cause large changes in the collector current I_C. This current passes through resistor R_4, producing a fluctuating DC voltage across it. The AC component of this voltage passes unhindered through capacitor C_3 to the output.

The circuit of Fig. 21-9 represents the basis for many amplifier systems at all commonly encountered signal frequencies. The common-emitter configuration can produce more gain than

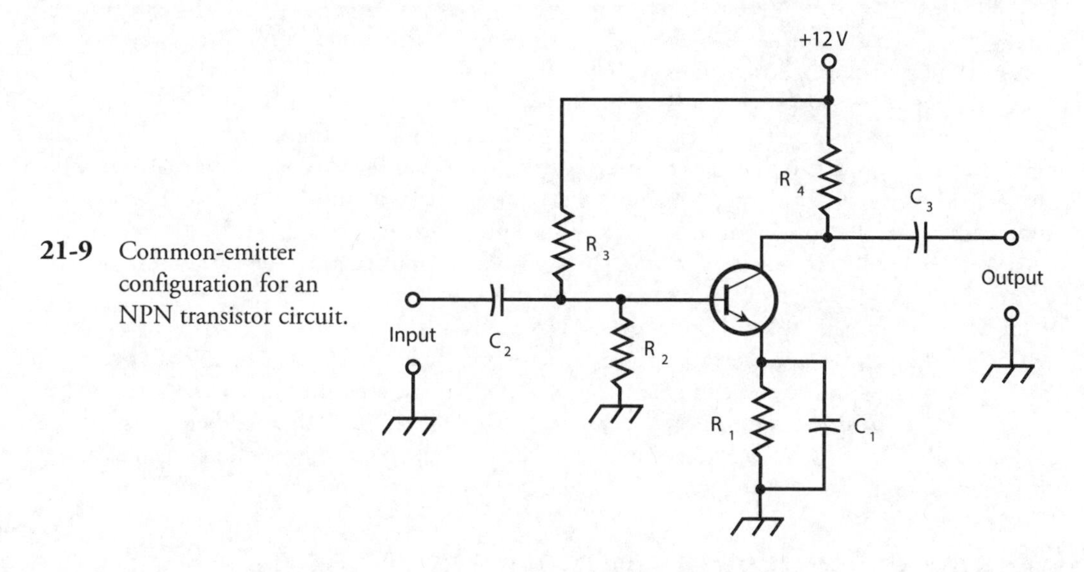

21-9 Common-emitter configuration for an NPN transistor circuit.

any other arrangement. The output wave appears inverted (in phase opposition) with respect to the input wave. If the input signal constitutes a pure sine wave, then the common-emitter circuit shifts the signal phase by 180°.

Common-Base Configuration

As its name implies, the *common-base circuit* (Fig. 21-10) has the base at signal ground. The DC bias is the same as that for the common emitter circuit, but we apply the input signal at the emitter instead of at the base. This arrangement gives rise to fluctuations in the voltage across resistor R_1, causing variations in I_B. These small current fluctuations produce large variations in the current through R_4. Therefore, amplification occurs. The output wave follows along in phase with the input wave.

The signal enters the transistor through capacitor C_1. Resistor R_1 keeps the input signal from "shorting out" to ground. Resistors R_2 and R_3 provide the base bias. Capacitor C_2 holds the base at

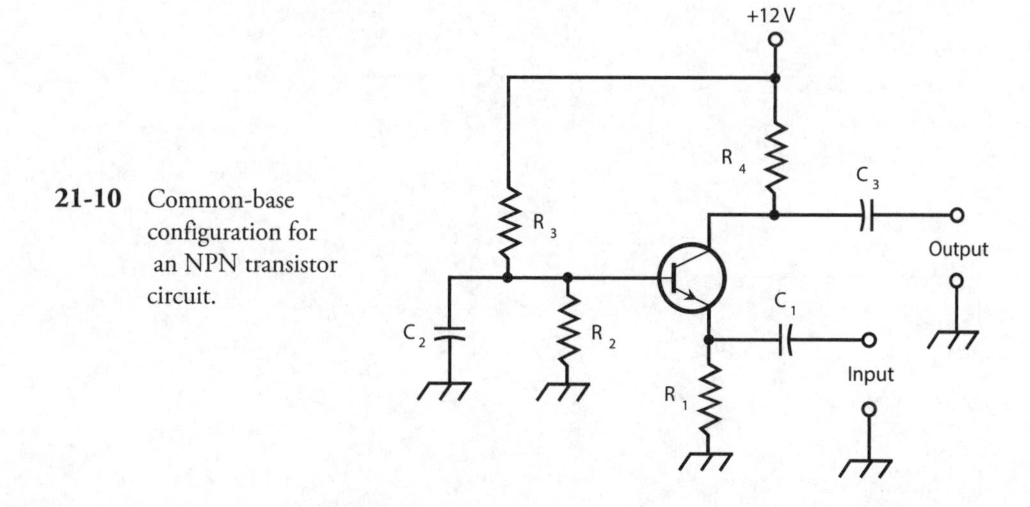

21-10 Common-base configuration for an NPN transistor circuit.

signal ground. Resistor R_4 keeps the output signal from "shorting out" through the power supply. We get the output through capacitor C_3. The common-base circuit exhibits a relatively low input impedance, and provides somewhat less gain than a common-emitter circuit.

A common-base amplifier offers better *stability* than most common-emitter circuits do. By "better stability," we mean that the common-base circuit is less likely to *break into oscillation* (generate a signal of its own) as a result of amplifying some of its own output. The main reason for this "good behavior" is the low input impedance of the common-base circuit, which requires the input signal to offer significant power to drive the system into amplification. The common-base circuit isn't sensitive enough to get out of control easily!

Sensitive amplifiers, such as optimally-biased common-emitter circuits, can pick up some of their own output as a result of stray capacitance between the input and output wires. This little bit of "signal leakage" can provide enough energy at the input to cause the whole circuit to "chase its own tail." When *positive* (in-phase) *feedback* gives rise to oscillation in an amplifier, we say that the amplifier suffers from *parasitic oscillation,* or *parasitics*, which can cause a radio transmitter to put out signals on unauthorized frequencies, or make a radio receiver stop working altogether.

Common-Collector Configuration

A *common-collector circuit* (Fig. 21-11) operates with the collector at signal ground. We apply the AC input signal at the transistor base, just as we do with the common-emitter circuit. The signal passes through C_2 onto the base. Resistors R_2 and R_3 provide the correct base bias. Resistor R_4 limits the current through the transistor. Capacitor C_3 keeps the collector at signal ground. Fluctuating DC flows through R_1, and a fluctuating voltage, therefore, appears across it. The AC part of this voltage passes through C_1 to the output. Because the output follows the emitter current, some engineers and technicians call this arrangement an *emitter-follower circuit.*

The output wave of a common-collector circuit appears exactly in phase with the input wave. The transistor exhibits a relatively high input impedance, while its output impedance remains low. For this reason, the common-collector circuit can take the place of a transformer when we want to match a high-impedance circuit to a low-impedance circuit or load. A well-designed emitter-follower circuit can function over a wider range of frequencies than a typical wirewound transformer can.

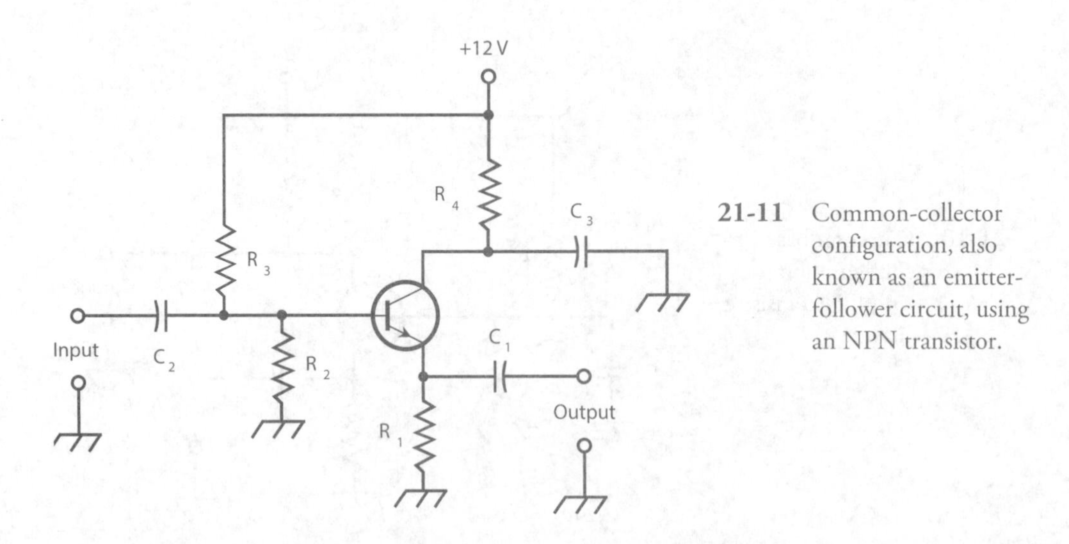

21-11 Common-collector configuration, also known as an emitter-follower circuit, using an NPN transistor.

Quiz

Refer to the text in this chapter if necessary. A good score is at least 18 correct. Answers are in the back of the book.

1. In a common-base transistor circuit, the output and input waves differ in phase by
   (a) ¼ of a cycle.
   (b) ⅓ of a cycle.
   (c) ½ of a cycle.
   (d) None of the above

2. Which of the following circuit configurations do engineers sometimes use in place of conventional wirewound transformers to match a high input impedance to a low output impedance?
   (a) Common emitter
   (b) Common base
   (c) Common collector
   (d) Any of the above

3. Current will *never* flow in the B-C junction of a grounded-emitter bipolar transistor when we
   (a) reverse-bias the E-B junction and apply no input signal.
   (b) forward-bias the E-B junction beyond forward breakover and apply no input signal.
   (c) zero-bias the E-B junction and apply a strong input signal.
   (d) forward-bias the E-B junction beyond forward breakover and apply a weak input signal.

4. Figure 21-12 illustrates a bipolar transistor and several other components in
   (a) a common-emitter configuration.
   (b) an emitter-follower configuration.
   (c) a common-base configuration.
   (d) a common-collector configuration.

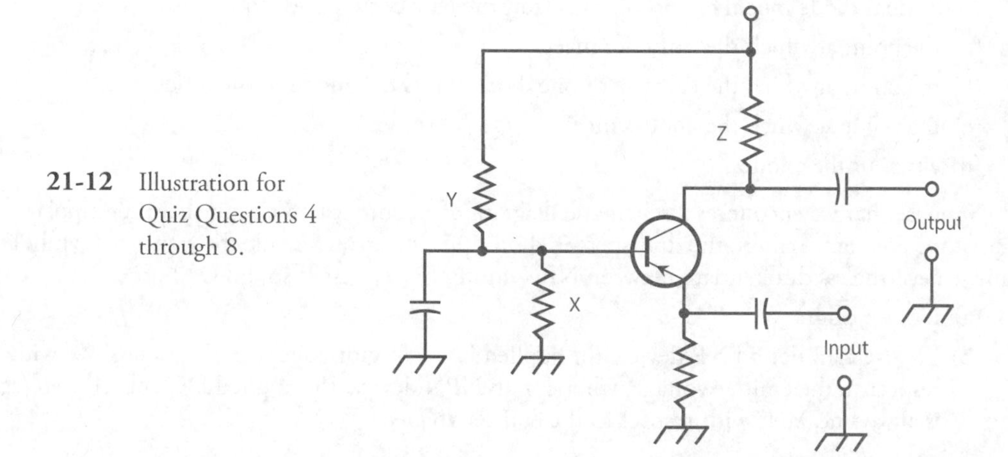

21-12 Illustration for Quiz Questions 4 through 8.

5. What, if any, major errors exist in the circuit of Fig. 21-12?
   (a) Nothing is wrong, assuming that we choose the component values properly.
   (b) We should use an NPN transistor, not a PNP transistor.
   (c) The power-supply polarity at the collector should be positive, not negative.
   (d) We should transpose the input and output terminals.
6. In the circuit of Fig. 21-12, what purpose does component X serve?
   (a) It keeps the signal from "shorting out" to ground.
   (b) It helps to establish the proper bias at the base.
   (c) It ensures that the circuit won't break into oscillation.
   (d) It keeps the base at signal ground.
7. In the circuit of Fig. 21-12, what purpose does component Y serve?
   (a) It keeps the input isolated from the output.
   (b) It keeps the output signal from "shorting out" through the power supply.
   (c) It ensures that the circuit won't break into oscillation.
   (d) It helps to establish the proper bias at the base.
8. In the circuit of Fig. 21-12, what purpose does component Z serve?
   (a) It helps the circuit to function as an oscillator by providing feedback.
   (b) It keeps the output signal from "shorting out" through the power supply.
   (c) It helps to establish the proper bias at the base.
   (d) It ensures that the output wave remains in phase opposition with respect to the input wave.
9. In an emitter-follower circuit, we apply the input signal between the
   (a) collector and ground.
   (b) emitter and collector.
   (c) base and ground.
   (d) base and collector.
10. In the dual-diode model of a PNP transistor, the base corresponds to
    (a) the point at which the cathodes meet.
    (b) the point at which the cathode of one diode meets the anode of the other.
    (c) the point at which the anodes meet.
    (d) either of the anodes.
11. Suppose that we encounter a schematic diagram of a complicated circuit that uses bipolar transistors. For some reason, the draftsperson didn't put the arrows inside the transistor symbols. Can we nevertheless differentiate between NPN and PNP devices? If so, how?
    (a) No, we can't.
    (b) Yes, we can. For a PNP device, the applied DC collector voltage is always positive with respect to the emitter voltage, while for an NPN device, the applied DC collector voltage is always negative with respect to the emitter voltage.

(c) Yes, we can. For a PNP device, the applied DC collector voltage is always negative with respect to the emitter voltage, while for an NPN device, the applied DC collector voltage is always positive with respect to the emitter voltage.

(d) Yes, we can. For a PNP device, the E-B junction is always forward-biased, while for an NPN device, the E-B junction is always reverse-biased.

12. With no signal input, a properly connected common-emitter NPN bipolar transistor would have the highest value of I_C when

(a) we forward-bias the E-B junction considerably beyond forward breakover.

(b) we connect the base directly to the negative power-supply terminal.

(c) we reverse-bias the E-B junction.

(d) we connect the base directly to electrical ground.

13. Suppose that for a certain transistor at a specific constant frequency, we find that the alpha equals 0.9315. What's the beta?

(a) We can't determine it because our figure for the alpha makes no sense. We must have made a mistake when we determined the alpha!

(b) 13.60

(c) 0.4823

(d) 1.075

14. Suppose that for a certain transistor at a certain frequency, we find that the beta equals 0.5572. What's the alpha?

(a) We can't determine it because our figure for the beta makes no sense. We must have made a mistake when we determined the beta!

(b) 1.258

(c) 0.3578

(d) 1.795

15. Suppose that for a certain transistor at a certain frequency, we find that the alpha equals exactly 1.00. What's the beta?

(a) We can't define it.

(b) 0.333

(c) 0.500

(d) 1.00

16. In a common-emitter circuit, we normally take the output from the

(a) emitter.

(b) base.

(c) collector.

(d) More than one of the above

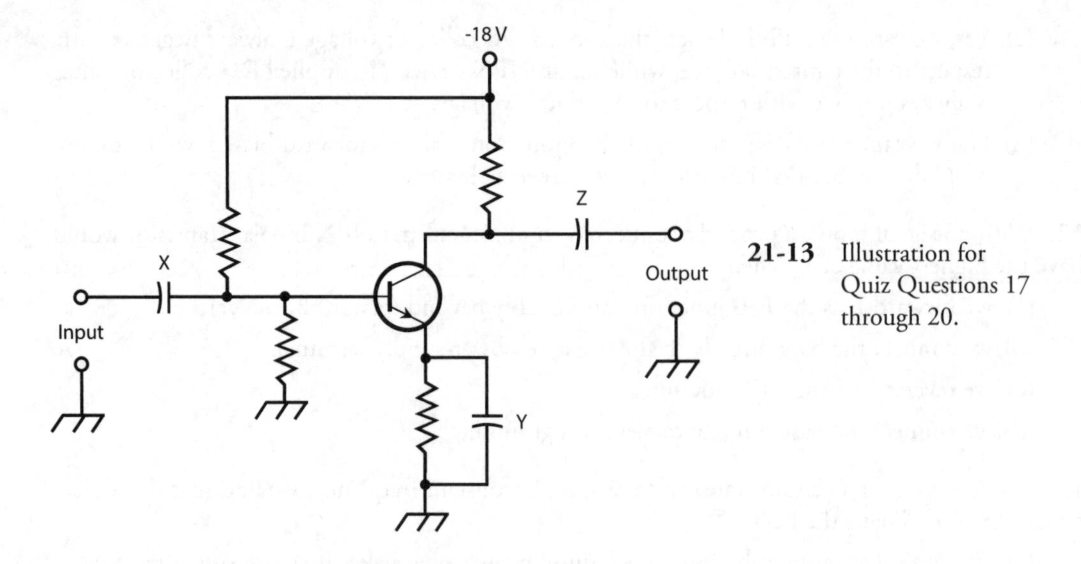

21-13 Illustration for Quiz Questions 17 through 20.

17. What major error exists in the circuit of Fig. 21-13?
    (a) The power supply voltage is too high for any bipolar transistor to handle.
    (b) We should use an NPN transistor, not a PNP transistor.
    (c) The power-supply polarity at the collector should be positive, not negative.
    (d) We should transpose the input and output terminals.
18. In the circuit of Fig. 21-13, what purpose does component X serve?
    (a) It keeps the signal from "shorting out" through the emitter.
    (b) It helps to establish the proper bias at the collector.
    (c) It keeps the signal from "feeding back" into the input device.
    (d) It blocks DC to or from the external input device, while letting the AC signal pass.
19. In the circuit of Fig. 21-13, what purpose does component Y serve?
    (a) It keeps the input isolated from the output.
    (b) It keeps the input signal from "shorting out" through the power supply.
    (c) It keeps the emitter at signal ground, while allowing a DC voltage to exist there.
    (d) It helps to establish the proper bias at the base.
20. In the circuit of Fig. 21-13, what purpose does component Z serve?
    (a) It keeps the circuit from breaking into oscillation.
    (b) It blocks DC to or from the external output device while letting the AC signal pass.
    (c) It matches the transistor's impedance to the impedance of the external output device, or load.
    (d) It ensures that the output wave remains in phase with the input wave.

22
CHAPTER

Field-Effect Transistors

THE BIPOLAR DEVICE ISN'T THE ONLY FORM OF TRANSISTOR THAT CAN SWITCH, AMPLIFY, OR OSCILLATE. A *field-effect transistor* (FET) can also do these things. Two main types of FET exist: the *junction FET* (JFET) and the *metal-oxide-semiconductor FET* (MOSFET).

Principle of the JFET

In a JFET, an *electric field* within the device affects the amount of current that can flow through it. Charge carriers (electrons or holes) move from the *source* (S) electrode to the *drain* (D) electrode to produce a *drain current* I_D that normally equals the *source current*, I_S. The rate of flow of charge carriers—that is, the current—depends on the voltage at a control electrode called the *gate* (G). Fluctuations in *gate voltage* E_G cause changes in the current through the *channel*, the path that the charge carriers follow between the source and the drain. The current through the channel normally equals I_D. Under the right conditions, small fluctuations in E_G can cause large variations in I_D. This fluctuating drain current can, in turn, produce significant fluctuations in the voltage across an output resistance.

N-Channel versus P-Channel

Figure 22-1 is a simplified drawing of an *N-channel JFET* (at A) and its schematic symbol (at B). The N type material forms the channel. The majority carriers in the channel are electrons, while the minority carriers are holes. We normally place the drain at a positive DC voltage with respect to the source, using an external power supply or battery.

In an N-channel device, the gate consists of P type material. Another section of P type material, called the *substrate*, forms a boundary on the side of the channel opposite the gate. The voltage on the gate produces an electric field that interferes with the flow of charge carriers through the channel. As E_G becomes more negative, the electric field chokes an increasing amount of the current though the channel, so I_D decreases.

A *P-channel JFET* (Figs. 22-2A and B) has a channel of P type semiconductor material. The majority charge carriers in the channel are holes, while the minority carriers are electrons. Using an

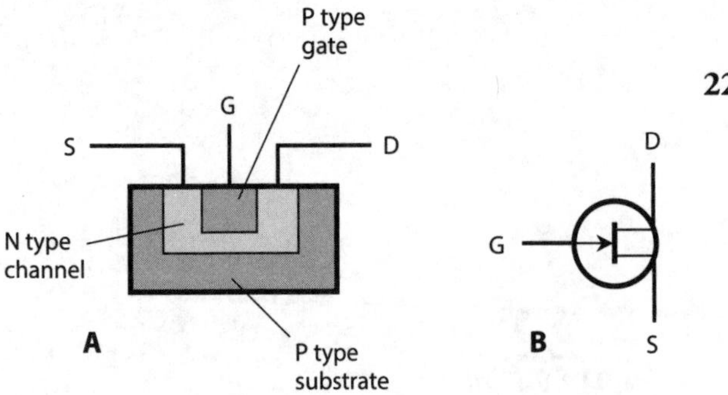

22-1 At A, a simplified structural drawing of an N-channel JFET. We identify the electrodes as S = source, G = gate, and D = drain. At B, the schematic symbol.

external power supply or battery, we place the drain at a negative DC voltage with respect to the source. The more positive E_G gets, the more the electric field chokes off the current through the channel, and the smaller I_D becomes.

You can recognize the N-channel JFET in schematic diagrams by the presence of a small arrow pointing inward at the gate. You can recognize the P-channel JFET by the arrow pointing outward at the gate. Alternatively, you can tell the N-channel device from the P-channel device (in case the symbols lack arrows) by looking at the power-supply polarity. A positive drain indicates an N-channel JFET, and a negative drain indicates a P-channel JFET.

In electronic circuits, N-channel and P-channel devices can do the same kinds of things. The main difference lies in the power-supply or battery polarity. We can almost always replace an N-channel JFET with a P-channel JFET, reverse the polarity, and expect the circuit to work the same—assuming that the new device has the right specifications. Just as we'll find different kinds of bipolar transistors, so will we encounter various types of JFETs, each suited to a particular application. Some JFETs work well as weak-signal amplifiers and oscillators; others find their niche in power amplification; some work ideally as high-speed switches.

Field-effect transistors have certain advantages over bipolar transistors. Perhaps the most important asset arises from the fact that JFETs, in general, create less *internal noise* than bipolar transistors do. This property makes JFETs excellent for use as weak-signal amplifiers at very high or ultra-high radio frequencies; in general, they do better than bipolar transistors in this respect. Field-effect transistors exhibit high input impedance values—in some cases so high that they draw virtually no current, while nevertheless providing significant signal output.

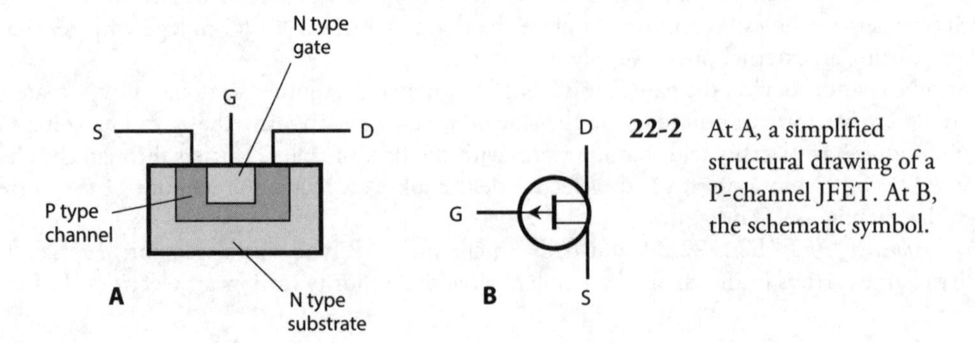

22-2 At A, a simplified structural drawing of a P-channel JFET. At B, the schematic symbol.

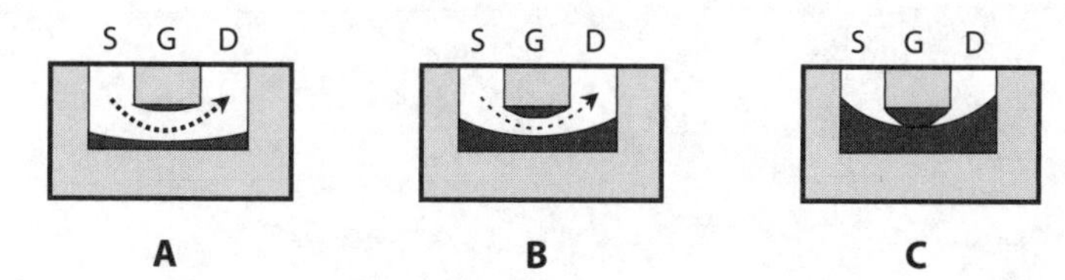

22-3 At A, the depletion region (darkest area) is narrow, the channel (white area) is wide, and many charge carriers (heavy dashed line) flow. At B, the depletion region is wider, the channel is narrower, and fewer charge carriers flow. At C, the depletion region obstructs the channel, and no charge carriers flow.

Depletion and Pinchoff

The JFET works because the voltage at the gate generates an electric field that interferes, more or less, with the flow of charge carriers along the channel. Figure 22-3 shows a simplified rendition of the situation for an N-channel device.

As the drain voltage E_D increases, so does the drain current I_D, up to a certain level-off value. This property holds true as long as the gate voltage E_G remains constant, and doesn't get too large (negatively). However, as E_G becomes increasingly negative (Fig. 22-3A), a *depletion region* (shown as a solid dark area) begins to form in the channel. Charge carriers can't flow in the depletion region, so they must pass through a narrowed channel.

As E_G becomes more negative still, the depletion region widens, as shown in Fig. 22-3B. The channel narrows further, and the current through it declines some more. Ultimately, if the gate voltage becomes negative enough, the depletion region completely obstructs the flow of charge carriers, and the channel current drops to zero under no-signal conditions (Fig. 22-3C). We call this condition *pinchoff.* It's the equivalent of cutoff in a bipolar transistor.

JFET Biasing

Figure 22-4 illustrates two biasing methods for N-channel JFET circuits. In Fig. 22-4A, we ground the gate through resistor R_2. The source resistor R_1 limits the current through the JFET. The drain current I_D flows through R_3, producing a voltage across R_3. That resistor also keeps the output signal from "shorting out" through the power supply or battery. The AC output signal passes through C_2 to the next circuit or load.

In Fig. 22-4B, we connect the gate through potentiometer R_2 to a source of voltage that's negative with respect to ground. Adjusting this potentiometer results in a variable negative gate voltage E_G between R_2 and R_3. Resistor R_1 limits the current through the JFET. The drain current I_D flows through R_4, producing a voltage across it. This resistor also keeps the output signal from "shorting out" through the power supply or battery. The AC output signal passes through C_2 to the next circuit or load.

In both circuits of Fig. 22-4, we provide the drain with a positive DC voltage relative to ground. For P-channel JFET circuits, simply reverse the polarities in Fig. 22-4, and replace the N-channel symbols with P-channel symbols.

Typical power-supply voltages in JFET circuits are comparable to those for bipolar transistor circuits. The voltage between the source and drain, abbreviated E_D, can range from about 3 V to 150 V DC; most often it's 6 to 12 V DC. The biasing arrangement in Fig. 22-4A works well for

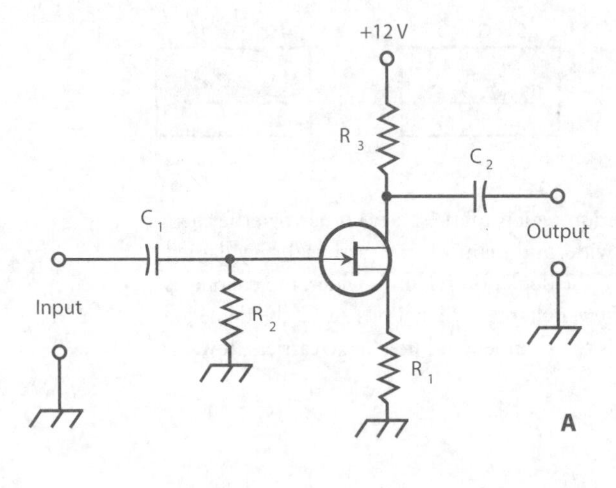

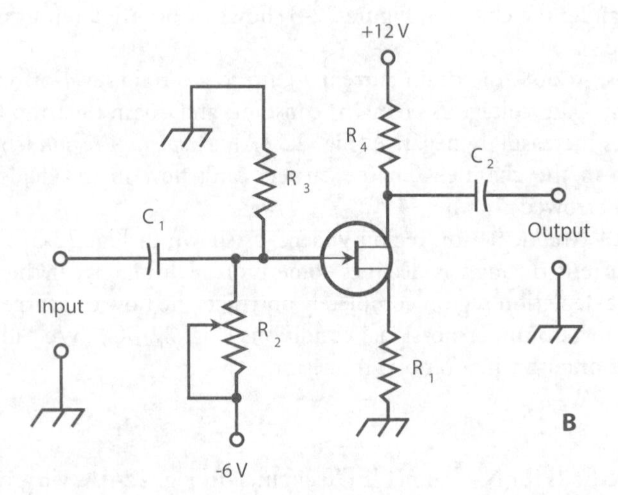

22-4 Two methods of biasing an N-channel JFET. At A, fixed gate bias; at B, variable gate bias.

weak-signal amplifiers, low-level amplifiers, and oscillators. The scheme at B works well in power amplifiers having substantial input signal amplitudes.

Amplification

Figure 22-5 shows a graph of I_D as a function of E_G for a hypothetical N-channel JFET. We assume that drain voltage E_D remains constant. When E_G is fairly large and negative, the JFET operates in a pinched-off state, so no current flows through the channel. As E_G gets less negative, the channel opens up, and I_D begins flowing. As E_G gets still less negative, the channel grows wider and I_D increases. As E_G approaches the point where the source-gate (S-G) junction (which constitutes a P-N junction) experiences forward breakover, the channel conducts as well as it possibly can; it's "wide open."

If E_G becomes positive enough so that the S-G junction goes past the forward-breakover point and conducts, some of the current in the channel "leaks out" through the gate. We rarely, if ever,

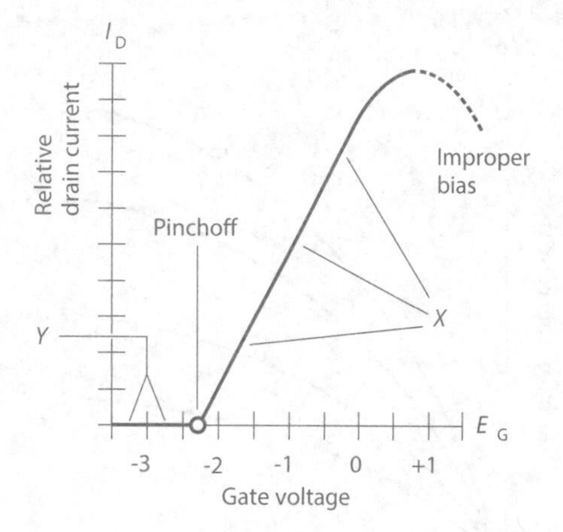

22-5 Relative drain current (I_D) as a function of gate voltage (E_G) for a hypothetical N-channel JFET.

want to see this state of affairs. We want the gate voltage to control the width of the channel and thereby control the current through it, but never to "suck" current out of it. Any current that flows out through the gate represents current that can't contribute to the output of the JFET.

Think of a JFET channel as a garden hose. When you want to reduce the flow of water at the output end of a hose, you can insert an adjustable valve somewhere along the length of the hose, or you can step down on the hose to pinch it narrower. You would not want to punch a hole in the hose to reduce the flow at the output because that action would let water go to waste.

The FET Amplifies Voltage

We can obtain the best amplification for weak signals when we set the no-signal gate voltage E_G so as to maximize the slope ("rise over run") of the drain-current versus gate-voltage (I_D versus E_G) curve. In Fig. 22-5, the range marked *X* shows the general region where this ideal condition exists. For power amplification, in which the JFET receives an input signal that's fairly strong to begin with, we'll often get the best results when we bias the JFET at or beyond pinchoff, in the range marked *Y*.

In either circuit shown in Fig. 22-4, I_D passes through the drain resistor. Small fluctuations in E_G cause large changes in I_D, and these variations in turn produce wide swings in the DC voltage across R_3 (in the circuit at A) or R_4 (in the circuit at B). The AC part of this voltage goes through capacitor C_2, and appears at the output as a signal of much greater AC voltage than that of the input signal at the gate. The JFET, therefore, acts as a *voltage amplifier*.

Drain Current versus Drain Voltage

Do you suspect that the drain current I_D, passing through the channel of a JFET, increases in a linear manner with increasing drain voltage E_D? This notion seems reasonable, but it's not what usually happens. Instead, I_D rises for a while as E_D increases steadily, and then I_D starts to level off as we increase E_D still more. We can plot I_D graphically as a function of E_D for various values of E_G under no-signal conditions. When we do that, we get a *family of characteristic curves*.

Figure 22-6 shows a family of characteristic curves for a hypothetical N-channel JFET. Engineers want to see graphs like this when choosing a JFET to serve in a specialized role, such as weak-signal amplification, oscillation, or power amplification. The graph of I_D versus E_G, one example of which appears in Fig. 22-5, is also an important specification that engineers consider. Characteristic curves portray DC behavior only; such curves are always derived under no-signal conditions.

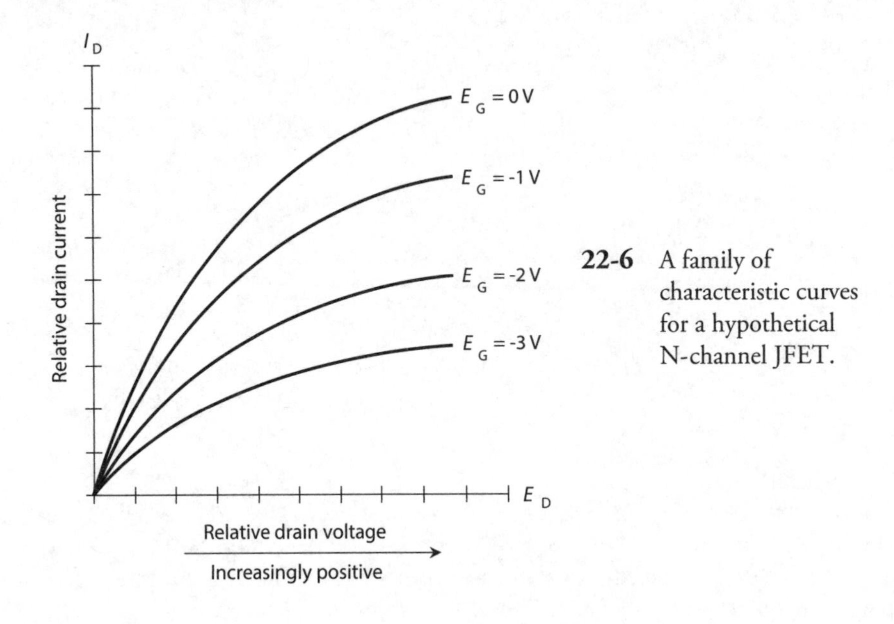

22-6 A family of characteristic curves for a hypothetical N-channel JFET.

Transconductance

We learned in Chap. 22 that the beta of a bipolar transistor tells us how well the device can amplify a signal in a practical circuit. We can also call the bipolar beta the *dynamic current amplification*. In a JFET, engineers call its equivalent the *dynamic mutual conductance,* or *transconductance.*

Refer again to Fig. 22-5. Suppose that we set the gate voltage E_G at a certain value, resulting in a drain current I_D. If the gate voltage changes by a small amount dE_G, then the drain current will change by an increment dI_D. The transconductance g_{FS} equals the ratio of the change in the drain current to the change in the gate voltage, as follows:

$$g_{FS} = dI_D / dE_G$$

The transconductance at any particular bias point translates to the slope ("rise over run") of a line tangent to the curve of Fig. 22-5 at that point.

As we can see from Fig. 22-5, the value of g_{FS} varies as we move along the curve. When we bias the JFET beyond pinchoff in the region marked *Y*, the slope of the curve equals zero; it's a horizontal line. In the range marked *Y*, we'll observe no fluctuation in I_D when E_G changes by small amounts. When there's a change in E_G, we'll see a change in I_D only when the channel conducts during at least part of the cycle of any input signal that we apply.

The region of greatest transconductance corresponds to the portion of the curve marked *X*, where the slope is the steepest. This region represents conditions under which we can derive the most gain from the device. Because this part of the curve constitutes a straight line, we can expect to get excellent linearity from any amplifier that we care to build with the JFET, provided that we keep the input signal from getting so strong that it drives the device outside of range *X* during any part of the cycle.

If we bias the JFET beyond the range marked *X*, the slope of the curve decreases, and we can't get as much amplification as we do in the range marked *X*. In addition, we can't expect the JFET to remain linear because the curve does not constitute a straight line in the "improper bias" range. If we keep on biasing the gate to greater and greater positive voltages, we get to the broken portion

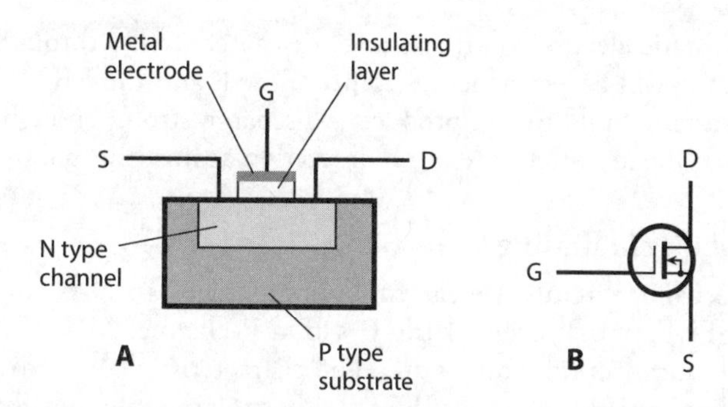

22-7 At A, the functional structure of an N-channel MOSFET. At B, the schematic symbol. Electrodes are S = source, G = gate, and D = drain.

of the curve, arriving at the zone where the S-G junction goes past forward breakover and draws current out of the channel.

The MOSFET

The acronym *MOSFET* (pronounced "*moss*-fet") stands for *metal-oxide-semiconductor field-effect transistor*. As with JFETs, two main types of MOSFET exist: The N-channel device and the P-channel device. Figure 22-7 shows a simplified cross-sectional drawing of an N-channel MOSFET along with the schematic symbol. The P-channel device and its symbol appear in Fig. 22-8.

Extremely High Input Impedance

When engineers originally developed the MOSFET, they called it an *insulated-gate field-effect transistor* or *IGFET*. That's still a pretty good term. The gate electrode is electrically insulated from the channel by a thin layer of dielectric material. As a result, the input impedance normally exceeds that of a JFET. In fact, the gate-to-source (G-S) resistance of a typical MOSFET compares favorably to the resistance of a capacitor of similar physical dimensions—it's "almost infinity"!

Because of its extreme G-S resistance, a MOSFET draws essentially no current, and therefore, essentially no power, from the input signal source. This property makes the MOSFET ideal for use in low-level and weak-signal amplifier circuits. Because the G-S "capacitor" is physically tiny, its capacitance is tiny as well, so the device can function quite well up to ultra-high radio frequencies (above 300 MHz).

Despite their obvious advantages in the impedance and frequency-response departments, MOS devices suffer from a big shortcoming: They're electrically fragile. When building or servicing circuits with MOS devices, technicians must use special equipment to ensure that their hands don't

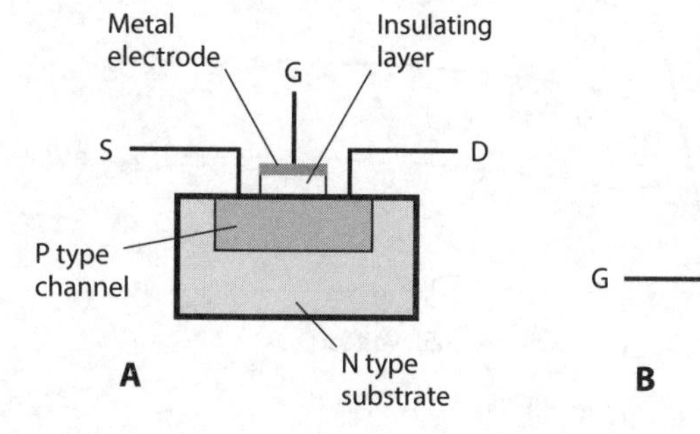

22-8 At A, the functional structure of a P-channel MOSFET. At B, the schematic symbol.

carry "static electricity." If a "static" discharge occurs through the dielectric of a MOS device, the dielectric will be permanently compromised, and the MOSFET must be discarded. Even a small electrostatic buildup can produce a discharge strong enough to destroy a MOSFET. Warm and humid climates offer little or no protection against this hazard.

Flexibility in Biasing

In electronic circuits, we can sometimes replace an N-channel JFET directly with an N-channel MOSFET, or a P-channel JFET with a P-channel MOSFET. However, in a few situations this simple substitution won't work. The characteristic curves for MOSFETs differ qualitatively from the characteristic curves for JFETs with similar amplifying characteristics.

The main difference between MOSFET behavior and JFET behavior arises from the fact that in a MOSFET, the source and the gate do not come together as a P-N junction. Instead, the two electrodes are physically separated by a "gap" of dielectric, so forward breakover can't occur. We can apply gate bias that's far more positive than +0.6 V to an N-channel MOSFET, or a lot more negative than −0.6 V to a P-channel MOSFET, and we'll never see current "leak" out through the gate as it would do in a JFET (unless, of course, we apply a voltage so great that *arcing* (sparking) takes place across the dielectric "gap").

Figure 22-9 illustrates a family of characteristic curves for a hypothetical N-channel MOSFET under no-signal conditions. The horizontal axis portrays DC drain voltage, while the vertical axis portrays drain current. For any specific gate voltage, the drain current increases rapidly at first as we increase the drain voltage. However, as we continue to increase the drain voltage, the drain current rises at a slower rate, eventually leveling off. Once a particular I_D-versus-E_D curve has "flattened out," we can't increase the drain current any further by applying more DC drain voltage.

Depletion Mode versus Enhancement Mode

In the FET devices we've discussed so far, the channel normally conducts; as the depletion region grows, choking off the channel, the charge carriers must pass through a narrower and narrower path. We call

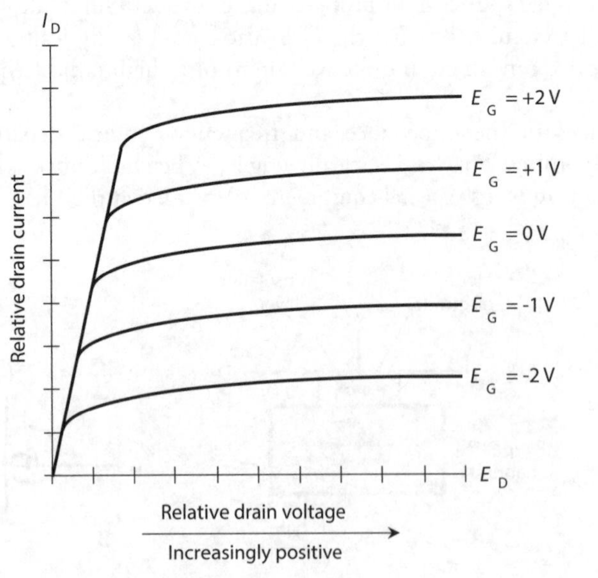

22-9 A family of characteristic curves for a hypothetical N-channel MOSFET.

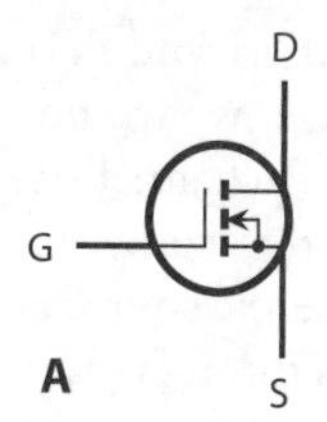

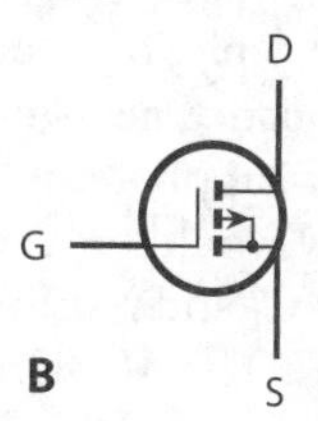

22-10 Schematic symbols for enhancement–mode MOSFETs. At A, the N-channel device; at B, the P-channel device.

this state of affairs the *depletion mode* for a field-effect transistor. A MOSFET can also function in the depletion mode. The drawings and schematic symbols of Figs. 22-7 and 22-8 show depletion-mode MOSFETs. The characteristic curves in Fig. 22-9 illustrate the behavior of a typical depletion-mode N-channel MOSFET. (To get the graphs for a P-channel device, reverse all the polarities.)

Metal-oxide semiconductor technology also allows an entirely different means of operation. An *enhancement-mode* MOSFET normally has a pinched-off channel. We must apply a bias voltage E_G to the gate so that a channel will form. If $E_G = 0$ in an enhancement-mode MOSFET, then $I_D = 0$ with no signal input. The enhancement-mode MOSFET, like its depletion-mode cousin, has an extremely low capacitance and high impedance between the gate and the channel.

In an N-channel enhancement-mode device, a *positive* voltage at the gate (with respect to the source) causes a conductive path to form in the channel. In the P-channel enhancement-mode device, we must apply a *negative* voltage at the gate in order to get the channel to conduct current. As the voltage increases, assuming the polarity is correct, the conductive channel grows wider, and the source-to-drain conductivity improves. This effect occurs, however, only up to a certain maximum current for a given constant DC drain voltage.

Figure 22-10 shows the schematic symbols for N-channel and P-channel enhancement-mode devices. Note that the vertical line is broken, rather than solid as it appears in the symbol for a depletion-mode MOSFET.

Common-Source Configuration

In a *common-source circuit,* we place the source at signal ground and apply the input signal to the gate. Figure 22-11 shows the general configuration for an N-channel JFET. The device could be

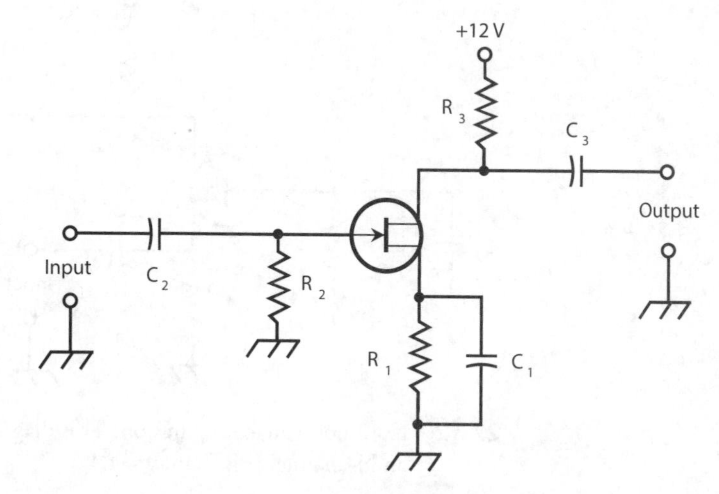

22-11 Common-source configuration. This diagram shows an N-channel JFET circuit.

an N-channel, depletion-mode MOSFET and the circuit arrangement would look the same. For an N-channel, enhancement-mode device, we would need an extra resistor between the gate and the positive power supply terminal. For P-channel devices, the schematics would be the same except that the supply would provide a negative, rather than a positive, voltage.

Capacitor C_1 and resistor R_1 place the source at signal ground, while elevating the source above ground for DC. The AC signal enters through C_2. Resistor R_2 adjusts the input impedance and provides bias for the gate. The AC signal passes out of the circuit through C_3. Resistor R_3 keeps the output signal from being shorted out through the power supply, while allowing for a positive voltage at the drain. This circuit can serve as the basic configuration for low-level RF amplifiers and oscillators.

The common-source arrangement provides the greatest gain of the three FET circuit configurations. The output wave appears in phase opposition with respect to the input wave.

Common-Gate Configuration

In the *common-gate circuit* (Fig. 22-12), we place the gate at signal ground and apply the input signal to the source. This diagram shows an N-channel JFET. For other types of FETs, the same considerations, as described above for the common-source circuit, apply. Enhancement-mode devices require an extra resistor between the gate and the positive supply terminal (or the negative terminal if the MOSFET is P-channel).

The DC bias for the common-gate circuit is basically the same as that for the common-source arrangement, but the signal follows a different path. The AC input signal enters through C_1. Resistor R_1 keeps the input signal from shorting out to ground. Gate bias is provided by R_1 and R_2. Capacitor C_2 places the gate at signal ground. (In some common-gate circuits, the gate goes directly to ground, and R_2 and C_2 are not used.) The output signal leaves the circuit through C_3. Resistor R_3 keeps the output signal from shorting out through the power supply, while still allowing the FET to receive the necessary DC voltage.

The common-gate arrangement produces less gain than its common-source counterpart, but it's far less likely to break into unwanted oscillation, making it a good choice for RF power-amplifier circuits. The output wave follows along in phase with the input wave.

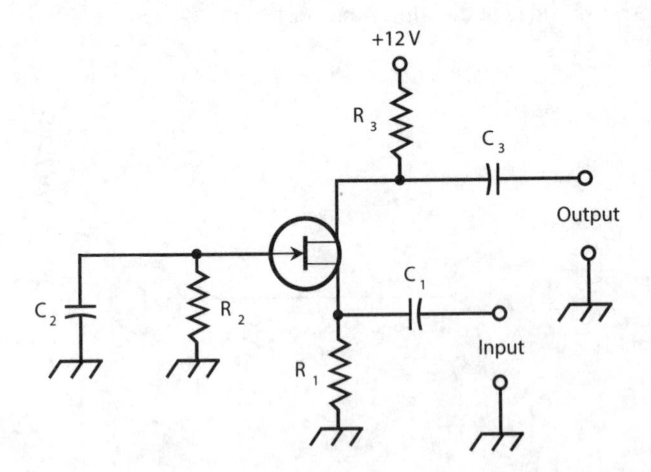

22-12 Common-gate configuration. This diagram shows an N-channel JFET circuit.

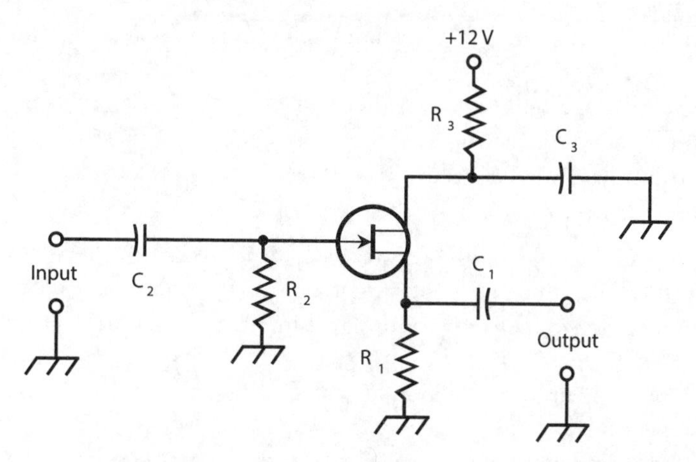

22-13 Common-drain configuration, also known as a source follower. This diagram shows an N-channel JFET circuit.

Common-Drain Configuration

Figure 22-13 shows a *common-drain circuit.* We place the drain at signal ground. It is sometimes called a *source follower* because the output waveform follows the signal at the source. The FET is biased in the same way as for the common-source and common-gate circuits. In Fig. 22-13, we see an N-channel JFET, but any other kind of FET could be used, reversing the polarity for P-channel devices. Enhancement-mode MOSFETs would need an extra resistor between the gate and the positive supply terminal (or the negative terminal if the MOSFET is P-channel).

The input signal passes through C_2 to the gate. Resistors R_1 and R_2 provide gate bias. Resistor R_3 limits the current. Capacitor C_3 keeps the drain at signal ground. Fluctuating DC (the channel current) flows through R_1 as a result of the input signal; this causes a fluctuating DC voltage to appear across R_1. We take the output from the source, and its AC component passes through C_1.

The output wave of the common-drain circuit is in phase with the input wave. This scheme constitutes the FET equivalent of the bipolar common-collector (emitter follower) arrangement. The output impedance is low, so this type of circuit works quite well for matching a high input impedance to a low output or load impedance over a wide range of frequencies.

Quiz

Refer to the text in this chapter if necessary. A good score is at least 18 correct. Answers are in the back of the book.

1. In a JFET, current through the channel varies because of the effects of
   (a) a magnetic field.
   (b) an electric field.
   (c) leakage current.
   (d) avalanche current.

2. In a P-channel JFET, assuming that the drain voltage remains constant, pinchoff occurs when we place the gate at a
   (a) tiny negative voltage with respect to the source.
   (b) significant negative voltage with respect to the source.
   (c) tiny positive voltage with respect to the source.
   (d) significant positive voltage with respect to the source.

3. When we bias a JFET at the point on its I_D versus E_G curve at which we can expect to derive the most amplification, the value of dI_D/dE_G, representing the slope of the curve at that point, is
   (a) zero ("running horizontally") with no signal input, and the curve appears as a straight line near the point.
   (b) positive ("ramping upward to the right") with no signal input, and the curve appears as a straight line near the point.
   (c) negative ("ramping downward to the right") with no signal input, and the curve bends downward near the point.
   (d) positive ("ramping upward to the right") with no signal input, and the curve bends downward near the point.

4. Under no-signal conditions, a zero-biased enhancement-mode MOSFET operates in a state of
   (a) pinchoff.
   (b) avalanche breakdown.
   (c) saturation.
   (d) forward breakover.

5. Figure 22-14 is a simplified cutaway diagram of
   (a) a P-channel JFET.
   (b) an N-channel JFET.
   (c) a P-channel MOSFET.
   (d) an N-channel MOSFET.

6. In Fig. 22-14, the item marked X constitutes
   (a) a thin wafer of N type material.
   (b) a thin layer of dielectric material.
   (c) a thin layer of highly conductive material.
   (d) a P-N junction.

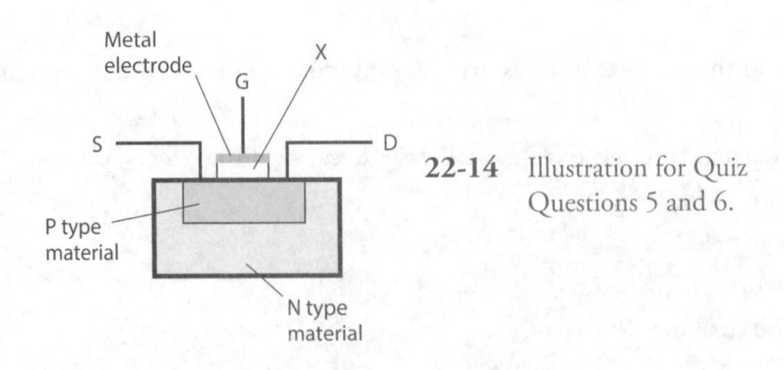

22-14 Illustration for Quiz Questions 5 and 6.

7. A bipolar transistor might work better than a JFET or MOSFET when we want an amplifier to
   (a) have good weak-signal performance.
   (b) have low input impedance.
   (c) produce high transconductance.
   (d) draw the most possible power from the input signal source.

8. A properly operating MOSFET presents a G-S resistance
   (a) of practically zero.
   (b) lower than that of a JFET.
   (c) comparable to that of a similar-sized capacitor.
   (d) lower than that of a bipolar transistor.

9. In a FET of any kind, we generally don't want to see DC flow between the
   (a) source and the drain.
   (b) source and the channel.
   (c) drain and the channel.
   (d) gate and the channel.

10. In a MOSFET, the majority carriers are electrons
    (a) if the device has an N-type channel.
    (b) when forward breakover occurs.
    (c) when avalanche breakdown occurs.
    (d) under no circumstances.

11. Which of the following circuits produces an output signal wave that's precisely in phase with the input signal wave?
    (a) The common-gate circuit
    (b) The common-drain circuit
    (c) The source follower
    (d) All of the above

12. A small electrostatic discharge, such as might appear on a technician's hands, can easily destroy a MOSFET by
    (a) fusing the source to the drain.
    (b) eliminating the charge carriers in the channel.
    (c) destroying the insulating properties of the dielectric.
    (d) causing forward breakover at the gate-drain junction.

13. A significant difference between MOSFETs and JFETs is the fact that
    (a) JFETs usually have lower input impedances.
    (b) JFETs are less electrically rugged.
    (c) JFETs are physically larger.
    (d) JFETs require far higher operating voltages.

14. We can recognize a depletion-mode MOSFET (as opposed to an enhancement-mode MOSFET) in schematic diagrams by the presence of
    (a) an arrow pointing inward.
    (b) a broken vertical line inside the circle.
    (c) an arrow pointing outward.
    (d) a solid vertical line inside the circle.

15. In a source follower, from which two points do we obtain the output signal?
    (a) The drain and ground
    (b) The drain and gate
    (c) The gate and source
    (d) The source and ground

16. Figure 22-15 illustrates a source follower with two major errors. To correct one of the errors, we must
    (a) transpose the input and output terminals.
    (b) replace the JFET with a depletion-mode MOSFET.
    (c) replace the JFET with an enhancement-mode MOSFET.
    (d) replace the gate resistor with a capacitor.

17. To correct the second error in Fig. 22-15, we must
    (a) connect the drain directly to the positive DC voltage source by replacing the drain resistor with a length of wire.
    (b) replace the gate capacitor with a resistor.
    (c) replace the source capacitor with a resistor.
    (d) reverse the DC power-supply polarity.

18. Figure 22-16 illustrates a family of characteristic curves for a hypothetical N-channel, depletion-mode MOSFET. The curves V through Z portray the behavior of the device for various
    (a) DC source voltages under no-signal conditions.
    (b) AC input signal voltages.

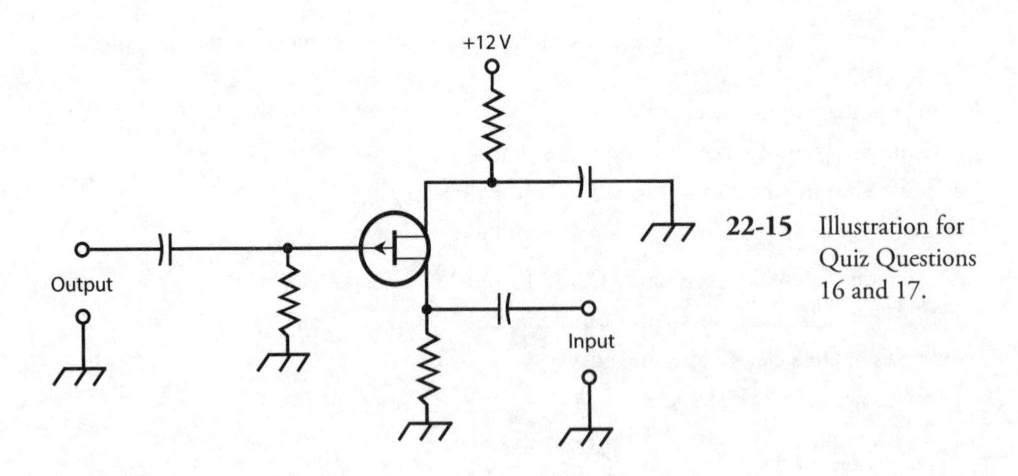

22-15 Illustration for Quiz Questions 16 and 17.

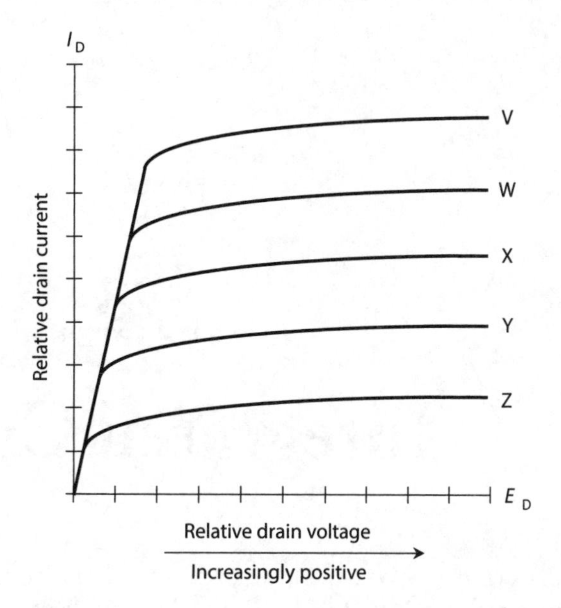

22-16 Illustration for Quiz Questions 18 through 20.

(c) DC gate voltages under no-signal conditions.
(d) AC output signal voltages.

19. What can we say about the relative voltages for curves V through Z in Fig. 22-16? Remember, we're dealing with an N-channel, depletion-mode MOSFET.
(a) They're DC voltages that get progressively less positive (or more negative) as we proceed down from V to Z.
(b) They're DC voltages that get progressively less negative (or more positive) as we proceed down from V to Z.
(c) They're AC voltages whose peak-to-peak values increase as we proceed down from V to Z.
(d) They're AC voltages whose peak-to-peak values decrease as we proceed down from V to Z.

20. Notice how all the curves in Fig. 22-16 tend to "level off" as we move toward the right in the coordinate grid. What does the "leveling-off" tell us about the general behavior of this particular N-channel, depletion-mode MOSFET?
(a) As we increase the positive DC drain voltage under no-signal conditions, the drain current increases slowly at first, then more and more rapidly.
(b) As we increase the positive DC drain voltage under no-signal conditions, the drain current increases rapidly at first, then more and more slowly.
(c) As we increase the peak-to-peak AC input signal voltage, the drain current increases slowly at first, then more and more rapidly.
(d) As we derive increasing peak-to-peak AC output signal voltage from the device, the drain current increases rapidly at first, then more and more slowly.

23
CHAPTER

Integrated Circuits

MOST *INTEGRATED CIRCUITS* (ICS), ALSO CALLED *CHIPS*, LOOK LIKE GRAY OR BLACK BOXES WITH protruding metal terminals called *pins*. In schematic diagrams, engineers represent ICs as triangles or rectangles, usually with component designators printed inside, and with emerging lines (representing the pins in the actual device) leading to external components.

Advantages of IC Technology

Integrated circuits have advantages over *discrete components* (individual transistors, diodes, capacitors, and resistors). The most important considerations follow.

Compactness

An obvious asset of IC design is economy of space. An IC is far more compact than an equivalent circuit made from discrete components. Integrated circuits allow for the construction of more sophisticated systems in smaller packages than discrete components do.

High Speed

The interconnections among internal IC components are physically tiny, making high switching speeds possible. As we increase the speed with which charge carriers can get from one component to another, we increase the number of computations that a system can do within a given span of time, and we reduce the time it takes for the system to perform complicated operations.

Low Power Consumption

Integrated circuits consume less power than equivalent discrete-component circuits. This advantage becomes a necessity in battery-operated systems. Because ICs use so little current, they produce less heat than their discrete-component equivalents, resulting in efficiency. The low-current feature also minimizes problems, such as frequency drift or intermittent failure, that can occur in equipment that gets hot with use.

Reliability

Integrated circuits fail less often, per component-hour of use, than systems built up from discrete components. The lower failure rate results from the fact that all component interconnections are sealed within the IC case, preventing the intrusion of dust, moisture, or corrosive gases. Therefore, ICs generally suffer less *downtime* (periods during which the equipment is out of service for repairs) than discrete-component systems do.

Ease of Maintenance

Integrated-circuit technology minimizes hardware maintenance costs and streamlines maintenance procedures. Many appliances use sockets for ICs, and replacement involves nothing more than finding the faulty IC, unplugging it, and plugging in a new one. Technicians use special tools to remove and replace ICs soldered directly onto circuit boards without sockets.

Modular Construction

Modern IC appliances use *modular construction*, in which individual ICs perform defined functions within a circuit board. The circuit board, or *card*, in turn, fits into a socket and has a specific purpose. Repair technicians, using computers programmed with customized software, locate the faulty card, remove it, and replace it with a new one, getting the appliance back to the consumer in the shortest possible time. Another technician troubleshoots the faulty card, getting it ready for use in an appliance that arrives in the future with a failure in the same card.

Limitations of IC Technology

No technological advancement ever comes without a downside. Integrated circuits have limitations that engineers must consider when designing an electronic device or system.

Inductors Impractical

While some components are easy to fabricate onto chips, other components defy the IC manufacturing process. Inductors, except for components with extremely low values (in the nanohenry range), constitute a prime example. Devices using ICs must generally be designed to work with discrete inductors (coils) external to the ICs themselves. This constraint need not pose a problem, however. Resistance-capacitance (*RC*) circuits can do most things that inductance-capacitance (*LC*) circuits can do, and *RC* circuits can be etched onto an IC chip with no difficulty.

High Power Impossible

The small size and low current consumption of ICs comes with an inherent limitation. In general, we cannot fabricate a high-power amplifier onto an IC chip. High-power operation necessitates a certain minimum physical mass and volume because the components generate a lot of heat. Effective removal of this heat requires hefty objects, such as *heatsinks* (which resemble old-fashioned steam radiators scaled down to circuit size), air blowers, or liquid cooling systems.

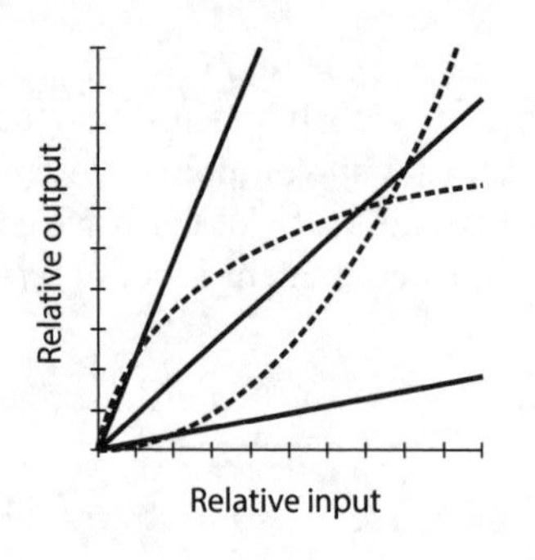

23-1 In a linear IC, the relative output is a linear (straight-line) function of the relative input. The solid lines show examples of linear IC characteristics. The dashed curves show functions that aren't characteristic of properly operating linear ICs.

Linear ICs

A *linear IC* processes *analog signals* such as voices and music. The term "linear" arises from the fact that the amplification factor remains constant as the input amplitude varies. In technical terms, the output signal strength constitutes a *linear function* of the input signal strength, as shown by any of the three solid, straight lines in the graph of Fig. 23-1.

Operational Amplifier

An *operational amplifier*, or *op amp*, is a specialized linear IC that consists of several bipolar transistors, resistors, diodes, and capacitors, all connected together so that high gain is possible over a wide range of frequencies. Some ICs contain two or more op amps, so you'll hear or read about *dual op amps* or *quad op amps*. Some ICs contain one or more op amps in addition to other circuits.

An op amp has two inputs, one *non-inverting* and one *inverting*, and one output. When a signal goes into the non-inverting input, the output wave emerges in phase coincidence with the input wave. When a signal goes into the inverting input, the output wave appears "upside-down" with respect to the input wave. An op amp has two power supply connections, one for the emitters of the internal bipolar transistors (V_{ee}) and one for the collectors (V_{cc}). The usual schematic symbol is a triangle (Fig. 23-2).

Op Amp Feedback and Gain

One or more external resistors determine the gain of an op amp. Normally, we place a resistor between the output and the inverting input to obtain a so-called *closed-loop configuration*. The feedback is negative (out of phase), causing the gain to remain lower than would hold true if no feedback existed. As we reduce the value of this resistor, the gain decreases because the negative feedback increases.

If we remove the feedback resistor, we get an *open-loop configuration*, in which the op amp produces its maximum rated gain. Figure 23-3 is a schematic diagram of a non-inverting closed-loop

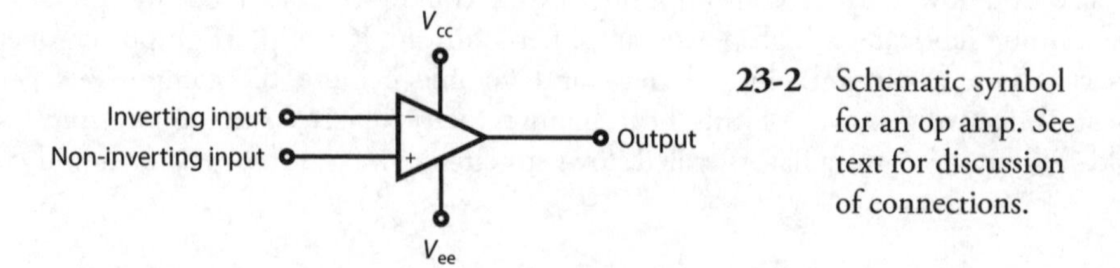

23-2 Schematic symbol for an op amp. See text for discussion of connections.

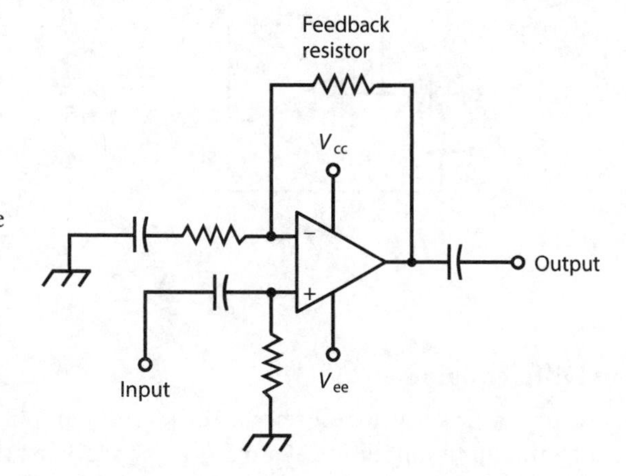

23-3 A closed-loop op amp circuit with negative feedback. If we remove the feedback resistor, we get an open-loop circuit.

amplifier. Open-loop op amps sometimes exhibit instability, especially at low frequencies, breaking unexpectedly into oscillation. Open-loop op-amp circuits also generate significant internal noise, which can cause trouble in some applications.

If we install an *RC* combination in the feedback loop of an op amp, the gain depends on the input-signal frequency. Using specific values of resistance and capacitance, we can make a frequency-sensitive filter that provides any of four different characteristics, as shown in Fig. 23-4:

1. A *lowpass response* that favors low frequencies
2. A *highpass response* that favors high frequencies
3. A *resonant peak* that produces maximum gain at and near a single frequency
4. A *resonant notch* that produces maximum loss at and near a single frequency

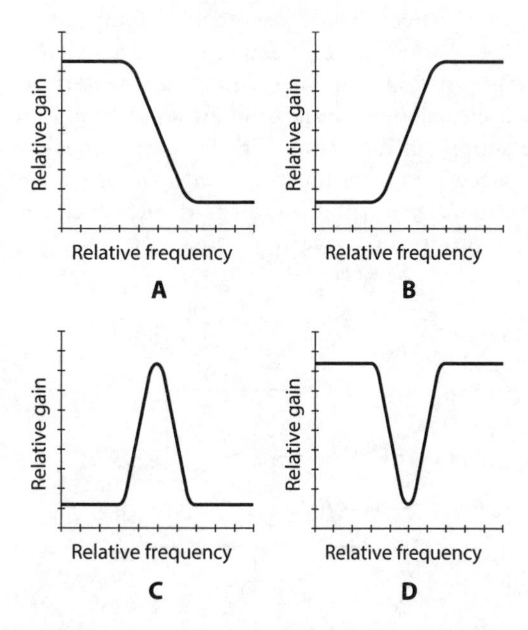

23-4 Gain-versus-frequency response curves. At A, lowpass; at B, highpass; at C, resonant peak; at D, resonant notch.

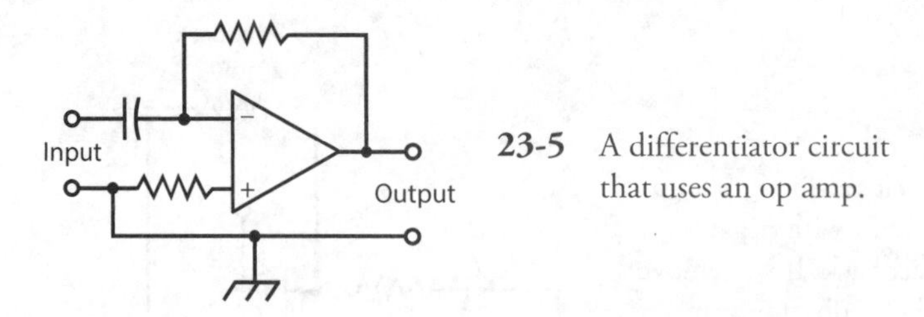

23-5 A differentiator circuit that uses an op amp.

Op Amp Differentiator

A *differentiator* is a circuit whose instantaneous output amplitude varies in direct proportion to the rate at which the input amplitude changes. In the mathematical sense, the circuit mathematically *differentiates* the input signal wave function. Op amps lend themselves well to use as differentiators. Figure 23-5 shows an example.

When the input to a differentiator is a constant DC voltage, the output equals zero (no signal). When the instantaneous input amplitude increases, the output is a positive DC voltage. When the instantaneous input amplitude decreases, the output is a negative DC voltage. If the input amplitude fluctuates periodically (a sine wave, for example), the output voltage varies according to the *instantaneous rate of change* (the mathematical *derivative*) of the instantaneous input-signal amplitude. We, therefore, observe an output signal with the same frequency as that of the input signal, although the waveform might differ.

In a differentiator circuit, a pure sine-wave input produces a pure sine-wave output, but the phase is shifted 90° to the left (¼ cycle earlier in time). Complicated input waveforms produce a wide variety of output waveforms.

Op Amp Integrator

An *integrator* is a circuit whose instantaneous output amplitude is proportional to the accumulated input signal amplitude as a function of time. The circuit mathematically *integrates* the input signal. In theory, the function of an integrator is the inverse, or opposite, of the function of a differentiator. Figure 23-6 shows how we can configure an op amp to obtain an integrator.

If we supply an integrator with an input signal waveform that fluctuates periodically, the output voltage varies according to the *integral*, or *antiderivative*, of the input voltage. This process yields an output signal with the same frequency as that of the input signal, although the waveform often differs. A pure sine-wave input produces a pure sine wave output, but the phase is shifted 90°

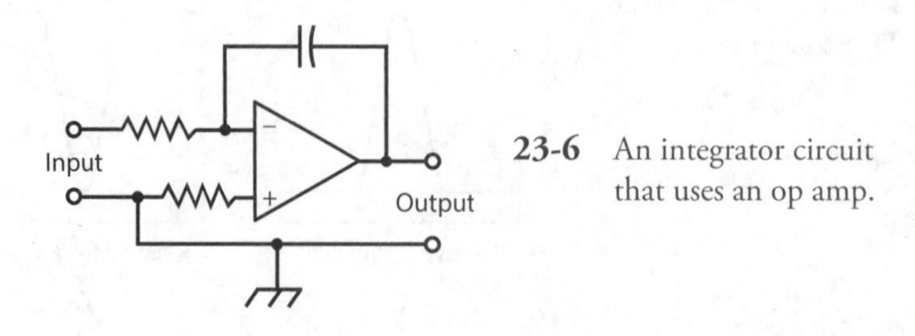

23-6 An integrator circuit that uses an op amp.

to the right (¼ cycle later in time). Complex input waveforms can produce many types of output waveforms.

The performance of a practical integrator differs in one important way from the operation of a theoretically ideal integrator. If the mathematical integral of an input function yields an endlessly increasing output function, the actual output voltage rises to a certain maximum, either positive or negative, and then stays there. Obviously, we can't get an output voltage that increases forever without limit! The maximum voltage at the output of a real-world integrator never exceeds the power-supply or battery voltage.

Voltage Regulator

A *voltage regulator IC* acts to control the output voltage of a power supply. This feature is important with precision electronic equipment. Voltage-regulator ICs exist with various voltage and current ratings. Typical voltage regulator ICs have three terminals, just as most transistors do. Because of this resemblance, these ICs are sometimes mistaken for power transistors!

Timer

A *timer IC* is a specialized oscillator that produces a delayed output. We can tailor the delay time to suit the needs of a particular device. The delay is generated by counting the number of oscillator pulses; the length of the delay can be adjusted by means of external resistors and capacitors. Timer ICs find widespread application in digital frequency counters, where a precise time interval or "window" must be provided.

Multiplexer

A *multiplexer IC* allows us to combine several different signals in a single channel by means of a process called *multiplexing*. An analog multiplexer can also be used in reverse; then it works as a *demultiplexer*. Therefore, you'll sometimes hear or read about a *multiplexer/demultiplexer IC*.

Comparator

A *comparator IC* has two inputs. It literally compares the voltages at the two inputs, which we call input A and input B. If the voltage at input A significantly exceeds the voltage at input B, the output equals about +5 V, giving us the logic 1, or high, state. If the voltage at input A is less than or equal to the voltage at input B, we get an output voltage of about +2 V or less, yielding the logic 0, or low, state.

Voltage comparators are available for a variety of applications. Some can switch between low and high states at a rapid rate, while others are slow. Some have low input impedance, and others exhibit high input impedance. Some are intended for AF or low-frequency RF use; others work well in video or high-frequency RF applications.

Voltage comparators can actuate, or *trigger*, other devices, such as relays, alarms, and electronic switching circuits.

Digital ICs

A *digital IC*, also sometimes called a *digital-logic IC*, operates using two discrete states: high (logic 1) and low (logic 0). Digital ICs contain massive arrays of logic *gates* that perform logical operations at high speed.

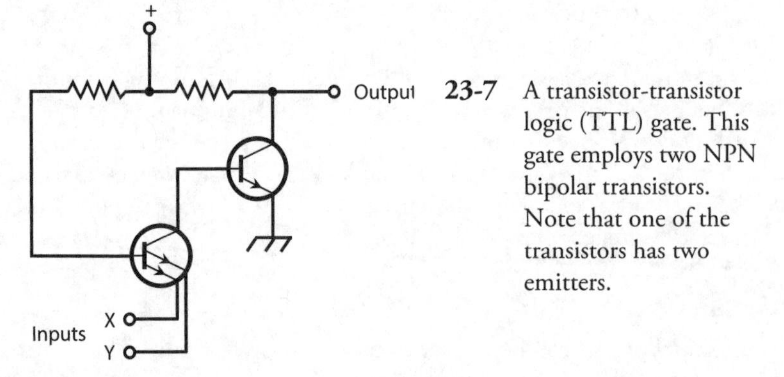

23-7 A transistor-transistor logic (TTL) gate. This gate employs two NPN bipolar transistors. Note that one of the transistors has two emitters.

Transistor-Transistor Logic

In *transistor-transistor logic* (TTL), arrays of bipolar transistors, some with multiple emitters, operate on DC pulses. Figure 23-7 illustrates the internal details of a basic TTL gate using two NPN bipolar transistors, one of which is a *dual-emitter* device. In TTL, the transistors are always either cut off or saturated. No analog amplification can occur; analog input signals are "swamped" by the digital states. Because of this well-defined duality, TTL systems offer excellent immunity to external noise, which generally is manifest in analog form.

Emitter-Coupled Logic

Emitter-coupled logic (ECL) constitutes another common bipolar-transistor scheme. In an ECL device, the transistors don't operate at saturation, as they do with TTL. This biasing scheme increases the speed of ECL relative to TTL, but it also increases vulnerability to external noise. "Unsaturated" transistors are sensitive to, and can actually amplify, analog disturbances. Figure 23-8 shows the internal details of a basic ECL gate using four NPN bipolar transistors.

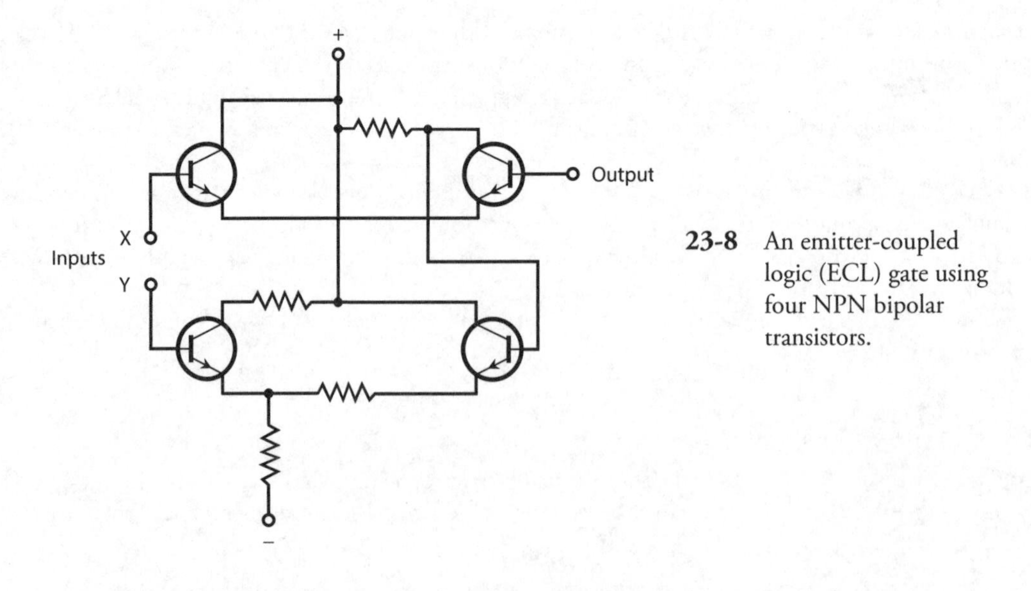

23-8 An emitter-coupled logic (ECL) gate using four NPN bipolar transistors.

Metal-Oxide-Semiconductor Logic

Digital ICs can be constructed using metal-oxide-semiconductor (MOS) devices. *N-channel MOS* (NMOS) *logic*, pronounced "EN-moss logic," offers simplicity of design, along with high operating speed. *P-channel MOS logic*, pronounced "PEA-moss logic," resembles NMOS logic, but usually exhibits a lower maximum operating speed. We can summarize the construction details of these two schemes as follows:

1. An NMOS IC is the counterpart of a discrete-component circuit that uses only N-channel MOSFETs.
2. A PMOS IC is the counterpart of a discrete-component circuit that uses only P-channel MOSFETs.

Complementary-metal-oxide-semiconductor (CMOS) *logic*, pronounced "SEA-moss logic," employs both N type and P type silicon on a single chip—analogous to using both N-channel and P-channel MOSFETs in a discrete-component circuit. The advantages of CMOS technology include extremely low current drain, high operating speed, and immunity to external noise.

All forms of MOS logic ICs require careful handling to prevent destruction by electrostatic discharges. The precautions are the same as those for handling discrete MOSFETs. All personnel who work with MOS ICs should "ground themselves" by wearing metal wrist straps connected to electrical ground, and by ensuring that the humidity in the lab does not get too low. When MOS ICs are stored, the pins should be pushed into conductive foam specifically manufactured for that purpose.

Component Density

In digital ICs, the number of transistors per chip is called the *component density*. It keeps increasing year by year. There's a practical limit to component density (imposed by the physical volumes of individual atoms), although some engineers think they'll be able to make single atoms perform multiple operations someday. In general, as IC component density increases, so does operating speed.

Small-Scale Integration

In *small-scale integration* (SSI), there are fewer than 10 transistors on a chip. These devices can carry the largest currents of any IC type because low component density translates into relatively large volume and mass per component. Small-scale integration finds application in voltage regulators and other moderate-power systems.

Medium-Scale Integration

In *medium-scale integration* (MSI), we have 10 to 100 transistors per chip. This density allows for considerable miniaturization, but does not constitute a high level of component density, relatively speaking, these days. An advantage of MSI is the fact that individual logic gates can carry fairly large currents in some applications. Both bipolar and MOS technologies can be adapted to MSI.

Large-Scale Integration

In *large-scale integration* (LSI), we have 100 to 1,000 transistors per semiconductor chip, a full *order of magnitude* (a factor of 10 times) more dense than MSI. Electronic wristwatches, single-chip calculators, *microcomputers*, and *microcontrollers* are examples of devices using LSI ICs.

Very-Large-Scale Integration

Very-large-scale integration (VLSI) devices have from 1,000 to 1,000,000 transistors per chip, up to three orders of magnitude more dense than LSI. High-end microcomputers, microcontrollers, and *memory chips* are made using VLSI.

Ultra-Large-Scale Integration

You might sometimes hear of *ultra-large-scale integration* (ULSI). Devices of this kind have more than 1,000,000 transistors per chip. The principal uses for ULSI technology include high-level computing, *supercomputing*, *robotics*, and *artificial intelligence* (AI). We should expect to see expanded use of ULSI technology in the future, particularly in medical devices.

IC Memory

Binary digital data, in the form of high and low states (logic 1 and 0), can be stored in memory chips that take a wide variety of physical forms. Some IC memory chips require a continuous source of backup voltage or they'll lose their data. Others can hold the data in the absence of backup voltage, in some cases for months or years. In electronic devices, we encounter two main types of memory: *random-access* and *read-only*.

Random-Access Memory

A *random-access memory* (RAM) chip stores binary data in *arrays*. The data can be *addressed* (selected) from anywhere in the matrix. Data is easily changed and stored back in RAM, in whole or in part. Engineers sometimes call a RAM chip *read/write memory*.

An example of the use of RAM is a word-processing computer file that you're actively working on. This paragraph, this chapter, and in fact, the whole text of this book was written in semiconductor RAM in small sections before being incrementally stored on the computer hard drive, and ultimately on external media.

There are two major categories of RAM: *dynamic RAM* (DRAM) and *static RAM* (SRAM). A DRAM chip contains transistors and capacitors, and data is stored as charges on the capacitors. The charge must be replenished frequently, or it will vanish through discharge. Replenishing is done automatically several hundred times per second. An SRAM chip uses a flip-flop to store the data. This arrangement gets rid of the need for constant replenishing of charge, but SRAM ICs require more elements than DRAM chips to store a given amount of data.

With any RAM chip, the data will vanish when we remove power unless we provide a means of *memory backup*. The most common memory-backup scheme involves the use of a small cell or battery with a long shelf life. Modern IC memories need so little current to store their data that a *backup battery* lasts as long in the circuit as it would last sitting on a shelf doing nothing!

Read-Only Memory

The data in a *read-only memory* (ROM) chip can be easily accessed, in whole or in part, but not written over. A standard ROM chip is *programmed* at the factory. We call such permanent programming *firmware*. Some ROM chips exist that you can program and reprogram yourself.

If the data in memory disappears when we remove all sources of power, we have so-called *volatile memory*. If the data is retained indefinitely after we remove the power, then we have *nonvolatile memory*. While the data in most RAM chips is volatile, the data in ROM chips is nonvolatile.

An *erasable programmable read-only memory* (EPROM) chip is a ROM device that we can reprogram by following a certain procedure. It's more difficult to rewrite data in EPROM than in RAM; the usual process for erasure involves exposure to ultraviolet (UV). We can recognize an EPROM chip by the presence of a transparent window with a removable cover, through which UV rays are focused to erase the data. The IC must be removed from the circuit board, exposed to UV for several minutes, reprogrammed, and finally returned to the circuit board.

The data in some EPROM chips can be erased by electrical means. Such an IC is called an *electrically erasable programmable read-only memory* (EEPROM) chip. We can reprogram a device of this type without removing it from its circuit card.

Quiz

Refer to the text in this chapter if necessary. A good score is at least 18 correct. Answers are in the back of the book.

1. We can keep an op amp from producing too much gain or minimize the risk of unwanted oscillation in an op amp by means of
   - (a) positive feedback.
   - (b) negative feedback.
   - (c) inductive feedback.
   - (d) capacitive feedback.

2. We find both N-channel and P-channel MOSFETs etched onto
   - (a) an NMOS chip.
   - (b) a PMOS chip.
   - (c) a CMOS chip.
   - (d) All of the above

3. If we want to build an analog audio amplifier for driving a headset or a small speaker, we might use
   - (a) a TTL IC.
   - (b) a linear IC.
   - (c) an EPROM IC.
   - (d) an ECL IC.

4. Which, if any, of the following component types is *least* practical for direct fabrication on an IC chip?
   - (a) Resistor
   - (b) Capacitor
   - (c) Inductor
   - (d) Transistor

5. Which of the following IC types would we most likely use to build a circuit that takes a pure sine-wave input and produces a pure sine-wave output at the same frequency but lagging the input by 90°?
   (a) Op amp
   (b) TTL
   (c) NMOS
   (d) PMOS

6. Fill in the blank to make the following sentence true: "In a DRAM chip, data is stored as ________."
   (a) sine waves at various frequencies
   (b) currents through tiny resistors
   (c) magnetic fields in tiny inductors
   (d) charges in tiny capacitors

7. Which of the following characteristics usually represents an advantage of a discrete-component circuit over the use of an equivalent IC?
   (a) Higher power-handling capacity
   (b) Streamlined maintenance
   (c) Improved reliability
   (d) All of the above

8. In which of the following types of memory chip can we overwrite data with the least amount of difficulty?
   (a) ROM
   (b) EPROM
   (c) EEPROM
   (d) RAM

9. Which of the following characteristics do we commonly see in CMOS chips?
   (a) Sensitivity to damage by electrostatic discharge
   (b) Suitability for analog use only
   (c) High current demand
   (d) Extremely limited speed

10. An integrator IC shifts the phase of a pure AC sine-wave input signal by
   (a) 180°.
   (b) 90°.
   (c) 45°.
   (d) an amount that depends on the frequency.

11. With respect to memory, which of the following expressions tells us that the data remains intact even if we remove all external sources of power?
    (a) Saturated
    (b) Static
    (c) Dynamic
    (d) Nonvolatile

12. In terms of the maximum number of transistors on a single chip, how many orders of magnitude larger is LSI than VLSI?
    (a) One
    (b) Two
    (c) Three
    (d) The premise of the question is wrong. The maximum number of transistors on an LSI chip is *smaller* than the maximum number of transistors on a VLSI chip.

13. In which of the following devices would you likely find an IC serving as the main component?
    (a) The filter in a high-voltage power supply
    (b) The final amplifier in a TV broadcast transmitter
    (c) An electronic calculator
    (d) All of the above

14. Which of the following ICs might we use to trigger a specific hardware device, such as an alarm bell?
    (a) A comparator
    (b) An EPROM
    (c) A notch filter
    (d) A multiplexer

15. The component density specification for an IC gives us a good idea of the
    (a) average diameter of the elements in the chip.
    (b) average surface area of the elements in the chip.
    (c) average mass-to-volume ratio of the elements in the chip.
    (d) total number of individual elements in the chip.

16. If we supply a differentiator IC with a constant input of +2 V DC, we observe an output of
    (a) +2 V DC.
    (b) 0 V DC, that is, nothing at all.
    (c) −2 V DC.
    (d) pure, sine-wave AC.

17. If we want to reprogram a typical EPROM chip, what must we do?
    (a) Simply input the new data
    (b) Expose the chip to UV radiation
    (c) Expose the chip to a magnetic field
    (d) Nothing; the data in an EPROM chip can't be altered

18. Which of the following characteristics represents a significant limitation of ECL compared with TTL?
    (a) Slower speed
    (b) Far higher current demand
    (c) Greater sensitivity to noise
    (d) Shorter working life

19. We can overwrite data, although perhaps with some difficulty, in any of the following types of memory except
    (a) ROM.
    (b) RAM.
    (c) EPROM.
    (d) EEPROM.

20. A differentiator IC shifts the phase of a pure AC sine-wave input signal by
    (a) 180°.
    (b) 90°.
    (c) 45°.
    (d) An amount that depends on the frequency.

24
CHAPTER

Electron Tubes

ELECTRON TUBES, CALLED *VALVES* IN ENGLAND AND KNOWN SIMPLY AS *TUBES* IN THE UNITED STATES, constituted the main active component in electronic equipment manufactured before the mid-1960s. In diodes, transistors, and ICs, the charge carriers are electrons and holes that "hop" from atom to atom in a solid medium. In a tube, the charge carriers are free electrons that fly through a vacuum or rarefied gas between charged metal electrodes.

The Main Advantage

Most electronics aficionados regard tubes as obsolete. Solid-state components have so many advantages over tubes, in so many applications, that this sentiment might seem completely justified. But despite their *physical* bulkiness, electron tubes have excellent *electrical* ruggedness. They're harder to "fry"!

Voltage spikes, or *transients*, on power lines can instantly destroy solid-state components. Some transients can get through a *power supply* to the diodes, transistors, and ICs in a system. As a result, sensitive solid-state components, especially ICs, burn out unless you place a transient suppressor in the AC line leading into the equipment.

Severe transients do not occur often, but the effects can be disastrous. An extreme voltage spike causes a massive burst of current. Usually, the cause is a lightning bolt striking a power line near the system. The *current surge* burns out the transient suppressor along with many of the transistors, diodes, and ICs beyond. In contrast, equipment built with vacuum tubes would probably keep working through the event unaffected.

Did You Know?

Some audiophiles insist that tube-based audio power amplifiers produce richer, fuller sound output than similar amplifiers that use power transistors.

Vacuum versus Gas-Filled

We can place most electron tubes into either of two categories: the *vacuum tube*, which has all or most of the internal gases removed, and the *gas-filled tube*, which contains low-pressure chemical vapor.

The electrons in a vacuum tube accelerate to considerable speed, resulting in current that can be guided in specific directions. An AC input signal can rapidly change the beam intensity and/or direction, producing most of the effects familiar to users of semiconductor devices, including:

- Rectification
- Amplification
- Oscillation
- Modulation
- Detection (demodulation)
- Mixing (heterodyning)
- Switching

In addition, specialized tubes can perform functions, such as the following:

- Waveform analysis
- Frequency-spectrum analysis
- Video image interception
- Video image display
- Radar image rendering
- Voltage regulation
- General lighting

As early as the 1800s, scientists knew that electrons could carry electric current through a vacuum. They also knew that hot electrodes emit electrons more easily than cool ones do. These phenomena were put to use in the first electron tubes, known as *diode tubes*, for the purpose of rectification. You'll rarely see diode tubes nowadays, although you might find them in power supplies designed to deliver several thousand volts at high current for long periods at a 100% *duty cycle* (continuous operation). Diode tubes have only two elements: a *cathode*, which usually receives the more negative voltage, and the *anode*, which usually receives the more positive voltage. Electrons can flow easily from the cathode to the anode, but not vice-versa.

A wide variety of gas-filled tubes find application in electricity and electronics. Some gas-filled tubes have a constant voltage drop, no matter how much or how little current they carry, making them useful as voltage regulators for high-voltage, high-current power supplies. They can withstand conditions that would destroy semiconductor regulating devices. Other gas-filled tubes emit infrared (IR), visible light, and/or ultraviolet (UV) radiant energy. This property can be put to use for practical and decorative lighting. We can use a small *neon bulb* (a form of gas-filled tube) to construct an audio *relaxation oscillator*, as shown in Fig. 24-1. Gas-filled tubes usually have only two elements, so technically we can call them diode tubes.

Electrode Configurations

In a tube, the cathode emits electrons. A wire *filament*, similar to the glowing element in an incandescent bulb, can heat the cathode to help drive electrons from it. The cathode of a tube is analogous to the source of an FET, or to the emitter of a bipolar transistor. The anode, also called the *plate*,

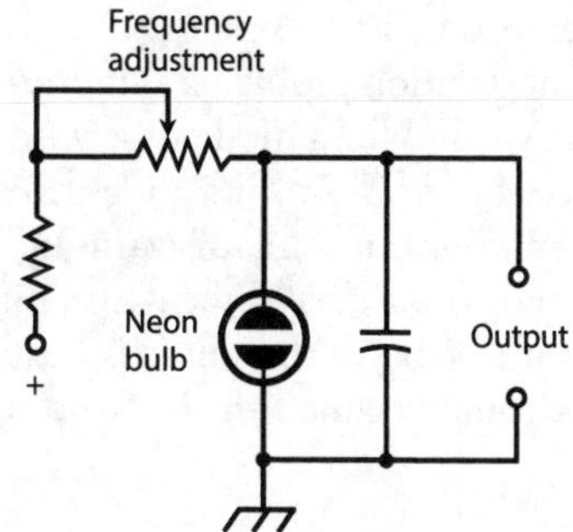

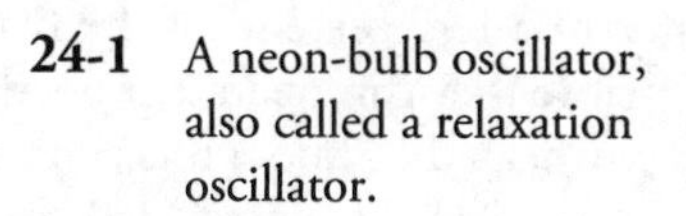

24-1 A neon-bulb oscillator, also called a relaxation oscillator.

collects the electrons. The plate serves as the tube counterpart of the drain of an FET or the collector of a bipolar transistor. In most tubes, intervening *grids* control the flow of electrons from the cathode to the plate. The grids are the counterparts of the gate of an FET or the base of a bipolar transistor.

Directly Heated Cathode

In some tubes, the filament also serves as the cathode. Engineers call this type of electrode a *directly heated cathode*. We apply the negative supply voltage directly to the filament. The filament voltage for most directly heated cathode tubes ranges from approximately 6 V to 12 V DC. We should use DC, not AC, to heat the filament in a tube with a directly heated cathode because AC modulates the output signal with a severe, unwanted AC "hum." Figure 24-2A shows the schematic symbol for a diode tube with a directly heated cathode.

Indirectly Heated Cathode

In many types of tubes, a cylindrical cathode surrounds the filament. The cathode gets hot as it absorbs IR energy radiated by the filament. We call this device an *indirectly heated cathode*. The fila-

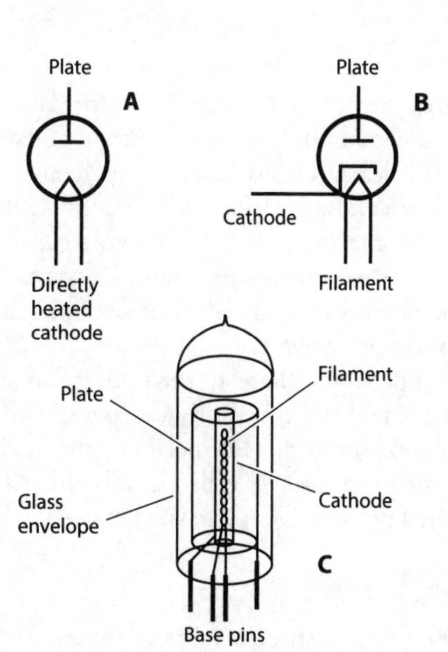

24-2 At A, the schematic symbol for a diode tube with a directly heated cathode. At B, the symbol for a diode tube with an indirectly heated cathode. At C, a simplified rendition of the construction of a diode tube.

ment normally receives 6 to 12 V AC or DC. In a tube with an indirectly heated cathode, filament AC does not cause modulation problems as it does in a directly heated cathode device. Figure 24-2B shows the schematic symbol for a diode tube with an indirectly heated cathode.

Because the electron emission in a tube depends to a large extent on the filament, or *heater*, tubes need a certain amount of time to "warm up" before they can operate properly. This time can vary from a few seconds (for a small tube with a directly heated cathode) to a couple of minutes (for massive power-amplifier tubes with indirectly heated cathodes). The warm-up time for an indirectly heated cathode tube roughly equals the boot-up time for a personal computer loaded up with a lot of software.

Cold Cathode

In a gas-filled tube, the electron-emitting element has no filament to heat it. We call such an electrode a *cold cathode*. Various chemical elements are used in gas-filled tubes. In fluorescent devices, we'll often find neon, argon, or xenon. In older gas-filled *voltage-regulator* (VR) tubes, mercury vapor was used. In a mercury-vapor VR tube, the warm-up period was determined by the length of time needed for the elemental mercury, which forms a liquid at room temperature, to vaporize (a couple of minutes). These tubes have become rare in recent years as engineers and system designers have learned about the toxic nature of mercury.

Plate

The plate, or anode, of a tube forms a metal cylinder concentric with the cathode and filament, as shown in Fig. 24-2C. We connect the plate to the positive DC supply voltage, sometimes through external inductors, transformer windings, and/or resistors. Tubes operate at dangerous—and in some cases deadly—plate voltages ranging from 50 V to more than 3 kV. Technicians unfamiliar with vacuum tubes should not attempt to service equipment that contains them. The output of a tube-type amplifier is almost always taken from the plate circuit. The plate exhibits high impedance for signal output, similar to the output impedance of a typical JFET.

Control Grid

In a vacuum tube, the flow of current (in the form of electrons moving from the cathode to the plate) can be controlled with an intervening electrode called the *control grid* (or simply the *grid*). It's a wire mesh or screen that lets electrons pass through to an extent that depends on the voltage that we apply. The grid impedes the flow of electrons if we provide it with a negative voltage relative to the cathode.

As we increase the negative *grid bias* voltage, the control grid obstructs the flow of electrons reaching the plate to an increasing extent. If we make the negative grid bias voltage large enough, the grid completely obstructs the flow of electrons, and we observe no current in the plate circuit. We call that condition *cutoff*.

If we apply a positive voltage to the grid, it can accelerate the flow of electrons to the plate, but only to a limited extent. Positive grid bias "bleeds off" some current that would otherwise reach the plate, a phenomenon somewhat like positive gate bias in an N-channel JFET. When the current from the cathode to the plate (usually called simply the *plate current*) reaches its maximum possible value for a given applied positive DC plate voltage, we say that the tube has reached a state of *saturation*.

Triode Tube

A tube with one grid constitutes a *triode*. Figure 24-3A shows its schematic symbol. The cathode appears at the bottom, the control grid is in the middle (dashed line), and the plane is at the top.

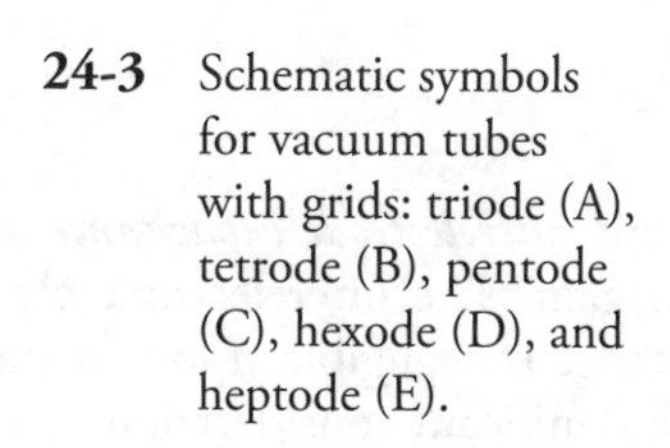
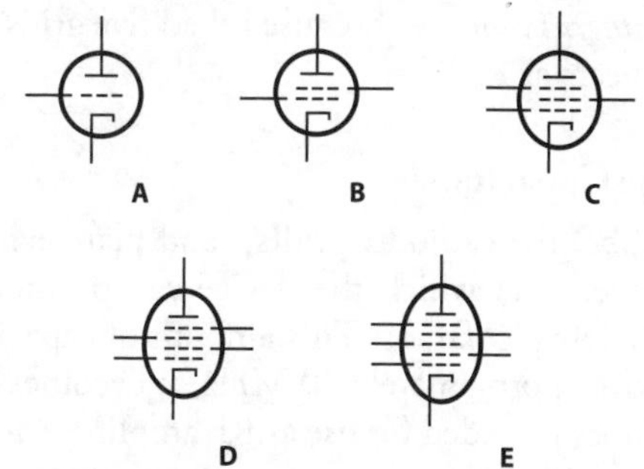

24-3 Schematic symbols for vacuum tubes with grids: triode (A), tetrode (B), pentode (C), hexode (D), and heptode (E).

In this particular example, we see an indirectly heated cathode; the filament is not shown. This omission is standard in schematics showing tubes with indirectly heated cathodes. (In a tube with a directly heated cathode, the filament symbol serves as the cathode symbol.) In most practical circuits, we'll want to bias the control grid with a negative DC voltage whose absolute value can range from zero to roughly half the DC plate voltage.

Tetrode Tube

Some vacuum tubes contain a second grid between the control grid and the plate. This grid takes the form of a wire spiral or coarse screen. We call it the *screen grid* or *screen.* It normally carries a positive DC voltage amounting to 25% to 35% of the plate voltage. The screen grid reduces the capacitance between the control grid and plate, minimizing the natural tendency of a vacuum-tube-based amplifier to oscillate (the *stability* improves, compared with a triode in the same application). The screen grid can also serve as a second control grid, allowing us to inject two signals into the tube. We might do that if we want to use the tube as a mixer or amplitude modulator. This type of tube has four elements, and is known as a *tetrode*. Figure 24-3B shows its schematic symbol, in which the screen grid appears immediately above the control grid and immediately below the plate.

Pentode Tube

The electrons in a tetrode, especially one provided with high DC plate voltage, sometimes bombard the plate with such force that some of them bounce back, or knock other electrons from the plate. This so-called *secondary emission* can hinder tube performance and, at high power levels, cause the screen current to rise to the point at which it physically destroys that electrode! We can eliminate this problem by placing another grid, called the *suppressor grid,* or *suppressor*, between the screen and the plate. The suppressor repels *secondary electrons* emanating from the plate, preventing most of them from reaching the screen. The suppressor also reduces the capacitance between the control grid and the plate more than a screen grid can do all by itself. We can obtain more gain and better stability with a *pentode*, or tube with five elements, than we can get with a tetrode or triode. Figure 24-3C shows the schematic symbol for a pentode. Occasionally, you'll find a pentode tube with the suppressor grid (topmost dashed line in the figure) internally shorted to the cathode.

Hexode and Heptode Tubes

In radio and TV receivers prior to the 1960s, tubes with four or five grids were sometimes used. These tubes had six and seven elements, respectively, and were called the *hexode* and the *heptode*. Such tubes usually formed the basis for signal mixers (frequency converters) or amplitude modulators. The schematic symbol for a hexode appears at D in Fig. 24-3; the symbol for a heptode, also known as a *pentagrid converter* because it had five grids and found its main application in frequency conversion, appears at E.

Interelectrode Capacitance

In a vacuum tube, the cathode, grid(s), and plate exhibit *interelectrode capacitance* that limits the maximum frequency at which the device can produce gain. The interelectrode capacitance in a typical tube is a few picofarads. This amount of capacitance is negligible at low frequencies, but at frequencies above approximately 30 MHz, it becomes a significant consideration.

Vacuum tubes intended for use as RF amplifiers are designed to minimize interelectrode capacitance. Interelectrode capacitance can cause a vacuum tube to oscillate when we want it to serve only as an amplifier. The oscillation problem tends to worsen as we increase the frequency; it's less likely (and less severe if it does occur) in tetrodes and pentodes than in triodes.

Circuit Arrangements

Signal amplification constitutes the main application for vacuum tubes, especially in radio and TV transmitters at power levels of more than 1 kW. Some high-fidelity audio systems also employ vacuum tubes. In recent years, tubes have gained favor with some popular musicians who insist that "tube amps" provide better sound quality than amplifiers using power transistors, even though no theoretical basis exists for such claims. If you ever do any work with tube-type amplifiers, you'll encounter two basic vacuum-tube amplifier circuit arrangements: the *grounded-cathode* circuit and the *grounded-grid* circuit.

Grounded-Cathode Configuration

Figure 24-4 illustrates a grounded-cathode circuit using a triode tube. Some tube-type RF power amplifiers and audio amplifiers employ this design. The circuit exhibits moderate input impedance and high output impedance. If necessary, we can match the tube output impedance to the load impedance by tapping a coil in the output circuit (as shown here), or by using a transformer. The output wave occurs in phase opposition with respect to the input wave.

Grounded-Grid Configuration

Figure 24-5 shows a basic grounded-grid RF amplifier circuit. The input impedance is low, and the output impedance is high. The output impedance can be matched to the load impedance by the same means as with the grounded cathode arrangement. A grounded-grid amplifier requires more driving (input) power than the grounded-cathode amplifier does, if we expect to get a certain amount of output power. A grounded-cathode amplifier might produce 1 kW of RF output with

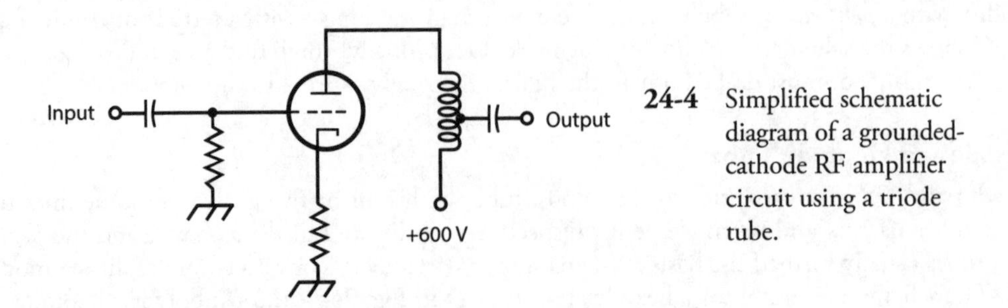

24-4 Simplified schematic diagram of a grounded-cathode RF amplifier circuit using a triode tube.

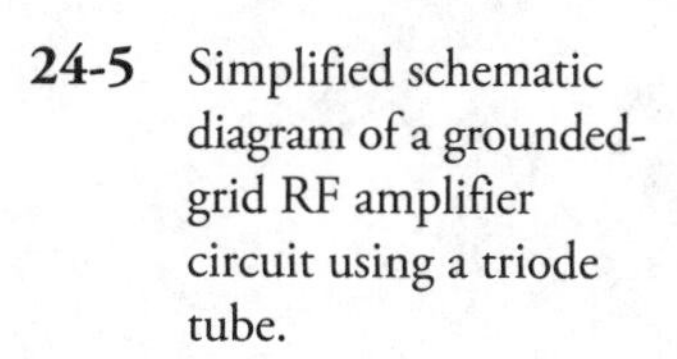
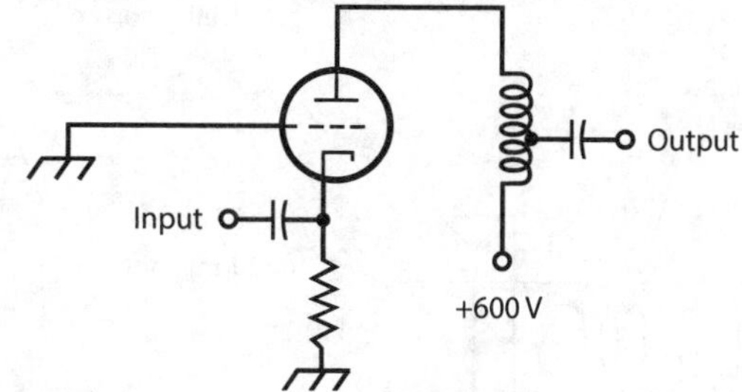

24-5 Simplified schematic diagram of a grounded-grid RF amplifier circuit using a triode tube.

only a 7-W input signal, but a grounded-grid amplifier might need 70 W of drive to produce 1 kW of RF output. A grounded grid amplifier has a significant advantage, however: it's less likely to break into unwanted oscillation than a grounded-cathode circuit. The output wave occurs in phase with the input wave.

Plate Voltage

In the circuits of Figs. 24-4 and 24-5, the plate voltages (+600 V DC) are given only as hypothetical values. With the plate voltages indicated in these examples, we could expect the circuits to produce 75 to 150 W of useful signal output, as long as they receive sufficient drive (input power) and we apply the proper grid bias. An amplifier rated at 1 kW output would require a plate voltage of +2 kV DC to +5 kV DC. In high-power radio and TV broadcast transmitters producing in excess of 50 kW RF output, we would need even higher DC plate voltages.

Cathode-Ray Tubes

"Ancient" TV receivers and computer monitors use *cathode-ray tubes* (CRTs). So do "ancient" oscilloscopes, spectrum analyzers, and radar sets. A few are still in service, however.

Electron Beam

In a CRT, a specialized cathode called an *electron gun* emits an electron beam that gains speed and definition as it passes through a series of positively charged *anodes*. The electron beam hits a glass screen whose inner surface has a *phosphor* coating. The phosphor glows visibly, as seen from the face of the CRT, wherever the electrons strike it. The beam *scanning* pattern is controlled by a pair of magnetic or electrostatic fields. One field causes the beam to scan rapidly across the screen in a horizontal direction. Another field moves the beam vertically. When we apply a signal to the electrodes that produce the deflection of the electron beam, an orderly pattern shows up on the screen. This pattern can portray the graph of a signal wave, a fixed image, an animated image, a computer text display, or any other type of visible image.

Electromagnetic CRT

Figure 24-6 shows a simplified cross-sectional view of an *electromagnetic CRT*. The device has two sets of *deflecting coils*, one for the horizontal plane and the other for the vertical plane. (To keep the illustration from getting too cluttered, we illustrate only one set of deflecting coils here.) As the instantaneous current through the coils increases, so does the intensity of the magnetic field surrounding them, causing increased deflection of the electron beam as it passes between the coils. The electron beam bends at right angles to the magnetic lines of flux between the coils.

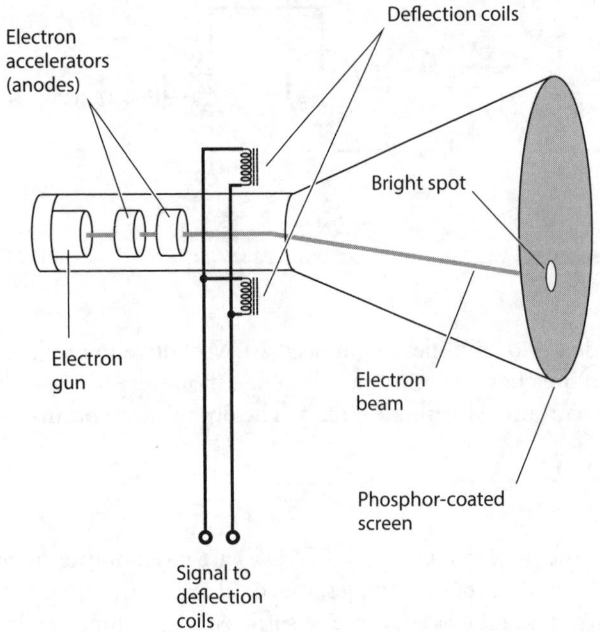

24-6 Simplified cross-sectional rendition of an electromagnetic CRT.

In an *oscilloscope*, the horizontal deflecting coils receive a sawtooth waveform, causing the electron beam to scan, or *sweep*, at a precise, adjustable speed across the screen from left to right as viewed from the front. After each timed left-to-right sweep, the beam returns, almost instantly, to the left side of the screen for the next sweep. The vertical deflecting coils receive the signal whose waveform we want to observe. The instantaneous signal current in the deflecting coils makes the electron beam move up and down. The combination of vertical and horizontal beam motion produces a display of the applied signal waveform as a function of time.

Electrostatic CRT

In an *electrostatic CRT*, charged metal electrodes, rather than current-carrying coils, cause deflection of the electron beam. When voltages appear on the *deflecting plates*, the beam is bent in the direction of the electric lines of flux between the plates. The greater the instantaneous voltage between a pair of deflecting plates, the stronger the electric field, and the greater the extent to which the beam deflects at that instant in time.

The principal advantage of an electrostatic CRT is the fact that it generates a far less intense fluctuating EM field than an electromagnetic CRT. Some people believe that such fields, called *extremely low frequency* (ELF) energy because they occur at frequencies far below those in the usual RF spectrum, might have adverse effects on people who use CRT-equipped devices such as desktop computers for extended periods of time. In recent years, with the evolution of *liquid crystal displays* (LCDs) and *plasma displays* as alternatives to the CRT type of display, ELF concerns have diminished.

Tubes Above 300 MHz

Specialized vacuum tubes are used for RF operation at frequencies above 300 MHz. Communications specialists call these ranges the *ultra high frequency* (UHF) *band*, which extends from 300 MHz to 3 GHz, and the *microwave band*, which extends from 3 GHz all the way up to the lower end of the infrared (IR) part of the EM spectrum. The *magnetron* and the *Klystron* are examples of tubes that can generate and amplify UHF and microwave signals.

Magnetron

A magnetron contains a cathode and a surrounding anode. The anode is divided by radial barriers into sections called *cavities*. The output is taken from an opening in the anode, and passes into a waveguide that serves as a transmission line for the RF output energy. A magnetron can generate more than 1 kW of RF power output at a frequency of 1 GHz. As the frequency increases, the realizable power output decreases.

The cathode is connected to the negative terminal of a high-voltage source, and the anode is connected to the positive terminal. Electrons flow radially outward from the cathode to the anode. A magnetic field is applied lengthwise through the cavities. As a result of this applied field, the electron paths bend into spirals. The electric field produced by the high voltage, interacting with the longitudinal magnetic field and the effects of the cavities, causes the electrons to bunch up into so-called *clouds*. The swirling movement of the electron clouds causes a fluctuating current in the anode. The frequency of this current depends on the shapes and sizes of the cavities. Small cavities result in the highest oscillation frequencies; larger cavities produce oscillation at relatively lower frequencies.

Klystron

A Klystron has an electron gun, one or more cavities, and a device that modulates the electron beam. Several different Klystron geometries exist. The most common are the *multi-cavity Klystron* and the *reflex Klystron*.

In a multi-cavity Klystron, the electron beam is *velocity-modulated* in the first cavity. This causes the density of electrons (the number of particles per unit of volume) in the beam to change as the beam moves through subsequent cavities. The electrons tend to "bunch up" in some regions and "spread out" in other regions. The intermediate cavities increase the magnitude of the electron beam modulation, resulting in amplification. Output is taken from the last cavity. Peak power levels in some multi-cavity Klystrons can exceed 1 MW (one megawatt or 10^6 W), although the average power is much less.

A reflex Klystron has a single cavity. A *retarding field* causes the electron beam to periodically reverse direction, producing a phase reversal that allows energy to be drawn from the electrons. A typical reflex Klystron can produce signals on the order of a few watts at frequencies of 300 MHz and above.

Quiz

Refer to the text in this chapter if necessary. A good score is at least 18 correct. Answers are in the back of the book.

1. We can accurately refer to the cathode in a CRT as the
   (a) plate.
   (b) dynode.
   (c) source.
   (d) electron gun.

2. The plate of a vacuum tube is the electrical counterpart of the
   (a) source of a MOSFET.
   (b) collector of a bipolar transistor.
   (c) cathode of a semiconductor diode.
   (d) gate of a JFET.

3. We might find mercury vapor in
   (a) a Klystron.
   (b) an old VR tube.
   (c) a pentagrid converter.
   (d) a CRT.

4. Specialized two-electrode, gas-filled tubes can serve as essential components in
   (a) dim-light detectors.
   (b) signal mixers.
   (c) audio oscillators.
   (d) All of the above

5. In a CRT, the anodes
   (a) accelerate the electrons.
   (b) emit the electrons.
   (c) deflect the electrons.
   (d) fluoresce when the electrons strike them.

6. In a tube with an *indirectly heated* cathode,
   (a) we can use AC to heat the filament.
   (b) the grid is connected to the filament.
   (c) the filament serves as the cathode.
   (d) no filament exists.

7. In a pentode tube, which electrode do we bias with a positive DC voltage relative to the cathode?
   (a) Control grid
   (b) Screen grid
   (c) Suppressor grid
   (d) All of the above

8. Which type of tube was once used primarily for rectification in power supplies?
   (a) Klystron
   (b) Magnetron
   (c) Pentagrid converter
   (d) Diode

9. A tube with two grids usually constitutes a
   (a) heptode.
   (b) hexode.
   (c) pentode.
   (d) tetrode.

10. We can reduce problems arising from secondary-electron emission in a tetrode vacuum-tube amplifier by
    (a) increasing the plate voltage.
    (b) switching to a pentode tube.
    (c) connecting the screen grid directly to the plate.
    (d) connecting the control grid directly to the cathode.

11. A pentode tube *always* has
    (a) a directly heated cathode.
    (b) an indirectly heated cathode.
    (c) a screen grid.
    (d) a dynode.

12. Which of the following effects can result directly from interelectrode capacitance in a vacuum-tube power amplifier?
    (a) Excessive UV emission
    (b) Excessive power gain
    (c) Excessive plate current
    (d) None of the above

13. In a magnetron, the cavity dimensions affect the
    (a) output frequency.
    (b) extent of electron-beam deflection.
    (c) brightness of the image.
    (d) image resolution.

14. In an electrostatic CRT, a voltage between the deflecting plates
    (a) bends the electron beam.
    (b) blocks the electron beam.
    (c) increases the electron-gun output.
    (d) reduces the gain.

15. Which of the following devices maintains a constant voltage between its cathode and its anode, independent of the current?
    (a) All Klystron vacuum tubes
    (b) All triode vacuum tubes
    (c) Certain gas-filled diode tubes
    (d) All of the above

16. In some pentode tubes, the suppressor is internally shorted to the
    (a) cathode.
    (b) control grid.
    (c) screen grid.
    (d) plate.

17. If we increase the negative DC control-grid bias voltage in a triode, tetrode, or pentode vacuum tube, we'll eventually reach a condition of
    (a) saturation.
    (b) oscillation.
    (c) cutoff.
    (d) negative resistance.

18. In some tetrode tubes, the screen is connected to
    (a) an external signal source.
    (b) the cathode.
    (c) the control grid.
    (d) the plate.

19. In an electromagnetic CRT, a current through the deflecting coils causes the electron beam to
    (a) speed up but not change direction.
    (b) bend toward the north magnetic pole and away from the south pole.
    (c) bend at right angles to the magnetic flux lines.
    (d) slow down but not change direction.

20. We normally apply DC, not AC, to the filament of a tube with a directly heated cathode because AC can
    (a) cause filament overheating.
    (b) cause unwanted modulation of the signal.
    (c) produce undesirable magnetic fields.
    (d) produce undesirable UV radiation.

25
CHAPTER
Power Supplies

A *POWER SUPPLY* CONVERTS UTILITY AC TO PURE DC OF THE SORT WE GET FROM AN ELECTROchemical or solar battery. In this chapter, we'll examine the components of a typical power supply.

Power Transformers

We can categorize power transformers in two general ways: step-down or step-up. As you remember, the output, or secondary, voltage of a step-down transformer is lower than the input, or primary, voltage. The reverse holds true for a step-up transformer, where the output voltage exceeds the input voltage.

Step-Down

Most electronic devices need only a few volts to function. The power supplies for such equipment use step-down power transformers, with the primary windings connected to the utility AC outlets. The transformer's physical size and mass depend on the amount of current that we expect it to deliver. Some devices need only a small current at a low voltage. The transformer in a radio receiver, for example, can be physically small. A large amateur radio transmitter or hi-fi amplifier needs more current. The secondary windings of a transformer intended for those applications must consist of heavy-gauge wire, and the cores must have enough bulk to contain the large amounts of magnetic flux that the coils generate.

Step-Up

Some circuits need high voltage. The cathode-ray tube (CRT) in an old-fashioned home television (TV) set needs several hundred volts, for example. Some amateur radio power amplifiers use vacuum tubes working at more than 1 kV DC. The transformers in these appliances are step-up types. They must have considerable bulk because of the number of turns in the secondary, and also because high voltages can spark, or *arc,* between wire turns if the windings aren't spaced far enough apart. If a step-up transformer needs to supply only a small amount of current, however, it can be fairly small and light.

Transformer Ratings

Engineers rate power transformers according to the maximum output voltage and current they deliver. For a given unit, we'll often read or hear about the *volt-ampere* (VA) capacity, which equals the product of the nominal output voltage and maximum deliverable current.

A transformer with 12-V output, capable of providing up to 10 A of current, has a VA capacity of 12 V × 10 A, or 120 VA. The nature of power-supply filtering, which we'll learn about later in this chapter, makes it necessary for the power-transformer VA rating to significantly exceed the actual power in watts that the load consumes.

A high-quality, rugged power transformer, capable of providing the necessary currents and/or voltages, constitutes an integral and critical part of a well-engineered power supply. The transformer is usually the most expensive power-supply component to replace if it burns out, so we must always choose a transformer with the appropriate specifications when designing and building a power supply.

Rectifier Diodes

Rectifier diodes are available in various sizes, intended for different purposes. Most rectifier diodes comprise silicon semiconductor materials, so we call them *silicon rectifiers*. Some rectifier diodes are fabricated from selenium, so we call them *selenium rectifiers*. When we work with power-supply diodes, we must pay close attention to two specifications: the *average forward current* (I_o) rating and the *peak inverse voltage* (PIV) rating.

Average Forward Current

Electric current always produces some heat because every material medium offers at least a little bit of resistance. If we drive too much current through a diode, the resulting heat will destroy the P-N junction. When designing a power supply, we must use diodes with an I_o rating of at least 1.5 times the expected average DC forward current. If this current is 4.0 A, for example, the rectifier diodes should be rated at $I_o = 6.0$ A or more.

In a power supply that uses one or more rectifier diodes, the I_o flows through each individual diode. The current drawn by the *load* can, and usually does, differ from I_o. Also, note that I_o represents an *average* figure. The *instantaneous* forward current is another thing entirely, and can range up to 15 or 20 times I_o depending on the nature of the filtering circuit.

Some diodes have *heatsinks* to help carry heat away from the P-N junction. We can recognize a selenium rectifier by the appearance of its heatsink, which looks like a miniature version of an old-fashioned baseboard radiator built around a steam pipe.

We can connect two or more identical rectifier diodes in parallel to increase the current rating over that of an individual diode. When we do this, we should connect a small-value resistor in series with each diode to equalize the current. Every one of these resistors should have a value such that the voltage drop across it equals roughly 1 V under normal operating conditions.

Peak Inverse Voltage

The PIV rating of a diode tells us the maximum instantaneous reverse-bias voltage that it can withstand without avalanche breakdown. A well-designed power supply has diodes whose PIV ratings significantly exceed the peak AC input voltage. If the PIV rating is not great enough, the diode or diodes in a supply conduct current for part of the reverse cycle. This conduction degrades the efficiency of the power supply because the reverse current "bucks" the forward current.

We can connect two or more identical rectifier diodes in series to obtain a higher PIV rating than a single diode alone can provide. We'll sometimes see engineers take advantage of this technique when designing high-voltage power supplies, such as those needed for tube-type power amplifiers. High-value resistors, of about 500 ohms for each peak-inverse volt, are placed in parallel with each diode to distribute the reverse bias equally among the diodes. In addition, each diode is shunted by (connected in parallel with) a capacitor of approximately 0.01 μF.

Half-Wave Circuit

The simplest rectifier circuit, called the *half-wave rectifier* (Fig. 25-1A), has a single diode that "chops off" half of the AC cycle. The effective output voltage from a power supply that uses a half-wave rectifier is much less than the peak transformer output voltage, as shown in Fig. 25-2A. The peak voltage across the diode in the reverse direction can range up to 2.8 times the applied RMS AC voltage.

Most engineers like to use diodes whose PIV ratings equal at least 1.5 times the maximum expected peak reverse voltage. Therefore, in a half-wave rectifier circuit, the diodes should be rated for at least 2.8 × 1.5, or 4.2, times the RMS AC voltage that appears across the secondary winding of the power transformer.

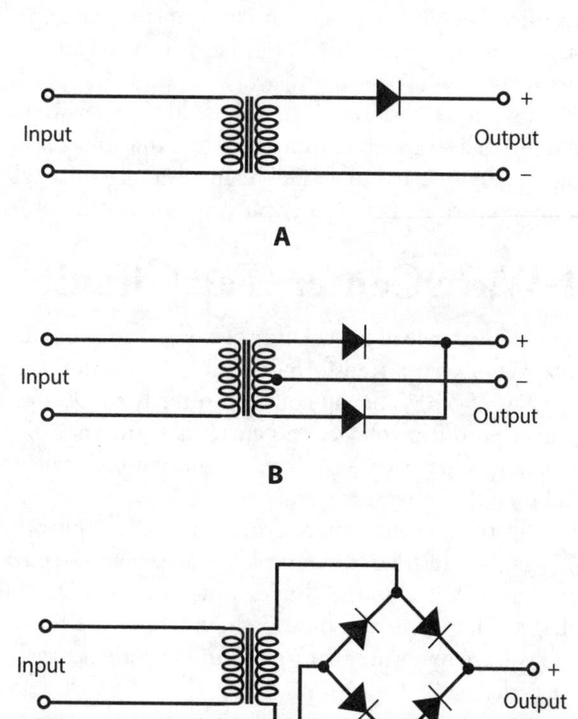

25-1 At A, a half-wave rectifier circuit. At B, a full-wave center-tap rectifier circuit. At C, a full-wave bridge rectifier circuit.

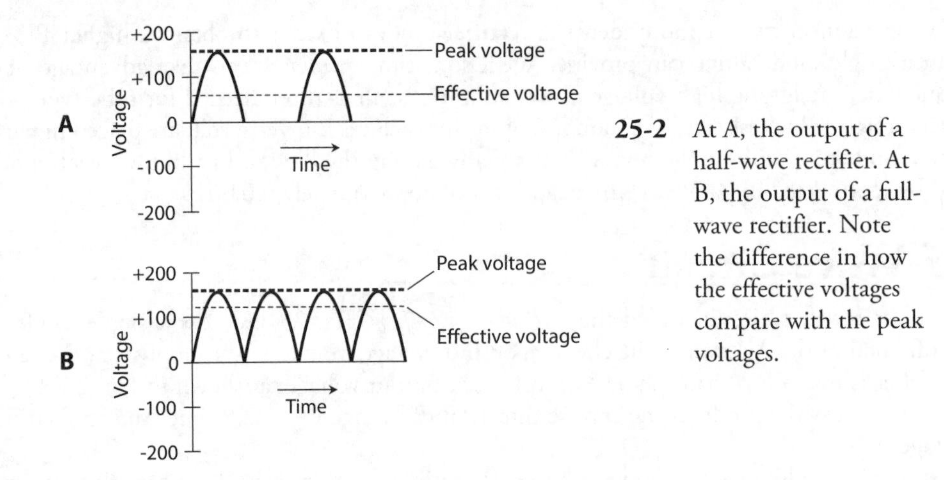

25-2 At A, the output of a half-wave rectifier. At B, the output of a full-wave rectifier. Note the difference in how the effective voltages compare with the peak voltages.

Half-wave rectification has shortcomings. First, the output is difficult to filter. Second, the output voltage can drop considerably when the supply must deliver high current. Third, half-wave rectification puts a strain on the transformer and diodes because it *pumps* them, meaning that the circuit works the diodes hard during half the AC cycle and lets them "loaf" during the other half.

Half-wave rectification will usually suffice when we want to design a power supply that will never have to deliver much current, or when the voltage can vary without affecting the behavior of the equipment connected to it. The main advantage of a half-wave circuit is the fact that it costs less than more sophisticated circuits because it contains fewer parts.

Full-Wave Center-Tap Circuit

We can take advantage of both halves of the AC cycle by means of *full-wave rectification*. A *full-wave center-tap rectifier* has a transformer with a connection called a *tap* at the center of the secondary winding (Fig. 25-1B). The tap connects directly to *electrical ground*, also called *chassis ground*. This arrangement produces voltages and currents at the ends of the secondary winding that oppose each other in phase. These two AC waves are individually half-wave rectified, cutting off one half of the cycle and then the other, repeatedly.

The effective output voltage from a power supply with a full-wave, center-tap rectifier (Fig. 25-2B) is greater, relative to the peak voltage, than the effective output voltage with the half-wave rectifier. The PIV across the diodes can, nevertheless, range up to 2.8 times the applied RMS AC voltage. Therefore, the diodes should have a PIV rating of at least 4.2 times the applied RMS AC voltage to ensure that they won't experience avalanche breakdown during any portion of the wave cycle.

The output of a full-wave center-tap rectifier is easier to filter than that of a half-wave rectifier, because the frequency of the pulsations in the DC (known as the *ripple frequency*) from a full-wave rectifier equals twice the ripple frequency of the pulsating DC from a half-wave rectifier, assuming identical AC input frequency in either situation. If you compare Fig. 25-2B with Fig. 25-2A, you will see that the full-wave-rectifier output is "closer to pure DC" than the half-wave rectifier output. Another advantage of a full-wave center-tap rectifier is the fact that it treats the transformer and diodes more "gently" than a half-wave rectifier does.

When we connect a load to the output of a power supply that uses a full-wave center-tap rectifier circuit, the voltage drops less than it does with the same load connected to a half-wave supply. However, because the transformer is more sophisticated, the full-wave center-tap circuit costs more than a half-wave circuit that delivers the same output voltage at the same rated maximum current.

Full-Wave Bridge Circuit

We can get full-wave rectification using a circuit known as a *full-wave bridge rectifier*, often called simply a *bridge*. Figure 25-1C shows a schematic diagram of a typical full-wave bridge circuit. The output waveform looks the same as the waveform that we get from the output of a full-wave center-tap circuit (Fig. 25-2B).

The effective output voltage from a power supply that uses a full-wave bridge rectifier is somewhat less than the peak transformer output voltage, as shown in Fig. 25-2B. The peak voltage across the diodes in the reverse direction equals about 1.4 times the applied RMS AC voltage. Therefore, each diode needs to have a PIV rating of at least 1.4×1.5, or 2.1, times the RMS AC voltage that appears at the transformer secondary in order to prevent avalanche breakdown from occurring during any part of the cycle.

The bridge circuit does not require a center-tapped transformer secondary. It uses the entire secondary winding on both halves of the wave cycle, so it makes more efficient use of the transformer than the full-wave center-tap circuit does. The bridge circuit also places less strain on the individual diodes than a half-wave or full-wave center-tap circuit does.

Voltage-Doubler Circuit

Diodes and capacitors can be interconnected to deliver a DC output of approximately twice the positive or negative peak AC input voltage. We call this arrangement a *voltage-doubler power supply*. It works well as long as the load draws low current. However, if we place a "heavy load" on a voltage-doubler power supply, the *voltage regulation* deteriorates. The voltage drops considerably when the current demand is significant; if the current demand fluctuates, so does the output voltage (the higher the current, the lower the voltage).

The best way to build a high-voltage power supply involves the use of a step-up transformer, not a voltage-doubling arrangement. Nevertheless, we can sometimes get away with a voltage-doubler power supply when we must minimize the overall cost, and when we never require it to deliver much current.

Figure 25-3 is a simplified diagram of a voltage-doubler power supply. This particular system takes advantage of the entire AC cycle, so it constitutes a *full-wave voltage doubler*. It subjects the diodes to reverse-voltage peaks of 2.8 times the applied RMS AC voltage. Therefore, the diodes should have a PIV rating of at least 4.2 times the RMS AC voltage that appears across the transformer secondary. When the current demand is low, the DC output voltage of this type of power supply equals approximately 2.8 times the RMS AC input voltage.

Proper operation of a voltage-doubler power supply depends on the ability of the capacitors to hold a charge under maximum load. The capacitors must have large values, as well as high *working-voltage* ratings. The capacitors serve two purposes: to boost the voltage and to filter the output.

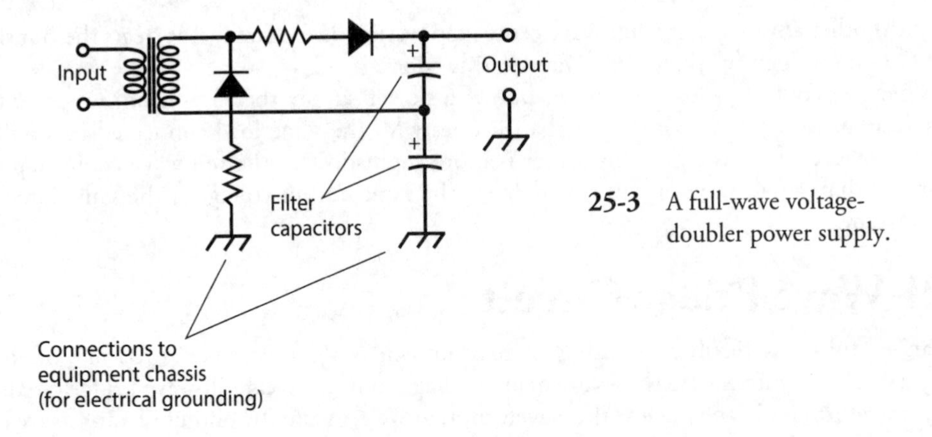

25-3 A full-wave voltage-doubler power supply.

The resistors, which have low ohmic values and appear in series with the diodes, protect the diodes against surge currents that occur when the power supply is first switched on.

Power-Supply Filtering

Most DC-powered devices need something more "pure" than the rough, pulsating DC that comes straight out of a rectifier circuit. We can eliminate, or at least minimize, the pulsations (ripple) in the rectifier output using a *power-supply filter*.

Capacitors Alone

The simplest power-supply filter consists of one or more large-value capacitors, connected in parallel with the rectifier output, as shown in Fig. 25-4. A good component for this purpose is an *electrolytic capacitor*. This type of capacitor is *polarized*, meaning that we must connect it in the correct direction. Any given electrolytic capacitor also has a certain maximum rated working voltage. Pay attention to these particulars if you ever work with electrolytic capacitors!

Filter capacitors function by "trying" to maintain the DC voltage at its peak level, as shown in Fig. 25-5. The output of a full-wave rectifier (drawing A) lends itself more readily to this process than the output of a half-wave rectifier (drawing B). With a full-wave rectifier receiving a 60-Hz AC electrical input, the ripple frequency equals 120 Hz, but with a half-wave rectifier it's only 60 Hz. The filter capacitor, therefore, gets recharged twice as often with a full-wave rectifier as it does with a half-wave rectifier.

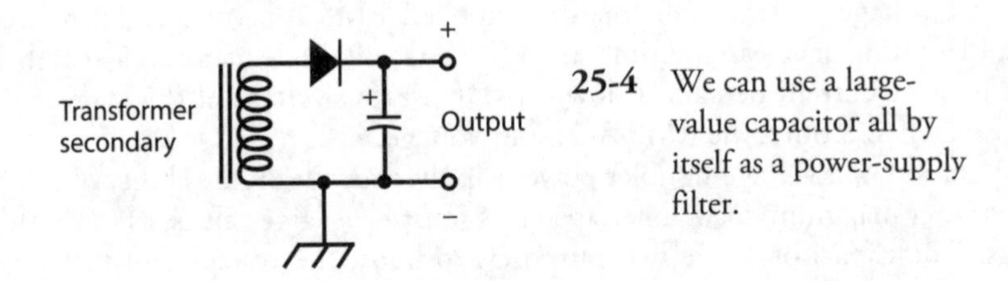

25-4 We can use a large-value capacitor all by itself as a power-supply filter.

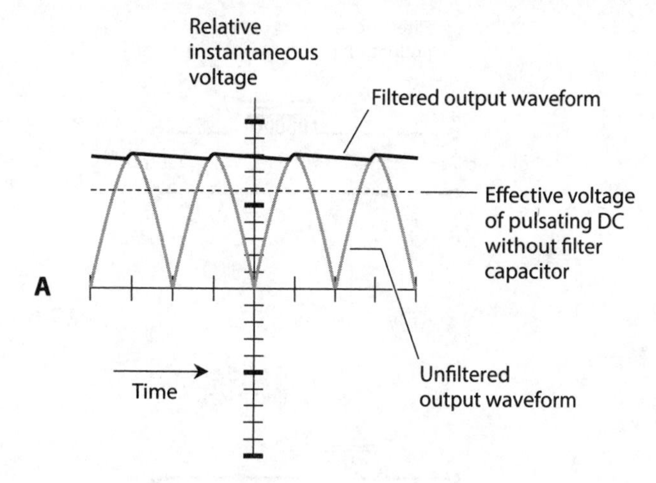

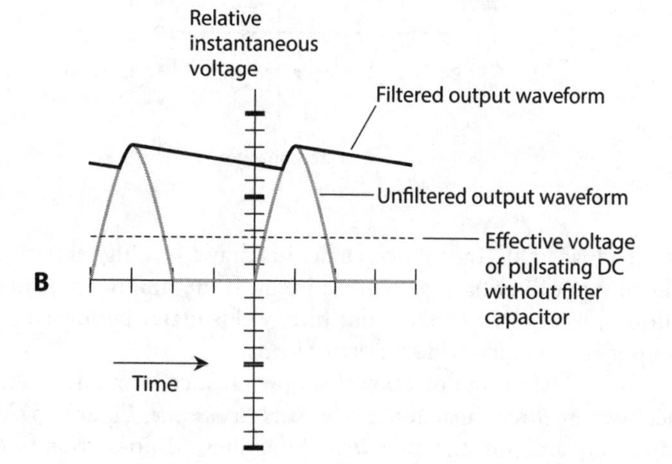

25-5 Filtering of ripple from a full-wave rectifier (A) and from a half-wave rectifier (B).

The two illustrations in Fig. 25-5 reveal the reason why a full-wave rectifier can produce more "pure" DC (for a given peak voltage and filter capacitance) than a half-wave rectifier can. The full-wave output gives the capacitor a "less bumpy ride," while the half-wave output lets the capacitor discharge more between each "refresh pulse."

Capacitors and Chokes

We can obtain enhanced ripple suppression by placing a large-value inductor in series with the rectifier output along with a large-value capacitor in parallel. When an inductor serves in this role, we call it a *filter choke.*

In a filter that uses a capacitor and an inductor, we can place the capacitor on the rectifier side of the choke to construct a *capacitor-input filter* (Fig. 25-6A). If we locate the filter choke on the rectifier side of the capacitor, we get a *choke-input filter* (Fig. 25-6B). Capacitor-input filtering works well when a power supply does not have to deliver much current. The output voltage, when

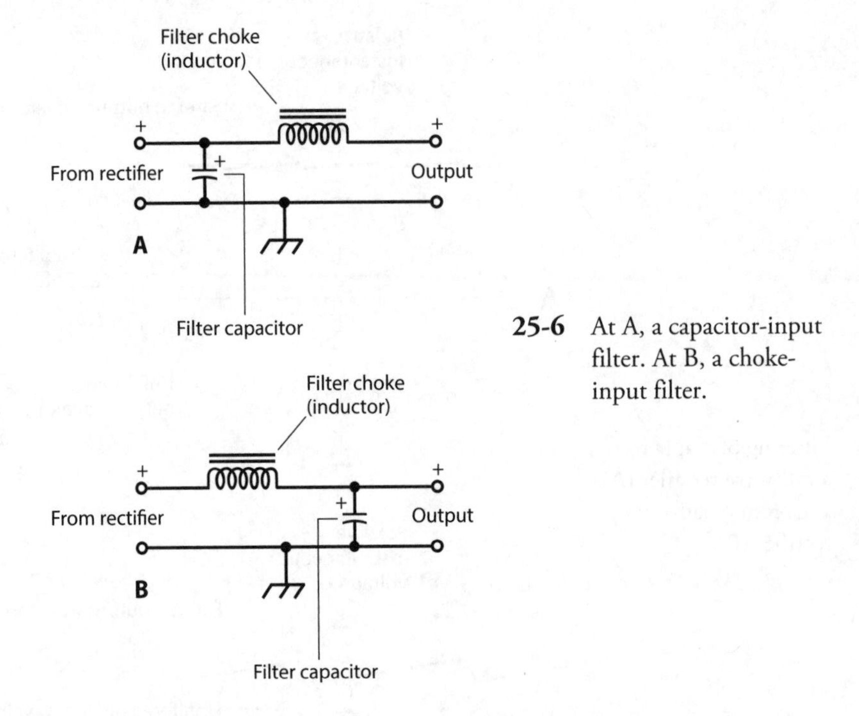

25-6 At A, a capacitor-input filter. At B, a choke-input filter.

the load is "light" (not much current is drawn), is higher with a capacitor-input filter than with a choke-input filter having identical input. If the supply needs to deliver large or variable amounts of current, however, a choke-input filter yields better performance because it produces a more stable output voltage for a wide variety of loads.

If the DC output of a power supply must contain an absolute minimum of ripple, we can connect two or three capacitor/choke pairs in *cascade*. Figure 25-7 shows an example. Each inductor/capacitor pair constitutes a *section* of the filter. Multi-section filters can consist of capacitor-input or choke-input sections; the two types should never be mixed in the same filter.

In the example of Fig. 25-7, both capacitor/choke pairs are called *L sections* (not because of inductance, but because of their geometric shapes in the schematic diagram). If we eliminate the

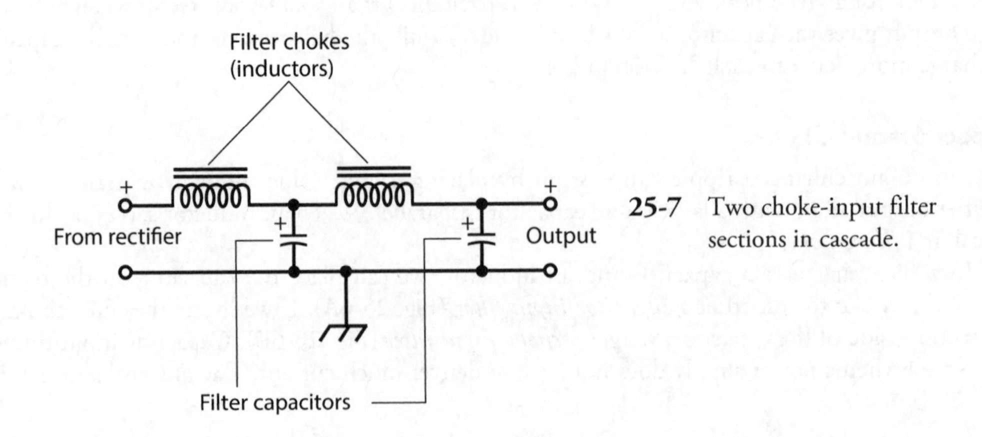

25-7 Two choke-input filter sections in cascade.

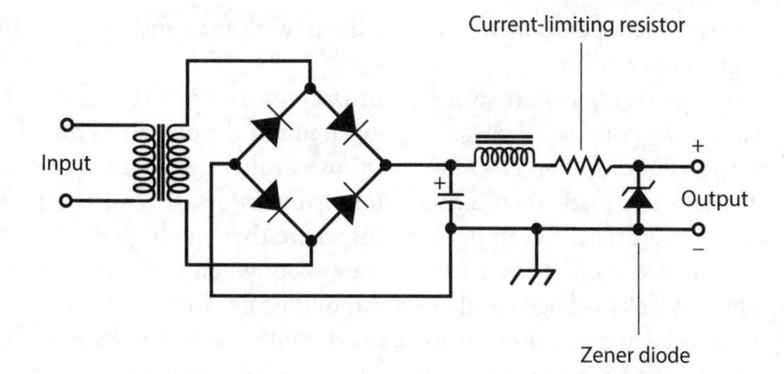

25-8 A power supply with a Zener-diode voltage regulator in the output.

second capacitor, the filter becomes a *T section* (the inductors form the top of the T, and the capacitor forms the stem). If we move the second capacitor to the input and remove the second choke, the filter becomes a *pi section* (the capacitors form the pillars of an uppercase Greek letter pi, and the inductor forms the top).

Voltage Regulation

If we connect a Zener diode in parallel with the output of a power supply so that the component receives a reverse bias, the diode limits the output voltage. The diode must have an adequate power rating to prevent it from burning out. In addition, we must connect a resistance in series with it to limit the current. The limiting voltage depends on the particular Zener diode used. Zener diodes are available for any reasonable power-supply voltage.

Figure 25-8 is a diagram of a full-wave bridge DC power supply including a Zener diode for voltage regulation. Note the direction in which we connect the Zener diode: the arrow points from minus to plus. This polarity goes contrary to the orientation of a rectifier diode. We must take care to connect a Zener diode with the correct polarity, or it will burn out as soon as we apply power to the circuit!

A simple Zener-diode voltage regulator, such as the one shown in Fig. 25-8, does not function effectively when we use the power supply with equipment that draws high current. The problem arises because the series resistor, essential to prevent destruction of the diode, creates a significant voltage drop when it carries more than a small amount of current. When we expect that a power supply will have to deliver a lot of current, we can employ a *power transistor* along with the Zener diode to obtain voltage regulation. Figure 25-9 shows an example. In this circuit, the resistor

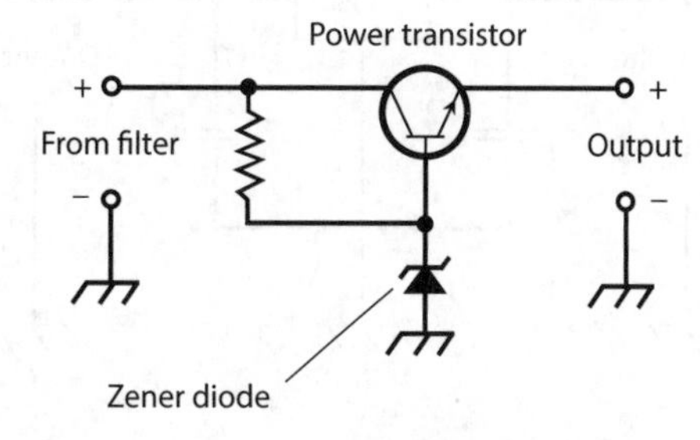

25-9 A voltage-regulator circuit using a Zener diode and a transistor.

ensures proper operation of the transistor without causing a drop in the output voltage under high-current conditions.

Voltage regulators are available in *integrated-circuit* (IC) form. The *regulator IC*, also called a *regulator chip*, goes in the power-supply circuit at the output of the filter. In high-voltage power supplies, specialized *electron tubes* can function as voltage regulators. Electron tubes can withstand higher momentary overloads than Zener diodes, transistors, or chips. However, some engineers consider *regulator tubes* archaic, even in scenarios where they might perform better than solid-state devices.

Some electronic devices can operate okay when connected to an *unregulated power supply* (a supply that lacks voltage regulator components or circuits). Other devices absolutely require a supply with voltage regulation. In any case, the rated output voltage of the power supply should always match the rated operating voltage of all the equipment that runs from it. Also, the supply should be capable of delivering the required amount of current, ideally with a "safety margin" of at least 10 percent.

Voltage Regulator ICs

As with many common circuit functions, the problem of voltage regulation is now mostly dealt with using special purpose voltage regulator Integrated Circuits (ICs).

The popular 78XX family of voltage regulator ICs are three-pin devices looking like a power transistor. The last two digits of the 78XX IC's name denotes the regulated output voltage that the IC produces, so the 7812 has a 12V output, the 7809 9V and the 7805 5V. Internally, these ICs operate in much the same way as the power transistor and Zener diode shown in Figure 25-9. However they also incorporate thermal protection so that if the IC package starts to get hot enough to damage the IC, it drops the output voltage and, hence, reduces the current until the device recovers.

Figure 25-10 shows how a 7805 might be used to provide a regulated 5-V output from an input of between 7 and 18 V. The IC is generally surrounded by a pair of capacitors. The capacitor at its input is needed only if the regulator is not situated close to the power supply filter. The output capacitor improves the stability of the regulation and the response of the regulator to sudden changes in current requirement. Typical values of input and output capacitors are 0.33μF and 0.1μF, respectively.

Switched-Mode Power Supplies (SMPS)

If you pick up your cellphone charger, you can tell just by its weight that there is not a regular transformer in there. A cellphone charger will contain a transformer, but it will be a small light-weight

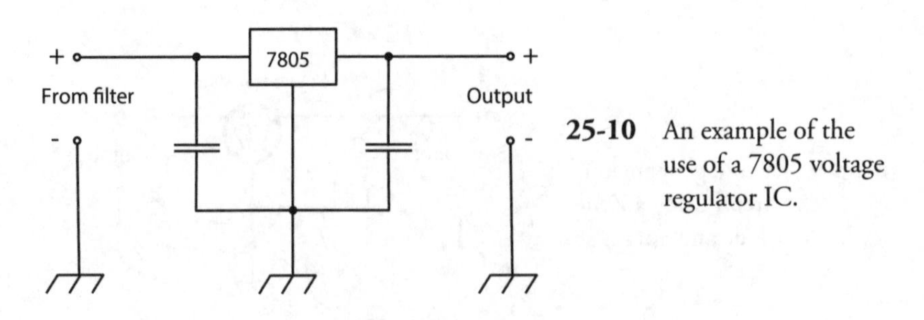

25-10 An example of the use of a 7805 voltage regulator IC.

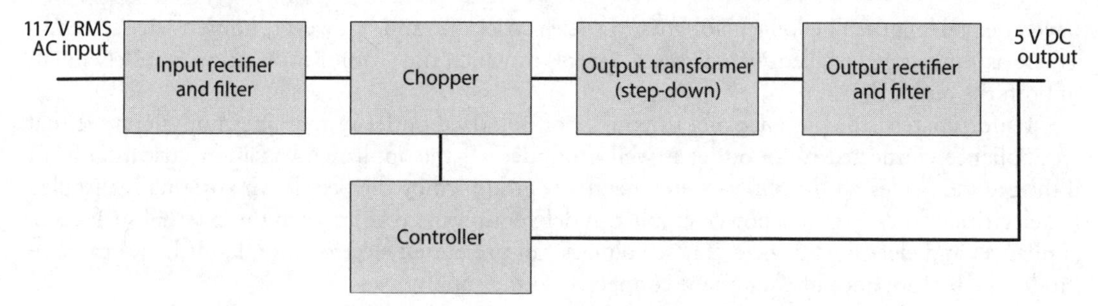

25-11 Block diagram for a SMPS.

high-frequency transformer. In fact, most modern power sources for consumer products use switched-mode power supplies or SMPSs.

Figure 25-11 shows how an SMPS operates. Instead of using a transformer to reduce the AC voltage, the high voltage AC input is rectified and filtered to produce high voltage DC. This DC is then "chopped," or switched, at high frequency into a series of short pulses that feed an output transformer that steps down the voltage. The resulting low voltage AC is then rectified and filtered into low voltage DC. Because the transformer operates at such a high frequency, it can be made much smaller and lighter than a 60-Hz transformer.

DC voltage regulation is achieved by feedback from the DC output to the controller that alters the width of the pulses being chopped that, in turn, alters the DC output voltage.

To keep the DC output isolated from the high voltage AC, the feedback path to the controller uses an optoisolator. This might seem like a lot of extra complexity, but the size, weight, and cost advantages of reducing the transformer size is enough to make SMPSs worthwhile. Generally an SMPS IC that includes the chopper and controller circuitry in a single package is used.

Equipment Protection

The output of a power supply should always remain free of sudden changes that can damage equipment or components, or interfere with their proper performance. To ensure the safety of personnel working around electrical and electronic systems, significant voltages must never appear on the external surfaces of a power supply, or on the external surfaces of any equipment connected to it.

Grounding

The best electrical ground for a power supply is the "third wire" ground provided in up-to-date AC utility circuits. When you examine a typical AC outlet in the United States, you'll see a "hole" shaped like an uppercase letter D turned on its side. That is—or should be—the electrical ground connection. The contacts inside this "hole" should go to a wire that ultimately terminates in a metal rod driven into the earth at the point where the electrical wiring enters the building. That connection constitutes an *earth ground*.

In older buildings, *two-wire AC systems* are common. You can recognize this type of system by noting the presence of two slots in the utility outlets without any ground "hole." Some of these systems obtain reasonable grounding by means of a scheme called *polarization*, in which the two slots have unequal length. The longer slot goes to the electrical ground. However, no two-wire system is as safe as a properly installed *three-wire AC system*, in which the ground connection is independent of both the outlet slots.

Unfortunately, the presence of a three-wire or polarized outlet system does not guarantee that an appliance connected to an outlet is well grounded. If the appliance has an electrical fault, or if the ground "holes" at the outlets weren't actually grounded by the people who originally installed the electrical utility system, a power supply can deliver unwanted voltages to the external surfaces of appliances and electronic devices. These voltages can present an electrocution hazard, and can also hinder the performance of equipment connected to the supply.

Warning!

All exposed metal surfaces of power supplies should be connected to the grounded wire of a three-wire electrical cord. Never defeat or remove the "third prong" of the plug. Always ensure that the electrical system in the building has been properly installed, so you don't work under the illusion that your system has a good ground when it actually does not. If you have any doubts about these matters, consult a professional electrician.

Surge Currents

At the instant we switch a power supply on, a surge of current occurs, even with nothing connected to the supply output. This surge takes place because the filter capacitors require an initial charge, forcing them to draw a large current for a short time. The surge current greatly exceeds the normal operating current. An extreme surge can destroy the rectifier diodes if they aren't sufficiently rated and/or protected. The phenomenon is most often seen in high-voltage supplies and voltage-multiplier circuits. We can minimize the risk of surge-related diode failure in at least three ways:

1. Use diodes with a current rating of many times the normal operating level.
2. Connect several diodes in parallel wherever a diode is called for in the circuit. *Current-equalizing resistors* ensure that no single diode "hoards" or "hogs" more than its share of the current (Fig. 25-12). The resistors should have small, identical ohmic values.
3. Use an *automatic switching* circuit in the transformer primary. Such a system applies a reduced AC voltage to the transformer for a couple of seconds immediately after power is first applied, and then delivers the full voltage once the filter capacitors have attained a significant charge.

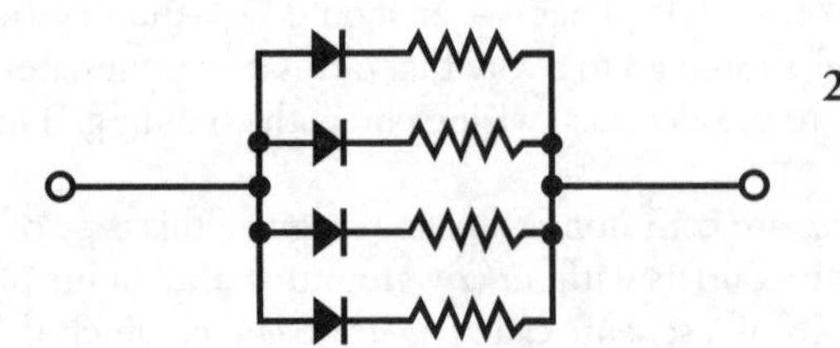

25-12 Diodes in parallel, with current-equalizing resistors in series with each diode.

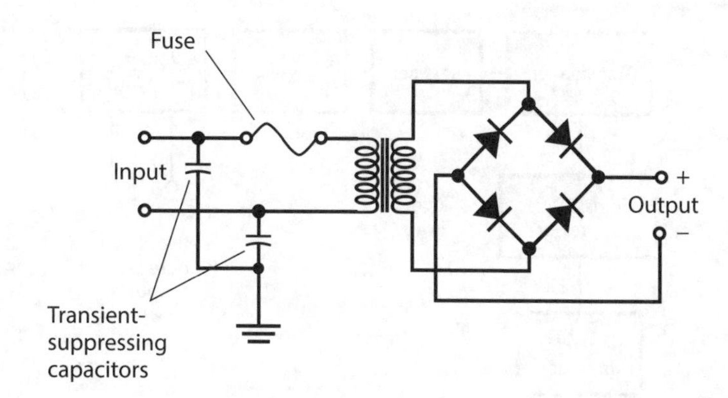

25-13 A full-wave bridge rectifier with transient-suppression capacitors and a fuse in the transformer primary circuit.

Transients

The AC that we observe at utility outlets is a sine wave with a constant voltage near 117 V RMS or 234 V RMS. However, in most household circuits, we occasionally observe voltage spikes, known as transients, that can attain positive or negative peak values of several thousand volts. Transients can result from sudden changes in the load in a utility circuit. A thundershower can produce transients throughout an entire town. Unless we take some measures to suppress them, transients can destroy the diodes in a power supply. Transients can also cause problems with sensitive electronic equipment, such as computers or microcomputer-controlled appliances.

We can get rid of most common transients by connecting a small capacitor of about 0.01 μF, rated for 600 V or more, between each side of the transformer primary and electrical ground, as shown in Fig. 25-13. A *disk ceramic capacitor* (not an electrolytic capacitor) works well for this purpose. Disk ceramic capacitors have no polarity issues, so we can connect them in either direction and expect them to work well.

Commercially made transient suppressors are available. These devices use sophisticated methods to prevent sudden voltage spikes from reaching levels at which they can cause problems. It's a good idea to use transient suppressors with all sensitive electronic devices, including computers, hi-fi stereo systems, and television sets. In the event of a thundershower, the best way to protect such equipment is to physically unplug it from the wall outlets until the storm has passed.

Fuses

A *fuse* comprises a piece of soft wire that melts, breaking a circuit if the current exceeds a certain level. We should connect the fuse in series with the transformer primary, as shown in Fig. 25-13. A short circuit or overload anywhere in the power supply, or in equipment connected to it, will burn the fuse out. If a fuse blows out, we must replace it with another fuse having the same specifications. Fuses are rated in amperes (A). Thus, a 5-A fuse will carry up to 5 A before blowing out, and a 20-A fuse will carry up to 20 A, regardless of the power-supply voltage.

Fuses are available in two types: the *quick-break fuse* and the *slow-blow fuse*. A quick-break fuse consists of a straight length of wire or a metal strip. A slow-blow fuse usually has a spring inside along with the wire or strip. Quick-break fuses in slow-blow situations can burn out needlessly, causing inconvenience. Slow-blow fuses in quick-break environments might not provide adequate protection to the equipment, letting excessive current flow for too long before blowing out.

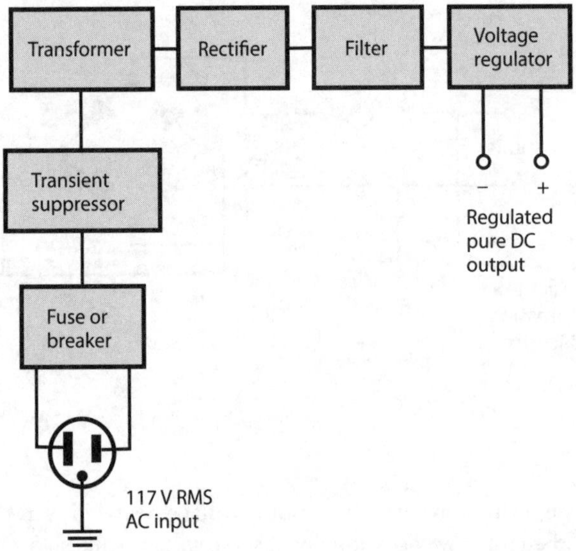

25-14 Block diagram of a complete power supply that can deliver high-quality DC output with AC input.

Circuit Breakers

A *circuit breaker* performs the same function as a fuse, except that we can reset a breaker by turning off the power supply, waiting a little while, and then pressing a button or flipping a switch. Some breakers reset automatically when the equipment has been shut off for a certain length of time. Circuit breakers, like fuses, are rated in amperes.

If a fuse or breaker repeatedly burns out or trips, or if it blows or trips immediately after we replace or reset it, then a serious problem exists with the power supply or with the equipment connected to it. Burned-out diodes, a faulty transformer, and shorted filter capacitors in the supply can all cause trouble. A short circuit in the equipment connected to the supply, or the connection of a device in the wrong direction (polarity), can cause repeated fuse blowing or circuit-breaker tripping.

Never replace a fuse or breaker with a larger-capacity unit to overcome the inconvenience of repeated fuse/breaker blowing/tripping. Find the cause of the trouble, and repair the equipment as needed. The "penny in the fuse box" scheme can endanger equipment and personnel, and it increases the risk of fire in the event of a short circuit. Such an action can also cause extensive damage to power-supply components including diodes, transformers, or filter chokes.

The Complete System

Figure 25-14 illustrates a complete power supply in block-diagram format. Note the sequence in which the portions of the system, called *stages*, follow one another.

Warning!

High-voltage power supplies can retain deadly voltages even after they've been switched off and unplugged. This danger exists because the filter capacitors retain their charge for a while, even in the absence of applied power. If you have any doubt about your ability to safely build or work with a power supply, leave it to a professional.

Quiz

Refer to the text in this chapter if necessary. A good score is at least 18 correct. Answers are in the back of the book.

1. Voltage regulation can be accomplished by means of a Zener diode across a power supply's filter output, reverse-biased, along with a series-connected
   (a) voltage-limiting capacitor.
   (b) power-limiting diode.
   (c) current-limiting resistor.
   (d) All of the above

2. Suppose that we apply a 60-Hz pure AC sine wave of 330 V pk-pk (peak-to-peak), having no DC component, to the input of a full-wave bridge rectifier circuit. The effective DC output voltage is
   (a) more than 330 V.
   (b) exactly 330 V.
   (c) slightly less than 330 V.
   (d) considerably less than 330 V.

3. The output of a rectifier circuit with excellent filtering compares favorably to the output of
   (a) a DC battery of the same voltage.
   (b) an AC transformer with the same RMS secondary voltage.
   (c) a DC battery with half the voltage.
   (d) an AC transformer with the same peak-to-peak secondary voltage.

4. Which of the following components is *not always* required in a power supply designed to produce 24 V pure DC output with 117 V RMS AC input?
   (a) A transformer
   (b) A rectifier circuit
   (c) A filtering circuit
   (d) A voltage-regulator circuit

5. Suppose that you see a fuse with a straight wire and a spring inside. You can assume that this fuse
   (a) has a low current rating.
   (b) has a high current rating.
   (c) is a slow-blow type.
   (d) is a quick-break type.

6. Transients can result from
   (a) intermittent diode failure in a power supply.
   (b) localized thundershowers.
   (c) improper installation of filter capacitors.
   (d) the use of improperly rated Zener diodes.

7. Suppose that a fuse blows out repeatedly. We get tired of the inconvenience, and replace the fuse with a unit having a higher current rating. This action can give rise to all of the following dangers except one. Which one?
   - (a) Serious damage (or further damage) might occur to electronic components in the supply.
   - (b) Personnel who work with equipment connected to the supply might receive deadly electrical shocks.
   - (c) One or more of the components in the supply might catch on fire.
   - (d) Voltage or current spikes might occur on the power lines outside the house.

8. Which of the following characteristics represents an advantage of a half-wave rectifier circuit in certain applications?
   - (a) It uses the whole transformer secondary for the full AC input cycle.
   - (b) The pulsating DC output is easier to filter than the output of a full-wave circuit.
   - (c) It costs less than other rectifier types because it uses fewer components.
   - (d) It offers superior voltage regulation compared to all other rectifier types.

9. If we want to build a power supply designed to provide well-filtered, high-voltage DC at low current levels without the need for good regulation, the cheapest option would be to use a
   - (a) harmonic-generator circuit.
   - (b) full-wave, center-tap circuit.
   - (c) full-wave bridge circuit.
   - (d) voltage-doubler circuit.

10. In a power supply designed to produce 800-V DC output from a 117-V RMS AC utility line, which of the following components does the incoming electricity encounter first?
    - (a) A diode
    - (b) A transformer
    - (c) A filter capacitor or choke
    - (d) A voltage regulator

11. Figure 25-15 illustrates a complete power supply. What, if anything, is wrong with this diagram?
    - (a) Two of the rectifier diodes are connected backwards.
    - (b) The Zener diode is connected backwards.
    - (c) The inductor should be connected directly across the capacitor.
    - (d) Nothing is wrong.

12. Assuming that we correct any errors that might exist in the circuit in Fig. 25-15, what sort of rectifier arrangement does it use?
    - (a) Quarter-wave
    - (b) Half-wave
    - (c) Full-wave
    - (d) Voltage-doubler

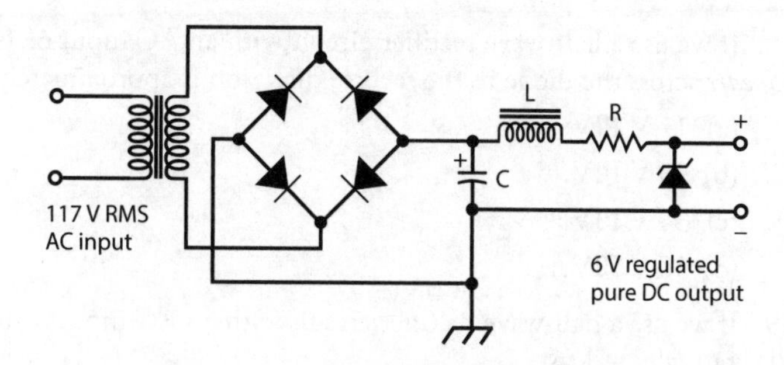

25-15 Illustration for Quiz Questions 11 through 15.

13. Assuming that we correct any errors that might exist in the circuit in Fig. 25-15, what sort of transformer will we need to build it?
    (a) Step-up
    (b) Step-down
    (c) Air-core
    (d) Center-tapped

14. Assuming that we correct any errors that might exist in the circuit in Fig. 25-15, what purpose does inductor L serve?
    (a) It helps to rectify the AC input to get pulsating DC output.
    (b) It limits the current through capacitor C.
    (c) It limits the voltage to protect the rectifier diodes.
    (d) It works with capacitor C to get rid of the ripple.

15. Assuming that we correct any errors that might exist in the circuit in Fig. 25-15, what purpose does resistor R serve?
    (a) It helps to filter the AC input.
    (b) It limits the current through the Zener diode.
    (c) It increases the voltage to charge up capacitor C.
    (d) It works with the Zener diode to maximize the current.

16. A voltage regulator IC looks a lot like
    (a) a large resistor.
    (b) an electrolytic capacitor.
    (c) a power transistor.
    (d) a Zener diode.

17. If we use a full-wave bridge rectifier circuit with an AC input of 12 V RMS, each diode should have a PIV rating of at least
    (a) 17 PIV.
    (b) 25 PIV.
    (c) 34 PIV.
    (d) 50 PIV.

18. If we use a half-wave rectifier circuit with an AC input of 12 V RMS, the PIV that *actually appears* across the diode in the reverse direction is approximately
    (a) 17 V PIV.
    (b) 25 V PIV.
    (c) 34 V PIV.
    (d) 50 V PIV.
19. If we use a half-wave rectifier circuit with an AC input of 12 V RMS, the diode should have a PIV rating of at least
    (a) 17 PIV.
    (b) 25 PIV.
    (c) 34 PIV.
    (d) 50 PIV.
20. An SMPS contains, among other components,
    (a) a chopper.
    (b) an optoisolator.
    (c) rectifiers.
    (d) All of the above

26
CHAPTER

Amplifiers and Oscillators

NOW THAT WE KNOW HOW TRANSISTORS WORK, WE'RE READY TO LEARN ABOUT AMPLIFIERS AND oscillators that use these devices. But first, let's take a closer look at a topic that we've encountered briefly already: the decibel (dB) as an expression of relative signal strength.

The Decibel Revisited

We can consider amplitude increases as *positive gain* and amplitude decreases as *negative gain*. For example, if a circuit's output signal amplitude equals +6 dB relative to the input signal amplitude, then the output exceeds the input. If the output signal amplitude is −14 dB relative to the input, then the output is weaker than the input. In the first case, we say that the circuit has a gain of 6 dB. In the second case, we say that the circuit has a gain of −14 dB or a loss of 14 dB.

For Voltage

Consider a circuit with an RMS AC input voltage of E_{in} and an RMS AC output voltage of E_{out}, with both voltages expressed in the same units (volts, millivolts, microvolts, or whatever). Also suppose that the input and output impedances both constitute pure resistances of the same ohmic value. We can calculate the *voltage gain* of the circuit, in decibels, with the formula

$$\text{Gain (dB)} = 20 \log (E_{out}/E_{in})$$

In this equation, "log" stands for the *base-10 logarithm* or *common logarithm*. Scientists and engineers write the base-10 logarithm of a quantity x as "log x" or sometimes as "$\log_{10} x$." We can consider the coefficient 20 in the above equation to be an exact value when we make calculations, no matter how many significant digits we need.

Logarithms can have other bases besides 10, the most common of which is the *exponential constant*, symbolized as e and equal to approximately 2.71828. Scientists and engineers express the *base-e logarithm* (also called the *natural logarithm*) of a quantity x as "$\log_e x$" or "ln x." Without knowing all of the mathematical particulars concerning how *logarithmic functions* behave, we can calculate the logarithms of specific numbers with the help of a good scientific calculator. From now on, let's agree that when we say "logarithm" or write "log," we mean the base-10 logarithm.

Problem 26-1

Suppose that a circuit has an RMS AC input of 1.00 V and an RMS AC output of 14.0 V. How much gain, in decibels, does this circuit produce?

Solution

First, you must find the ratio E_{out}/E_{in}. Because $E_{out} = 14.0$ V RMS and $E_{in} = 1.00$ V RMS, the ratio equals 14.0/1.00, or 14.0. Next, find the logarithm of 14.0. A calculator tells you that this quantity is quite close to 1.146128036. Finally, multiply this number by 20 and then round off to three significant figures, getting 22.9 dB.

Problem 26-2

Suppose that a circuit has an RMS AC input voltage of 24.2 V and an RMS AC output voltage of 19.9 V. What's the gain in decibels?

Solution

Find the ratio $E_{out}/E_{in} = 19.9/24.2 = 0.822314\ldots$. (The sequence of three dots, called an *ellipsis*, indicates extra digits introduced by the calculator. You can leave them in until the final round-off.) When you use a calculator to find the logarithm of this quantity, you get $\log 0.822314\ldots = -0.0849622\ldots$. The gain equals $20 \times (-0.0849622)$, which rounds off to -1.70 dB.

For Current

We can calculate current gain or loss figures in decibels in the same way as we calculate voltage gain or loss figures. If I_{in} represents the RMS AC input current and I_{out} represents the RMS AC output current (in the same units as I_{in}, such as amperes, milliamperes, microamperes, or whatever), then

$$\text{Gain (dB)} = 20 \log (I_{out}/I_{in})$$

For this formula to work, the input and output impedances must both comprise pure resistances, and the ohmic values must be identical.

For Power

We can calculate the *power gain* of a circuit, in decibels, by cutting the coefficient of the formula in half, from exactly 20 to exactly 10, accurate to as many significant digits as we want. If P_{in} represents the input signal power and P_{out} represents the output signal power (in the same units as P_{in}, such as watts, milliwatts, microwatts, kilowatts, or whatever), then

$$\text{Gain (dB)} = 10 \log (P_{out}/P_{in})$$

For this formula to work, the input and output impedances should both show up as pure resistances, but their ohmic values can differ.

Problem 26-3

Suppose that a power amplifier has an input of 5.72 W and an output of 125 W. What's the gain in decibels?

Solution

Find the ratio $P_{out}/P_{in} = 125/5.72 = 21.853146\ldots$. Then find the logarithm, obtaining log $21.853146\ldots = 1.339513\ldots$. Finally, multiply by 10 and round off to obtain a gain figure of $10 \times 1.339513\ldots = 13.4$ dB.

Problem 26-4

Suppose that an *attenuator* (a circuit designed deliberately to produce power loss) provides 10 dB power reduction. You supply this circuit with an input signal at 94 W. What's the output power?

Solution

An attenuation of 10 dB represents a gain of -10 dB. You know that $P_{in} = 94$ W, so P_{out} constitutes the unknown in the power gain formula. You must, therefore, solve for P_{out} in the equation

$$-10 = 10 \log (P_{out}/94)$$

First, divide each side by 10, getting

$$-1 = \log (P_{out}/94)$$

To solve this equation, you must take the *base-10 antilogarithm*, also known as the *antilog* or *inverse log*, of each side. The antilog function "undoes" the work of the log function. The antilog of a value x can be abbreviated as "antilog x." Pure mathematicians, as well as some scientists and engineers, denote it as "$\log^{-1} x$." Antilogarithms of specific numbers, like logarithms, can be determined with any good scientific calculator. (Function keys for the antilogarithm vary, depending on the particular calculator you use. You might have to enter the value and then hit an "Inv" key followed by a "log" key; you might enter the value and then hit a "10^x" key.) When you take the antilogarithm of both sides of the above equation, you get

$$\text{antilog } (-1) = \text{antilog } [\log (P_{out}/94)]$$

Working out the value on the left-hand side with a calculator, and noting on the right-hand side that the antilog "undoes" the work of the log, you get

$$0.1 = P_{out}/94$$

Multiplying each side by 94 tells you that

$$94 \times 0.1 = P_{out}$$

Finally, when you multiply out the left-hand side and transpose the left-hand and right-hand sides of the equation, you get the answer as

$$P_{out} = 9.4 \text{ W}$$

Decibels and Impedance

When determining the voltage gain (or loss) and the current gain (or loss) for a circuit in decibels, you can expect to get the same figure for both parameters only when the complex input impedance is *identical* to the complex output impedance. If the input and output impedances differ (either reactance-wise or resistance-wise, or both), then the voltage gain or loss generally differs from the current gain or loss.

Consider how transformers work. A step-up transformer, in theory, has voltage gain, but this voltage increase alone doesn't make a signal more powerful. A step-down transformer can exhibit theoretical current gain, but again, this current increase alone doesn't make a signal more powerful. In order to make a signal more powerful, a circuit must increase the signal *power*—the *product* of the voltage and the current!

When determining power gain (or loss) for a particular circuit in decibels, the input and output impedances *do not* matter, as long as they're both free of reactance. In this sense, positive power gain always represents a real-world increase in signal strength. Similarly, negative power gain (or power loss) always represents a true decrease in signal strength.

Whenever we want to work out decibel values for voltage, current, or power, we should strive to get rid of all the reactance in a circuit, so that our impedances constitute pure resistances. Reactance "artificially" increases or decreases current and voltage levels; and while reactance theoretically consumes no power, it can have a profound effect on the behavior of instruments that measure power, and it can cause power to go to waste as unwanted heat dissipation.

Basic Bipolar-Transistor Amplifier

In the previous chapters, you saw some circuits that use transistors. We can apply an input signal to some control point (the base, gate, emitter, or source), causing a much greater signal to appear at the output (usually the collector or drain). Figure 26-1 shows a circuit in which we've wired up an NPN bipolar transistor to serve as a *common-emitter amplifier*. The input signal passes through C_2 to the base. Resistors R_2 and R_3 provide the base bias. Resistor R_1 and capacitor C_1 allow for the emitter to maintain a constant DC voltage relative to ground, while keeping it grounded for the AC signal. Resistor R_1 also limits the current through the transistor. The AC output signal goes through capacitor C_3. Resistor R_4 keeps the AC output signal from shorting through the power supply.

In this amplifier, the optimum capacitance values depend on the design frequency of the amplifier, and also on the impedances at the input and output. In general, as the frequency and/or circuit impedance increase, we need less capacitance. At audio frequencies (abbreviated AF and representing frequencies from approximately 20 Hz to 20 kHz) and low impedances, the capacitors might

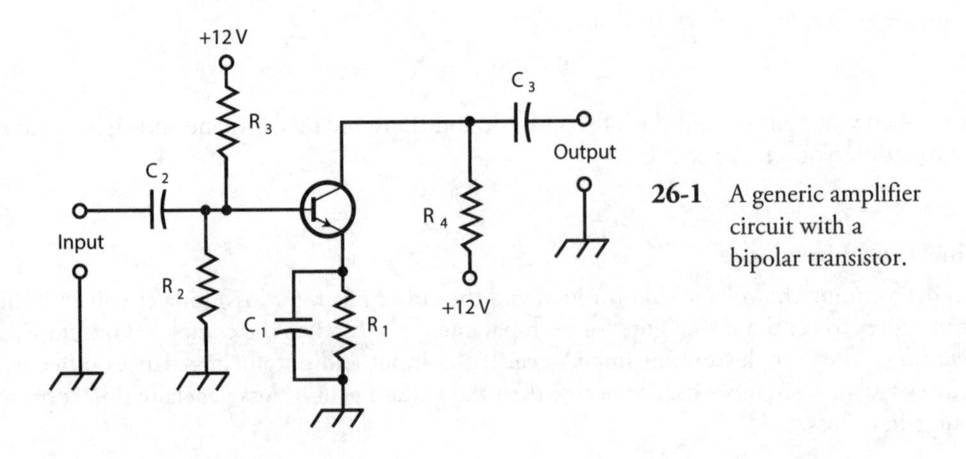

26-1 A generic amplifier circuit with a bipolar transistor.

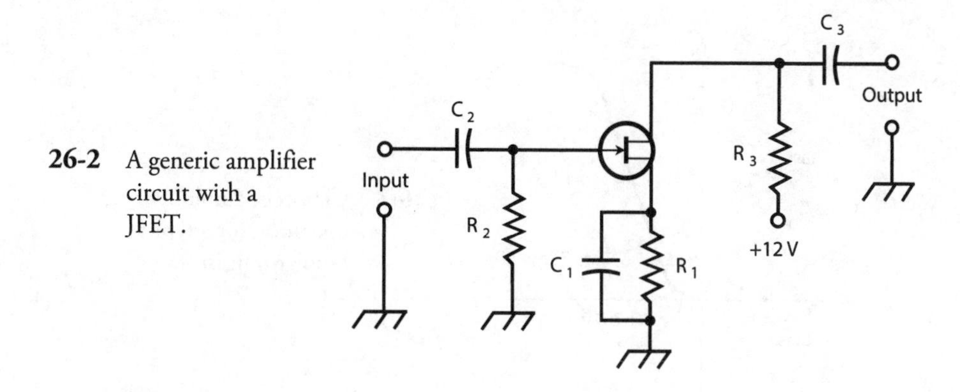

26-2 A generic amplifier circuit with a JFET.

have values as large as 100 μF. At radio frequencies (RF) and high impedances, values will normally equal only a fraction of a microfarad, down to picofarads at the highest frequencies and impedances. The optimum resistor values also depend on the application. In the case of a weak-signal amplifier, typical values are 470 ohms for R_1, 4.7 k for R_2,10 k for R_3, and 4.7 k for R_4.

Basic FET Amplifier

Figure 26-2 shows an N-channel JFET hooked up as a *common-source amplifier*. The input signal passes through C_2 to the gate. Resistor R_2 provides the gate bias. Resistor R_1 and capacitor C_1 give the source a DC voltage relative to ground, while grounding it for signals. The output signal goes through C_3. Resistor R_3 keeps the output signal from shorting through the power supply.

A JFET exhibits high input impedance, so the value of C_2 should be small. If we use a MOSFET rather than a JFET, we'll get a higher input impedance, so C_2 will be smaller yet, sometimes 1 pF or less. The resistor values depend on the application. In some instances, we won't need R_1 and C_1 at all, and we can connect the source directly to ground. If we use resistor R_1, its optimum value depends on the input impedance and the bias needed for the FET. For a weak-signal amplifier, typical values are 680 ohms for R_1, 10 k for R_2, and 100 ohms for R_3.

Amplifier Classes

Engineers classify analog amplifier circuits according to the bias arrangement as *class A*, *class AB*, *class B*, and *class C*. Each class has its own special characteristics, and works best in its own unique set of circumstances. A specialized amplifier type, called *class D*, can also be used in certain situations.

The Class-A Amplifier

With the previously mentioned component values, the amplifier circuits in Figs. 26-1 and 26-2 operate in the class-A mode. This type of amplifier is *linear*, meaning that the output waveform has the same shape as (although a greater amplitude than) the input waveform.

When we want to obtain class-A operation with a bipolar transistor, we must bias the device so that, with no signal input, it operates near the middle of the straight-line portion of the I_C versus I_B

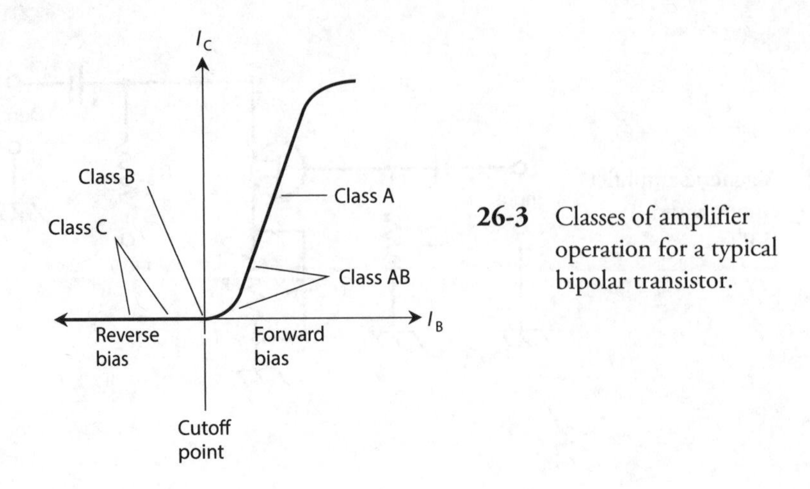

26-3 Classes of amplifier operation for a typical bipolar transistor.

(collector current versus base current) curve. Figure 26-3 shows this situation for a bipolar transistor. With a JFET or MOSFET, the bias must be such that, with no signal input, the device operates near the middle of the straight-line part of the I_D versus E_G (drain current versus gate voltage) curve as shown in Fig. 26-4.

When we use a class-A amplifier in the "real world," we must never allow the input signal to get too strong. An excessively strong input signal will drive the device out of the straight-line part of the characteristic curve during part of the cycle. When this phenomenon occurs, the output waveform will no longer represent a faithful reproduction of the input waveform, and the amplifier will become *nonlinear*. In some types of amplifiers, we can tolerate (or even encourage) nonlinearity, but we'll want a class-A amplifier to operate in a linear fashion at all times.

Class-A amplifiers suffer from one outstanding limitation: The transistor draws current whether or not any input signal exists. The transistor must "work hard" even when no signal comes in. For weak-signal applications, the "extra work load" doesn't matter much; we must concern ourselves entirely with getting plenty of gain out of the circuit.

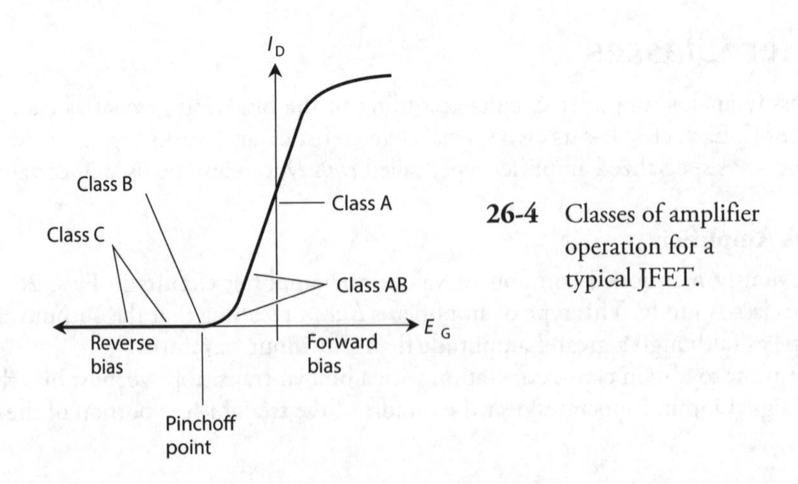

26-4 Classes of amplifier operation for a typical JFET.

The Class-AB Amplifier

When we bias a bipolar transistor close to cutoff under no-signal conditions, or when we bias a JFET or MOSFET near pinchoff, the input signal always drives the device into the nonlinear part of the *operating curve*. (Figures 26-3 and 26-4 are examples of operating curves.) Then, by definition, we have a class-AB amplifier. Figures 26-3 and 26-4 show typical bias zones for class-AB amplifier operation. A small collector or drain current flows when no input signal exists, but it's less than the no-signal current in a class-A amplifier.

Engineers sometimes specify two distinct modes of class-AB amplification. If the bipolar transistor or FET never goes into cutoff or pinchoff during any part of the input signal cycle, the amplifier works in *class* AB_1. If the device goes into cutoff or pinchoff for any part of the cycle (up to almost half), the amplifier operates in *class* AB_2.

In a class-AB amplifier, the output-signal wave doesn't have the same shape as the input-signal wave. But if the signal wave is *modulated*, such as in a voice radio transmitter, the waveform of the *modulating signal* comes out undistorted anyway. Therefore, class-AB operation can work very well for RF power amplifiers.

The Class-B Amplifier

When we bias a bipolar transistor exactly at cutoff or an FET exactly at pinchoff under zero-input-signal conditions, we cause an amplifier to function in the *class-B* mode. The operating points for class-B operation are labeled on the curves in Figs. 26-3 and 26-4. This scheme, like the class-AB mode, lends itself well to RF power amplification.

In class-B operation, no collector or drain current flows under no-signal conditions. Therefore, the circuit does not consume significant power unless a signal goes into it. (Class-A and class-AB amplifiers consume some DC power under no-signal conditions.) When we provide an input signal, current flows in the device during half of the cycle. The output signal waveform differs greatly from the input waveform. In fact, the wave comes out "half-wave rectified" as well as amplified.

You'll sometimes hear of class-AB or class-B "linear amplifiers," especially when you converse with amateur (ham) radio operators. In this context, the term "linear" refers to the fact that the amplifier doesn't distort the *modulation waveform* (sometimes called the *modulation envelope*), even though the *carrier waveform* is distorted because the transistor is not biased in the straight-line part of the operating curve under no-signal conditions.

Class-AB_2 and class-B amplifiers draw power from the input signal source. Engineers say that such amplifiers require a certain amount of *drive* or *driving power* to function. Class-A and class-AB_1 amplifiers theoretically need no driving power, although the input signal must provide a certain amount of voltage to influence the behavior of the transistor.

The Class-B Push-Pull Amplifier

We can connect two bipolar transistors or FETs in a pair of class-B circuits that operate in tandem, one for the positive half of the input wave cycle and the other for the negative half. In this way, we can get rid of waveform distortion while retaining all the benefits of class-B operation. We call this type of circuit a *class-B push-pull amplifier*. Figure 26-5 shows an example using two NPN bipolar transistors.

Resistor R_1 limits the current through the transistors. Capacitor C_1 keeps the input transformer center tap at signal ground, while allowing for some DC base bias. Resistors R_2 and R_3 bias the transistors precisely at their cutoff points. For best results, the two transistors must be identical. Not only should their part numbers match, but we should pick them out *by experiment* for each amplifier circuit that we build, to ensure that the characteristics coincide as closely as possible.

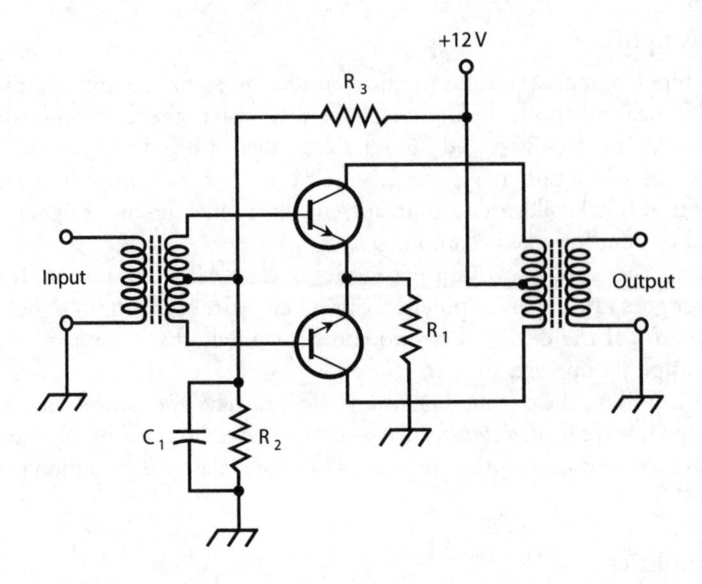

26-5 A class-B push-pull amplifier using NPN bipolar transistors.

Class-B push-pull circuits provide a popular arrangement for audio-frequency (AF) power amplification. Push-pull design offers the easy transistor workload of the class-B mode with the low-distortion, linear amplification characteristics of the class-A mode. However, a push-pull amplifier needs two center-tapped transformers, one at the input and the other at the output, a requirement that makes push-pull amplifiers more bulky and expensive than other types.

All push-pull amplifiers share a unique and interesting quality: They "cancel out" the *even-numbered* harmonics in the output. This property offers an advantage in the design and operation of wireless transmitters because it gets rid of concerns about the second harmonic, which usually causes more trouble than any other harmonic. A push-pull circuit doesn't suppress *odd-numbered* harmonics any more than a "single-ended" circuit (one that employs a single bipolar transistor or FET) does.

The Class-C Amplifier

We can bias a bipolar transistor or FET past cutoff or pinchoff, and it will still work as a power amplifier (PA), provided that the drive is sufficient to overcome the bias during part of the cycle. We call this mode *class-C* operation. Figures 26-3 and 26-4 show no-signal bias points for class-C amplification.

Class-C amplifiers are nonlinear, even for amplitude-modulation (AM) envelopes. Therefore, engineers generally use class-C circuits only with input signals that are either "full-on" or "full-off." Such signals include old-fashioned *Morse code*, along with digital modulation schemes in which the frequency or phase (but not the amplitude) of the signal can vary, but the amplitude is always either zero or maximum.

A class-C amplifier needs a lot of driving power. It does not produce as much gain as other amplification modes. For example, we might need 300 W of signal drive to get 1 kW of signal power output from a class-C power amplifier. However, we get more "signal bang for the buck" in class C than we do with an amplifier working in any other mode. That is to say, we get optimum *efficiency* from the class-C scheme.

The Class-D Amplifier

A class-D amplifier differs radically from traditional amplifiers; its output transistors (generally MOSFETs) act in a digital way, always either on or off, as opposed to analog modes for the other classes. Class-D amplifiers use pulse-width modulation (PWM), which you'll learn about in Chap. 27 (Wireless Transmitters and Receivers) and Chap. 29 (Microcontrollers). Figure 26-6 shows the principle of class-D operation.

A class-D amplifier uses a comparator, which you learned about in Chap. 23 (Integrated Circuits) and a triangular-wave oscillator to convert an analog input signal to a train of pulses having different lengths. The pulse length varies in proportion to the instantaneous input-signal voltage. These pulses get amplified by a class-C-like push-pull stage. Then a lowpass filter, playing the role of a digital-to-analog converter, transforms the pulses back into an analog signal. (For low-quality audio applications with a loudspeaker output, the speaker can't respond fast enough to follow the pulse frequency, so it converts the digital signal to analog form without the need for a separate lowpass filter.)

To visualize the digitization process, imagine the input signal slowly changing compared to the output of the triangular-wave oscillator. Let's say that the input signal is, at first, stronger than the signal from the triangular-wave oscillator, so the comparator output is high. At some point, the input signal gets weaker than the triangular signal, and the comparator output goes low. The time that this transition takes depends on the voltage of the input signal. The higher the voltage, the longer the pulse. The proportion of the time that the pulse remains high is called the *duty cycle*. The higher the duty cycle, the stronger the output signal.

When built with MOSFETs that have input impedances on the order of a few megohms, a class-D amplifier can supply several tens of watts of power without overheating. Class-D amplifiers almost always use ICs rather than discrete components, and have largely replaced analog amplifiers in consumer devices, such as cell phones, notebook computers, and tablet computers.

On the downside, class-D amplifiers cause some distortion. For hi-fi audio amplification, therefore, analog designs remain superior; class-A amplifiers are ideal for that purpose.

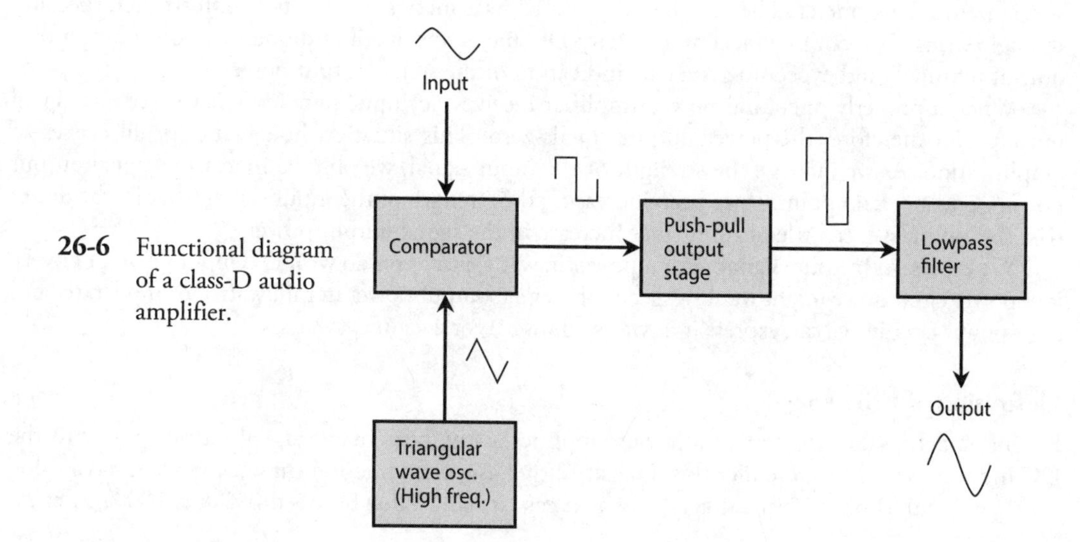

26-6 Functional diagram of a class-D audio amplifier.

Efficiency in Power Amplifiers

An efficient power amplifier not only provides optimum output power with minimum heat generation and minimum strain on the transistors, but it conserves energy as well. These factors translate into reduced cost, reduced size and weight, and longer equipment life as compared with inefficient power amplifiers.

The DC Input Power

Suppose that you connect an ammeter or milliammeter in series with the collector or drain of an amplifier and the power supply. While the amplifier operates, the meter will show a certain reading. The reading might appear constant, or it might fluctuate with changes in the input signal level. The DC *collector input power* to a bipolar-transistor amplifier circuit equals the product of the collector current I_C and the collector voltage E_C. Similarly, for an FET, the *DC drain input power* equals the product of the drain current I_D and the drain voltage E_D. We can categorize DC input power figures as *average* or *peak* values. The following discussion involves only average power.

We can observe significant DC collector or drain input power even when an amplifier receives no input signal. A class-A circuit operates this way. In fact, when we apply an input signal to a class-A amplifier, the average DC collector or drain input power *does not change* compared to the value under no-signal conditions! In class AB_1 or class AB_2, we observe low current (and, therefore, low DC collector or drain input power) with zero input signal, and higher current (and, therefore, higher DC input power) when we apply a signal to the input. In classes B and C, we see no current (and, therefore, zero DC collector or drain input power) when no input signal exists. The current, and therefore, the DC input power, increase with increasing signal input.

We usually express or measure the DC collector or drain input power in watts, the product of amperes and volts, as long as the circuit has no reactance. We can express it in milliwatts for low-power amplifiers, or kilowatts (and, in extreme cases, megawatts) for high-power amplifiers.

The Signal Output Power

If we want to accurately determine the *signal output power* from an amplifier, we must employ a specialized AC wattmeter. The design of AF and RF wattmeters represents a sophisticated specialty in engineering. We can't connect an ordinary DC meter and rectifier diode to a power amplifier's output terminals and expect to get a true indication of the signal output power.

When a properly operating power amplifier receives no input signal, we never see any signal output, and therefore, the power output equals zero. This situation holds true for all classes of amplification. As we increase the strength of the input signal, we observe increasing signal output power up to a certain point. If we keep increasing the strength of the input signal (that is, the drive) past this point, we see little or no further increase in the signal output power.

We express or measure signal output power in watts, just as we do with DC input power. For very low-power circuits, we might want to specify the signal output power in milliwatts; for moderate- and high-power circuits we can express it in watts, kilowatts, or even megawatts.

Definition of Efficiency

Engineers define the *efficiency* of a power amplifier as the ratio of the signal output power to the DC input power. We can render this quantity either as a plain fraction (in which case it has a value between 0 and 1) or as a percentage (in which case it has a value between 0% and 100%). Let P_{in}

represent the DC input power to a power amplifier. Let P_{out} represent the signal output power. We can quantify the efficiency (eff) as the ratio

$$\text{eff} = P_{out} / P_{in}$$

or as the percentage

$$\text{eff}_{\%} = 100\ P_{out} / P_{in}$$

Problem 26-5

Suppose that a bipolar-transistor amplifier has a DC input power of 120 W and a signal output power of 84 W. What's the efficiency in percent?

Solution

We should use the formula for amplifier efficiency $\text{eff}_{\%}$ expressed as a percentage. When we go through the arithmetic, we obtain

$$\text{eff}_{\%} = 100\ P_{out} / P_{in}$$
$$= 100 \times 84 / 120 = 100 \times 0.70 = 70\%$$

Problem 26-6

Suppose that the efficiency of an FET amplifier equals 0.600. If we observe 3.50 W of signal output power, what's the DC input power?

Solution

Let's plug in the given values to the formula for the efficiency of an amplifier expressed as a ratio, and then use simple algebra to solve the problem. We get

$$0.600 = 3.50 / P_{in}$$

which solves to

$$P_{in} = 3.50 / 0.600 = 5.83\ \text{W}$$

Efficiency versus Class

Class-A amplifiers have efficiency figures from 25% to 40%, depending on the nature of the input signal and the type of transistor used. A good class-AB_1 amplifier has an efficiency rating somewhere between 35% and 45%. A class-AB_2 amplifier, if well designed and properly operated, can exhibit an efficiency figure up to about 50%. Class-B amplifiers are typically 50% to 65% efficient. Class-C amplifiers can have efficiency levels as high as 75%.

Drive and Overdrive

In theory, class-A and class-AB_1 power amplifiers don't draw any power from the signal source to produce significant output power. This property constitutes one of the main advantages of these two modes. We need provide only a certain AC signal voltage at the control electrode (the base, gate, emitter, or source) for class-A or class-AB_1 circuits to produce useful output signal power. Class-AB_2 amplifiers need some driving power to produce AC power output. Class-B

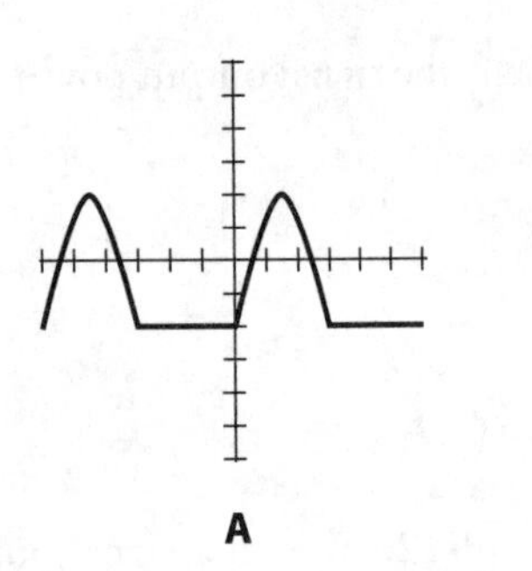

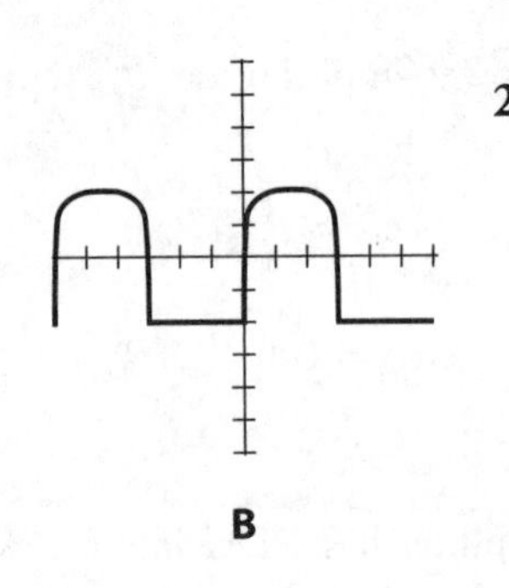

26-7 At A, an oscilloscope display of the signal output waveform from a properly operating class-B power amplifier. At B, a display showing distortion in the waveform caused by overdrive.

amplifiers require more drive than class-AB_2 circuits do, and class-C amplifiers need the most driving power of all.

Whatever type of PA that we employ in a given situation, we must always make certain that we don't allow the driving signal to get too strong because it will produce an undesirable condition known as *overdrive*. When we force an amplifier to operate in a state of overdrive, we get excessive distortion in the output signal waveform. This distortion can adversely affect the modulation envelope. We can use an *oscilloscope* (or *scope*) to determine whether or not this type of distortion is taking place in a particular situation. The scope gives us an instant-to-instant graphical display of signal amplitude as a function of time. We connect the scope to the amplifier output terminals to scrutinize the signal waveform. The output waveform for a particular class of amplifier always has a characteristic shape; overdrive causes a form of distortion known as *flat topping*.

Figure 26-7A illustrates the output signal waveform for a properly operating class-B amplifier. Figure 26-7B shows the output waveform for an overdriven class-B amplifier. Note that in drawing B, the waveform peaks appear blunted. This unwanted phenomenon shows up as distortion in the modulation on a radio signal, and also an excessive amount of signal output at harmonic frequencies. The efficiency of the circuit can be degraded, as well. The blunted wave peaks cause higher-than-normal DC input power but no increase in useful output power, resulting in below-par efficiency.

Audio Amplification

The circuits that we've examined so far have been generic but not application-specific. With capacitors of several microfarads, and when biased for class-A operation, these circuits offer good representations of audio amplifiers.

Frequency Response

High-fidelity (or hi-fi) audio amplifiers, of the kind used in music systems, must have more or less constant gain from 20 Hz to 20 kHz (at least), and preferably over a wider range such as 5 Hz to 50 kHz. Audio amplifiers for voice communications need to function well only between approximately 300 Hz and 3 kHz. In digital communications, audio amplifiers are designed to work over a narrow range of frequencies, sometimes less than 100 Hz wide.

Transistorized hi-fi amplifiers usually contain resistance-capacitance (*RC*) networks that tailor the frequency response. These networks constitute *tone controls*, also called *bass* and *treble* controls. The simplest hi-fi amplifiers use a single control to provide adjustment of the tone. Sophisticated amplifiers have separate controls, one for bass and the other for treble. The most advanced hi-fi systems make use of *graphic equalizers*, having controls that affect the amplifier gain over several different frequency spans.

Volume Control

Audio amplifier systems usually consist of two or more *stages*. A stage has one bipolar transistor or FET (or a push-pull combination) along with resistors and capacitors. We can cascade two or more stages to get high gain. In one of the stages, we incorporate a *volume control*. The simplest volume control is a potentiometer that allows us to adjust the amplifier system gain without affecting its linearity.

Figure 26-8 illustrates a basic volume control. In this amplifier, the gain through the transistor remains constant even as the input signal strength varies. The AC output signal passes through C_1 and appears across potentiometer R_1, the volume control. The *wiper* (indicated by the arrow) of the potentiometer "picks off" more or less of the AC output signal, depending on the position of the control shaft. Capacitor C_2 isolates the potentiometer from the DC bias of the following stage.

We should always place an audio volume control in a low-power stage. If we put the potentiometer at a high-power point, it will have to dissipate considerable power when we set the volume at a low level. High-power potentiometers, as you might guess, cost more than low-power ones, and they're harder to find. Even if we manage to obtain a potentiometer that can handle the strain, placing the volume control at a high-power point will cause the amplifier system to suffer from poor efficiency.

Transformer Coupling

We can use transformers to transfer (or *couple*) signals from one stage to the next in a cascaded amplifier system (also called an *amplifier chain*). Figure 26-9 illustrates *transformer coupling* between two amplifier circuits. Capacitors C_1 and C_2 keep the lower ends of the transformer windings at signal ground. Resistor R_1 limits the current through the first transistor Q_1. Resistors R_2 and R_3 provide the base bias for transistor Q_2.

Transformer coupling costs more per stage than *capacitive coupling* does. However, transformer coupling can provide optimum signal transfer between amplifier stages. By selecting a transformer with the correct turns ratio, the output impedance of the first stage can be perfectly matched to the input impedance of the second stage, assuming that no reactance exists in either circuit.

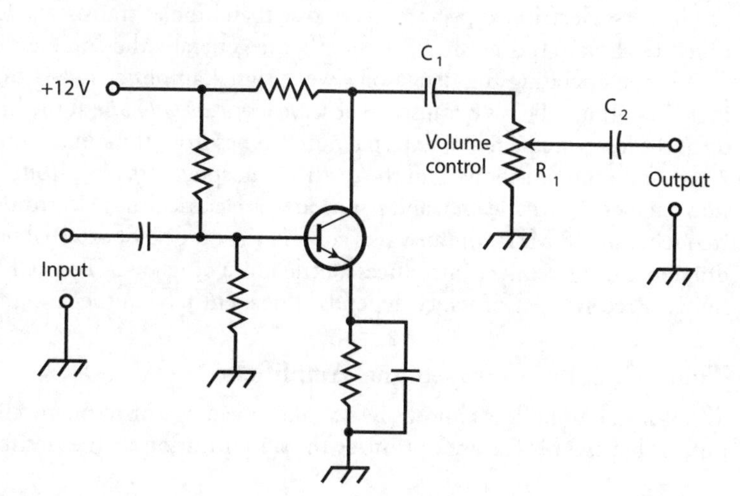

26-8 We can use a basic volume control (potentiometer R_1) to vary the gain in a low-power audio amplifier.

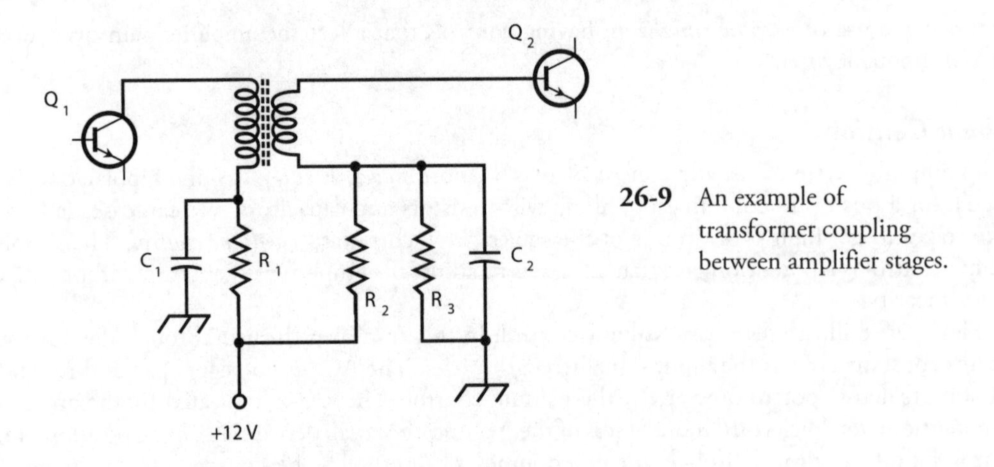

26-9 An example of transformer coupling between amplifier stages.

Radio-Frequency Amplification

The *RF spectrum* extends upward in frequency to well over 300 GHz. Sources disagree as to the exact low-frequency limit. Some texts put it at 3 kHz, some at 9 kHz, some at 10 kHz, and a few at the upper end of the AF range, usually defined as 20 kHz.

Weak-Signal versus Power Amplifiers

The *front end,* or first amplifying stage, of a radio receiver requires the most sensitive possible amplifier. The sensitivity depends on two factors: the gain (or amplification factor) and the *noise figure,* a measure of how well a circuit can amplify desired signals while generating a minimum of *internal noise.*

All semiconductor devices create internal noise as a consequence of charge-carrier movement. We call this phenomenon *electrical noise.* Internal noise can also result from the inherent motion of the molecules that comprise the semiconductor material; that's known as *thermal noise.* Random, rapid fluctuations in current can give rise to a third effect called *shot-effect noise.* In general, JFETs produce less electrical and shot-effect noise than bipolar transistors do. Gallium arsenide FETs, also called *GaAsFETs* (pronounced "gasfets"), are generally the "quietest" of all semiconductor devices.

As the operating frequency of a weak-signal amplifier increases, the noise figure gets increasingly important. That's because we observe less *external noise* at the higher radio frequencies than we do at the lower frequencies. External noise comes from the sun (*solar noise*), from outer space (*cosmic noise*), from thundershowers in the earth's atmosphere (*sferics*), from human-made internal combustion engines (*ignition noise*), and from various electrical and electronic devices (*appliance noise*). At a frequency of 1.8 MHz, the airwaves contain a great deal of external noise, and it doesn't make much difference if the receiver introduces a little noise of its own. But at 1.8 GHz, we see far less external noise, so receiver performance depends almost entirely on the amount of internally generated noise.

Tuned Circuits in Weak-Signal Amplifiers

Weak-signal amplifiers almost always take advantage of resonant circuits in the input, at the output, or both. This feature optimizes the amplification at the desired frequency, while helping to minimize noise on unwanted frequencies. Figure 26-10 is a schematic diagram of a typical tuned GaAsFET weak-signal RF amplifier designed for operation at about 10 MHz.

In some weak-signal RF amplifier systems, engineers use transformer coupling between stages and connect capacitors across the primary and/or secondary windings of the transformers. This tactic produces resonance at a frequency determined by the capacitance and the transformer winding inductance. If the set of amplifiers is intended for use at only one frequency, this method of coupling, called *tuned-circuit coupling*, enhances the system efficiency, but increases the risk that the stages will break into oscillation at the resonant frequency.

Broadband PAs

We can design RF power amplifiers to operate in either the *broadband* mode or the *tuned* mode. As these terms suggest, broadband amplifiers work over a wide range of frequencies without adjustment for frequency variations, while tuned amplifiers require adjustment of the resonant frequencies of internal circuits.

A broadband PA offers convenience because it does not require tuning within its design frequency range. The operator need not worry about critical adjustments, or bother altering the circuit parameters when changing the frequency. However, broadband PAs are, as a whole, less efficient than tuned PAs. Another disadvantage of broadband PAs is the fact that they will amplify any signal in the design frequency range, whether or not the operator wants that amplification to occur. For example, if some earlier stage in a radio transmitter oscillates at a frequency different from the intended signal frequency, and if this undesired signal falls within the design frequency range of the broadband PA, then that signal will undergo amplification along with the desired signal, producing unintended *spurious emission* from the transmitter.

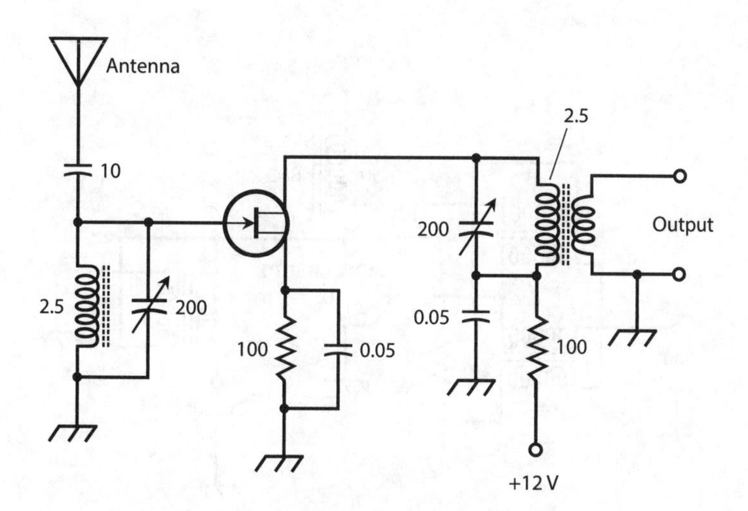

26-10 A tuned RF amplifier for use at approximately 10 MHz. Resistances are in ohms. Capacitances are in microfarads (μF) if less than 1, and in picofarads (pF) if more than 1. Inductances are in microhenrys (μH).

Figure 26-11 is a schematic diagram of a typical broadband PA using an NPN power transistor. This circuit can provide several watts of continuous RF power output over a range of frequencies from 1.5 MHz through 15 MHz. The transformers constitute a critical part of this circuit. They must work efficiently over a 10:1 range of frequencies. The 50-μH component labeled "RFC" is an *RF choke*, which passes DC and low-frequency AC while blocking high-frequency AC (that is, RF signals).

Tuned PAs

A tuned RF PA offers improved efficiency compared with broadband designs. The tuned circuits minimize the risk of spurious signals from earlier stages being amplified and transmitted over the air. Tuned PAs can work into a wide range of load impedances. In addition to a *tuning control*, or resonant circuit that adjusts the output of the amplifier to the operating frequency, the tuned amplifier incorporates a *loading control* that optimizes the signal transfer between the amplifier and the load (usually an antenna).

Tuned PAs have one significant limitation: The "tune-up" procedure (usually the adjustment of variable capacitors and/or variable inductors) takes time, and improper adjustment can result in damage to the transistor. If the tuning and/or loading controls aren't properly set, the amplifier efficiency will drop to near zero, while the DC power input remains high. Solid-state devices overheat quickly under these conditions because the excess power "has nowhere to go" except to dissipate itself as heat in the amplifier's components.

Figure 26-12 illustrates a tuned RF PA that can provide a few watts of useful power output at about 10 MHz. The transistor is the same type as the one used in the broadband amplifier of Fig. 26-11. The operator should adjust the tuning and loading controls (left-hand and right-hand variable capacitors, respectively) for maximum power output as indicated by an RF wattmeter.

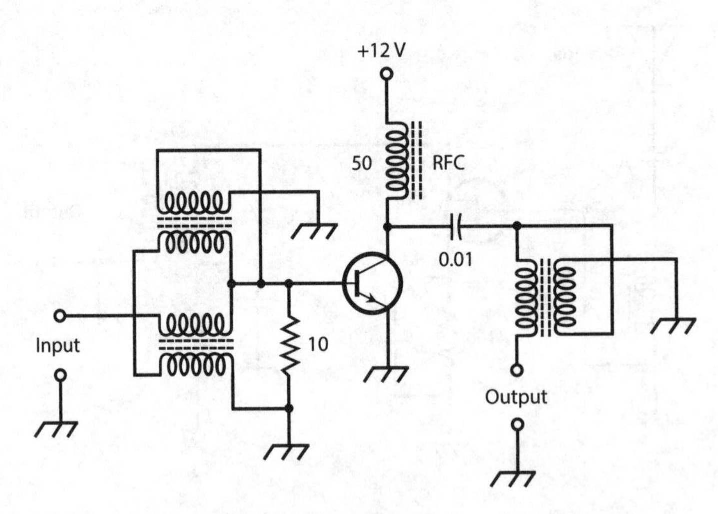

26-11 A broadband RF power amplifier, capable of producing a few watts output. Resistances are in ohms. Capacitances are in microfarads (μF). Inductances are in microhenrys (μH). The 50-μH component labeled "RFC" is an RF choke.

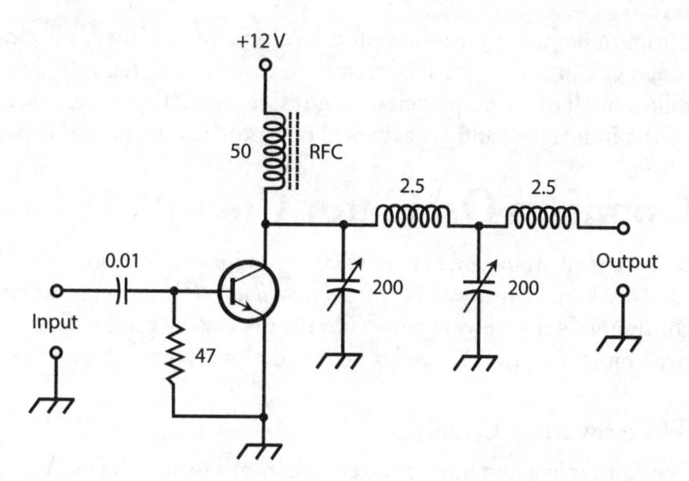

26-12 A tuned RF power amplifier, capable of producing a few watts of output. Resistances are in ohms. Capacitances are in microfarads (μF) if less than 1, and in picofarads (pF) if more than 1. Inductances are in microhenrys (μH).

How Oscillators Work

An oscillator is a specialized amplifier with positive feedback. Radio-frequency oscillators generate signals in a wireless broadcast or communications system. Audio-frequency oscillators find applications in hi-fi systems, music synthesizers, electronic sirens, security alarms, and electronic toys.

Positive Feedback

Feedback normally occurs either in phase with the input signal, or in phase opposition relative to the input signal. If we want to make an amplifier circuit oscillate, we must introduce some of its output signal back to the input in phase (that's called *positive feedback*). If we introduce some of the output signal back to the input in phase opposition, we have *negative feedback* that reduces the overall gain of the amplifier. Negative feedback is not always bad; engineers deliberately use it in some amplifiers to prevent unwanted oscillation.

The AC output signal wave from a common-emitter or common-source amplifier occurs in phase opposition with respect to the input signal wave. If you couple the collector to the base through a capacitor, you won't get oscillation. You must invert the phase in the feedback process if you want oscillation to occur. In addition, the amplifier must exhibit a certain minimum amount of gain, and the coupling from the output to the input must be substantial. The positive feedback path must be easy for a signal to follow. Most oscillators comprise common-emitter or common-source amplifier circuits with positive feedback.

The AC output signal wave from a common-base or common-gate amplifier is in phase with the input signal wave. You might, therefore, suppose that such circuits would make ideal candidates for oscillators. However, the common-base and common-gate circuits produce less gain than their common-emitter and common-source counterparts, so it's more difficult to make them oscillate. Common-collector and common-drain circuits are even worse in this respect because they have negative gain!

Feedback at a Single Frequency

We can control the frequency of an oscillator using tuned, or resonant, circuits, usually consisting of inductance-capacitance (LC) or resistance-capacitance (RC) combinations. The LC scheme is common in radio transmitters and receivers; the *RC* method is more often used in audio work. The tuned circuit makes the feedback path easy for a signal to follow at one frequency, but difficult to follow at all other frequencies. As a result, oscillation takes place at a stable frequency, determined by the inductance and capacitance, or by the resistance and capacitance.

Common Oscillator Circuits

Many circuit arrangements can reliably produce oscillation. The following several circuits all fall into a category known as *variable-frequency oscillators* (VFOs) because we can adjust their signal frequencies continuously over a wide range. Oscillators usually produce less than 1 W of RF power output. If we need more power, we'll need to follow the oscillator with one or more stages of amplification.

The Armstrong Circuit

We can force a common-emitter or common-source class-A amplifier to oscillate by coupling the output back to the input through a transformer that reverses (inverts) the phase of the fed-back signal. The schematic diagram of Fig. 26-13 shows a common-source amplifier whose drain circuit is coupled to the gate circuit by means of a transformer. We control the frequency by adjusting a capacitor connected in series with the transformer secondary winding. The inductance of the transformer secondary, along with the capacitance, forms a resonant circuit that passes energy easily at one frequency, while attenuating (suppressing) the energy at other frequencies. Engineers call this type of circuit an *Armstrong oscillator*. We can substitute a bipolar transistor for the JFET, as long as we bias the device for class-A amplification.

The Hartley Circuit

Figure 26-14 illustrates another method of obtaining controlled RF feedback. In this example, we use a PNP bipolar transistor. The circuit has a single coil with a tap on the winding. A variable capacitor in parallel with the coil determines the oscillating frequency, and allows for frequency adjustment. This circuit is called a *Hartley oscillator*.

In the Hartley circuit, as well as in most other RF oscillator circuits, we must always use the minimum amount of feedback necessary to obtain reliable, continuous oscillation. The location of

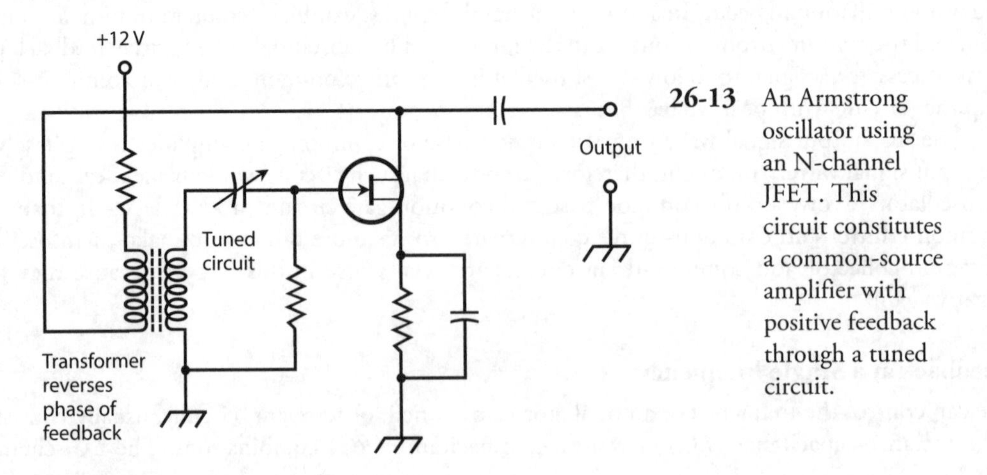

26-13 An Armstrong oscillator using an N-channel JFET. This circuit constitutes a common-source amplifier with positive feedback through a tuned circuit.

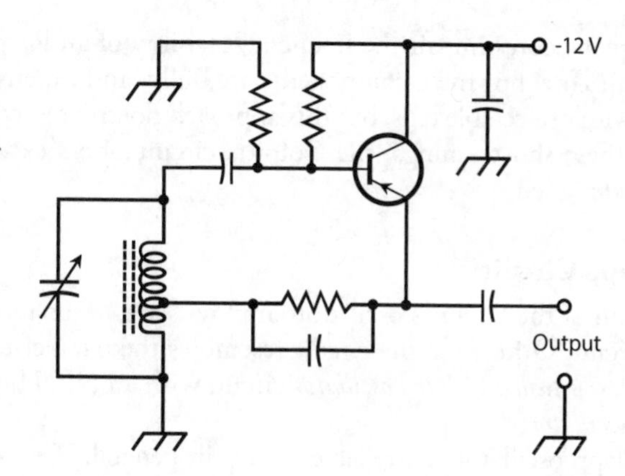

26-14 A Hartley oscillator using a PNP bipolar transistor. We can recognize the Hartley circuit by the tapped inductor in the tuned *LC* circuit.

the coil tap determines the amount of feedback. The circuit shown in Fig. 26-14 takes only about 25% of its amplifier power to produce the feedback. We can, therefore, use the other 75% of the power as useful signal output.

The Colpitts Circuit

We can tap the capacitance, instead of the inductance, in the tuned circuit of an RF oscillator. This arrangement gives us a *Colpitts oscillator*. Figure 26-15 is a schematic diagram of a P-channel JFET wired up to function as a Colpitts oscillator. We control the amount of feedback by "tweaking" the ratio of the two capacitances connected in parallel with the variable inductor, which provides for the frequency adjustment. The two capacitors across the variable inductor are fixed, not variable. This feature offers convenience and saves money because we'll likely have trouble finding a dual variable capacitor that maintains the correct ratio of capacitances throughout its tuning range (as we would need if the inductor in the tuned circuit weren't variable).

Unfortunately, finding a variable inductor for use in a Colpitts oscillator can prove almost as difficult as obtaining a suitable dual variable capacitor. We can use a *permeability-tuned* coil, but

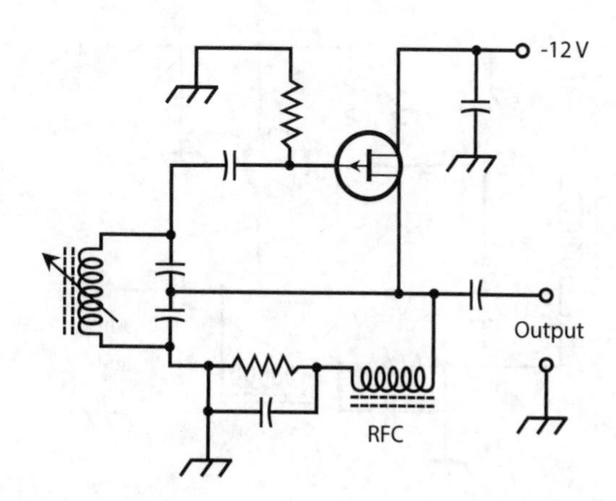

26-15 A Colpitts oscillator using a P-channel JFET. The Colpitts circuit can be recognized by the split capacitance in the tuned *LC* circuit.

ferromagnetic cores impair the frequency stability of an RF oscillator. We can use a *roller inductor* with an air core, but these components are bulky and expensive. We can use a fixed inductor with several switch-selectable taps, but this approach doesn't allow for continuous frequency adjustment. Despite these shortcomings, the Colpitts circuit offers exceptional stability and reliability when properly designed.

The Clapp Circuit

A variation of the Colpitts oscillator employs series resonance, instead of parallel resonance, in the tuned circuit. Otherwise, the circuit resembles the parallel-tuned Colpitts oscillator. Figure 26-16 shows a *series-tuned Colpitts oscillator* circuit with an NPN bipolar transistor. Some engineers call it a *Clapp oscillator.*

A Clapp oscillator is a reliable circuit in general. We can easily get it to oscillate and keep it going. Its frequency won't change much if we build it with high-quality components. The Clapp design allows us to employ a variable capacitor for frequency control, while accomplishing feedback through a capacitive voltage divider.

Getting the Output

Have you noticed something strange about the Hartley, Colpitts, and Clapp oscillators diagrammed in Figs. 26-14 through 26-16? If not, look again, and compare these circuits with class-A common-emitter and common-source amplifiers. In these oscillators, we take the output from the emitter or source, not from the collector or drain as we would normally do with an amplifier. Why, you ask, would we want to take this approach in oscillator design?

Theoretically, we can get the output of an oscillator from the collector or drain to get maximum gain. But in an oscillator, stability and reliability are more important than gain. We can get all the gain we want in amplifiers following an oscillator. We obtain better stability in an oscillator when we take the output signal from the emitter or source, as compared with taking it from the collector or drain. In that arrangement, variations in the load impedance have less effect on the frequency of oscillation, and a sudden decrease in load impedance is less likely to cause the oscillator to fail outright.

To prevent the output signal from shorting to ground, we can connect an RF choke (RFC) in series with the emitter or source in the Colpitts and Clapp oscillator circuits. The choke, which comprises a high-value inductor, allows DC to pass while blocking high-frequency AC (exactly the

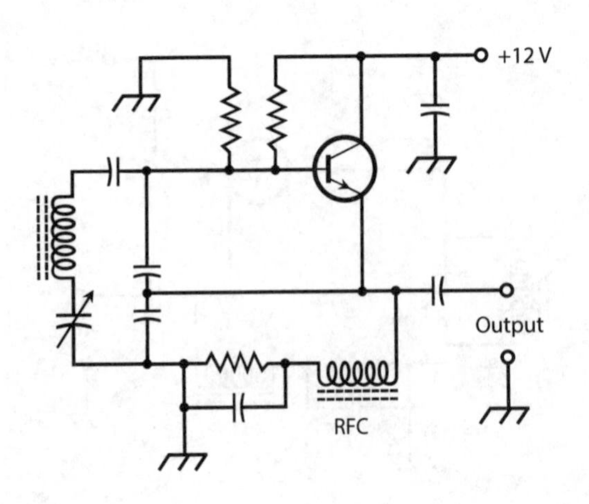

26-16 A series-tuned Colpitts oscillator, also known as a Clapp oscillator. This circuit uses an NPN bipolar transistor.

opposite behavior from that of a *blocking capacitor*). Typical values for RF chokes range from about 100 μH at high frequencies such as 15 MHz, to 10 mH at low frequencies, such as 150 kHz.

The Voltage-Controlled Oscillator

We can adjust the frequency of a VFO to some extent by connecting a varactor diode in the tuned *LC* circuit. Recall that a varactor, also called a *varicap*, is a semiconductor diode that functions as a variable capacitor when reverse-biased. As the reverse-bias voltage increases, the junction capacitance decreases, provided that we don't apply so much voltage that avalanche breakdown occurs.

The Hartley and Clapp oscillator circuits lend themselves to varactor-diode frequency control. We can connect the varactor in series or in parallel with the main tuning capacitor. We must isolate the varactor for DC with blocking capacitors. If you look back to Chap. 20 for a moment and check Fig. 20-9, you'll see an effective method of connecting a varactor in a tuned *LC* circuit. We call the resulting limited-range VFO a *voltage-controlled oscillator* (VCO).

Varactors cost less, weigh less, and take up less physical space than variable capacitors or inductors. These factors constitute the chief advantages of a VCO over an old-fashioned VFO that employs only a variable capacitor and a fixed inductor, or only a variable inductor and a fixed capacitor.

Diode-Based Oscillators

At ultra-high frequencies (UHF) and microwave radio frequencies, certain types of diodes can function as oscillators. In Chap. 20, you learned about these components, which include the Gunn diode, the IMPATT diode, and the tunnel diode.

Crystal-Controlled Oscillators

In an RF oscillator, we can use a *quartz crystal* in place of a tuned *LC* circuit, as long as we don't have to change the frequency often. *Crystal-controlled* oscillators offer frequency stability superior to that of *LC* tuned VFOs.

Several schemes exist for the connection of quartz crystals into bipolar or FET circuits for the purpose of obtaining oscillation. One common circuit is the *Pierce oscillator*. We can obtain a Pierce oscillator by connecting a JFET and a quartz crystal, as shown in Fig. 26-17. This circuit takes advantage of an N-channel JFET, but we could just as well use an N-channel MOSFET, a P-channel JFET, or a P-channel MOSFET. The crystal frequency can be varied by about plus-or-minus a tenth

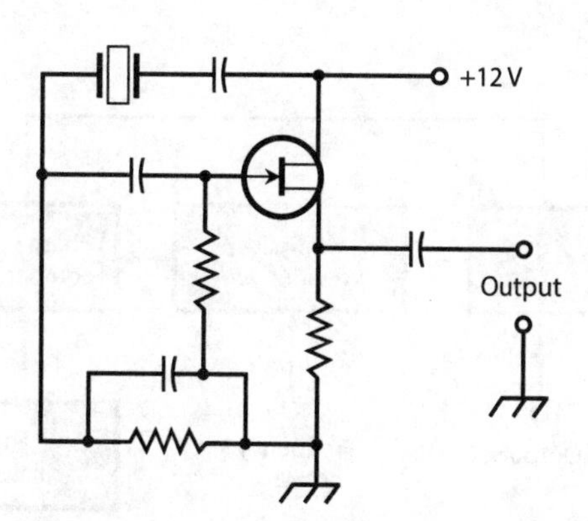

26-17 A Pierce oscillator circuit using an N-channel JFET.

of one percent (±0.1%, or one part in 1000) by means of an inductor or capacitor in parallel with the crystal. However, the oscillation frequency is determined mainly by the thickness of the quartz wafer, and by the angle at which it was cut from the original mineral sample.

Crystals change in frequency as the temperature changes. But they're far more stable than *LC* circuits, most of the time. If an engineer needs a crystal oscillator with exceptional frequency stability, the crystal can be enclosed in a temperature-controlled chamber called a *crystal oven*. In this environment, the crystal maintains its rated frequency so well that it can function as a *frequency standard* against which other frequency-dependent circuits, including oscillators, are calibrated.

The Phase-Locked Loop

One type of oscillator that combines the flexibility of a VFO with the stability of a crystal oscillator is the *phase-locked loop* (PLL). The PLL makes use of a circuit called a *frequency synthesizer*. The output of a VCO passes through a *programmable multiplier/divider*, a digital circuit that divides and/or multiplies the VCO frequency by integral (whole-number) values that we can freely choose. As a result, the output frequency can equal any rational-number multiple of the crystal frequency. We can, therefore, adjust a well-designed PLL circuit in small digital increments over a wide range of frequencies. Figure 26-18 is a block diagram of a PLL.

The output frequency of the multiplier/divider remains "locked," by means of a *phase comparator*, to the signal from a crystal-controlled *reference oscillator*. As long as the output from the multiplier/divider stays exactly on the reference oscillator frequency, the two signals remain exactly in phase, and the output of the phase comparator equals zero (that is, 0 V DC). If the VCO frequency begins to gradually increase or decrease (a phenomenon known as *oscillator drift*), the output frequency of the multiplier/divider also drifts, although at a different rate. Even a frequency change of less than 1 Hz causes the phase comparator to produce a *DC error voltage*. This error voltage is either positive or negative, depending on whether the VCO has drifted higher or lower in frequency. We apply the error voltage to a varactor, causing the VCO frequency to change in a direction opposite to that of the drift, creating a *DC feedback* circuit that maintains the VCO frequency at a precise value. Engineers call it a *loop* circuit that *locks* the VCO onto a particular frequency by means of *phase* sensing—hence the expression *phase-locked*.

The key to the stability of the PLL lies in the fact that the reference oscillator employs crystal control. When you hear that a radio receiver, transmitter, or transceiver is *synthesized*, you can have reasonable confidence that a PLL determines its operating frequency. The stability of a synthesizer can be enhanced by using an amplified signal from the shortwave time-and-frequency broadcast station WWV at 2.5, 5, 10, or 15 MHz, directly as the reference oscillator. These

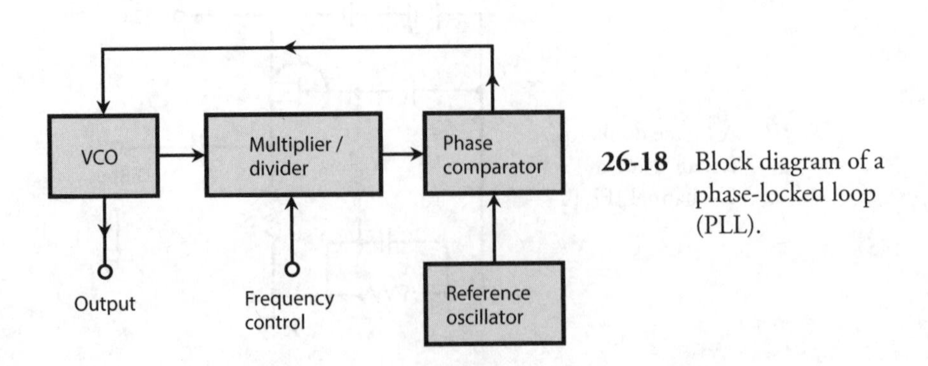

26-18 Block diagram of a phase-locked loop (PLL).

signals remain frequency-exact to a minuscule fraction of 1 Hz because they're controlled by atomic clocks. Most people don't need precision of this caliber, so you won't see consumer devices like ham radios and shortwave receivers with *primary-standard* PLL frequency synthesis. But some corporations and government agencies use the primary-standard method to ensure that their systems stay "on frequency."

Oscillator Stability

In an oscillator, the term *stability* can have either of two distinct meanings: constancy of frequency (or minimal frequency drift), and reliability of performance.

Frequency Stability

In the design and construction of a VFO of any kind, the components—especially the capacitors and inductors—must, to the greatest extent possible, maintain constant values under all anticipated conditions.

Some types of capacitors hold their values better than others as the temperature rises or falls. Polystyrene capacitors behave very well in this respect. Silver-mica capacitors can work when polystyrene units aren't readily available. Air-core coils exhibit the best temperature stability of all inductor configurations. They should be wound, when possible, from stiff wire with strips of plastic to keep the windings in place. Some air-core coils are wound on hollow cylindrical cores, made of ceramic or phenolic material. Ferromagnetic solenoidal or toroidal cores aren't very good for use in VFO coils because these materials change permeability as the temperature varies. This variation alters the inductance, in turn affecting the oscillator frequency.

The best oscillators, in terms of frequency stability, are crystal-controlled. This category includes circuits that oscillate at the fundamental frequency of the quartz crystal, circuits that oscillate at one of the crystal harmonic frequencies, or PLL circuits that oscillate at frequencies derived from the crystal frequency by means of programmable multiplier/dividers.

Reliability

An oscillator should always start working as soon as we apply DC power. It should keep oscillating under all normal conditions. The failure of a single oscillator can cause an entire receiver, transmitter, or transceiver to stop working.

When an engineer builds an oscillator and puts it to use in a radio receiver, transmitter, or audio device, *debugging* is always necessary. Debugging comprises a trial-and-error process—often quite tedious—of getting the flaws or "bugs" out of the circuit so that it will work well enough to to be mass-produced. Rarely can an engineer build something "straight from the drawing board" and have it work perfectly on the first test! In fact, if you build two oscillators from the same diagram, with the same component types and values, you shouldn't be surprised if one circuit works okay and the other one doesn't work at all. Problems of this sort usually happen because of differences in the quality of components that don't show up until you conduct "real-world" circuit tests.

Oscillators are designed to work into relatively high load impedances. If we connect an oscillator to a load that has a low impedance, that load will "try" to draw a lot of power from the oscillator. Under such conditions, even a well-designed oscillator might stop working or not start up, when we first switch it on. Oscillators aren't meant to produce powerful signals; we can use amplifiers for that purpose! You need never worry that an oscillator's load impedance might get too high. In general, as we increase the load impedance for an oscillator, its overall performance improves.

Audio Oscillators

Audio oscillators appear in myriad electronic devices including doorbells, ambulance sirens, electronic games, telephone sets, and toys that play musical tunes. All AF oscillators are, in effect, AF amplifiers with positive feedback.

Audio Waveforms

At AF, oscillators can use *RC* or *LC* combinations to determine the frequency, but *RC* circuits are generally preferred. If we want to build an audio oscillator using *LC* circuits, we'll need large inductances requiring the use of ferromagnetic cores.

At RF, oscillators are usually designed to produce a sine-wave output. A pure sine wave represents energy at one and only one frequency. Audio oscillators, in contrast, don't always concentrate all their energy at a single frequency. (A pure AF sine wave, especially if continuous and frequency-constant, can give you a headache!) The various musical instruments in a band or orchestra all sound different from each other, even when they all play a note at the same frequency. This difference in sound *timbre* (the "character" of the sound) arises from the fact that each instrument generates its own unique AF waveform. As you know, a clarinet sounds different than a trumpet, which in turn sounds different from a cello or piano, even if they all play the same note such as middle C.

Imagine that you use a time-domain laboratory display to scrutinize the waveforms of musical instruments. You can build an arrangement for this purpose with a high-fidelity microphone, a sensitive, low-distortion audio amplifier, and an oscilloscope. You'll see that each instrument has its own "signature." Therefore, each instrument's unique sound qualities can be reproduced using AF oscillators whose waveform outputs match those of the instrument. Electronic music synthesizers use audio oscillators to generate the notes that you hear.

The Twin-T Oscillator

Figure 26-19 shows a popular AF circuit called a *Twin-T oscillator* that can serve well for general-purpose use. The frequency depends on the values of the resistors R and capacitors C. The output note constitutes a near-perfect sine wave, although not quite perfect. (The small amount of distortion helps to alleviate the "ear-brain irritation" typically produced by an absolutely pure AF sinusoid.) The circuit shown in this example uses two PNP bipolar transistors biased for class-A amplification.

The Multivibrator

Another popular AF oscillator circuit makes use of two identical common-emitter or common-source amplifier circuits, hooked up so that the signal goes around and around between them. Lay people sometimes call this arrangement a "multivibrator," although the technical term *multivibrator* more appropriately applies to various *digital* signal-generating circuits.

In the example of Fig. 26-20, two N-channel JFETs are connected to form a "multivibrator" for use at AF. Each of the two transistors amplifies the signal in class-A mode and inverts the phase. Every time the signal goes from any particular point all the way around the circuit, it arrives back at that point inverted twice so it's in phase with its "former self," producing positive feedback.

We can set the frequency of the oscillator of Fig. 26-20 by means of an *LC* circuit. The coil can have a ferromagnetic core because stability is not of great concern and because we need such a core to obtain enough inductance to produce resonance at AF. Toroidal or pot cores work well in this application. The value of *L* can range from about 10 mH to as much as 1 H. We choose the

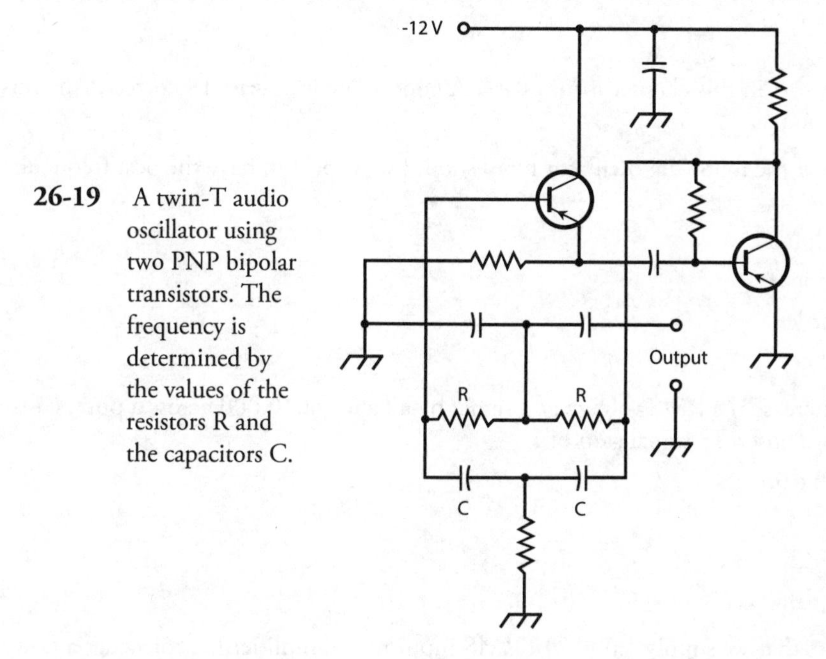

26-19 A twin-T audio oscillator using two PNP bipolar transistors. The frequency is determined by the values of the resistors R and the capacitors C.

capacitance according to the formula for resonant circuits (which you learned earlier in this course), to obtain an AF output note at the frequency desired.

Integrated-Circuit Oscillators

In recent years, solid-state technology has advanced to the point that entire electronic systems can be *etched* onto silicon chips. Such devices have become known as *integrated circuits* (ICs). The *operational amplifier*, also called an *op amp*, is a type of IC that performs exceptionally well as an AF oscillator because it has high gain, and it can easily be connected to produce positive feedback. You learned about op amps in Chap. 23.

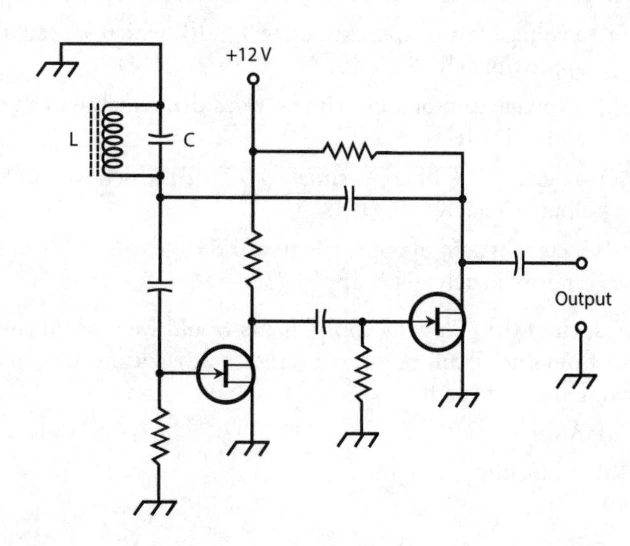

26-20 A "multivibrator" audio oscillator using two N-channel JFETs. The frequency is determined by the values of the inductor L and the capacitor C.

Quiz

Refer to the text in this chapter if necessary. A good score is at least 18 correct. Answers are in the back of the book.

1. Which of the following oscillator types should we expect to have the best frequency stability?
   (a) Colpitts
   (b) Clapp
   (c) Hartley
   (d) Pierce

2. If we increase the RMS voltage of a signal by a factor of 10,000 across a pure, constant resistance, we observe a signal gain of
   (a) 100 dB.
   (b) 80 dB.
   (c) 40 dB.
   (d) 20 dB.

3. Suppose that we supply 1.00 W of RMS input to an amplifier that provides a power gain of 33.0 dB. What's the output, assuming that no reactance exists in the system?
   (a) 2.00 kW RMS
   (b) 330 W RMS
   (c) 200 W RMS
   (d) 50.0 W RMS

4. Imagine that we apply a signal of 30 V RMS to the primary winding of a perfectly efficient impedance-matching transformer (it dissipates no power as heat in its core or windings), obtaining 10 V RMS across the secondary. Also suppose that no reactance exists in the circuits connected to the primary and secondary. This transformer technically introduces
   (a) a voltage loss of approximately 9.5 dB, which we can also call a voltage gain of approximately −9.5 dB.
   (b) a voltage gain of approximately 4.8 dB, which we can also call a voltage loss of approximately −4.8 dB.
   (c) a current loss of approximately 9.5 dB, which we can also call a current gain of approximately −9.5 dB.
   (d) a current gain of approximately 4.8 dB, which we can also call a current loss of approximately −4.8 dB.

5. Which of the following components would we most likely choose if we want to allow DC to pass from one circuit point to another, but we want to keep high-frequency AC signals from following the same path?
   (a) A varactor
   (b) A blocking capacitor

(c) An RF choke
(d) A Gunn diode

6. Figure 26-21 is a diagram of
   (a) a Pierce oscillator.
   (b) a class-B push-pull amplifier.
   (c) an Armstrong oscillator.
   (d) a broadband power amplifier.
7. What, if any, significant technical errors exist in the circuit in Fig. 26-21?
   (a) No significant technical errors exist.
   (b) The power supply polarity is wrong.
   (c) We must use an enhancement-mode MOSFET, not a depletion-mode MOSFET.
   (d) The transformer must have an air core, not a powdered-iron core.
8. What precaution must we observe if we expect the circuit of Fig. 26-21 to perform correctly?
   (a) We must connect the transformer windings to ensure that the drain-to-gate feedback occurs in the proper phase.
   (b) We must set the variable capacitor to obtain the maximum possible amplification factor.
   (c) We must not allow the output load impedance to exceed the value of the resistor between the transformer primary and the source of DC voltage.
   (d) All of the above

9. In which of the following bipolar-transistor amplifier types does collector current flow for less than half of the signal cycle?
   (a) Class-C
   (b) Class-B
   (c) Class-AB_2
   (d) Class-AB_1

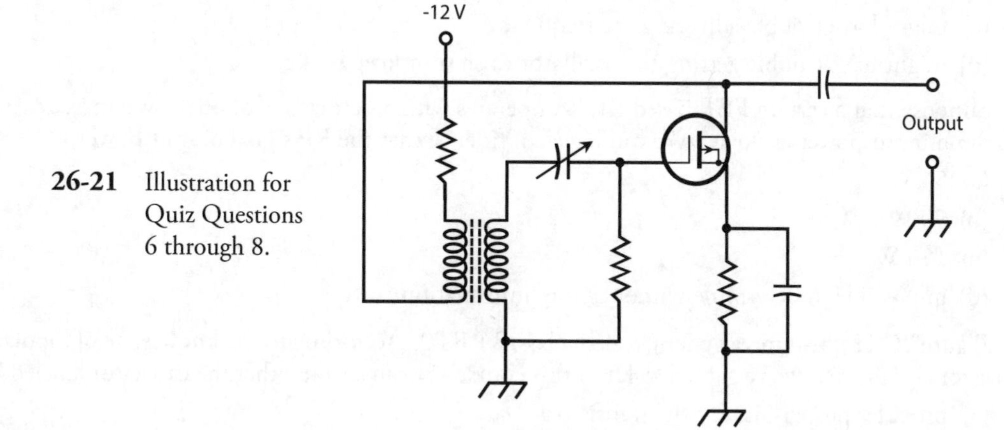

26-21 Illustration for Quiz Questions 6 through 8.

10. When designing and testing a tuned class-B push-pull RF power amplifier, we must
    (a) bias the transistors to ensure that collector or drain current flows in both devices during the entire AC input signal cycle.
    (b) select the capacitors so as to allow the system to work over a wide range of frequencies without adjustment.
    (c) set the output tuned circuit to resonate at an even harmonic of the input frequency.
    (d) select two bipolar or field-effect transistors whose characteristics are as nearly identical as possible.

11. In a Hartley oscillator, the output-signal frequency depends on the
    (a) gain of the transistor.
    (b) tuned-circuit inductance and capacitance.
    (c) dimensions of a quartz crystal.
    (d) feedback path in a phase-locked loop (PLL).

12. Which FET amplifier type introduces little or no distortion into the AC signal wave, with drain current during the entire signal cycle?
    (a) Class A
    (b) Class AB_1 or AB_2
    (c) Class B
    (d) Class C

13. We can make a class-B amplifier linear for the AC signal waveform by
    (a) minimizing the output impedance.
    (b) biasing the transistor considerably past cutoff or pinchoff.
    (c) connecting two transistors in a push-pull arrangement.
    (d) no known means.

14. If we connect a load with very low, purely resistive impedance to the output of an oscillator, we
    (a) maximize the output power.
    (b) enhance the linearity, while allowing for harmonics.
    (c) might have trouble adjusting the frequency.
    (d) might have trouble getting the oscillator to start or keep going.

15. Suppose that a certain FET-based RF PA operates with an efficiency of 60%. We measure the DC drain input power as 90 W. We can have confidence that the RF signal output power is
    (a) 54 W
    (b) 90 W
    (c) 150 W
    (d) impossible to determine without more information.

16. Figure 26-22 illustrates a generic tuned, class-B RF PA. According to the knowledge of bipolar-transistor circuits that we've gained so far in this course, we can surmise that the capacitor labeled V
    (a) provides proper bias for the transistor.
    (b) allows the AC input signal to enter but provides DC isolation.

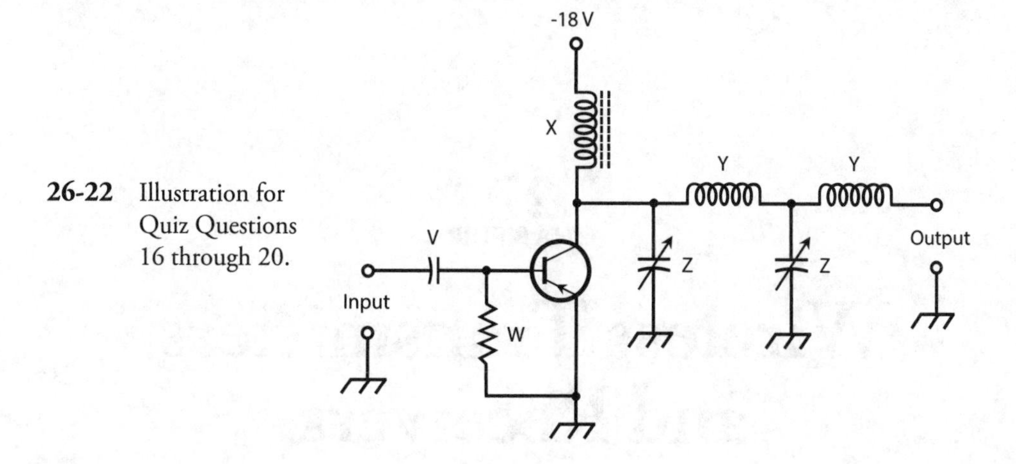

26-22 Illustration for Quiz Questions 16 through 20.

(c) determines the resonant frequency of the input circuit.
(d) keeps the signal from shorting through the power supply.

17. According to the knowledge of bipolar-transistor circuits that we've gained so far in this course, we can surmise that the resistor labeled W in Fig. 26-22
(a) provides proper bias for the transistor.
(b) allows the AC input signal to enter but provides DC isolation.
(c) determines the resonant frequency of the input circuit.
(d) keeps the signal from shorting through the power supply.

18. According to the knowledge of bipolar-transistor circuits that we've gained so far in this course, we can surmise that the RF choke labeled X in Fig. 26-22
(a) prevents excessive current from flowing in the collector.
(b) allows the AC output signal to leave but provides DC isolation.
(c) determines the resonant frequency of the output circuit.
(d) keeps the signal from shorting through the power supply.

19. According to the knowledge of bipolar-transistor circuits that we've gained so far in this course, we can surmise that the inductors labeled Y in Fig. 26-22
(a) help to optimize the signal transfer to the output.
(b) ensure that the output signal remains in phase with the input signal.
(c) provide enough feedback to keep the circuit from oscillating.
(d) keep the transistor from operating in a state of overdrive.

20. According to the knowledge of bipolar-transistor circuits that we've gained so far in this course, we can surmise that the capacitors labeled Z in Fig. 26-22
(a) help to optimize the signal transfer to the output.
(b) ensure that the output signal remains in phase with the input signal.
(c) provide enough feedback to keep the circuit from oscillating.
(d) keep the transistor from operating in a state of overdrive.

27 CHAPTER

Wireless Transmitters and Receivers

IN WIRELESS COMMUNICATIONS, A *TRANSMITTER* CONVERTS DATA INTO *ELECTROMAGNETIC* (EM) *WAVES* intended for recovery by one or more *receivers*. In this chapter, we'll learn how to convert data to an *EM field*, and then learn how we can intercept and decode that field at remote points.

Modulation

When we *modulate* a wireless signal, we "write" data onto an EM wave. We can carry out this process by varying the amplitude, the frequency, or the phase of the wave. We can also obtain a modulated signal by generating a series of multiple-wave pulses and varying their duration, amplitude, or timing. The heart of a wireless signal comprises a sine wave called the *carrier* whose frequency can range from a few kilohertz (kHz) to many gigahertz (GHz). If we expect effective data transfer, the carrier frequency must be at least 10 times the highest frequency of the modulating signal.

On/Off Keying

The simplest form of modulation involves *on/off keying* of the carrier. We can *key* the oscillator of a radio transmitter to send *Morse code*, one of the simplest known *binary digital* modulation modes. The duration of a Morse-code *dot* equals the duration of one *binary digit*, more often called a *bit*. (A binary digit is the smallest or shortest possible unit of data in a system whose only two states are "on" and "off.") A *dash* measures three bits in duration. The space between dots and dashes within a *character* equals one bit; the space between characters in a *word* equals three bits; the space between words equals seven bits. Some technicians refer to the *key-down* (full-carrier) condition as *mark* and the *key-up* (no-signal) condition as *space*. Amateur radio operators who enjoy using the Morse code send and receive it at speeds ranging from about 5 words per minute (wpm) to around 60 wpm.

Frequency-Shift Keying

We can send digital data faster and with fewer errors than Morse code allows if we use *frequency-shift keying* (FSK). In some FSK systems, the carrier frequency shifts between mark and space conditions, usually by a few hundred hertz or less. In other systems, a two-tone audio-frequency (AF) sine wave

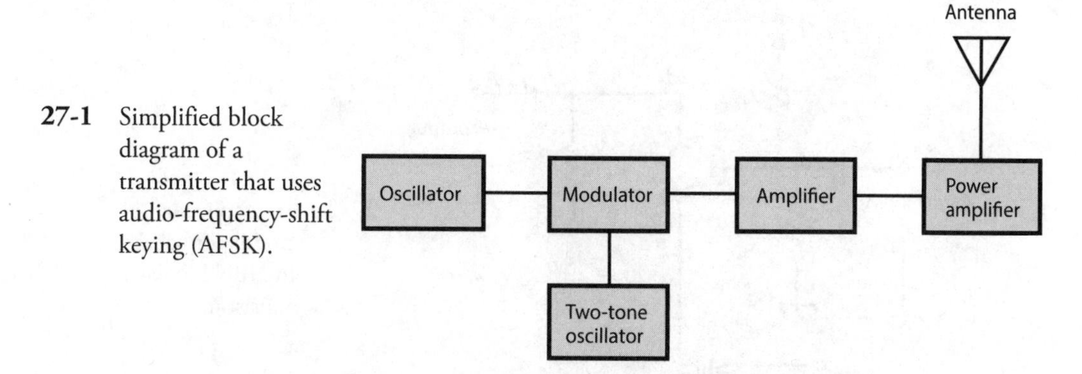

27-1 Simplified block diagram of a transmitter that uses audio-frequency-shift keying (AFSK).

modulates the carrier, a mode known as *audio-frequency-shift keying* (AFSK). The two most common codes used with FSK and AFSK are *Baudot* (pronounced "baw-DOE") and *ASCII* (pronounced "ASK-ee"). The acronym ASCII stands for *American Standard Code for Information Interchange.*

In *radioteletype* (RTTY), FSK, and AFSK systems, a *terminal unit* (TU) converts the digital signals into electrical impulses that operate a teleprinter or display characters on a computer screen. The TU also generates the signals necessary to send RTTY when an operator types on a keyboard. A device that sends and receives AFSK is sometimes called a *modem*, an acronym that stands for *modulator/demodulator*. A modem is basically the same as a TU. Figure 27-1 is a block diagram of an AFSK transmitter.

The main reason why FSK or AFSK work better than on/off keying is the fact that the space signals are identified as such, rather than existing as mere gaps in the data. A sudden noise burst in an on/off keyed signal can "confuse" a receiver into falsely reading a space as a mark, but when the space is positively represented by its own signal, this type of error happens less often at any given data speed.

Amplitude Modulation

An AF voice signal has frequencies mostly in the range between 300 Hz and 3 kHz. We can modulate some characteristic of an RF carrier with an AF voice waveform, thereby transmitting the voice information over the airwaves. Figure 27-2 shows a simple circuit for obtaining *amplitude modulation* (AM). We can imagine this circuit as an RF amplifier for the carrier, with the instantaneous gain dependent on the instantaneous audio input amplitude. We can also think of the circuit as a mixer that combines the RF carrier and the audio signals to produce sum and difference signals above and below the carrier frequency.

The circuit shown in Fig. 27-2 performs well as long as we don't let the AF input amplitude get too great. If we inject too much audio, we get *distortion* (nonlinearity) in the transistor resulting in degraded *intelligibility* (understandability), reduced *circuit efficiency* (ratio of DC power input to useful power output), and excessive output signal *bandwidth* (the difference between the highest and lowest component frequency). We can express the modulation extent as a percentage ranging from 0%, representing an *unmodulated carrier*, to 100%, representing the maximum possible modulation we can get without distortion. If we increase the modulation beyond 100%, we observe the same problems as we do when we apply excessive AF input to a modulator circuit, such as the one shown in Fig. 27-2. In an AM signal that's modulated at 100%, we find that 1/3 of the signal power conveys the data, while the carrier wave consumes the other 2/3 of the power.

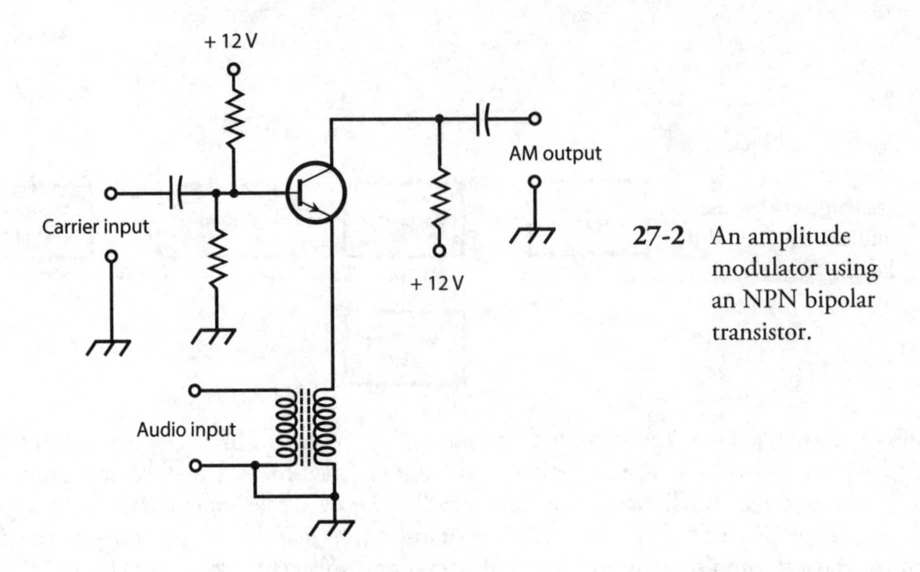

27-2 An amplitude modulator using an NPN bipolar transistor.

Figure 27-3 shows a *spectral display* of an AM voice radio signal. The horizontal scale is calibrated in increments of 1 kHz per division. Each vertical division represents 3 dB of change in signal strength. The maximum (reference) amplitude equals 0 dB relative to 1 mW (abbreviated as 0 dBm). The data exists in *sidebands* above and below the carrier frequency. These sidebands constitute sum and difference signals produced by mixing in the modulator circuit between the audio and the carrier. The RF energy between −3 kHz and the carrier frequency is called the *lower sideband* (LSB); the RF energy from the carrier frequency to +3 kHz is called the *upper sideband* (USB).

The signal bandwidth equals the difference between the maximum and minimum sideband frequencies. In an AM signal, the bandwidth equals twice the highest audio modulating frequency. In

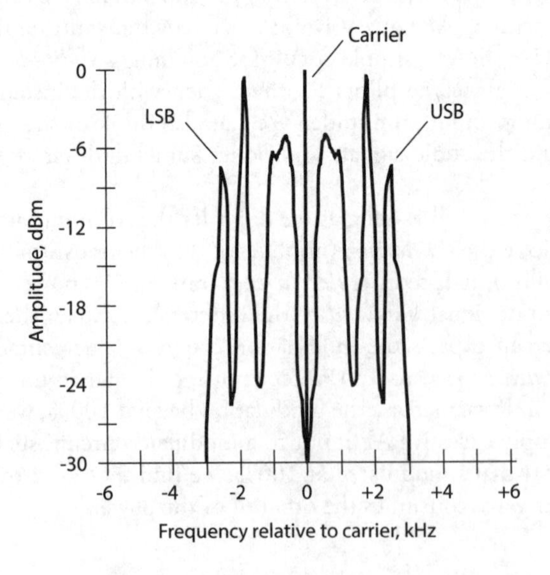

27-3 Spectral display of a typical amplitude-modulated (AM) voice communications signal.

the example of Fig. 27-3, all the AF voice energy exists at or below 3 kHz, so the signal bandwidth equals 6 kHz, typical of AM voice communications. In standard AM broadcasting in which music is transmitted along with voices, the AF energy is spread over a wider bandwidth, nominally 10 kHz to 20 kHz. The increased bandwidth provides for better *fidelity* (sound quality).

Single Sideband

In AM at 100% modulation, the carrier wave consumes ⅔ of the signal power, and the sidebands exist as mirror-image duplicates that, combined, employ only ⅓ of the signal power. These properties make AM inefficient and needlessly redundant.

If we could get rid of the carrier and one of the sidebands, we'd still convey all the information we want while consuming far less power. Alternatively, we could get a stronger signal for a given amount of RF power. We could also reduce the signal bandwidth to a little less than half that of an AM signal modulated with the same data. The resulting *spectrum savings* would allow us to fit more than twice as many signals into a specific range, or *band*, of frequencies. During the early twentieth century, communications engineers perfected a way to modify AM signals in this way. They called the resulting mode *single sideband* (SSB), a term which endures to this day.

When we remove the carrier and one of the sidebands from an AM signal, the remaining energy has a spectral display resembling Fig. 27-4. In this case, we eliminate the USB along with the carrier, leaving only the LSB. We could just as well remove the LSB along with the carrier, leaving only the USB.

Balanced Modulator

We can almost completely suppress the carrier in an AM signal using a *balanced modulator*—an amplitude modulator/amplifier using two transistors with the inputs connected in push-pull and the outputs connected in parallel, as shown in Fig. 27-5. This arrangement "cancels" the carrier wave in the output signals, leaving only LSB and USB energy. The balanced modulator produces a *double-sideband suppressed-carrier* (DSBSC) signal, often called simply *double sideband* (DSB). One of the sidebands can be suppressed in a subsequent circuit by a *bandpass filter* to obtain an SSB signal.

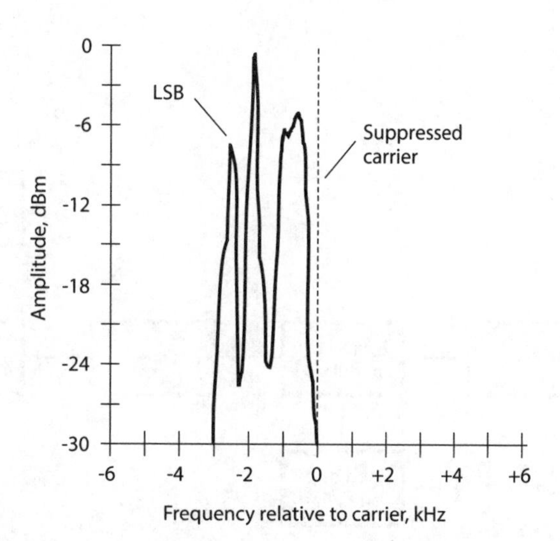

27-4 Spectral display of a typical single-sideband (SSB) voice communications signal, in this case lower sideband (LSB).

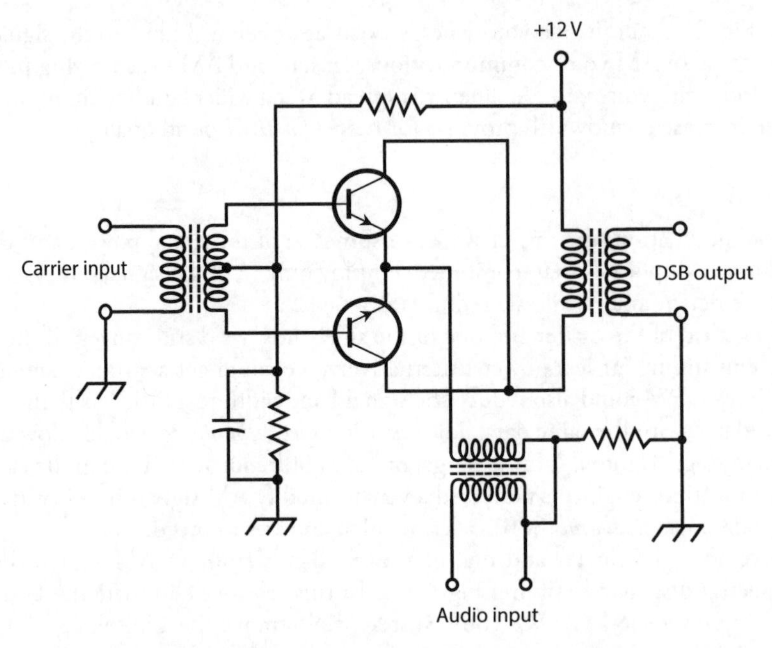

27-5 A balanced modulator using two NPN bipolar transistors. We connect the bases in push-pull and the collectors in parallel.

Basic SSB Transmitter

Figure 27-6 is a block diagram of a simple SSB transmitter. The RF amplifiers that follow any type of amplitude modulator, including a balanced modulator, must all operate in a linear manner to prevent distortion and unnecessary spreading of the signal bandwidth, a condition that some engineers and radio operators call *splatter*. These amplifiers generally work in class A except for the PA, which

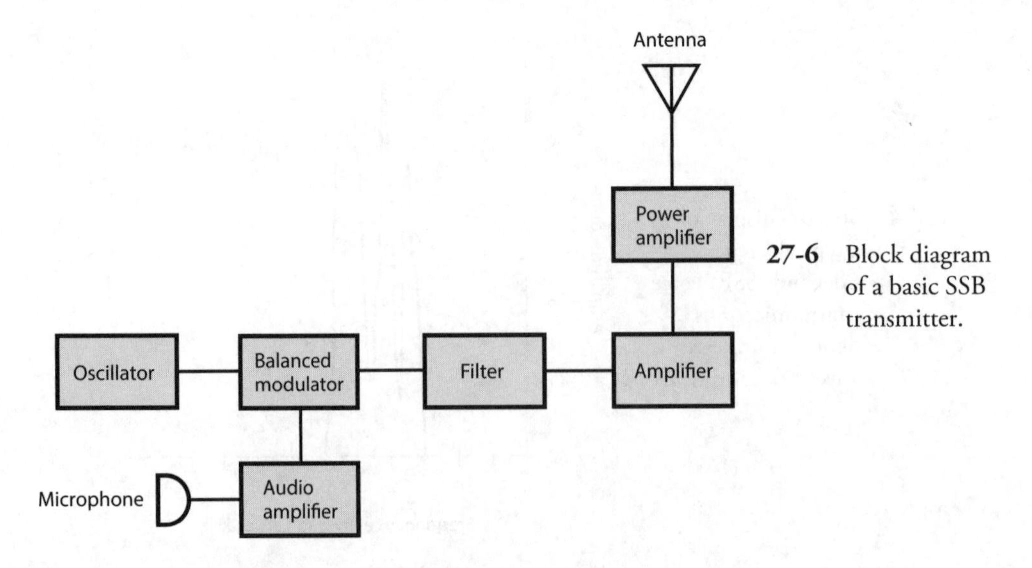

27-6 Block diagram of a basic SSB transmitter.

operates in class AB or class B. We'll never see a class-C amplifier as the PA in an SSB transmitter because class-C operation distorts any signal whose amplitude varies over a continuous range.

Frequency Modulation

In *frequency modulation* (FM), the instantaneous signal amplitude remains constant; the instantaneous frequency varies instead. We can get FM by applying an audio signal to the varactor diode in a voltage-controlled oscillator (VCO). Figure 27-7 shows an example of this scheme, known as *reactance modulation*. This circuit employs a Colpitts oscillator, but we could use any other type of oscillator and get similar results. The varying voltage across the varactor causes its capacitance to change in accordance with the audio waveform. The fluctuating capacitance causes variations in the resonant frequency of the inductance-capacitance (*LC*) tuned circuit, causing small fluctuations in the frequency generated by the oscillator.

Phase Modulation

We can indirectly obtain FM if we modulate the phase of the oscillator signal. When we vary the phase from instant to instant, we inevitably provoke small variations in the frequency as well. Any instantaneous phase change shows up as an instantaneous frequency change (and vice-versa). When we employ *phase modulation* (PM), we must process the audio signal before we apply it to the modulator, adjusting the *frequency response* of the audio amplifiers. Otherwise the signal will sound muffled when we listen to it in a receiver designed for ordinary FM.

Deviation for FM and PM

In an FM or PM signal, we can quantify the maximum extent to which the instantaneous carrier frequency differs from the unmodulated-carrier frequency in terms of a parameter called *deviation*. For most FM and PM voice transmitters, the deviation is standardized at ±5 kHz. We call this mode *narrowband FM* (NBFM). The bandwidth of an NBFM signal roughly equals that of an AM signal containing the same modulating information. In FM hi-fi music broadcasting, and in some other applications, the deviation exceeds ±5 kHz, giving us a mode called *wideband FM* (WBFM).

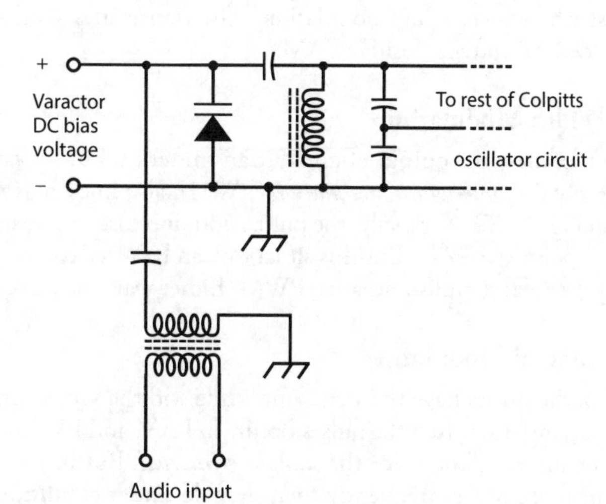

27-7 Generation of a frequency-modulated (FM) signal by employing reactance modulation in a Colpitts oscillator. We can modify other oscillator types in a similar way.

The deviation obtainable with FM is greater, for a given oscillator frequency, than the deviation that we get with PM. However, we can increase the deviation of any FM or PM signal with the help of a *frequency multiplier*. When the signal passes through a frequency multiplier, the deviation gets multiplied along with the carrier frequency. The deviation in the final output should equal the highest modulating audio frequency if we expect optimum audio fidelity. Therefore, ±5 kHz is more than enough deviation for voice communications. For music, a deviation of ±15 kHz or ±20 kHz is required for good reproduction.

Modulation Index for FM and PM

In any FM or PM signal, the ratio of the frequency deviation to the highest modulating audio frequency is called the *modulation index*. Ideally, this figure should be somewhere between 1:1 and 2:1. If it's less than 1:1, the signal sounds muffled or distorted, and efficiency is sacrificed. Increasing the modulation index much beyond 2:1 broadens the bandwidth without providing significant improvement in intelligibility or fidelity.

Power Amplification for FM and PM

A class-C PA can function in an FM or PM transmitter without causing distortion because the signal amplitude remains constant. Nonlinearity (characteristic of class-C operation) has no meaning, let alone any adverse effects, when the signal amplitude never changes! For this reason, we'll often find class-C PAs in FM and PM transmitters, especially those that have high output power. Remember: Class C offers the best efficiency of any PA mode!

Pulse-Amplitude Modulation

We can modulate a signal by varying some aspect of a constant stream of signal pulses. In *pulse-amplitude modulation* (PAM), the strength of each individual pulse varies according to the modulating waveform. In this respect, PAM resembles AM. Figure 27-8A shows an amplitude-versus-time graph of a hypothetical PAM signal. The modulating waveform appears as a dashed curve, and the pulses appear as vertical gray bars. Normally, the pulse amplitude increases as the instantaneous modulating-signal level increases (*positive PAM*). But this situation can be reversed, so higher audio levels cause the pulse amplitude to go down (*negative PAM*). Then the signal pulses are at their strongest when there is no modulation. The transmitter works harder to produce negative PAM than it does to produce positive PAM.

Pulse-Width Modulation

We can modulate the output of an RF transmitter output by varying the width (duration) of signal pulses to obtain *pulse-width modulation* (PWM), also known as *pulse duration modulation* (PDM) as shown in Fig. 27-8B. Normally, the pulse width increases as the instantaneous modulating-signal level increases (*positive PWM*). But this situation can be reversed (negative *PWM*). The transmitter must work harder to accomplish negative PWM. Either way, the peak pulse amplitude remains constant.

Pulse-Interval Modulation

Even if all the pulses have the same amplitude and the same duration, we can obtain pulse modulation by varying how often the pulses occur. In PAM and PWM, we always transmit the pulses at the same time interval, known as the *sampling interval*. But in *pulse interval-modulation* (PIM), pulses can occur more or less frequently than they do under conditions of no modulation. Figure 27-8C

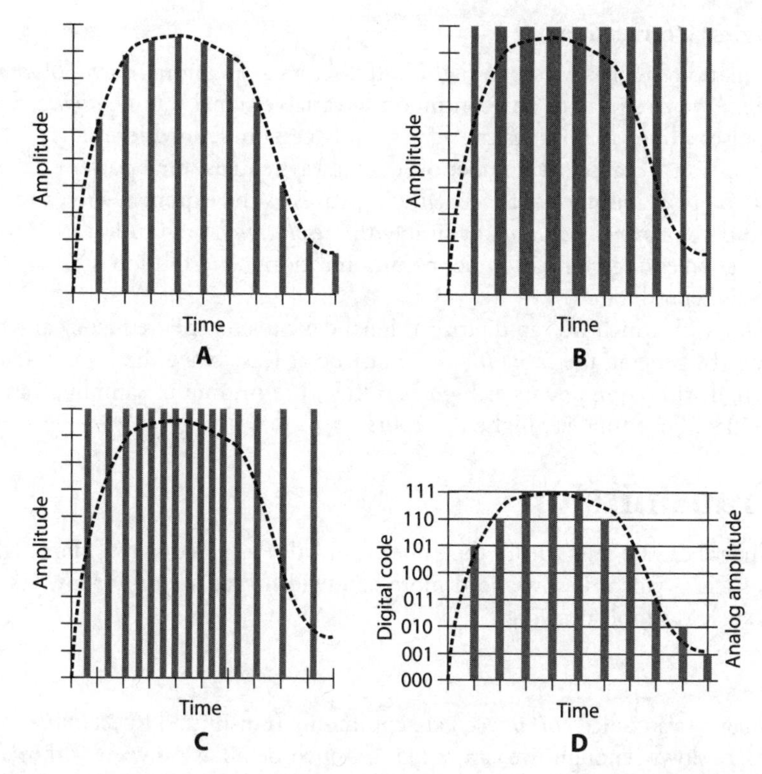

27-8 Time-domain graphs of various modes of pulse modulation. At A, pulse-amplitude modulation (PAM); at B, pulse-width modulation (PWM), also called pulse-duration modulation (PDM); at C, pulse-interval modulation (PIM); at D, pulse-code modulation (PCM).

shows a hypothetical PIM signal. Every pulse has the same amplitude and the same duration, but the time interval between them changes. When there is no modulation, the pulses emerge from the transmitter evenly spaced with respect to time. An increase in the instantaneous data amplitude might cause pulses to be sent more often, as is the case in Fig. 27-8C (*positive PIM*). Alternatively, an increase in instantaneous data level might slow down the rate at which the pulses emerge (*negative PIM*).

Pulse-Code Modulation

In *digital communications*, the modulating data attains only certain defined states, rather than continuously varying. Compared with old-fashioned *analog communications* (where the state is always continuously variable), digital modes offer improved *signal-to-noise* (S/N) *ratio*, narrower signal bandwidth, better accuracy, and superior reliability. In *pulse-code modulation* (PCM), any of the above-described aspects—amplitude, width, or interval—of a pulse sequence (or *pulse train*) can be varied. But rather than having infinitely many possible states, the number of states equals some power of 2, such as 2^2 (four states), 2^3 (eight states), 2^4 (16 states), 2^5 (32 states), 2^6 (64 states), and so on. As we increase the number of states, the fidelity and data transmission speed improve, but the signal gets more complicated. Figure 27-8D shows an example of eight-level PCM.

Analog-to-Digital Conversion

Pulse-code modulation, such as we see in Fig. 27-8D, serves as a common form of *analog-to-digital* (A/D) *conversion.* A voice signal, or any continuously variable signal, can be *digitized*, or converted into a train of pulses whose amplitudes can achieve only certain defined levels.

In A/D conversion, because the number of states always equals some power of 2, we can represent the signal as a binary-number code. Fidelity improves as the exponent increases. The number of states is called the *sampling resolution*, or simply the *resolution.* A resolution of $2^3 = 8$ (as shown in Fig. 27-8D) is good enough for basic voice communications. A resolution of $2^4 = 16$ can provide fairly decent music reproduction.

The efficiency with which we can digitize a signal depends on the frequency at which we carry out the sampling. In general, the *sampling rate* must be at least twice the highest data frequency. For an audio signal with components as high as 3 kHz, the minimum sampling rate for effective digitization is 6 kHz. For music it's higher, of course.

Image Transmission

Non-moving images can be sent within the same bandwidth as voice signals. For high-resolution, moving images such as video files, we need more bandwidth than we do for low-resolution, non-moving images such as simple drawings.

Facsimile

Non-moving images (also called *still images*) are commonly transmitted by *facsimile*, also called *fax*. If we send the data slowly enough, we can render as much detail as we want within a 3-kHz-wide band, the standard for voice communications. This flexibility explains why detailed fax images can be sent over a *plain old telephone service* (POTS) line.

In an electromechanical fax transmitter, a paper document is wrapped around a *drum.* The drum rotates at a slow, controlled rate. A spot of light scans laterally across the document. The drum moves the document so that a single line is scanned with each pass of the light spot. This process continues, line by line, until the device has scanned the complete *frame* (image). A *photodetector* picks up the light rays reflected from the paper. Dark portions of the image reflect less light than bright parts, so the current through the photodetector varies as the light beam passes over various regions. This current modulates a carrier in one of the modes described earlier, such as AM, FM, or SSB. Typically, the data for pure black corresponds to a 1.5-kHz audio sine wave, and the data for pure white corresponds to a 2.3-kHz audio sine wave. Gray shades produce audio sine waves having frequencies between these extremes.

A fax receiver duplicates the transmitter's scanning rate and pattern, and a display or printer reproduces the image in *grayscale* (shades of gray ranging from black to white, without color).

Slow-Scan Television

We can think of *slow-scan television* (SSTV) as a fast, repetitive form of fax. An SSTV signal, like a fax signal, propagates within a band of frequencies as narrow as that of a human voice. And, like fax, SSTV transmission reproduces only still images, not moving ones. However, an SSTV system scans and transmits a full image in much less time than a fax machine does. The time required to send a complete frame (image or scene) is only eight seconds, rather than a minute or more. Therefore, we can send multiple images in a reasonable amount of time, giving our viewers some sense of motion in a scene. This speed bonus comes with a tradeoff: in SSTV, we get lower *resolution*, meaning less image detail, than we get with fax.

We can program a personal computer so that its monitor will act as an SSTV display. We can also find converters that allow us to look at SSTV signals on a conventional TV set.

An SSTV frame contains 120 lines. The black and white frequencies are the same as those for fax transmission; the darkest parts of the picture are sent at 1.5 kHz and the brightest parts are sent at 2.3 kHz. *Synchronization (sync) pulses,* which keep the receiving apparatus in step with the transmitter, are sent at 1.2 kHz. A *vertical sync pulse* tells the receiver that it's time to begin a new frame; this pulse lasts for 30 milliseconds (ms). A *horizontal sync pulse* tells the receiver when it's time to start a new line in a frame; its duration equals 5 ms. These pulses prevent *rolling* (haphazard vertical image motion) or *tearing* (lack of horizontal synchronization).

Fast-Scan Television

Old-fashioned analog television is also known as *fast-scan TV* (FSTV). While broadcasters no longer use this mode, some radio amateurs still do. The frames are transmitted at the rate of 30 per second. There are 525 lines per frame. The quick frame time, and the increased resolution, of FSTV make it necessary to use a much wider frequency band than is the case with fax or SSTV. A typical video FSTV signal takes up 6 MHz of spectrum space, or 2000 times the bandwidth of a fax or SSTV signal. Fast-scan TV is usually sent using AM or wideband FM. With AM, one of the sidebands can be filtered out, leaving only the carrier and the other sideband. Engineers call this mode *vestigial sideband* (VSB) transmission. It cuts the bandwidth of an FSTV signal down to about 3 MHz.

Because of the large amount of spectrum space needed to send FSTV, this mode isn't practical for use at frequencies below about 30 MHz. Even at 30 MHz, a vestigial sideband signal consumes 10 percent of the entire RF spectrum below its own frequency! In the "olden days of TV," all commercial FSTV transmission was carried out above 50 MHz, with the great majority of channels having frequencies far higher than this. Channels 2 through 13 on your TV receiver are sometimes called the *very high frequency* (VHF) *channels*; the higher channels are called the *ultra high frequency* (UHF) *channels*.

Figure 27-9 portrays a time-domain graph of the waveform of a single line in an FSTV video signal. This graph shows us 1/525 of a complete frame. The highest instantaneous signal amplitude corresponds to the blackest shade, and the lowest amplitude corresponds to the lightest shade.

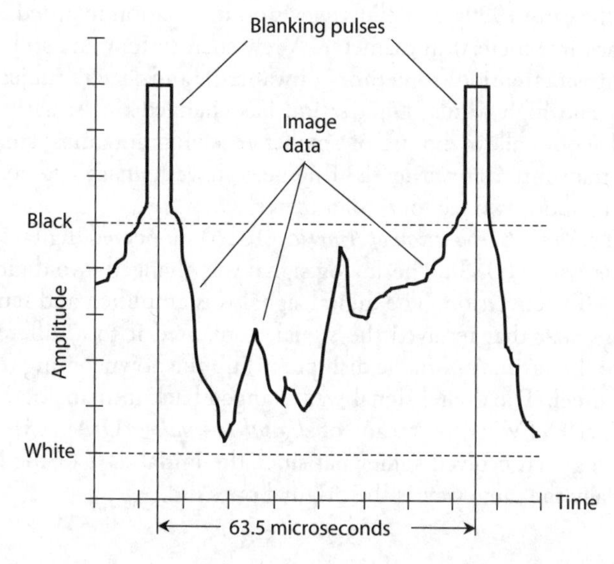

27-9 Time-domain graph of a single line in a conventional fast-scan television (FSTV) video frame.

Therefore, the FSTV signal is sent "negatively." This convention allows *retracing* (moving from the end of one line to the beginning of the next) to be synchronized between the transmitter and the receiver. A well-defined, strong *blanking pulse* tells the receiver when to retrace, and it also shuts off the beam while the receiver display is retracing. If you watched much commercial analog TV using rooftop antennas or "rabbit ears," did you notice that weak TV signals had poor *contrast*? Weakened blanking pulses result in incomplete retrace blanking. But this little problem is better than having the TV receiver completely lose track of when it should retrace, as might happen if the highest instantaneous signal amplitude went with the lightest image shade.

Color FSTV works by sending three separate monochromatic signals, corresponding to the *primary colors* red, blue, and green. The signals are black-and-red, black-and-blue, and black-and-green. The receiver recombines these signals and displays the resulting video as a fine, interwoven matrix of red, blue, and green dots. When viewed from a distance, the dots are too small to be individually discernible. Various combinations of red, blue, and green intensities can yield any color that the human eye can perceive.

High-Definition Television

The term *high-definition television* (HDTV) refers to any of several methods for getting more detail into a TV picture than could ever be done with FSTV. The HDTV mode also offers superior sound quality, making for a more satisfying home TV and home theater experience.

A standard FSTV picture has 525 lines per frame, but HDTV systems have between 787 and 1125 lines per frame. The image is scanned about 60 times per second. High-definition TV is usually sent in a digital mode; this offers another advantage over conventional FSTV. Digital signals propagate better, are easier to deal with when they are weak, and can be processed in ways that analog signals defy.

Some HDTV systems take advantage of a technique called *interlacing* that "meshes" two *rasters* (complete image frames) together. Interlacing effectively doubles the image resolution without doubling the cost of the hardware. However, interlaced images can exhibit annoying *jitter* in fast-moving or fast-changing scenes.

Digital Satellite TV

Until the early 1990s, a satellite television installation required a dish antenna roughly six to 10 feet (two or three meters) in diameter. A few such systems are still in use. The antennas are expensive, they attract attention (sometimes unwanted), and they're subject to damage from ice storms, heavy snows, and high winds. Digitization has changed this situation. In any communications system, digital modes allow the use of smaller receiving antennas, smaller transmitting antennas, and/or lower transmitter power levels. Engineers have managed to get the diameter of the receiving dish down to about two feet or ⅔ of a meter.

The *Radio Corporation of America* (RCA) pioneered digital satellite TV with its so-called *Digital Satellite System* (DSS). The analog signal was changed into digital pulses at the transmitting station using A/D conversion. The digital signal was amplified and sent up to a satellite. The satellite had a *transponder* that received the signal, converted it to a different frequency, and retransmitted it back to the earth. A portable dish picked up the downcoming (or *downlink*) signal. A *tuner* selected the channel. The digital signal was changed back into analog form, suitable for viewing on a conventional FSTV set, by means of *digital-to-analog* (D/A) *conversion*. Although digital satellite TV technology has evolved somewhat since the initial days of the RCA DSS, today's systems work in essentially the same way as the original ones did.

The Electromagnetic Field

In a radio or television transmitting antenna, electrons constantly move back and forth. Their velocity constantly changes as they speed up in one direction, slow down, reverse direction, speed up again, and so on. Any change of velocity (speed and/or direction) constitutes *acceleration.* When charged particles accelerate in a certain way, they produce an *electromagnetic* (EM) *field.*

How It Happens

When electrons move, they generate a magnetic (M) field. When electrons accelerate, they generate a *changing* M field. When electrons accelerate back and forth, they generate an *alternating* M field at the same frequency as that of the electron motion.

An alternating M field gives rise to an alternating electric (E) field, which, in turn, spawns another alternating M field. This process repeats indefinitely in the form of an EM field that *propagates* (travels) through space at the speed of light. The E and M fields expand alternately outward from the source in spherical wavefronts. At any given point in space, the lines of E flux run perpendicular to the lines of M flux. The waves propagate in a direction perpendicular to both the E and M flux lines.

Frequency versus Wavelength

All EM fields have two important properties: the *frequency* and the *wavelength.* When we quantify them, we find that they exhibit an *inverse relation:* as one increases, the other decreases. We've already learned about AC frequency. We can express EM wavelength as the physical distance between any two adjacent points at which either the E field or the M field has identical amplitude and direction.

An EM field can have any conceivable frequency, ranging from centuries per cycle to quadrillions of cycles per second (or hertz). The sun has a magnetic field that oscillates with a 22-year cycle. Radio waves oscillate at thousands, millions, or billions of hertz. Infrared (IR), visible light, ultraviolet (UV), X rays, and gamma rays comprise EM fields that alternate at many trillions (million millions) of hertz. The wavelength of an EM field can likewise vary over the widest imaginable range, from many trillions of miles to a tiny fraction of a millimeter.

Let f_{MHz} represent the frequency of an EM wave in megahertz as it travels through free space. (Technically, free space constitutes a vacuum, but we can consider the air at the earth's surface equivalent to free space for most "real-world" applications.) Let L_{ft} represent the wavelength of the same wave in feet. Then

$$L_{\text{ft}} = 984/f_{\text{MHz}}$$

If we want to express the wavelength L_{m} in meters, then

$$L_{\text{m}} = 300/f_{\text{MHz}}$$

The inverses of these formulas are

$$f_{\text{MHz}} = 984/L_{\text{ft}}$$

and

$$f_{\text{MHz}} = 300/L_{\text{m}}$$

Velocity Factor

In media other than free space, EM fields propagate at less than the speed of light. As a result, the wavelength grows shorter according to a quantity called the *velocity factor*, symbolized v. The value of v can range from 0 (representing no movement at all) to 1 (representing the speed of propagation in free space, which equals approximately 186,000 mi/s or 300,000 km/s). We can also express the velocity factor as a percentage $v_{\%}$. In that case, the smallest possible value is 0%, and the largest is 100%. The velocity factor in practical situations rarely falls below 0.50 or 50%, and it usually exceeds 0.60 or 60%.

Velocity factor constitutes a crucial parameter in the design of RF transmission lines and antenna systems, when sections of cable, wire, or metal tubing must be cut to specific lengths measured in wavelengths or fractions of a wavelength. Taking the velocity factor v, expressed as a ratio, into account, we can modify the above-mentioned four formulas as follows:

$$L_{\text{ft}} = 984v/f_{\text{MHz}}$$
$$L_{\text{m}} = 300v/f_{\text{MHz}}$$
$$f_{\text{MHz}} = 984v/L_{\text{ft}}$$
$$f_{\text{MHz}} = 300v/L_{\text{m}}$$

The EM and RF Spectra

Physicists, astronomers, and engineers refer to the entire range of EM wavelengths as the *electromagnetic* (EM) *spectrum*. Scientists use logarithmic scales to depict the EM spectrum according to the wavelength in meters, as shown in Fig. 27-10A. The radio-frequency (RF) *spectrum*, which includes radio, television, and microwaves, appears expanded in Fig. 27-10B, where we label the axis according to frequency. The RF spectrum is categorized in *bands* from *very low frequency* (VLF) through *extremely high frequency* (EHF), according to the breakdown in Table 27-1. The exact lower limit of the VLF range is a matter of disagreement in the literature. Here, we define it as 3 kHz.

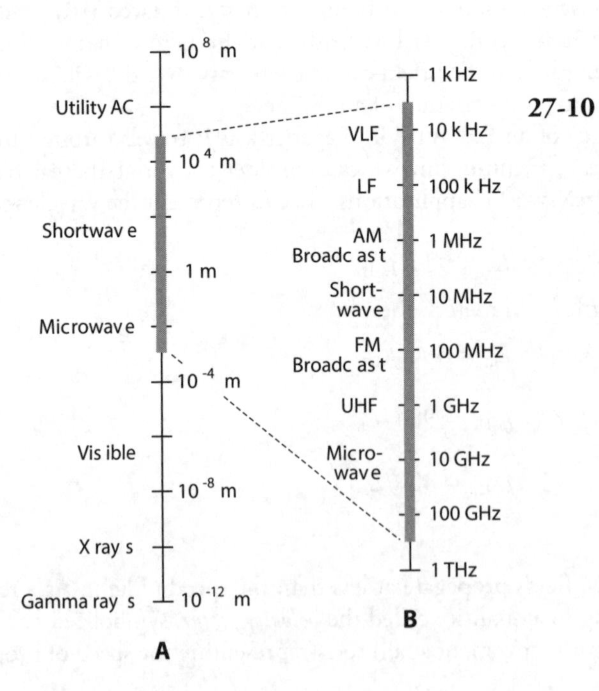

27-10 At A, the electromagnetic (EM) spectrum from 10^8 m to 10^{-12} m. Each vertical division represents two orders of magnitude (a 100-fold increase or decrease in the wavelength). At B, the radio-frequency (RF) portion of the EM spectrum, with each vertical division representing one order of magnitude (a 10-fold increase or decrease in the wavelength).

Table 27-1. Bands in the RF Spectrum.

Frequency Designation	Frequency Range	Wavelength Range
Very Low (VLF)	3 kHz–30 kHz	100 km–10 km
Low (LF)	30 kHz–300 kHz	10 km–1 km
Medium (MF)	300 kHz–3 MHz	1 km–100 m
High (HF)	3 MHz–30 MHz	100 m–10 m
Very High (VHF)	30 MHz–300 MHz	10 m–1 m
Ultra High (UHF)	300 MHz–3 GHz	1 m–100 mm
Super High (SHF)	3 GHz–30 GHz	100 mm–10 mm
Extremely High (EHF)	30 GHz–300 GHz	10 mm–1 mm

Wave Propagation

Radio-wave propagation has fascinated scientists ever since Marconi and Tesla discovered, around the year 1900, that EM fields can travel over long distances without any supporting infrastructure. Let's examine some wave-propagation behaviors that affect wireless communications at radio frequencies.

Polarization

We can define the orientation of E-field *lines of flux* as the *polarization* of an EM wave. If the E-field flux lines run parallel to the earth's surface, we have *horizontal polarization.* If the E-field flux lines run perpendicular to the surface, we have *vertical polarization.* Polarization can also have a "slant," of course.

In some situations, the E-field flux lines rotate as the wave travels through space. In that case we have *circular polarization* if the E-field intensity remains constant. If the E-field intensity is more intense in some planes than in others, we have *elliptical polarization.* A circularly or elliptically polarized wave can rotate either *clockwise* or *counterclockwise* as we watch the wavefronts come toward us. The rotational direction is called the *sense* of polarization. Some engineers use the term *right-hand* instead of clockwise and the term *left-hand* instead of counterclockwise.

Line-of-Sight Wave

Electromagnetic waves follow straight lines unless something makes them bend. *Line-of-sight* propagation can often take place when the receiving antenna can't be seen visually from the transmitting antenna because radio waves penetrate nonconducting opaque objects, such as trees and frame houses, to some extent. The line-of-sight wave consists of two components called the *direct wave* and the *reflected wave*, as follows:

1. In the direct wave, the longest wavelengths are least affected by obstructions. At very low, low and medium frequencies, direct waves can *diffract* around things. As the frequency rises, especially above about 3 MHz, obstructions have a greater and greater blocking effect on direct waves.
2. In the reflected wave, the EM energy reflects from the earth's surface and from conducting objects like wires and steel beams. The reflected wave always travels farther than the direct wave. The two waves might arrive at the receiving antenna in perfect phase coincidence, but usually they don't.

If the direct and reflected waves arrive at the receiving antenna with equal strength but 180° out of phase, we observe a *dead spot*. The same effect occurs if the two waves arrive inverted in phase with respect to each other (that is, in phase opposition). The dead-spot phenomenon is most noticeable at the highest frequencies. At VHF and UHF, an improvement in reception can sometimes result from moving the transmitting or receiving antenna only a few inches or centimeters! In mobile operation, when the transmitter and/or receiver are moving, multiple dead spots produce rapid, repeated interruptions in the received signal, a phenomenon called *picket fencing*.

Surface Wave

At frequencies below about 10 MHz, the earth's surface conducts AC quite well, so vertically polarized EM waves can follow the surface for hundreds or thousands of miles, with the earth helping to transmit the signals. As we reduce the frequency and increase the wavelength, we observe decreasing *ground loss*, and the waves can travel progressively greater distances by means of *surface-wave propagation*. Horizontally polarized waves don't travel well in this mode because the conductive surface of the earth "shorts out" horizontal E flux. At frequencies above about 10 MHz (corresponding to wavelengths shorter than roughly 30 m), the earth becomes lossy, and surface-wave propagation rarely occurs for distances greater than a few miles.

Sky-Wave EM Propagation

Ionization in the upper atmosphere, caused by solar radiation, can return EM waves to the earth at certain frequencies. The so-called *ionosphere* has several dense zones of ionization that occur at fairly constant, predictable altitudes.

The *E layer*, which lies about 50 mi (roughly 80 km) above the surface, exists mainly during the day, although nighttime ionization is sometimes observed. The E layer can provide medium-range radio communication at certain frequencies.

At higher altitudes, we find the *F_1 layer* and the *F_2 layer*. The F_1 layer, normally present only on the daylight side of the earth, forms at about 125 mi (roughly 200 km) altitude; the F_2 layer exists at about 180 mi (roughly 300 km) over most, or all, of the earth, the dark side as well as the light side. Sometimes the distinction between the F_1 and F_2 layers is ignored, and they are spoken of together as the *F layer*. Communication by means of F-layer propagation can usually be accomplished between any two points on the earth at some frequencies between 5 MHz and 30 MHz.

The lowest ionized region is called the *D layer*. It exists at an altitude of about 30 mi (roughly 50 km), and is ordinarily present only on the daylight side of the planet. This layer absorbs radio waves at some frequencies, impeding long-distance ionospheric propagation.

Tropospheric Propagation

At frequencies above about 30 MHz (wavelengths shorter than about 10 m), the lower atmosphere bends radio waves towards the surface. *Tropospheric bending* occurs because the *index of refraction* of air, with respect to EM waves, decreases with altitude. Tropospheric bending makes it possible to communicate for hundreds of miles, even when the ionosphere will not return waves to the earth.

Ducting is tropospheric propagation that occurs somewhat less often than bending, but offers more dramatic effects. Ducting takes place when EM waves get "trapped" within a layer of cool, dense air sandwiched between two layers of warmer air. Like bending, ducting occurs almost entirely at frequencies above 30 MHz.

Still another tropospheric-propagation mode is called *tropospheric scatter*, or *troposcatter*. This phenomenon takes place because air molecules, dust grains, and water droplets scatter some of the

EM field. We observe troposcatter most commonly at VHF and UHF. Troposcatter always occurs to some extent, regardless of weather conditions.

Tropospheric propagation in general, without mention of the specific mode, is sometimes called *tropo*.

Auroral Propagation

In the presence of unusual solar activity, the *aurora* (northern lights or southern lights) can return radio waves to the earth, facilitating *auroral propagation*. The aurora occurs at altitudes of about 40 to 250 mi (roughly 65 to 400 km). Theoretically, auroral propagation is possible, when the aurora are active, between any two points on the earth's surface from which the same part of the aurora lie on a line of sight. Auroral propagation seldom occurs when either the transmitting station or the receiving station is located at a latitude less than 35° north or south of the equator.

Auroral propagation causes rapid and deep signal fading, which nearly always renders analog voice and video signals unintelligible. Digital modes work somewhat better, but the carrier frequency gets "spread out" or "smeared" over a band several hundred hertz wide as a result of phase modulation induced by auroral motion. This "spectral spreading" limits the maximum data transfer rate. Auroral propagation commonly takes place along with poor ionospheric propagation resulting from sudden eruptions called *solar flares* on the sun's surface.

Meteor-Scatter Propagation

Meteors produce ionized trails that persist for a fraction of a second up to several seconds. The exact duration of the trail depends on the size of the meteor, its speed, and the angle at which it enters the atmosphere. A single meteor trail rarely lasts long enough to allow transmission of much data. However, during a *meteor shower*, multiple trails can produce almost continuous ionization for a period of hours. Ionized regions of this type can reflect radio waves at certain frequencies. Communications engineers call this effect *meteor-scatter propagation*, or sometimes simply *meteor scatter*. It can take place at frequencies far above 30 MHz and over distances ranging from just beyond the horizon up to about 1500 mi (roughly 2400 km). The maximum communications range depends on the altitude of the ionized trail, and also on the relative positions of the trail, the transmitting station, and the receiving station.

Moonbounce Propagation

Earth-moon-earth (EME) communications, also called *moonbounce*, is routinely carried on by amateur radio operators at VHF and UHF. This mode requires a sensitive receiver using a low-noise preamplifier, a large, directional antenna, and a high-power transmitter. Digital modes work far better than analog modes for moonbounce.

Signal *path loss* presents the main difficulty for anyone who contemplates EME communications. Received EME signals are always weak. High-gain directional antennas must remain constantly aimed at the moon, a requirement that dictates the use of steerable antenna arrays. The EME *path loss* increases with increasing frequency, but this effect is offset by the more manageable size of high-gain antennas as the wavelength decreases.

Solar noise can pose a problem; EME communications becomes most difficult near the time of the new moon, when the moon lies near a line between the earth and the sun. The sun constitutes a massive broadband generator of EM energy! Problems can also occur with *cosmic noise* when the moon passes near "noisy" regions in the so-called *radio sky*. The constellation *Sagittarius* lies in the direction of the center of the Milky Way galaxy, and EME performance suffers when the moon passes in front of that part of the stellar background.

The moon keeps the same face more or less toward the earth at all times, but some back-and-forth "wobbling" occurs. This motion, called *libration* (not "liberation" or "libation"!), produces rapid, deep fluctuations in signal strength, a phenomenon known as *libration fading*. The fading becomes more pronounced as the operating frequency increases. It occurs as multiple transmitted EM wavefronts reflect from various "lunagraphical" features, such as craters and mountains on the moon's surface, whose relative distances constantly change because of libration. The reflected waves recombine in constantly shifting phase at the receiving antenna, sometimes reinforcing, and at other times canceling.

Transmission Media

Data can be transmitted over various *media*, which include *cable*, *radio* (also called *wireless*), *satellite links* (a specialized form of wireless), and *fiberoptics*. Cable, radio/TV, and satellite communications use the RF spectrum. Fiberoptics uses IR or visible light energy.

Cable

The earliest cables comprised plain wires that carried DC. Nowadays, data-transmission cables more often carry RF signals that can be amplified at intervals on a long span. The use of such amplifiers, called *repeaters*, greatly increases the distances over which data can be sent by cable. Another advantage of using RF is the fact that numerous signals can travel over a single cable, with each signal on a different frequency.

Cables can consist of pairs of wires, somewhat akin to lamp cords. But more often coaxial cable, of the type described and illustrated at the end of Chap. 10, is used. This type of cable has a center conductor that carries the signals, surrounded by a cylindrical, grounded shield that keeps signals confined to the cable, and also keeps external EM fields from interfering with the signals.

Radio

All radio and TV signals consist of EM waves traveling through the earth's atmosphere or outer space. In a radio transmitting station, the RF output goes into an *antenna system* located at some distance from the transmitter. To get from the transmitter's final amplifier to the antenna, the EM energy follows a *transmission line*, also called a *feed line*.

Most radio antenna transmission lines consist of coaxial cable. Other types of cable exist for special applications. At microwave frequencies, hollow tubes called *waveguides* can transfer the energy. A waveguide works more efficiently than coaxial cable at the shortest radio wavelengths.

Radio amateurs sometimes use a *parallel-wire* transmission line in which the RF currents in the two conductors are in phase opposition so their EM fields cancel each other out. This phase cancellation keeps the transmission line from radiating, guiding the EM field along toward the antenna.

Satellite Systems

At very high frequencies (VHF) and above, some communications circuits use satellites that follow *geostationary orbits* around the earth. If a satellite orbits directly over the equator at an altitude of approximately 35,800 km (22,200 mi) and travels from west to east, it follows the earth's rotation, thereby staying in the same spot in the sky as seen from the surface. That's why we call it a *geostationary satellite*.

A single geostationary satellite lies on a line of sight with a large set of locations that covers about 40% of the earth's surface. Three such satellites, placed at 120° (⅓-circle) intervals around the earth, allow coverage of all human-developed regions. Only the extreme polar regions lie "out of range." We can aim a *dish antenna* at a geostationary satellite, and once we've fixed the antenna in the correct position, we can leave it alone.

Another form of satellite system uses multiple "birds" in relatively low-altitude orbits that take them over, or nearly over, the earth's poles. These satellites exhibit continuous, rapid motion with respect to the earth's surface. If enough satellites of this type exist, the entire "flock" can work together, maintaining reliable communications between any two points on the surface at all times. Directional antennas aren't necessary in these systems, which engineers call *low earth orbit* (LEO) networks.

Fiberoptics

We can modulate beams of IR or visible light, just as we can modulate RF carriers. An IR or visible light beam has a frequency far higher than that of any RF signal, allowing modulation by data at rates faster than anything possible with radio.

Fiberoptic technology offers several advantages over wire cables (which are sometimes called *copper* because the conductors usually comprise that metallic element). A fiberoptic cable doesn't cost much, doesn't weigh much, and remains immune to interference from outside EM fields. A fiberoptic cable doesn't corrode as metallic wires do. Fiberoptic cables are inexpensive to maintain and easy to repair. An optical fiber can carry far more signals than a cable because the frequency bands are far wider in terms of megahertz or gigahertz.

In theory we can "imprint" the entire RF spectrum, from VLF through EHF, onto a single beam of visible light and transmit it through an optical fiber no thicker than a strand of human hair!

Receiver Fundamentals

A wireless receiver converts EM waves into the original messages sent by a distant transmitter. Let's define a few important criteria for receiver operation, and then we'll look at two common receiver designs.

Specifications

The *specifications* of a receiver quantify how well the hardware can actually do what we design and build it to do.

Sensitivity: The most common way to express receiver sensitivity is to state the number of microvolts that must exist at the antenna terminals to produce a certain *signal-to-noise ratio* (S/N) or *signal-plus-noise-to-noise ratio* (S+N/N) in decibels (dB). The sensitivity depends on the gain of the *front end* (the amplifier or amplifiers connected to the antenna). The amount of noise that the front end generates also matters because subsequent stages amplify its noise output as well as its signal output.

Selectivity: The *passband*, or bandwidth that the receiver can "hear," is established by a wideband *preselector* in the early RF amplification stages, and is honed to precision by narrowband filters in later amplifier stages. The preselector makes the receiver most sensitive within about plus-or-minus 10 percent (±10%) of the desired signal frequency. The narrowband filter responds only to the frequency or channel of a specific signal that we want to hear; the filter rejects signals in nearby channels.

Dynamic range: The signals at a receiver input can vary over several orders of magnitude (powers of 10) in terms of absolute voltage. We define dynamic range as the ability of a receiver to maintain a fairly constant output, and yet to keep its rated sensitivity, in the presence of signals ranging from extremely weak to extremely strong. A good receiver exhibits dynamic range in excess of 100 dB. Engineers can conduct experiments to determine the dynamic range of any receiver; commercial receiver manufacturers publish this specification as a selling point.

Noise figure: The less internal noise a receiver produces, in general, the better the S/N ratio will be. We can expect an excellent S/N ratio in the presence of weak signals only when our receiver has a low noise figure, a measure of internally generated circuit noise. The noise figure matters most at VHF, UHF, and microwave frequencies. Gallium-arsenide field-effect transistors (GaAsFETs) are known for the low levels of noise they generate, even at very high frequencies. We can get away with other types of FETs at lower frequencies. Bipolar transistors, which carry higher currents than FETs, generate more circuit noise than FETs do.

Direct-Conversion Receiver

A *direct-conversion receiver* derives its output by mixing incoming signals with the output of a tunable (variable frequency) *local oscillator* (LO). The received signal goes into a mixer along with the output of the LO. Figure 27-11 is a block diagram of a direct-conversion receiver.

For reception of on/off keyed Morse code, also called *radiotelegraphy* or *continuous-wave* (CW) mode, the LO, also called a *beat-frequency oscillator* (BFO), is set a few hundred hertz above or below the signal frequency. We can also use this scheme to receive FSK signals. The audio output has a frequency equal to the difference between the LO frequency and the incoming carrier frequency. For reception of AM or SSB signals, we adjust the LO to precisely the same frequency as that of the signal carrier, a condition called *zero beat* because the *beat frequency*, or difference frequency, between the LO and the signal carrier equals zero.

A direct-conversion receiver provides rather poor *selectivity*, meaning that it can't always separate incoming signals when they lie close together in frequency. In a direct-conversion receiver, we can hear signals on either side of the LO frequency at the same time. A *selective filter* can theoretically eliminate this problem. Such a filter must be designed for a fixed frequency if we expect it to work well. However, in a direct-conversion receiver, the RF amplifier must operate over a wide range of frequencies, making effective filter design an extreme challenge.

Superheterodyne Receiver

A *superheterodyne receiver*, also called a *superhet*, uses one or more local oscillators and mixers to obtain a constant-frequency signal. We can more easily filter a fixed-frequency signal than we can filter a signal that changes in frequency (as it does in a direct-conversion receiver).

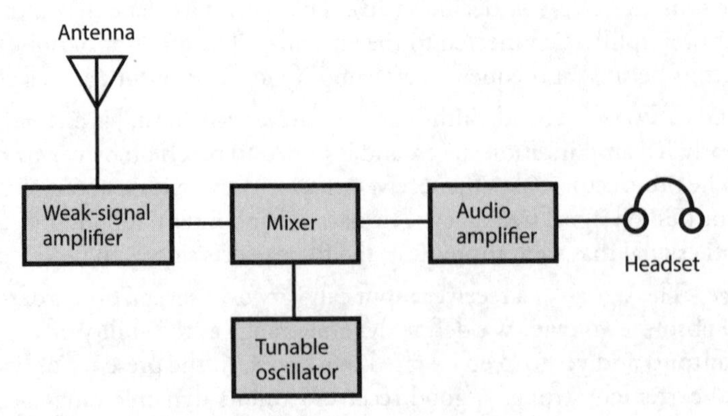

27-11 Block diagram of a direct-conversion receiver.

In a superhet, the incoming signal goes from the antenna through a tunable, sensitive front end, which is a precision weak-signal amplifier. The output of the front end mixes (heterodynes) with the signal from a tunable, unmodulated LO. We can choose the sum signal or the difference signal for subsequent amplification. We call this signal the *first intermediate frequency* (IF), which can be filtered to obtain selectivity.

If the first IF signal passes straight into the detector, we call our system a *single-conversion receiver*. Some receivers use a second mixer and second LO, converting the first IF to a lower-frequency *second IF*. Then we have a *double-conversion receiver*. The IF bandpass filter can be constructed for use on a fixed frequency, allowing superior selectivity and facilitating adjustable bandwidth. The sensitivity is enhanced because fixed IF amplifiers are easy to keep in tune.

Unfortunately, even the best superheterodyne receiver can intercept or generate unwanted signals. We call external false signals *images*; we call internally generated false signals *birdies*. If we carefully choose the LO frequency (or frequencies) when we design our system, images and birdies will rarely cause problems during ordinary operation.

Stages of a Single-Conversion Superhet

Figure 27-12 shows a block diagram of a generic single-conversion superheterodyne receiver. Individual receiver designs vary somewhat, but we can consider this example representative. The various stages break down as follows:

- The front end consists of the first RF amplifier, and often includes *LC* bandpass filters between the amplifier and the antenna. The dynamic range and sensitivity of a receiver are determined by the performance of the front end.
- The mixer stage, in conjunction with the tunable local oscillator (LO), converts the variable signal frequency to a constant IF. The output occurs at either the sum or the difference of the signal frequency and the tunable LO frequency.

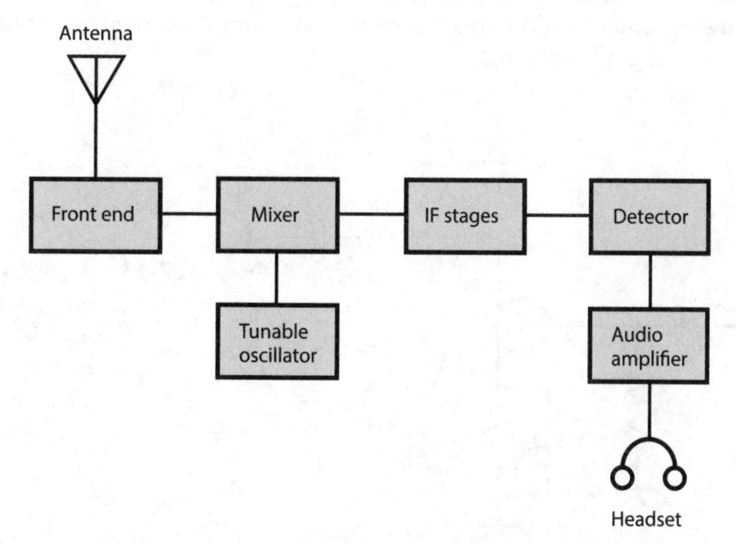

27-12 Block diagram of a single-conversion superheterodyne receiver.

- The IF stages produce most of the gain. We also get most of the selectivity here, filtering out unwanted signals and noise, while allowing the desired signal to pass.
- The detector extracts the information from the signal. Common circuits include the envelope detector for AM, the product detector for SSB, FSK, and CW, and the ratio detector for FM.
- One or two stages of audio amplification boost the demodulated signal to a level suitable for a speaker or headset. Alternatively, we can feed the signal to a printer, facsimile machine, or computer.

Predetector Stages

When we design and build a superheterodyne receiver, we must ensure that the stages preceding the first mixer provide reasonable gain but generate minimal internal noise. They must also be capable of handling strong signals without *desensitization* (losing gain), a phenomenon also known as *overloading*.

Preamplifier

All preamplifiers operate in the class-A mode, and most employ FETs. An FET has a high input impedance ideally suited to weak-signal work. Figure 27-13 shows a simple generic RF preamplifier circuit. Input tuning reduces noise and provides some selectivity. This circuit produces a 5 dB to 10 dB gain, depending on the frequency and the choice of FET.

We must ensure that a preamplifier remains linear in the presence of strong input signals. Nonlinearity can cause unwanted mixing among multiple incoming signals. These so-called *mixing products* produce *intermodulation distortion* (IMD), or *intermod*, that can spawn numerous false signals inside the receiver. Intermod can also degrade the S/N ratio by generating *hash*, a form of wideband noise.

Front End

At low and medium frequencies, considerable atmospheric noise exists, and the design of a front-end circuit is simple because we don't have to worry much about internally generated noise. (Conditions are bad enough in the antenna!)

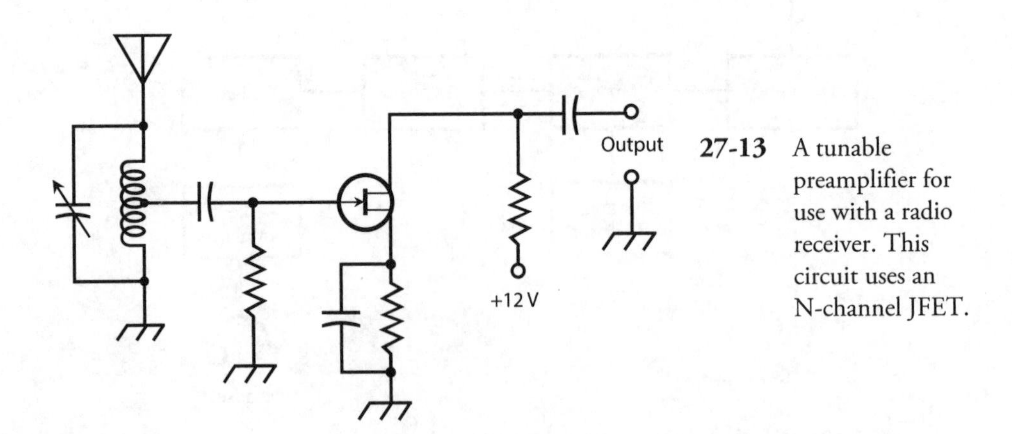

27-13 A tunable preamplifier for use with a radio receiver. This circuit uses an N-channel JFET.

Atmospheric noise diminishes as we get above 30 MHz or so. Then the main sensitivity-limiting factor becomes noise generated within the receiver. For this reason, front-end design grows in importance as the frequency rises through the VHF, UHF, and microwave spectra.

The front end, like a preamplifier, must remain as linear as possible. The greater the degree of nonlinearity, the more susceptible the circuit becomes to the generation of mixing products and intermod. The front end should also have the greatest possible dynamic range.

Preselector

The preselector provides a bandpass response that improves the S/N ratio, and reduces the likelihood of overloading by a strong signal that's far removed from the operating frequency. The preselector also provides *image rejection* in a superheterodyne circuit.

We can tune a preselector by means of *tracking* with the receiver's main tuning control, but this technique requires careful design and alignment. Some older receivers incorporate preselectors that must be adjusted independently of the receiver tuning.

IF Chains

A high IF (several megahertz) works better than a low IF (less than 1 MHz) for image rejection. However, a low IF allows us to obtain superior selectivity. Double-conversion receivers have a comparatively high first IF and a low second IF to get the "best of both worlds." We can cascade multiple IF amplifiers with tuned-transformer coupling. The amplifiers follow the mixer and precede the detector. Double-conversion receivers have two series, called *chains,* of IF amplifiers. The *first IF chain* follows the first mixer and precedes the second mixer, and the *second IF chain* follows the second mixer and precedes the detector.

Engineers sometimes express IF-chain selectivity by comparing the bandwidths for two power-attenuation values, usually −3 dB and −30 dB, also called *3 dB down* and *30 dB down.* This specification offers a good description of the bandpass response. We call the ratio of the bandwidth at −30 dB to the bandwidth at −3 dB the *shape factor.* In general, small shape factors are more desirable than large ones, but small factors can prove difficult to attain in practice. When we have a small shape factor and graph the system gain as a function of frequency, we get a curve that resembles a rectangle, so we can say that the receiver has a *rectangular response.*

Detectors

Detection, also called *demodulation,* allows a wireless receiver to recover the modulating information, such as audio, images, or printed data, from an incoming signal.

Detection of AM

A radio receiver can extract the information from an AM signal by half-wave rectifying the carrier wave and then filtering the output waveform just enough to smooth out the RF pulsations. Figure 27-14A shows a simplified time-domain view of how this process works. The rapid pulsations (solid curves) occur at the RF carrier frequency; the slower fluctuation (dashed curve) portrays the modulating data. The carrier pulsations are smoothed out by passing the output through a capacitor that's large enough to hold the charge for one carrier current cycle, but not so large that it dampens or obliterates the fluctuations in the modulating signal. We call this technique *envelope detection.*

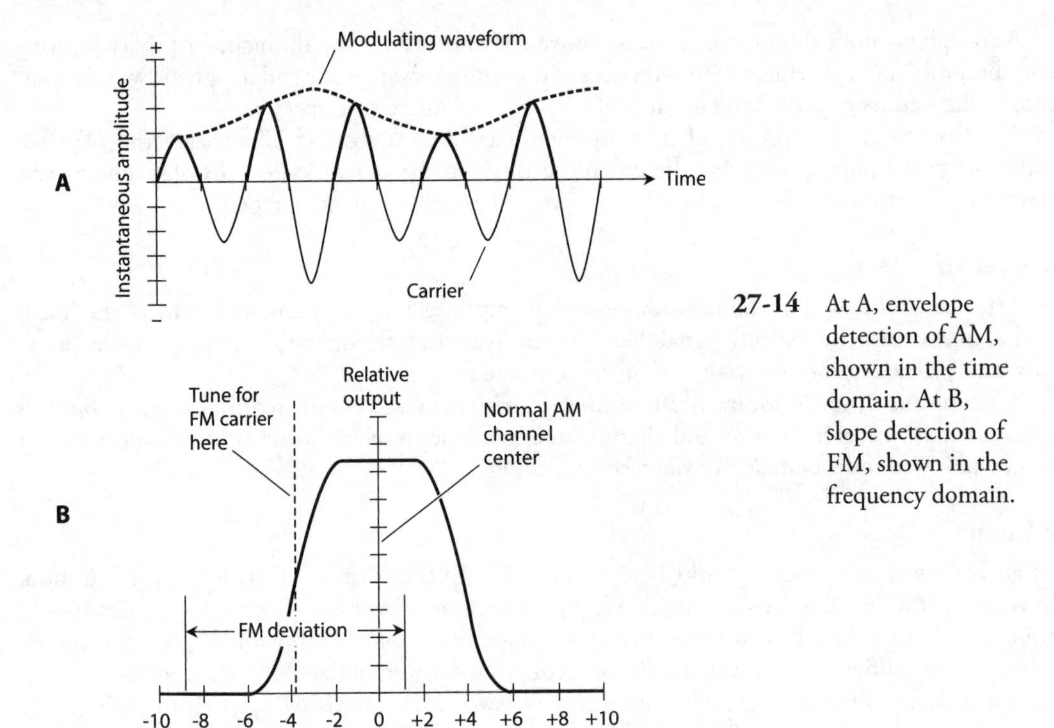

27-14 At A, envelope detection of AM, shown in the time domain. At B, slope detection of FM, shown in the frequency domain.

Detection of CW and FSK

If we want a receiver to detect CW, we must inject a constant-frequency, unmodulated carrier a few hundred hertz above or below the signal frequency. The local carrier is produced by a tunable *beat-frequency oscillator* (BFO). The BFO signal and the incoming CW signal heterodyne in a mixer to produce audio output at the sum or difference frequency. We can tune the BFO to obtain an audio "note" or "tone" at a comfortable listening pitch, usually 500 to 1000 Hz. This process is called *heterodyne detection.*

We can detect FSK signals using the same method as CW detection. The carrier beats against the BFO in the mixer, producing an audio tone that alternates between two different pitches. With FSK, the BFO frequency is set a few hundred hertz above or below both the *mark frequency* and the *space frequency*. The *frequency offset*, or difference between the BFO and the signal frequencies, determines the audio output frequencies. We adjust the frequency offset to get specific standard AF notes (such as 2125 Hz and 2295 Hz in the case of 170-Hz shift).

Slope Detection of FM and PM

We can use an AM receiver to detect FM or PM by setting the receiver frequency near, but not exactly at, the unmodulated-carrier frequency. An AM receiver has a filter with a passband of a few kilohertz and a selectivity curve such as that shown in Fig. 27-14B. If we tune the receiver so that the FM unmodulated-carrier frequency lies near either edge, or *skirt*, of the filter response, frequency variations in the incoming signal cause its carrier to "swing" in and out of the receiver passband. As a result, the instantaneous receiver output amplitude varies along with the modulating data on the FM or PM signal. In this system, known as *slope detection*, the relationship between the instantaneous deviation and the instantaneous output amplitude is nonlinear because the skirt of the passband is not a straight line (as we can see in Fig. 27-14B). Therefore, slope detection does not provide an optimum method of detecting FM or PM signals. The process can usually yield an intelligible voice, but it will ruin the quality of music.

Using a PLL to Detect FM or PM

If we inject an FM or PM signal into a PLL circuit, the loop produces an error voltage that constitutes a precise duplicate of the modulating waveform. A *limiter*, which keeps the signal amplitude from varying, can be placed ahead of the PLL so that the receiver doesn't respond to changes in the signal amplitude. Weak signals tend to abruptly appear and disappear, rather than fading, in an FM or PM receiver that employs limiting.

Discriminator for FM or PM

A *discriminator* produces an output voltage that depends on the instantaneous signal frequency. When the signal frequency lies at the center of the receiver passband, the output voltage equals zero. When the instantaneous signal frequency falls below the passband center, the output voltage becomes positive. When the instantaneous signal frequency rises above center, the output voltage becomes negative. The relationship between the instantaneous FM deviation (which, as we remember, can result indirectly from PM) and the instantaneous output amplitude is linear. Therefore, the detector output represents a faithful reproduction of the incoming signal data. A discriminator is sensitive to amplitude variations, but we can use a limiter to get rid of this problem, just as we do in a PLL detector.

Ratio Detector for FM or PM

A *ratio detector* comprises a discriminator with a built-in limiter. The original design was developed by RCA (Radio Corporation of America), and works well in high-fidelity receivers and in the audio portions of old-fashioned analog TV receivers. Figure 27-14C illustrates a simple ratio detector

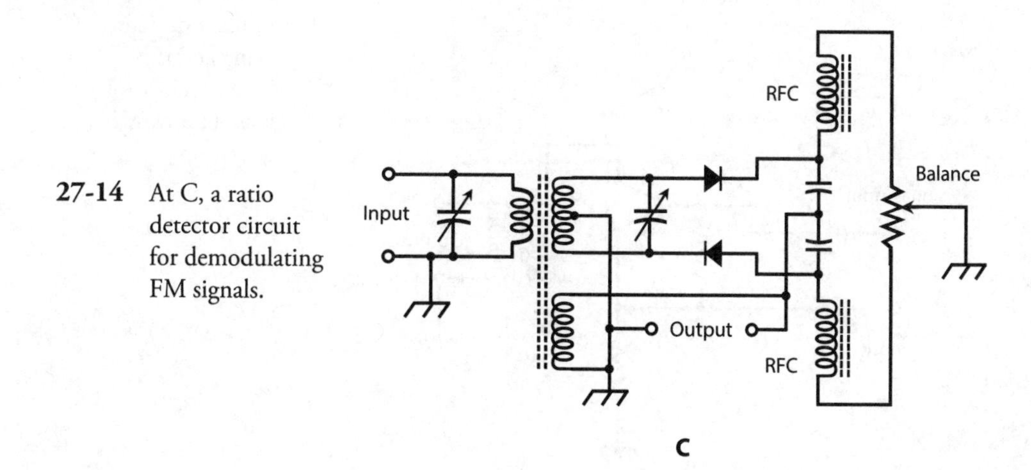

27-14 At C, a ratio detector circuit for demodulating FM signals.

circuit. The potentiometer marked "balance" should be adjusted experimentally to get optimum received-signal audio quality.

Detection of SSB

For reception of SSB signals, most communications engineers prefer to use a *product detector*, although a direct-conversion receiver can do the job. A product detector also facilitates reception of CW and FSK. The incoming signal combines with the output of an unmodulated LO, reproducing the original modulating signal data. Product detection occurs at a single frequency, rather than at a variable frequency, as in direct-conversion reception. The single, constant frequency results from mixing of the incoming signal with the output of the LO.

Figures 27-14D and 27-14E are schematic diagrams of product-detector circuits, which can also serve as mixers in superhet receivers. In the circuit shown at D, diodes are used, so we do not get any amplification. The circuit shown at E employs a bipolar transistor biased for class-B mode, providing some gain if the incoming signal has been sufficiently amplified by the front end before it arrives at the detector input. The effectiveness of the circuits shown in Figs. 27-14D or 27-14E

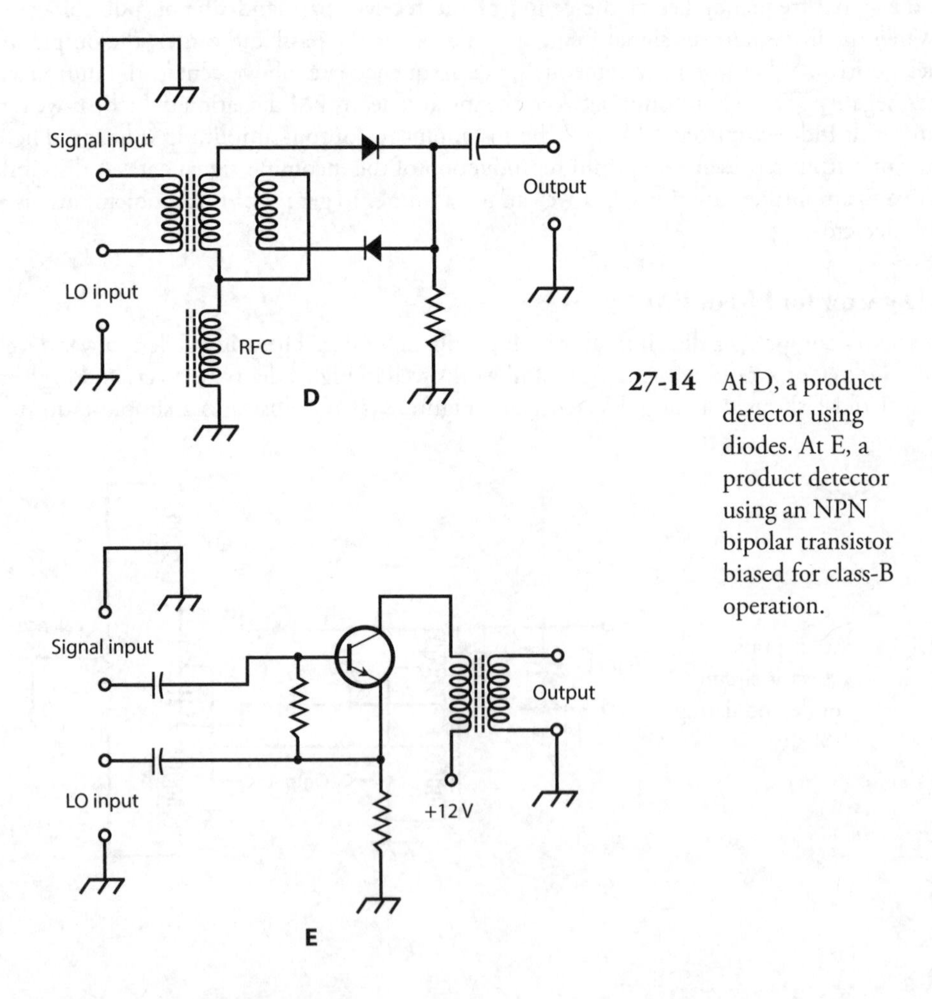

27-14 At D, a product detector using diodes. At E, a product detector using an NPN bipolar transistor biased for class-B operation.

lies in the nonlinearity of the semiconductor devices. This nonlinearity facilitates the heterodyning necessary to obtain sum and difference frequency signals that result in data output.

Postdetector Stages

We can obtain selectivity in a receiver by tailoring the frequency response in the AF amplifier stages following the detector, in addition to optimizing the RF selectivity in the IF stages preceding the detector.

Filtering

In a communications system, a human voice signal requires a band ranging from about 300 Hz to 3000 Hz for a listener to easily understand the content. An *audio bandpass filter*, with a passband of 300 Hz to 3000 Hz, can improve the intelligibility in some voice receivers. An ideal voice audio bandpass filter has little or no attenuation within the passband range but high attenuation outside the passband range, along with a near-rectangular response curve.

A CW or FSK signal requires only a few hundred hertz of bandwidth. Audio CW filters can narrow the response bandwidth to 100 Hz or less, but passbands narrower than about 100 Hz produce *ringing*, degrading the quality of reception at high data speeds. With FSK, the bandwidth of the filter must be at least as large as the difference (shift) between mark and space, but it need not (and shouldn't) greatly exceed the frequency shift.

An *audio notch filter* is a *band-rejection filter* with a sharp, narrow response. Band-rejection filters pass signals only below a certain lower cutoff frequency or above a certain upper cutoff frequency. Between those limits, in the so-called *bandstop range*, signals are blocked. A notch filter can "mute" an interfering unmodulated carrier or CW signal that produces a constant-frequency tone in the receiver output. Audio notch filters are tunable from at least 300 Hz to 3000 Hz. Some AF notch filters work automatically; when an interfering AF tone appears, the notch finds and "mutes" it within a few tenths of a second.

Squelching

A *squelch* silences a receiver when no incoming signals exist, allowing reception of signals when they appear. Most FM communications receivers use squelching systems. The squelch is normally *closed*, cutting off all audio output (especially receiver hiss, which annoys some communications operators) when no signal is present. The squelch *opens*, allowing everything to be heard, if the signal amplitude exceeds a *squelch threshold* that the operator can adjust.

In some systems, the squelch does not open unless an incoming signal has certain pre-determined characteristics. This feature is called *selective squelching*. The most common way to achieve selective squelching is the use of a *subaudible* (below 300 Hz) *tone generator* or an AF *tone-burst generator* in the transmitter. The squelch opens only in the presence of signals modulated by a tone, or sequence of tones, having the proper characteristics. Some radio operators use selective squelching to prevent unwanted transmissions from "coming in."

Fast-Scan TV

An analog fast-scan television (FSTV) receiver can follow a D/A converter for reception of advanced digital TV signals. This type of receiver has a tunable front end, an oscillator and mixer, a set of IF amplifiers, a video demodulator, an audio demodulator and amplifier chain, a picture CRT or display with associated peripheral circuitry, and a loudspeaker. Figure 27-15 is a block diagram of a receiver for FSTV.

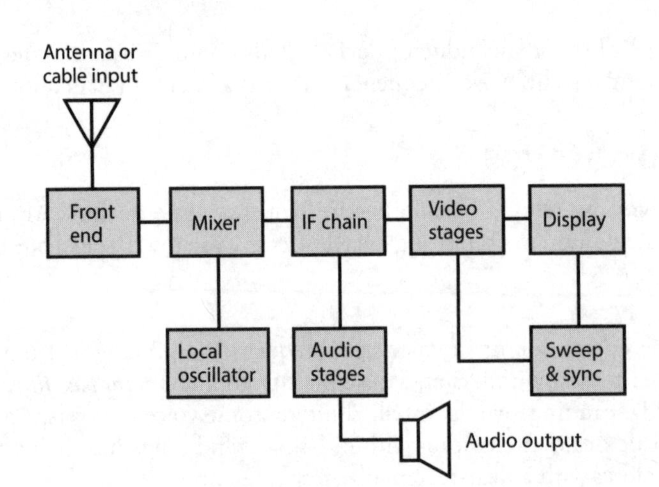

27-15 Block diagram of a conventional FSTV receiver.

In the United States, conventional FSTV broadcasts were once made on channels numbered from 2 through 69. Each channel was 6 MHz wide, including video and audio information. Channels 2 through 13 were called the *VHF TV broadcast channels*. Channels 14 through 69 constituted the *UHF TV broadcast channels*. In digital television these days, the D/A "converter box" serves as the program selector and outputs all signals on a single analog channel.

Slow-Scan TV

A *slow-scan television* (SSTV) communications station needs a transceiver with SSB capability, a standard FSTV set or personal computer, and a *scan converter* that translates between the SSTV signal and either FSTV imagery or computer video data. The scan converter contains two data converters (one for receiving and the other for transmitting), some digital *memory*, a *tone generator*, and a TV detector. Scan converters are commercially available. Computers can be programmed to perform this function. Some amateur radio operators build their own scan converters.

Specialized Wireless Modes

Communications engineers have a long history of innovation, developing numerous exotic wireless modes. In recent years, new modes have emerged; we can expect more to come, each of which offers specific advantages under strange or difficult conditions. Four common examples follow.

Dual-Diversity Reception

A *dual-diversity receiver* can reduce fading in radio reception at high frequencies (approximately 3 to 30 MHz) when signals propagate through the ionosphere and return to earth's surface. The system comprises two identical receivers tuned to the same signal and having separate antennas spaced several wavelengths apart. The outputs of the receiver detectors go into a single audio amplifier as shown in Fig. 27-16.

Dual-diversity receiver tuning is a sophisticated technology—we might call it an art—and good equipment for this purpose costs a lot of money. Some advanced diversity-reception installations

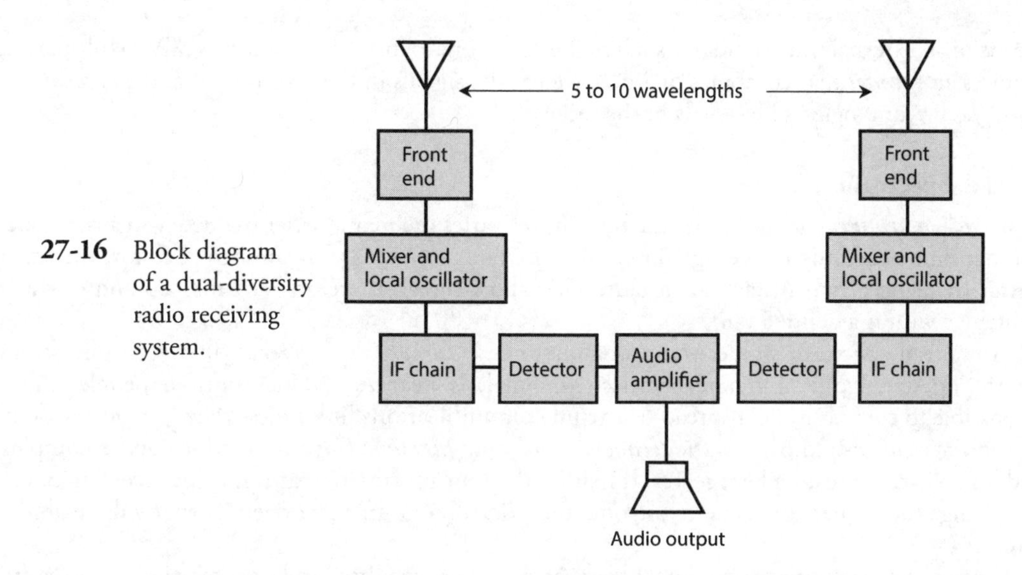

27-16 Block diagram of a dual-diversity radio receiving system.

employ three or more antennas and receivers, providing superior immunity to fading, but further compounding the tuning difficulty and further increasing the expense.

Synchronized Communications

Digital signals require less bandwidth than analog signals to convey a given amount of information per unit of time. The term *synchronized communications* refers to any of several specialized digital modes in which the transmitter and receiver operate from a common frequency-and-time standard to optimize the amount of data that can be sent in a communications channel or band.

In synchronized digital communications, also called *coherent communications*, the receiver and transmitter operate in lock-step. The receiver evaluates each transmitted data bit for a block of time lasting for the specified duration of a single bit. This process makes it possible to use a receiving filter having extremely narrow bandwidth. The synchronization requires the use of an external frequency-and-time standard, such as that provided by the National Institute of Standards and Technology (NIST) radio station WWV in the United States. Frequency dividers generate the necessary synchronizing signals from the frequency-standard signal. A tone or pulse appears in the receiver output for a particular bit if, but only if, the average signal voltage exceeds a certain value over the duration of that bit. False signals caused by filter ringing, sferics, or ignition noise are generally ignored because they rarely produce sufficient average bit voltage.

Experiments with synchronized communications have shown that the improvement in S/N ratio, compared with nonsynchronized systems, is several decibels at low to moderate data speeds.

Multiplexing

Signals in a communications channel or band can be intertwined, or *multiplexed*, in various ways. The most common methods are *frequency-division multiplexing* (FDM) and *time-division multiplexing* (TDM). In FDM, the channel is broken down into subchannels. The carrier frequencies of the signals are spaced so that they don't overlap. Each signal remains independent of all the others. A TDM system breaks signals down into segments of specific time duration, and then the segments are transferred in a rotating sequence. The receiver stays synchronized with the transmitter by means of an external time standard, such as the data from "shortwave" station WWV. Multiplexing requires an *encoder* that combines or "intertwines" the signals in the transmitter, and a *decoder* that separates or "untangles" the signals in the receiver.

Spread-Spectrum

In *spread-spectrum communications*, the transmitter varies the main carrier frequency in a controlled manner, independently of the signal modulation. The receiver is programmed to follow the transmitter frequency from instant to instant. The whole signal, therefore, "roams" up and down in frequency within a defined range.

In spread-spectrum mode, the probability of *catastrophic interference*, in which one strong interfering signal can obliterate the desired signal, is near zero. Unauthorized people find it impossible to eavesdrop on a spread-spectrum communications link unless they gain access to the *sequencing code*, also known as the *frequency-spreading function*. Such a function can be complex, and, of course, it must be kept secret. If neither the transmitting operator nor the receiving operator divulge the sequencing code to anyone, then (ideally) no unauthorized listener will be able to intercept it.

During a spread-spectrum contact between a given transmitter and receiver, the operating frequency can fluctuate over a range of several kilohertz, megahertz, or tens of megahertz. As a band becomes occupied with an increasing number of spread-spectrum signals, the overall noise level in the band appears to increase. Therefore, a practical limit exists to the number of spread-spectrum contacts that a band can handle. This limit is roughly the same as it would be if all the signals were constant in frequency, and had their own discrete channels. The main difference between fixed-frequency communications and spread-spectrum communications, when the band gets crowded, lies in the *nature* of the mutual interference.

A common method of generating spread-spectrum signals involves so-called *frequency hopping*. The transmitter has a list of channels that it follows in a certain order. The transmitter "jumps" or "hops" from one frequency to another in the list. The receiver must be programmed with this same list, in the same order, and must be synchronized with the transmitter. The *dwell time* equals the length of time that the signal remains on any given frequency; it's the same as the time interval at which frequency changes occur. In a well-designed frequency-hopping system, the dwell time is short enough so that a signal will not be noticed by an unauthorized listener using a receiver set to a constant frequency, and also will not cause interference on any frequency. The sequence contains numerous *dwell frequencies*, so the signal energy is diluted to the extent that, if someone tunes to any particular frequency in the sequence, they won't notice the signal.

Another way to obtain spread spectrum, called *frequency sweeping*, requires frequency-modulating the main transmitted carrier with a waveform that guides it "smoothly" up and down over the assigned band. The "sweeping FM" remains entirely independent of the actual data that the signal conveys. A receiver can intercept the signal if, but only if, its instantaneous frequency varies according to the same waveform, over the same band, at the same rate, and in the same phase as that of the transmitter. The transmitter and receiver in effect "roam all over the band," following each other from moment to moment according to a "secret map" that only they know.

Quiz

Refer to the text in this chapter if necessary. A good score is at least 18 correct. Answers are in the back of the book.

1. Which of the following communications modes has a mark component at one carrier frequency and a space component at a different carrier frequency?
   (a) CW
   (b) FSK
   (c) FSTV
   (d) FM

2. Which of the following communications modes has a mark component in which the carrier is "full-on" and a space component in which the carrier is entirely absent?
   (a) CW
   (b) FSK
   (c) FSTV
   (d) FM

3. We can demodulate FM signals with
   (a) a discriminator.
   (b) a ratio detector.
   (c) an envelope detector.
   (d) All of the above

4. If we want to demodulate an AM signal, we'll get the best results with
   (a) a discriminator.
   (b) a ratio detector.
   (c) an envelope detector.
   (d) a product detector.

5. We can accomplish spread-spectrum communications by means of
   (a) ratio detection.
   (b) product detection.
   (c) frequency hopping.
   (d) All of the above

6. Birdies can occur in a
   (a) direct-conversion receiver.
   (b) superheterodyne receiver.
   (c) ratio detector.
   (d) front end with poor dynamic range.

7. The dynamic range specification in a receiver tells us how well the system can handle signals
   (a) in diverse modulation modes.
   (b) over a wide range of frequencies.
   (c) in the presence of high noise levels.
   (d) from extremely weak to extremely strong.

8. Which of the following modes involves breaking a signal down into "pieces" of specific time duration, transmitting the "pieces" in a repeating sequence, and reassembling them back into the original signal at the receiver?
   (a) FSK
   (b) TDM
   (c) SSTV
   (d) CW

9. Suppose that an RF carrier has a frequency of 830 kHz. We can effectively modulate this carrier with information containing frequency components up to about
   (a) 83.0 Hz.
   (b) 830 Hz.
   (c) 8.30 kHz.
   (d) 83.0 kHz.

10. We might expect to observe a dead spot in communications reception when the direct and reflected waves arrive at the receiving antenna
   (a) with a 0° phase difference.
   (b) with a 180° phase difference.
   (c) with a 90° phase difference.
   (d) in any condition other than phase coincidence.

11. The earth's atmosphere generally exhibits a decreasing index of refraction, with respect to radio waves, as the altitude above the surface increases. At some radio frequencies, this property gives rise to
   (a) ionospheric reflection.
   (b) troposcatter.
   (c) tropospheric bending.
   (d) auroral propagation.

12. Suppose that we modulate a VHF carrier in the SSB mode with AF data having frequency components up to 20 kHz. What's the approximate bandwidth of the SSB signal?
   (a) 10 kHz
   (b) 20 kHz
   (c) 40 kHz
   (d) 80 kHz

13. In a LEO satellite network, each individual satellite follows an orbit
   (a) approximately 22,200 miles above the earth's surface.
   (b) that keeps it in a single spot above the earth's surface.
   (c) that keeps it constantly between the earth and the moon.
   (d) that takes it over, or nearly over, the earth's poles at low altitude.

14. We can generate a DSB, suppressed-carrier signal with a
    (a) frequency modulator.
    (b) phase modulator.
    (c) balanced modulator.
    (d) slope modulator.

15. Figure 27-17 illustrates a circuit designed to perform
    (a) ratio detection.
    (b) modulation.
    (c) oscillation.
    (d) product detection.

16. Based on the general knowledge of electronics that we've gained so far in this course, we know that the components marked X in Fig. 27-17
    (a) pass signals but not DC.
    (b) pass DC but not signals.
    (c) ensure that the transistor receives pure DC from the battery.
    (d) limit the current through the transistor E-B and B-C junctions.

17. What appears at the output terminals of the circuit of Fig. 27-17, assuming that we choose the correct component values and operate the system properly?
    (a) An FM signal
    (b) An SSB signal
    (c) A DSB, suppressed-carrier signal
    (d) An AM signal

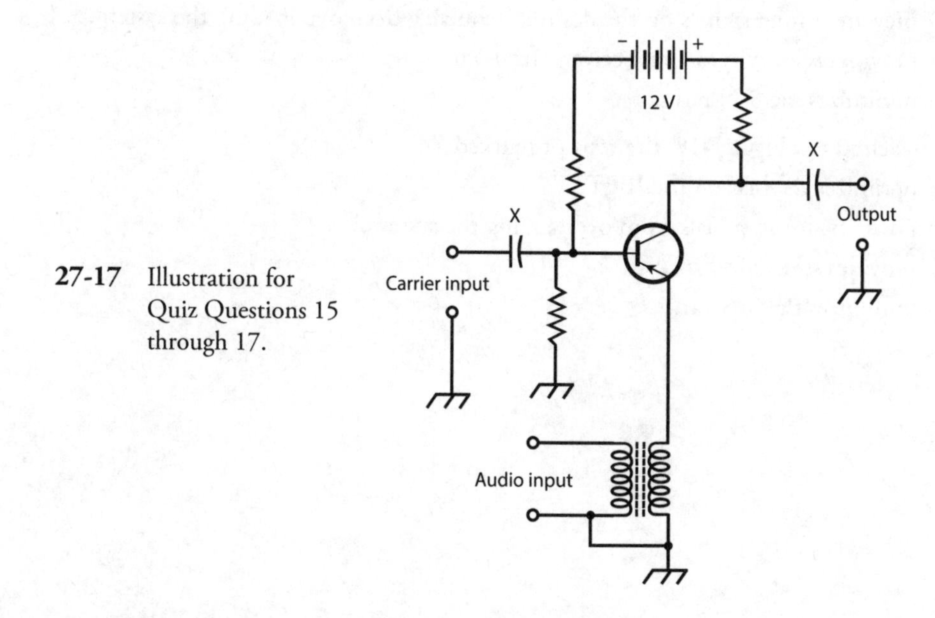

27-17 Illustration for Quiz Questions 15 through 17.

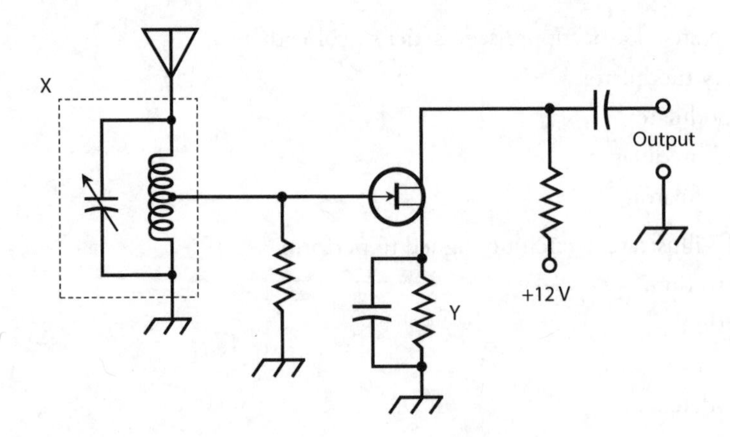

27-18 Illustration for Quiz Questions 18 through 20.

18. Figure 27-18 is a schematic diagram for a weak-signal amplifier that we might find in the front end of a radio communications receiver. What's wrong?
    (a) We should use a P-channel JFET, not an N-channel JFET.
    (b) We should install a blocking capacitor between the gate resistor and the center tap of the inductor.
    (c) We should replace the capacitor at the output terminals with an RF choke.
    (d) We should remove the capacitor between the source and ground.

19. In the circuit of Fig. 27-18, the *LC* circuit marked X
    (a) optimizes the bias on the JFET.
    (b) prevents strong signals on the desired frequency from overloading the system.
    (c) provides selectivity at the receiver's front end.
    (d) minimizes the S/N ratio.

20. In the circuit of Fig. 27-18, the resistor marked Y
    (a) optimizes the bias on the JFET.
    (b) prevents strong signals from overloading the system.
    (c) provides selectivity.
    (d) minimizes the S/N ratio.

28 CHAPTER

Digital Basics

ENGINEERS AND TECHNICIANS CALL AN ELECTRONIC SIGNAL *DIGITAL* WHEN IT CAN ATTAIN A LIMITED number of well-defined states. Digital signals contrast with *analog* signals, which vary over a continuous range and, therefore, can, in theory, attain infinitely many different instantaneous states. Figure 28-1 shows an example of an analog signal (at A) and a digital signal (at B).

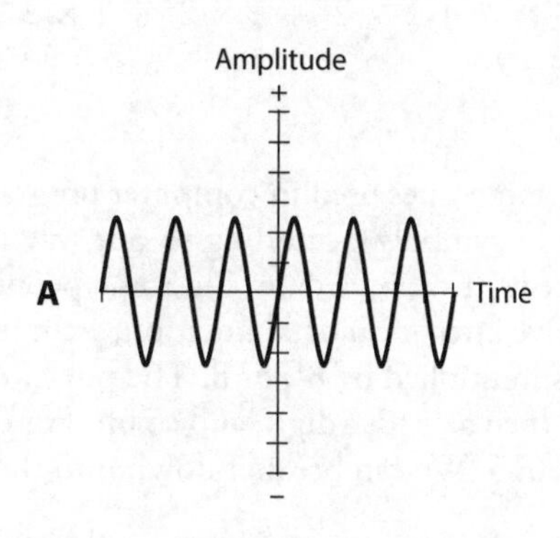

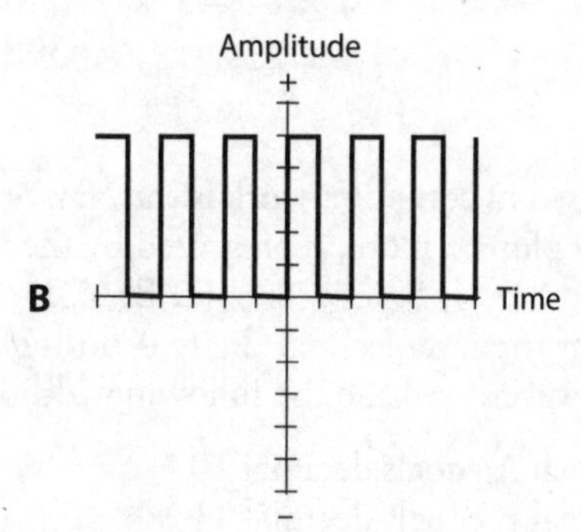

28-1 An analog wave (A) and a digital rendition of the same wave (B).

Numeration Systems

In everyday life, most of us deal with the *decimal number system*, which makes use of digits from the set {0, 1, 2, 3, 4, 5, 6, 7, 8, 9}. Machines, such as computers and communications devices, offer other numeration schemes.

Decimal

The familiar decimal number system is also called *base 10* or *radix 10*. When we express nonnegative integers in this system, we multiply the right-most digit by 10^0, or 1. We multiply the next digit to the left by 10^1, or 10. The power of 10 increases as we move to the left. Once we've multiplied the digits, we add up all the resulting values. For example:

$$8 \times 10^0 + 5 \times 10^1 + 0 \times 10^2 + 2 \times 10^3 + 6 \times 10^4 + 8 \times 10^5 = 862{,}058$$

Binary

The *binary number system* denotes numbers using only the digits 0 and 1. We'll sometimes hear this system called *base 2* or *radix 2*. When we express nonnegative integers in binary notation, we multiply the rightmost digit by 2^0, or 1. The next digit to the left is multiplied by 2^1, or 2. The power of 2 increases as we continue to the left, so we get a "fours" digit, then an "eights" digit, then a "16s" digit, and so on. For example, consider the decimal number 94. In the binary system, we would write this quantity as 1011110. It breaks down into the sum

$$0 \times 2^0 + 1 \times 2^1 + 1 \times 2^2 + 1 \times 2^3 + 1 \times 2^4 + 0 \times 2^5 + 1 \times 2^6 = 94$$

Octal

Another scheme, sometimes used in computer programming, goes by the name *octal number system* because it has eight symbols (according to our way of thinking), or 2^3. Every digit constitutes an element of the set {0, 1, 2, 3, 4, 5, 6, 7}. Some people call this system *base 8* or *radix 8*. When we express nonnegative integers in octal notation, we multiply the rightmost digit by 8^0, or 1. The next digit to the left is multiplied by 8^1, or 8. The power of 8 increases as we continue to the left, so we get a "64s" digit, then a "512s" digit, and so on. For example, we render the decimal quantity 3085 in octal form as 6015. We can break it down into the sum

$$5 \times 8^0 + 1 \times 8^1 + 0 \times 8^2 + 6 \times 8^3 = 3085$$

Hexadecimal

Another system used in computer work is the *hexadecimal number system*. It has 16 (2^4) symbols: the usual 0 through 9 plus six more, represented by the uppercase English letters A through F, yielding the digit set {0, 1, 2, 3, 4, 5, 6, 7, 8, 9, A, B, C, D, E, F}. This system is sometimes called *base 16* or *radix 16*. All of the hexadecimal digits 0 through 9 represent the same values as their decimal counterparts. However, we have the following additional digits:

- Hexadecimal A equals decimal 10
- Hexadecimal B equals decimal 11
- Hexadecimal C equals decimal 12

- Hexadecimal D equals decimal 13
- Hexadecimal E equals decimal 14
- Hexadecimal F equals decimal 15

When we express nonnegative integers in hexadecimal notation, we multiply the rightmost digit by 16^0, or 1. We multiply the next digit to the left by 16^1, or 16. The power of 16 increases as we continue to the left, so we get a "256s" digit, then a "4096s" digit, and so on. For example, we write the decimal quantity 35,898 in hexadecimal form as 8C3A. Remembering that C = 12 and A = 10, we can break the hexadecimal number down into the following sum:

$$A \times 16^0 + 3 \times 16^1 + C \times 16^2 + 8 \times 16^3$$
$$= 35{,}898$$

Digital Logic

Digital logic, also called simply *logic*, is the form of "reasoning" used by electronic machines. Engineers also use the term in reference to the circuits that make up digital devices and systems.

Boolean Algebra

Boolean algebra constitutes a system of logic using the numbers 0 and 1 with the operations AND (multiplication), OR (addition), and NOT (negation). Combinations of these operations give us two more, called NAND (NOT AND) and NOR (NOT OR). This system, which gets its name from the nineteenth-century British mathematician *George Boole*, plays a vital role in the design of digital electronic circuits.

- The AND operation, also called *logical conjunction*, operates on two or more quantities. Let's denote it using an asterisk, for example X * Y.
- The NOT operation, also called *logical inversion* or *logical negation*, operates on a single quantity. Let's denote it using a minus sign (−), for example −X.
- The OR operation, also called *logical disjunction*, operates on two or more quantities. Let's denote it using a plus sign (+), for example X + Y.

Table 28-1A breaks down all the possible input and output values for the above-described Boolean operations, where 0 indicates "falsity" and 1 indicates "truth." In mathematics and philosophy courses involving logic, you can expect to see other symbols used for conjunction and disjunction.

Theorems

Table 28-1B shows several logic equations that hold true under all circumstances, that is, for all values of the *logical variables* X, Y, and Z. We call such facts *theorems*. Statements on either side of the equals

Table 28-1A Boolean Operations

X	Y	−X	X * Y	X + Y
0	0	1	0	0
0	1	1	0	1
1	0	0	0	1
1	1	0	1	1

Table 28-1B Common Theorems in Boolean Algebra

Theorem (Logic Equation)	What It's Called
$X + 0 = X$	OR identity
$X * 1 = X$	AND identity
$X + 1 = 1$	
$X * 0 = 0$	
$X + X = X$	
$X * X = X$	
$-(-X) = X$	Double negation
$X + (-X) = 1$	
$X * (-X) = 0$	Contradiction
$X + Y = Y + X$	Commutative property of OR
$X * Y = Y * X$	Commutative property of AND
$X + (X * Y) = X$	
$X * (-Y) + Y = X + Y$	
$(X + Y) + Z = X + (Y + Z)$	Associative property of OR
$(X * Y) * Z = X * (Y * Z)$	Associative property of AND
$X * (Y + Z) = (X * Y) + (X * Z)$	Distributive property
$-(X + Y) = (-X) * (-Y)$	DeMorgan's Theorem
$-(X * Y) = (-X) + (-Y)$	DeMorgan's Theorem

(=) sign in each case are *logically equivalent*, meaning that one is true *if and only if (iff)* the other is true. For example, the statement $X = Y$ means "If X then Y, and if Y then X." Boolean theorems allow us to simplify complicated *logic functions*, facilitating the construction of a circuit to perform a specific digital operation using the smallest possible number of switches.

Positive versus Negative Logic

In so-called *positive logic*, a circuit represents the binary digit 1 with an *electrical potential* of approximately +5 V DC (called the *high state* or simply *high*), while the binary digit 0 appears as little or no DC voltage (called the *low state* or simply *low*). Some circuits employ *negative logic*, in which little or no DC voltage (low) represents logic 1, while +5 V DC (high) represents logic 0. In another form of negative logic, the digit 1 appears as a negative voltage (such as −5 V DC, constituting the low state) and the digit 0 appears as little or no DC voltage (the high state because it has the more positive voltage). To avoid confusion, let's stay with positive logic for the rest of this chapter!

Logic Gates

All digital electronic devices employ switches that perform specific logical operations. These switches, called *logic gates*, can have anywhere from one to several inputs and (usually) a single output.

- A *logical inverter*, also called a *NOT gate*, has one input and one output. It reverses, or inverts, the state of the input. If the input equals 1, then the output equals 0. If the input equals 0, then the output equals 1.

- An *OR gate* can have two or more inputs (although it usually has only two). If both, or all, of the inputs equal 0, then the output equals 0. If any of the inputs equal 1, then the output equals 1. Mathematical logicians would tell us that such a gate performs an *inclusive-OR operation* because it "includes" the case where both variables are high.
- An *AND gate* can have two or more inputs (although it usually has only two). If both, or all, of the inputs equal 1, then the output equals 1. If any of the inputs equal 0, then the output equals 0.
- An OR gate can be followed by a NOT gate. This combination gives us a *NOT-OR* gate, more often called a *NOR gate*. If both, or all, of the inputs equal 0, then the output equals 1. If any of the inputs equal 1, then the output equals 0.
- An AND gate can be followed by a NOT gate. This combination gives us a *NOT-AND* gate, more often called a *NAND gate*. If both, or all, of the inputs equal 1, then the output equals 0. If any of the inputs equal 0, then the output equals 1.
- An *exclusive OR gate*, also called an *XOR gate*, has two inputs and one output. If the two inputs have the same state (either both 1 or both 0), then the output equals 0. If the two inputs have different states, then the output equals 1. Mathematicians use the term *exclusive-OR operation* because it doesn't "include" the case where both variables are high.

Table 28-2 summarizes the functions of the above-defined logic gates, assuming a single input for the NOT gate and two inputs for the others. Figure 28-2 illustrates the schematic symbols that engineers and technicians use to represent these gates in circuit diagrams.

Clocks

In electronics, the term *clock* refers to a circuit that generates pulses at high speed and at precise, constant time intervals. The clock sets the tempo for the operation of digital devices. In a computer, the clock acts like a metronome for the *microprocessor*. We express or measure clock speeds

Table 28-2 Logic Gates and Their Characteristics

Gate Type	Number of Inputs	Remarks
NOT	1	Changes state of input.
OR	2 or more	Output high if any inputs are high. Output low if all inputs are low.
AND	2 or more	Output low if any inputs are low. Output high if all inputs are high.
NOR	2 or more	Output low if any inputs are high. Output high if all inputs are low.
NAND	2 or more	Output high if any inputs are low. Output low if all inputs are high.
XOR	2	Output high if inputs differ. Output low if inputs are the same.

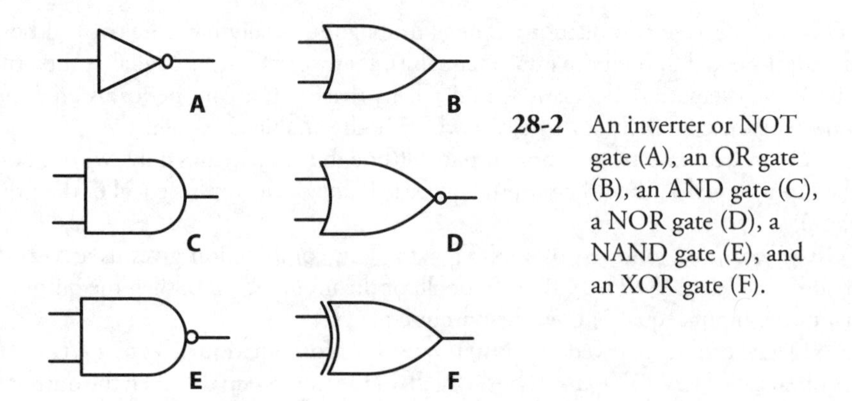

28-2 An inverter or NOT gate (A), an OR gate (B), an AND gate (C), a NOR gate (D), a NAND gate (E), and an XOR gate (F).

as frequencies in *hertz* (Hz). One hertz equals one pulse per second. Higher-frequency units work out as follows, just as they do with analog wave signals:

- A *kilohertz* (kHz) equals 1000 or 10^3 pulses per second
- A *megahertz* (MHz) equals 1,000,000 or 10^6 pulses per second
- A *gigahertz* (GHz) equals 1,000,000,000 or 10^9 pulses per second
- A *terahertz* (THz) equals 1,000,000,000,000 or 10^{12} pulses per second

In positive logic, a clock generates brief high pulses at regular intervals. The normal state is low.

Flip-Flops

A *flip-flop* is a specialized circuit constructed from logic gates, known collectively as a *sequential gate*. In a sequential gate, the output state depends on both the inputs and the outputs. The term "sequential" comes from the fact that the output depends not only on the states of the circuit at any given instant in time, but also on the states immediately preceding. A flip-flop has two states, called *set* and *reset*. Usually, the set state corresponds to logic 1 (high), and the reset state corresponds to logic 0 (low). In schematic diagrams, a flip-flop is usually shown as a rectangle with two or more inputs and two outputs. If the rectangle symbol is used, the letters FF, for "flip-flop," are printed or written at the top of the rectangle, either inside or outside. Several different types of flip-flop exist, as follows.

- In an *R-S flip-flop*, the inputs are labeled R (reset) and S (set). Engineers call the outputs Q and −Q. (Often, rather than −Q, we'll see Q′, or perhaps Q with a line over it.) Table 28-3A shows the input and output states. If R = 0 and S = 0, the output states remain at the values they've attained for the moment. If R = 0 and S = 1, then Q = 1 and −Q = 0. If R = 1 and S = 0, then Q = 0 and −Q = 1. When S = 1 and R = 1, the circuit becomes unpredictable.
- In a *synchronous flip-flop*, the states change when triggered by the signal from an external clock. In *static triggering*, the outputs change state only when the clock signal is either high or low. This type of circuit is sometimes called a *gated flip-flop*. In *positive-edge triggering*, the outputs change state at the instant the clock pulse is positive-going. In *negative-edge triggering*, the outputs change state at the instant the clock pulse is negative-going. The abrupt rise or fall of a pulse looks like the edge of a cliff (Fig. 28-3).

Table 28-3 Flip-Flop States

A: **R-S Flip-Flop** **R**	**S**	**Q**	**−Q**
0	0	Q	−Q
0	1	1	0
1	0	0	1
1	1	?	?

B: **J-K Flip-Flop** **J**	**K**	**Q**	**−Q**
0	0	Q	−Q
0	1	1	0
1	0	0	1
1	1	−Q	Q

- In a *master/slave (M/S) flip-flop*, the inputs are stored before the outputs can change state. This device comprises essentially two R-S flip-flops in series. We call the first flip-flop the *master* and the second flip-flop the *slave*. The master functions when the clock output is high, and the slave acts during the next ensuing low portion of the clock output. The time delay prevents confusion between the input and output.
- The operation of a *J-K flip-flop* resembles the functioning of an R-S flip-flop, except that the J-K device has a predictable output when the inputs both equal 1. Table 28-3B shows the input and output states for this type of flip-flop. The output changes only when a triggering pulse is received.
- The operation of an *R-S-T flip-flop* resembles that of an R-S flip-flop, except that a high pulse at an additional T input causes the circuit to change state.
- A circuit called a *T flip-flop* has only one input. Each time a high pulse appears at the input, the output reverses its state, either from 1 to 0 or else from 0 to 1. Note the difference between this type of circuit and a simple inverter (NOT gate)!

Counters

A *counter* literally counts digital pulses one by one. Each time the counter receives a high pulse, the binary number in its *memory* increases by 1. A *frequency counter* can measure the frequency of an AC wave or signal by tallying up the cycles over a precisely known period of time. The circuit consists

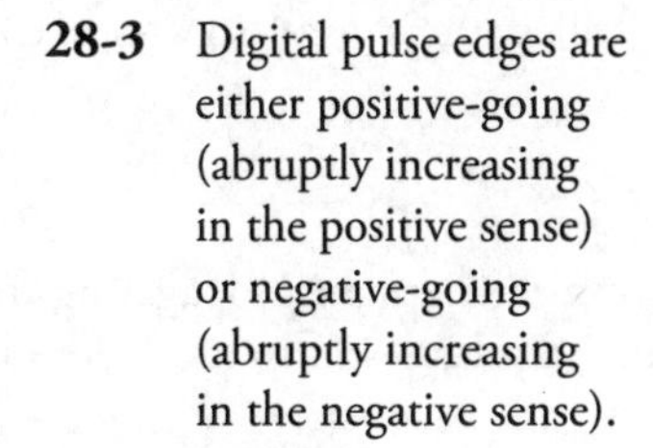

28-3 Digital pulse edges are either positive-going (abruptly increasing in the positive sense) or negative-going (abruptly increasing in the negative sense).

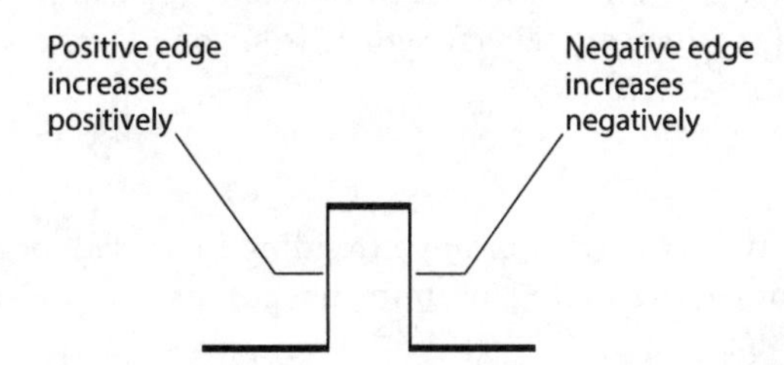

of a *gate*, which begins and ends each counting cycle at defined intervals. (Don't confuse this type of gate with the logic gates described a few moments ago.) The counter's accuracy depends on the *gate time*, or how long the gate remains open to accept pulses for counting. As we increase the gate time in a frequency counter, the accuracy improves. Although the counter tallies up the pulses as binary numbers, the display shows base-10 digital numerals.

Binary Communications

If we want to attain *multilevel signaling* (digital transmission with more than two states), we can represent each different signal level with a unique group of binary digits, representing a specific binary number. A group of three binary digits can represent up to 2^3, or eight, levels. A group of four binary digits can represent up to 2^4, or 16, levels. The term "binary digit" is commonly contracted to *bit*. A bit is represented by either logic 0 or logic 1. Some engineers call a group of eight bits an *octet*, and in many systems an octet corresponds to a unit called a *byte*.

Forms of Binary Signaling

Engineers have invented numerous forms, or *modes*, of binary communication. Three classical examples follow.

1. *Morse code* is the oldest binary mode. The logic states are called *mark* (key-closed or on) and *space* (key-open or off). Morse code is obsolete in modern systems, but amateur radio operators consider it a backup mode that can work in case digital signal-reading machines fail.
2. *Baudot*, also called the *Murray code*, is a five-unit digital code not widely used by today's digital equipment, except in a few antiquated *teleprinter* systems. There exist 2^5, or 32, possible representations.
3. The *American National Standard Code for Information Interchange* (ASCII) is a seven-unit code for the transmission of text and simple computer programs. There exist 2^7, or 128, possible representations.

Bits

We can represent large quantities of data according to powers of 2 or according to powers of 10. This duality can cause some confusion. Here's how the units build up when we talk about bits:

- A *kilobit* (kb) equals 10^3 or 1000 bits
- A *megabit* (Mb) equals 10^6 or 1,000,000 bits
- A *gigabit* (Gb) equals 10^9 or 1,000,000,000 bits
- A *terabit* (Tb) equals 10^{12} bits or 1000 Gb
- A *petabit* (Pb) equals 10^{15} bits or 1000 Tb
- An *exabit* (Eb) equals 10^{18} bits or 1000 Pb

We use power-of-10 multiples of *bits per second* (bps, kbps, Mbps, Gbps, and so on) to express data speed when we transmit digital signals from one location (called the *source*) to another location (called the *destination*).

Bytes

Data quantity in *storage* or *memory* (residing in a fixed location rather than propagating from one place to another) is specified in units that comprise power-of-2 multiples of bytes. Here's how the units build up:

- A *kilobyte* (KB) equals 2^{10} or 1024 bytes
- A *megabyte* (MB) equals 2^{20} or 1,048,576 bytes
- A *gigabyte* (GB) equals 2^{30} or 1,073,741,824 bytes
- A *terabyte* (TB) equals 2^{40} bytes or 1024 GB
- A *petabyte* (PB) equals 2^{50} bytes or 1024 TB
- An *exabyte* (EB) equals 2^{60} bytes or 1024 PB

Note also the following conventions concerning abbreviations for the units and prefix multipliers:

- The lowercase b stands for "bits."
- The uppercase B stands for "bytes."
- The lowercase k stands for 10^3 or 1000.
- The uppercase K stands for 2^{10} or 1024.
- We always denote the prefix multipliers M, G, T, P, and E in uppercase.

Baud

The term *baud* refers to the number of times per second that a signal changes state. We'll read about baud (sometimes called *baud rate*) only in texts and papers dated before about 1980. Bits per second (bps) and baud represent *qualitatively* different parameters, even though they might come out *quantitatively* close to each other for a particular digital signal. Some engineers used to speak and write about bps and baud as if they meant the same thing. They don't!

Examples of Data Speed

When we link multiple computers in a *network*, each computer has a *modem* (modulator/demodulator) connecting it to the communications medium. The slowest modem determines the speed at which the machines communicate. Table 28-4 shows common data speeds and the approximate time periods required to send one page, 10 pages, and 100 pages of double-spaced, typewritten text at each speed.

Table 28-4 Time Needed to Send Data at Various Speeds

Abbreviations: s = second, ms = millisecond (0.001 s), μs = microsecond (0.000001 s)

A:

Speed, kbps	Time for One Page	Time for 10 Pages	Time for 100 Pages
28.8	380 ms	3.8 s	38 s
38.4	280 ms	2.8 s	28 s
57.6	190 ms	1.9 s	19 s
100	110 ms	1.1 s	11 s
250	44 ms	440 ms	4.4 s
500	22 ms	220 ms	2.2 s

B:

Speed, Mbps	Time for One Page	Time for 10 Pages	Time for 100 Pages
1.00	11 ms	110 ms	1.1 s
2.50	4.4 ms	44 ms	440 ms
10.0	1.1 ms	11 ms	110 ms
100	110 μs	1.1 ms	11 ms

Data Conversion

We can convert an analog signal into a string of pulses whose amplitudes have a finite number of states, usually some power of 2. This scheme constitutes *analog-to-digital* (A/D) *conversion*; the reverse of digital-to-analog (D/A) conversion.

Figure 28-4 shows the functional difference between analog and digital signals. Imagine sampling the curve to obtain a sequence, or *train,* of pulses (A/D conversion), or smoothing out the pulses to obtain the curve (D/A conversion).

We can transmit and receive binary data one bit at a time along a single line or channel. This mode constitutes *serial data transmission.* Higher data speeds can be obtained by using multiple lines or a wideband channel, sending independent sequences of bits along each line or subchannel. Then we have *parallel data transmission.*

Parallel-to-serial (P/S) *conversion* involves the reception of bits from multiple lines or channels, and their retransmission one by one along a single line or channel. A *buffer* stores the bits from the parallel lines or channels while they await transmission along the serial line or channel. *Serial-to-parallel* (S/P) *conversion* involves the reception of bits one by one from a serial line or channel, and their retransmission in batches along several lines or channels. The output of an S/P converter cannot go any faster than the input, but we can find such a system useful when we want to interface between a serial-data device and a parallel-data device.

Figure 28-5 illustrates a circuit that employs a P/S converter at the source and an S/P converter at the destination. In this example, the words comprise eight-bit bytes. However, the words could have 16, 32, 64, or even 128 bits, depending on the communications scheme.

Data Compression

Data compression provides us with a way to maximize the amount of digital information that a machine can store in a given space, or that we can send within a certain period of time.

Text files can be compressed by replacing often-used words and phrases with symbols such as =, #, &, $, and @, as long as none of these symbols occurs in the uncompressed file. As the data is received, the machine can *decompress* it by substituting the original words and phrases for the symbols.

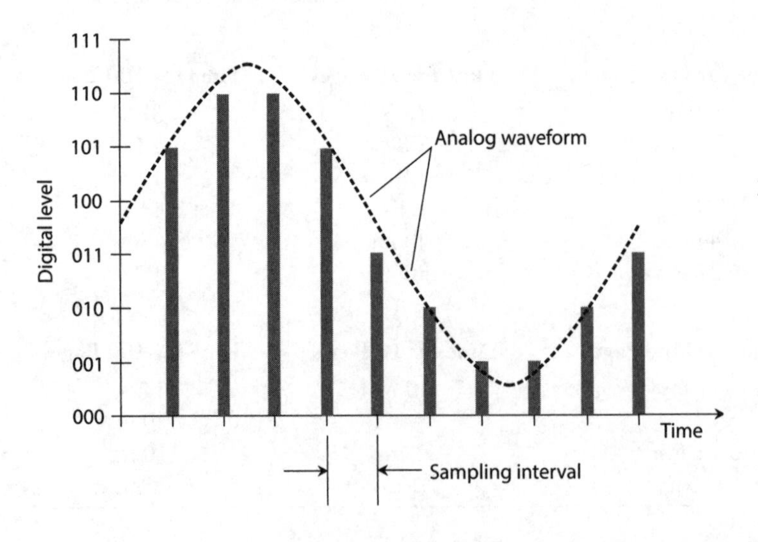

28-4 An analog waveform (dashed curve) and an 8-level digital representation of the same curve (vertical bars).

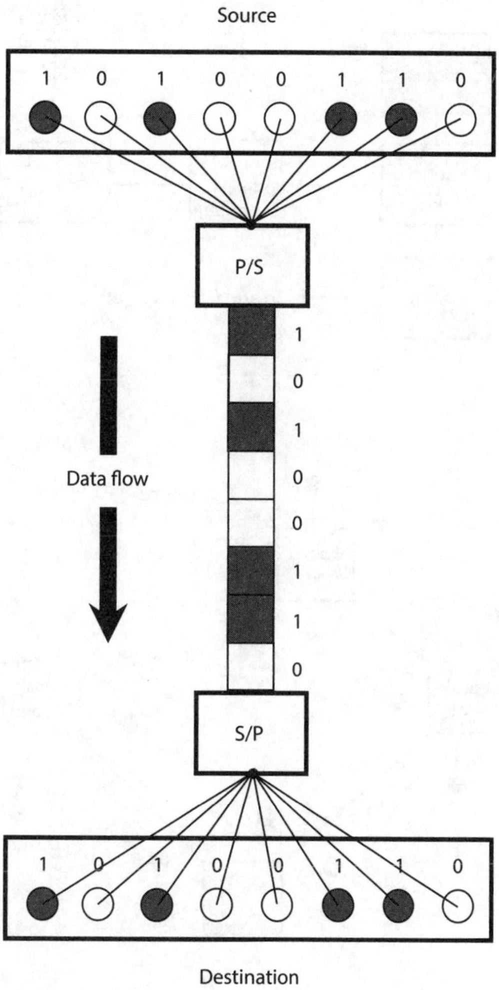

28-5 A communications circuit employing parallel-to-serial (P/S) conversion at the source and serial-to-parallel (S/P) conversion at the destination.

Digital image files can be compressed in either of two ways. In *lossless image compression*, detail is not sacrificed; only the redundant bits are eliminated. In *lossy image compression*, we lose some detail, although the loss is rarely severe enough to degrade the quality of the image to an objectionable extent.

Packet Wireless

In *packet wireless*, we connect a computer to a radio transceiver using a *terminal node controller* (TNC), which resembles a computer modem. Figure 28-6A shows an example. The computer has a modem as well as a TNC, so we can send and receive messages using conventional online services as well as radio.

Figure 28-6B shows how a packet-wireless message is routed. Black dots represent *subscribers*. Rectangles represent local nodes, each of which serves subscribers by means of short-range links at very-high, ultra-high, or microwave radio frequencies. The nodes are interconnected by terrestrial radio links if they are relatively near each other. If the nodes are widely separated, satellite links interconnect them.

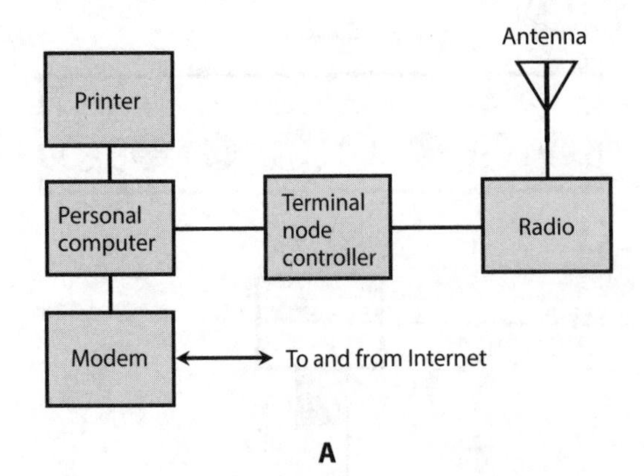

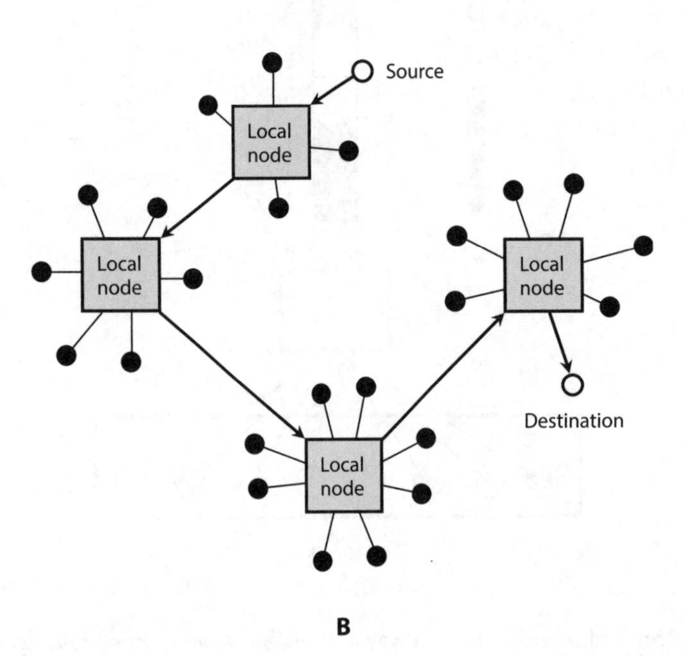

28-6 At A, a packet-wireless station. At B, passage of a packet through nodes in a wireless communications circuit.

Digital Signal Processing

Digital signal processing (DSP), a method of improving the precision of digital data, was originally used with radio and television receivers to produce a clear voice and/or picture from a marginal signal. Amateur radio operators did some of the earliest experiments with DSP. In analog modes, such as voice or video, the signals are first changed into digital form by A/D conversion. Then the digital data is "tidied up" so that the pulse timing and amplitude adhere strictly to the *protocol* (set of standards) for the type of digital data in use. Noise is greatly reduced or eliminated. Finally, the digital signal is changed back to the original voice or video by a D/A converter.

Digital signal processing can extend the workable range of almost any communications circuit because it allows reception under worse conditions than would be possible without it. Digital signal processing also improves the quality of marginal signals, so the receiving equipment or operator makes fewer errors. In circuits that use only digital modes, A/D and D/A conversion aren't required, but DSP can nevertheless "clean up" the signal, thereby improving the accuracy of the system and making it possible to copy data over and over many times (produce multi-generation copies).

Digital signals have discrete, well-defined states. A machine finds it easier to process a digital signal than to process an analog signal, which has a theoretically infinite number of possible states. In particular, digital signals have well-defined patterns that are easy for computers and microprocessors to recognize and clarify. Binary (two-state) signals are the simplest for machines to work with.

Even the most sophisticated digital computers can't deal directly with the complex curves of an analog function. Although the modulation envelope of a single-sideband (SSB) signal isn't difficult for an analog communications receiver to process, a digital machine sees such a waveform as an "alien code." When we convert analog data to binary digital format, a sophisticated digital electronic circuit can understand it and modify it. The most powerful mainstream microprocessors comprise binary digital devices.

The DSP circuit works by eliminating confusion between digital states. Figure 28-7A shows a hypothetical signal before DSP. Figure 28-7B shows the same signal after DSP. The electronic circuit makes its digital "decision" (high or low) for defined time intervals. If the incoming signal remains above a certain level for an interval of time, the DSP output equals logic 1 (the high state). If the level remains below the critical point for a time interval, then the output equals logic 0 (the low state). A sudden burst of noise, such as *sferics* (atmospheric "static") from a nearby thunderstorm, might fool a DSP circuit into thinking that a signal is high when it's really low. But overall, errors occur less often with DSP than without it.

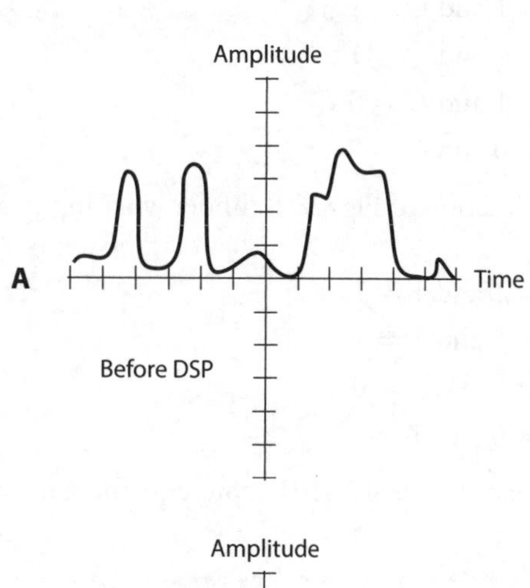

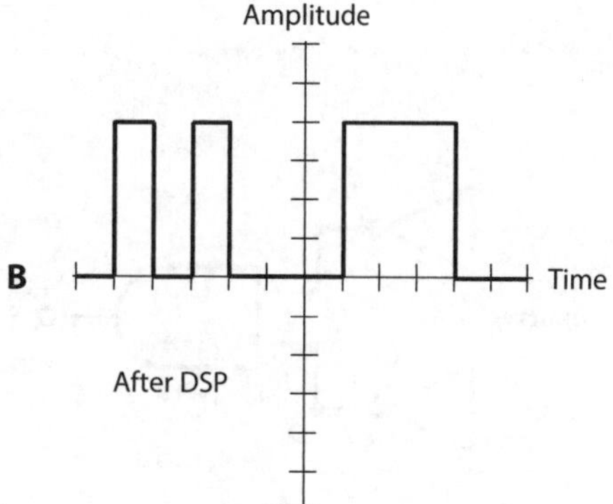

28-7 At A, a "dirty" binary signal before digital signal processing (DSP). At B, the same signal after DSP.

Quiz

Refer to the text in this chapter if necessary. A good score is at least 18 correct. Answers are in the back of the book.

1. In Boolean algebra, we represent the inclusive logic OR operation as
   (a) multiplication.
   (b) division.
   (c) addition.
   (d) subtraction.

2. The hexadecimal numeral C7 represents the same quantity as the decimal numeral
   (a) 127.
   (b) 199.
   (c) 212.
   (d) 263.

3. Suppose that we place a NOT gate in *cascade* (that is, in series) with each of the two inputs X and Y of an AND gate, as shown in Fig. 28-8. If $X = 1$ and $Y = 0$, how can we describe the states of the signals at points P and Q?
   (a) $P = 1$ and $Q = 1$
   (b) $P = 0$ and $Q = 1$
   (c) $P = 1$ and $Q = 0$
   (d) $P = 0$ and $Q = 0$

4. In the situation of Fig. 28-8, under what input conditions will we get logic 1 at the output point R?
   (a) $X = 1$ and $Y = 1$
   (b) $X = 0$ and $Y = 1$
   (c) $X = 1$ and $Y = 0$
   (d) $X = 0$ and $Y = 0$

5. The binary numeral 10101 represents the same quantity as the decimal numeral
   (a) 18.
   (b) 21.

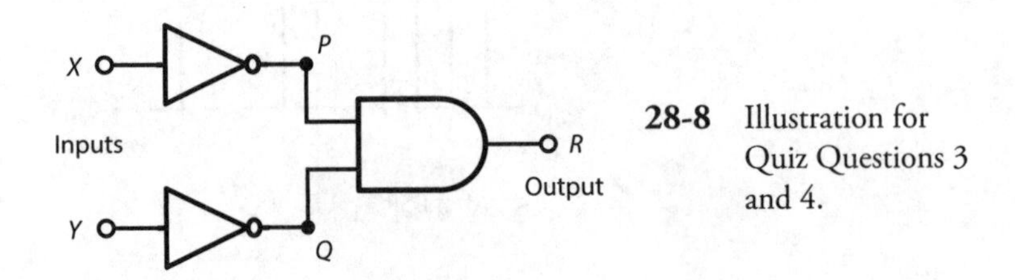

28-8 Illustration for Quiz Questions 3 and 4.

(c) 29.
(d) 57.

6. The octal numeral 425 represents the same quantity as the decimal numeral
   (a) 1034.
   (b) 517.
   (c) 277.
   (d) 194.

7. The decimal numeral 104 translates into the octal numeral
   (a) 173.
   (b) 161.
   (c) 150.
   (d) 137.

8. Suppose that we place a NOT gate at the output of an XOR gate, as shown in Fig. 28-9, and then we operate the combination of gates as a "black box." Under what input conditions X and Y will we get a low state at the output point Q?
   (a) Only when $X = 0$ and $Y = 0$
   (b) Only when $X = 1$ and $Y = 1$
   (c) Whenever $X = Y$ (X and Y have the same state)
   (d) Whenever $X \neq Y$ (X and Y have opposite states)

9. Under what input conditions X and Y will we get a high state at the output point Q in the "black box" of Fig. 28-9?
   (a) Only when $X = 0$ and $Y = 0$
   (b) Only when $X = 1$ and $Y = 1$
   (c) Whenever $X = Y$
   (d) Whenever $X \neq Y$

10. The second digit from the left in an eight-digit binary numeral carries a decimal value equal to a multiple of
    (a) 64.
    (b) 128.
    (c) 256.
    (d) 512.

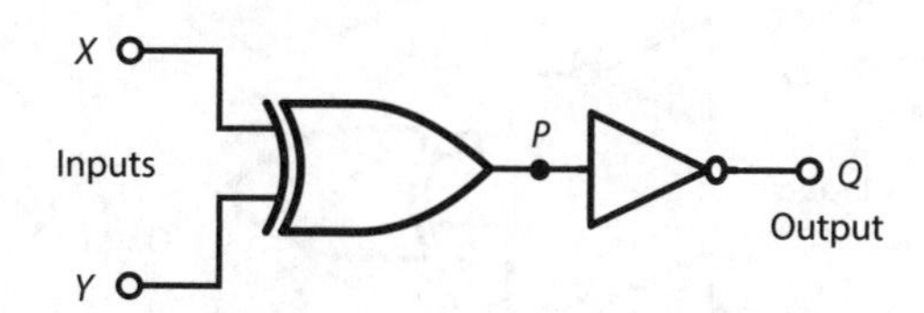

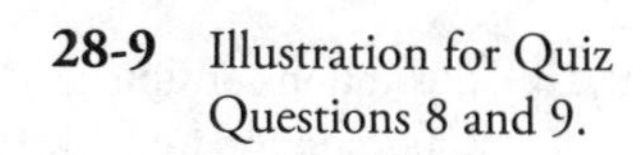
28-9 Illustration for Quiz Questions 8 and 9.

11. The decimal numeral 35 translates into the hexadecimal numeral
    (a) B2.
    (b) 2A.
    (c) AB.
    (d) 23.

12. When both inputs of an R-S flip-flop are high, the output states are
    (a) unpredictable.
    (b) both high.
    (c) both low.
    (d) opposite.

13. Suppose that we place a NOT gate in cascade with each of the two inputs X and Y of an XOR gate, as shown in Fig. 28-10. Under what input conditions will we get logic 1 at the output point R?
    (a) Only when $X = 1$ and $Y = 1$
    (b) Only when $X = 0$ and $Y = 0$
    (c) Whenever $X = Y$
    (d) Whenever $X \neq Y$

14. In the circuit of Fig. 28-10, under what input conditions will we get logic 0 at the output point R?
    (a) Only when $X = 1$ and $Y = 1$
    (b) Only when $X = 0$ and $Y = 0$
    (c) Whenever $X = Y$
    (d) Whenever $X \neq Y$

15. In the hexadecimal numeration system, what follows 999?
    (a) 1000
    (b) 99A
    (c) A000
    (d) A99

16. Digital signal processing can
    (a) enhance the distinction between high and low logic states.
    (b) minimize the number of errors per unit of time in a digital communications system.

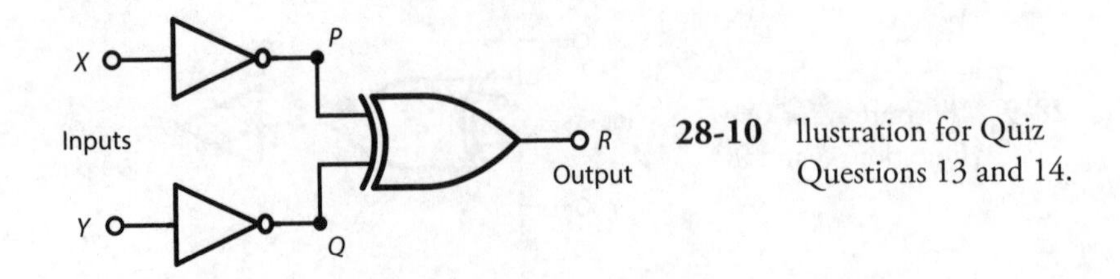

28-10 llustration for Quiz Questions 13 and 14.

(c) allow us to make accurate multigenerational copies of digital data.

(d) All of the above

17. The largest decimal value that we can represent as an eight-digit binary numeral is

(a) 511.

(b) 255.

(c) 127.

(d) 63.

18. To maximize the amount of useful information conveyed in a given time without increasing the actual number of characters or symbols sent, we can employ

(a) D/A conversion.

(b) A/D conversion.

(c) data compression.

(d) data acceleration.

19. If both of the inputs to a two-input NAND gate equal logic 0, the output state is

(a) low.

(b) high.

(c) unpredictable and unstable.

(d) a function of the previous states.

20. What do we call the number of times per second that a digital signal changes from high to low or vice versa?

(a) Baud

(b) Bits per second

(c) Characters per second

(d) Conversion rate

Test: Part 3

DO NOT REFER TO THE TEXT WHEN TAKING THIS TEST. A GOOD SCORE IS AT LEAST 37 CORRECT. Answers are in the back of the book. It's best to have a friend check your score the first time, so you won't memorize the answers if you want to take the test again.

1. When you design a power supply that uses rectifier diodes, you must make sure that each diode can handle
   (a) at least the expected average DC forward current.
   (b) at least half again the expected average DC forward current.
   (c) no more than 60 Hz.
   (d) at least 100 V RMS.
   (e) transients caused by lightning striking a nearby power line.

2. A crystal set radio receiver comprises an antenna, a detector, a headset shunted by a capacitor, and
   (a) an *LC* circuit.
   (b) an amplifier.
   (c) an oscillator.
   (d) a mixer.
   (e) all of the above

3. If you want current to flow between the source and drain of an N-channel JFET for the entire input cycle, you must make sure that
   (a) the gate-source junction remains reverse-biased, but at a voltage that never exceeds the pinchoff value during any part of the cycle.
   (b) the gate-source junction remains forward-biased, but at a voltage that never exceeds the pinchoff value during any part of the cycle.
   (c) the gate-source junction remains reverse-biased, but at a voltage that never falls below the pinchoff value during any part of the cycle.

(d) the gate-source junction remains forward-biased, but at a voltage that never falls below the pinchoff value during any part of the cycle.

(e) the gate-source junction remains zero-biased.

4. As you increase the negative DC gate voltage to a P-channel JFET while holding the negative DC drain voltage constant, what happens to the drain current, assuming that you apply no input signal?

(a) The drain current rises and eventually levels off when the gate-drain junction approaches a state of saturation.

(b) The drain current decreases because the channel narrows, reducing its conductance.

(c) The drain current increases because the source current increases, increasing the effective conductance of the channel.

(d) The drain current remains the same because the channel width remains constant.

(e) The drain current decreases because current bleeds off into the gate from the channel.

5. In the scenario of Question 4, suppose that you keep increasing the negative DC gate voltage while leaving the negative DC drain voltage constant. What will happen to the drain current in the absence of an input signal?

(a) The drain current will keep going down and approach zero because more and more of the channel current will bleed off through the gate.

(b) The drain current will remain constant until, at a certain point, the extreme negative gate voltage will destroy the device.

(c) The drain current will increase as the channel conductance improves, all the way up to the point at which the extreme negative gate voltage destroys the device.

(d) The channel current will drop to zero and then reverse direction.

(e) The drain current will attain its maximum value for the particular DC drain voltage that you apply.

6. If you increase by tenfold the *power* dissipated by a load that has no reactance, you get

(a) 3 dB gain.

(b) 6 dB gain.

(c) 10 dB gain.

(d) 20 dB gain.

(e) 100 dB gain.

7. If you increase by tenfold the *voltage* across a load that has no reactance, you get

(a) 3 dB gain.

(b) 6 dB gain.

(c) 10 dB gain.

(d) 20 dB gain.

(e) 100 dB gain.

8. In a depletion-mode JFET, DC gate bias of the appropriate polarity
   (a) keeps the input signal from leaking through to the source.
   (b) prevents destruction of the device by electrostatic discharge.
   (c) keeps the device from oscillating.
   (d) constricts the channel between the source and drain.
   (e) forms a channel between the source and drain.

9. If you double the *RMS current* through a load with a purely resistive impedance, you get a gain of
   (a) 3 dB.
   (b) 4 dB.
   (c) 6 dB.
   (d) 8 dB.
   (e) 12 dB.

10. How does the phase of the AC output signal compare with the phase of the AC input signal in a class-A, common-emitter, bipolar-transistor amplifier?
   (a) The output and input waves coincide in phase.
   (b) The output wave leads the input wave by 90°.
   (c) The output wave lags the input wave by 90°.
   (d) The output and input waves oppose each other in phase.
   (e) In an NPN device, the output wave leads the input wave by 90°; in a PNP device, the output wave lags the input wave by 90°.

11. Figure Test 3-1 shows the situation inside a semiconductor diode under certain conditions. The small black dots represent majority carriers on either side of the P-N junction. The presence of a depletion region suggests
   (a) forward bias.
   (b) reverse bias.
   (c) zero bias.
   (d) that the diode is intended for switching.
   (e) that the diode is intended for rectification.

12. In the scenario of Fig. Test 3-1, the particles moving toward point X are
   (a) protons.
   (b) neutrons.
   (c) electrons.
   (d) positrons.
   (e) holes.

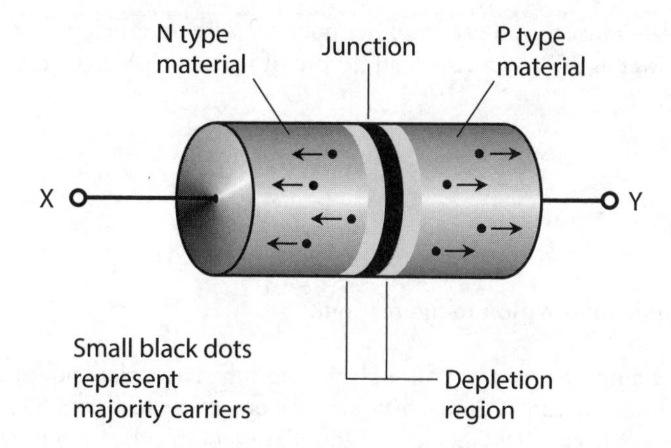

Test 3-1 Illustration for Part 3 Test Questions 11 and 12.

13. A PIN diode works best as a
    (a) rectifier.
    (b) detector.
    (c) mixer.
    (d) frequency multiplier.
    (e) high-speed switch.

14. Figure Test 3-2 shows a circuit that's intended for the output of an AC-to-DC power supply. The resistor
    (a) limits the output voltage.
    (b) limits the current that the Zener diode must handle.
    (c) maximizes the output voltage.
    (d) maximizes the current that the supply can deliver.
    (e) keeps the output power under control.

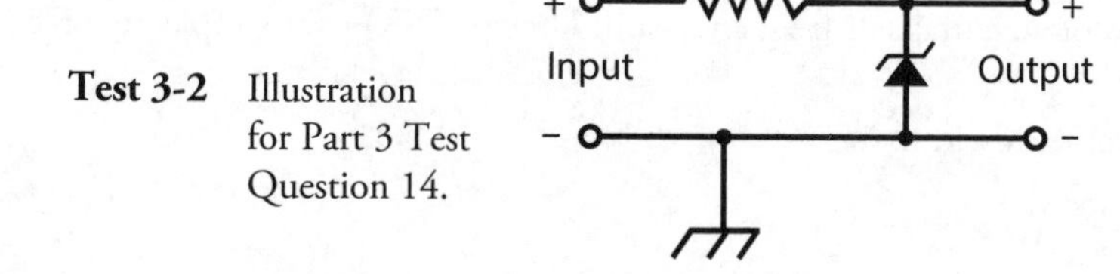

Test 3-2 Illustration for Part 3 Test Question 14.

15. A class-B bipolar-transistor power amplifier operates with an efficiency of 50%. We measure the RMS output power as 70 W across a load devoid of reactance. What's the DC collector input power?

    (a) 35 W

    (b) 49 W

    (c) 99 W

    (d) 140 W

    (e) We need more information to figure it out.

16. A class-C power amplifier using a 250-TH vacuum tube has a plate power input of 300 W and an RF power output of 225 W into a 50-ohm, reactance-free load at 3.550 MHz. (Have you ever seen one of things? It's a relic! Google on "250-TH vacuum tube" as a phrase.) What's the efficiency of this amplifier?

    (a) 66%

    (b) 75%

    (c) 80%

    (d) 88%

    (e) 92%

17. In an IC, component density correlates directly with

    (a) operating speed.

    (b) power output.

    (c) voltage rating.

    (d) current drain.

    (e) All of the above

18. If you input signals at 6.0 MHz and 2.5 MHz to a mixer, at which of the following frequencies should you *not* expect to get any output?

    (a) 8.5 MHz

    (b) 3.5 MHz

    (c) 6.0 MHz

    (d) 15.0 MHz

    (e) 2.5 MHz

19. What's a major advantage of frequency-shift keying (FSK) over simple on/off keying such as Morse code?

    (a) Narrower bandwidth

    (b) Higher speed

(c) Greater accuracy
(d) Improved frequency stability
(e) Reduced harmonic emission

20. In which of the following systems could you use a vacuum tube instead of a transistor?
(a) Hi-fi audio power amplifier
(b) Radio-frequency (RF) oscillator
(c) RF power amplifier
(d) Weak-signal RF amplifier
(e) All of the above

21. Figure Test 3-3 is a diagram of a power supply designed to produce high-voltage DC. It lacks an essential circuit. What is it, and what should it do?
(a) A filter to get pure DC at the output
(b) A capacitor to regulate the output voltage
(c) A resistor to limit the output current
(d) A diode to double the ripple frequency
(e) A coil to stabilize the output power

22. What's the purpose of the components marked X in Fig. Test 3-3?
(a) Stabilize the input voltage
(b) Limit the power that the device consumes
(c) Keep the input terminals from shorting to ground
(d) Ensure that the input is a pure AC sine wave
(e) None of the above

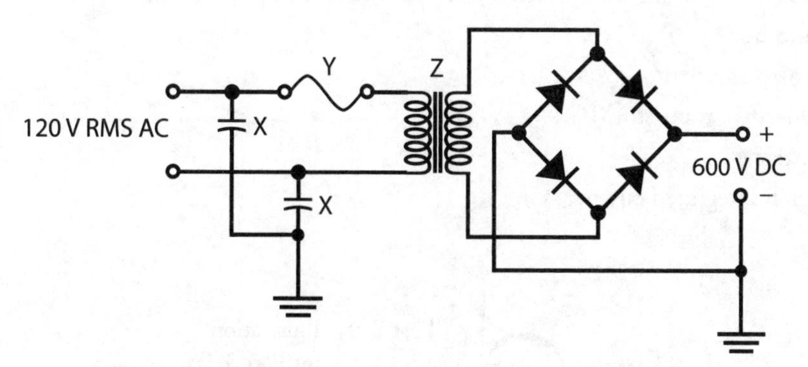

Test 3-3 Illustration for Part 3 Test Questions 21 through 24.

23. In Fig. Test 3-3, what's the component represented by the symbol marked Y?
    (a) A current limiter
    (b) A rectifier
    (c) An AC line filter
    (d) An incandescent lamp
    (e) A fuse

24. In Fig. Test 3-3, what's the component represented by the symbol marked Z?
    (a) An isolation transformer
    (b) A step-up transformer
    (c) A wave-smoothing coil
    (d) A ripple filter
    (e) A transient suppressor

25. Figure Test 3-4 shows the schematic symbol for
    (a) an NPN bipolar transistor.
    (b) a PNP bipolar transistor.
    (c) an N-channel JFET.
    (d) a P-channel JFET.
    (e) an enhancement-mode MOSFET.

26. In Fig. Test 3-4, what's the electrode represented by the line with the arrow?
    (a) The gate
    (b) The emitter
    (c) The source
    (d) The collector
    (e) The base

27. Which of the following components will you find in an optoisolator?
    (a) Photodiode
    (b) Bipolar transistor
    (c) Field-effect transistor (FET)
    (d) Capacitor
    (e) Digital integrated circuit (IC)

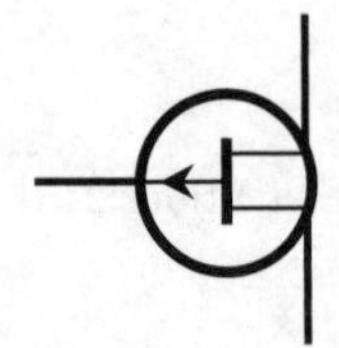

Test 3-4 Illustration for Part 3 Test Questions 25 and 26.

28. You bias a JFET beyond pinchoff (under no-input-signal conditions) for use as an RF power amplifier. It takes a fairly strong input signal to produce any output. Even then, channel current flows for less than half the cycle. You have a
    (a) class-A amplifier.
    (b) class-AB_1 amplifier.
    (c) class-AB_2 amplifier.
    (d) class-B amplifier.
    (e) class-C amplifier.

29. Imagine that over a small range, an increase in the voltage across a component causes the current to *decrease*. This is an example of
    (a) negative resistance.
    (b) parasitic conduction.
    (c) current-reversal effect.
    (d) avalanche effect.
    (e) None of the above, because such a situation can never occur.

30. Figure Test 3-5 shows how the collector current varies as a function of the base current for a bipolar transistor under no-signal conditions. If you want to use the device as a class-B amplifier, at which point should you bias it?
    (a) Point V
    (b) Point W
    (c) Point X
    (d) Point Y
    (e) Point Z

31. In Fig. Test 3-5, point V portrays
    (a) linear bias.
    (b) cutoff.
    (c) zero bias.
    (d) saturation.
    (e) pinchoff.

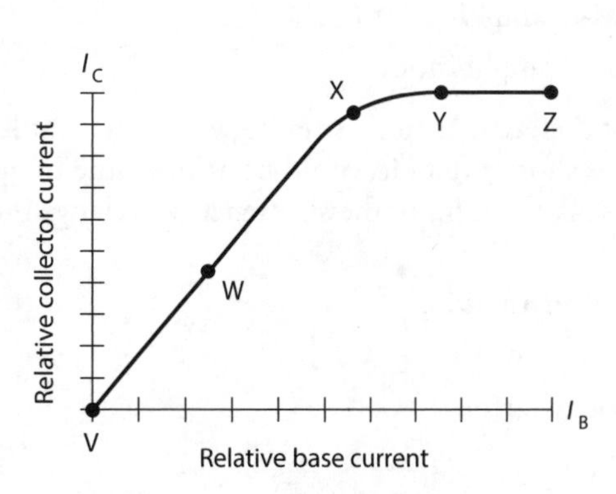

Test 3-5 Illustration for Part 3 Test Questions 30 and 31.

32. You can construct a push-pull amplifier with two identical JFETs, connecting the gates in phase opposition and connecting the drains
    (a) in phase coincidence.
    (b) 30° out of phase.
    (c) 60° out of phase.
    (d) 90° out of phase.
    (e) in phase opposition as well.

33. In an FM signal, the extent to which the instantaneous signal amplitude varies is called
    (a) zero, because it remains constant!
    (b) the modulation index.
    (c) the amplitude deviation.
    (d) the amplitude fluctuation.
    (e) the amplitude differential.

34. If we want a diode to completely block current in the reverse direction, we must ensure that
    (a) the N type material has a more positive voltage than the P type material, and the peak applied voltage remains below the avalanche threshold.
    (b) the N type material has a more negative voltage than the P type material, and the peak applied voltage remains below the avalanche threshold.
    (c) the N type material has a more positive voltage than the P type material, and the peak applied voltage exceeds the avalanche threshold.
    (d) the N type material has a more negative voltage than the P type material, and the peak applied voltage exceeds the avalanche threshold.
    (e) the peak applied voltage never exceeds the alpha cutoff value.

35. Figure Test 3-6 shows how the gate voltage affects the drain current in a typical JFET. If you bias the device near the middle of the range marked X without inputting any signal, you have conditions good for
    (a) class-A amplification.
    (b) class-B amplification.
    (c) class-C amplification.
    (d) class-D amplification.
    (e) class-E amplification.

36. In most dielectric materials, radio waves travel at less than the speed of light, so the wavelength is shorter than it would be at the same frequency in air or a vacuum. For any given dielectric, the ratio of the shortened wavelength to the wavelength in air or a vacuum is called its
    (a) wavelength ratio.
    (b) speed ratio.
    (c) velocity factor.

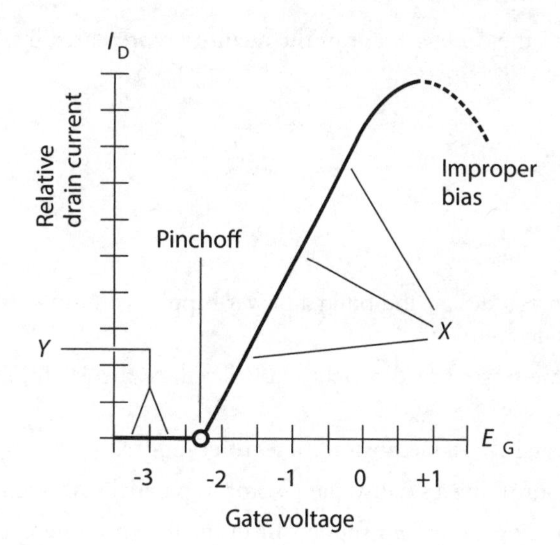

Test 3-6 Illustration for Part 3 Test Question 35.

(d) propagation factor.
(e) frequency factor.

37. Figure Test 3-7 shows the schematic symbols for
    (a) N-channel (at A) and P-channel (at B) depletion-mode JFETs.
    (b) N-channel and P-channel enhancement-mode JFETs.
    (c) N-channel and P-channel depletion-mode MOSFETs.
    (d) N-channel and P-channel enhancement-mode MOSFETs.
    (e) nothing that we've discussed in this book, at least not yet.
38. The devices symbolized in Fig. Test 3-7 have
    (a) extremely high input impedances.
    (b) extremely low input impedances.
    (c) excellent linearity but no way to produce any gain.
    (d) high bias voltage requirements.
    (e) no known characteristic; neither has appeared in this book, at least not yet.

Test 3-7 IIllustration for Part 3 Test Questions 37 and 38.

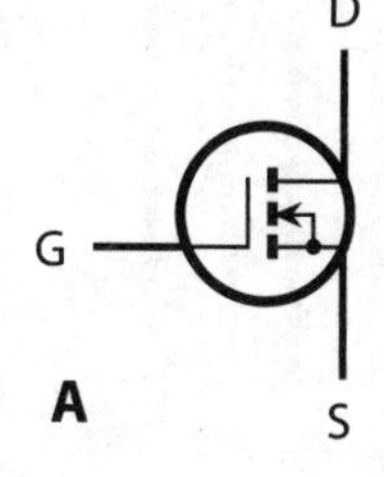

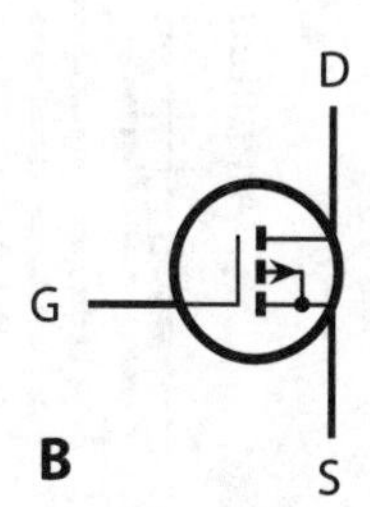

39. What's the decimal value of the quantity represented by the binary numeral 11011?
    (a) 18
    (b) 22
    (c) 27
    (d) 31
    (e) 37

40. When you design and build a power supply, you can best minimize the risk of transient-related diode failure by
    (a) inserting a timed switch that will reduce the AC input voltage for a few seconds after the transient occurs.
    (b) using RF diodes instead of rectifier diodes.
    (c) connecting a small-value resistor in parallel with each diode.
    (d) placing a transient suppressor in the AC input line.
    (e) using a transformer that can handle extremely high current.

41. How can we render the decimal quantity 45 in binary terms?
    (a) 101101
    (b) 100100
    (c) 110001
    (d) 100011
    (e) 111011

42. Figure Test 3-8 is a time-domain graph illustrating the principle of
    (a) pulse-code modulation.
    (b) pulse-amplitude modulation.
    (c) pulse-duration modulation.

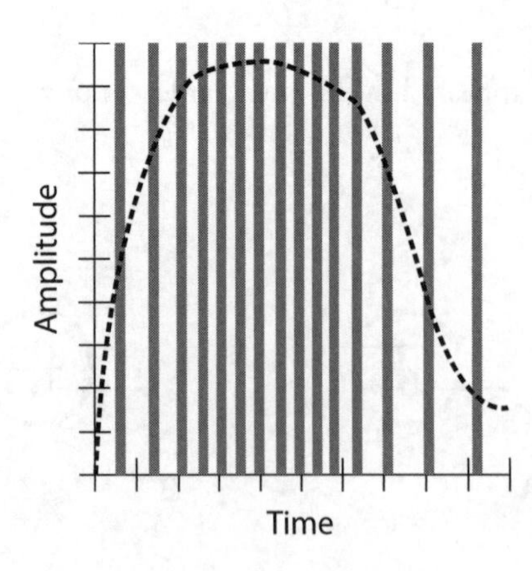

Test 3-8 Illustration for Part 3 Test Question 42.

(d) pulse-interval modulation.
(e) pulse-width modulation.

43. You can increase the sensitivity of your HF radio receiver by
(a) using a squelch to silence the receiver when no signal comes in.
(b) placing a preamplifier between the antenna and the front end.
(c) reducing the voltage of the receiver's power supply or battery.
(d) incorporating a balanced modulator to cancel out noise.
(e) adding a notch filter to increase the IF bandwidth.

44. For class-C power amplification, you should, under no-signal conditions, bias a bipolar transistor
(a) so drain current never flows, no matter how strong the input signal gets.
(b) somewhat beyond cutoff.
(c) exactly at cutoff.
(d) such that steady drain current flows.
(e) to saturation.

45. How many grids does a tetrode tube have?
(a) None
(b) One
(c) Two
(d) Three
(e) Four

46. Inside some pentode tubes, one of the grids is directly connected to the cathode. Which grid?
(a) The collector grid
(b) The control grid
(c) The screen grid
(d) The suppressor grid
(e) The gate grid

47. In a class-A JFET amplifier, a small change in the instantaneous gate voltage should produce
(a) a large change in the instantaneous drain voltage.
(b) an even smaller change in the instantaneous source voltage.
(c) a large change in the instantaneous channel current.
(d) an equivalent change in the instantaneous channel voltage.
(e) no change in the instantaneous channel current.

48. A grounded-grid amplifier using a triode vacuum tube
    (a) is less likely to oscillate than a grounded-cathode amplifier.
    (b) needs less driving power than a grounded-cathode amplifier.
    (c) has high input impedance, making it ideal for weak-signal RF amplification.
    (d) can operate with far lower voltage than a grounded-cathode amplifier.
    (e) All of the above

49. In the hexadecimal numeration system, what follows A as you count upward?
    (a) Nothing; that system has no digit A.
    (b) B
    (c) C
    (d) D
    (e) 10

50. The charge carriers in a vacuum tube are
    (a) protons.
    (b) neutrons.
    (c) holes.
    (d) electrons.
    (e) ions.

4
PART

Specialized Devices and Systems

29

CHAPTER

Microcontrollers

A *MICROCONTROLLER* IS A SINGLE IC THAT CONTAINS WHAT WOULD HAVE PASSED FOR MOST OF A home computer in the 1980s. It includes a simple 8-bit (sometimes more) microprocessor, nonvolatile flash memory that holds a program to be run, and *random access memory* (RAM) to hold temporary data and values. Many microcontrollers also have some *electrically erasable programmable read-only memory* (EEPROM) that is used for nonvolatile storage of program data. That is, unlike normal RAM, the contents are not lost when the microcontroller is not powered. A microcontroller also has numerous *general-purpose input/output* (GPIO) pins that you can use to interface with sensors, switches, LEDs, displays, etc. It is essentially a computer on a chip.

Benefits

The benefit of having a single microcontroller chip instead of many separate logic chips is cost savings. You will find microcontrollers in almost any item of consumer electronics that you care to mention. You will find them in everything from an electric toothbrush to a car (which probably has tens of microcontrollers in it).

Figure 29-1 shows how a typical microcontroller chip is organized. The *central processing unit* (CPU) fetches instructions one at a time from the flash memory. These instructions together form the program running on the microcontroller and include things such as:

- Add numbers together
- Compare numbers
- Jump to another part of the program depending on a comparison
- Read a digital input
- Write a digital output

Microcontrollers perform things one step at a time and use a *clock* to trigger each step. The clock is an oscillator of generally 1 MHz to 20 MHz for a low-power microcontroller. Each time the clock ticks over, another instruction is performed.

Microcontrollers interact with the world through their general-purpose input/output (GPIO) pins. These can be configured to be inputs or outputs, just like the inputs and outputs to logic gates

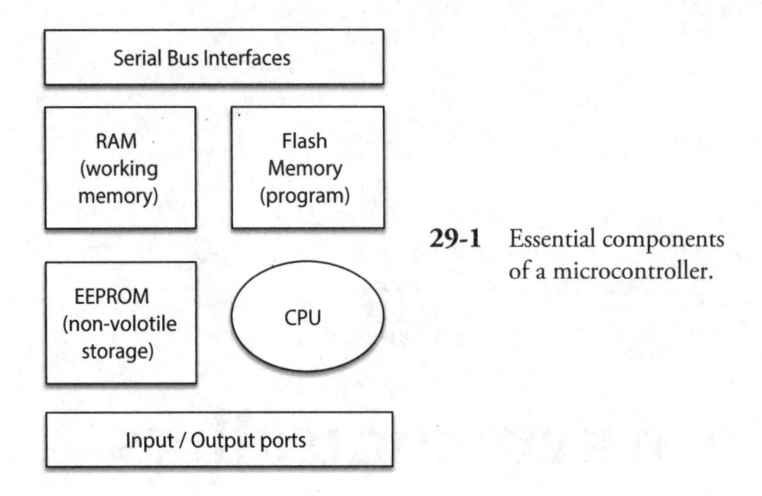

29-1 Essential components of a microcontroller.

described in Chap. 28 "Digital Basics." However, because what links the pins is software rather than, say, a hardware AND gate, the microcontroller is almost infinitely flexible, limited only by its speed, memory, and the ability of you, the programmer, to write the program that controls it.

Microcontrollers generally have a method of converting between continuous analog signals and the on/off world of digital electronics. This means that their use is not restricted to just the digital, but they can also be pressed into service for some analog electronic applications.

A special subspecies of microcontroller called the *digital signal processing* (DSP) microcontroller is optimized for such analog activities. An echo effects box for a guitar or even tone controls for audio amplifiers are often implemented by digitizing an analog signal, doing some math, and then converting the digital value back into an analog signal again.

The downside of all this flexibility is that the microcontroller is not made knowing how to perform all this magic; it must be programmed, and so the person wanting to use it has to learn a little computer programming as well as electronics. The programs for a microcontroller are generally short and simple, and you will learn a lot more about this in the next chapter in which we look at the extremely popular microcontroller board, the Arduino.

Microcontroller boards, such as the Arduino, combine a microcontroller chip with supporting components like voltage regulators and a USB programming interface so that you can get programs onto your microcontroller without having to have separate programming hardware.

All Shapes and Sizes

Microcontrollers are available in all sorts of package sizes, from 3 pins to hundreds of pins, so you can pick a device with the features, performance, price, and number of GPIO pins that you need for your design. In fact, there are so many different devices that it is not easy to navigate the vast array of devices on offer. To simplify things a little, devices are often grouped into families of microcontrollers that have the same basic structure and programming instructions, but have different numbers of pins and different quantities of flash memory for program storage.

The ATtiny family of microcontrollers from the Atmel chip manufacturer are typical of low-end microcontrollers. Table 29-1 shows the features of these common microcontrollers. The guide prices are somewhat arbitrary but are based on the price for one chip at the time of writing.

Table 29-1. ATtiny Microcontrollers

Microcontroller	Package Pins	GPIO Pins	Flash Memory (kB)	Guide Price (c)
ATtiny13	8	6	1	60
ATtiny45	8	6	4	90
ATtiny44	14	12	4	90
ATtiny2313	20	18	2	120

General-Purpose Input/Output (GPIO) Pins

The clever part of a GPIO pin is the "general purpose" part. This means that your program running on the microcontroller can decide whether the pin should be an input or an output and also, if it is an input, whether or not to enable a built-in resistor to pull the input up to the positive supply of the microcontroller.

Figure 29-2 is a simplified schematic diagram for a typical GPIO pin. The output drive is *push-pull*. This means that when configured as an output, the pin can either *sink* (receive) or source current, although generally only between 10 mA and 40 mA in each direction. In addition, control logic allows the whole output to be disabled so that the pin can act as a digital input.

Microcontrollers often have some pins that can be used as analog inputs. This requires the pin to be connected to a comparator; by comparing the voltage at the input to a series of different voltages generated within the microcontroller, the pin can be used to measure a voltage.

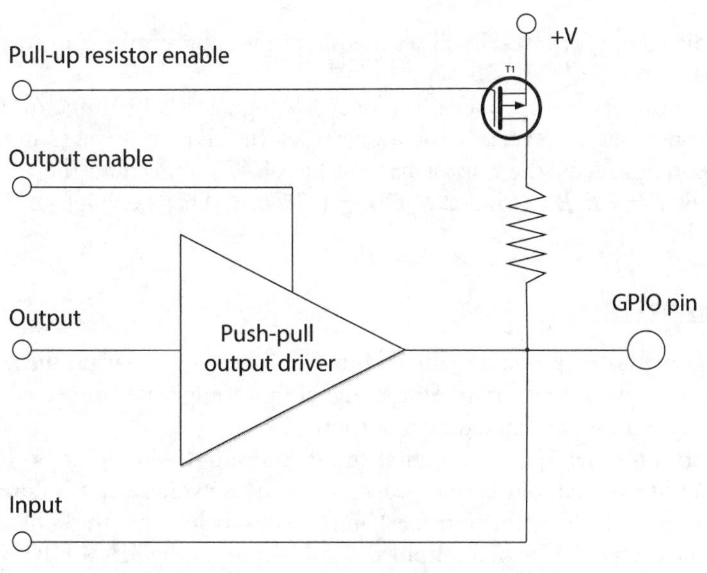

29-2 Simplified schematic of a GPIO pin.

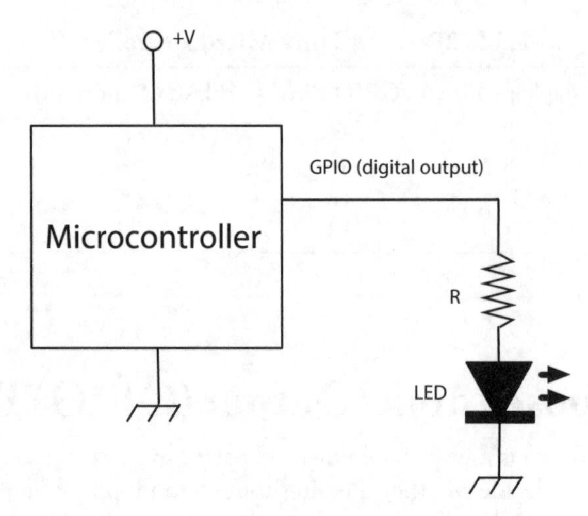

29-3 Lighting an LED with a digital output.

Digital Outputs

The most common thing to do with a GPIO pin is to have it act as a digital output. That is, to turn something on and off. This could be something connected directly to the GPIO pin, like an LED, or it could use a transistor or relay to provide more current or voltage, or both, to whatever is being controlled.

Figure 29-3 shows how you would typically connect the GPIO pin of a microcontroller to an LED. The series resistor is necessary because GPIO pins can typically supply only a few tens of milliamps.

Setting the digital output HIGH will set the pin at the supply voltage for the microcontroller (5 V or 3.3 V) and setting it LOW will set it to 0 V.

Most microcontrollers can operate at 5-V or 3.3-V logic levels (5-V micros will nearly always work with 3.3-V ones, but the reverse is not always true). In this case, if you assume that the microcontroller is operating at 5 V, the resistor has a value of 470 ohms, and the LED has a forward voltage of 2 V, then $I = E/R = (5 - 2)/470 = 6.28$ mA. That's enough for the LED to light reasonably brightly.

Digital Inputs

When you configure a GPIO pin to be a digital input, it is because you want the microcontroller to do something in response to that input. So the digital input might be connected to a push switch or a motion detector sensor that has a digital output.

When a microcontroller "reads" a digital input, the input will either be HIGH or LOW. HIGH usually means over half of the microcontroller's supply voltage (2.5 V for a 5-V microcontroller and 1.65 V for a 3.3-V microcontroller) and LOW is below that threshold.

It's fine to connect a 3.3-V digital output to a 5-V input, as the 3.3-V HIGH voltage will still be above the 2.5-V input threshold, but you must not connect a 5-V digital output to a 3.3-V digital input unless the input is described as "5-V tolerant." It is not uncommon for a 3.3-V microcontroller to have some 5-V-tolerant inputs just to make it easier to interface them in systems with multiple supply voltages.

Because microcontrollers can implement all sorts of complex logic, there is usually no need to connect a switch to a microcontroller that is any more complex than a simple momentary action, push-to-make switch. Such switches are arranged to switch to ground, using a pull-up resistor that keeps the input high until it is connected to ground by the switch. Figure 29-4A shows a switch connected to a digital input using the internal pull-up resistor of the microcontroller, and Fig. 29-4B shows a similar switch using an external pull-up resistor.

To avoid the need for an external pull-up resistor, the pin logic for a microcontroller normally includes a built-in pull-up resistor that can be enabled or disabled from the program running on your microcontroller. The value of this resistor is often specified as a range rather than an exact value, and this might typically be 30 kilohms to 50 kilohms. Under most circumstances this will work just fine, but if you have a long lead to your switch, then you may need to supply your own external and much smaller value of resistor, say, 270 ohms. In this case, the resistor should be connected as shown in Fig. 29-4B.

Switch contacts tend to bounce. That is, when you press the button, you do not get a single clean closure of the contacts, but rather the contacts close and open a number of times in quick succession. For some applications, this does not matter, but if, say, alternate presses of the switch button are used by the microcontroller to turn something else on and off, then if the digital input receives an even number of bounces in rapid succession, it could seem like nothing happened when the button was pressed.

Debouncing switches for use in digital circuitry without a microprocessor requires hardware debouncing; however, if the switch is connected to a microcontroller, then it can be debounced using software without the need for any extra components. You will see an example of how to do this with an Arduino microcontroller board in Chap. 30.

An ordinary digital input normally requires the program running on the microcontroller to repeatedly read the value of the digital input until it changes. This means that some very short pulses on a digital input might be missed because they went high then low again before the program on the microcontroller could register the change. This won't matter for pressing a button, but for such very short-pulsed inputs (microseconds), then some of the microcontroller's GPIO pins can be

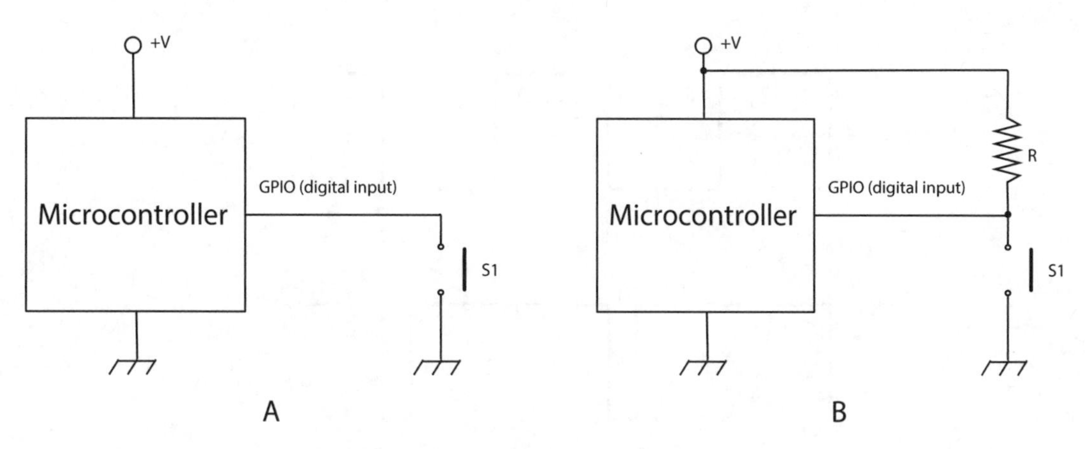

29-4 Connecting a switch to a digital input.

designated as "interrupts." These are guaranteed to be registered by the microcontroller and cause your program to stop whatever it was doing and instead run an interrupt service routine.

PWM Outputs

A few microcontrollers provide true analog outputs. These allow the output to be varied in a large number of steps (often 1024 or 4096) between 0 and the microcontroller supply voltage. These microcontrollers are the exception though, and in general, microcontrollers use a technique called PWM (see the section "The Class D Amplifier" in Chap. 26 and the section "Pulse-Width Modulation" in Chap. 27) to provide something that approximates an analog output.

Rather than using PWM to transmit a signal, as described in Chap. 27, or to amplify an audio signal, as described in Chap. 26, a microcontroller uses low-frequency PWM to control the power to, say, a motor or LED, controlling its speed or brightness by altering the duration of the pulses powering the output device.

Figure 29-5 shows how PWM works. If the pulses are short (say, high for just 5% of the time), then only a small amount of energy is delivered with each pulse. The longer the pulse, the more energy is supplied to the load. When powering a motor, this will control the speed at which the motor rotates. When driving an LED using PWM, the brightness appears to change. In fact, an LED can turn on and off millions of times per second, and so the PWM pulses will become pulses of light. The human eye and brain do the averaging trick for us, making the brightness of the LED vary with the pulse length.

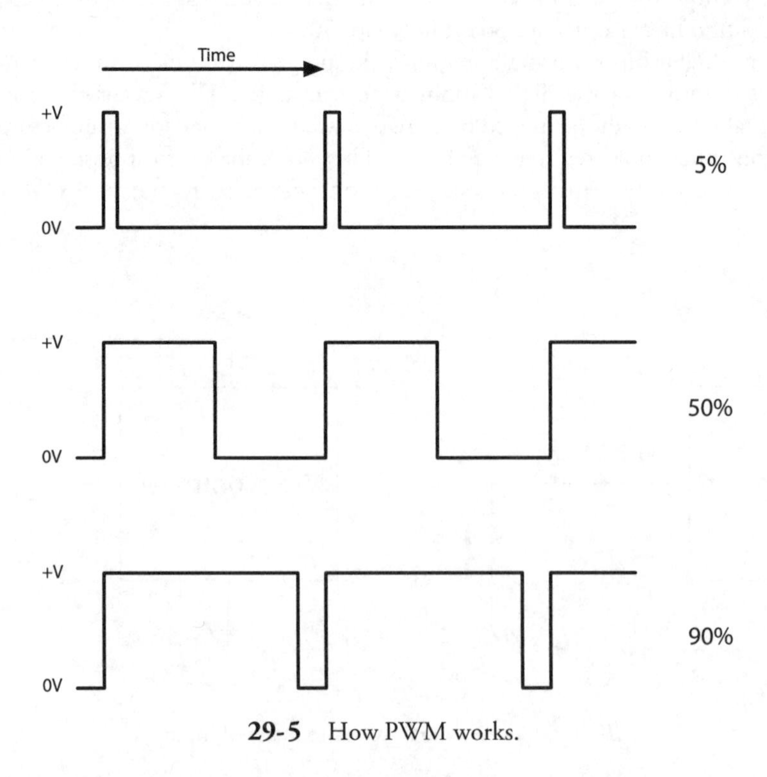

29-5 How PWM works.

Most microcontrollers have hardware PWM available on some, or all, of their pins. If they do not, then PWM can be implemented in software quite easily. The pulse frequency of the PWM outputs is generally around 500 Hz but is also configurable to be higher or lower, depending on the application.

Analog Inputs

Converting a voltage (between 0 and the microcontroller supply voltage) to a number for use by the program running on the microcontroller requires the use of an *A/D converter* (analog-to-digital converter), perhaps more often called an ADC these days.

Nearly all microcontrollers use a technique called *successive approximation* to convert the analog voltage to a digital value. This technique uses a *digital-to-analog converter* (DAC), which converts a digital value to an analog voltage, and a comparator, as shown in Fig. 29-6.

The GPIO pin being used as an analog input is connected to one input of a voltage comparator. The other input of the comparator is connected to the output of a DAC. The DAC generates an analog voltage depending on the binary number on D0 to D9 supplied by the microcontroller. So, if this value is 0 (D0 to D9 all low) then the output voltage of the DAC will be 0. If, on the other hand, D0 to D9 are all high (the maximum value), then the output voltage from the DAC is the microcontroller supply voltage.

It's the comparator's job to provide feedback to the microcontroller in a game of "higher/lower" to gradually home in on a number input to the DAC that just balances the input from the GPIO pin. This takes a number of steps. In fact, the number of steps is equal to the number of bits input to the DAC (10 in Fig. 29-6).

The trick is to start with the highest value bit (D9) in Fig. 29-6. Assuming a 5-V microcontroller, then if bit D9 is high and all the other bits are low, then the output voltage of the DAC will be 2.5 V (half of 5 V). If the GPIO input voltage is above this, then the comparator output will be high. If it is less, then the output of the comparator will be low. So, if the output of the comparator (Result in Fig. 29-6) is high, the microcontroller leaves bit D9 set and moves on to the next bit D8, and applies the same test using the comparator. This is repeated for the other bits of the DAC.

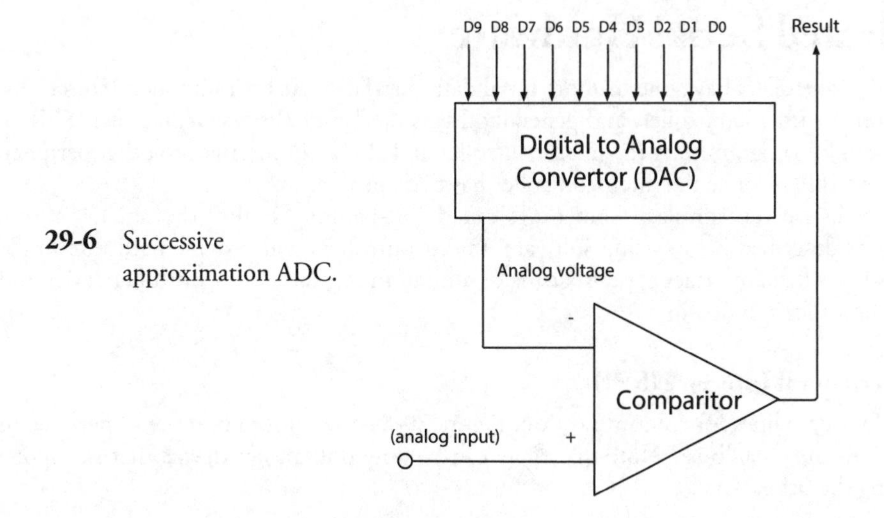

29-6 Successive approximation ADC.

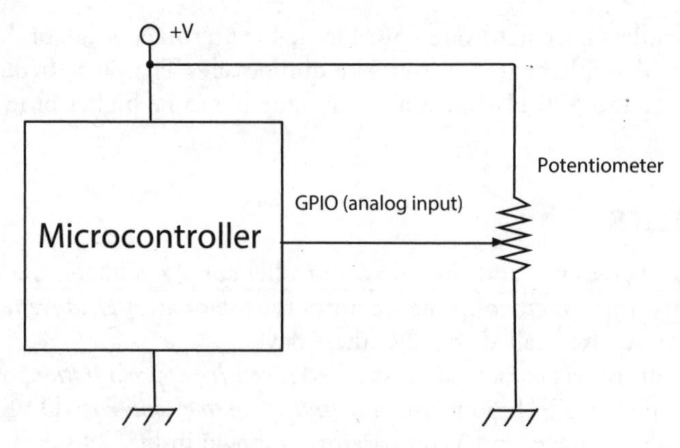

29-7 Attaching a potentiometer to the analog input of a microcontroller.

Microcontrollers have different bit resolutions of ADC. The example here is 10 bits, which is typical of many microcontrollers and is what is used in the ATmega328 microcontroller used by the Arduino. Some microcontrollers have 8- or 12-bit resolutions. The more bits of resolution, the more precise the reading, but the longer it takes to make.

An 8-bit resolution ADC gives a value between 0 and 255 ($2^8 - 1$). So, a voltage of 3 V at the GPIO pin of a 5-V microcontroller would result in a reading of $3/5 \times 255 = 153$. For a microcontroller with 10-bit ADC, the range of value would be 0 to 1023 ($2^{10} - 1$).

Figure 29-7 shows how you can attach a potentiometer to the analog input of a microcontroller so that the position of the potentiometer's knob can be read by the program running on the microcontroller.

The resistance of the potentiometer needs to be much lower than the input impedance of the comparator. Most microcontroller comparators will have an input impedance in the megohm range, so a potentiometer of perhaps 10 k should be fine.

Dedicated Serial Hardware

Most microcontrollers have one or more serial data interfaces. At least one such interface is needed to program the microcontroller, and generally this is the *Serial Peripheral Interface* (SPI). This can be used both for programming the microcontroller and also as an interface to other peripheral chips or microcontrollers, once the microcontroller has been programmed.

Although you can simulate (sometimes called "bit-banging") all of the serial communication mechanisms described below using software, microcontrollers will provide hardware implementations of some of the interface types that make programming the microcontroller easier and higher speed communication possible.

Serial Peripheral Interface (SPI)

SPI uses four data lines for communication. Figure 29-8 shows how a number of peripherals can be connected to the data "bus." Note that there can be only one master device that is responsible for controlling the other devices.

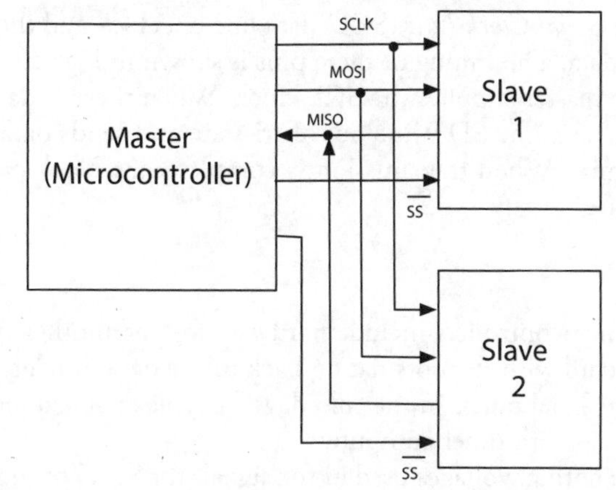

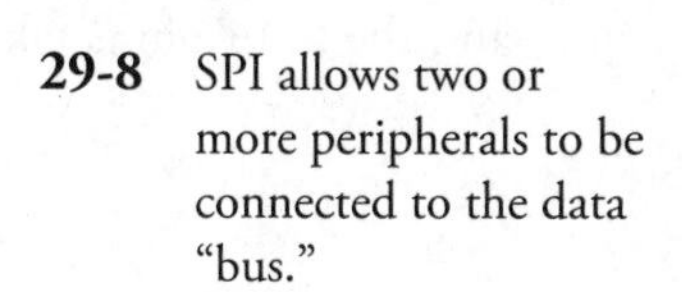
29-8 SPI allows two or more peripherals to be connected to the data "bus."

Each of the slave devices has a dedicated *slave select* (SS) line, and the master enables just the device it wants to communicate with. Two data lines are needed because separate lines are used for each direction of communication. *Master Out Slave In* (MOSI) carries the data from the master to the slave device, and *Master In Slave Out* (MISO) the reverse. A separate clock signal synchronizes the data transmission.

I²C

Inter-Integrated Circuit (I²C), also sometimes known as *two-wire interface* (TWI), serves much the same purpose as SPI, although it has two wires for data rather than the four of SPI. I²C is often used by displays and other peripheral modules that are designed to be connected to a microcontroller.

The two data lines of I²C are open-drain connections that operate as both inputs and outputs at the microcontroller. They need pull-up resistors so that when not being driven low as an output, they are high. The protocol that controls the direction of data flow also ensures that the situation never occurs in which one end of the bus is connected to a digital output that is low and the other end to a digital output that is high.

Figure 29-9 shows how two microcontrollers might communicate with each other using I²C.

I²C devices are either masters or slaves, and there can be more than one master device per bus. In fact, devices are allowed to change roles, although this is not usually done.

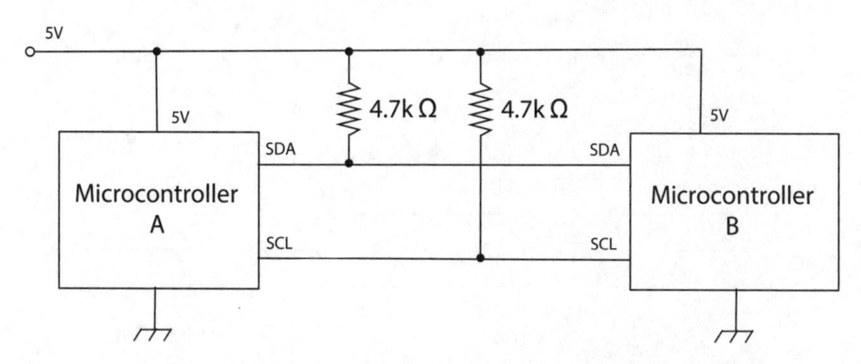

29-9 Microcontroller-to-microcontroller communication using I²C.

The *Serial Clock Line* (SCL) data line is a clock and the other, the *Serial Data Line* (SDA), carries the data. The timing of these pins is shown in Fig. 29-10.

The master supplies the SCL clock. When there is data to be transmitted, the sender (master or slave) takes the SDA line out of tri-state and sends data as logic highs or lows in time with the clock signal. When transmission is complete, the clock can stop, and the SDA pin is taken back to tri-state.

Serial

Many microcontrollers include hardware for yet another type of interface called *Serial.* This is an old standard with its roots dating back to the days of teletypes. Some computers can still be found that have Serial ports. In the "old days," people attached modems to them for communicating over phone lines with other computers.

The normal voltages used in the signals for Serial ports conform to the RS232 standard and use voltages that swing both positive and negative with respect to GND. This is not terribly convenient when using microcontrollers. For this reason, microcontrollers often use the same communication protocol, but at logic levels. This is called *TTL Serial,* or often just *Serial.* Let's use a capital S to distinguish this type of communication from general serial communication.

Electrically, TTL Serial uses two data pins, Transmit (Tx) and Receive (Rx). It is not a bus, and the connection is point to point, so there are no problems of addressing different devices. If you want to connect more than one Serial device to your microcontroller, then the microcontroller will need more than one Serial port.

Another remnant from early computer history is the nomenclature around the bandwidth of serial connections. A serial connection has to be set to the same baud rate at both ends of the connection. The *baud rate* (see also Chap. 28) is the number of times that the signal can change state per second. This is taken to be the same as the bits transmitted per second of the data payload, but that does include start, stop, or possible parity bits, so the actual transmission of data in bits per second is a little slower than the baud rate. To simplify matching up the baud rates at each end of the connection, a set of standard baud rates is used: 110, 300, 600, 1200, 2400, 4800, 9600, 14,400, 19,200, 38,400, 57,600, 115,200, 128,000, and 256,000.

Of these, 1200 is probably the slowest baud rate commonly in use and many TTL Serial devices will not go as high as 115,200. Perhaps the most common rate is 9600 baud. Devices often default to this rate, but can be configurable to other rates.

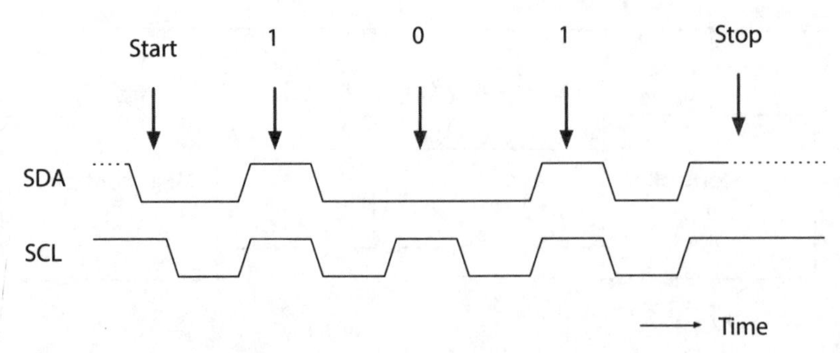

29-10 Timing diagram for I^{2}C.

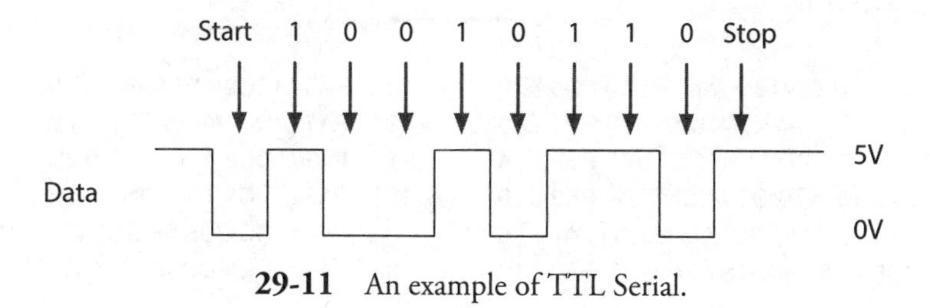

29-11 An example of TTL Serial.

As well as the baud rate, other parameters that define a Serial connection are the number of bits per word, the type of parity bit, and the number of start and stop bits. Almost universally, these are defined as 8, none, and 1 respectively, which often gets abbreviated to 8N1.

Bits are simply sent as high or low logic levels (Fig. 29-11). As there is no separate clock signal, timing is critical, so after the start bit, the receiver will sample at the appropriate rate until it has read the eight data bits and the one stop bit. The least significant bit of the data is sent first.

When connecting two Serial devices together, you connect the Tx pin of one device to the Rx pin of the other and vice-versa. For some devices, such as GPS receivers, only one-way communication is needed, as the device just repeatedly transmits data. In this case, the link needs to be established in only one direction.

Most microcontrollers have built-in hardware for TTL Serial called a *Universal Asynchronous Receiver Transmitter* (UART).

USB

While electrically similar to RS232, *Universal Serial Bus* (USB) is one of the most complex interfaces around in terms of software and communication protocols. This complexity is because it can be used with full-sized computers with a vast array of peripherals, from keyboards and mice to printers and cameras. Fortunately, its use with microcontrollers is generally simplified.

While microcontrollers generally do not have USB hardware built in, most families of microcontrollers will have a member with USB hardware that allows the microcontroller to communicate with a PC or emulate a mouse or keyboard. In fact, even if the microcontroller does not have hardware USB support, a Serial interface and some slick software will normally allow the microcontroller to "talk" USB.

Microcontrollers are often used to implement USB peripherals, such as keyboards and mice. The USB connector has two power pins (GND and 5 V as well as two data pins D+ and D−). The USB standard dictates that a USB host (usually your computer) must supply 5 V and at least 500 mA on its supply terminals.

An Example—The ATtiny44

Figure 29-12 shows the pin connections of the ATtiny44 microcontroller taken from its datasheet. As you can see, each pin has a variety of roles. This is typical of a modern microcontroller in which flexibility is paramount.

Looking at Fig. 29-12, you can see that the only two pins that do not have multiple roles are pin 14 (GND) and pin 1 (VCC, the positive supply). The other pins are theoretically all available to be used as GPIO pins. The supply voltage can be anything between 2.7 V and 5.5 V.

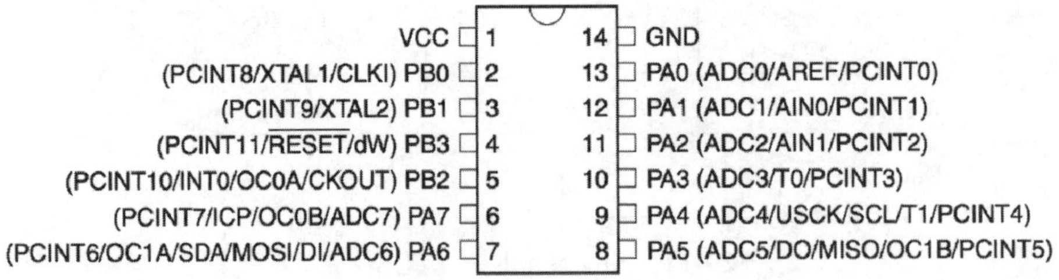

Left function	Pin	Pin	Right function
VCC	1	14	GND
(PCINT8/XTAL1/CLKI) PB0	2	13	PA0 (ADC0/AREF/PCINT0)
(PCINT9/XTAL2) PB1	3	12	PA1 (ADC1/AIN0/PCINT1)
(PCINT11/$\overline{\text{RESET}}$/dW) PB3	4	11	PA2 (ADC2/AIN1/PCINT2)
(PCINT10/INT0/OC0A/CKOUT) PB2	5	10	PA3 (ADC3/T0/PCINT3)
(PCINT7/ICP/OC0B/ADC7) PA7	6	9	PA4 (ADC4/USCK/SCL/T1/PCINT4)
(PCINT6/OC1A/SDA/MOSI/DI/ADC6) PA6	7	8	PA5 (ADC5/DO/MISO/OC1B/PCINT5)

29-12 The pinout of an ATtiny44.

The GPIO pins on the ATtiny44 are grouped into two ports, port A (PA) and port B (PB). Starting with the right side of the chip, pins 13 to 8 are labelled PA0 to PA5. These pins are all available for use as GPIO pins on port A. The other functions for those pins are indicated in parentheses. The pins PA0 to PA5 are also marked as ADC0 to ADC5. This indicates that all of those pins can be used as analog inputs.

Note that the SPI pins needed for programming (MOSI, MISO and USCK) can also be configured for use as GPIO pins.

On the left of the IC, you can see that pins 2 and 3 have the second function of XTAL1 and XTAL2. This allows you to connect a crystal between those pins to set the clock frequency of the microcontroller, which produces a much more accurate clock frequency than that provided by the internal resonator circuit of the microcontroller. This particular microcontroller will work with clock frequencies up to 20 MHz.

The choice between using the internal resonator circuit of the microcontroller for the clock frequency or sacrificing two GPIO pins for a more accurate clock frequency is made during the programming of the microcontroller chip.

Programming Languages

The machine code of a microcontroller is a series of instructions that tell it what to do. The instructions are not intended to be used directly by humans. Instead, the human programmer creates a text file containing a program written in a so-called *high-level language* on his/her main computer. This text file is then fed into a program called a *compiler*, which converts the high-level language into a machine code file that is installed into the flash memory of the microcontroller.

There are many programming languages that are used to program microcontrollers. The most common of these is the C programming language. C has the advantage that it is fairly easy to use, while still being relatively "close to the metal," so that programs written with it generally compile down to be compact and efficient. This is something that you need when programming a microcontroller with its limited memory.

In the next chapter, you will learn a little C programming on a microcontroller.

Programming a Microcontroller

There are various ways of getting a program onto a microcontroller, and they all involve the use of a regular computer, and often specialized programming hardware, such as the AVR Dragon programmer shown in Fig. 29-13.

29-13 The AVR Dragon programmer.

This board has a *zero insertion force* (ZIF) socket on the left that holds the microcontroller IC while it is being programmed. On the right of Fig. 29-13, you can see the USB interface that is used to connect the programmer to your computer.

During programming, the microcontroller's SPI pins are used to allow access to the microcontroller's flash memory.

Programming hardware, like the AVR Dragon, can also be used to program microcontrollers that are already soldered into place on a PCB, using a method called *In-Circuit Serial Programming* (ICSP). Here, the SPI pins of the microcontroller are connected to header pins on the PCB so that they can be connected to programming hardware using a "programming cable."

Another approach to programming microcontrollers is to program the microcontroller with a *bootloader,* using one of the methods described above. The bootloader runs each time that the microcontroller is restarted. It pauses for a fraction of a second to see if its Serial interface is being used to send it a new program. If it is, then it copies the received data into its flash memory and then restarts itself. Note that the bootloader is not destroyed in this process, so it can be used the next time a program needs to be uploaded.

Allowing the microcontroller to be programmed over Serial rather than SPI means that it can be programmed using hardware no more complex than a USB or Serial adaptor. The Arduino actually includes such a converter on board, so that all the hardware that you need to program an Arduino is a USB cable.

Quiz

Refer to the text in this chapter if necessary. A good score is 18 correct. Answers are in the back of the book.

1. What's the main purpose of flash memory in a microcontroller?
   (a) To store program variables
   (b) To store program code
   (c) To store charge
   (d) To interface with GPIO pins

2. What's the main purpose of RAM in a microcontroller?
   (a) To store program variables
   (b) To store program code
   (c) To store charge
   (d) To interface with GPIO pins

3. Which of the following statements about GPIO pins is true?
   (a) Once a GPIO pin has been set to be an input or output, it cannot subsequently be changed.
   (b) GPIO pins can usually source and sink large currents.
   (c) GPIO pins can be switched between acting as an input to acting as an output while a program is running on the microcontroller.
   (d) GPIO pins normally operate at 10 V.

4. Which of the following connections is *not* allowed because it could damage a microcontroller?
   (a) Connecting a 3.3-V digital output to a 5-V input
   (b) Connecting a 5-V digital output to a 3.3-V input
   (c) Connecting one digital output pin of a microcontroller to a digital input on the same microcontroller
   (d) Connecting a 1.5-V cell between GND and a digital input of a microcontroller

5. Under which of the following circumstances would you use an external pull-up resistor on a digital input?
   (a) Every time you connect a switch to a digital input.
   (b) Whenever the lead between the digital input and the switch is long.
   (c) Whenever you connect a digital output from one microcontroller to the digital input of another.
   (d) When the digital input is on a 3.3V microcontroller.

6. Which of the following voltages at a digital input is most likely to result in a logical HIGH being registered by the microcontroller?
   (a) 2 V on a 5-V microcontroller
   (b) 0 V on a 5-V microcontroller
   (c) 2 V on a 3.3-V microcontroller
   (d) None of the above

7. What current would flow through an LED with a forward voltage of 2 V and a series resistor of 1k (1000 ohms) driven by a 5-V digital output?
   (a) 1 mA
   (b) 2 mA
   (c) 3 mA
   (d) 4 mA

8. Debouncing of switches is
   (a) only possible using hardware.
   (b) usually done using software.
   (c) never necessary.
   (d) None of the above

9. How are interrupt inputs different from regular inputs?
   (a) They aren't; they're the same thing!
   (b) They use faster logic.
   (c) They use slower logic.
   (d) When interrupt inputs change signal level, they trigger an interrupt service routine.

10. Which of the following statements about PWM is true?
    (a) PWM is the only method for producing an analog output from a microcontroller.
    (b) Microcontrollers generally use high frequencies for PWM (over 100 kHz).
    (c) A duty cycle of 100% in a PWM output from a 5-V microcontroller will produce an output voltage of 0 V.
    (d) A duty cycle of 100% in a PWM output from a 5-V microcontroller will produce an output voltage of 5 V.

11. PWM control of an LED
    (a) changes the actual brightness of the LED.
    (b) changes the peak voltage across the LED.
    (c) changes the peak current through the LED.
    (d) changes the apparent brightness of the LED as interpreted by the human eye/brain.

12. An 8-bit analog input to a microcontroller will be digitized into a number in the range
    (a) 0 to 256.
    (b) 0 to 1023.
    (c) 0 to 4096.
    (d) 0 to 255.

13. What would be the digitized value of a 1-V input to an analog input of a 5-V microcontroller with an 8-bit ADC?
    (a) 1
    (b) 51
    (c) 48
    (d) 127

14. Which of the following statements about SPI is true?
    (a) SPI is used as the programming interface for most microcontrollers.
    (b) SPI can only be used to connect one device to a microcontroller.
    (c) SPI is not clocked and requires accurate time for data transmission.
    (d) SPI is never used as the programming interface for a microcontroller.

15. Which of the following statements about I^2C is true?
    (a) I^2C is used as the programming interface for most microcontrollers.
    (b) I^2C can only be used to connect one device to a microcontroller.
    (c) I^2C is not clocked and requires accurate time for data transmission.
    (d) I^2C is never used as the programming interface for a microcontroller.
16. Which of the following statements about TTL Serial is true?
    (a) TTL Serial is always used as the programming interface for microcontrollers.
    (b) A single TTL Serial connection can be connected to only one device at a time.
    (c) TTL Serial uses voltages that swing above and below 0 V.
    (d) TTL Serial is never used as the programming interface for a microcontroller.
17. Which of the following baud rates is *not* commonly used with serial interfaces?
    (a) 100
    (b) 300
    (c) 1200
    (d) 9600
18. Which of the following statements about USB is true?
    (a) Some microcontrollers include a hardware USB interface.
    (b) USB host devices provide 3-V power.
    (c) The USB standard states that USB peripherals can draw a maximum of 100 mA from the USB power connections.
    (d) None of the above
19. The C programs that you write for a microcontroller are
    (a) directly copied into the flash memory of a microcontroller.
    (b) converted into a form suitable for storing in the microcontroller's flash memory using software called an interpreter.
    (c) converted into a form suitable for storing in the microcontroller's flash memory using software called a compiler.
    (d) stored in the microcontroller's RAM.
20. What's the purpose of a bootloader?
    (a) To allow a microcontroller to be programmed using SPI
    (b) To allow a microcontroller to be programmed using I^2C
    (c) To allow a microcontroller to be programmed using Serial and USB
    (d) None of the above

30
CHAPTER

Arduino

ELECTRONICS HOBBYISTS HAVE BEEN USING MICROCONTROLLERS AS LONG AS THOSE DEVICES HAVE existed. But they have become extremely popular on the back of the runaway success of the Arduino microcontroller boards. Here are some of the reasons for the success of the Arduino platform:

- Built-in USB interface and programmer (no extra hardware required)
- Simple to use *Integrated Development Environment* (IDE) with which to write your programs
- Works with Microsoft Windows, Mac, and Linux computers
- Simple programming language
- Standard GPIO socket arrangement for the addition of plug-in "shields"
- Open-source hardware design, so you get to see the schematic diagram of the board and understand exactly how it works

The Arduino Uno/Genuino

Although there is now a wide range of different Arduino models, the "classic" Arduino model is the Arduino Uno, more specifically the Arduino Uno revision 3. This is the device that most people think of as an Arduino. After legal difficulties, the originators of the Arduino have had to change the brand name to Genuino rather than Arduino. The boards are electrically identical, but cosmetically, look slightly different. Figure 30-1 shows an Arduino Uno with key components annotated.

The Arduino Uno's USB port is used when programming the Arduino, but it can also be used to communicate with the Arduino from your PC and to provide 5 V DC to the device.

The Arduino Uno actually has two microcontroller chips on the board. The main microcontroller is the ATmega328 chip fitted into an IC socket. The second microcontroller is just to the right of the USB socket. This second microcontroller has a built-in USB interface and serves the single purpose of acting as a USB interface to the ATmega328. The USB interface microcontroller has its own *In-Circuit Serial Programming* (ICSP) header that allows it to be programmed during manufacture.

In the top left of the board, there is a reset switch. Pressing this takes the RESET pin of the ATmega328 low, causing the microcontroller to reset.

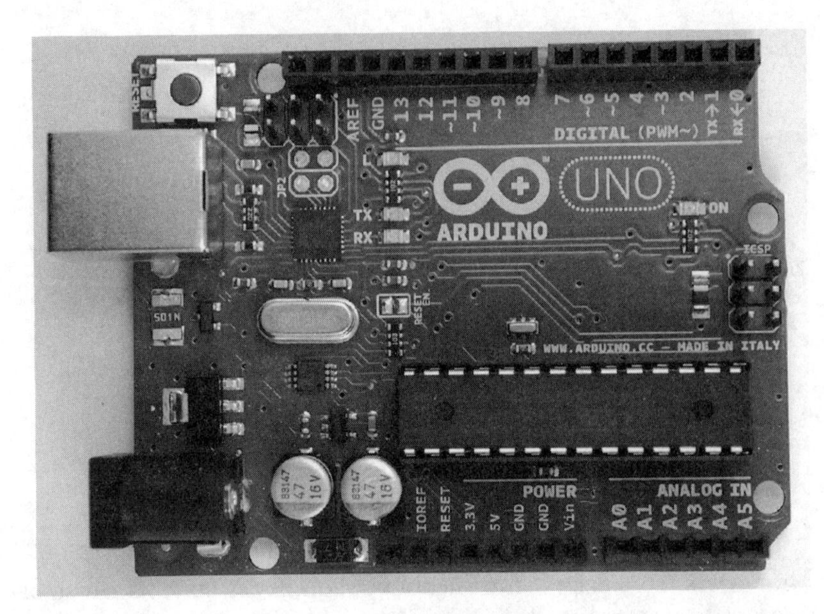

30-1 The Arduino Uno.

The top side of the board has two sections of female header pins connected to the GPIO pins of the ATmega328 labeled 0 to 13. Pin 13 is actually wired to an LED built onto the board, known as the "L" LED. This is useful because it allows you to experiment with the Arduino GPIO pins as a digital output without having to attach any external components. The pins 0 and 1 double as a TTL Serial connection that is also used for communication and programming over USB, so it is best not to use pins 0 and 1 for anything else.

After pin 13, there is a ground pin (GND) and a pin labeled AREF. The AREF pin can be connected to a voltage lower than 5 V to compress the range of the analog inputs, although this feature is not often used. The Arduino Uno operates at 5 V, and so all its digital outputs will provide 5 V when HIGH.

The two unlabeled sockets to the left of AREF are the SDA and SCK pins of an I^2C interface.

The ICSP header pins on the right of the board are for factory programming of the ATmega328 with the bootloader that allows all subsequent programming of the ATmega328 to take place over USB rather than over ICSP, thus avoiding the need for special programming hardware.

The ATmega328 itself is mounted in an IC socket. This allows easy replacement of the IC should it be accidentally damaged by, say, shorting out an output pin. A socket also means that you can use the Arduino to program an ATmega328 and then remove the programmed IC and install it in an electronics project without having to commit the entire Arduino to the project. You can buy ATmega328 ICs already programmed with the Arduino bootloader to replace the chip.

At the bottom right of the board, you have six pins labeled A0 to A5 that are primarily intended for use as analog inputs, although they can all also be used as digital inputs or outputs.

The final section of the socket header is mostly concerned with power. The V_{in} pin can be used as an alternative to the DC Power Socket or the USB connector to power the Arduino from an unregulated 7-V to 12-V input. There are also regulated 5-V and 3.3-V power supplies from the voltage regulators built onto the board, or if being powered over USB, 5 V DC from the USB socket.

To allow for projects that require accurate timing, the Arduino uses an external 16-MHz crystal to provide the frequency for the clock to the ATmega328.

Setting Up the Arduino IDE

The Arduino IDE is a program that you run on your Windows, Mac, or Linux PC that lets you type in your program and then upload it to your Arduino board over USB. To install the latest version of the Arduino IDE, follow the instructions on the following website:

https://www.arduino.cc/en/Main/Software

When you have the software installed, run the Arduino IDE; then from the "File" menu, select the option "Examples," "01 Basic," and then "Blink," and you should see a window that looks something like Figure 30-2.

The "Compile/Check" button compiles the Arduino C code without actually attempting to upload it to the board itself. The "Upload" button first compiles the Arduino C code and then uploads it to the Arduino over USB.

In the Arduino world, programs are called "sketches." The next three buttons allow you to start a new sketch, open an existing sketch, or save the current sketch. Each sketch is saved as a file with the extension .iso. Each sketch file will automatically be placed inside a folder of the same name as the sketch file, but without the extension.

The top-right icon on the Arduino IDE's toolbar opens a separate window called the "Serial Monitor" that allows you to communicate with the Arduino from your computer.

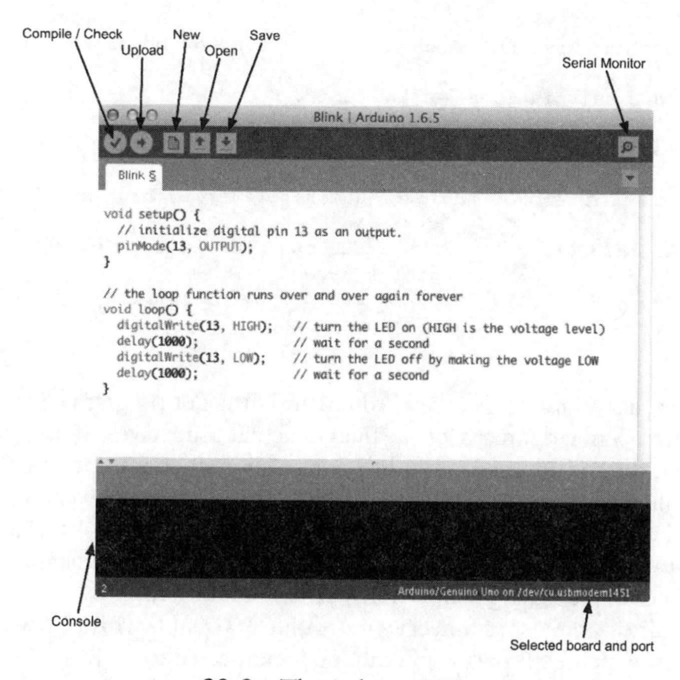

30-2 The Arduino IDE.

The console area of the IDE is where any error messages will appear when you try to compile a sketch or upload it to an Arduino.

Programming "Blink"

The Arduino Uno has a tiny LED attached to pin 13, labeled "L." Your first step with an Arduino should be to make this LED blink. You may find that when you plug the LED lead into your Arduino, the "L" LED is already blinking. This happens because Arduinos are generally sold with the LED Blink sketch preinstalled. To prove that you are making the LED blink, you can make it blink faster.

When you start up the IDE and open the Blink sketch, the text of the sketch appears as shown below:

```
/*
  Blink
  Turns on an LED on for one second, then off for one second, repeatedly.

  Most Arduinos have an on-board LED you can control. On the Uno and
  Leonardo, it is attached to digital pin 13. If you're unsure which
  pin the on-board LED is connected to on your Arduino model, check
  the documentation at http://www.arduino.cc

  This example code is in the public domain.

  modified 8 May 2014
  by Scott Fitzgerald
 */

// the setup function runs once when you press reset or power the board
void setup() {
  // initialize digital pin 13 as an output.
  pinMode(13, OUTPUT);
}

// the loop function runs over and over again forever
void loop() {
  digitalWrite(13, HIGH);   // turn the LED on (HIGH is the voltage level)
  delay(200);              // wait for a second
  digitalWrite(13, LOW);    // turn the LED off by making the voltage LOW
  delay(200);              // wait for a second
}
```

The text between `/*` and `*/` is called "comment." It is not program code; it's just comments to explain the sketch. Similarly, many of the lines of actual code have a // after them followed by a description of what the code does. This is also "comment" rather than program code. When using the // style of comment, the comment begins with the // and ends at the end of the line.

Edit the two lines that say: `delay(1000)` to read `delay(200)` and then save the sketch. When you come to save the sketch, the IDE will ask you to choose a new name for it because it is a read-only, built-in example. Save the sketch in a different location.

If you haven't already done so, connect your Arduino to your PC with a USB cable. You then need to let the IDE know which USB port your Arduino is connected to, so from the Tools menu, select the "Port:" option, as shown in Figure 30-3, and select the option that has (Arduino/Genuino Uno)

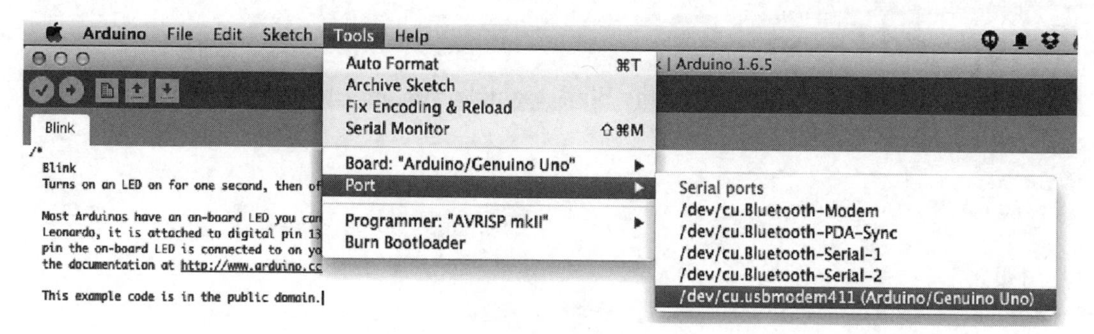

30-3 Selecting the serial port.

after it. If you have a Mac or Linux computer, then the name will be something like the one shown in Figure 30-3; if you are using Windows, then it will be called COM followed by a number.

You also need to tell the IDE what type of board is connected, so from the Tools menu, select "Board:" and then make sure that "Arduino/Genuino Uno" is selected.

Click on the upload button, and after the sketch has compiled, you should see the LEDs labeled TX and RX flicker while the sketch is uploaded into the flash memory of the ATmega328. This should take only a few seconds, after which the "L" LED should blink fairly fast.

Programming Fundamentals

Before taking apart the Blink sketch and adding some bells and whistles, let's explore the fundamental concepts of programming.

A program, or in Arduino terminology a "sketch," is a text file that contains a series of commands to be executed by the computer. In the case of the Arduino, these commands are in a programming language called C.

The commands will run (or "be executed") in order, one after the other. A microcontroller can do only one thing at a time. For example the lines below will be executed one after the other to make the LED blink.

```
digitalWrite(13, HIGH);    // turn the LED on (HIGH is the voltage level)
delay(200);                // wait for a second
digitalWrite(13, LOW);     // turn the LED off by making the voltage LOW
delay(200);
```

These commands are pretty self-explanatory, or if nothing else, explained by the comments that follow them. Note that the comment, "wait for a second," is now wrong because we changed the delay. It should now say, "wait for 200 milliseconds (⅕ of a second)."

There is a little more to the structure of this sketch that we will come to in the next section. But for now, the important thing to understand is that these programming commands will be executed in turn, one after the other.

In fact, it is not the text of the above program that is copied into the flash memory of the ATmega328 microcontroller, but rather the compiled version of this code. In other words, the Arduino IDE "compiles" the text that we typed into a compact machine code form that is then copied into the flash memory of the ATmega328 by the bootloader program on the ATmega328.

Setup and Loop

If you strip away the comments from the Blink example, you are left with the following lines of actual program code.

```
void setup() {
  pinMode(13, OUTPUT);
}

void loop() {
  digitalWrite(13, HIGH);
  delay(200);
  digitalWrite(13, LOW);
  delay(200);

}
```

You can see that the code is divided into two blocks or "functions." The first block starts with the line `void setup {` and the end of `setup` is marked by a `}`. In this case, there is just a single line between { and } that specifies that pin 13 should be set to be an output. But in other situations, there may be multiple lines of code in `setup`. The `setup` function runs once when the Arduino resets.

The second function `loop` will be run over and over. As soon as it finishes the last line between its { and }, it will start over with the first line. Notice how each of the lines of code within `loop` ends in a semicolon.

The first line inside `loop` uses the Arduino command `digitalWrite` to set pin 13 high. The next line `delay` instructs the Arduino processor to do nothing for 200 milliseconds or ⅕ of a second. The next two lines set pin 13 low and then delay for another 200 milliseconds before the cycle starts again.

Variables and Constants

One of the most important ideas in programming is that of *variables*. A variable is a way of giving a name to a value. For example, the Blink sketch makes pin 13 toggle on and off, but if you wanted to use pin 10 instead, you would have to change every occurrence of "13" to "10" (three places) in the sketch. That's not so bad in this sketch, but in a more complicated sketch, there might be lots of places where the pin number would need to be replaced, and missing one of them when making a change to the program could result in a bug that takes awhile to track down and fix.

To avoid problems like this, it is a good idea to use a variable to give the pin a name. This has the added advantage that rather than just appearing as a number that could mean anything, the pin now has a logical name that will tell someone looking at the code what it is for. In the code below, Blink has been improved by the addition of a variable to define the LED pin. The changes are highlighted in bold. Because typing in code is tedious and error prone, all the example sketches are available as a download from the GitHub page for the book at the website

https://github.com/simonmonk/tyee6

To download them, click on the "Download ZIP" button on the GitHub page. Extract the ZIP file to some convenient location. Each listing in this chapter has a comment at the top that identifies the sketch file in the downloads.

```
// blink_variable
int ledPin = 13;

void setup() {
  pinMode(ledPin, OUTPUT);
}

void loop() {
  digitalWrite(ledPin, HIGH);
  delay(200);
  digitalWrite(ledPin, LOW);
  delay(200);
}
```

You could take this a stage further and use a variable for the delay value like this:

```
int ledPin = 13;

int blinkDelay = 200;
void setup() {
  pinMode(ledPin, OUTPUT);
}
void loop() {
  digitalWrite(ledPin, HIGH);
  delay(blinkDelay);
  digitalWrite(ledPin, LOW);
  delay(blinkDelay);
}
```

Variable names must be a single word (no spaces), and by convention, they start with a lowercase letter and use an uppercase letter to separate the different parts of the variable name so that the variable name can be usefully descriptive. In this example, the most logical name for the variable (ignoring all syntax rules and conventions for a moment) would be `LED pin`, but we are discouraged from starting the variable name with an uppercase letter, and we cannot have a space in the variable name. So we end up with a variable name of `ledPin` that satisfies the requirements of starting with a lowercase letter and being all one word. The uppercase `P` of `Pin` helps us to read the variable name.

Variables are defined right at the start of the program before `setup`. The word `int` is short for *integer* and specifies that the variable will contain a whole number. Later on we will meet other types of variables.

In the example above, once `ledPin` and `blinkDelay` are defined, there are no instructions later on in the program code that would change them. This means that they can be called *constants*. You can, optionally, tell the Arduino compiler that these variables are constants by prefixing them with the word `const`, as shown below:

```
const int ledPin = 13;
const int blinkDelay = 200;
```

The code will work whether you insert `const` or not, but if you do include `const`, it tells someone reading the code that the variable isn't going to change later in the program, and it also allows the compiler to make slightly more efficient and compact code, saving you a few bytes of program size and RAM usage while the program runs.

The Serial Monitor

It can sometimes be a little difficult to know what the Arduino is doing. Yes, it can blink its built-in LED, but apart from that, if you have a problem with the software, it can be tricky to work out just what is going wrong. The same USB interface that allows you to upload a sketch from your computer to your Arduino can also be used to allow the Arduino to communicate with your computer and provide valuable feedback on the value of variables and what the program is doing.

To try out the Serial Monitor, modify your Blink sketch so that it appears as below.

```
// blink_serial_monitor
void setup() {
  pinMode(13, OUTPUT);
  Serial.begin(9600);
}

void loop() {
  Serial.println("on");
  digitalWrite(13, HIGH);
  delay(200);
  Serial.println("off");
  digitalWrite(13, LOW);
  delay(200);
}
```

The line that has been added to `setup` starts serial communication with the Arduino IDE running on your computer at the baud rate specified. A baud rate of 9600 is the default. You can change this to a higher or lower value, but you will have to also change the drop-down list on the Serial Monitor to match the value that you use in the sketch.

Now, inside the `loop` function, two new lines have been added that send a text message of `on` and `off` over the serial connection to the Serial Monitor.

Upload the sketch to your Arduino, and you should find that it behaves just like it did before you modified it. However, if you click on the Serial Monitor icon on the Arduino IDE, then the Serial Monitor window will open, as shown in Figure 30-4.

Each time the LED is turned on and off, you will see the message `on` or `off` appear in the Serial Monitor.

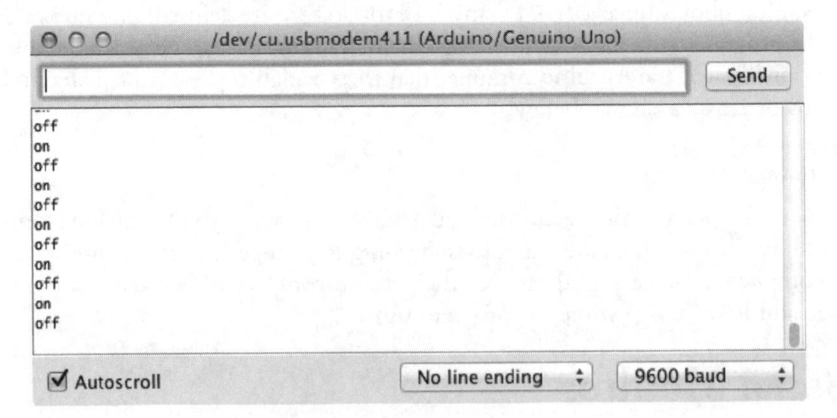

30-4 The Serial Monitor.

The values on and off inside the Serial.println commands are called *strings*. They are the C language's way of representing text. If you are from a conventional programming background, then you are probably familiar with the concept of strings and use them extensively in your programming. Strings are not commonly used when writing programs for an Arduino, as an Arduino is often used in applications that do not have any means of displaying text. The exceptions to this are situations in which the Arduino is communicating with a device like your PC that does have a means of displaying text, or when you have attached display hardware to the Arduino.

Ifs

As a program runs, the normal sequence of events is to run one command after the other. However, sometimes you will need to run only some of the commands if a condition is true. For example, you might want to run only certain commands when a switch, connected to a digital input, has been pressed. Another example might be that you want to run the command to turn a digital output on only if the temperature, measured using a temperature sensor, is greater than a certain value. The mechanism for doing this is to use the C if command. Assuming that you already have a variable called temperature that contains the temperature, you could write:

```
if (temperature > 90) {
  Serial.println("Its hot!");
}
```

Don't worry for now where temperature gets its value from. The important point is the structure of the if command. After the word if, there is an expression in parentheses called the "condition," and in this case, the condition is that the value in the variable temperature is greater than (>) 90. There then follows an { to indicate the start of a block of code. All the lines of code between this { and the corresponding } will be run only if the temperature indeed exceeds 90. The line Serial.println("Its hot!"); is tabbed right to show that it belongs to the if command.

Iteration

Another common departure from simply executing a series of commands one after the other is to repeat those commands a number of times. Although the loop function in an Arduino sketch will repeat the commands it contains indefinitely, sometimes you need more control over how many times some lines of code are executed.

The main C language command for repeating things a number of times is the for command. The following sketch uses a for command to send the numbers 1 to 10 to the Serial Monitor.

```
// count_to_ten_once
void setup() {
  Serial.begin(9600);
  for (int i = 1; i <= 10; i++) {
    Serial.print(i);
  }
}

void loop() {
}
```

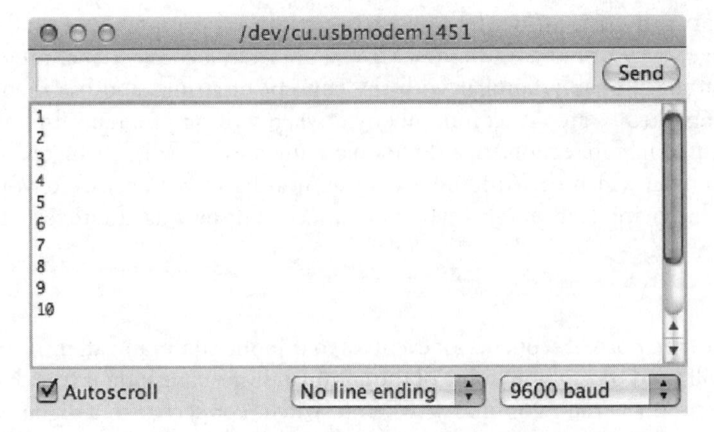

30-5 The Serial Monitor output counting to 10.

Because we want the Arduino to count to 10 only once, the `for` loop is in `setup`. If we wanted the sketch to count to 10 over and over, then we would move the three lines of the `for` loop to the currently empty `loop` function so that the sketch looked like this:

```
// count_to_ten_repeat
void setup() {
  Serial.begin(9600);
}

void loop() {
  for (int i = 1; i <= 10; i++) {
    Serial.println(i);
  }
}
```

The parenthesized expression after the word `for` contains three little snippets of code separated by semicolons. The first defines a counter variable called `i`, and the second section is the condition for staying in the loop. In this case, that means that the code will stay in the loop as long as `i` is less than or equal to (<=) 10. The final section (`i++`) means that 1 will be added to the value of `i` each time around the loop. The lines of code in between the { and } of the `for` loop will be run repeatedly until `i` is no longer less than or equal to 10.

The end result of this process is that the numbers between 1 and 10 get displayed on the Serial Monitor, as shown in Fig. 30-5.

There is a second type of `loop` command in C called a `while` loop that you will find useful from time to time.

The `while` command looks rather like an `if` command at first glance, but rather than just doing the things inside its { and } if the condition is true, it repeatedly executes the commands inside the { and } while the condition is true, stopping only when the condition is not true. At this point, the program continues to run the lines of code (if there are any) after the `while` loop.

If you rewrite the `for` example above to use `while` it would look like this:

```
// count_to_ten_once_while
void setup() {
  Serial.begin(9600);
  int i = 1;
```

```
  while (i <= 10) {
    Serial.println(i);
    i++;
  }
}

void loop() {
}
```

The variable `i` is now defined before the loop starts.

Functions

Functions are named blocks of code. Every sketch has to contain a `setup` and a `loop` function, but you can also define your own functions. This tends to happen if there are some lines of code that you want to use in several places in your sketch. Rather than repeat them in the code, you define them as a function. Writing you own functions helps to make your programs easier to understand. In the following example sketch, there is a user-defined function called `blink` that makes the "L" LED pin 13 blink 10 times.

```
// blink_function_broken
const int ledPin = 13;
void setup() {
  pinMode(ledPin, OUTPUT);
}

void loop() {
}

void blink() {
  for (int i = 1; i <= 10; i++) {
    digitalWrite(ledPin, HIGH);
    delay(1000);
    digitalWrite(ledPin, LOW);
    delay(1000);
  }
}
```

If you were to run this sketch, then the LED would not actually blink at all. That is because, although you have defined the `blink` function saying what the code must do and giving it a name, nowhere in the sketch do you actually "call" the function, telling it to run.

This separation of *defining* a function versus *running* a function is a very important distinction. In defining the function, we have created a named piece of code that knows how to "blink," but nowhere have we actually told it to go ahead and do some blinking. It's a bit like giving someone written instructions for making a cup of coffee but not actually telling them to go ahead and make a cup.

To fix this so that you actually get some blinking of the LED, you can put the line highlighted in bold into `setup`.

```
// blink_function
const int ledPin = 13;
void setup() {
  pinMode(ledPin, OUTPUT);
  blink();
}
```

```
void loop() {
}

void blink() {
  for (int i = 1; i <= 10; i++) {
    digitalWrite(ledPin, HIGH);
    delay(1000);
    digitalWrite(ledPin, LOW);
    delay(1000);
  }
}
```

It does not matter where in the sketch you define the function, although it is most common to put your own functions after `setup` and `loop`. To execute or "call" the `blink` function, use the name of the function followed by `()`.

The current `blink` function is pretty inflexible: it can blink only 10 times, it always blinks `ledPin`, and the fixed delay value means it can only blink at one speed. You can make this function much more flexible and general purpose by "parameterizing" the function so it reads

```
// blink_function_params
const int ledPin = 13;

void setup() {
  pinMode(ledPin, OUTPUT);
  blink(ledPin, 20, 200);
}

void loop() {
}

void blink(int pin, int times, int period) {
  for (int i = 1; i <= times; i++) {
    digitalWrite(pin, HIGH);
    delay(period);
    digitalWrite(pin, LOW);
    delay(period);
  }
}
```

The changes are highlighted in bold. The first change is that the call to `blink` inside `setup` now has three things in the parentheses separated by commas. These are called *parameters* and will be "passed to" the function when it is called. The first is the pin to blink (in this case `ledPin`). The second is the number of times to flash, and the last is the delay period between turning the pin on and off.

The parameters inside the function are called *local variables* because they apply only inside the function. So when the function is called and the first parameter supplied is `ledPin`, the value of `ledPin` will be transferred to the local variable `pin` inside the function. Such local variables are accessible only within the function itself, whereas the other variables that we have met so far, such as `ledPin`, are called *global variables* because they are accessible throughout the sketch.

Data Types

A variable of type `int` in Arduino C uses two bytes of data. The type `int` is used for most variables. An exception is when the `int` range of −32768 to 32767 is not enough because you want to represent a number greater than 32767 or less than −32768, in which case a `long` using four bytes of data will give you big numbers.

Another situation in which an `int` won't work arises when you want to represent real numbers that have digits after the decimal place. The `float` data type uses the binary equivalent of scientific notation. That is, the number is split into a mantissa and exponent. This gives an enormous range of values but limited precision.

Using 0.0 rather than just 0 helps to emphasize that the number is a real number and not just an integer. Table 30-1 breaks down the data types available.

So far our variables have all been `ints` and declared like

```
int x = 0;
```

Setting the initial value for the variable by following the declaration with = and then a value is optional, but considered good practice, as it removes any ambiguity about the value of the variable.

To declare a `float` you write something like

```
float x = 0.0;
```

Generally speaking, when performing calculations that involve a mixture of different types (say `ints` and `floats`), the compiler does a pretty good job of automatically converting types,

Table 30-1. Data Types in Arduino C

Type	Memory (bytes)	Range	Notes
`boolean`	1	true or false (0 or 1)	
`char`	1	−128 to +128	Used to represent an ASCII character code, e.g., A is represented as 65. Its negative numbers are not normally used.
`byte`	1	0 to 255	Often used for communicating serial data, as a single unit of data.
`int`	2	−32768 to +32767	
unsigned `int`	2	0 to 65536	Can be used for extra precision where negative numbers are not needed. Use with caution as arithmetic with `ints` may cause unexpected results.
`long`	4	−2,147,483,648 to 2,147,483,647	Needed only for representing very big numbers.
unsigned `long`	4	0 to 4,294,967,295	See unsigned `int`
`float`	4	−3.4028235E+38 to + 3.4028235E+38	
`double`	4	same as `float`	Normally this would be 8 bytes and higher precision than `float` with a greater range. However on Arduino it is the same as `float`.

as you would expect. For example, the following code would produce the expected result of 25,000,000.00:

```
// calc_1
void setup() {
  Serial.begin(9600);
  float x = 5000.0;
  int y = 5000;
  float result = x * y;
  Serial.println(result);
}

void loop() {
}
```

The dangerous situations in which you may not get the result you expect arise when the combined number gets too big for the `int` data type, for example:

```
// calc_2
void setup() {
  Serial.begin(9600);
  int x = 500;
  int y = 500;
  int result = (y * x) / 1000;
  Serial.println(result);
}

void loop() {
}
```

In this case, you would expect `result` to be 250, or 250,000/1000. However, if you run this sketch, the result will actually be −12. That happens because the first step of the calculation is to multiply 500 by 500, which gives a result of 250,000, which is above the limit for an `int`. In C, once the number exceeds the limit, the value "wraps-around" to negative numbers, giving meaningless results like the one here. If you try changing `x` and `y` to be `longs` the result will be as expected.

In summary, whenever you are doing arithmetic, always think through the maximum values that might arise, even for intermediate values in the calculations, and use a data type that has a big enough range for them.

Interfacing with GPIO Pins

When it comes to using the Arduino's GPIO pins, there are a number of built-in functions that you use first to define whether the pin is to act as an input or an output, and then, either to read the value of an input into the sketch or write a value to an output.

Setting the Pin Mode

Unless otherwise specified, an Arduino pin is an input without the pin's pull-up resistor being active. The `pinMode` command allows you to choose whether the pin is to act as an input or output and whether the pull-up resistor should be enabled or not.

Normally the mode of a pin is set in the `setup` function, but you can change the mode of a pin at any time as the sketch runs.

The `pinMode` built-in function takes two parameters. The first is the pin whose mode is to be set, and the second is the *mode*. The mode must be INPUT, INPUT_PULLUP, or OUTPUT. These are constants defined in Arduino C.

Digital Read

To read the digital value of an input pin, the built-in function `digitalRead` is used. This takes the pin to be read as its parameter and *returns a value* of 0 or 1. Returning a value means that the result of reading the digital input can be assigned to a variable, as shown in the following example sketch:

```
// digital_read
const int inputPin = 7;

void setup() {
  Serial.begin(9600);
  pinMode(inputPin, INPUT);
}

void loop() {
  int x = digitalRead(inputPin);
  Serial.println(x);
  delay(1000);
}
```

The variable `x` is defined inside the `loop` function, making it a local variable accessible only within the `loop` function. The result of the `digitalRead` function call will be assigned to the variable and will be 1 if the pin is high and 0 if the pin is low. Two special constants called `HIGH` and `LOW` are defined in the Arduino code so that you can use them instead of 1 and 0 respectively.

If you wanted to make a somewhat over-engineered light switch, you could attach a switch to pin 7 of the Arduino, and then install the following sketch to turn the built-in "L" LED on when the switch is pressed.

```
// digital_read_switch
const int switchPin = 7;
const int ledPin = 13;

void setup() {
  pinMode(switchPin, INPUT_PULLUP);
  pinMode(ledPin, OUTPUT);
}

void loop() {
  if (digitalRead(switchPin) == LOW) {
    digitalWrite(ledPin, HIGH);
  }
  else {
    digitalWrite(ledPin, LOW);
  }
}
```

Rather than use a variable to store the result of the `digitalRead` (as we did in the previous example), the `digitalRead` function is used directly in the condition part of the `if` command. To compare two things to see if they are equal, you use a double equals sign (==) rather than the single sign (=) that you use to assign a value to a variable. The condition is that the result of calling digitalRead be LOW because the input is normally HIGH due to the pull-up resistor and only goes LOW when the switch is closed.

In this case, the if statement has an else counterpart that is run if the condition is not true.

The above example could, of course, easily be made without an Arduino at all, simply by putting a switch, an LED, and a current limiting resistor in series and connected to a voltage source.

If you wanted to change things so that pressing a push switch toggled the LED between on and off, then you would need the sketch to use a variable to keep track of whether the LED was last on or off (called its *state*). This is what such a sketch would look like:

```
// digital_read_toggle
const int switchPin = 7;
const int ledPin = 13;

int ledState = LOW;

void setup() {
  pinMode(switchPin, INPUT_PULLUP);
  pinMode(ledPin, OUTPUT);
}

void loop() {
  if (digitalRead(switchPin) == LOW) {
    ledState = ! ledState;
    digitalWrite(ledPin, ledState);
    delay(100);
    while (digitalRead(switchPin) == LOW) {}
  }
}
```

Now, when digitalRead detects that the switch button has been pressed, the variable ledState is toggled using the not (!) command. This sets ledState to be the inverse of its current setting. That is, if ledState is HIGH, it sets it LOW and vice-versa.

The digitalWrite function is then used to set the output to the new state. After that, there is a delay of 100 milliseconds that allows time for the switch contacts to settle, as they often "bounce" between high and low during the pressing of the switch.

You don't want ledState to be immediately toggled again, so the while loop effectively waits until the switch has been released.

Digital Write

You have already used the digitalWrite command to turn the "L" LED on the Arduino board on and off. The command takes two parameters, the first being the pin to control and the second being 1 or 0 for HIGH and LOW, respectively.

The command will control the pin only if the pin has already been set to be a digital output using the pinMode command.

An Arduino pin can source or sink 40mA without any risk of damaging the ATmega328 microcontroller, which is fine for controlling an LED directly, but other devices, such as a relay or DC motor, will require some current amplification.

Because we are concerned only with turning things on and off, the amplification does not need to be linear, so a simple control with a single transistor is all that is required in most circumstances.

Figure 30-6 shows the use of an NPN bipolar transistor to provide the 50 to 100mA (or so) needed to drive the coil of a typical relay. The resistor should limit the current to less than 40mA,

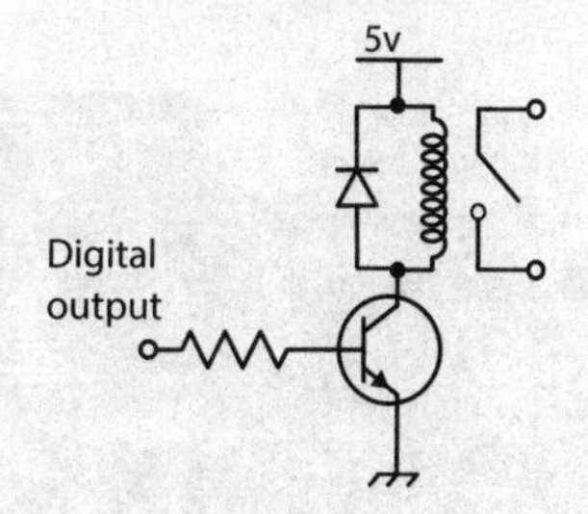

30-6 Controlling a relay from a digital output.

so 150 ohms would be ideal. A simple low-cost transistor, like the 2N3904, is suitable. The diode across the relay coil is necessary to snub (that is, suppress) any pulses of voltage resulting from driving the inductive load of the relay coil.

Analog Input

The Arduino has six pins labeled A0 to A5 that can provide 10-bit analog inputs. The built-in function `analogRead` returns a number between 0 and 1023 corresponding to the voltage at the analog input. The following sketch illustrates this, along with the math necessary to convert the reading into a voltage and print it out at the Serial Monitor.

```
// analog_read
const int analogPin = A0;

void setup() {
  Serial.begin(9600);
}

void loop() {
  int reading = analogRead(analogPin);
  float volts = reading * 5.0 / 1023.0;
  Serial.println(volts);
  delay(1000);
}
```

The `int reading` is multiplied by 5.0 and divided by 1023.0. That is, both 5.0 and 1023.0 have a decimal point so that C knows they are `floats` rather than `ints`.

The Serial Monitor will display the voltage at pin A0 once a second, and you can actually use this scheme to measure some of the voltages on the Arduino board itself. Figure 30-7 shows a jumper wire linking A0 to GND.

While this is in place, the readings should be 0 V. Change the jumper wire so that it now links A0 and the "3-V" pin. The Serial Monitor should report a voltage of around 3.3 V. Finally connect A0 to the Arduino's "5-V" pin. Figure 30-8 shows the output of the Serial Monitor.

Although you need to be careful not to exceed the maximum input voltage to an analog input of 5 V, you can, of course, use a pair of resistors as a voltage divider, if you need to measure higher voltages.

Analog Write

An Arduino Uno does not have true analog outputs, but rather uses *pulse width modulation* (PWM) as described in Chap. 29. Only the pins on an Arduino Uno marked with a ~ (3, 5, 6, 9, 10, and 11) can provide hardware supported PWM.

30-7 An Arduino measuring its own voltages.

The command to set the duty cycle of such an output is the `analogWrite` command. This takes two parameters: the pin to control and the duty cycle as an `int` between 0 and 255. A value of 0 means the duty cycle will be 0 and the output will be fully off. A value of 255 and it will be fully on. In fact it's just like `digitalWrite` except that instead of the value being 0 or 1 (LOW or HIGH) the value is between 0 and 255.

To illustrate, we can use an Arduino with a potentiometer attached to control the brightness of an LED, as shown in Fig. 30-9.

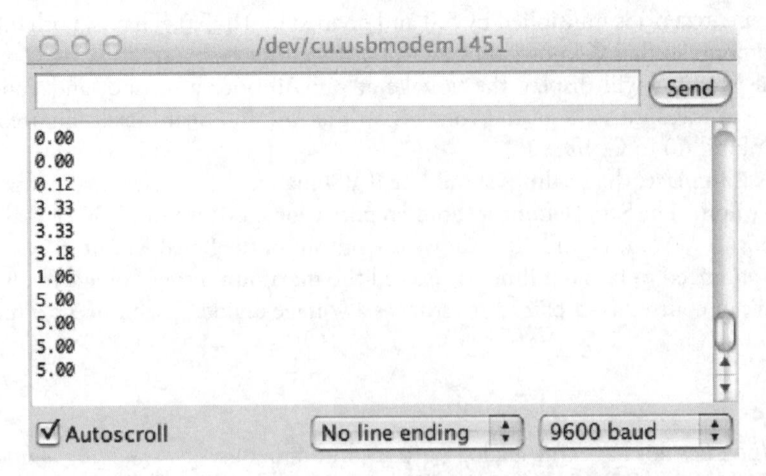

30-8 The Serial Monitor reporting analog readings.

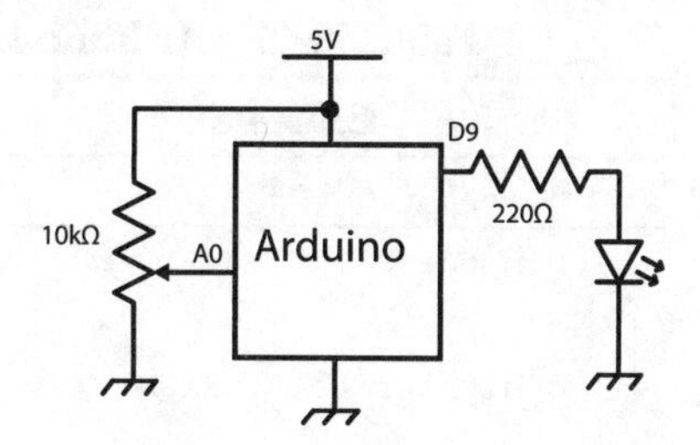

30-9 Controlling LED brightness with a potentiometer.

The sketch for this project simply takes the analog reading from the potentiometer (0 to 1023) and divides it by 4 to put it in the range 0 to 255 needed by analogWrite.

```
// analog_write
const int potPin = A0;
const int ledPin = 9;

void setup() {
  pinMode(ledPin, OUTPUT);
}

void loop() {
  int reading = analogRead(potPin);
  analogWrite(ledPin, reading / 4);
}
```

As you turn the potentiometer knob, the apparent brightness of the LED will increase as the duration of the pulses arriving at it increases. This way of controlling the apparent brightness of an LED actually works much better than controlling the voltage to the LED because, until the voltage reaches the working forward voltage of the LED (typically at least 1.6 V), the LED will not emit any light at all.

The Arduino C Library

There are a large number of commands available in the Arduino library, some of which you have already met. The most commonly used commands are listed in Table 30-2. For a full reference to all the Arduino commands, see the official Arduino documentation at the website

http://www.arduino.cc

Libraries

The Arduino IDE uses the concept of libraries to organize code that you might want to use in your sketches. These libraries contain program code that others have written and shared. This makes it easier to interface with certain types of hardware.

For example, the Arduino IDE comes supplied with a load of libraries pre-installed that you can make use of. It is a convention that libraries should include some example sketches that will get you started quickly when using the libraries. You can get an idea of the libraries included with the Arduino IDE from the "File" and then the "Examples" menu (see Fig. 30-10).

Table 30-2. Arduino Library Functions

Command	Example	Description
Digital IO		
pinMode	pinMode(8, OUTPUT);	Sets pin 8 to be an output. The alternative is to set it to be INPUT or INPUT_PULLUP.
digitalWrite	digitalWrite(8, HIGH);	Sets pin 8 high. To set it low, use the constant LOW instead of HIGH
digitalRead	int i; i = digitalRead(8);	This will set the value of i to HIGH or LOW depending on the voltage at the pin specified (in this case pin 8).
pulseIn	i = pulseIn(8, HIGH)	Returns the duration in microseconds of the next HIGH pulse on pin 8
tone	tone(8, 440);	Make pin 8 oscillate at 440 Hz.
noTone	noTone(8);	Cut short the playing of any tone that was in progress on pin 8.
Analog IO		
analogRead	int r; r = analogRead(A0);	Assigns a value to r of between 0 and 1023: 0 for 0 V, 1023 if the pin A0 is 5 V.
analogWrite	analogWrite(9, 127);	This command outputs a PWM signal. The duty cycle is a number between 0 and 255, 255 being 100 %. This must be used by one of the pins marked as PWM on the Arduino board (3, 5, 6, 9, 10 and 11).
Time Commands		
millis	unsigned long l; l = millis();	The variable type long in Arduino is represented in 32 bits. The value returned by millis() will be the number of milliseconds since the last reset. The number will wrap around after approximately 50 days.
micros	long l; l = micros();	See millis, except this is microseconds since the last reset. It will wrap after approximately 70 minutes.
delay	delay(1000);	Delay for 1000 milliseconds or 1 second.
delayMicroseconds	delayMicroseconds(100000);	Delay for 100,000 microseconds. Note the minimum delay is 3 microseconds, the maximum is around 16 milliseconds.
Interrupts		
attachInterrupt	attachInterrupt(1, myFunction, RISING);	Associates the function myFunction with a rising transition on interrupt 1 (D3 on an Uno).
detachInterrupt	detachInterrupt(1);	Disables any interrupt on interrupt 1.

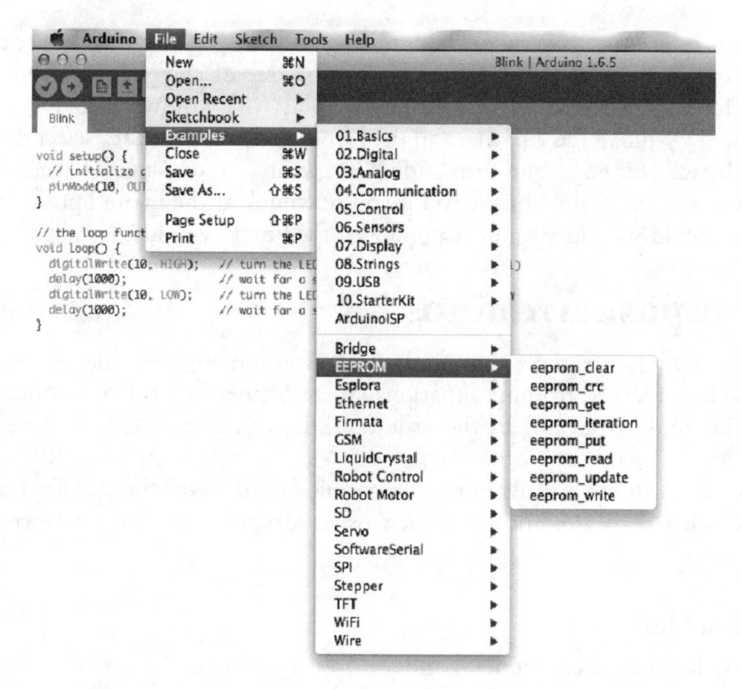

30-10 Library example sketches.

The top half of the example sketches list is not related to libraries, but below the line, all the sketches are contributed from libraries. For example, to store values persistently so that they are not lost when the Arduino is reset, you can use the EEPROM library. The "eeprom_clear" sketch is shown below with some of the comments removed for brevity.

```
#include <EEPROM.h>
void setup()
{
  for ( int i = 0 ; i < EEPROM.length() ; i++ ) {
    EEPROM.write(i, 0);
  }

  // turn the LED on when we're done
  digitalWrite(13, HIGH);
}
void loop(){ /** Empty loop. **/ }
```

To show the Arduino IDE that a library is required the `#include` command is used, followed by the name of the library's header file. The best way to get the command right for your own sketch is just to copy it from one of the example sketches.

Other libraries that are included with the IDE are for network programming with network hardware (Ethernet), liquid-crystal displays (LCDs), reading and writing SD cards (SD) controlling servomotors (Servo), and stepper motors (Stepper).

If you have a piece of hardware that you want to control from your Arduino, then chances are there will already be a library for it included in the Arduino IDE, or there will be a library written by someone else that you can install in your Arduino IDE.

The Arduino community is very good at creating and sharing libraries. In the spirit of open-source cooperation, libraries are almost always provided free of charge and without any kind of licensing restrictions on their use. When you find a library that you want to use, it will be in the form of a ZIP file. Download the ZIP file, and then from the Arduino IDE, select the menu option "Sketch," then "Include Library," and then "Add ZIP Library," and navigate to the ZIP file you just downloaded. This will install the library, and if you now look at the menu option "File" and then "Examples," you should see a new set of examples for the library you just installed.

Special Purpose Arduinos

The Arduino Uno is by far the most commonly used Arduino; however, there are other models of Arduino that are better suited to some situations. Some of these are official Arduino boards, and others are made by third parties and use the Arduino IDE or a different IDE, but the same Arduino C language.

There are many Arduino and Arduino-compatible boards. New boards are being developed all the time, so rather than attempt a comprehensive survey, let's focus on three representative Arduino boards.

The Arduino Pro Mini

The Arduino Uno has a lot of components and features. This can make it a little wasteful to embed an Arduino Uno into a project. For example, the Arduino has a built-in USB interface, and if you need this only while programming the board, it is unnecessary to have that USB interface permanently attached to your project.

The Arduino Pro Mini (Fig. 30-11) separates the USB interface from the rest of the Arduino so that the Arduino itself is smaller and lower-cost. This makes it more reasonable to embed an Arduino permanently into a project once you have perfected it. If you need to update the Arduino's sketch, you can just reattach the USB interface and program it again.

Programming an Arduino Pro Mini is just like programming an Arduino Uno. You just have to select a board type of "Arduino Pro or Pro Mini" in the "Board" section of the "Tools" menu.

30-11 The Arduino Pro Mini (left) and USB interface (right).

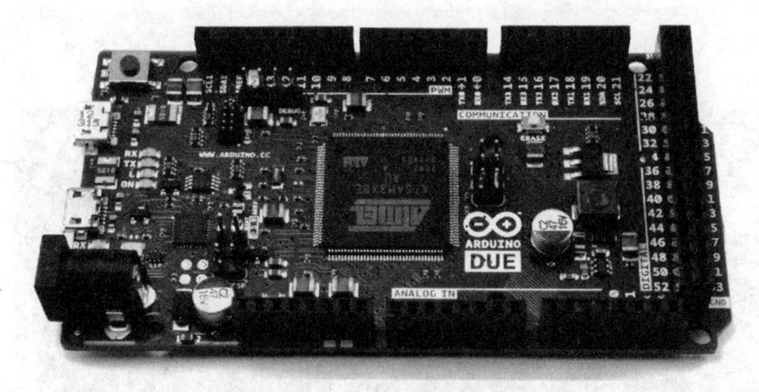

30-12 The Arduino Due.

Arduino Due

At the other extreme, sometimes an Arduino Uno does not have enough GPIO pins or is not fast enough for some demanding task. The Arduino Due is designed for such applications. (Fig. 30-12).

The Due has 54 GPIO pins, although these operate at 3.3 V rather than the 5 V of an Arduino Uno. Besides having more pins, the Due also has a much faster processor (80-MHz clock) and a 32-bit architecture. You can find full specifications for the board at the website

https://www.arduino.cc/en/Main/ArduinoBoardDue

Arduino Particle Photon

One of the more interesting unofficial Arduino-type boards is the Photon (Fig. 30-13). This board looks similar to the Arduino Pro Mini, but the big difference is that it has built-in Wi-Fi hardware making it ideal for *Internet of Things* (IoT) projects.

30-13 The Arduino Particle Photon.

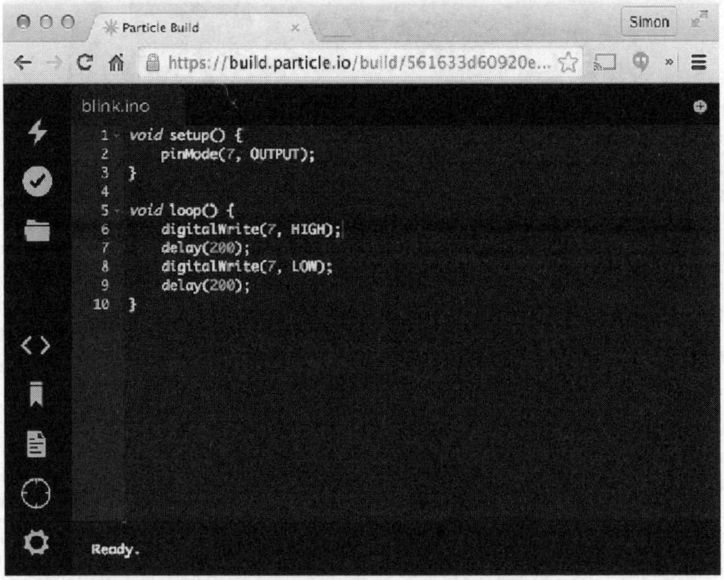

30-14 The Arduino Particle IDE

The board connects to your home Wi-Fi network using a smartphone app for first-time configuration. After that, you do not need any physical connection to the Photon to program it, but rather you program it over the Internet using a cloud service provided by Particle.

The Particle IDE (Fig. 30-14) looks rather like the Arduino IDE, and the language you use is the same as Arduino C with a few extensions to make network communication easier.

Shields

The Arduino header sockets can be used to attach so-called *shields*, which stack on top of the Arduino and are available for all sorts of purposes. An Internet search will find you no end of different shields for all sorts of purposes, including:

- Motor control
- Relays
- Ethernet and Wi-Fi
- Various types of display
- Sensors

Shields often have an accompanying Arduino library that makes them easy to use.

Quiz

Refer to the text in this chapter if necessary. A good score is 18 correct. Answers are in the back of the book.

1. The Arduino Uno has two microcontroller chips because
    (a) that way it can perform twice as fast.
    (b) one of the microcontroller chips is dedicated to providing a USB interface.

(c) one of the microcontroller chips provides a video interface.

(d) None of the above

2. The "L" LED of the Arduino Uno is connected to digital pin
   (a) 10.
   (b) 11.
   (c) 12.
   (d) 13.
3. The Arduino Uno clock frequency is
   (a) 4 MHz.
   (b) 8 MHz.
   (c) 16 MHz.
   (d) 20 MHz.
4. The Arduino Uno has regulated voltage outputs of
   (a) 3.3 V and 5 V.
   (b) 3.3 V only.
   (c) 5 V only.
   (d) None of the above
5. An Arduino can be powered
   (a) only through its USB port.
   (b) through its USB port or the DC barrel jack socket.
   (c) through its DC barrel jack socket or "V_{in}" pin.
   (d) through its USB port or the DC barrel jack socket or the "V_{in}" pin.
6. To use the Arduino IDE, you need a PC running
   (a) Windows.
   (b) Linux.
   (c) Mac OS X.
   (d) Any of the above
7. The USB interface of an Arduino can be used to
   (a) program the Arduino.
   (b) allow communication between the Arduino and your PC.
   (c) power the Arduino.
   (d) All of the above
8. The Arduino's six pins labeled A0 to A5 can be used
   (a) only as analog inputs.
   (b) as analog inputs or analog outputs.
   (c) as analog or digital inputs.
   (d) as analog inputs, digital inputs, or digital outputs.

9. The Arduino `delay` function pauses execution in time units of
   (a) seconds.
   (b) milliseconds.
   (c) microseconds.
   (d) clock cycles.

10. Why would the "L" LED not turn on in the following sketch?

```
void setup() {
  digitalWrite(13, HIGH);
}
void loop() {}
```

   (a) Because pin 13 is not set to be an output
   (b) Because the "L" LED is not connected to pin 13
   (c) Because there is nothing in the loop function
   (d) Because pin 13 is not usable as an output

11. In the context of programming, a variable is
   (a) a numeric value such as 123.
   (b) a way of giving a name to a value so that you can refer to that value by name and later change it if you want to.
   (c) a named value that cannot subsequently be changed by the program.
   (d) None of the above

12. A constant
   (a) is a variable that cannot be changed while the program is running.
   (b) is something known with certainty about what the program should do.
   (c) is the same thing as a variable.
   (d) uses more memory than a variable.

13. A "Sketch" is
   (a) a rough design on paper for how your program is to work.
   (b) a schematic for the electronic components that you are planning to connect to an Arduino.
   (c) code to be "included" into your program that provides access to specialized hardware.
   (d) the Arduino terminology for a program.

14. What output would you see on the Serial Monitor after running this sketch?

```
void setup() {
  Serial.begin(9600);
  int x = 12;
  x++;
  Serial.println(x);
}
void loop() {}
```

   (a) 0
   (b) 12

(c) 13

(d) 9600

15. What will happen when you run the following sketch?

```
void setup() {
  pinMode(13, OUTPUT);
}

void loop() {
  digitalWrite(13, HIGH);
  delay(200);
  digitalWrite(13, LOW);
}
```

(a) The "L" LED will blink 200 times a second.

(b) The "L" LED will blink 5 times a second.

(c) The "L" LED will appear to be on all the time.

(d) The "L" LED will appear to be off all the time.

16. Every Arduino sketch must contain the function or functions:

(a) `setup`.

(b) `setup` and `loop`.

(c) `loop`.

(d) No functions are mandatory.

17. Which of the following statements best describes the Serial Monitor?

(a) It is a USB hardware interface on the Arduino board.

(b) It monitors network traffic on your computer.

(c) It is used for programming an Arduino.

(d) It is part of the Arduino IDE that allows communication with an Arduino over USB.

18. Which of the following `for` constructs will count from 5 to 10 using the variable `i`?

(a) `for (int i =5; i <= 10; i++)`

(b) `for (int j =5; i <=10; j++)`

(c) `for (int i = 5; i < 10; i++)`

(d) `for (int i = 1; i <= 5; i++)`

19. What is the maximum positive value that you can hold in a variable of type `int`?

(a) 255

(b) 256

(c) 65535

(d) 32767

20. Why do you need to use a reverse-biased diode across the coils of a relay?

(a) To boost the current

(b) To boost the voltage

(c) To "snub" voltage spikes as a result of switching the current to the coil

(d) None of the above

31
CHAPTER

Transducers, Sensors, Location, and Navigation

IN THIS CHAPTER, YOU'LL LEARN ABOUT ELECTRONIC DEVICES THAT CONVERT ENERGY FROM ONE form to another, devices that can detect phenomena and measure their intensity, systems that can help you determine your location (or the location of some other object), and devices that facilitate navigation for vessels, such as ships, aircraft, and robots.

Wave Transducers

In electronics, *wave transducers* convert AC or DC into acoustic or electromagnetic (EM) waves. They can also convert these waves into AC or DC signals.

Dynamic Transducer for Sound

A *dynamic transducer* comprises a coil and magnet that translates mechanical vibration into varying electrical current or vice-versa. The most common examples are the *dynamic microphone* and the *dynamic speaker*.

Figure 31-1 is a functional diagram of a dynamic transducer. A diaphragm is attached to a coil that can move back and forth rapidly along its axis. A permanent magnet rests inside the coil. Sound waves cause the diaphragm and coil to move together, producing fluctuations in the magnetic field

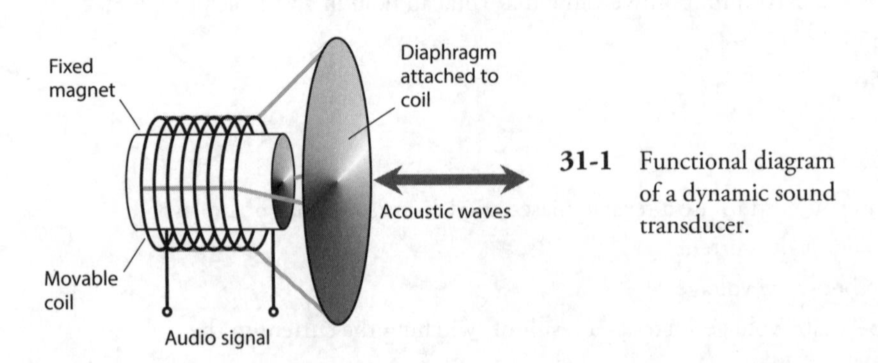

31-1 Functional diagram of a dynamic sound transducer.

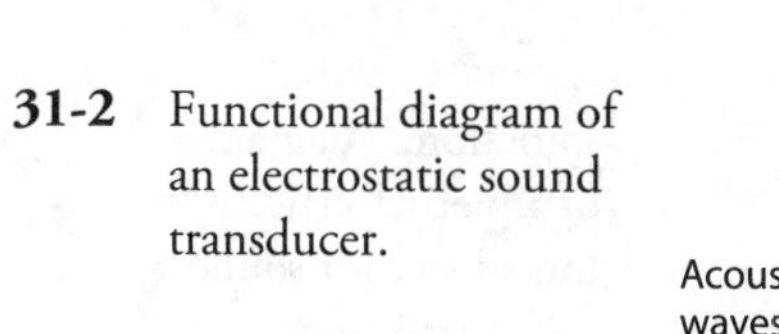
31-2 Functional diagram of an electrostatic sound transducer.

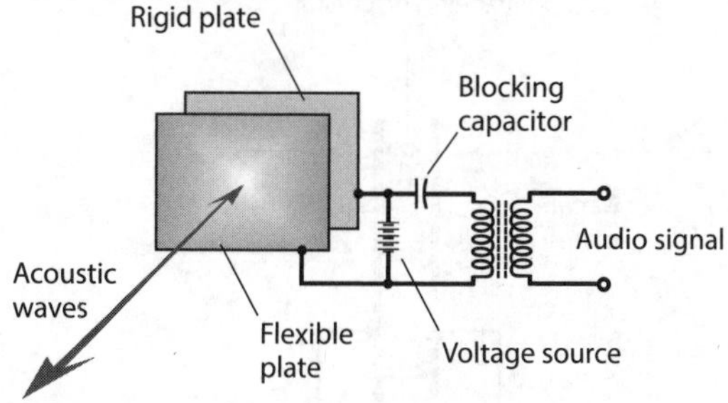

within the coil. As a result, audio AC flows in the coil, having the same waveform as the sound that strikes the diaphragm.

If we apply an audio signal to the coil, the AC in the wire generates a magnetic field that produces forces on the coil. These forces cause the coil to move, pushing the diaphragm back and forth to create acoustic waves in the surrounding air.

Electrostatic Transducer for Sound

An *electrostatic transducer* takes advantage of the forces produced by electric fields. Two metal plates, one flexible and the other rigid, are placed parallel to each other and close together, as shown in Fig. 31-2.

In an *electrostatic pickup*, incoming sound waves vibrate the flexible plate, producing small, rapid changes in the spacing, and therefore, the capacitance, between the plates. We apply a constant DC voltage between the plates. As the plate-to-plate capacitance varies, the electric field intensity between them fluctuates, causing variations in the current through the transformer primary winding. Audio signals appear across the secondary winding.

In an *electrostatic emitter*, fluctuating currents in the transformer produce changes in the voltage between the plates. This AC voltage results in electrostatic field variations, producing forces that push and pull the flexible plate in and out. The motion of the flexible plate produces sound waves in the air.

We can use electrostatic transducers in most applications in which dynamic transducers will work. Advantages of electrostatic transducers include light weight and good sensitivity. The relative absence of magnetic fields can also constitute an asset in certain situations.

Piezoelectric Transducer for Sound and Ultrasound

Figure 31-3 shows a *piezoelectric transducer* that consists of a slab-like *crystal* of quartz or ceramic material sandwiched between two metal plates. Piezoelectric transducers can function at higher frequencies than dynamic or electrostatic transducers can, so piezoelectric transducers are favored in ultrasonic applications such as intrusion detectors.

When acoustic waves strike one or both of the plates, the metal vibrates. This vibration transfers to the crystal by mechanical contact. The crystal generates weak electric currents when subjected to the mechanical stress. Therefore, an AC voltage develops between the two metal plates, with a waveform identical to that of the acoustic disturbance.

If we apply an electrical audio signal to the plates, the fluctuating current causes the crystal to vibrate in sync with the current. The metal plates vibrate also, producing an acoustic disturbance in the surrounding medium.

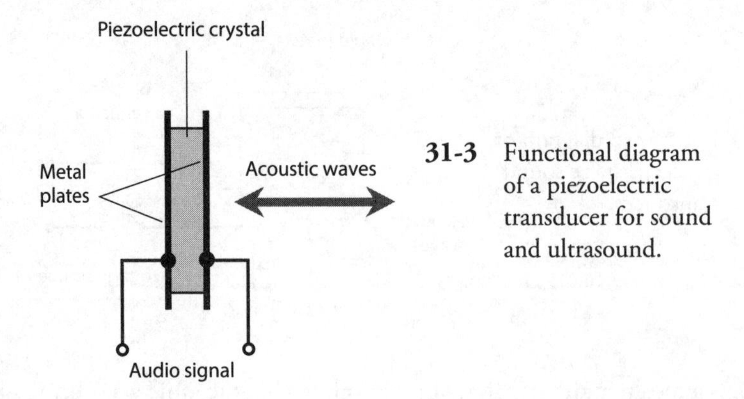

31-3 Functional diagram of a piezoelectric transducer for sound and ultrasound.

Transducers for RF Energy

The term *radio-frequency* (*RF*) *transducer* is a fancy expression for an *antenna*. Two basic types exist: the *receiving antenna* and the *transmitting antenna*.

Transducers for IR and Visible Light

Many wireless devices transmit and receive energy in the infrared (IR) spectrum, which spans frequencies higher than those of radio waves but lower than those of visible light. Some wireless devices transmit and receive EM signals in the visible range, although we will encounter them less often than we find IR devices.

The most common IR transmitting transducer is the infrared-emitting diode (IRED). When you apply fluctuating DC to the device, it emits IR rays. The fluctuations in the current constitute modulation, causing variations in the intensity of the rays emitted by the semiconductor P-N junction. The modulated IR carries information such as which channel you want your television (TV) set to "see" or whether you want to raise or lower the volume. You can focus an IR beam using optical lenses or mirrors to *collimate* the rays (make them parallel) for line-of-sight transmission through clear air over distances of up to several hundred meters.

Infrared receiving transducers resemble photodiodes or photovoltaic cells. The fluctuating IR energy from the transmitter strikes the P-N junction of the receiving diode. If the receiving device is a photodiode, you apply a current to it. This current varies rapidly in accordance with the signal waveform on the IR beam from the transmitter. If the receiving device is a photovoltaic cell, it produces the fluctuating current all by itself, without the need for an external power supply. In either case, the current fluctuations are weak, and you must amplify them before sending them to the equipment (TV set, garage-door opener, oven, security system, or whatever) controlled by the wireless system.

Displacement Transducers

A *displacement transducer* measures a distance or angle traversed, or the distance or angle separating two points. Conversely, a displacement transducer can convert an electrical signal into mechanical movement over a certain distance or angle. A device that measures or produces movement in a straight line constitutes a *linear displacement transducer*. If it measures or produces movement through an angle, we call it an *angular displacement transducer*.

Pointing and Control Devices

A *joystick* can produce movement or control variable quantities in two dimensions. The device has a movable lever attached to a ball bearing within a control box. You can manipulate the lever by hand up and down, or to the right and left. Some joysticks allow you to rotate the lever for control in a third dimension. Joysticks are used in computer games, for entering coordinates into a computer, and for the remote-control of robots.

A *mouse* is a peripheral device commonly used with personal computers. By sliding the mouse around on a flat surface, you can position a cursor or arrow on the computer display. Push-button switches on the top of the unit actuate the computer to perform whatever function the cursor or arrow shows. These actions are called *clicks*.

A *trackball* resembles an inverted mouse, or a two-dimensional joystick without the lever. Instead of pushing the device around on a flat surface, you manipulate a ball bearing with the index finger of one hand, causing the display cursor to move vertically and horizontally. Push-button switches on a computer keyboard, or on the trackball box itself, actuate the functions.

An *eraser-head pointer* is a rubber button approximately five millimeters (5 mm) in diameter, usually placed in the center of a computer keyboard. You move the cursor on the display by pushing against the button. Clicking is done with button switches on the keyboard.

A *touch pad* is a sensitive plate approximately the size and shape of a credit card. You place your index finger on the plate and move your finger around, producing movement of the display cursor or arrow. You can do clicks just as you do with a trackball or eraser-head pointer.

Electric Motor

An *electric motor* converts electrical energy into angular (and in some cases linear) mechanical energy. Motors can operate from either AC or DC, and range in size from tiny devices used in microscopic robots to huge machines that pull passenger trains. You learned the basic principle of the DC motor in Chap. 8. In a motor designed to work with AC, no commutator exists. Instead, the alternations in the current keep the polarity correct at all times, so the shaft does not "lock up." The rotational speed of an AC motor depends on the frequency of the applied AC. With 60 Hz AC, for example, the rotational speed equals 60 revolutions per second (60 r/sec) or 3600 revolutions per minute (3600 r/min). When you connect a motor to a load, the rotational force required to turn the shaft increases and the motor draws increasing power from the source.

Stepper Motor

A *stepper motor* turns in small increments, rather than continuously. The *step angle*, or extent of each turn, varies depending on the particular motor. It can range from less than 1° of arc to a quarter of a circle (90°). A stepper motor turns through its designated step angle and then stops, even if the coil current continues. When the shaft of a stepper motor has come to rest with current going through its coils, the shaft resists external rotational force; it "tries to stay in place."

Conventional motors run at hundreds or thousands of revolutions per minute. A stepper motor usually runs at far lower speeds, almost always less than 180 r/min. A stepper motor has the most turning power when operated at its slowest speeds, and the least turning power when operated at its highest speeds.

When we supply current pulses to a stepper motor at a constant frequency, the shaft rotates in increments, one step for each pulse. In this way, the device can maintain a precise speed. Because of the braking effect, this speed holds constant over a wide range of mechanical turning resistances.

Stepper motors work well in applications requiring point-to-point motion. Specialized robots can perform intricate tasks with the help of microcomputer-controlled stepper motors.

Selsyn and Synchro

A *selsyn* is an indicating device that shows the direction in which an object points. The selsyn consists of a transmitting unit and a receiving (or indicator) unit. As the shaft of the transmitting unit rotates, the shaft of the receiving unit, which forms part of a stepper motor, follows along exactly. Figure 31-4 shows how we can employ a selsyn to serve as a remote direction indicator for a wind vane. As the vane rotates, the indicator unit shaft moves through the same number of angular degrees as the transmitting unit shaft.

A pair of selsyns can display the orientation of a space-communications antenna that has two separate rotators, one for the azimuth (compass bearing) and another for the elevation. The selsyn for azimuth bearings has a range of 0° to 360°. The selsyn for elevation bearings has a range of 0° to 90°.

A *synchro* is a two-way selsyn that can control certain mechanical devices and also display their status. Synchros work well for robotic *teleoperation*, or remote control. Some synchros are programmable. The operator inputs a number into a microcomputer that controls the generator unit, and the receiver unit changes position accordingly. Synchros are also used for precision control of directional communications antennas, such as the Yagi, corner reflector, or dish.

Electric Generator

An *electric generator* is constructed in much the same way as an AC motor, although it functions in the opposite sense. Some generators can also function as motors; we call dual-purpose devices of this type *motor/generators*.

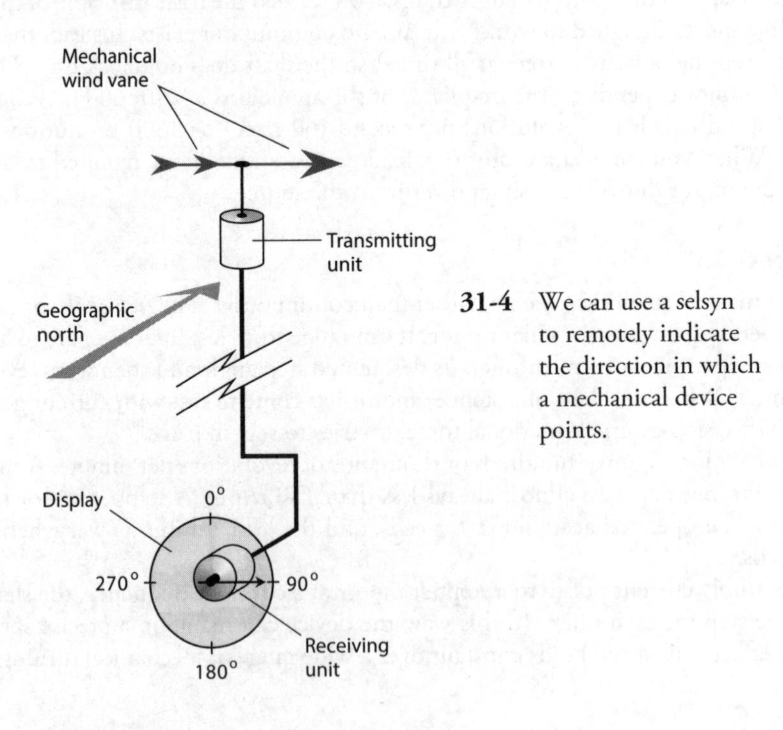

31-4 We can use a selsyn to remotely indicate the direction in which a mechanical device points.

A typical generator produces AC from the mechanical rotation of a coil in a strong magnetic field. Alternatively, a permanent magnet can be rotated within a coil of wire. We can drive the rotating shaft with a gasoline engine, a steam turbine, a water turbine, a wind turbine, or any other source of mechanical power. A commutator can work with a generator to produce pulsating DC output, which we can filter, if desired, to obtain pure DC to operate electronic equipment.

Small portable gasoline-powered generators, capable of delivering a few kilowatts, can be purchased in department stores or home-and-garden stores. Larger generators, which usually burn propane or methane ("natural gas"), allow homes or businesses to retain a continuous supply of electrical power during a utility disruption. The largest generators are found in power plants, and can produce many kilowatts.

Small generators can function in synchro systems. These specialized generators allow remote control of robotic devices. A generator can be used to measure the speed at which a vehicle or rolling robot moves. The shaft of the generator is connected to one of the wheels, and the generator output voltage and frequency vary directly with the angular speed of the wheel. In this case we have a *tachometer*.

Optical Encoder

When we use digital radios, we adjust the frequency in discrete steps, not continuously. A typical frequency increment is 10 Hz for "shortwave" radios and 200 kHz for FM broadcast radios. An *optical encoder*, also called an *optical shaft encoder*, offers an alternative to mechanical switches or gear-driven devices that wear out with time.

An optical encoder comprises two LEDs, two photodetectors, and a device called a *chopping wheel*. The LEDs shine on the photodetectors through the wheel. The wheel, which has alternately transparent and opaque radial bands (Fig. 31-5), is attached to a rotatable shaft and a control knob. As we rotate the knob, the light beams are interrupted. Each interruption causes the frequency to change by a specified increment.

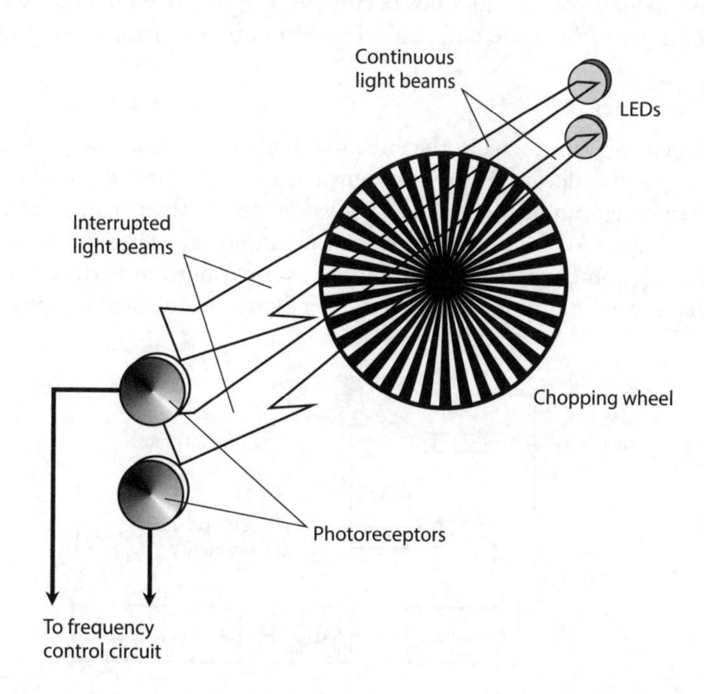

31-5 An optical encoder uses LEDs and photodetectors to sense the direction and extent of shaft rotation.

When used for frequency control, an optical shaft encoder can sense the difference between the "frequency-up" command (clockwise shaft rotation) and the "frequency-down" command (counterclockwise shaft rotation) according to which photodetector senses each sequential beam interruption first.

Detection and Measurement

A *sensor* employs one or more transducers to detect or measure parameters, such as temperature, humidity, barometric pressure, pressure, texture, proximity, and the presence of certain substances.

Capacitive Pressure Sensor

Figure 31-6 portrays the functional details of a *capacitive pressure sensor*. Two metal plates, separated by a layer of compressible dielectric foam, create a variable capacitor that we connect in parallel with an inductor. The resulting inductance/capacitance (*LC*) circuit determines the frequency of an oscillator. If an object strikes or pushes against the sensor, the plate spacing momentarily decreases, causing an increase in the capacitance and, therefore, a decrease in the oscillator frequency. When the object moves away from the transducer, the foam layer springs back to its original thickness, the plates return to their original spacing, and the oscillator frequency returns to normal.

We can convert the output of a capacitive pressure sensor to digital data using an *analog-to-digital converter* (ADC). This signal can go to a microcomputer, such as a robot controller. We can mount pressure sensors in various places on a mobile robot, such as the front, back, and sides. Then, for example, physical pressure on the sensor in the front of the robot can send a signal to the controller, which tells the machine to move backward.

A capacitive pressure sensor can be fooled by massive conducting or semiconducting objects in its vicinity. If such a mass comes near the transducer, the capacitance can change even if direct mechanical contact does not occur. We call this phenomenon *body capacitance*. In some applications, we can tolerate body capacitance; in other situations, we can't.

Elastomer

When we want to avoid the effects of body capacitance, we can use an *elastomer* device instead of a capacitive device for pressure sensing. An elastomer is a flexible substance resembling rubber or plastic that can be used to detect the presence or absence of mechanical pressure.

Figure 31-7 illustrates how we can employ an elastomer to detect and locate a pressure point. The elastomer conducts electricity fairly well, but not perfectly. It has a foam-like consistency, so that it can be compressed. Conductive plates are attached to opposite faces of the elastomer pad.

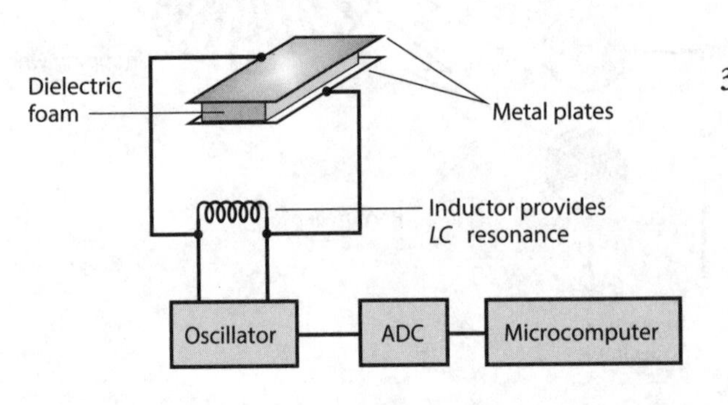

31-6 A capacitive pressure sensor. When force is applied, the spacing between the plates decreases, causing the capacitance to increase and the oscillator frequency to go down.

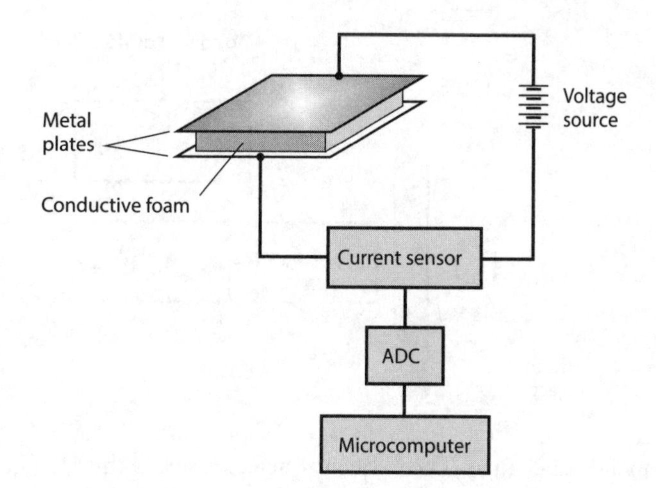

31-7 An elastomer pressure sensor detects applied force without unwanted capacitive effects.

When pressure appears at some point in the pad, the material compresses, and its electrical resistance goes down. The drop in resistance produces an increase in the current between the plates. As the applied pressure increases, the elastomer grows thinner, and the current goes up more. The current-change data can be sent to a microcomputer, such as a robot controller.

Back-Pressure Sensor

A motor produces a measurable pressure that depends on the *torque* ("turning force") that we apply. A *back-pressure sensor* detects and measures the torque that the motor exerts at any given instant in time. The sensor produces a signal, usually a variable voltage, that increases as the torque increases. Figure 31-8 is a functional diagram of the system.

Robotics engineers use back-pressure sensors to limit the forces applied by robot grippers, arms, drills, hammers, or other so-called *end effectors*. The *back voltage*, or signal produced by the sensor, reduces the torque applied by the motor, preventing damage to objects that the robot handles, and also ensuring the safety of people working around the robot.

Capacitive Proximity Sensor

A *capacitive proximity sensor* uses an RF oscillator, a frequency detector, and a metal plate connected into the oscillator, as shown in Fig. 31-9. The resulting device *takes advantage* of the body capacitance effects that can confound capacitive pressure sensors. We design the oscillator so that any variation in the capacitance of the plate, with respect to the environment, causes the oscillator

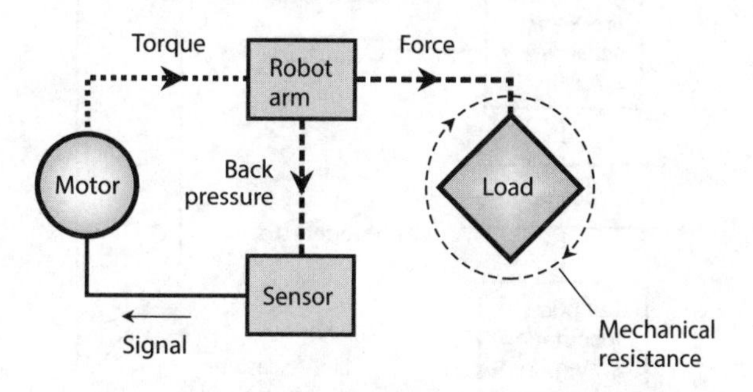

31-8 A back-pressure sensor governs the force applied by a robot arm or other mechanical device.

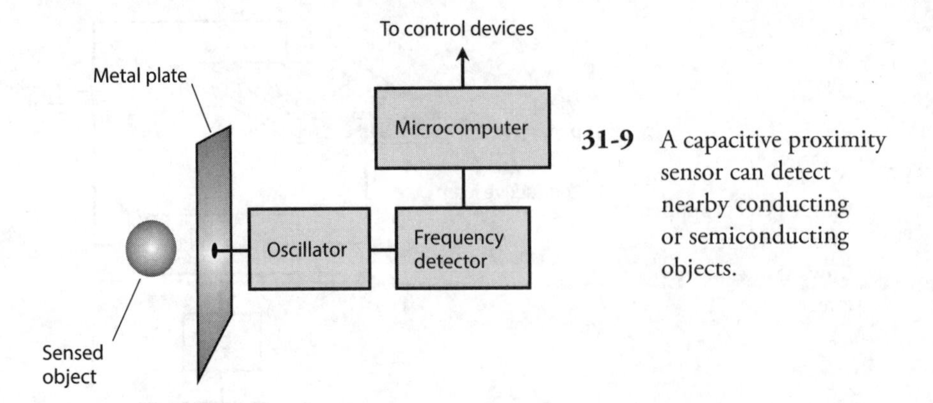

31-9 A capacitive proximity sensor can detect nearby conducting or semiconducting objects.

frequency to change. The frequency detector senses this change and transmits a signal to a microcomputer or robot controller.

Substances that conduct electricity to some extent (such as metal, salt water, and living tissue) are sensed more easily by capacitive transducers than are materials that do not conduct, such as dry wood, plastic, glass, or dry fabric. For this reason, capacitive proximity sensors work poorly, if at all, in environments that lack conductive objects. A machine shop would present a better venue for a robot with capacitive proximity sensing than, say, a child's bedroom.

Photoelectric Proximity Sensor

Reflected light can help a robot "know" when it's approaching a physical barrier. A *photoelectric proximity sensor* contains a light-beam generator, a photodetector, a frequency-sensitive amplifier, and a microcomputer, interconnected and operated as shown in Fig. 31-10.

The light beam reflects from the object, and the photodetector picks up some of the reflected light. The tone generator modulates the light beam at a certain frequency, say, 1000 Hz. The

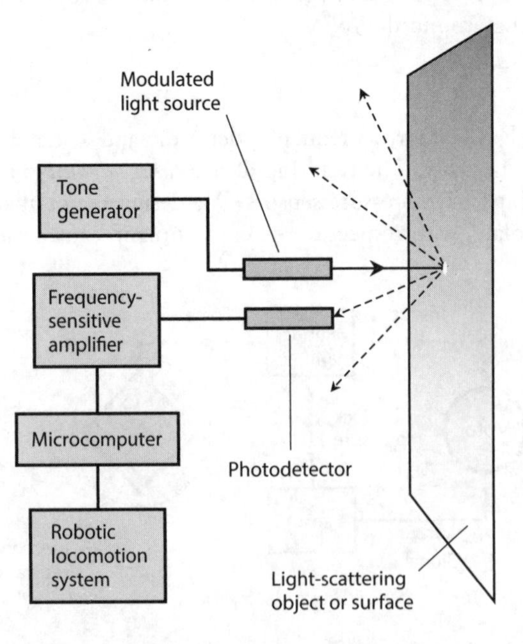

31-10 A photoelectric proximity sensor. Modulation of the light beam allows the device to distinguish between sensor-generated light and background illumination.

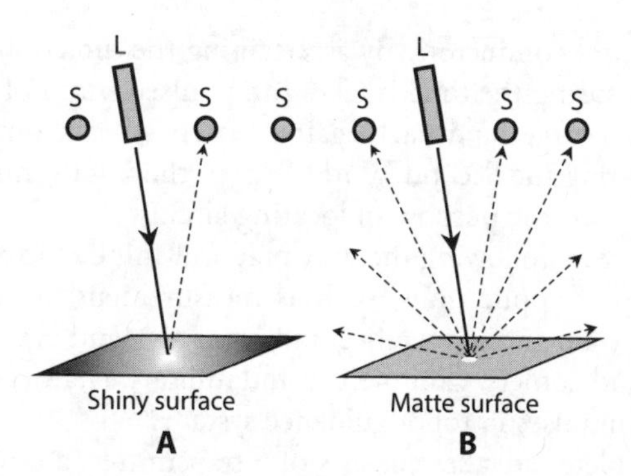

31-11 In texture sensing, lasers (L) and sensors (S) analyze a shiny surface (at A) and a matte surface (at B). Solid lines represent incident light; dashed lines represent reflected light.

photodetector's amplifier responds only to light modulated at that frequency. This modulation scheme prevents false imaging that could otherwise result from stray illumination from flashlights or sunlight. (Such light sources are unmodulated, and won't actuate a sensor designed to respond only to modulated light.) As the robot approaches an object, the robot controller senses the increasing intensity of the reflected, modulated beam. The robot can then steer clear of the obstruction.

Photoelectric proximity sensing doesn't work for objects that don't reflect light, or for shiny objects, such as glass windows or mirrors, approached at a sharp angle. In these scenarios, none of the light beam reflects back toward the photodetector, so the object remains "invisible" to the robot.

Texture Sensor

Texture sensing is the ability of a machine to determine whether an object has a shiny surface or a rough (matte) surface. A simple texture sensor contains a laser and several light-sensitive receptors.

Figure 31-11 shows how a combination of laser (L) and sensors (S) can tell the difference between a flat shiny surface (drawing A) and a flat matte (roughened) surface (drawing B). The shiny surface, such as a flat glass mirror or flat pane of glass, reflects light at the incidence angle only. But the matte surface, such as a sheet of paper, scatters light in all directions. The shiny surface reflects the beam back entirely to the sensor in the path of the beam whose reflection angle equals its incidence angle. The matte surface reflects the beam back to all the sensors. We can program a microcomputer to tell the difference.

Certain types of surfaces can confuse a texture sensor of the type portrayed in Fig. 31-11. For example, we would define a pile of small ice cubes as shiny on a microscopic scale but rough on a large scale. Depending on the diameter of the laser beam, the texture sensor might interpret such a surface as either shiny or matte. The determination can also be affected by the motion of the sensor relative to the surface. A surface interpreted as shiny when standing still relative to the sensor might be interpreted as matte when moving relative to the sensor.

Location Systems

The devices described in the previous paragraphs function best in short-range applications (with the exception of RF antennas). In this section, we'll examine a few medium-range and long-range applications of transducers and sensors. These applications fall into the broad category of *location systems*.

Radar

The term *radar* derives from the words *radio detection and ranging*. Electromagnetic (EM) waves having certain frequencies reflect from various objects, especially if those objects contain metals or other electrical conductors. By ascertaining the direction(s) from which radio signals are returned, and by measuring the time it takes for a pulsed beam of EM energy to travel from the transmitter location to a target and back again, a radar set can pinpoint the geographic positions of distant objects. During the Second World War in the 1940s, military personnel put this property of radio waves to use for the purpose of locating aircraft.

In the years following the war, practical-minded experimenters discovered that radar can serve in a variety of applications, such as measurement of automobile speed (by the police), weather forecasting (rain and snow reflect radar signals), and even the mapping of planets, planetary moons, asteroids, and comets. Commercial and military aviators use radar extensively. In recent years, radar has also found uses in robot guidance systems.

A complete radar set consists of a transmitter, a directional antenna with a narrow main lobe and high gain, a receiver, and an indicator or display. The transmitter produces intense pulses of RF microwaves at short intervals. The pulses propagate outward in a sharply defined beam from the antenna, and the wavefronts strike objects at various distances. Shortly after the pulse transmission, the receiving antenna picks up the reflected signals, known as *echoes*. As the distance to a reflecting object (often called a *target*) increases, so does the delay time between pulse transmission and echo reception. The transmitting antenna rotates in a horizontal plane at constant angular speed, allowing observation in all *azimuth bearings* (compass directions).

A typical circular radar display comprises a CRT or LCD, equipped to show scans of azimuth and *range* (distance to the target). Figure 31-12 shows the basic display configuration. The observing station's location corresponds to the center of the display. Azimuth bearings are indicated in degrees clockwise from true north, and are marked around the perimeter of the screen. The range is indicated by the radial displacement of the echo; the farther away the target, the farther from the display center the echo or blip. The radar display constitutes a dynamic set of *plane polar coordinates*. In Fig. 31-12, a target appears at an azimuth of about 124° (east-southeast). Its range is near the maximum for the display, known as the *radar horizon*.

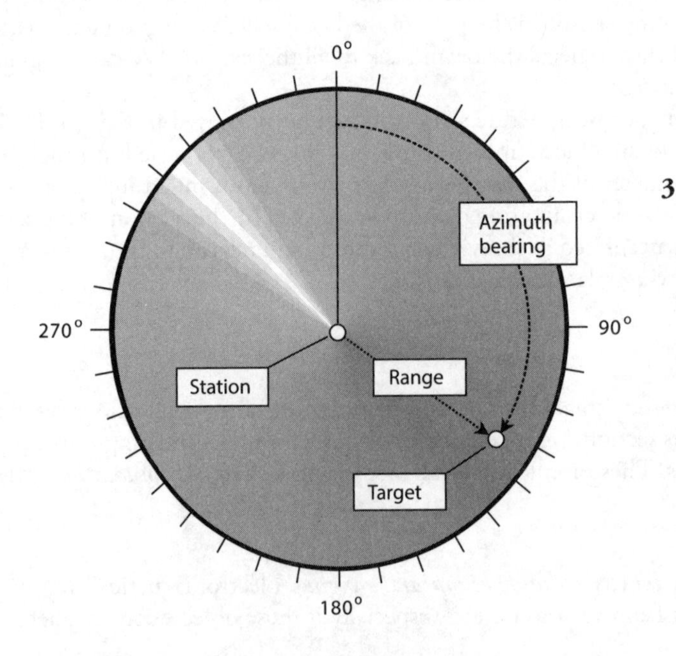

31-12 A radar display. The light radial band shows the azimuth direction in which the microwave beam is currently transmitted and received. (Not all radar displays show this band.)

For any particular installation, the distance to the radar horizon depends on the height of the antenna above the ground, the nature of the terrain in the area, the transmitter output power and antenna gain, the receiver sensitivity, and the weather conditions in the vicinity. Airborne long-range radar can detect echoes from several hundred kilometers (km) away under ideal conditions. A low-power system with the antenna at a low height might have a radar horizon of only 50 to 70 km.

The fact that precipitation reflects radar echoes creates an ongoing nuisance to aviation personnel, but it's invaluable for weather forecasting and observation. Radar allows aviators and meteorologists to detect and track severe thunderstorms and hurricanes. A *mesocyclone*, which is a severe thunderstorm likely to produce tornadoes, generates a characteristic "hook-shaped" echo on a radar display. The eyewall and rainbands surrounding the center of a hurricane show up as rings, arcs, or broken-up spirals on the display.

Some radar sets can detect changes in the frequency of the returned pulse, thereby allowing measurement of wind speeds in hurricanes and tornadoes, or the speeds and acceleration rates of approaching and receding targets. A system of this type is called *Doppler radar* because the frequency change results from the *Doppler effect* on EM waves.

Sonar

Sonar is a medium-range method of proximity sensing. The acronym derives from the words *sonic navigation and ranging*. The principle is simple: Bounce acoustic waves off of objects, and measure the time it takes for the echoes to return.

An elementary sonar system consists of an AC pulse generator, an acoustic emitter, an acoustic pickup, a receiver, a delay timer, and an indicating device, such as a numeric display, CRT, LCD, or pen recorder. The transmitter sends out acoustic waves through the medium, usually water or air. These waves are reflected by objects, and the echoes are picked up by the receiver. The distance to an object is determined on the basis of the echo delay, assuming that the speed of the acoustic waves in the medium is known.

Figure 31-13A shows a simple sonar system. The microcomputer can generate a *computer map* on the basis of sounds returned from various directions in two or three dimensions. This map can help a mobile robot or vessel navigate in its environment. However, the system can be "fooled" if the echo delay equals or exceeds the time interval between pulses, as shown in Fig. 31-13B. To overcome this conundrum, the microcomputer can instruct the pulse generator to send pulses of various frequencies in a defined, rotating sequence. The microcomputer keeps track of which echo corresponds to which pulse.

Acoustic waves travel faster in water than they do in the air. The amount of salt in water makes a difference in the propagation speed when sonar is used on boats (in depth finding, for example). The density of water can vary because of temperature differences as well. If the true speed of the acoustic waves is not accurately known, false readings will result. In fresh water, acoustic waves travel at about 1400 meters per second (m/s), or 4600 feet per second (ft/s). In salt water, acoustic waves travel at about 1500 m/s (4900 ft/s). In air, acoustic waves travel at approximately 335 m/s (1100 ft/s).

In the atmosphere, sonar can operate with audible sound waves, but ultrasound is often used instead. Ultrasound has a frequency too high to hear, ranging from about 20 kHz to more than 100 kHz. The most significant advantage of ultrasound is the fact that people who work around the sonar devices can't hear the signals, and will, therefore, not experience distraction, headaches, or other ill effects from them. In addition, ultrasonic sonar is less likely than audible sonar to be confused by people talking, heavy equipment, loud music, and other common noise sources. At frequencies higher than the range of human hearing, acoustical disturbances don't occur as often, or with as much intensity, as they do within the hearing range.

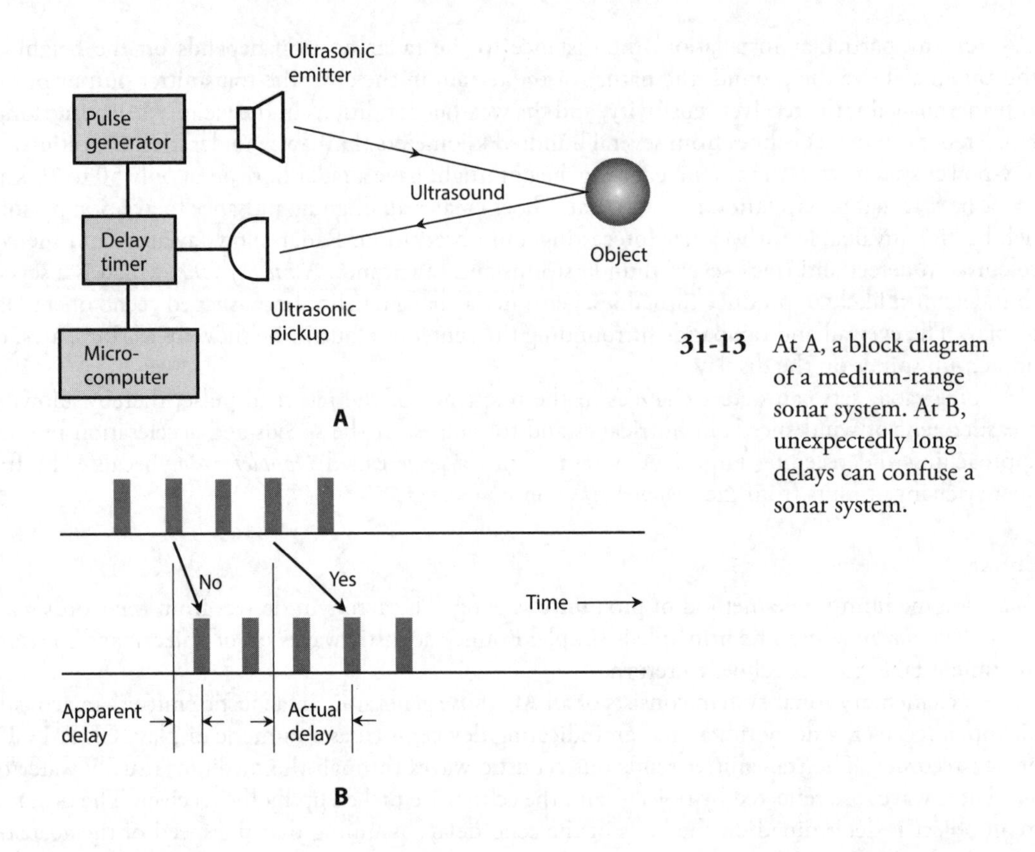

31-13 At A, a block diagram of a medium-range sonar system. At B, unexpectedly long delays can confuse a sonar system.

In its most advanced forms, sonar can rival *vision systems* (also called *machine vision*) as a means of mapping the environment for a mobile robot or vessel at close range. However, sonar has an outstanding limitation: All acoustic waves, including sound and ultrasound, require a material medium (usually a gas or liquid) in order to propagate. Therefore, sonar won't work in outer space, which practically constitutes a vacuum. Another significant limitation of sonar results from the fact that acoustic waves travel far more slowly than EM waves, such as radar pulses or visible-light beams, severely limiting the practical range (or *sonar horizon*).

Signal Comparison

A machine or vessel can find its geographic position by comparing the signals from two fixed stations at known positions, as shown in Fig. 31-14A. By adding 180° to the bearings of the sources X and Y, the machine or vessel (small shaded rectangle) obtains its bearings as "seen" from the sources (small spheres). The machine or vessel can determine its direction and speed by taking two readings separated by a certain amount of time. In the old days, the navigators of aircraft and oceangoing vessels physically plotted diagrams resembling Fig. 31-14A on paper maps using a pen, a straight-edged ruler, and a drafting compass, a process known as *triangulation*. Nowadays, computers do that work, with speedier and more accurate results.

Figure 31-14B is a block diagram of an *acoustic direction finder* such as a mobile robot might employ. The receiver has a signal-strength indicator and a servo that turns a directional ultrasonic transducer. The system has two remotely located signal sources, called *beacons*, that operate at

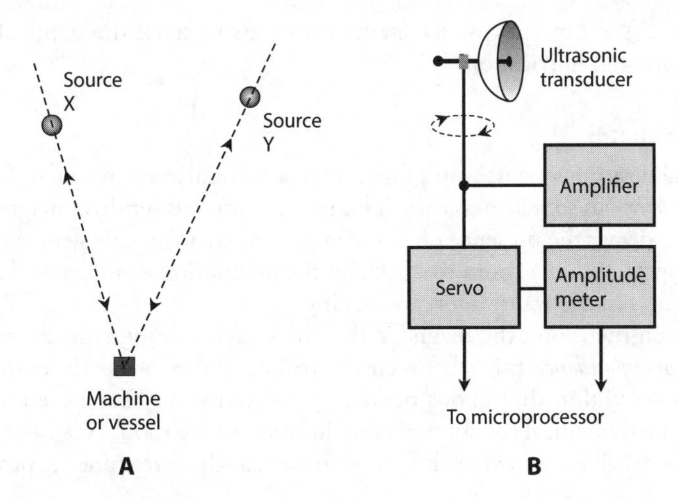

31-14 A simple direction-finding scheme (A) and an ultrasonic direction finder (B).

different frequencies. When the transducer rotates so the signal from one beacon reaches its maximum amplitude, a bearing is obtained by comparing the orientation of the transducer with some known standard such as a magnetic compass. The same is done for the other beacon. A computer determines the precise location of the robot based on this data.

Radio Direction Finding (RDF)

We can use a radio receiver, equipped with a signal-strength indicator and connected to a rotatable, directional antenna, to determine the direction from which RF signals arrive. *Radio-direction-finding* (RDF) equipment aboard a mobile vehicle facilitates determining the location of an RF transmitter. An RDF receiver can also serve to find one's own position with respect to two or more RF transmitters operating on different frequencies.

At frequencies below approximately 300 MHz, an RDF receiver employs a small loop or loopstick antenna shielded against the electric component of radio waves, so it picks up only the magnetic part of the EM field. The loop rotates until a well-defined minimum, or *null*, occurs in the received signal strength, indicating that the axis of the loop lies along a line toward the transmitter. When we take signal-strength readings from two or more locations separated by a sufficient distance, we can pinpoint the location of the transmitter by finding the intersection point of the azimuth bearing lines on a map. A computer, obviously, can do the same.

At frequencies above approximately 300 MHz, a directional transmitting and receiving antenna, such as a Yagi, quad, dish, or helical type, gives better results than a small loop. When we use such an antenna for RDF, the azimuth bearing corresponds to a signal maximum, or *peak*, rather than to a signal null.

Navigational Methods

Navigation involves the use of location devices over a continuous period, thereby deriving a function of position versus time. We can use this technique to determine whether or not a vessel follows its prescribed course. We can also use navigation methods to track the paths of military targets, severe thunderstorms, and hurricanes.

Fluxgate Magnetometer

When conventional position sensors won't function in a particular environment for a mobile robot, a *fluxgate magnetometer* can sometimes work. This system employs sensitive magnetic receptors and a microcomputer to detect the presence of, and changes in, an artificially generated magnetic field. The robot can navigate within a room by checking the orientation of magnetic lines of flux generated by electromagnets in the walls, floor, and ceiling.

For each point in the room, the magnetic flux lines have a unique direction and intensity, so that a *one-to-one correspondence* exists between the magnetic flux *vector* (intensity-direction composite) and the points within the robot's operating environment. This correspondence translates into a two-variable mathematical function of every location in the room. We can program the robot controller with this function, allowing the robot to constantly determine its position to within a few millimeters.

Epipolar Navigation

Epipolar navigation works by evaluating how an image changes as viewed from a single, moving perspective. The system needs only one observation point (that is, one observer location) at any given moment in time.

Imagine that you pilot an aircraft over the ocean. The only land in sight is a small island. The on-board computer "sees" an image of the island that constantly changes shape. Figure 31-15 shows three sample sighting positions (A, B, C) and the size/shape of the island as "seen" by a machine vision system in each case. The computer has the map data, so it "knows" the true size, shape, and location of the island.

The computer compares the shape and size of the image it "sees" at each point in time, from the vantage point of the aircraft, with the actual shape and size of the island from the map data. From this information, the computer can ascertain the aircraft's altitude, ground speed, ground direction, geographic latitude, and geographic longitude.

Loran

The acronym *loran* derives from the words *long-range navigation.* Loran is one of the oldest electronic navigation schemes. The system employs RF pulse transmission at low and medium frequencies (typically below 2 MHz) from multiple transmitters at specific geographic locations. A computer on board an oceangoing vessel can determine the ship's location by comparing the time difference in the arrival of the signals from two different transmitters at known locations. Based on the fact that radio waves propagate at the speed of light in free space (approximately 299,792 km/s or 186,282 mi/s), the computer can determine the distance to each transmitter, and from this data, calculate the location of the ship relative to the transmitters. In recent years, loran has been largely supplanted by the *Global Positioning System* (GPS).

Global Positioning System (GPS)

The Global Positioning System (GPS) comprises a network of radiolocation and radionavigation units that operate on a worldwide basis. The system employs several satellites, and allows determination of latitude, longitude, and altitude.

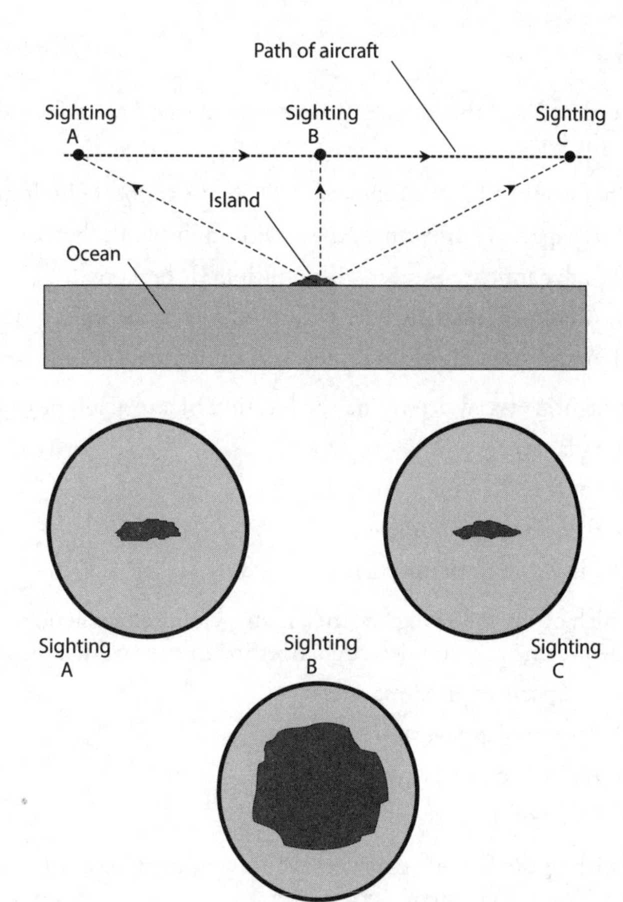

31-15 Epipolar navigation is the opto-electronic counterpart of human spatial perception.

All GPS satellites transmit signals in the microwave part of the radio spectrum. The signals are modulated with codes that contain timing information used by the receiving apparatus to make measurements. A GPS receiver determines its location by measuring the distances to several different satellites. This task is carried out by precisely timing the signals as they travel between the satellites and the receiver. The process resembles old-fashioned navigators' triangulation, except that it takes place in three dimensions (space) rather than in two dimensions (on the surface of the earth).

Electromagnetic waves propagate a little more slowly in the earth's ionosphere than they do in outer space. The extent of this reduction depends on the signal frequency, the ionization density in the upper atmosphere, and the angle at which the signal beams must work their way through the ionized layers. The GPS employs dual-frequency transmission to compensate for this effect. A GPS receiver uses a computer to process the information received from the satellites, giving the user an indication of position accurate to within a few meters.

Increasingly, automobiles, trucks, and pleasure boats are sold with GPS receivers installed as standard equipment. If you're driving your car through a remote area and you get lost, you can use GPS to locate your position. Using a cell phone, wireless Internet connection, Citizens Band (CB) radio transceiver, or amateur ("ham") radio transceiver, you can call for help and inform authorities of your exact position, which appears on an easy-to-read map.

Quiz

Refer to the text in this chapter if necessary. A good score is at least 18 correct. Answers are in the back of the book.

1. What would a mobile robot most likely use to locate itself in a large vacant parking lot?
   (a) A capacitive proximity sensor with multiple RF beacons
   (b) A dynamic transducer with multiple IR beacons
   (c) An optical shaft encoder with multiple visible-light beacons
   (d) An acoustic direction finder with multiple ultrasonic beacons

2. At night, you can determine the location of a tornado near you using a
   (a) radar set.
   (b) sonar set.
   (c) fluxgate magnetometer.
   (d) capacitive proximity sensor.

3. Which of the following systems, if any, can use magnetic receptors and a microcomputer to sense the presence of, and detect changes in, an external magnetic field?
   (a) A capacitive proximity sensor
   (b) An epipolar navigator
   (c) A synchro or selsyn
   (d) None of the above

4. Which of the following devices would you use to control and display the compass direction toward which a dish antenna points?
   (a) A fluxgate magnetometer
   (b) The GPS
   (c) A synchro
   (d) A capacitive proximity sensor

5. An elastomer pad, considered as an object all by itself, has variable
   (a) conductance.
   (b) magnetic-field intensity.
   (c) sensitivity to ultrasound.
   (d) inductive reactance.

6. An azimuth-range radar display has a geometric configuration similar to that of
   (a) an analog meter.
   (b) a digital bar graph meter.
   (c) an analog TV display.
   (d) a polar-coordinate system.

7. You can use sonar to measure sound
   (a) loudness.
   (b) frequency.
   (c) speed.
   (d) None of the above

8. A fluxgate magnetometer allows a robot to locate itself on the basis of
   (a) body capacitance with nearby objects.
   (b) the orientations of external magnetic fields.
   (c) the phase relationships among multiple ultrasonic beacon signals.
   (d) optical delays to and from stationary reflectors.

9. Which of the following devices is *not* an electromechanical transducer?
   (a) An electric motor
   (b) An electric generator
   (c) A selsyn
   (d) A fluxgate magnetometer

10. At low radio frequencies, an RDF system typically employs
    (a) a Yagi antenna.
    (b) a small loop or loopstick antenna.
    (c) a dish antenna.
    (d) a half-wave dipole antenna.

11. A robot can use a back-pressure sensor to
    (a) measure the intensity of visible light.
    (b) regulate the torque produced by an electric drill.
    (c) determine the orientation of magnetic lines of flux.
    (d) regulate the output of an acoustic transducer.

12. When you apply AF voltage to the plates of an electrostatic transducer, the electric field between the plates fluctuates in intensity, causing
    (a) a magnetic field that moves the coil, producing sound waves.
    (b) the capacitance to vary, thereby generating a phase shift.
    (c) forces that move the flexible plate, producing sound waves.
    (d) the plates to charge and discharge, producing AF current.

13. Which of the following devices is a form of RF transducer?
    (a) A small loop antenna
    (b) A back-pressure sensor
    (c) An elastomer
    (d) An epipolar navigation sensor

14. In a dynamic pickup or microphone, incident sound waves cause AF current to flow in the coil as a result of
    (a) the coil's motion within a magnetic field.
    (b) the coil's motion within an electrostatic field.
    (c) a DC voltage applied to the coil.
    (d) changes in the capacitive reactance across the coil.

15. A motor whose shaft rotates in discrete increments (not continuously) is called
    (a) an incremental motor.
    (b) a stepper motor.
    (c) a fractional motor.
    (d) a selsyn.

16. A photoelectric proximity sensor might have trouble detecting
    (a) a gray curtain.
    (b) a black wall.
    (c) a red ball.
    (d) All of the above

17. How many simultaneous observing locations does an epipolar navigation system require?
    (a) One
    (b) Two
    (c) Three
    (d) Four

18. Which of the following devices or systems operates at the lowest frequency?
    (a) A radar set
    (b) The GPS
    (c) A photoelectric proximity sensor
    (d) A loran system

19. Which of the following devices or systems operates on the basis of fluctuating magnetic forces between physical objects?
    (a) A sonar system
    (b) A loran system
    (c) A dynamic transducer
    (d) A capacitive proximity sensor

20. You would use a piezoelectric transducer to detect
    (a) magnetic fields.
    (b) ultrasonic waves.
    (c) microwave RF signals.
    (d) torque produced by a motor.

32
CHAPTER
Acoustics and Audio

IN SOUND RECORDING AND REPRODUCTION, ESPECIALLY WITH MUSIC, FIDELITY SUPERSEDES ALL other considerations. Audio enthusiasts consider low distortion and esthetics—and in some cases massive output power—more important than sheer efficiency.

Acoustics

Acoustics is the science of sound waves. Sound consists of molecular vibrations at audio frequencies (AF), ranging from about 20 Hz to 20 kHz. Young people can hear the full range of AF sound. As they age, people lose hearing sensitivity at the upper and lower frequency extremes.

Audio Frequencies

Musicians divide the AF range into three broad, vaguely defined parts, called *bass* (pronounced "base"), *midrange*, and *treble*. The bass frequencies start at 20 Hz and extend to 150 or 200 Hz. Midrange begins at this point, and extends up to 2 or 3 kHz. Treble consists of the audio frequencies higher than midrange. As the frequency increases, the wavelength, in any particular medium, grows shorter.

In air, sound travels at about 1100 feet per second (ft/s) or 335 meters per second (m/s). The relationship between the frequency f of a sound wave in hertz and the wavelength λ_{ft} in feet is therefore

$$\lambda_{ft} = 1100/f$$

The relationship between f in hertz and λ_m in meters is

$$\lambda_m = 335/f$$

These formulas also hold for frequencies in kilohertz and wavelengths in millimeters.

A sound disturbance traveling in sea-level air at 20 Hz has a wavelength of 55 ft (17 m). A sound of 1.0 kHz produces a wave 1.1 ft (34 cm) long. At 20 kHz, a sound wave in the air measures only 0.055 ft (17 mm) long. In other media such as air at extreme altitudes, fresh water, salt water, or metals, the foregoing formulas do not apply.

Waveforms

The frequency, or *pitch*, of a sound disturbance represents one of several variables that acoustic waves can possess. Another important factor is the shape of the wave. The *acoustic waveform* determines the *timbre* (sometimes erroneously called "tone"). The simplest acoustic waveform comprises a sine wave (or *sinusoid*), in which all of the energy exists at a single frequency. Sinusoidal sound waves rarely occur in nature, but we can synthesize them with specialized audio oscillators. A good artificial example of an audio sinusoid is the *beat note* or *heterodyne* produced by a steady, unmodulated carrier in a communications receiver using a product detector.

In music, most of the notes produced by "real-world" instruments have complex waveforms containing energy at a specific fundamental frequency and its harmonics. The simplest examples include sawtooth, square, and triangular waves. The shape of the waveform depends on the distribution of energy among the fundamental frequency and the harmonics. In theory, a sound wave can have infinitely many different shapes at a single frequency such as 1 kHz. As a result, a note at any particular frequency can exhibit unlimited variations in timbre.

Path Effects

A flute, clarinet, guitar, and piano can each produce a sound wave at 1 kHz, but the timbre differs for each instrument. The waveform affects the way that acoustic disturbances reflect from physical objects. Acoustics engineers must consider this variability when designing sound systems and concert halls to ensure that all of the instruments sound realistic everywhere in the room.

Suppose that you have a sound system set up in your living room, and that, for the particular placement of speakers with respect to your ears, sound waves propagate well at 1, 3, and 5 kHz, but poorly at 2, 4, and 6 kHz. This performance-versus-frequency variability will affect how musical instruments sound to you, distorting the sounds from some instruments more than the sounds from other instruments. Unless all sounds, at all frequencies, reach your ears in the same proportions as they come from the speakers, you will not hear the music as it originally came from the instruments.

Figure 32-1 shows a listener, a speaker, and three sound reflectors called *baffles*. The waves following paths X, Y, and Z, when reflected by the baffles along with the wave following the direct path D, add up to something different, at the listener's ears, for each frequency of sound. This phenomenon is impossible to prevent. That's why it's difficult to design an acoustical room, such as a concert auditorium, that will propagate sound waves effectively at all frequencies for every listener.

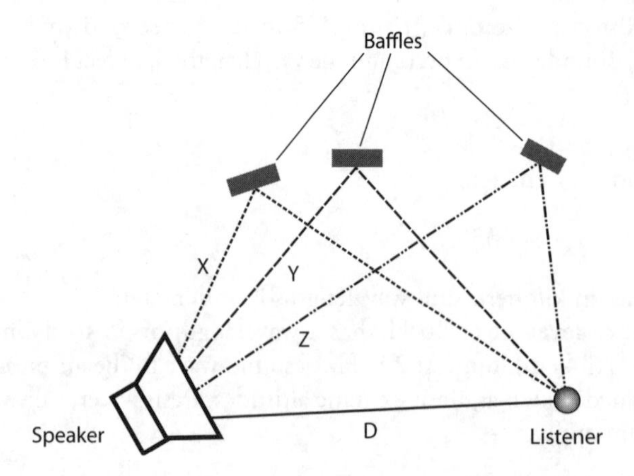

32-1 Acoustic waves that follow reflected paths X, Y, and Z combine with acoustic waves that follow the direct path D to produce the sound that a listener hears.

Loudness and Phase

You do not perceive the *loudness* (also called *volume*) of sound in direct proportion to the power contained in the disturbance. Your ears and brain sense sound levels according to the logarithm of the actual intensity. Another variable is the *phase* with which waves arrive at your ears. Phase allows you to perceive the direction from which a sound arrives. Phase can also affect the perceived sound volume.

The Decibel in Acoustics

You've learned about the decibel as a unit of relative signal voltage, current, and power. Decibels are also used in acoustics to express relative sound power. If you abruptly change the volume setting in a hi-fi (high fidelity) amplifier so that you can just barely detect the difference in sound loudness, the increment equals plus-or-minus one decibel (±1 dB). If you use the volume control to abruptly double or halve the actual acoustic-wave power coming from a set of speakers, you perceive a change of ±3 dB.

If you expect decibels to have meaning in acoustics, you must have a reference level against which you can measure all other sound power levels. Have you read that a household vacuum cleaner produces 80 dB of sound? This figure represents a sound-power comparison with respect to the *threshold of hearing*, the faintest sound that a person with good hearing can detect in a *quiet room* specially designed to have a minimum of background noise.

Phase in Acoustics

Even if only one sound source exists, acoustic waves reflect from the walls, ceiling, and floor of a room. In the scenario of Fig. 32-1, imagine the baffles as two walls and the ceiling. The three sound paths X, Y, and Z will almost certainly differ in length, so the sound waves reflected from these surfaces won't arrive in the same phase at the listener's ears. The direct path D (a straight line from the speaker to the listener) always represents the shortest possible route. In this situation, there exist at least four different paths by which sound waves can propagate from the speaker to the listener. In some practical scenarios, there are dozens, hundreds, or thousands of different paths. In theory, there might be infinitely many!

Suppose that, at a certain frequency, the acoustic waves for all paths happen to arrive in exactly the same phase in the listener's ears. Sounds at that frequency will seem exaggerated in volume. The same phase coincidence might also occur at harmonics of this frequency. This situation can cause problems because *acoustic peaks*, called *antinodes*, distort the original sound. At certain other frequencies, the waves might mix in phase opposition, yielding *acoustic nulls* called *nodes* or *dead zones*. If the listener moves a few meters (or even, in some cases, a few centimeters), the volume at any affected frequency will change. A new antinode or node might then present itself at another set of frequencies.

One of the biggest challenges in acoustical design is the avoidance of significant antinodes and nodes. In a home hi-fi system, this task might comprise nothing more than minimizing the extent to which sound waves reflect from the ceiling, the walls, the floor, and the furniture. Acoustical tile can be installed on the ceiling, the walls can be papered or covered with cork tile, the floor can be carpeted, and the furniture can be upholstered with cloth. In large auditoriums and music halls, the problem becomes more complex because of the larger sound propagation distances involved, and also because of the fact that some sound waves, especially at higher audio frequencies, reflect from balconies, chairs, lighting fixtures, and even people in the audience.

Technical Considerations

Regardless of its size, a good hi-fi sound system must have certain characteristics. Let's look at two of the most important technical considerations: *linearity* and *dynamic range*. In acoustics, these terms have almost, but not quite, the same meanings as they do in communications.

Linearity

In acoustics, we define linearity as the extent to which the output waveform of an amplifier constitutes a faithful reproduction of the input waveform. In hi-fi equipment, all the amplifiers must be as linear as the state of the art allows.

If you connect a *dual-trace oscilloscope* (an oscilloscope that lets you observe two waveforms at the same time) to the input and output terminals of a hi-fi audio amplifier with good linearity, the output waveform shows up as a vertically magnified duplicate of the input waveform. When you apply an input signal to the horizontal scope input and the resulting output signal to the vertical scope input, the display shows a straight, but slanted, line. In an amplifier with poor linearity, the instantaneous output-versus-input function is not a straight line. The output waveform does not represent a faithful reproduction of the input, and distortion occurs. In some RF amplifiers, this state of affairs can be tolerated, but it's unacceptable in a hi-fi audio system.

Engineers design hi-fi amplifiers to work with input signals up to a certain peak-to-peak amplitude. If the signal amplitude exceeds this limit, the system's *active components* (usually transistors) become nonlinear, and distortion takes place. In a hi-fi system equipped with VU or distortion meters, excessive input causes the needles to "kick up" into the red ranges of the scales during peaks.

Dynamic Range

In a sound system, we can define the dynamic range as the ratio of the maximum power output to the minimum power output that the system can deliver while maintaining acceptable performance with low distortion. As the dynamic range increases, the sound quality improves for music or programming having a wide range of volume levels. We express dynamic range in decibels (dB).

At low volume levels, *background noise* limits the dynamic range of a hi-fi system. In an analog system, most of this noise comes from the audio amplification stages. In a tape recording, we also observe some *tape hiss*. A scheme called *Dolby* (a trademark of Dolby Laboratories) can minimize the effects of background noise in analog recording. However, digital recording and reproduction systems produce less internal noise than analog systems do.

At high volume levels, the power-handling capability of an audio amplifier limits the dynamic range. If we hold all other factors constant, we can expect a 100-W audio system to have greater dynamic range than a 50-W system. The speaker size is also important. As speakers get physically larger, their ability to handle high power improves, resulting in increased dynamic range. That's why serious audio enthusiasts sometimes purchase sound systems with amplifiers and speakers that seem unnecessarily large.

Components

You can set up a hi-fi system in myriad ways. A true *audiophile* ("sound lover" or serious hi-fi enthusiast) assembles a complex system over a period of time. Following are some basic considerations that can serve as guidelines when choosing system components.

Configurations

The simplest type of home hi-fi system resides in a single box, with an AM/FM radio receiver and a *compact disk* (CD) player. The speakers are generally external, with short connecting cables. The assets of this so-called *compact hi-fi system* include small size and low cost. The main limitation is, as you might expect, limited audio output.

More sophisticated hi-fi systems have separate boxes containing individual devices such as:

- An AM tuner
- An FM tuner
- An amplifier or pair of amplifiers
- A CD player
- A computer and its peripherals (optional)

The computer facilitates downloading music files or *streaming audio* from the Internet, creating ("burning") CDs, and composing and editing electronic music. A high-end system can also include a *satellite radio* receiver, a *tape player*, a *turntable*, or other nonstandard peripheral. The individual hardware units in this type of system, known as a *component hi-fi system*, are interconnected with shielded cables. A component system costs more than a compact system, but it offers superior sound quality, more audio power, and greater versatility than a compact system. You can tailor the system to your preferences.

Some hi-fi manufacturers build all their equipment cabinets to a standard width and then mount the cabinets, one above the other, in a metal framework called a *rack*. A *rack-mounted hi-fi system* saves floor space and makes a system look professional. The rack can be equipped with wheels so that the whole system, except for the external speakers, can be rolled from place to place.

Figure 32-2 is a block diagram of a typical home stereo hi-fi system. You should ground the amplifier chassis to minimize hum and noise, and to minimize susceptibility to interference from

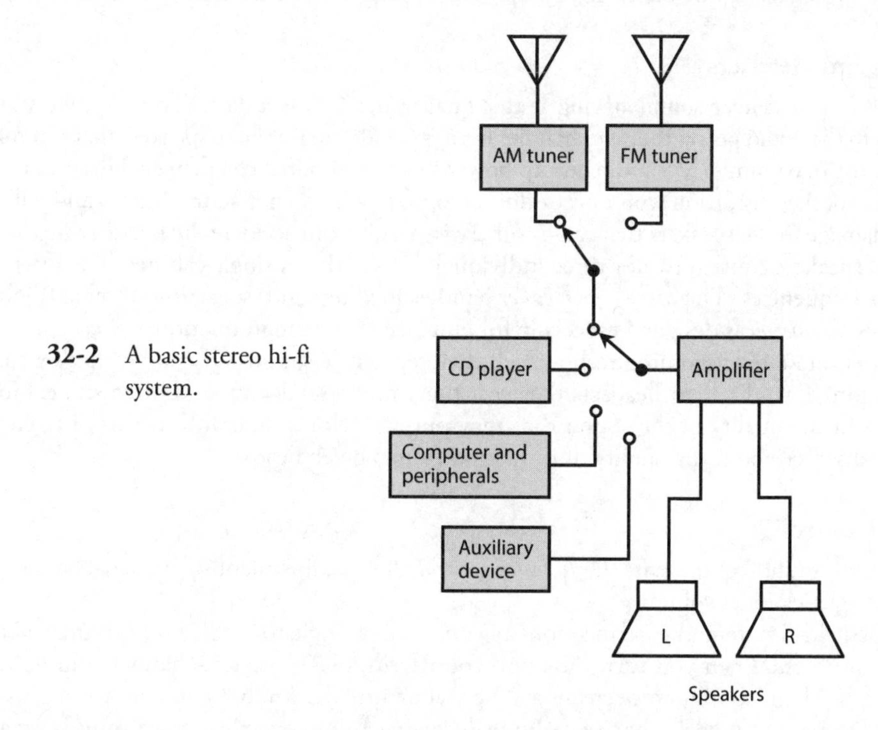

32-2 A basic stereo hi-fi system.

external sources. The AM antenna usually comprises a loopstick built into the cabinet or mounted on the rear panel. The FM antenna can be an indoor type, such as television "rabbit ears," or a directional outdoor antenna equipped with lightning-protection hardware.

The Tuner

A *tuner* is a radio receiver capable of receiving signals in the standard AM broadcast band (535 to 1605 kHz) and/or the standard FM broadcast band (88 to 108 MHz). Tuners don't have built-in amplifiers. A tuner can provide enough power to drive a headset, but it usually takes an external amplifier to provide enough power for speakers.

Modern hi-fi tuners employ frequency synthesizers and digital readouts. Most tuners have numerous programmable *memory channels* that allow you to select your favorite stations with a push of a single button, no matter where the stations fall within the frequency band. Most tuners also have *seek* and/or *scan* modes that allow the radio to automatically search the band for any station with reasonable signal strength.

The Amplifier

In a hi-fi system, an amplifier delivers medium or high audio power to a set of speakers. An amplifier always has at least one input, but more often three or more: one for a CD player, another for a tuner, and still others for auxiliary devices such as a tape player, turntable, or computer. A few milliwatts of input power can produce audio output up to the amplifier's limit, in some cases hundreds of watts.

Amplifier prices rise with increasing power output and improvements in the dynamic range. A simplified hi-fi amplifier forms the basis for a *public-address* (PA) system. Popular music bands use massive amplifiers, some of which employ *vacuum tubes* that offer electrical ruggedness and excellent linearity. Tube-type amplifiers require bulky, massive power supplies that deliver dangerous DC voltages.

Speakers and Headsets

No amplifier can deliver sound having better quality than the speakers allow. Speakers are rated according to the audio power they can handle. It's a good idea to purchase speakers that can tolerate at least twice the maximum RMS audio output power that the amplifier can deliver. This precaution will ensure that speaker distortion won't occur during loud, low-frequency sound bursts, and will prevent physical damage to the speakers that might otherwise result from accidentally overdriving them.

Good speakers contain two or three individual units within a single cabinet. The *woofer* reproduces bass frequencies. The *midrange speaker* handles medium and, sometimes, treble (high) audio frequencies. A *tweeter* is designed especially for enhanced treble reproduction.

Headsets are rated according to how well they reproduce sound. Of course, that's a subjective consideration. Two different headsets that cost the same at retail can, and often do, exhibit huge differences in the quality of the sound that they put out. Not only that, but two different people will likely disagree about the quality of the sound from a given headset.

Balance Control

In hi-fi stereo sound equipment, the *balance control* allows adjustment of the left-channel volume versus the right-channel volume.

In a basic hi-fi system, the balance control consists of a single rotatable knob connected to a pair of potentiometers. When you turn the knob counterclockwise, the left-channel volume increases and the right-channel volume decreases. When you turn the knob clockwise, the right-channel volume increases and the left-channel volume decreases. In more sophisticated sound systems, you can adjust the balance with two independent volume controls, one for the left channel and the other for the right channel.

Proper balance is important in stereo hi-fi. A balance control can compensate for such factors as variations in speaker placement, relative loudness in the channels, and the acoustical characteristics of the room in which the equipment is installed.

Tone Control

You can adjust the amplitude-versus-frequency characteristic of a hi-fi sound system with a so-called *tone control.* In its simplest form, a tone control consists of a single knob or slide device. The counterclockwise, lower, or left-hand settings of this control result in strong bass and weak treble audio output. The clockwise, upper, or right-hand settings result in weak bass and strong treble. When you set the control to mid-position, the amplifier exhibits a more or less flat audio response; the bass, midrange, and treble sounds emerge in roughly the same proportions as they existed in the original recorded or received signal.

Figure 32-3A shows how a single-knob tone control can be incorporated into an audio amplifier. The amplifier is designed to exhibit an exaggerated treble response in the absence of the tone control. The potentiometer attenuates the treble to a variable extent.

A more versatile tone control has two capacitors and two potentiometers, as shown in Fig. 32-3B. We insert the series resistance-capacitance (*RC*) circuit in parallel with the audio output; it attenuates the treble to a variable extent. We insert the parallel *RC* circuit in series with the audio path; it attenuates the bass to a variable extent. We can adjust the two potentiometers separately, although some interaction occurs.

Audio Mixer

If you simply connect two or more audio sources to the same input terminals of a single amplifier, you can't expect good results. Different signal sources (such as a computer, a tuner, and a CD player) will likely have different impedances. When connected together, the impedances appear in parallel, causing impedance mismatches for most or all of the sources, as well as at the amplifier input. As a result, you'll get degradation of system efficiency and poor overall performance.

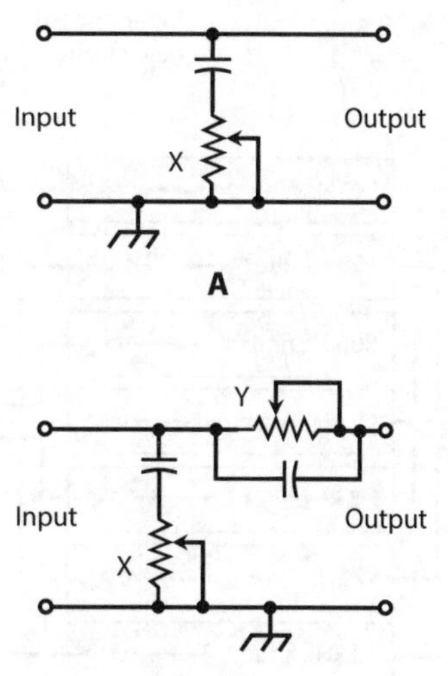

32-3 Methods of tone control. At A, a single potentiometer/capacitor combination (X) provides treble attenuation only. At B, one potentiometer/capacitor combination (X) attenuates the treble, and the other (Y) attenuates the bass.

Another problem arises from the fact that the signal amplitudes from various sources almost always differ. A microphone produces almost no AF power all by itself, whereas some tuners produce enough to drive a pair of small speakers. Connecting both of these components directly together will cause the tuner signal to obliterate the microphone signal. In addition, the tuner output terminals might "see" the microphone as a miniature speaker or headset and damage it by forcing audio energy into it.

An *audio mixer* eliminates the problems involved with connecting multiple devices to a single channel. First, it isolates the inputs from each other, so that no impedance mismatch or competition exists among the sources. Second, it allows you to independently control the gain at each input, facilitating adjustment of individual amplitudes so that the signals blend in the relative proportions you want.

Graphic Equalizer

A *graphic equalizer* allows for adjustment of the relative loudness of audio signals at various frequencies so that you can fine-tune the amplitude-versus-frequency output characteristic of hi-fi sound equipment. Serious hi-fi stereo enthusiasts and recording engineers employ these devices, which act as high-level, precision tone controls.

A typical graphic equalizer comprises several independent gain controls, each one affecting a different part of the audible spectrum. The controls are usually slide potentiometers with calibrated scales. The slides move up and down, or, in some cases, left to right. When you set the potentiometers so that the slides are all at the same level, the audio output or response is flat, meaning that no particular range is amplified or attenuated with respect to the whole AF spectrum. By moving any one of the controls, you can adjust the gain within a certain frequency range without affecting the gain outside that range. The positions of the controls on the front panel provide an intuitive graph of the output or response curve.

Figure 32-4 is a block diagram of a hypothetical graphic equalizer with seven gain controls. (Seven isn't a "magic number"; we use it here only as an example.) The input goes to an *audio splitter* that breaks the signal into several paths of equal impedance, and prevents interaction among the circuits. The individual signals are fed to audio bandpass filters, each one having its own gain control. In this schematic, the slide potentiometers appear as variable resistors following the filters. Finally, the signals pass through an audio mixer, and the composite goes to the output.

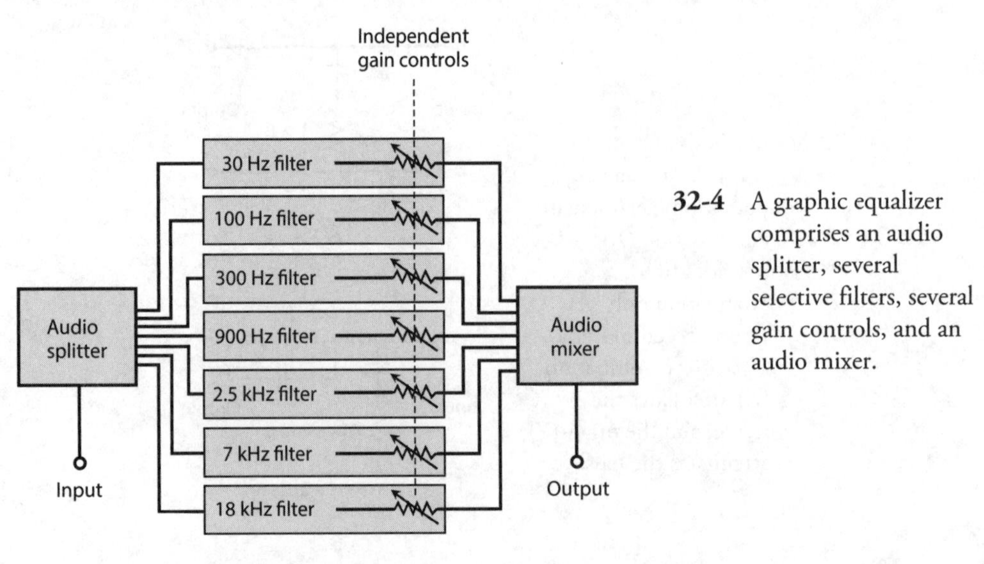

32-4 A graphic equalizer comprises an audio splitter, several selective filters, several gain controls, and an audio mixer.

Engineers face multiple challenges in the design and proper use of graphic equalizers. Each filter gain control must operate independently of all the others. Judicious choice of filter frequencies and responses is important. The filters must not introduce distortion. The active devices, if any, must not generate significant audio noise. Graphic equalizers aren't designed or built to handle high power, so they must be placed at low-level points in an audio amplifier chain. In a multichannel circuit such as a stereo sound system, a separate graphic equalizer can be installed in each channel path.

Specialized Systems

Mobile and portable hi-fi systems operate at low DC voltages. Typical audio power levels are much lower than in home hi-fi systems. Speakers have smaller sizes as well. In portable systems, headsets often replace speakers.

Mobile Systems

Mobile hi-fi systems, designed for cars and trucks, usually have four speakers. The left stereo channel drives the left front and left rear speakers; the right stereo channel drives the right front and right rear speakers. The balance control adjusts the ratio of sound volume between the left and right channels for both the front and rear speaker sets. Another control adjusts the ratio of sound volume between the front and rear sets.

A mobile hi-fi system has an AM/FM receiver and the capability to reproduce music from recorded or downloaded media. Some older vehicles have systems with cassette tape players. Some new cars and trucks have satellite radio receivers in addition to all the other standard media. One note of caution: Tapes and CDs are heat-sensitive, so you shouldn't store them in a car or truck and then leave the vehicle out in the sun.

Portable Systems

Portable hi-fi systems operate from batteries. The most well-known is the *headphone radio* or *Walkman*. Dozens of designs exist. Some include only an FM radio; some have AM/FM reception capability. Some have a small box with a cord that runs to the headset; others are entirely contained in the headset. Some have multiple media players and satellite radio receivers. The sound quality is usually excellent, although you'll need a good headset if you want to get the best from them.

Another form of portable hi-fi set, sometimes called a *boom box*, can produce several watts of audio output, and delivers the sound to a pair of speakers built into the box. A typical boom box has about the same dimensions as a desktop computer tower. It includes an AM/FM radio and various media players. The system gets its name from the loud bass acoustic energy peaks that its speakers can deliver.

Quadraphonic Sound

Quadraphonic sound refers to four-channel audio recording and reproduction. Audiophiles often call it *quad stereo* or *four-channel stereo*. Each of the four channels operates independently of the other three. In a well designed quad stereo system, the speakers should be level with the listener, equidistant from the listener, and separated by angles of 90° from the listener's point of view. If the listener faces north, for example, the left front speaker will lie to the northwest, the right front speaker will lie to the northeast, the left rear speaker will lie to the southwest, and the right rear speaker will lie to the southeast. This geometry provides optimum balance, and also facilitates the greatest possible directional sound contrast.

Hard Recording Media

Methods of recording sound, particularly music, have evolved dramatically since the ascent of digital technology. Existing and historical *hard recording media* include the *compact disk, analog audio tape, digital audio tape,* and *vinyl disk.* Computer flash drives and variants thereof can store digital audio data as well, but they're more appropriately called *soft recording media.*

Compact Disk

A compact disk (sometimes spelled "disc"), also called a CD, is a plastic disk with a diameter of 4.72 in (12.0 cm), capable of storing sound, images, computer programs, and computer files. Digital sound, when recorded on the surface of a CD, suffers little from the hiss and crackle that bedevil recordings on analog media because the information on the disk exists in binary form. A bit (binary digit) has either the 1 (high) or the 0 (low) logic state. The distinction between these two states is more clear-cut than the subtle fluctuations in the level of an analog signal.

When an engineer records a CD for hi-fi use, the sound initially undergoes *analog-to-digital* (A/D) *conversion,* changing the continuously variable AF waves into digital logic bits. Microscopic craters called *pits* are then physically burned into the surface of the disk, one pit for each bit. The pits are arranged in a spiral path called a *track* that would measure several kilometers long if straightened out. *Digital signal processing* (*DSP*) minimizes the noise introduced by environmental factors, such as microscopic particles on the disk or random electronic impulses in circuit hardware. A *scrambling* process can "smear" recordings throughout the disk, rather than burning the pits in a direct sequence.

Compact-disk players recover the sound from a disk without any hardware physically touching the surface. A laser beam scans the disk. The pits scatter the incident beam, but the unpitted plastic surface reflects the beam like a mirror. A digitally modulated beam, therefore, emerges from the disk; a sensor picks up the beam and converts it into electrical currents. These currents proceed to a *descrambling* circuit, then to a *digital-to-analog* (D/A) *converter,* then to a DSP circuit, and finally to the audio amplifiers. Speakers or headphones convert the AF currents into sound waves.

With a CD player, the track location processes are entirely electronic, and they can all be done quickly. Tracks are assigned numbers that you select by pressing buttons. You cannot damage the CD, no matter how much you skip around among the songs. You can move instantly to any desired point within an individual track. You can program the system to play only those tracks you want, ignoring the others.

Analog Audio Tape

We can classify analog audio tape recorders and playback devices as *cassette type* or *reel-to-reel type.* Although these systems are obsolete in hi-fi applications, we'll still encounter them once in a while. A typical audio cassette plays for 30 minutes on each side; longer-playing cassettes allow recording for as much as 60 minutes per side. The longer tapes are thinner and more subject to stretching than the shorter tapes.

A reel-to-reel tape feed system resembles an old-fashioned movie projector. The tape resides on two flat spools called the *supply reel* and the *take-up reel.* The reels rotate counterclockwise as the tape passes through the recording/playback mechanism. When the take-up reel fills up and the supply reel empties out, both reels can be flipped over and interchanged for recording or playback on the "other side" of the tape. (Actually, the process takes place on the same side of the tape, but on different paths called *tracks.*) The speed is usually 1.875, 3.75, or 7.5 inches per second (in/s).

In the *record mode,* the tape moves past the *erase head* before anything is recorded on it. If magnetic impulses already exist on the tape, the erase head removes them before recording anything else. This process prevents *doubling*, or the simultaneous presence of two programs on the tape. In sophisticated tape recorders, you can disable the erase head if you want doubling to occur. The *recording head* comprises a small electromagnet that generates a fluctuating magnetic field, whose instantaneous flux density varies in direct proportion to the instantaneous level of the audio input signal, thereby magnetizing the tape in a pattern that duplicates the waveform of the signal. The *playback head* is normally deactivated in the record mode. However, you can use the playback head while recording to create an echo effect.

In the *playback mode*, neither the erase head nor the recording are activated. The playback head acts as a sensitive magnetic-field detector. As the tape moves past it, the playback head is exposed to a fluctuating magnetic field whose waveform duplicates that produced by the recording head when the audio was originally recorded on the tape. This magnetic field induces weak AF currents in the playback head. These currents are amplified and delivered to a speaker, headset, or other output device.

Digital Audio Tape

Digital audio tape (DAT) is magnetic recording tape on which binary digital data can be recorded. In digital audio recording, tape noise remains minimal because it has an analog nature, and the digital system doesn't respond very much to analog fluctuations. Some electronic noise arises in the analog amplification stages following D/A conversion, but to a far lesser extent than the noise generated in older, fully analog systems. The reduced noise in DAT equipment provides more true-to-life reproduction than analog methods could ever offer.

With DAT, you can make multi-generation copies with practically no degradation in audio fidelity. (For the same reason, a computer can repeatedly read and overwrite data on a magnetic hard drive.) On DAT, distinct magnetized regions on the tape represent the data bits. While analog signals appear "fuzzy" in the sense that they vary continuously, digital signals come out "crisp"; they're either totally present or else totally absent. Imperfections in the recording apparatus, the tape, and the pickup head affect digital signals less than they affect analog signals. A well-designed DSP system can eliminate the minute flaws that creep into a digital signal when you record it or play it back.

Vinyl Disk

Vinyl disks were superseded years ago by CDs and Internet downloads, but some audiophiles still harbor a "vinyl love affair"! Some vinyl disks, and the turntables that can play them, have attained value as collectors' items. The main trouble with vinyl is its susceptibility to physical damage. Even if you try hard to preserve a vinyl disk, it will gradually acquire imperfections that lead to a "scratchy" sound when you play the contents back. In addition, electrostatic effects can produce "crackling" noises when the atmospheric humidity drops. Friction between the *stylus*, or "playback needle," which physically rides along a groove in the disk, gives rise to charge buildup and subsequent small discharges that produce the noise.

Vinyl disks require a *turntable* that spins at selectable speeds of 33 and 45 revolutions per minute (r/min). Common systems include *rim drive*, *belt drive*, and *direct drive*. In the rim drive, a small wheel rotates while in contact with the turntable, thereby causing the turntable to spin at a slower, regulated speed. A belt drive works in much the same way as the fan belt in a motor vehicle functions; the motor drives the belt, which in turn drives the turntable. In a direct drive system, the motor shaft goes straight to the turntable shaft without any intervening gears, wheels, or belts.

Electromagnetic Interference

The term *electromagnetic interference* (EMI) refers to unwanted phenomena in which appliances, circuits, devices, and systems upset each other's operation because of EM fields they produce or pick up.

When EMI results from an electronic system's improper response to nearby RF transmitters, engineers call it *radio-frequency interference* (RFI). Audio systems are particularly vulnerable to EMI effects.

EMI from Computers

A computer will produce wideband EM energy, especially if it has a CRT monitor. The digital pulses in the *central processing unit* (CPU) can also cause problems in some cases. Hi-fi tuners in close proximity to computers or their peripherals can pick up EM energy from these sources. The EM fields escape the computer through the interconnecting cables and power cords, which act as miniature transmitting antennas.

When you place a hi-fi system and a computer next to each other, such as when you want to amplify streaming audio from the Internet or you want all your electronic systems in one place, you should anticipate EMI when you set the tuner to certain frequencies. Some Internet connection devices, such as *cable modems* or *wireless routers*, can also cause EMI. Even cordless telephone sets can sometimes cause trouble.

RFI from Radio and Television Transmitters

Hi-fi sound equipment can malfunction because of strong RF fields from a nearby radio or television broadcast transmitter, even when the transmitter functions according to its specifications and in compliance with the standards set by the Federal Communications Commission (FCC). In these cases, and also in cases involving Citizens Band (CB) radios and amateur ("ham") radios, the root cause of the problem is rarely a transmitter malfunction. Instead, the trouble usually arises from inferior home-entertainment-equipment design. The EM energy can enter through speaker wires, power cords, the tuner antenna, and cables between an amplifier and externals, such as a CD player or tape deck.

As the number of connecting cables in a home entertainment system increases, the likelihood of interference from an RF field of a given intensity and frequency also increases. In addition, the risk of RFI increases as interconnecting cables grow longer. Good engineering principles dictate minimizing the number of connecting cables, and keeping them as short as possible. If you have excess cable for a given interconnection and you don't want to cut it shorter, you can coil it up and tape the coil in place. You should also ensure that the entire system has a good electrical ground connection.

Amateur Radio and RFI

If you're an amateur radio operator ("radio ham") with a sophisticated or high-powered station, you might find yourself taking the blame for interference to home entertainment equipment, whether the problem is technically your station's fault or not. In situations of this kind, you should use the minimum amount of transmitter output power necessary to maintain the desired communication. (That's the law according to the FCC, anyway!) You should make certain that your transmitters are aligned properly, so that they radiate signals only at the frequencies intended. Antenna systems should be located and installed so as to radiate as little energy as possible into nearby homes and other buildings.

Unfortunately, a large proportion of RFI cases result from inadequate or nonexistent built-in protection for home entertainment equipment, particularly a lack of EM shielding. Therefore, you'll likely have difficulty solving an RFI problem solely on the basis of modifications to your radio station. You may nevertheless mitigate the trouble by reducing your transmitter power, switching to another frequency band, or operating only when the home entertainment equipment is not in use. A compromising attitude can help to secure the cooperation of a neighbor who is experiencing RFI from your amateur radio station.

As an amateur radio operator—even a highly qualified one—you should never attempt to modify a neighbor's home-entertainment equipment. If something goes wrong with the neighbor's hardware later, he or she might blame you. Once in a while, a manufacturer of home-entertainment equipment will offer technical support in RFI cases, but as most of us already know, technical-support departments often leave a lot to be desired even when equipment completely breaks down. If you're contemplating the purchase of a large home-entertainment system from a particular vendor, you might do well to do some research on the Internet and find out how previous users have rated the vendor's technical support department.

EMI from Appliances and Power Lines

Hi-fi tuners can pick up EMI from appliances, such as vacuum cleaners, light dimmers, heating pads, electric blankets, hair dryers, and television sets. All of these devices contain components that generate electric sparks and/or produce harmonics of the AC utility wave because of nonlinear operation. Utility lines can also radiate considerable EM energy. These fields rarely get strong enough to interfere with consumer electronic systems directly, although they often cause trouble for shortwave radio listeners and amateur radio operators.

Power-line interference arises from high-current electric sparks between points that lie in close proximity on the power line, and that differ greatly in voltage. Engineers call this phenomenon *arcing*. A malfunctioning transformer, a failing street light, or a defective insulator can arc, generating EMI that can prove difficult to locate and eradicate. Sometimes, the utility company will offer their help. If you live near an amateur radio operator and you suspect that power-line interference is causing trouble with your hi-fi system, chances are good that the "radio ham" is also having trouble with it. The radio amateur's technical expertise may help you track down the source of the power-line noise if the utility company won't cooperate.

Gasoline-powered *internal combustion engines* used in lawn mowers, weed trimmers, snow blowers, cars, trucks, farm implements, and road construction equipment occasionally cause EMI to hi-fi systems. This type of interference resembles power-line EMI, but usually constitutes a less severe problem because of its intermittent or infrequent nature. In most situations of this kind, the offending device is easy to locate.

Other Potential Problems

Unwanted *RF mixing* can occur in the most unsuspected places. (This type of mixing, also called *heterodyning*, is not the same thing as the process that takes place in an "audio mixing" console.) At RF, *mixing products* arise at the sum and difference frequencies of other signals when they combine in a nonlinear device. Heterodyning can give rise to a species of RFI that affects radio receivers and hi-fi tuners, and that can be almost impossible to track down and correct. Poor electrical connections in house wiring, plumbing, and exterior metallic structures, such as fences and rain gutters, can generate mixing products and harmonics in the presence of RF fields from multiple radio transmitters in the vicinity.

Intermodulation, a particularly obnoxious form of RF mixing, sometimes occurs in the downtown areas of large cities, where many powerful wireless transmitters operate simultaneously. The number of mixing products and harmonics in these areas can become so great that they constitute *broadband RF noise*. Intermodulation (sometimes informally called "intermod") causes false signals in radio receivers, often sounding "hashy" or broken-up. In the worst cases, intermod can completely ruin FM stereo reception.

Interference to hi-fi tuners can sometimes result from harmonic emissions or *spurious emissions* (output signals at frequencies other than the design frequency, but not harmonically related) from a nearby broadcast, CB, or amateur-radio transmitter. Emissions of this type don't occur often with well-designed CB and amateur-radio transmitters because these systems employ relatively low power. However, if you live in the shadow of a broadcast or cellular communications tower, you might experience trouble from the harmonics and/or spurious signals generated by the attendant transmitters.

Preventive Measures

You can employ various tactics in your quest to keep stray RF energy out of home entertainment equipment. You can install multiple RF chokes and bypass capacitors in power cords and interconnecting cables. However, you must ensure that these components won't interfere with the transmission of power, signals, or data through cables. For advice, consult a competent engineer or the manufacturer of the equipment in question. Here's a word of caution: If you install RF chokes or bypass capacitors, you'll probably void the equipment warranty if the installation involves internal modification or the cutting of built-in cords or wires.

RF shielding can help to prevent sensitive electronic apparatus from picking up stray RF fields. The simplest way to provide RF shielding for a circuit or device is to surround it with a metal sheet, mesh, or screen and connect the metal to a good electrical ground. Because metals constitute electrical conductors, an external RF field induces electric currents in them. These currents oppose the currents in the RF field. If the metal enclosure (known as a *Faraday cage*) is well-grounded, it electrically short-circuits the RF energy. Obviously you won't wrap all your hi-fi gear in aluminum foil or window screening, but if you're shopping for a new system, particularly a tuner or amplifier, you might also want to buy a metal cabinet for the hardware.

In addition to using metallic enclosures, you should also shield all of your interconnecting cords if you want your system to have optimum protection against RFI. In a shielded cable, all the signal-carrying conductors are surrounded by a tubular copper braid that's electrically grounded through the connectors at the ends of the cable. The most popular form of shielded cable is coaxial cable. It can replace two-wire cords between an amplifier and other parts of a system.

More Help

The *American Radio Relay League* (ARRL), 225 Main Street, Newington, CT, publishes books about EMI and RFI phenomena, their causes, and ways to deal with problems when they occur. These publications are intended mainly for amateur radio operators, but high-end audiophiles might also find them useful.

Quiz

Refer to the text in this chapter if necessary. A good score is 18 correct. Answers are in the back of the book.

1. You measure the maximum output power P_{max} that a hi-fi system can deliver while maintaining acceptable performance, and then you measure the minimum output power P_{min} that the system can deliver with good audio audibility and quality. What does the ratio P_{max}/P_{min} represent?
    - (a) Distortion tolerance
    - (b) Dynamic range

(c) Linearity factor

(d) Signal-to-noise ratio

2. You have a pair of hi-fi speakers that can handle up to 150 W RMS at all audio frequencies without producing unacceptable distortion, even on the loudest sound bursts. You want to buy a hi-fi system. What's the maximum RMS power that your system can produce without overstressing the speakers?

(a) 50 W

(b) 75 W

(c) 100 W

(d) 150 W

3. Despite their nostalgic value to collectors, vinyl disks

(a) cause intermodulation distortion.

(b) can be easily damaged.

(c) produce unreliable audio.

(d) All of the above

4. On a compact disk (CD), sound is recorded in the form of

(a) tiny pits in the plastic.

(b) grooves like those on a vinyl disk.

(c) variations in the color of the plastic.

(d) tiny bumps on the plastic.

5. In air at sea level, the wavelength of a pure audio tone at 335 Hz is approximately

(a) 2.0 meters.

(b) 1.4 meters.

(c) 1.0 meter.

(d) 50 centimeters.

6. What do you call the extent to which the output wave from an audio amplifier "looks like" the input wave (as seen on an oscilloscope, for example)?

(a) Similarity

(b) Shape factor

(c) Linearity

(d) Amplitude

7. In a hi-fi amplifier, you can adjust the relative loudness of the bass, midrange, and treble using the

(a) tone control.

(b) balance control.

(c) linearity control.

(d) audio mixer.

8. You can minimize a hi-fi power amplifier's susceptibility to interference from external sources by
   (a) avoiding the use of batteries for power.
   (b) using a graphic equalizer.
   (c) placing a capacitor in series with the input line.
   (d) grounding its chassis.

9. Most digital AM/FM tuners have
   (a) frequency synthesizers.
   (b) programmable memory channels.
   (c) seek and/or scan modes.
   (d) All of the above

10. Most children can hear sounds ranging between frequencies of roughly
   (a) 40 Hz and 10 kHz.
   (b) 20 Hz and 20 kHz.
   (c) 10 Hz and 40 kHz.
   (d) 10 Hz and 60 kHz.

11. In a vinyl disk player, which of the following objects, if any, physically touches the disk during sound playback?
   (a) Stylus
   (b) Laser
   (c) Tweeter
   (d) Nothing physically touches the disk.

12. An audio mixer allows you to
   (a) connect two or more devices (such as a tuner and CD player) to an amplifier that has only one input.
   (b) combine multiple audio outputs to work with a single pair of speakers or a single headset.
   (c) control the tone levels independently for several different AF ranges.
   (d) connect a radio transmitter to the output of an audio amplifier to broadcast AM or FM programs.

13. When direct and reflected sound waves arrive at a certain point in phase coincidence, then a listener at that point is
   (a) at an antinode.
   (b) in a dead zone.
   (c) at the threshold of hearing.
   (d) at a nonlinear point.

14. Which of the following devices will convert low-frequency AF current to bass sound?
   (a) Woofer
   (b) Mixer

(c) Tweeter
(d) Inverter

15. What do some people call a portable hi-fi receiver designed for private listening through headphones?
(a) Stroller
(b) Sound box
(c) Walkman
(d) Boom box

16. In air at sea level, sound waves travel at roughly
(a) 1100 feet per second (ft/sec).
(b) 550 ft/sec.
(c) 335 ft/sec.
(d) 670 ft/sec.

17. Shielded connecting cables between the amplifier and speakers help to protect a hi-fi system against
(a) variations in AC line voltage.
(b) transients on the AC line.
(c) ripple in the power supply output.
(d) EMI from nearby wireless devices.

18. A sophisticated tone control that allows you to adjust loudness within specific AF ranges is called
(a) an audio mixer.
(b) a linearity control.
(c) a graphic equalizer.
(d) a level adjuster.

19. If you halve the frequency of a pure audio tone in air at sea level, then its wavelength
(a) quadruples.
(b) doubles.
(c) stays the same.
(d) becomes half as great.

20. If you quadruple the wavelength of a pure audio tone in air at sea level, then its propagation speed
(a) quadruples.
(b) doubles.
(c) stays the same.
(d) becomes half as great.

33
CHAPTER

Lasers

THE WORD LASER IS AN ACRONYM THAT STANDS FOR THE TECHNICAL TERM *LIGHT AMPLIFICATION* by *stimulated emission of radiation*. Lasers have diverse applications in electronic systems.

How a Laser Works

In theory, laser radiation differs from ordinary EM radiation in two ways. First, all of the energy exists at a single wavelength, and therefore, at a single frequency. At visible-light wavelengths, this property gives the light a vivid hue ("color"). Second, the "wave packets" (known as *photons*) in a laser beam all propagate in phase coincidence, so all the wave peaks and troughs "line up" with one another. When we talk about lasers, we usually refer to devices that produce coherent rays in the ultraviolet (UV), visible-light, or infrared (IR) portions of the spectrum, expressing wavelengths in *nanometers* (nm), where 1 nm = 10^{-9} m.

Spectral Distribution and Coherence

The visible light from a conventional source, such as the sun or a house lamp, is not a clean, sine-wave EM disturbance. Components exist in the red, or longest, visible wavelength (upwards from about 770 nm) through the violet, or shortest, visible wavelength (up to about 390 nm). Energy components usually exist outside this range as well. If we look at a graph of light intensity as a function of wavelength for a particular light source, we get a *spectral distribution* for that source.

The spectral distribution for an *incandescent lamp* skews toward the red end of the visible spectrum, as shown in Fig. 33-1A. A common *fluorescent tube* produces energy oriented toward the middle visible wavelengths (Fig. 33-1B). Some lamps emit light at several discrete wavelengths, producing a distribution that shows multiple well-defined *spectral lines*, as shown in Fig. 33-1C.

Even a lamp that emits light at discrete wavelengths, such as a *mercury-vapor lamp*, *sodium-vapor lamp*, or *neon-gas lamp*, produces myriad waves in random phase for each specific wavelength (Fig. 33-2A). Because of the random phase, we can call such light *incoherent*, whether it's *monochromatic* (single-colored, existing at a single wavelength) or not. We use the term *coherent* to describe the visible light or other EM energy from a source that radiates at a single wavelength (Fig. 33-2B) and for which all of the wavefronts line up with each other. No ordinary lamp produces coherent light, but lasers do.

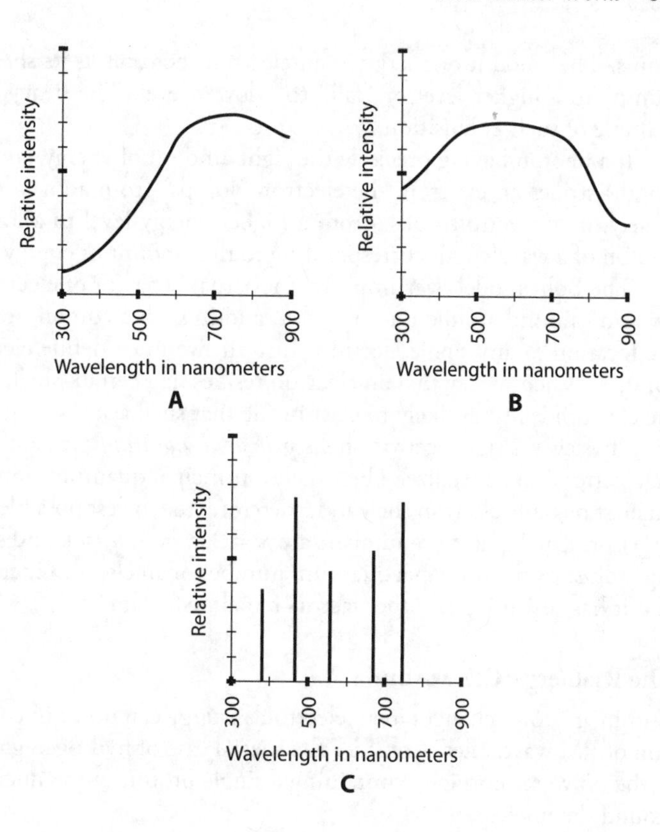

33-1 At A, the spectral distribution of light from an incandescent lamp. At B, the spectral distribution of light from a fluorescent lamp. At C, the spectral distribution of light from a hypothetical lamp that emits energy at multiple discrete wavelengths.

The Bohr Atom

When an electron is bound by an atomic nucleus, that electron can normally exist only at certain discrete energy levels. All the way back in Chap. 1 (Fig. 1-2), we saw a simplified model of an atom, known as the *Bohr atom* after its inventor, physicist *Niels Bohr*, who hypothesized it in the early 1900s. Look at that figure again now. The light dashed circles represent cross sections of spheres called *electron shells*. The electrons orbit around the nucleus in the shells, each of which has a fixed

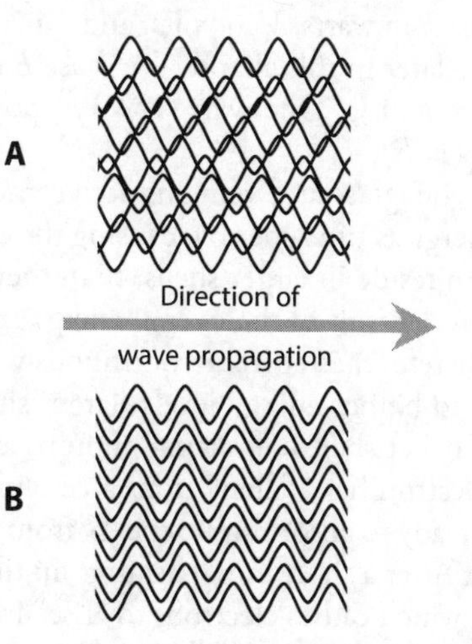

33-2 At A, waves of monochromatic light with waveforms aligned at random. At B, waves of coherent monochromatic light.

radius. The amount of energy in an electron increases as its shell radius increases. An electron can "jump" to a higher level or "fall" to a lower level. The heavy dashed path shows a hypothetical example of such a transition.

If a photon having precisely the right amount of energy (and therefore, exactly the right wavelength) strikes an electron, the electron "jumps" from a given energy level to a higher one. Conversely, if an electron "falls" from a higher-energy level to a lower-energy level, the atom emits a photon of a wavelength corresponding to the amount of energy that the electron loses.

The Bohr model oversimplifies the actual behavior of electrons. In the modern view, electrons "swarm" around atomic nuclei so fast, and in such a complicated way, that we can never pinpoint the location of any single electron. Instead, we must define electron movement in terms of probabilities. When we say that an electron resides in a certain shell, we mean that at any point in time, the electron is just as likely to exist inside that shell as outside it.

Physicists assign electron shells *principal quantum numbers* within atoms, using the lowercase, italic letter n to generalize. The smallest principal quantum number is $n = 1$, which represents the smallest possible electron shell and, therefore, the lowest possible electron energy level. Progressively larger principal quantum numbers are $n = 2$, $n = 3$, $n = 4$, and so on, representing shells of increasing radius. As the principal quantum number of an electron increases, so does the amount of energy that it has, assuming all other factors remain constant.

The Rydberg-Ritz Formula

Within an atom, changes in an electron's energy can occur in conjunction with absorption or emission of EM waves. Let's consider the special case of hydrogen gas. Hydrogen atoms are the simplest in the universe, normally containing a single proton in the nucleus and a single electron in "orbit" around the nucleus.

Imagine a glass cylinder from which all the air has been pumped out. Suppose that we introduce a small amount of hydrogen gas into the cylinder. Then we connect a DC source to electrodes at either end of the cylinder, obtaining a device called a *gas-discharge tube*. If we apply a high enough DC source voltage V, the gas inside the tube ionizes, and an electrical current I flows. As a result, the gas dissipates a certain amount of power P. The power, voltage, and current relate according to the formula

$$P = VI$$

where we express P in watts, V in volts, and I in amperes. We use V, rather than E, to represent voltage here because later in this chapter we'll use E to represent energy. The standard unit of energy is the *joule* (abbreviated J). One watt (1 W) of power represents energy dissipated at the rate of one *joule per second* (1 J/s).

If the DC voltage source remains active and stays connected to the electrodes, the gas in the tube absorbs energy as time goes by, causing the electrons in the atoms to attain higher energy levels, and therefore, to reside in larger shells, than they would do if no voltage source were connected to the electrodes. As a result of the overall energy gain, the gas gets hot. This process cannot continue indefinitely. Therefore, as current continuously flows through the ionized hydrogen gas, electrons not only "rise" to higher energy levels (larger shells), but they also constantly "fall" back to lower energy levels (smaller shells), emitting photons at various EM wavelengths.

Once an electron has "fallen" from a certain shell to another shell having a lower energy level, that electron is ready to absorb some energy from the DC source again, "rise" again, and then eventually "fall" again. After a short initial heating-up time, the amount of energy absorbed by the gas from the DC source, which causes electrons to "rise" from smaller shells to larger ones, balances the energy radiated by the gas as electrons "fall" from larger shells back to smaller ones.

A hydrogen atom has several electron shells, so an electron can gain or lose any one of numerous *specific* energy quantities as it moves from one shell to another. Energy, therefore, radiates from ionized hydrogen gas in the form of photons having multiple discrete wavelengths λ as follows:

$$\lambda = 1/[R_{\text{H}} (n_1^{-2} - n_2^{-2})]$$

where we express λ in meters, we let R_{H} represent a constant called the *Rydberg constant*, n_1 represents the principal quantum number of the smaller shell to which a particular electron "falls," and n_2 represents the principal quantum number of the larger shell from which that same electron has just "fallen." Physicists call the above equation the *Rydberg-Ritz formula*, named after *Johannes Rydberg* and *Walter Ritz* who first published an academic paper about it in 1888. The value of the Rydberg constant, accurate to six significant figures, is

$$R_{\text{H}} = 1.09737 \times 10^7$$

This figure is sometimes rounded off to three significant figures as 1.10×10^7. We define it in units of *per meter* (/m or m^{-1}). Some texts denote the constant as R with an infinity-symbol subscript (R_∞). Once in a while you'll see it symbolized simply as R, but in that case you must not confuse it with the *universal gas constant.*

We've examined how electrons relate to EM energy for hydrogen gas. Other gases, as well as liquids and solids, can behave in a similar way, but their Rydberg constants all differ. Every element, mixture, and compound has a unique Rydberg constant, and a unique set of wavelengths at which it can absorb or emit energy under certain conditions. With some materials, we can exploit this phenomenon to generate coherent light.

Wave Amplification

In a *gas laser*, the atoms behave somewhat differently than they do in a simple gas-discharge tube because we keep the gas in a super-energized condition in which an abnormally large number of electrons exist in high-energy shells. This state of affairs, called a *population inversion*, constitutes the critical factor responsible for the *stimulated emission* of photons that makes a laser work.

When a photon strikes an electron in a gas laser tube, that photon is not absorbed. Instead, it continues to travel in the same direction, and with the same energy, as it did before the collision. The electron, instead of "rising" to a higher shell, actually "falls" to a lower shell, as shown in Fig. 33-3.

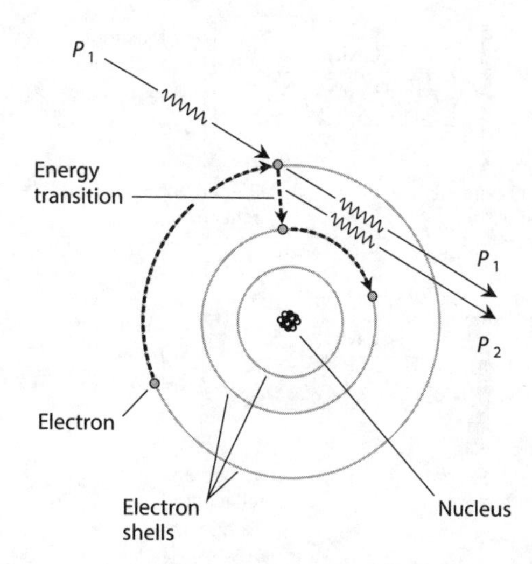

33-3 Amplification of EM energy by means of a photon interacting with an atom.

When this transition occurs, the electron loses energy instead of gaining energy. Therefore, a new photon P_2 emanates from the electron, leaving the atom along with the original photon P_1.

If the original photon P_1 has exactly the right amount of energy, the new photon P_2 emerges at the same wavelength, and going in the same direction, as the original photon P_1. Photon P_1 then travels along with P_2 in such a way that the two wave disturbances exist in phase coincidence. When transitions of this sort occur in large numbers, the super-energized gas, in effect, boosts the amplitude of the incoming EM energy, a phenomenon called *wave amplification*. That's how the gas laser generates its intense energy beam. A similar phenomenon can also occur in certain liquids and solids.

The Cavity Laser

One of the most common types of lasers, the *cavity laser*, has a *resonant cavity*, usually in the shape of a cylinder or prism, and an external source of energy. The cavity is designed so that EM waves bounce back and forth inside it, reinforcing each other in phase coincidence.

Basic Configuration

The cavity contains (or consists of) a gas, liquid, or solid substance called the *lasing medium*. We place mirrors at each end so the EM waves reflect back and forth many times between them. One mirror reflects all the energy that strikes it, and the other mirror reflects approximately 95 percent of the incident energy. The mirrors can be flat or concave. The ends of the cavity can be perpendicular to the lengthwise cavity axis or oriented at a slant.

Figure 33-4 illustrates two common cavity laser designs. These drawings are "squashed horizontally" for clarity. The waves represent the EM energy, usually IR or visible light, resonating between the reflectors. Only a few wave cycles appear in this simplified functional drawing. In a practical cavity laser, thousands or millions of wave cycles occur between each reflection.

Pumping

Energy is supplied to the lasing medium by means of a process called *pumping*, so that the intensity of the beam builds up as the internal beam repeatedly reflects from the mirrors. The waves emerge

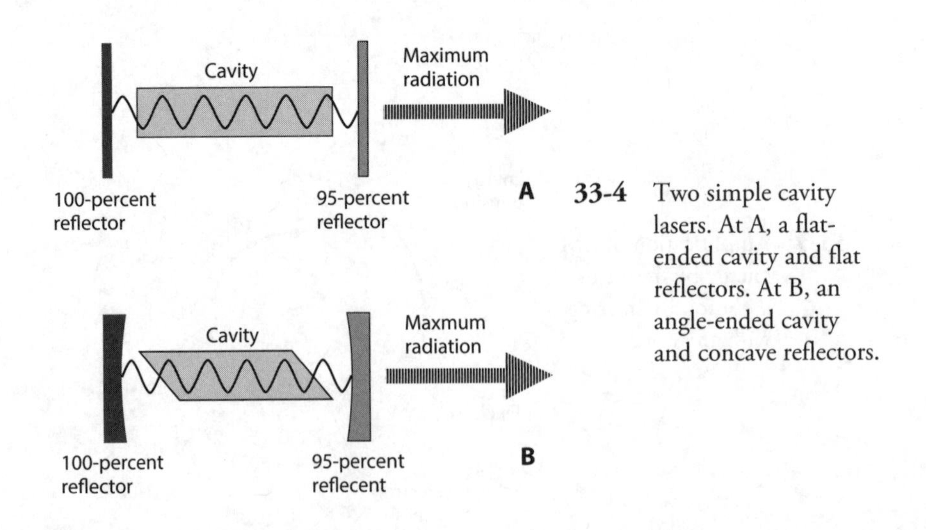

33-4 Two simple cavity lasers. At A, a flat-ended cavity and flat reflectors. At B, an angle-ended cavity and concave reflectors.

from the partially silvered (95-percent reflective) mirror in the form of coherent radiation, which concentrates the wave energy into the most intense possible beam.

Beam Radius

The *beam radius* depends on the physical radius of the cavity and also on its length. Generally, small-radius cavities produce narrower beams at close range than large-radius cavities do. However, large-diameter cavities usually have less *beam divergence*, which quantifies the extent to which the beam radius increases with increasing distance from the source, once the beam gets several cavity lengths away from the partially silvered mirror. At a great distance in free space (a vacuum), we can expect a large-radius laser to produce a narrower beam than a small-radius laser of the same type. Beam divergence occurs as a result of imperfections in the hardware, and also because of scattering and diffusion in the air or other medium through which the laser beam travels.

Output Wavelength

The *output wavelength* from a cavity laser depends on the length of the cavity and on the atomic resonant wavelengths of the lasing medium. Many different sizes of cavity lasers can produce an output of a given wavelength. Some cavity lasers are continuously tunable (adjustable) over a range of wavelengths. Most operate in the IR or visible part of the EM spectrum.

Beam Energy Distribution

We can portray the *beam energy distribution* at a given distance from a laser device by generating a cross-sectional graph of beam intensity versus the radius from the beam center in a plane perpendicular to the beam axis. Figures 33-5 and 33-6 show hypothetical examples for so-called "spot" and "ring" distributions, respectively. In all four of these graphs, the relative intensity appears on the horizontal axis, and the relative distance from the beam center is shown on the vertical axis, with the intersection of the two axes representing the beam center. Figure 33-7 illustrates typical examples of "spot" and "ring" beams as they would look if we projected them onto a white screen in a dark room.

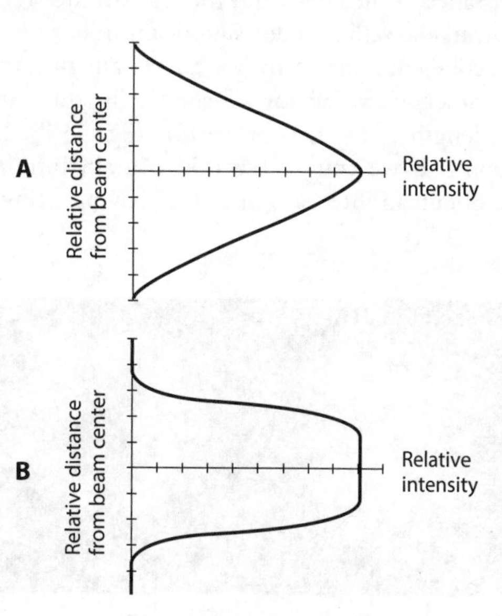

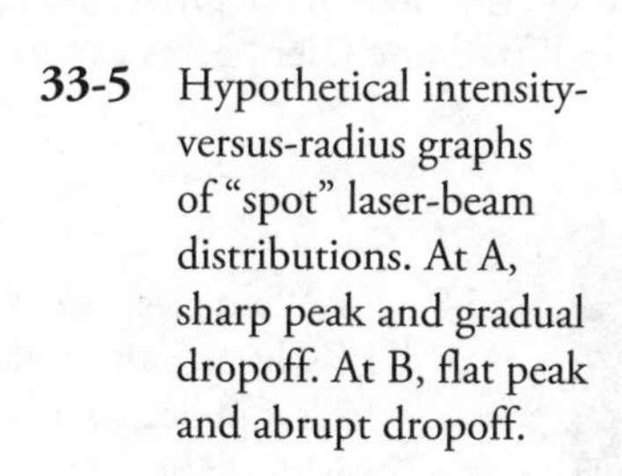

33-5 Hypothetical intensity-versus-radius graphs of "spot" laser-beam distributions. At A, sharp peak and gradual dropoff. At B, flat peak and abrupt dropoff.

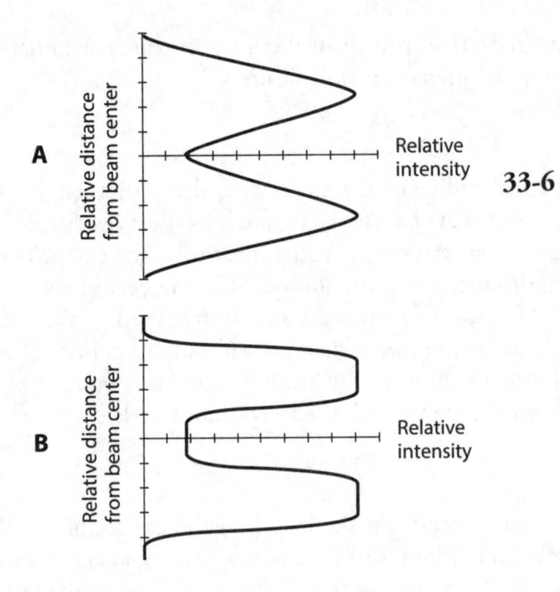

33-6 Hypothetical intensity-versus-radius graphs of "ring" laser-beam distributions. At A, sharp peaks and gradual dropoffs. At B, flat peaks and abrupt dropoffs.

Focusing and Collimating the Beam

Because the rays from a distant light source follow essentially parallel paths within any small zone in space, we can bring the rays to a sharp focus so that they cover a minuscule area. The radius of the focused beam image depends on the distance to the source, the size of the source, and the focal length of the lens or mirror. Have you ever used a convex lens to focus sunlight and cause something to catch on fire? As the diameter of the lens or mirror increases, it gathers more light and the focused beam becomes more intense, assuming that the focal length remains constant.

When we want to focus a laser beam, the area of the lens or mirror need only be large enough so that it captures the whole beam. In theory, this minimum capture area never changes, no matter what the distance of the laser from the lens. In the "real world," the laser beam spreads a little, but the beam-divergence effect is not significant unless the distance from the laser is great.

With a coherent light beam, we can obtain an extremely small region of light at the focus of a convex lens or a concave mirror. Theoretically, this region constitutes a geometric point regardless of the focal length of the lens or mirror (Fig. 33-8). In practice, of course, we never actually get a perfect point. If we could, it would contain infinite energy density, a physical impossibility. Instead, we obtain an intense spot or ring having a tiny radius. Some laser beams can be focused to

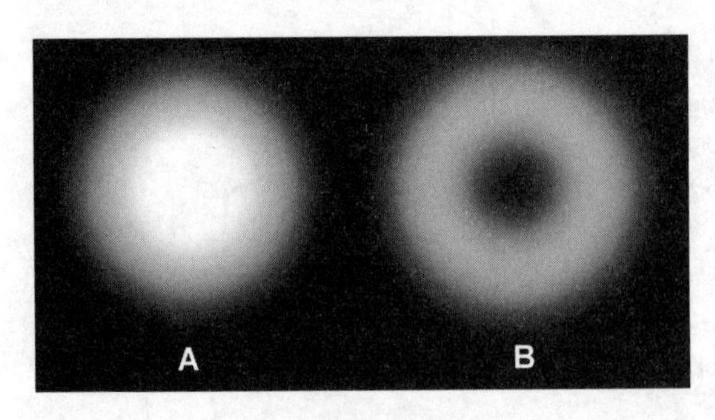

33-7 At A, a typical "spot" laser beam as it would look if projected onto a white screen. At B, the projection of a "ring" beam on the same screen.

Geometric point

Convex lens

Parallel rays from laser

A

33-8 In theory, a coherent light beam can be focused to a geometric point by a convex lens (A) or by a concave mirror (B).

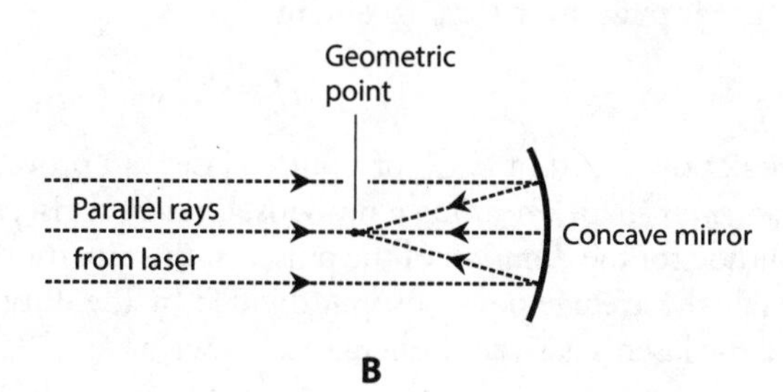

microscopic precision, forming energy spots or rings so small that they can alter the genes in living biological cells.

Continuous versus Pulsed

Some lasers deliver output power at a constant rate as time passes, and therefore, have equal peak and average power output. We call such devices *continuous-wave* (CW) *lasers*. Other lasers produce brief bursts having extreme peak power output but much lower average power output. We call them *pulsed lasers*. Pulsed-laser output can be achieved in a variety of ways.

A pulsed-laser technology known as *Q-switching* involves allowing the intensity of the light in the cavity to build up, while the cavity remains in a state unfavorable for lasing. When the intensity reaches a certain level, the cavity conditions are adjusted, producing resonance so that we get an intense, brief burst of coherent radiation. The process is repeated indefinitely, producing energy bursts at regular intervals.

Pulsed lasing can also be obtained by supplying the cavity with bursts of energy, rather than with a continuous supply. This technique is useful with lasing media that cannot maintain CW operation. A brilliant flash lamp can serve as the pumping source in this mode.

Energy, Efficiency, and Power

The *output energy* and *average output power* obtainable from a cavity laser depend on the physical size of the cavity, and on the amount of energy or power supplied by the pumping source.

We can determine the output energy by measuring the total radiant energy E_{out} of the beam, in joules, over a period of time. We might carry out the measurement by allowing the laser to strike a perfectly absorbing target (a so-called *black body*) that converts all the incident radiant energy to heat, and then measuring the amount of heat produced with a *calorimeter* or *joule meter*.

The efficiency *eff* of the laser can be determined by measuring the total pumping energy E_{in} over a certain period, measuring E_{out} for the same period, and then finding the ratio

$$eff = E_{out} / E_{in}$$

The average output power $P_{avg\text{-}out}$ in watts is determined by measuring the output energy E_{out} in joules produced over a certain period of time t in seconds, and then dividing the energy in joules by the number of seconds over which the energy measurement has been made, giving us

$$P_{avg\text{-}out} = E_{out} / t$$

In terms of power, we can calculate the efficiency *eff* by dividing the average output power $P_{avg\text{-}out}$ by the average input power $P_{avg\text{-}in}$ to obtain

$$eff = P_{avg\text{-}out} / P_{avg\text{-}in}$$

The peak power output $P_{pk\text{-}out}$ of a pulsed laser can prove difficult to determine unless the pulses have a *rectangular* shape, meaning a near-instantaneous rise, a near-instantaneous decay, and a constant amplitude for the duration of the pulse. In this situation (the theoretical ideal), the peak power output equals the average power output divided by the duty cycle D, which tells us the proportion of the time the laser is active. Mathematically, we have

$$P_{pk\text{-}out} = P_{avg\text{-}out} / D$$

In a pulsed laser, D is always smaller than 1 (usually much smaller), so P_{pk} always exceeds P_{avg} (usually by a large factor). If we know the peak output power and the duty cycle, then we can determine the average power using the formula

$$P_{avg\text{-}out} = P_{pk\text{-}out} D$$

If we know the peak power and the average power, then we can calculate the duty cycle using the formula

$$D = P_{avg\text{-}out} / P_{pk\text{-}out}$$

Semiconductor Lasers

As we've learned, some semiconductor diodes emit radiant energy when forward-biased so that current flows. This phenomenon, called *photoemission,* occurs as electrons in the atoms, driven to higher shells from the energy provided by a DC source, "fall" back down to lower energy states. The wavelength depends on the mixture of semiconductors. Most diodes emit energy in the visible or IR range.

Laser Diode

A *laser diode,* also called an *injection laser,* is a special LED or IRED with a relatively large and flat P-N junction. If the forward current through the junction remains below a certain threshold level, the device behaves like an ordinary LED or IRED, producing incoherent light or IR radiation. When the threshold current is reached, the emissions become coherent. Laser diodes, as well as conventional LEDs and IREDs, are designed to work properly only when forward-biased. They operate like ordinary semiconductor diodes when reverse-biased; they fail to radiate whether or not the reverse bias exceeds the avalanche voltage.

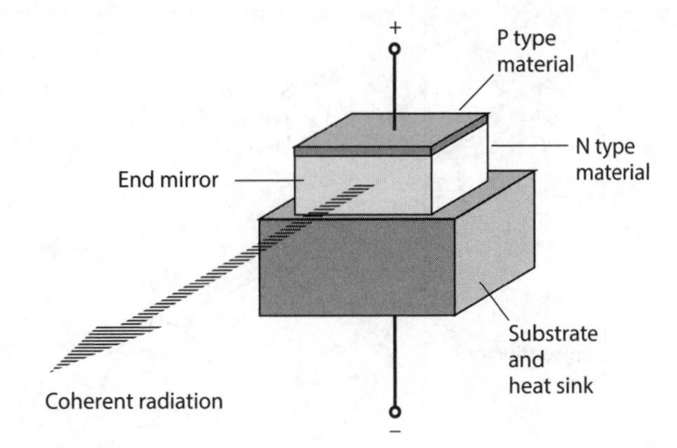

33-9 Simplified functional diagram of a laser diode. Coherent radiation is emitted in the plane of the P-N junction.

Gallium arsenide (GaAs) is a compound used in the manufacture of various semiconductor devices, including diodes, LEDs, IREDs, FETs, and ICs. Pure GaAs constitutes an N type semiconductor material. The charge carriers move fast and easily, so engineers say that GaAs has excellent, or high, *carrier mobility*.

GaAs LEDs are used in numeric displays for consumer items, such as electronic clocks. We'll also find them in sophisticated systems, such as radio transceivers, test instruments, calculators, and computers. GaAs IREDs, which have a primary emission wavelength of approximately 900 nm, find applications in fiber-optic systems, robotic proximity sensors, handheld remote-control boxes for home appliances, and motion detectors for security systems.

All GaAs LEDs and IREDs can be rapidly switched or modulated because of the high carrier mobility. We can modulate some *GaAs laser* diodes at frequencies in excess of 1 GHz (10^9 Hz), making them ideal for use in broadband or high-speed optical communications systems. Figure 33-9 illustrates the construction of a typical GaAs laser diode. The P and N type semiconductor materials are built up on top of a base called the *substrate*, which also serves to carry away excess heat by means of thermal conduction.

Hybrid Silicon Laser

In order to meet the demand for inexpensive, mass-producible silicon-based LEDs and IREDs, engineers developed the *hybrid silicon laser* (HSL) by combining silicon in layers with certain other semiconductor compounds called *III-V materials*. These compounds, which include *indium phosphide* (InP) and GaAs, allow for energy emission when excited by either an electrical current or an external source of coherent light. The lasing medium is surrounded by mirrors to form an enclosed medium called a *waveguide*, which produces coherent light by creating a resonant state in which the waves reinforce each other.

Quantum Cascade Laser

In some semiconductor lasers, electrons make "cascading" jumps among multiple atoms. Figure 33-10 shows, in simplified form, how this process occurs. This type of device, called a *quantum cascade laser* (QCL), produces more output than a conventional semiconductor laser because of a "chain reaction" of energy transitions. Figure 33-10 shows only two semiconductor layers, but a typical QCL has many layers, producing a massive cascade effect. The emission wavelength, which

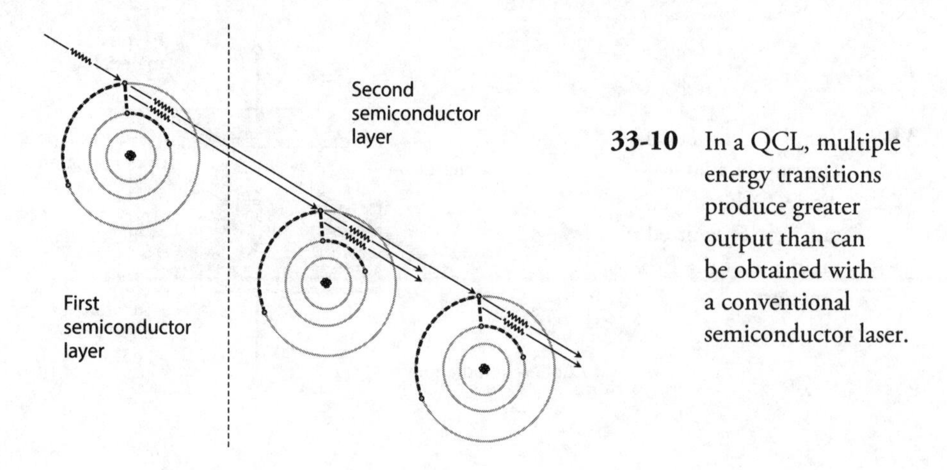

33-10 In a QCL, multiple energy transitions produce greater output than can be obtained with a conventional semiconductor laser.

depends on the thickness of the semiconductor layers, lies in the middle-IR to far-IR range, considerably longer than the waves of visible red light, but shorter than radio microwaves.

Vertical-Cavity Surface-Emitting Laser

In a *vertical-cavity surface-emitting laser* (VCSEL), a coherent visible or IR beam emerges perpendicular to the plane of the P-N junction, in contrast to the usual laser diode in which the beam emerges in the plane of the junction. Lasing takes place inside a region called the *quantum well*, surrounded on either side by semiconductor layers that also serve as end reflectors (Fig. 33-11).

In practice, the VCSEL offers a couple of advantages over a conventional laser diode. First, technicians can test a VCSEL at multiple stages prior to production, a procedure not always practical with other types of semiconductor devices. Second, a VCSEL generally emits a narrower beam than a conventional laser diode does.

Vertical-cavity devices based on GaAs operate at wavelengths from the visible red (about 650 nm) to the near-IR (about 1300 nm). Devices based on InP can function at wavelengths up to approximately 2000 nm. The wavelength can be adjusted by varying the thicknesses of the reflecting layers.

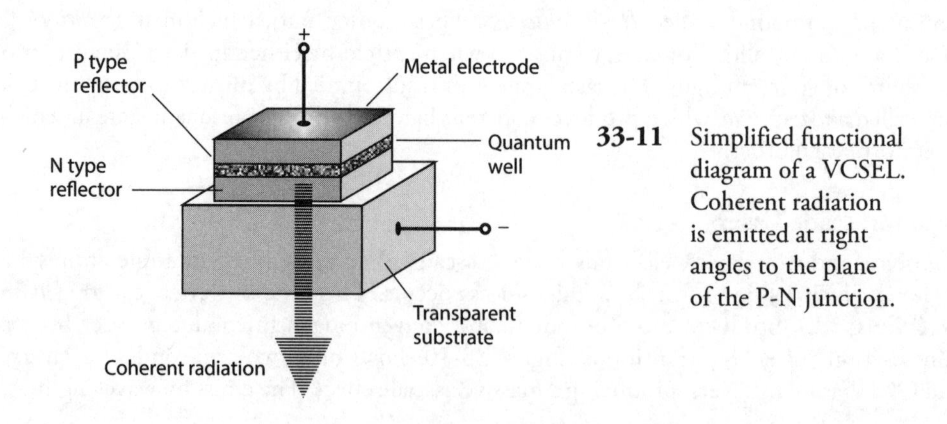

33-11 Simplified functional diagram of a VCSEL. Coherent radiation is emitted at right angles to the plane of the P-N junction.

Solid-State Lasers

Solid-state lasers include the *ruby laser*, the *crystal laser*, and various lasers using a solid combination of *yttrium*, *aluminum*, and *garnet* (called YAG). A solid-state laser is optically pumped by a *flash tube* that surrounds the lasing medium.

Ruby Laser

The lasing medium in a ruby laser comprises *aluminum oxide* with a trace of *chromium* added. The ruby crystal is a cylindrical piece of this material, which appears as a reddish solid with reflective surfaces at each end. One reflector is totally silvered to reflect all of the incident light. The other is partially silvered so that it reflects about 95 percent of the incident light. The laser beam emerges from the partially silvered end.

Figure 33-12 is a simplified diagram of a ruby laser. The flash tube emits brilliant, brief pulses of visible light that cause energy transitions in the electrons of the ruby crystal. Coherent emission occurs in pulses at a wavelength of approximately 694 nm, at the extreme red end of the visible spectrum.

With 1 W of power input to the flash tube, a typical ruby laser produces a few milliwatts of coherent light output. The rest of the input power dissipates as heat or radiates as stray light from the flash tube. The output pulses have duration on the order of 0.5 to 1.0 ms. In some arrangements, a small optical ruby laser pumps a large one. The large laser acts as an amplifier to obtain peak power output far greater than a single device can produce.

Crystal Laser

A solid-state laser having multiple compounds in the lasing medium is called a *mixed crystal laser*. A mixture of gallium arsenide and *gallium antimonide*, termed *gallium-arsenide-antimonide* (GaAsSb), constitutes the cavity material in some such devices. Other materials used in mixed crystal lasers include *gallium-arsenide-phosphide* (GaAsP) and *aluminum-gallium-arsenide* (AlGaAs).

Engineers "grow" solid-state lasing crystals in a liquid solution, following a process similar to that used to "grow" conventional semiconductor diode and transistor crystals. The semiconductor materials are *doped* (mixed with trace amounts of impurities) in various concentrations to obtain N and P properties. The emission wavelength can be adjusted by varying the relative concentrations of the substances in the cavity medium. In a GaAsSb laser, the wavelength can be tuned continuously from roughly 900 to 1200 nm in the near-IR range.

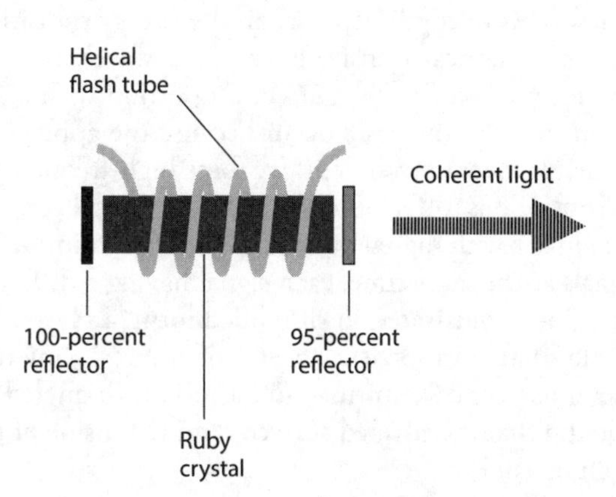

33-12 Simplified functional diagram of an optically pumped ruby laser.

An *insulating crystal laser* consists of a slightly doped sample of a *rare-earth element.* The resulting laser emission usually occurs at near-IR or visible red wavelengths. The insulating crystal laser is designed to operate at a cool temperature, so the device needs an elaborate heat-dissipation system. Overheating will result in deterioration of the lasing properties of the crystal. Wavelength adjustment is difficult, so the insulating crystal laser is generally used at only a single frequency.

Neodymium Laser

The *neodymium-yttrium-aluminum-garnet* (or *neodymium-YAG*) laser is a specialized solid-state laser that combines liquid and solid materials. The elemental neodymium rests in a liquid solution and remains confined within a solid YAG crystal. Energy is delivered into the laser medium by *arc lamps* or other bright sources of light focused on the crystal.

The wavelength of the neodymium-YAG laser is about 1065 nm in the near-IR portion of the spectrum. The efficiency is on the order of 1 percent. High-power neodymium-YAG lasers, like insulating crystal lasers, require a method of cooling to prevent damage to the crystal. The neodymium-YAG laser can operate in the pulsed mode or in the continuous mode.

Alternatively, a neodymium-based laser can take the form of an insulating crystal device. Glass can be doped with neodymium to obtain laser operation with intense optical pumping. When maintained at 20°C or lower, the efficiency of a *neodymium-glass laser* is about 3 percent when glass of high optical quality is used. The peak output power can range up to several hundred gigawatts. The device operates at an extremely low duty cycle; the average output power is a minuscule fraction of the peak output power.

Other Noteworthy Lasers

Some of the earliest lasers employed gas and liquid media, and their basic designs have not changed for decades. A few examples follow.

Helium-Neon Laser

Small lasers using gas-filled tubes containing helium and neon are available through scientific hobby companies. The *helium-neon* (He-Ne) *laser* emits vivid red light. The gas is excited by an electric current, typically producing emissions at 633 nm. Some designs produce IR output at 1150 nm or 3390 nm. The output power typically ranges from 10 mW to 100 mW, and the efficiency is approximately 5 percent at the visible red wavelength.

The lasing wavelength depends on the energy transitions shared by the two gases. Helium and neon undergo identical transitions at some wavelengths when their concentration levels have a certain value and occur in the correct proportion with respect to each other. Energy transitions in the electron shells of the gases occur because the applied current ionizes the gases, producing free electrons that frequently collide. If we force high-frequency AC through the ionized gas, the output becomes continuous, offering advantages for optical communications because a continuous laser is easy to modulate with digital or analog data. A continuous-mode He-Ne laser can be modulated by many signals at the same time, each signal having a different frequency.

Figure 33-13 illustrates, in simplified form, a He-Ne laser excited by RF energy. Some He-Ne lasers employ flat mirrors at each end of the resonant tube, and other designs (such as the one shown here) use concave mirrors and a tube with angled ends. The mirrors are of the *first-surface* type, meaning that the silvered surface is on the inside of the mirror glass, facing the interior of the gas-containing tube.

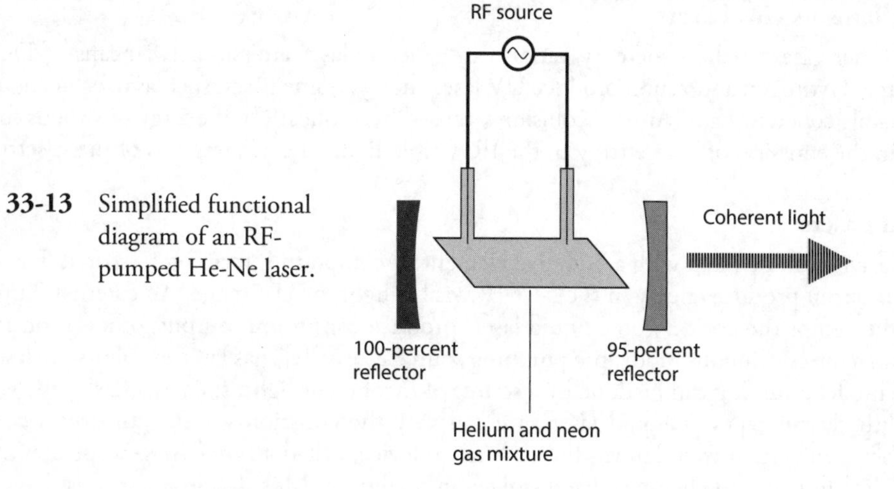

33-13 Simplified functional diagram of an RF-pumped He-Ne laser.

Nitrogen-Carbon-Dioxide-Helium Laser

A mixture of nitrogen, carbon dioxide, and helium can be used to make a laser that operates at a wavelength of about 1060 nm in the IR spectrum. Energy at this wavelength propagates with exceptionally low loss over long distances through clear air. The *nitrogen-carbon-dioxide-helium* (N-CO_2-He) *laser* can generate several kilowatts of continuous output power with pulse peaks in the gigawatt range. The gases can be excited by DC, as shown in Fig. 33-14.

Argon-Ion Laser

The *argon-ion laser* is a gas laser operating at about 480 nm, producing blue visible light. The ionized argon is kept at low pressure. The ionization and energy input are produced by passing an electric current through the gas. The power output is moderate but the efficiency is low, mandating the use of a cooling system. The pictorial representation of the argon-ion laser is essentially the same as that for the N-CO_2-He laser (Fig. 33-14).

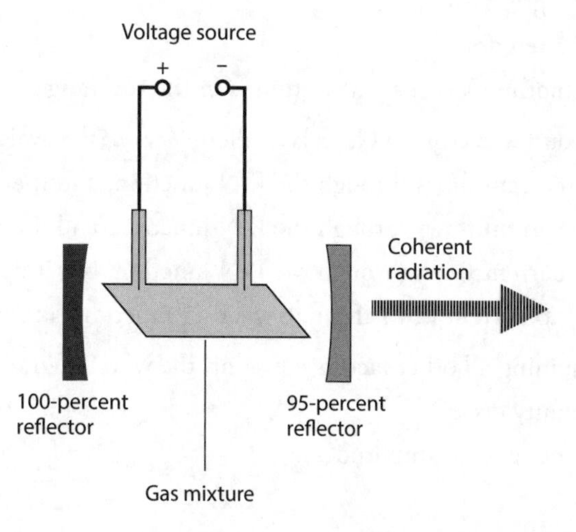

33-14 Simplified functional diagram of a gas laser excited by DC from an external voltage source.

Miscellaneous Gas Lasers

Many other gases, such as mercury vapor, can generate laser emissions by means of DC electric discharge. Hydrogen and xenon produce UV laser energy. Some gases, such as oxygen and chlorine, emit visible coherent light. Atomic collisions, caused by application of energy in various forms, can result in the emission of laser energy in the IR, visible-light, and UV regions of the spectrum.

Liquid Lasers

A laser cavity can be filled with a liquefied element or compound, forming a colored dye. A typical *liquid laser* can produce energy in the near-IR, visible light, or UV range. An external source of visible light pumps the device. Some liquid lasers produce continuous output; others produce pulsed output. In the continuous mode, the pumping source is usually a gas laser or solid-state laser. In the pulsed mode, pumping can be done by a source of incoherent light, such as a flash tube.

With certain types of liquid lasers, we can vary the emission wavelength over a continuous range between certain well-defined limits. Such a device, called a *tunable laser*, operates at a wavelength determined by a chemical dye dissolved in the liquid. Most liquid lasers must have effective cooling systems. Otherwise, overheated liquid can damage or destroy the cavity enclosure.

Quiz

Refer to the text in this chapter if necessary. A good score is 18 correct. Answers are in the back of the book.

1. In a pulsed laser, the duty cycle correlates with the ratio of
   (a) average to peak output power.
   (b) highest to lowest frequency.
   (c) longest to shortest wavelength.
   (d) minimum to maximum brilliance.
2. If you shine the beam from an argon-ion laser at a white wall, you'll see
   (a) a blue dot.
   (b) a green dot.
   (c) a red dot.
   (d) nothing because the output is in the UV range.
3. If you reverse-bias a GaAs laser diode *beyond* the avalanche voltage,
   (a) current flows through the P-N junction, and it emits coherent radiation.
   (b) current flows through the P-N junction, and it emits incoherent radiation.
   (c) current flows through the P-N junction, but it does not radiate anything.
   (d) no current flows through the P-N junction, and it does not radiate anything.
4. Assuming all other factors constant, the wavelength of a pulsed laser depends on the
   (a) duty cycle.
   (b) peak pulse amplitude.

(c) power output.

(d) None of the above

5. If you shine the beam from a xenon gas laser at a white wall, you'll see

(a) a blue dot.

(b) a green dot.

(c) a red dot.

(d) nothing because the output is in the UV range.

6. When an electron moves from a certain shell in an atom to another shell having a larger radius, the electron

(a) gains energy.

(b) loses energy.

(c) emits a photon.

(d) attains a lower frequency.

7. Suppose that the peak output power from a neodymium-glass laser is 500 GW. If the average output power is 50 W and the output occurs as rectangular pulses, what's the duty cycle?

(a) 10^{-12}

(b) 10^{-11}

(c) 10^{-10}

(d) 10^{-9}

8. If the laser described in Question 7 has an average input power of 1.0 kW, what's its efficiency?

(a) 1 percent

(b) 5 percent

(c) 10 percent

(d) 20 percent

9. The light from a laser in the visible range has

(a) varying hue with photons in phase coincidence.

(b) varying hue with photons in random phase.

(c) vivid hue with photons in phase coincidence.

(d) vivid hue with photons in random phase.

10. In a VCSEL, the energy beam

(a) comes off at a right angle to the plane of the P-N junction.

(b) is more brilliant than the beam from any other type of laser.

(c) has longer wavelength than the beam from any other type of laser.

(d) All of the above

11. A laser produces rectangular pulses of constant amplitude. The duty cycle is 5.00 percent and the average output power is 25.0 W. What's the peak output power?
    (a) 100 W
    (b) 250 W
    (c) 500 W
    (d) More information is needed to answer this question.

12. What's the efficiency of the laser described in Question 11 if the average input power is 40.0 W?
    (a) 50.0 percent
    (b) 62.5 percent
    (c) 66.7 percent
    (d) 75.0 percent

13. You connect a power source that provides 6.600 kV DC to a gas-discharge tube, ionizing the gas and causing 3.300 mA to flow. How much power does the ionized gas dissipate?
    (a) 2.200 W
    (b) 21.78 W
    (c) 220.0 W
    (d) 217.8 W

14. What type of laser produces more output power than a conventional semiconductor laser as a result of a "chain reaction" of energy transitions?
    (a) GaAsSb
    (b) Neodymium YAG
    (c) Ruby
    (d) Quantum cascade

15. What type of laser emits pulsed visible light at about 694 nm?
    (a) GaAsSb
    (b) Neodymium YAG
    (c) Ruby
    (d) Quantum cascade

16. If you reverse-bias a laser diode *below* the avalanche voltage,
    (a) current flows through the P-N junction, and it emits coherent radiation.
    (b) current flows through the P-N junction, and it emits incoherent radiation.
    (c) current flows through the P-N junction, but it does not radiate anything.
    (d) no current flows through the P-N junction, and it does not radiate anything.

17. A GaAs LED can be modulated or switched rapidly because it has
    (a) high carrier mobility.
    (b) long-wavelength emissions.

(c) high avalanche thresholds.

(d) high forward-breakover thresholds.

18. You modify the pulsed laser described all the way back in Question 11 so the duty cycle doubles to 10.0 percent, but you take pains to keep the average input and output power the same. What's the peak output power of the modified laser?

(a) 100 W

(b) 250 W

(c) 500 W

(d) More information is needed to answer this question.

19. What happens to the efficiency of the laser described in Question 11 if you modify it as described in Question 18?

(a) It becomes ¼ as great.

(b) It becomes half as great.

(c) It stays the same.

(d) It doubles.

20. You connect an IR laser diode in series with a current-limiting resistor and a source of variable DC voltage to reverse-bias the P-N junction. Initially, you set the voltage far below the avalanche threshold. Then you gradually increase the reverse voltage until it reaches and passes the avalanche point. How does the P-N junction behave during this process?

(a) At first, the P-N junction does not conduct, and the device radiates nothing. When the avalanche threshold is reached and passed, current flows and the diode emits coherent IR.

(b) At first, the P-N junction does not conduct, and the device radiates nothing. When the avalanche threshold is reached and passed, current flows and the diode emits incoherent IR. However, no laser diode can emit coherent radiation when reverse-biased. If you want lasing to occur, you must forward-bias the junction considerably beyond the forward-breakover threshold.

(c) At first, the P-N junction does not conduct, and the device radiates nothing. When the avalanche threshold is reached and passed, current flows but the diode still radiates nothing. If you want radiation to occur at all, you must forward-bias the junction at or beyond the forward-breakover threshold.

(d) At first, the P-N junction does not conduct, and the device radiates nothing. When the avalanche threshold is reached and passed, current flows and the diode emits incoherent IR. When the voltage rises further and reaches a certain point, the radiation becomes coherent.

34
CHAPTER

Advanced Communications Systems

THE TERM *WIRELESS* AROSE IN THE EARLY 1900S WHEN INVENTORS AND EXPERIMENTERS BEGAN sending and receiving messages using EM fields. Before long, wireless acquired more specific terminology, such as *radio*, *television*, and *EM communications*. In the 1980s and 1990s, the term *wireless* emerged again, this time in the consumer context.

Cellular Communications

Wireless telephone sets operate in a specialized communications system called *cellular*. Originally, the cellular communications network served mainly traveling business people. Nowadays, many (if not most) ordinary folks regard *cell phones* as necessities, and most cell-phone sets have non-voice features, such as text messaging, Web browsing, video displays, and digital cameras.

How Cellular Systems Work

A cell phone looks like a hybrid between a cordless telephone receiver and a walkie-talkie, but smaller. Some cell phones have dimensions so tiny that an unsuspecting person might mistake them for packs of chewing gum or candy. A cell-phone unit contains a radio transmitter and receiver combination called a *transceiver*. Transmission and reception take place on different frequencies, so you can talk and listen at the same time, and easily interrupt the other party if necessary, a communications capability known as *full duplex*.

In an ideal cellular network, every phone set constantly lies within range of at least one base station (also called a *repeater*), which picks up transmissions from the portable units and retransmits the signals to the telephone network, to the Internet, and to other portable units. A so-called *cell* encompasses the region of coverage for any particular repeater.

When a cell phone operates in motion, say, while the user sits in a car or on a boat, the set can move around in the network, as shown in Fig. 34-1. The dashed curve represents a hypothetical vehicle path. Base stations (dots) transfer access to the cell phone among themselves, a process called *handoff*. The hexagons show the limits of the transmission/reception range (or *cell*) for each base station. All the base stations are thereby connected to the regional telephone system, making it possible

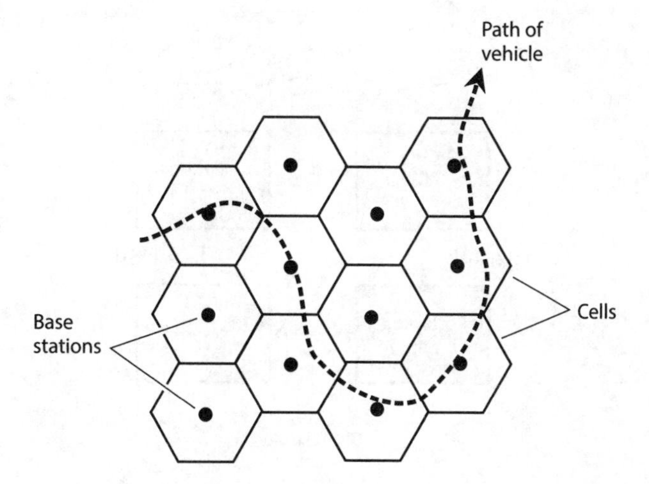

34-1 In an ideal cellular system, a moving cell-phone set (dashed line) always remains within range of at least one base station.

for the user of any portable unit to place calls to, or receive calls from, anyone else in the system, whether those other subscribers have cell phones or old-fashioned *landline* phones.

Cellular connections sometimes suffer from frequency call loss or *breakup* when signals transfer from one repeater to another. A technology called *code-division multiple access* (*CDMA*) reduces the frequency and severity of this problem compared to the early years of cellular technology. In CDMA, the repeater coverage zones overlap significantly, but signals don't interfere with each other because every phone set possesses a unique signal code. Rather than abruptly switching from one base-station zone to the next, the signal goes through a region in which it flows through more than one base station at a time. This *make-before-break* scheme helps to mitigate, but not eliminate, one of the most annoying problems inherent in cellular communication.

Caveats and Cautions

Unfortunately, call loss and breakup can occur even in a well-designed and constructed network when a user makes or receives calls from a physical location that suffers from poor reception. In a digital communications system, signals don't fade in and out as they do with analog radios. Instead, the signals tend to appear and disappear completely, sometimes off-again and on-again in a maddening flurry. Who among us has not experienced this phenomenon while using a cell phone? In some cities, one often sees shattered cell-phone sets lying in gutters or parking lots—victims of *chucking*, in which a furious user has hurled a phone set to the pavement with major-league-baseball speed. (I do not recommend nor condone this practice.)

In order to use a cellular network, you must purchase or rent a phone set and pay a monthly fee, or else buy time in advance. When using the system, you must never forget that your communications are not necessarily private. In this age of personal-security concerns and corporate espionage, you had better assume that *every word you say, every character you type, every photograph you take, and every video you capture can be intercepted by anyone in the world.* It's easier for unauthorized people to eavesdrop on wireless communications than to intercept wire or cable communications, and such people have no scruples about posting "hacked" conversations, texts, pictures, or videos on the Internet.

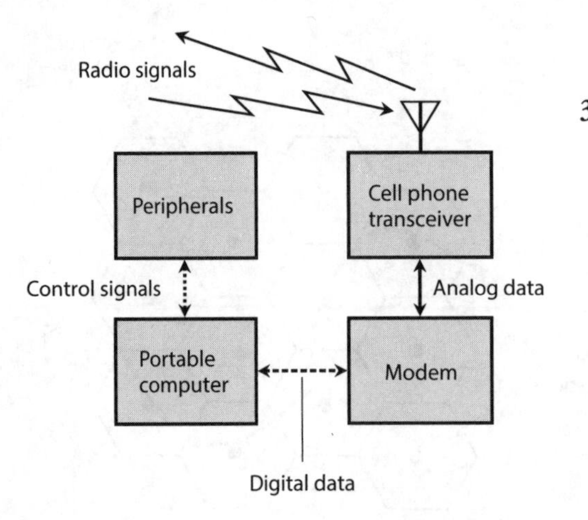

34-2 You can equip a cell phone with a modem (if it doesn't already have a modem built-in), allowing portable or mobile access to online networks with the full power of your personal computer.

Cell Phones and Computers

You can connect a *personal computer* (PC) to an older cell-phone set (one that doesn't have Internet access built-in) with a portable *modem* that converts incoming computer data from analog to digital form, and also converts outgoing data from digital to analog form. In this way, you can access the Internet from anywhere within range of a cellular base station, and have all the power of your home computer's Web browser at your disposal. Figure 34-2 is a block diagram of this scheme.

In recent years, high-end cell-phone sets have become available with modems built-in, but they generally have less sophisticated Web browsers than personal computers do.

Most commercial aircraft have telephones at each row of seats, along with jacks into which you can plug a modem or portable computer. If you plan to access the Internet from an aircraft, you must generally use the phones provided by the airline, not your own cell phone because radio transceivers can cause interference with flight instruments. You must also observe the airline's restrictions concerning the operation of electronic equipment while in flight.

Satellites and Networks

A *satellite communications* system resembles a huge cellular network with the repeaters in space rather than on the earth's surface. The zones of coverage are large, and they constantly change in size and shape if the satellite moves relative to the earth's surface.

Geostationary-Orbit Satellites

Geostationary satellites serve important roles in television (TV) broadcasting, telephone and data communication, weather and environmental data-gathering, radiolocation, and radionavigation.

In geostationary satellite networks, earth-based stations can communicate through a single "bird" only when the stations both lie on a line of sight with the satellite. If two stations are nearly on opposite sides of the planet, say, in Australia and Wisconsin, they must operate through two satellites to obtain a link (Fig. 34-3). In this situation, signals are relayed between the two satellites, as well as between either satellite and its respective earth-based station.

A potential problem with geostationary satellite links results from the fact that the signal must follow such a long path that perceptible propagation delays occur. Engineers call this delay, and

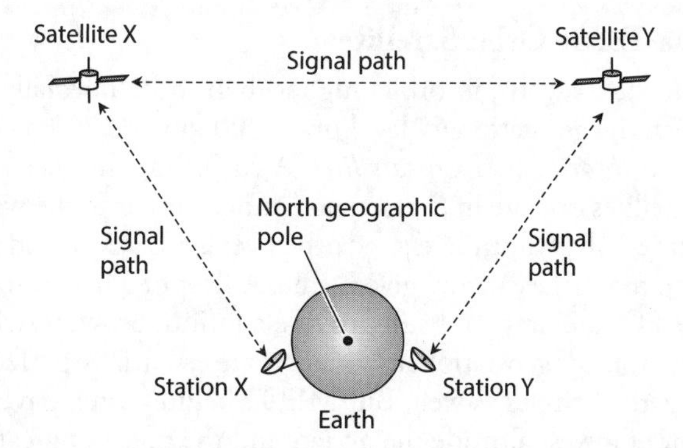

34-3 A communications link involving two geostationary satellites.

its observed effects, *latency*. The delay doesn't cause much trouble with casual communications or Web browsing, but it slows things down when computers are linked to combine their processing power. It also makes high-end, Internet-based *computer-game playing* between two or more users difficult or impossible because one user can't quickly respond to the others' moves in fast-paced action scenes.

Low-Earth-Orbit Satellites

The earliest communications satellites orbited only a few hundred kilometers above the earth. They were *low-earth-orbit* (LEO) *satellites*. Because of their low-altitude orbits, LEO satellites took only about 90 minutes to complete one revolution. Communication was intermittent at best, because a satellite remained within range of any given ground station for only a few minutes at a time. This limitation was the main reason why geostationary satellites became predominant once rocket technology progressed to the point at which satellites could reach the necessary altitude and acquire the necessary orbital precision.

Despite their advantages, geostationary satellites have certain limitations. A geostationary orbit requires constant adjustment because a tiny change in altitude will cause the satellite to get out of sync with the earth's rotation. Geostationary satellites are expensive to launch and maintain. When communicating through them, significant latency exists because of the path's length. Reliable communications demand fairly high transmitter RF output power, and a sophisticated, precisely aimed antenna, usually a helix or dish. These problems with geostationary satellites have brought about a revival of the LEO paradigm. Instead of one single satellite, the new concept dictates a large fleet of them, in effect building a "cellular network in reverse."

In a well-designed and well-implemented LEO-satellite system, at least one satellite always lies in direct line-of-sight range for every user on earth. The satellites can relay messages throughout the fleet. Therefore, any two points on the surface can always make, and maintain, contact through the satellites. The satellites follow *polar orbits* (routes that pass over or near the earth's geographic poles) to optimize the geographical coverage. Even if you work in the Arctic or Antarctic, you can use a LEO-satellite system. This flexibility does not hold true with geostationary satellite networks, where the regions immediately around the geographic poles remain invisible to the satellites.

A LEO-satellite communications link is easier to access and use than a geostationary-satellite link. A small, simple, non-direction antenna will suffice; you don't have to aim it in any particular direction. The transmitter can reach the network using only a few watts of RF output power. The latency rarely exceeds 100 milliseconds (ms), compared with as much as 400 ms for geostationary-satellite links.

Medium-Earth-Orbit Satellites

Some satellites revolve in orbits higher than those normally considered low-earth, but at altitudes lower than the geostationary level of 35,800 km (22,200 mi). These intermediate "birds" are called *medium-earth-orbit* (MEO) *satellites*. A MEO satellite takes several hours to complete each orbit. MEO satellites operate in fleets, in a manner similar to the way LEO satellites are deployed. Because the average MEO altitude exceeds the average LEO altitude, each MEO "bird" can cover a larger region on the surface at any given time. A fleet of MEO satellites can be smaller than a comparable fleet of LEO satellites, and still provide continuous, worldwide communications.

The orbits of geostationary satellites are essentially perfect circles, and most LEO satellites orbit in near-perfect circles as well. But MEO satellites often have elongated, or *elliptical*, orbits. We call the point of lowest altitude the *perigee* and the point of greatest altitude the *apogee*. They can differ considerably. The MEO satellite orbits at a speed that depends on its altitude. The lower the altitude, the faster the satellite moves. A satellite with an elliptical orbit crosses the sky rapidly when it "swoops low" near perigee, and slowly when it "flies high" near apogee. Users find it easiest to use an MEO satellite when its apogee is high above the horizon, as seen from the earth surface; under these conditions, the "bird" stays in the visible sky for a long time.

Every time a MEO satellite completes one orbit, the earth rotates beneath it. The rotation of the earth rarely coincides with the orbital period of the satellite. Therefore, successive apogees for a MEO satellite occur over different points on the earth's surface. This so-called *apogee drift* complicates satellite tracking, necessitating computers programmed with accurate orbital data. For a MEO system to effectively provide worldwide coverage without localized periodic blackouts, the various satellites must follow diverse yet coordinated orbits. In addition, the network must have enough satellites so that each point on the earth always lies on a line of sight with one or more of them. Ideally, every user should always observe at least one in-sight "bird" near its apogee.

Wireless Local-Area Networks

A *local-area network* (LAN) comprises a group of computers linked together within a building, campus, or other small region. The interconnections in early LANs were made with wire cables, but wireless links have become standard today. A *wireless LAN* offers flexibility because the computer users can move around without having to bother with plugging and unplugging cables. This arrangement works especially well when most, or all, of the users have portable computers. The geometric arrangement of major system components is called the *LAN topology*. Two major wireless LAN arrangements dominate: the *client-server topology* and the *peer-to-peer topology*.

A client-server wireless LAN (Fig. 34-4A) has one large, powerful, central computer called a *file server*, to which all the smaller personal computers are linked. The file server has enormous computing power, high speed, and large storage capacity. It can contain all the data for every user. End users do not communicate directly. All the data must pass through the file server.

In a peer-to-peer wireless LAN (Fig. 34-4B), all of the computers have more or less equal computing power, speed, and storage capacity. Each user maintains his or her own data. Subscribers can communicate directly without the data having to pass through any intermediary. This mode offers greater privacy and individuality than the client-server topology, but it tends to slow down when many users need to share data all at once.

Large institutions favor client-server LANs, while small businesses and schools, or departments within a larger corporation or university, prefer cheaper and more user-friendly peer-to-peer LANs. In the illustrations of Fig. 34-4, only three personal computers are shown in each network. However, any LAN can have as few as two, or as many as several dozen, computers.

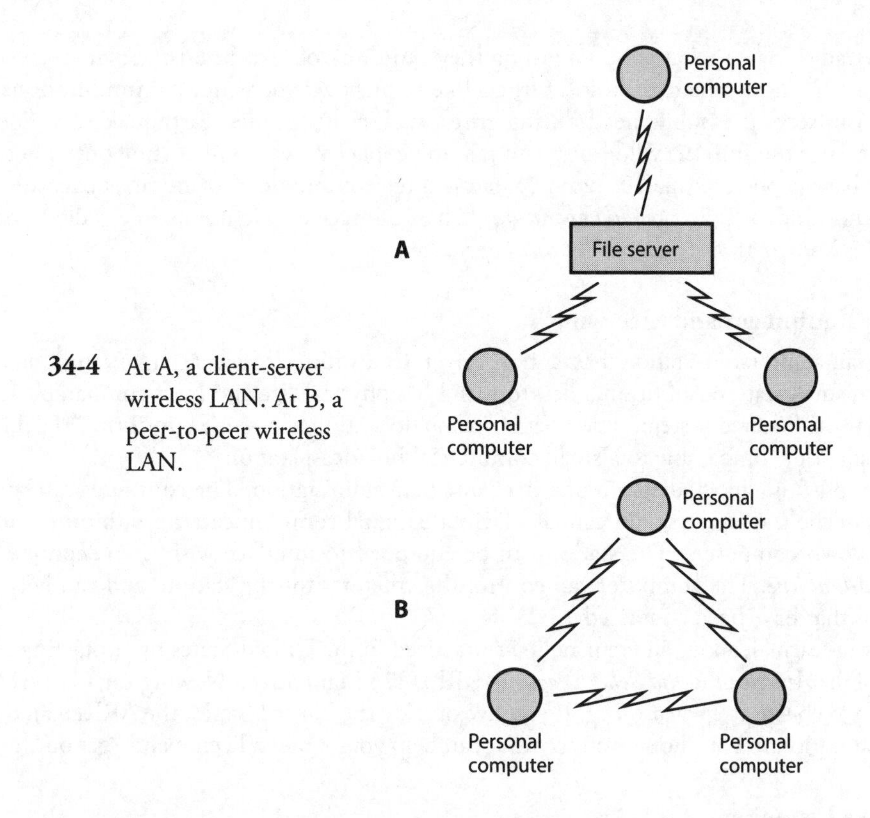

34-4 At A, a client-server wireless LAN. At B, a peer-to-peer wireless LAN.

Home Internet users sometimes employ a modified version of the arrangement shown in Fig. 34-4A. In place of the file server, a device called a *wireless router* provides a hub through which the computers can communicate. The router connects to the Internet through a high-speed interface such as a *cable modem*, allowing several computers in a household to have Internet access at the same time.

Amateur and Shortwave Radio

In most countries of the world, people can obtain government-issued licenses to send and receive messages by radio for nonprofessional purposes. Americans call this hobby *amateur radio*, or *ham radio*. If you want only to listen to communications and broadcasting, and not to transmit signals, you do not need a license in the United States (although you do need one in some countries).

Who Uses Amateur Radio?

Anyone can use ham radio provided they can pass the tests necessary to obtain a license. Amateur radio operators (or "hams") communicate in diverse modes including voice, Morse code, television, and various forms of digital text messaging. Text messaging can be done in real time, or by posting short passages similar to *electronic mail* (e-mail). Radio amateurs have set up their own radio networks to take advantage of a mode known as *packet communications*, or *packet radio*. Most packet networks these days have Internet gateways, but a few amateur radio operators store digital information at their stations, in effect creating a "virtual Internet" independent of the conventional communications infrastructure.

Some radio "hams" chat about anything they can think of (except business matters, which are illegal to discuss using amateur radio). Others like to practice emergency communications skills so that they can serve the public need during crises, such as hurricanes, earthquakes, or floods. Still others like to go out into the wilderness and talk to people far away while sitting out under the stars and using battery power. Amateur radio operators often communicate from cars, boats, aircraft, and bicycles; this mode is called *mobile operation*. When transceivers are used while walking or hiking, the mode is known as *portable,* or *handheld operation.*

Amateur Equipment and Licensing

A simple amateur radio station has a transceiver (transmitter/receiver), a microphone, and an antenna. A small station can fit on a desktop, and has physical size and mass comparable to a component type hi-fi stereo system. The operator can add station accessories until the "rig" becomes a large installation, comparable to a small commercial broadcast station.

Figure 34-5 is a block diagram of a fixed amateur radio station. The computer can control the functions of the transceivers, and can also facilitate digital communications with other radio amateurs who own computers. The station can be equipped to interface with the telephone services, also called *landline.* The computer can control the antennas for the station, and can keep a log of all stations that have been contacted.

You can learn all about amateur radio as practiced in the United States by contacting the headquarters of the *American Radio Relay League* (ARRL), 225 Main Street, Newington, CT 06111. They maintain a Web site at www.arrl.org. If you live outside the United States, the ARRL can direct you to an organization in your home country that can help you obtain a license and "get on the air."

Shortwave Listening

The high-frequency (HF) portion of the radio spectrum, at frequencies between 3 and 30 MHz, is sometimes called the *shortwave band.* This term constitutes a misnomer by contemporary and technical standards. The wavelengths greatly exceed those of EM fields at ultra high frequencies (UHF), microwaves, and infrared (IR), at which wireless devices commonly operate these days. In free space, a frequency of 3 MHz corresponds to a wavelength of 100 m, while a frequency of 30 MHz

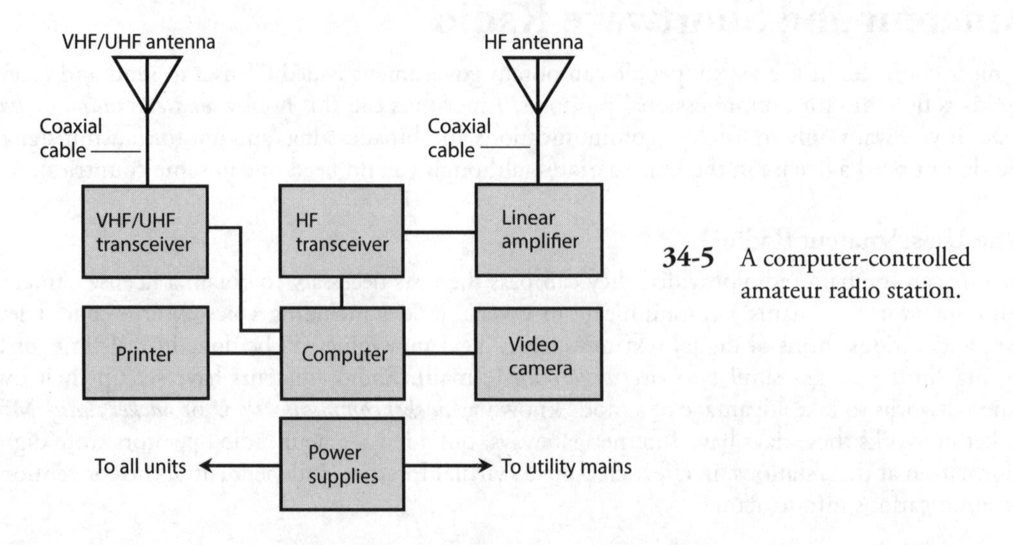

34-5 A computer-controlled amateur radio station.

corresponds to a wavelength of 10 m. Around the year 1920 when the shortwave band got its name, the wavelengths between 100 m and 10 m were indeed short in relative terms; most broadcast and communications signals had wavelengths in the multi-kilometer range!

Anyone can build or obtain a shortwave or general-coverage radio receiver, install a modest outdoor antenna, and listen to signals from all around the world. This hobby is called *shortwave listening*, or *SWLing*. In the United States, the proliferation of computers and on-line communications has, to some extent, overshadowed SWLing. These days, most young people grow up ignorant of shortwave broadcasting and communications, although this mode still predominates in much of the world. Nevertheless, some folks find never-ending fascination in the fact that people can contact each other using "old-fashioned radios" over vast distances using wireless devices alone, without the need for any human-made infrastructure other than an antenna at the source and another antenna at the destination. The ionosphere returns shortwave signals to the earth's surface, allowing reliable global broadcasting and communication to take place today, exactly as it has since the first days of radio during the early 1900s.

You can find commercially manufactured shortwave receivers if the prospect of SWLing interests you. Some electronics stores carry one or more models, along with antenna equipment, for a complete installation. Amateur-radio conventions, also called "hamfests," can serve as sources of shortwave receiving equipment at bargain prices. For information about events of this sort in your area, you can contact the American Radio Relay League at www.arrl.org, or visit your local amateur radio club.

Security and Privacy

In recent years, people have grown increasingly concerned about the security and privacy of electronic communications, particularly wireless. When a wireless system is compromised, even the most expert engineers might not detect the intrusion until the system or its subscribers have suffered irreparable harm. In some cases, the intrusion escapes detection altogether, and the victims never find out why strange and recurrent personal security problems continue to plague them.

Wireless versus Wired

Wireless eavesdropping differs from conventional *wiretapping* in two fundamental ways. First, eavesdropping is easier to do in wireless systems than in hard-wired systems. "Antique" hard-wired phone sets might seem inconvenient, but when you use them, your privacy will generally remain more secure than it will if you use a wireless device. Second, eavesdropping of a wireless link can be carried out secretly, but a good engineer can usually detect and locate a *leak* or *tap* in a hard-wired system.

If you use wireless devices to perform any part of a communications link, then a spy can place an eavesdropping receiver within range of your transmitting antenna (Fig. 34-6) and intercept your transmissions. The existence of a *wireless tap* has no effect on the electronic characteristics of any equipment in your system. A wary engineer might detect the presence of the eavesdropping receiver by noticing the spurious output caused by its IF oscillators, but in today's "RF-saturated" world, the presence of "one more signal" makes virtually no difference in the EM environment.

Levels of Security

We can categorize telecommunications security in terms of four levels, ranging from 0 (no security) to 3 (the most secure connections technology allows).

No Security (Level 0) In a communications system with level 0 security, anyone can eavesdrop on a connection at any time, as long as they're willing to spend the money and time to obtain the

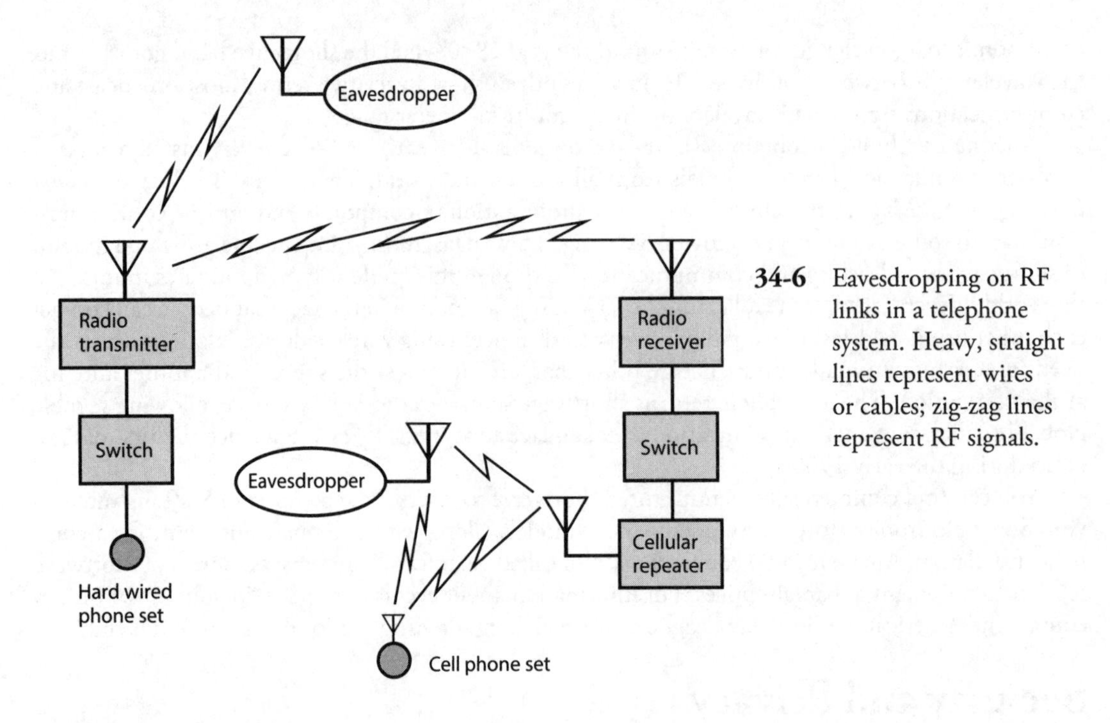

34-6 Eavesdropping on RF links in a telephone system. Heavy, straight lines represent wires or cables; zig-zag lines represent RF signals.

necessary equipment. Two examples of level 0 links are amateur radio and Citizens' Band (CB) voice communications.

Wire-equivalent Security (Level 1) An end-to-end hard-wired connection requires considerable effort to tap, and sensitive detection apparatus can usually reveal the existence of a wiretap. A communications system with *level-1 security* must have certain characteristics for optimum effectiveness:

- The cost must stay affordable
- The system must be reasonably safe for personal financial transactions
- When network usage is heavy, the degree of privacy afforded to each subscriber should not decrease, relative to the case when network usage is light
- Ciphers, if used, should remain unbreakable for at least 12 months, and preferably for 24 months or more
- *Encryption* ("secret-cipher") *technology*, if used, should be updated at least once every month

Security for Commercial Transactions (Level 2) Some financial and business data demand protection beyond the wire-equivalent level. Some companies and individuals refuse to transfer money by electronic means because they fear criminals will gain access to an account. In a communications system with *level-2 security*, the encryption used in commercial transactions should be sufficiently strong so that a potential intruder (also called a *hacker*) would need at least 10 years, and preferably 20 years or more, to break the cipher. Users should update the technology at least once a year.

Military Level Security (Level 3) Security to *military specifications* (also called *mil spec*) involves the most sophisticated encryption available. Technologically advanced countries, and entities with economic power, have an advantage here. However, as technology gains ever more (and arguably too much) power and influence over human activities, aggressor nations and terrorists might injure powerful nations by seeking out, and striking at, the weak points in communications infrastructures. In a communications system with *level-3 security*, the encryption scheme should be such that engineers believe it would take a hacker at least 20 years, and preferably 40 years or more, to break the cipher. The technology should be updated as often as economics allow.

Extent of Encryption

In a wireless system, we can achieve reasonable security and privacy with *digital encryption* that renders signals readable only to receivers with the necessary *decryption key*. This practice makes it difficult for unauthorized people to access or disrupt the system. The best decryption keys are complicated and obscure, making it hard (hopefully impossible) for hackers to figure it out.

For level-1 security, encryption is required only for the wireless portion(s) of the circuit. We should change the cipher at regular intervals to keep it "fresh." The block diagram of Fig. 34-7A shows *wireless-only encryption* for a hypothetical cellular telephone connection.

If we want to attain level-2 or level-3 security, we must use *end-to-end encryption*, in which we encrypt the signal at all intermediate points, even those parts of the link for which the signals propagate by wire or cable. Figure 34-7B shows this scheme for a hypothetical cellular connection.

Security with Cordless Phones

Most cordless phones are designed to make it difficult for unauthorized people to "hijack" a telephone line. Prevention of eavesdropping enjoys a lower level of priority, except in expensive cordless systems. If you have any concerns about using a cordless phone in a particular situation, use a hard-wired phone set instead.

If someone knows the frequencies at which a cordless handset and base unit operate, and if that person wants to eavesdrop on conversations that take place using that system, then that individual can

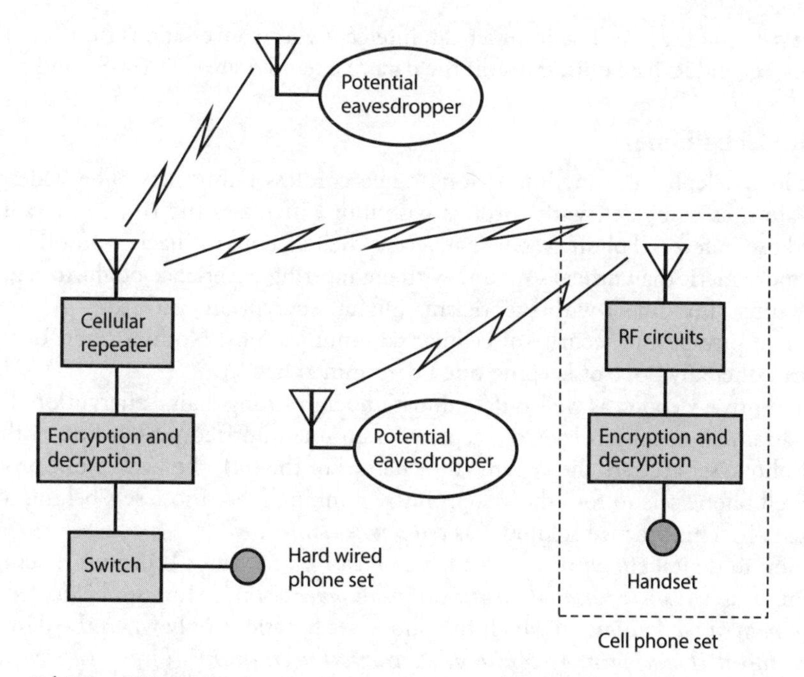

34-7A Wireless-only encryption. Heavy, straight lines represent wires or cables; zig-zag lines represent RF signals.

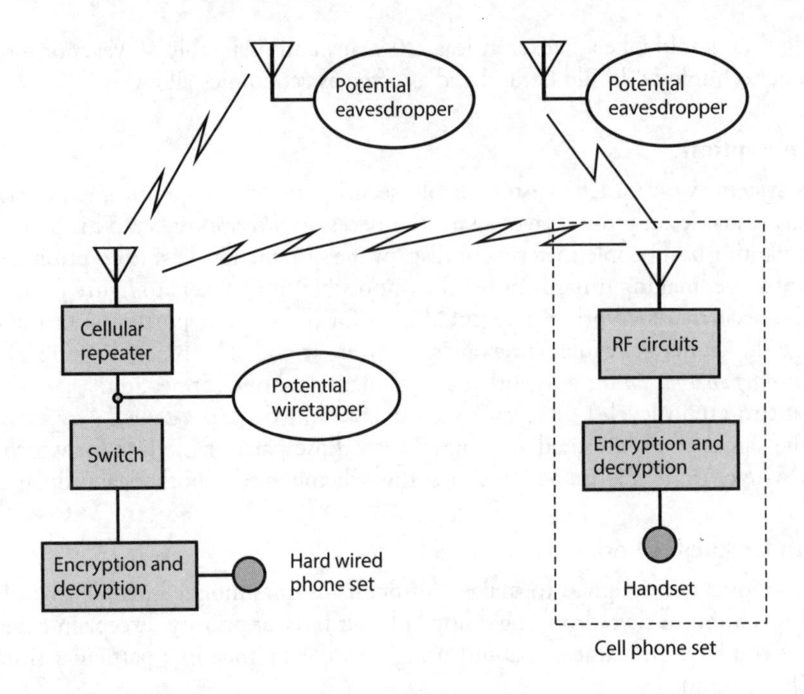

34-7B End-to-end encryption. Heavy, straight lines represent wires or cables; zig-zag lines represent RF signals.

place a wireless tap on the line. The intruder can intercept your conversations from a point near your cordless phone set and its base unit, transmit the data to a remote site (Fig. 34-8), and record it there.

Security with Cell Phones

In effect, cellular telephones function as long-range cordless phone sets. The wider coverage of cellular repeaters, as compared with cordless base units, increases the risk of eavesdropping and unauthorized use. A few cell phone vendors advertise their systems as "hacker-proof." Some of these claims have more merit than others. Anyone with engineering experience ought to regard the word "proof" (meaning "immune") with skepticism. Digital encryption constitutes the most effective way to maintain privacy and security of cellular communications. Nothing short of strong digital encryption can offer any hope of keeping out a determined hacker.

Access and privacy codes, as well as data, must undergo strong digital encryption if a cell-phone system is to attain an optimum level of security. If an unauthorized person knows the code with which a cell phone set accesses the system (the "name" of the set), the hacker can program one or more rogue cell phone sets to fool the system into "thinking" that those sets belong to the user of the authorized set. This practice is known as *cell phone cloning*.

In addition to digital encryption of data, *user identification* (user ID) must be employed. The simplest form of user ID is a *personal identification number* (PIN). More sophisticated systems can employ *voice-pattern recognition*, in which the phone set functions only when the designated user's voice speaks into it. *Hand-print recognition*, *electronic fingerprinting*, or *iris-pattern recognition*—which, along with voice-pattern recognition, constitute examples of *biometric security* measures—can further protect against malicious eavesdroppers or hackers.

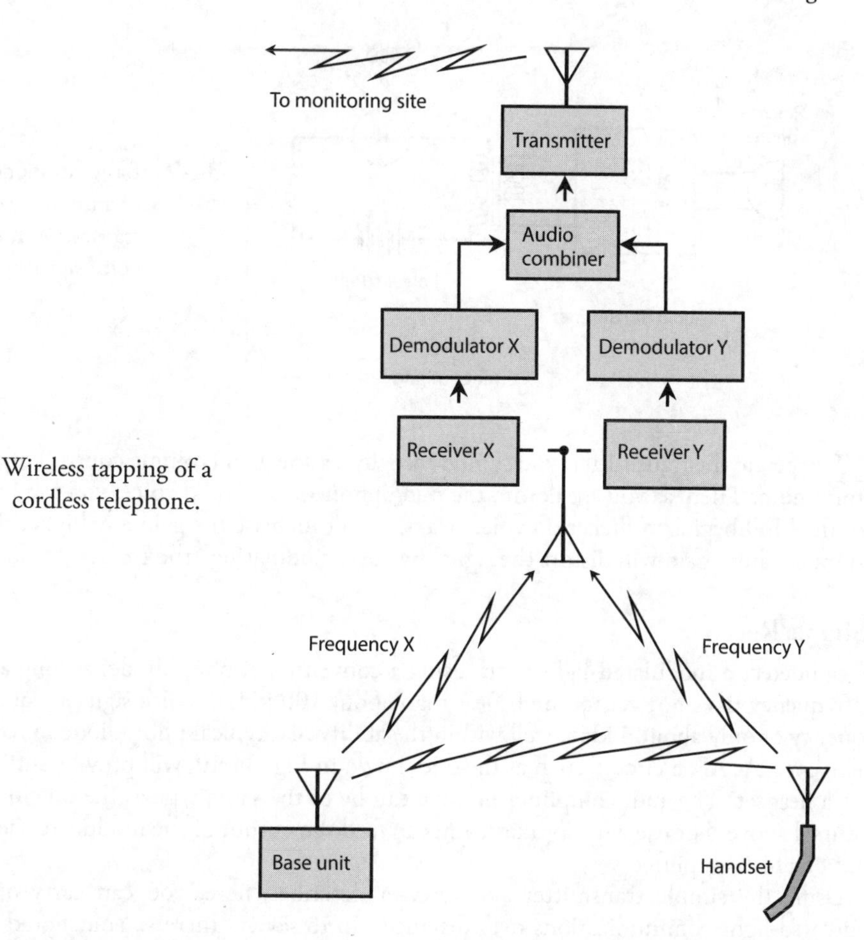

34-8 Wireless tapping of a cordless telephone.

Modulated Light

Electromagnetic energy of any frequency can undergo modulation for the purpose of transmitting information. *Modulated light* has recently become a significant means of conveying data, although the concept has existed for more than a century.

A Simple Transmitter

Figure 34-9 illustrates the components of a simple modulated-light voice transmitter that you can build for a few dollars using parts available from electronics supply stores. This circuit uses a microphone, an audio amplifier, a transformer, a current-limiting resistor, an LED, and a source of DC power such as a battery. Some hobby electronics stores sell complete audio amplifier modules. Alternatively, you can construct a simple audio amplifier circuit using a rugged bipolar transistor or *operational amplifier* ("op amp") chip. Ideally, the module or amplifier should have a built-in gain control.

The DC from the battery or power supply provides constant illumination for the LED. The current-limiting resistance should be low enough to allow the LED to glow even when no output comes from the audio amplifier, but high enough to prevent damage to the LED by the DC power source. When someone speaks into the microphone, the amplifier produces AF output. In the LED, the AF waveform appears superimposed on the DC from the power supply.

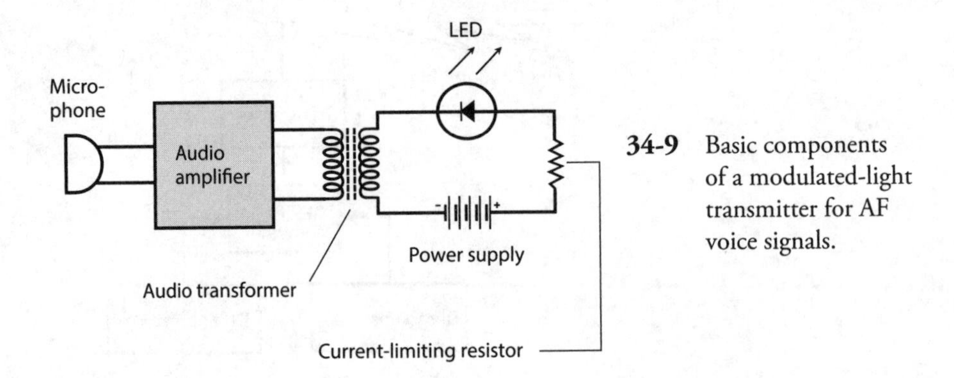

34-9 Basic components of a modulated-light transmitter for AF voice signals.

To operate the transmitter, you should initially set the gain (volume control) of the amplifier at its minimum. Then, as you speak into the microphone in a normal voice, you can increase the gain until the LED begins to flicker on voice peaks. Don't increase the gain any higher than that point. Too much audio gain will distort the signal by "overmodulating" the LED.

A Simple Receiver

We can detect a modulated-light beam using a conventional photodiode, as long as the modulating frequency does not get too high (less than about 100 kHz). Voice signals require a maximum frequency of only about 3 kHz, well within the ability of a typical photodiode to recover the signal intelligence. At AF, a circuit, such as the one shown in Fig. 34-10, will provide sufficient output to drive a headset. The audio amplifier module can be of the same type as the one in the transmitter described above. For use with speakers, you can feed the output of the module to the external audio input of a hi-fi amplifier.

Using the simple transmitter and receiver described here, you can carry out line-of-sight modulated-light communications over distances up to several meters. You should not expose the photodiode to a bright external source of light because such light can bias or even saturate the photodiode, making it insensitive to the tiny fluctuations in illumination from the transmitter. You should also avoid exposing the receiver to the light output from lamps that operate from 60-Hz utility AC power sources. The AC utility current modulates the light from all such lamps, producing severe interference to AF modulated-light signals.

Using this receiver, you can observe some interesting human-made and natural phenomena. For example, the visible image from a cathode-ray tube (CRT) screen, such as the ones used in older television sets and analog computer monitors, produces bizarre sounds when demodulated by this

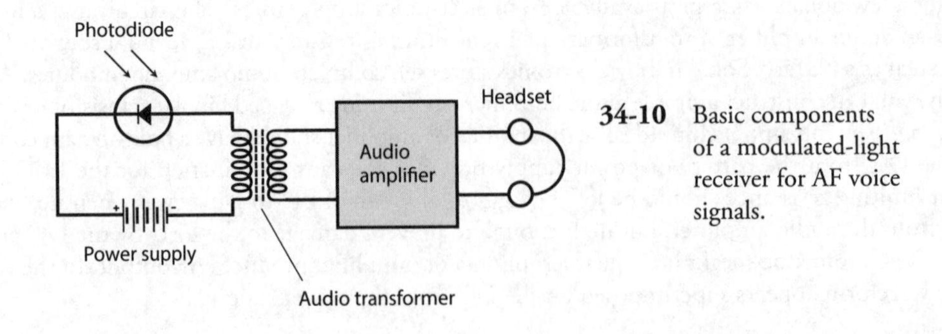

34-10 Basic components of a modulated-light receiver for AF voice signals.

circuit. You can have fun "listening" to sunlight shining through the leaves of a tree on a windy day, or reflecting from the pond or puddle when a light breeze produces ripples on the surface.

Extending the Range

Engineers use various methods to increase the communications range of line-of-sight modulated-light systems, including *collimation* of the transmitted beam (making the rays parallel), maximizing the receiver *aperture* (light-gathering area), and increasing the power of the transmitted signal.

A paraboloidal or spherical reflector can collimate the light beam from the transmitter LED. As the reflector diameter increases, the beam becomes narrower, allowing for propagation over longer distances. Large reflectors are expensive and can prove difficult to find; as an alternative, you can use a *Fresnel lens* of the type used in overhead projectors. These lenses consist of flat plastic etched with circular grooves that cause the material to behave like a convex lens. Fresnel lenses typically measure several centimeters square. You can get them from specialty hobby stores and catalogs. You should place the LED on one side of the lens, at a distance from the lens surface equal to the focal length. The collimated light beam will emerge from the other side of the lens.

You can use a second Fresnel lens to increase the receiver aperture. You should keep the receiver's "field of view" narrow to allow reception of the transmitted beam, while minimizing interference from sources in other directions. If you have a telescope, you can insert the photodiode into the eyepiece holder (instead of a regular telescope eyepiece). Then you can use the telescope's "finder" to visually aim it at the transmitted light source. A Fresnel lens can provide a large receiver aperture, but provides a less precise field of view than a full-sized telescope. If you use a Fresnel lens to gather the light, you can mount it in one side of an opaque box with a window cut out to fit the lens, and place the photodiode inside the box opposite the lens. Ideally, the distance across the box (from the lens to the photodiode) should equal the focal length of the Fresnel lens.

Inverse-Square Law

The light intensity from a point source, or from any relatively distant source, varies inversely in proportion to the square of the distance. If the source is directional (say, the beam from a lantern or flashlight), this relation holds true as long as we stay in the center of the beam. If d_1 and d_2 represent distances from a point source, and if P_1 and P_2 represent the intensities (power densities) as observed from these distances, respectively, and if we make all measurements in the same units, then

$$P_2/P_1 = d_1^2/d_2^2$$

In practical applications, the inverse-square law applies at great distances to collimated light rays, just as it does to an uncollimated point source. At a certain distance from a collimated source, the rays effectively diverge because the illuminating object (such as an LED) at the collimator's focus does not form a perfect geometric point. The collimating lens or reflector casts a magnified real (focused) LED image that gets larger as the distance increases, doubling in height and width as the distance doubles, tripling in height and width as the distance triples, and so on. If we multiply the distance by n, the area of the image grows by a factor of n^2, while the amount of radiant power remains the same. If our light receptor is small compared to the magnified real image, increasing the distance to the source causes the intercepted power to diminish according to the same inverse-square relation as it would if the source were a point without a collimator.

In the earth's atmosphere, air molecules absorb some visible light, especially at certain wavelengths. Most of the absorption occurs in the blue and violet parts of the visible-light spectrum. Dust and water vapor, rain, fog, particulate pollutants, ozone, and carbon monoxide all increase the absorption of light by the lower atmosphere. Absorption and scattering cause light-beam attenuation at a more rapid rate than would hold true in a vacuum.

An ideal laser device creates a beam of light with rays that all run perfectly parallel to each other. In theory, therefore, the beam intensity does not diminish with increasing distance from the source. In practice, we can approach this ideal but never attain it. A sophisticated laser device concentrates its energy into a beam that remains narrow for many kilometers, explaining why lasers lend themselves to line-of-sight, long-distance communication.

Fiber Optics

In 1970, *Robert Maurer* of Corning Glass Works demonstrated the practicality of *fiber optics* for high-speed, high-volume communications after an exceptional grade of glass became available from Standard Telecommunication Laboratories. Since that time, fiber-optic communications systems have gained widespread popularity throughout the world.

Advantages

Besides allowing for the transmission of signals, optical fibers offer immunity from electromagnetic interference (EMI). A strong radio signal, thunderstorm, solar storm, or nearby high-voltage power line does not affect visible light or IR rays traveling along an optical fiber. Conversely, the signals in an optical fiber do not cause EMI to external devices or systems. All of the signal energy stays within the fiber. Potential eavesdroppers find the data contained in the visible light or IR rays traveling along an optical fiber more difficult to intercept than the data from a conventional wire, cable, or wireless system. The minerals that compose glass fibers are cheap and plentiful, and can be "mined" with minimal environmental impact. Fiber-optic cables can be submerged or buried, and they do not corrode as metal conductors do. Optical fibers last longer than wires or cables, and they require less frequent maintenance.

Light Sources

Two types of light source customarily serve in fiber-optic communications systems: the laser diode and the conventional LED or IRED. The laser diode produces a beam that spreads out more rapidly with distance than the beams from large lasers such as the cavity type, but beam divergence does not pose a problem in fiber optics because the fiber keeps the beam confined. A collimating lens at the input end of the fiber keeps the light rays nearly parallel, so that none of the light energy escapes from the fiber.

A laser diode, LED, or IRED emits energy when sufficient current passes through it. If the no-signal current remains within a certain range, the instantaneous emitted power varies in proportion to the instantaneous applied current. The response rate is rapid, so the beam can be easily modulated by varying the current in sync with the data to be transmitted. Efficient beam amplitude modulation (AM) results.

As the modulated beam passes along the optical fiber, the relative intensity varies in the same proportion all along the fiber, even though the absolute intensity decreases because of loss. Therefore, the signal at the receiving end of the fiber, although weaker than the signal at the transmitting end, has identical modulation characteristics.

Multimode Fiber Designs

Two basic types of optical fiber exist, known as *multimode* and *single-mode*. In a multimode fiber, the transmission medium has a diameter of at least 10 times the longest wavelength to be carried.

For visible-light systems, that's approximately 7.5 micrometers (μm), where 1 μm equals 1000 nm, or 10^{-6} m. In an IR system, the minimum required fiber diameter is somewhat larger. In a single-mode fiber, the transmission-medium diameter can range down to approximately one full wavelength. The following discussions concern only multimode fibers.

Multimode optical fibers are made from glass or plastic with certain impurities added. The impurities affect the *refractive index*, or the extent to which light rays bend when they pass into or out of the medium at a fixed angle. A typical multimode optical fiber has a *core* surrounded by a tubular *cladding*. The cladding has a lower refractive index than the core, which constitutes the transmission medium. Two basic multimode designs have gained common usage: the *step-index* optical fiber and the *graded-index* optical fiber.

In a step-index fiber (Fig. 34-11A), the core has a uniform index of refraction and the cladding has a lower index of refraction, also uniform. At the boundary, the refractive index changes abruptly. Ray *X* enters the core parallel to the fiber axis and travels without striking the boundary unless a bend occurs in the fiber. If a bend occurs, ray *X* veers off-center and behaves like ray *Y*, striking the boundary between the core and the cladding. Each time ray *Y* encounters the boundary, the ray undergoes *total internal reflection* because of the difference in refractive indices. Therefore, ray *Y* stays within the core.

In a graded-index optical fiber (Fig. 34-11B), the core has a refractive index that's at its maximum along the central axis and steadily decreases outward from the center. At the boundary, an abrupt drop occurs in the refractive index, so the cladding has a lower index of refraction than any part of the core. Ray *X* enters the core parallel to the fiber axis and travels without striking the boundary unless a bend occurs in the fiber. In case of a bend, ray *X* veers off center and behaves like ray *Y*. As ray *Y* moves away from the center of the core, the index of refraction decreases, causing the ray to veer back toward the center. If ray *Y* encounters an especially sharp bend in the fiber, the ray strikes the boundary between the core and the cladding, and the ray undergoes total internal reflection.

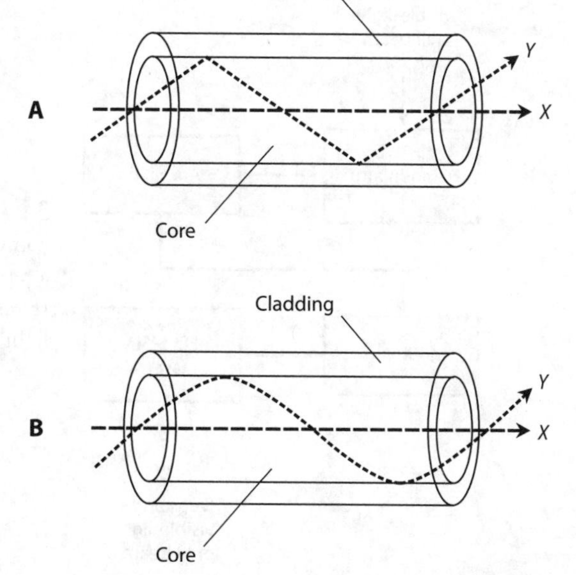

34-11 At A, a step-index optical fiber. At B, a graded-index fiber. In both designs, rays *X* and *Y* stay inside the core.

Fiber Bundling

Two or more optical fibers can be bundled to form a complex cable. Layers of durable, waterproof material, such as polyethylene, protect the individual fibers from damage and moisture intrusion. A tough outer jacket protects the bundle from the elements. Each fiber can carry numerous rays of visible-light or IR energy, each ray having a different wavelength. Every individual ray can be modulated with thousands of RF signals, each signal having a different carrier frequency. Because the frequencies of visible-light and IR rays are much higher than the frequencies of the signals that modulate them, the attainable bandwidth in a fiber-optic cable link vastly exceeds the attainable bandwidth of any wire-cable or wireless link, allowing much higher data transmission speed.

Another notable advantage of fiber bundles over multiple-wire bundles results from the fact that, because the fibers don't carry electric currents, we never have to contend with *crosstalk* (mutual signal interference) among them. Current-carrying wires, unless individually shielded, suffer from crosstalk because the AC signal in any one of the wires can "leak" into the other wires nearby. Providing an EM shield for each wire in a large bundle makes a cable expensive, bulky, and heavy. Fiber-optic cables cost less, have smaller size, and have lower mass than electrical cables because EM shielding is not necessary.

Repeaters

A long-distance fiber-optic system must incorporate *repeaters* at intervals along the length of the cable. The repeaters comprise *opto-electronic transceivers* that intercept, demodulate, and amplify the signals and then re-modulate and retransmit them, usually (but not necessarily) using the same type of visible-light or IR source as the original optical transmitter.

Figure 34-12 is a block diagram of a modulated-light repeater comprising a demodulator that separates the signals from the faint incoming visible light or IR carrier, an amplifier to boost the signal strength, and a modulated LED, IRED, or laser that transmits a visible or IR beam containing all of the original signal data to the destination or to another repeater.

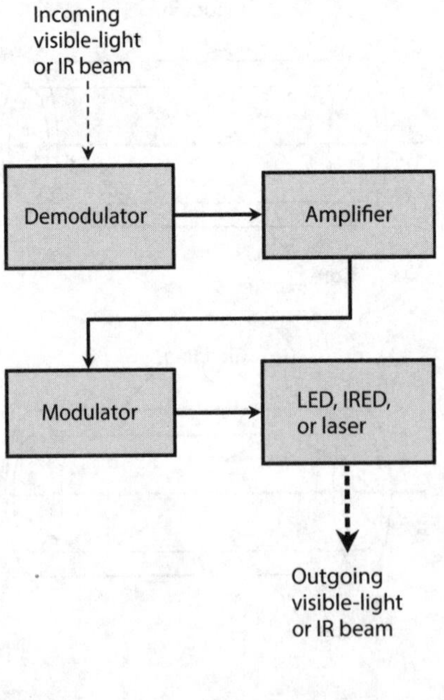

Incoming visible-light or IR beam
Demodulator
Amplifier
Modulator
LED, IRED, or laser
Outgoing visible-light or IR beam

34-12 In a fiber-optic communications system, a repeater amplifies and retransmits modulated light or IR beams.

Quiz

Refer to the text in this chapter if necessary. A good score is 18 correct. Answers are in the back of the book.

1. In a line-of-sight modulated-light communications system, you can maximize the range by using
   (a) an LED at the receiving end to pick up the light.
   (b) a Fresnel lens at the receiving end to capture and focus the light.
   (c) the highest possible modulating frequency at the transmitting end.
   (d) a blue light source at the transmitting end.

2. Optical fibers are not susceptible to
   (a) crosstalk when bundled into cables.
   (b) external electromagnetic (EM) fields.
   (c) corrosion such as occurs with metal wire.
   (d) All of the above

3. Hand-print recognition, electronic fingerprinting, and iris-pattern recognition are examples of
   (a) data conversion.
   (b) biometric security measures.
   (c) multimode data recognition.
   (d) optical encoding.

4. In a step-index multimode optical fiber, the refractive index of the core is
   (a) lower than the refractive index of the cladding, with a gradual transition.
   (b) lower than the refractive index of the cladding, with an abrupt transition.
   (c) higher than the refractive index of the cladding, with a gradual transition.
   (d) higher than the refractive index of the cladding, with an abrupt transition.

5. In a low-earth-orbit (LEO) satellite network,
   (a) all the satellites are within range of each other all the time.
   (b) every satellite's orbit takes it over the earth's poles.
   (c) at least one satellite appears stationary in the sky from at least one fixed point on the surface.
   (d) All of the above

6. You can maximize the range of a modulated-light transmitter for line-of-sight communication by
   (a) collimating the transmitted light rays.
   (b) using the shortest possible light wavelength.
   (c) using the highest possible modulation frequency.
   (d) All of the above

7. What do you call a local area network (LAN) in which every user's computer connects directly to a single, powerful central computer?
   (a) Peer-to-peer LAN
   (b) Repeater-based LAN
   (c) Client-server LAN
   (d) Distributed LAN

8. The most sophisticated data encryption systems employ
   (a) high-gain op amps.
   (b) high-speed microcontrollers.
   (c) complicated decryption keys.
   (d) pulse-width modulation (PWM).

9. Which of the following characteristics is an advantage of optical fiber over copper wire as a communications medium?
   (a) Optical fibers last longer than copper wires do.
   (b) You can submerge or bury optical fibers and they won't corrode.
   (c) The minerals that compose optical fibers don't cost much.
   (d) All of the above

10. In a cellular telephone network, base stations transfer connections to moving cell phones by means of a process called
   (a) handoff.
   (b) cell transfer.
   (c) zone switching.
   (d) multiple access.

11. If you increase your distance from a light source by a factor of 9, assuming that the total output from the source remains constant and you stay in the center of the beam, the brilliance will decrease by a factor of
   (a) 81.
   (b) 27.
   (c) 9.
   (d) 3.

12. What is *chucking* in the context of cellular telephone networks?
   (a) A cell phone using two or more base stations at the same time
   (b) A connection failing for no apparent reason
   (c) Interference to a cell phone from an unknown source
   (d) An angry user violently discarding a bad cell phone

13. In a geostationary-satellite link, latency can pose a problem for people who want to
   (a) communicate using modulated light.
   (b) casually browse the Internet.
   (c) combine the processing power of multiple computers.
   (d) use cordless phones in rural areas.

14. The light-emitting component in a fiber-optic transmitter can be
    (a) a photodiode.
    (b) an IRED.
    (c) a PIN diode.
    (d) a Gunn diode.

15. It's easiest to track a satellite in an elongated elliptical orbit when the satellite
    (a) is at and near apogee.
    (b) passes over the earth's poles.
    (c) is at and near perigee.
    (d) orbits over the earth's equator.

16. What's the technical term for programming an unauthorized cell phone so that a cellular network will accept it as part of the system?
    (a) Cloning
    (b) Cracking
    (c) Miscoding
    (d) Collimation

17. In a well-engineered cellular network, every phone set is always within range of at least one
    (a) other phone set.
    (b) modem.
    (c) repeater.
    (d) satellite.

18. Amateur radio is a hobby that
    (a) allows people to record commercial radio programs and then retransmit them, but not for profit.
    (b) allows people to send and receive messages by radio for nonprofessional purposes.
    (c) allows people to build radio receivers but not transmitters.
    (d) allowed nonprofessional radio communications until the start of the Vietnam War, but has remained illegal since then.

19. In effect, a communications satellite is an earth-orbiting
    (a) repeater.
    (b) data converter.
    (c) optical amplifier.
    (d) frequency multiplier.

20. The "shortwave radio" band encompasses frequencies of roughly
    (a) 0.3 to 3 MHz.
    (b) 3 to 30 MHz.
    (c) 30 to 300 MHz.
    (d) 300 MHz and higher.

35
CHAPTER

Antennas for RF Communications

WE CAN CATEGORIZE RF ANTENNAS INTO RECEIVING TYPES AND TRANSMITTING TYPES. NEARLY ALL transmitting antennas can receive signals quite well within the design frequency range. Some, but not all, receiving antennas can transmit RF signals with reasonable efficiency.

Radiation Resistance

When RF current flows in an electrical conductor, such as a wire or a length of metal tubing, some EM energy radiates into space. Imagine that we connect a transmitter to an antenna and test the whole system. Then we replace the antenna with a resistance/capacitance (*RC*) or resistance/inductance (*RL*) circuit and adjust the component values until the transmitter behaves exactly as it did when connected to the real antenna. For any antenna operating at a specific frequency, there exists a unique resistance R_R, in ohms, for which we can make a transmitter "think" that an *RC* or *RL* circuit is, in fact, that antenna. We call R_R the *radiation resistance* of the antenna.

Determining Factors

Suppose that we place a thin, straight, lossless vertical wire over flat, horizontal, perfectly conducting ground with no other objects in the vicinity, and feed the wire with RF energy at the bottom. In this situation, the radiation resistance R_R of the wire is a function of its height in wavelengths. If we graph the function, we get Fig. 35-1A.

Now imagine that we string up a thin, straight, lossless wire in free space (such as a vacuum with no other objects anywhere nearby) and feed it with RF energy at the center. In this case, R_R is a function of the overall conductor length in wavelengths. If we graph the function, we obtain Fig. 35-1B.

Antenna Efficiency

We rarely have to worry about *antenna efficiency* in a receiving system, but in a transmitting antenna system, efficiency always matters. The efficiency expresses the extent to which a transmitting antenna converts the applied RF power to actual radiated EM power. In an antenna, radiation resistance R_R

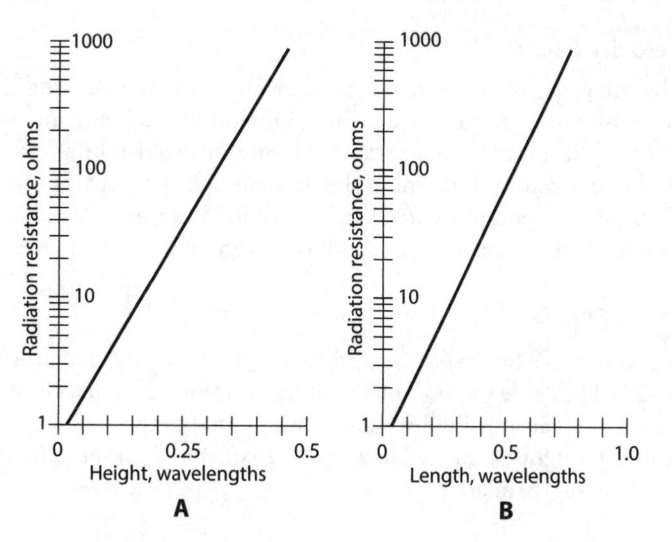

35-1 Approximate values of radiation resistance for vertical antennas over perfectly conducting ground (A) and for center-fed antennas in free space (B).

always appears in series with a certain *loss resistance* (R_L). We can calculate the antenna efficiency, *Eff,* as a ratio with the formula

$$Eff = R_R / (R_R + R_L)$$

As a percentage, we have

$$Eff_{\%} = 100\ R_R / (R_R + R_L)$$

We can obtain high efficiency in a transmitting antenna only when the radiation resistance *greatly exceeds* the loss resistance. In that case, most of the applied RF power "goes into" useful EM radiation, and relatively little power gets wasted as heat in the earth and in objects surrounding the antenna. When the radiation resistance is comparable to or smaller than the loss resistance, a transmitting antenna behaves in an inefficient manner. This situation often exists for extremely short antenna radiators because they exhibit low radiation resistance. If we want reasonable efficiency in an antenna with a low R_R value, we must do everything we can to minimize R_L. Even the most concerted efforts rarely reduce R_L to less than a few ohms.

If an antenna system has a high loss resistance, it can work efficiently if we design it to exhibit an extremely high radiation resistance. When an antenna radiator measures a certain height or length at a given frequency, and if we construct it of low-loss wire or metal tubing, we can get its radiation resistance to exceed 1000 ohms. Then we can construct an efficient antenna even in the presence of substantial loss resistance.

Half-Wave Antennas

We can calculate the physical span of an EM-field half wavelength in free space using the formula

$$L_{ft} = 492 / f_o$$

where L_{ft} represents the straight-line distance in feet, and f_o represents the frequency in megahertz. We can calculate the physical span of a half wavelength in meters, L_m, using the formula

$$L_m = 150 / f_o$$

Velocity Factor

The foregoing formulas represent theoretical ideals, assuming infinitely thin conductors that have no resistance at any EM frequency. Obviously, such antenna conductors do not exist in physical reality. The atomic characteristics of wire or metal tubing cause EM fields to travel along a real-world conductor a little more slowly than light propagates through free space. For ordinary wire, we must incorporate a *velocity factor* v of 0.95 (95 percent) to account for this effect. For tubing or large-diameter wire, v can range down to about 0.90 (90 percent).

Open Dipole

An *open dipole* or *doublet* comprises a half-wavelength radiator fed at the center, as shown in Fig. 35-2A. Each side or "leg" of the antenna measures a quarter wavelength long from the *feed point* (where the transmission line joins the antenna) to the end of the conductor. For a straight wire radiator, the length L_{ft}, in feet, at a design frequency f_o, in megahertz, for a center-fed, half-wavelength dipole is approximately

$$L_{ft} = 467/f_o$$

The length is meters is approximately

$$L_m = 143/f_o$$

These values assume that $v = 0.95$, as we would normally expect with copper wire of reasonable diameter. In free space, the impedance at the feed point of a center-fed, half-wave, open dipole constitutes a pure resistance of approximately 73 ohms. This resistance represents R_R alone. No reactance exists because a half-wavelength open dipole exhibits resonance, just as a tuned *RLC* circuit would if made with a discrete resistor, inductor, and capacitor. In fact, a dipole antenna (as well as many other types) constitutes an *RLC* circuit of a specialized, unique sort.

Folded Dipole

A *folded dipole antenna* consists of a half-wavelength, center-fed antenna constructed of two parallel wires with their ends connected together as shown in Fig. 35-2B. The feed-point impedance of the folded dipole is a pure resistance of approximately 290 ohms, or four times the feed-point resis-

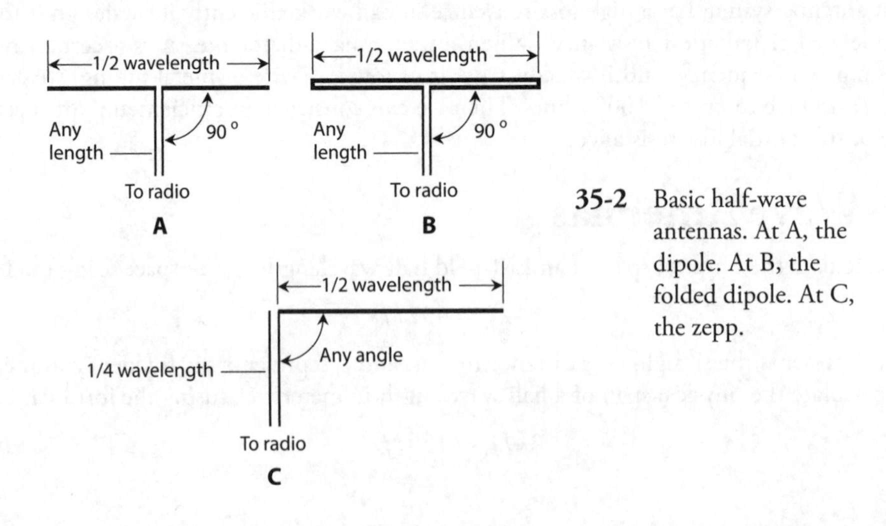

35-2 Basic half-wave antennas. At A, the dipole. At B, the folded dipole. At C, the zepp.

tance of a half-wave open dipole made from a single wire. This "resistance-multiplication" property makes the folded dipole ideal for use in parallel-wire transmission lines, which usually have high characteristic impedance (Z_o) values.

Half-Wave Vertical

Imagine that we stand a half-wave radiator "on its end" and feed it at the *base* (the bottom end) against an earth ground, coupling the transmission line to the antenna through an *LC* circuit called an *antenna tuner* or *transmatch* designed to cancel out reactances over a wide range of values. Then we connect the other end of the feed line to a radio transmitter. This type of antenna works as an efficient radiator even in the presence of considerable loss resistance R_L in the conductors, the surrounding earth, and nearby objects, because the radiation resistance R_R is high.

Zepp

A *zeppelin antenna*, also called a *zepp*, comprises a half-wave radiator fed at one end with a quarter-wave section of parallel-wire transmission line, as shown in Fig. 35-2C. The impedance at the feed point is an extremely high, pure resistance. Because of the specific length of the transmission line, the transmitter "sees" a low, pure resistance at the operating frequency. A zeppelin antenna can function well at all harmonics of the design frequency. If we use a transmatch to "tune out" reactance, we can use any convenient length of transmission line.

Because of its non-symmetrical geometry, the zepp antenna allows some EM radiation from the feed line as well as from the antenna. That phenomenon can sometimes present a problem in radio transmitting applications. Amateur radio operators have an expression for it: "RF in the shack." We can minimize problems of this sort by carefully cutting the antenna radiator to a half wavelength at the fundamental frequency, and by using the antenna only at (or extremely near) the fundamental frequency or one of its harmonics.

J Pole

We can orient a zepp antenna vertically, and position the feed line so that it lies in the same line as the radiating element. The resulting antenna, called a *J pole*, radiates equally well in all horizontal directions. The J pole offers a low-cost alternative to metal-tubing vertical antennas at frequencies from approximately 10 MHz up through 300 MHz. In effect, the J-pole is a half-wavelength vertical antenna fed with an *impedance matching section* comprising a quarter-wavelength section of transmission line. The J pole does not require any electrical ground system, a feature that makes it convenient in locations with limited real estate.

Some radio amateurs "hang" long J poles, cut for 3.5 MHz or 1.8 MHz, from kites or helium-filled balloons. Such an antenna offers amazing performance, but it presents a danger if not tethered to prevent it from breaking off and flying away with the kite or balloon. Such an antenna must never be "flown" where it might fall on a utility or power lines. Long, kite- or balloon-supported wire antennas can acquire massive electrostatic charges, even in clear weather—and they *literally* attract lightning. "Flying" such an antenna in unstable weather invites deadly disaster.

Quarter-Wave Verticals

The physical span of a quarter wavelength antenna is related to frequency according to the formula

$$L_{ft} = 246v/f_o$$

where L_{ft} represents a quarter wavelength in feet, f_o represents the frequency in megahertz, and v represents the velocity factor. If we express the length in meters as L_m, then the formula becomes

$$L_m = 75v/f_o$$

For a typical wire conductor, $v = 0.95$ (95 percent); for metal tubing, v can range down to approximately 0.90 (90 percent).

A quarter-wavelength vertical antenna must be operated against a low-loss RF ground if we want reasonable efficiency. The feed-point value of R_R over perfectly conducting ground is approximately 37 ohms, half the radiation resistance of a center-fed half-wave open dipole in free space. This figure represents radiation resistance in the absence of reactance, and provides a reasonable impedance match to most coaxial-cable type transmission lines.

Ground-Mounted Vertical

The simplest vertical antenna comprises a quarter-wavelength radiator mounted at ground level. The radiator is fed at the base with coaxial cable. The cable's center conductor is connected to the base of the radiator, and the cable's shield is connected to ground.

Unless we install an extensive *ground radial* system with a quarter-wave vertical antenna, it will exhibit poor efficiency unless the earth's surface in the vicinity is an excellent electrical conductor (salt water or a salt marsh, for example). In receiving applications, vertically oriented antennas "pick up" more human-made noise than horizontal antennas do. The EM fields from ground-mounted transmitting antennas are more likely to interfere with nearby electronic devices than are the EM fields from antennas installed high above the ground.

Ground Plane

A *ground-plane antenna* is a vertical radiator, usually ¼ wavelength tall, operated against a system of ¼-wavelength conductors called *radials*. The feed point, where the transmission line joins the radiator and the hub of the radial system, is elevated. When we place the feed point at least ¼ wavelength above the earth, we need only three or four radials to obtain low loss resistance for high efficiency. We extend the radials straight out from the feed point at an angle between 0° (horizontal) and 45° below the horizon. Figure 35-3A illustrates a typical ground-plane antenna.

A ground-plane antenna works best when fed with coaxial cable. The feed-point impedance of a ground-plane antenna having a quarter-wavelength radiator is about 37 ohms if the radials are horizontal; the impedance increases as the radials *droop*, reaching about 50 ohms at a *droop angle* of 45°. You've seen ground-plane antennas if you've spent much time around *Citizens Band* (CB)

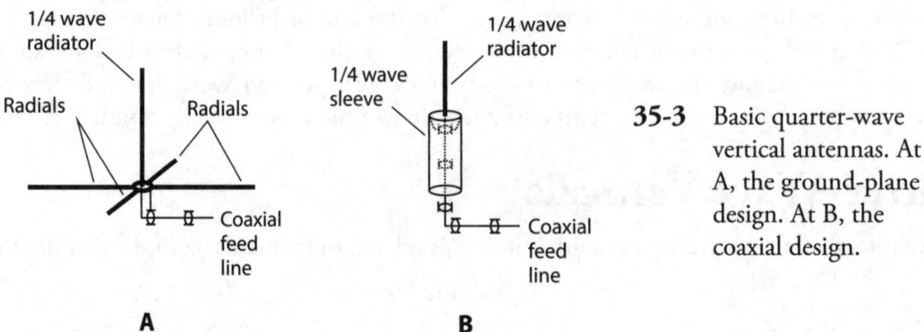

35-3 Basic quarter-wave vertical antennas. At A, the ground-plane design. At B, the coaxial design.

fixed radio installations that operate near 27 MHz, or if you've done much amateur-radio activity in the very-high-frequency (VHF) bands at 50 or 144 MHz.

Coaxial Antenna

We can extend the radials in a ground-plane antenna straight downward, and then merge them into a quarter-wavelength-long cylinder or sleeve concentric with the coaxial-cable transmission line. We run the feed line inside the radial sleeve, feeding the antenna "through the end" as shown in Fig. 35-3B. The feed-point radiation resistance equals approximately 73 ohms, the same as that of a half-wave open dipole. This type of antenna is sometimes called a *coaxial antenna*, a term that arises because of its feed-system geometry, not merely because it's fed with coaxial cable.

Loops

Any receiving or transmitting antenna made up of one or more turns of wire or metal tubing constitutes a *loop antenna*.

Small loop

A *small loop antenna* has a circumference of less than 0.1 wavelength (for each turn) and can function effectively for receiving RF signals. However, because of its small physical size, this type of loop exhibits low radiation resistance, a fact that makes RF transmission inefficient unless the conductors have minimal loss. While small transmitting loops exist, you won't encounter them often.

Even if a small loop has many turns of wire and contains an overall conductor length equal to a large fraction of a wavelength or more, the radiation resistance of a practical antenna is a function of its *actual circumference* in space, not the total length of its conductors. Therefore, for example, if we wind 100 turns of wire around a hoop that's only 0.1 wavelength in circumference, we have a low radiation resistance even though the total length of wire equals 10 full wavelengths!

A small loop exhibits the poorest response to signals coming from along its axis, and the best response to signals arriving in the plane perpendicular to its axis. A variable capacitor can be connected in series or parallel with the loop and adjusted until its capacitance, along with the inherent inductance of the loop, produces resonance at the desired receiving frequency. Figure 35-4 shows an example.

Communications engineers and radio amateurs sometimes use small loops for *radio direction finding* (RDF) at frequencies up to about 20 MHz, and also for reducing interference caused by

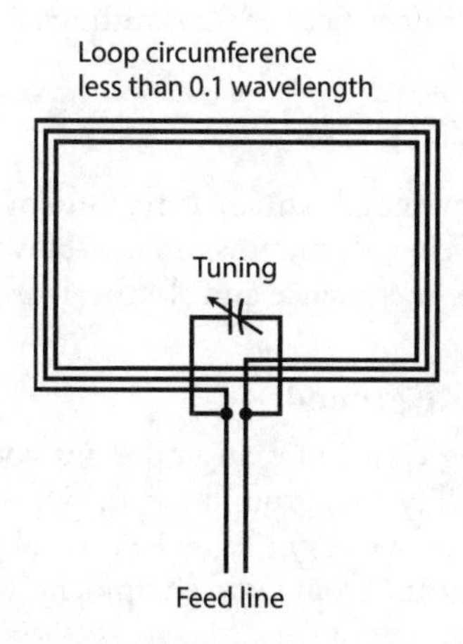

35-4 A small loop antenna with a capacitor for adjusting the resonant frequency.

human-made noise or strong local signals. A small loop exhibits a sharp, deep *null* along its axis (perpendicular to the plane in which the conductor lies). When the loop is oriented so that the null points in the direction of an offending signal or noise source, the unwanted energy can be attenuated by upwards of 20 dB.

Loopstick

For receiving applications at frequencies up to approximately 20 MHz, a *loopstick antenna* can function in place of a small loop. This device consists of a coil wound on a solenoidal (rod-shaped), powdered-iron core. A series or parallel capacitor, in conjunction with the coil, forms a tuned circuit. A loopstick displays directional characteristics similar to those of the small loop antenna shown in Fig. 35-4. The sensitivity is maximum off the sides of the coil (in the plane perpendicular to the coil axis), and a sharp null occurs off the ends (along the coil axis).

Large Loop

A *large loop antenna* usually has a circumference of either a half wavelength or a full wavelength, forms a circle, hexagon, or square in space, and lies entirely in a single plane. It can work well for transmitting or receiving.

A half-wavelength loop presents a high radiation resistance at the feed point. Maximum radiation/response occurs in the plane of the loop, and a shallow, rather broad null exists along the axis. A full-wavelength loop presents a radiation resistance (and zero reactance, forming a purely resistive impedance) of about 100 ohms at the feed point. The maximum radiation/response occurs along the axis, and minimum radiation/response (though not a true null) exists in the plane containing the loop.

The half-wavelength loop exhibits a slight power loss relative to a half-wave open or folded dipole in its *favored directions* (the physical directions in which it offers the best performance). The full-wavelength loop shows a slight gain over a dipole in its favored directions. These properties hold for loops up to several percent larger or smaller than exact half-wavelength or full-wavelength circumferences. Resonance can be obtained by means of a transmatch at the feed point, even if the loop itself does not exhibit resonance at the frequency of interest.

An extremely large loop antenna, measuring several wavelengths in circumference, can be installed horizontally among multiple supports such as communications towers, trees, or wooden poles. We might call this type of antenna a *giant loop*. The gain and directional characteristics of giant loops are hard to predict. If fed with open wire line using a transmatch at the transmitter end of the line, and if placed at least a quarter wavelength above the earth's surface, such an antenna can offer exceptional performance for transmitting and receiving.

Ground Systems

End-fed quarter-wavelength antennas require low-loss RF ground systems to perform efficiently. Center-fed half-wavelength antennas do not. However, good grounding is advisable for any antenna system to minimize interference and electrical hazards.

Electrical versus RF ground

Electrical grounding constitutes an important consideration for personal safety. A good electrical (that is, DC and utility AC) ground can help protect communications equipment from damage if lightning strikes in the vicinity. A good electrical ground also minimizes the risk of *electromagnetic interference* (EMI) to and from radio equipment. In a three-wire electrical utility system, the ground

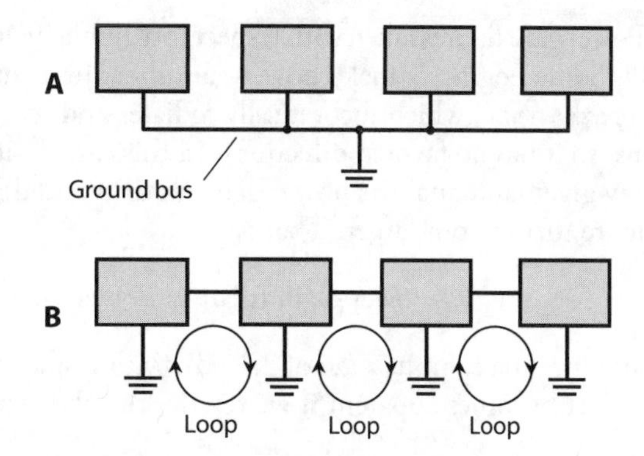

35-5 At A, the correct method for grounding multiple hardware equipment. At B, an incorrect method for grounding multiple equipment, creating RF ground loops.

prong on the plug should never be defeated because such modification can result in dangerous voltages appearing on exposed metal surfaces.

A good RF ground system can help minimize EMI, even if it's not necessary for efficient antenna operation. Figure 35-5 shows a proper RF ground scheme (at A) and an improper one (at B). In a good RF ground system, each device is connected to a common *ground bus*, which, in turn, runs to the earth ground through a single conductor. This conductor should have the smallest possible physical length. A poor ground system contains *ground loops* that can act like loop antennas and increase the risk of EMI.

Radials and the Counterpoise

A surface-mounted vertical antenna should employ as many grounded radial conductors as possible, and they should be as long as possible. The radials can lie on the earth's surface or be buried a few inches underground. In general, as the number of radials of a given length increases, the overall efficiency of any vertical antenna improves if all other factors remain constant. As the radial length increases and all other factors remain constant, vertical-antenna efficiency improves. The radials should all converge toward, and connect directly to, a ground rod at the feed point.

A specialized conductor network called a *counterpoise* can provide an RF ground without a direct earth-ground connection. A grid of wires, screen, or metal sheet is placed above the earth's surface and oriented horizontally to obtain *capacitive coupling* to the earth's conductive mass. A vertical antenna is located at the center of the counterpoise. This arrangement minimizes RF ground loss, although the counterpoise won't provide a good electrical ground unless connected to a *ground rod* driven into the earth, or to the utility system ground. Ideally, a counterpoise should have a radius of at least a quarter wavelength at the lowest anticipated operating frequency.

Gain and Directivity

The *power gain* of a transmitting antenna equals the ratio of the maximum *effective radiated power* (ERP) to the actual RF power applied at the feed point. Power gain is expressed in decibels (dB). It's usually expressed in an antenna's favored direction or directions.

Suppose that the ERP, in watts, for a given antenna equals P_{ERP}, and the applied power, also in watts, equals P. We can calculate the antenna power gain using the formula

$$\text{Power Gain (dB)} = 10 \log_{10} (P_{\text{ERP}} / P)$$

In order to define power gain, we must use a *reference antenna* with a gain that we define as 0 dB in its favored direction(s). A half-wavelength open or folded dipole in free space provides a useful reference antenna. Power gain figures taken with respect to a dipole (in its favored directions) are expressed in units called dBd. Some engineers make power-gain measurements relative to a specialized system known as an *isotropic antenna*, which theoretically radiates and receives equally well in all directions in three dimensions (so it has no favored direction). In this case, units of power gain are called dBi.

For any given antenna, the power gains in dBd and dBi differ by approximately 2.15 dB, with the dBi figure turning out larger. That is,

$$\text{Power Gain (dBi)} = \text{Power Gain (dBd)} + 2.15$$

An isotropic antenna exhibits a *loss* of 2.15 dB with respect to a half-wave dipole in its favored directions. This fact becomes apparent if we rewrite the above formula as

$$\text{Power Gain (dBd)} = \text{Power Gain (dBi)} - 2.15$$

Directivity Plots

We can portray antenna *radiation patterns* (for signal transmission) and response patterns (for reception) using graphical plots, such as those in Fig. 35-6. We assume, in all such plots, that the antenna occupies the center (or *origin*) of a *polar coordinate system*. The greater the radiation or reception capability of the antenna in a certain direction, the farther from the center we plot the corresponding point.

A dipole antenna, oriented horizontally so that its conductor runs in a north-south direction, has a *horizontal plane* (or *H-plane*) pattern similar to Fig. 35-6A. The *elevation plane* (or *E-plane*) pattern depends on the height of the antenna above *effective ground* at the viewing angle. In most locations, the effective ground is an imaginary plane or contoured surface slightly below the actual surface of the earth. With the dipole oriented so that its conductor runs perpendicular to the page, and the antenna ¼ wavelength above effective ground, the E-plane pattern for a half-wave dipole resembles Fig. 35-6B.

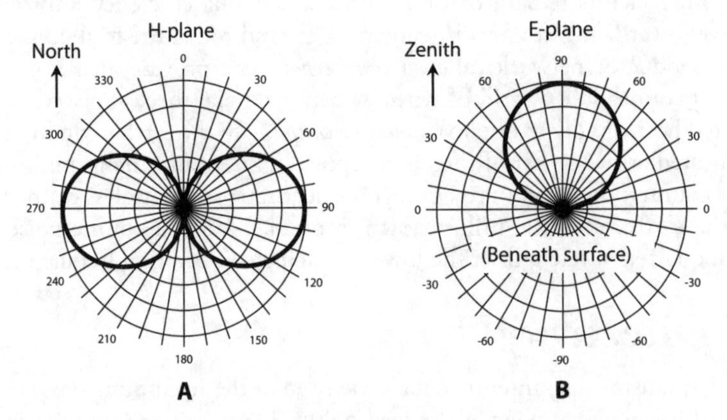

35-6 Directivity plots for a dipole. At A, the H-plane (horizontal plane) plot as viewed from high above the antenna. Coordinate numbers indicate compass-bearing (azimuth) angles in degrees. At B, the E-plane (elevation plane) plot as viewed from a point on the earth's surface far from the antenna. Coordinate numbers indicate elevation angles in degrees above or below the plane of the earth's surface.

Forward Gain

Forward gain is expressed in terms of the ERP in the *main lobe* (favored direction) of a *unidirectional* (one-directional) antenna compared with the ERP from a reference antenna, usually a half-wave dipole, in its favored directions. This gain is calculated and defined in dBd. In general, as the wavelength decreases (the frequency gets higher), we find it easier to obtain high forward gain figures.

Front-to-Back Ratio

The *front-to-back* (f/b) *ratio* of a unidirectional antenna quantifies the concentration of radiation/response *in the center* of the main lobe, relative to the direction *opposite the center* of the main lobe. Figure 35-7 shows a hypothetical directivity plot for a unidirectional antenna pointed north. The outer circle depicts the RF *field strength* in the direction of the center of the main lobe, and represents 0 dB (relative to the main lobe, not a dipole). The next smaller circle represents a field strength 5 dB down (a radiation /response level of −5 dB) with respect to the main lobe. Continuing inward, circles represent 10 dB down (−10 dB), 15 dB down (−15 dB), and 20 dB down (−20 dB). The origin represents 25 dB down (−25 dB) with respect to the main lobe, and also shows the location of the antenna. In this particular example, we can determine the f/b ratio by comparing the signal levels between north (azimuth 0°) and south (azimuth 180°). It appears to be 15 dB in Fig. 35-7.

Front-to-Side Ratio

The *front-to-side* (f/s) *ratio* provides us with another useful expression for the directivity of an antenna system. The specification applies to unidirectional antennas, and also to *bidirectional* antennas that have two favored directions, one opposite the other in space. We express f/s ratios in decibels (dB), just as we do with f/b ratios. We compare the EM field strength *in the favored direction* with the field strength *at right angles* to the favored direction. Figure 35-7 shows an example. In this situation, we can define two separate f/s ratios: one by comparing the signal level between north and east (the right-hand f/s ratio), and the other by comparing the signal level between north and west (the left-hand f/s ratio). In most directional antenna systems, the right-hand and left-hand f/s ratios theoretically equal each other. However, they sometimes differ in practice because of physical imperfections in the antenna structure, and also because of the effects of conducting objects or an irregular earth surface near the antenna. In the situation of Fig. 35-7, both the left-hand and right-hand f/s ratios appear to be roughly 17 dB.

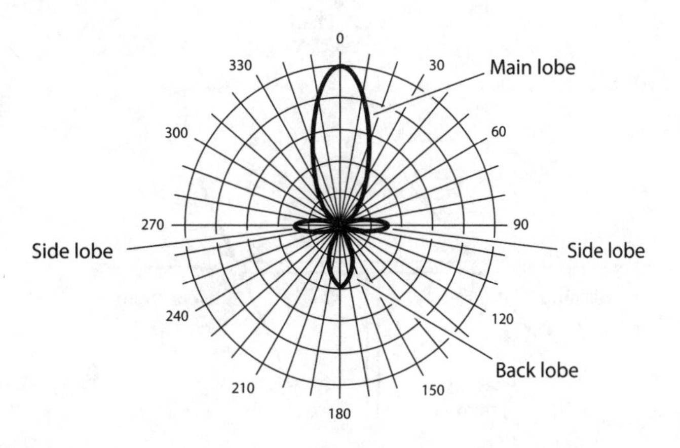

35-7 Directivity plot for a hypothetical antenna in the H (horizontal) plane. We can determine the front-to-back and front-to-side ratios from such a graph. Coordinate numbers indicate compass-bearing angles in degrees.

Phased Arrays

A *phased antenna array* uses two or more *driven elements* (radiators connected directly to the feed line) to produce power gain in some directions at the expense of other directions.

End-Fire Array

A typical *end-fire array* consists of two parallel half-wave open dipoles fed 90° out of phase and spaced ¼ wavelength apart, as shown in Fig. 35-8A. This geometry produces a unidirectional radiation pattern. Alternatively, the two elements can be driven in phase and spaced at a separation of one full wavelength, as shown in Fig. 35-8B, producing a bidirectional radiation pattern. When we design the *phasing system* (also called the *phasing harness*), we must cut the branches of the transmission line to precisely the correct lengths, taking the velocity factor of the line into account.

Longwire

A wire antenna measuring a full wavelength or more, and fed at a high-current point or at one end, constitutes a *longwire antenna.* A longwire antenna offers gain over a half-wave dipole. As we increase the length of the wire, making sure that it runs along a straight line for its entire length, the main lobes get more and more nearly in line with the antenna, and their magnitudes increase. The power gain in the *major lobes* (that is, the strongest lobes, representing the favored directions) in a straight longwire depends on the overall length of the antenna; the longer the wire, the greater

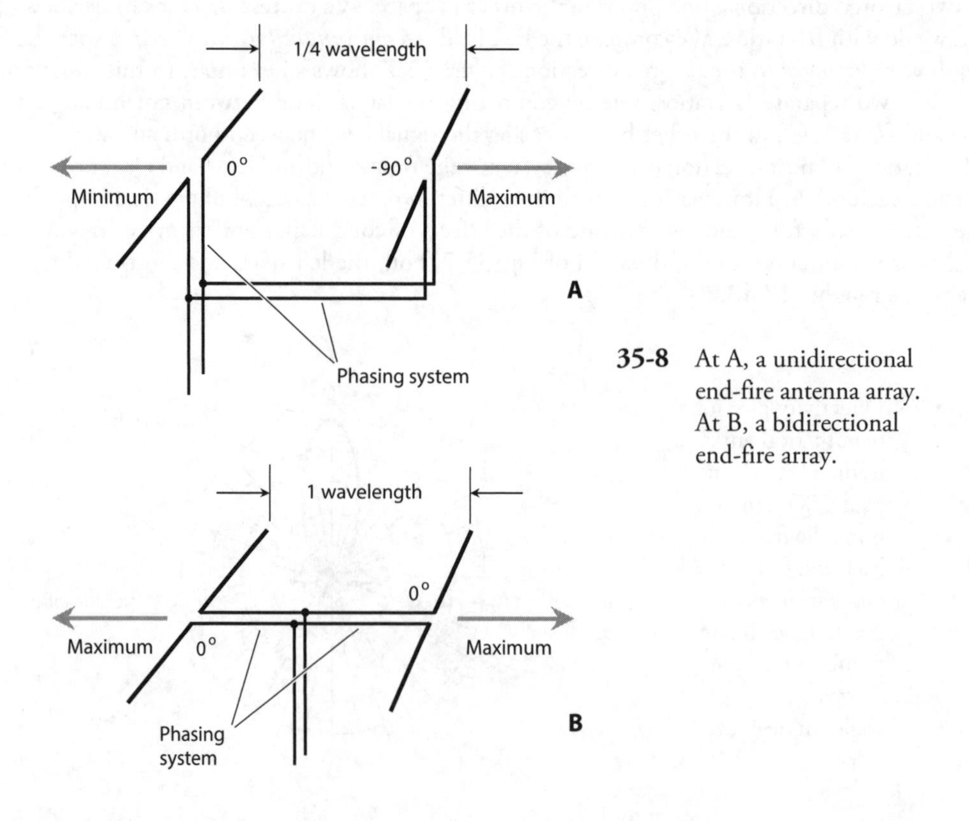

35-8 At A, a unidirectional end-fire antenna array. At B, a bidirectional end-fire array.

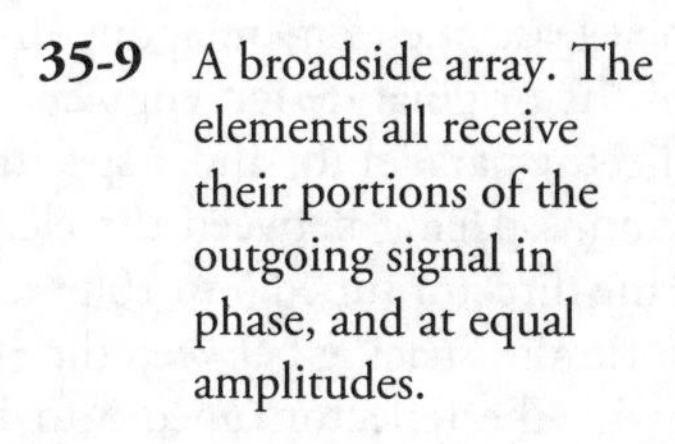
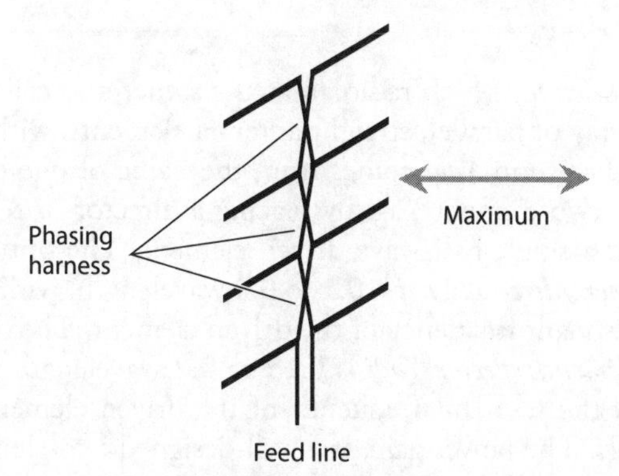

35-9 A broadside array. The elements all receive their portions of the outgoing signal in phase, and at equal amplitudes.

the gain. We can consider a longwire as a phased array because it contains multiple points along its span where the RF current attains maximum values. Each of these *current loops* acts like the center of a half-wave dipole or zepp antenna, so the whole longwire, in effect, constitutes a set of two or more half-wave antennas placed end-to-end, with each section phased in opposition relative to the adjacent section or sections.

Broadside Array

Figure 35-9 shows the geometric arrangement of a *broadside array*. The driven elements can each consist of a single radiator, as shown in this illustration, or they can consist of more complex antennas with directive properties. In the design shown here, all the driven elements are identical, and are spaced at ½-wavelength intervals along the phasing harness, which comprises a parallel-wire transmission line with a half-twist between each element to ensure that all the elements operate in phase coincidence. If we place a flat reflecting screen behind the array of dipoles in Fig. 35-9, we obtain a system known as a *billboard antenna*. The directional properties of any broadside array depend on the number of elements, on whether or not the elements have gain themselves, on the spacing among the elements, and on whether or not a reflecting screen is employed. In general, as we increase the number of elements, the forward gain, the f/b ratio, and the f/s ratios all increase.

Parasitic Arrays

Communications engineers use so-called *parasitic arrays* at frequencies ranging from approximately 5 MHz into the microwave range for obtaining directivity and forward gain. Examples include the *Yagi antenna* and the *quad antenna*. In the context of an antenna array, the term *parasitic* describes the characteristics of certain antenna elements. (It has nothing to do with parasitic oscillation, a phenomenon that can take place in malfunctioning RF power amplifiers.)

Concept

A *parasitic element* is an electrical conductor that forms an important part of an antenna system, but that we don't connect directly to the feed line. Parasitic elements operate by means of *EM coupling* to the driven element or elements. When power gain occurs in the direction of the parasitic element, we call that element a *director*. When power gain occurs in the direction opposite the parasitic element, we call that element a *reflector*. Directors normally measure a few percent shorter than the driven element(s). Reflectors are normally a few percent longer than the driven element(s).

Yagi

The *Yagi antenna*, which radio amateurs sometimes call a *beam antenna*, or simply a *beam*, comprises an array of parallel, straight antenna elements, with at least one element acting in a parasitic capacity. (The term *Yagi* comes from the name of one of the original design engineers.) We can construct a two-element Yagi by placing a director or reflector parallel to, and a specific distance away from, a single half-wave driven element. The optimum spacing between the elements of a *driven-element/director Yagi* is 0.1 to 0.2 wavelength, with the director tuned 5 to 10 percent higher than the resonant frequency of the driven element. The optimum spacing between the elements of a *driven-element/reflector Yagi* is 0.15 to 0.2 wavelength, with the reflector tuned 5 to 10 percent lower than the resonant frequency of the driven element. Either of these designs give us a *two-element Yagi*. The power gain of a well-designed two-element Yagi in its single favored direction is approximately 5 dBd.

A Yagi with one director and one reflector, along with the driven element, forms a *three-element Yagi*. This design scheme increases the gain and f/b ratio as compared with a two-element Yagi. An optimally designed three-element Yagi exhibits approximately 7 dBd gain in its favored direction. Figure 35-10 is a generic example of the relative dimensions of a three-element Yagi. Although this illustration can serve as a crude "drawing-board" engineering blueprint, the optimum dimensions of a three-element Yagi in practice will vary slightly from the figures shown here because of imperfections in real-world hardware, and also because of conducting objects or terrain irregularities near the system.

The gain, f/b ratio, and f/s ratios of a properly designed Yagi antenna all increase as we add elements to the array. We can obtain four-element, five-element, or larger Yagis by placing extra directors in front of a three-element Yagi. When multiple directors exist, each one should be cut slightly shorter than its predecessor. Some commercially manufactured Yagis have upwards of a dozen elements. As you can imagine, engineers must spend a lot of time "tweaking" the dimensions of such antennas to optimize their performance.

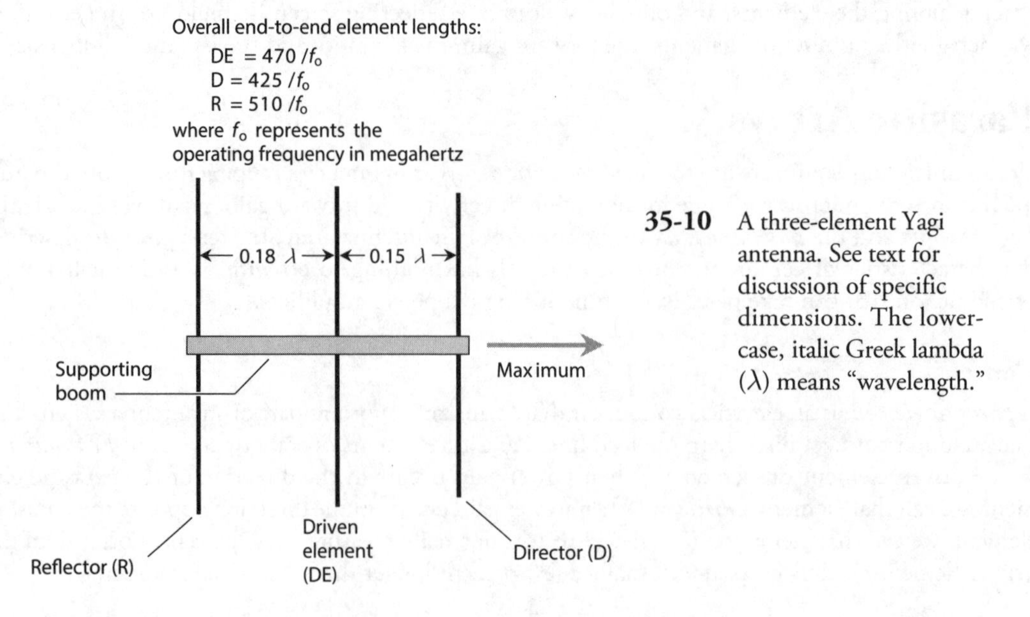

35-10 A three-element Yagi antenna. See text for discussion of specific dimensions. The lowercase, italic Greek lambda (λ) means "wavelength."

Quad

A *quad antenna* operates according to the same principles as the Yagi, except that full-wavelength loops replace the half-wavelength elements.

A *two-element quad* can consist of a driven element and a reflector, or it can have a driven element and a director. A *three-element quad* has one driven element, one director, and one reflector. The director has a perimeter of 0.95 to 0.97 wavelength, the driven element has a perimeter of exactly one wavelength, and the reflector has a perimeter of 1.03 to 1.05 wavelength. These figures represent electrical dimensions (taking the velocity factor of wire or tubing into account), and not free-space dimensions.

Additional directors can be added to the basic three-element quad design to form quads having any desired numbers of elements. The gain increases as the number of elements increases. Each succeeding director is slightly shorter than its predecessor. Long quad antennas are practical at frequencies above 100 MHz. At frequencies below approximately 10 MHz, quad antennas become physically large and unwieldy, although some radio amateurs have constructed quads at frequencies down to 3.5 MHz.

Antennas for Ultra-High and Microwave Frequencies

At ultra high frequencies (UHF) and microwave frequencies, high-gain, many-element antennas have reasonable physical dimensions and mass because the wavelengths are short.

Waveguides

A *waveguide* is a specialized RF transmission line comprising a hollow metal "pipe" or "duct" with a circular or rectangular cross section. The EM field travels down the waveguide quite efficiently, provided that the wavelength is short enough (or the cross-sectional dimensions of the pipe are large enough). In order to efficiently propagate an EM field, a *rectangular waveguide* must have height and width that both measure at least 0.5 wavelength, and preferably more than 0.7 wavelength. A *circular waveguide* should measure at least 0.6 wavelength in diameter, and preferably 0.7 wavelength or more.

The characteristic impedance (Z_o) of a waveguide varies with the frequency. In this sense, a waveguide behaves differently than a coaxial or parallel-wire RF transmission line, whose Z_o value remains independent of the frequency over the entire range of wavelengths for which the line is designed.

A properly installed and maintained waveguide acts as an exceptional RF transmission line because dry air has essentially zero loss, even at UHF and microwave frequencies. However, if we expect a waveguide to work properly, we must keep its interior free from dirt, dust, insects, spider webs, and condensation. Even a small obstruction can seriously degrade the performance and cause significant power loss.

The main limitation of a waveguide, from a practical standpoint, is its relative inflexibility, both figuratively and literally. We can't run a waveguide from one point to another in a haphazard fashion, as we can do with coaxial cable. Bends or turns in a waveguide present a particular problem, because we must make them gradually. We can't simply "turn a corner" with a waveguide! Another limitation involves the usable frequency range. Waveguides are impractical for use at frequencies below approximately 300 MHz because the required cross-sectional dimensions become prohibitively large.

Horn

The *horn antenna* has a characteristic shape like a squared-off trumpet horn. It provides a unidirectional radiation and response pattern, with the favored direction coincident with the opening of the horn. The feed line is a waveguide that joins the antenna at the narrowest point (throat) of the horn. Horns are sometimes used all by themselves, but they can also feed large *dish antennas* at UHF and microwave frequencies. The horn design optimizes the f/s ratio by minimizing extraneous radiation and response that occurs if a dipole is used as the driven element for the dish.

Dish

Most people are familiar with dish antennas because of their widespread use in consumer satellite TV and Internet services. Although the geometry looks simple to the casual observer, a dish antenna must be precisely shaped and aligned if we want it to function as intended. The most efficient dish, especially at the shortest wavelengths, comprises a *paraboloidal reflector*, so named because it's a section of a *paraboloid* (the three-dimensional figure that we get when we rotate a *parabola* around its axis). However, a *spherical reflector*, having the shape of a section of a sphere, can also work in most dish-antenna system designs.

A dish-antenna feed system consists of a coaxial line or waveguide from the receiver and/or transmitter along with a horn or helical driven element at the focal point of the reflector. Figure 35-11A shows an example of *conventional dish feed.* Figure 35-11B shows an alternative scheme known as *Cassegrain dish feed.* The term *Cassegrain* comes from the resemblance of this antenna design to that of a *Schmidt-Cassegrain reflector telescope.* All dish antennas work well at UHF and microwave frequencies for transmitting and receiving. For hobby use, they're physically impractical at frequencies much below 300 MHz.

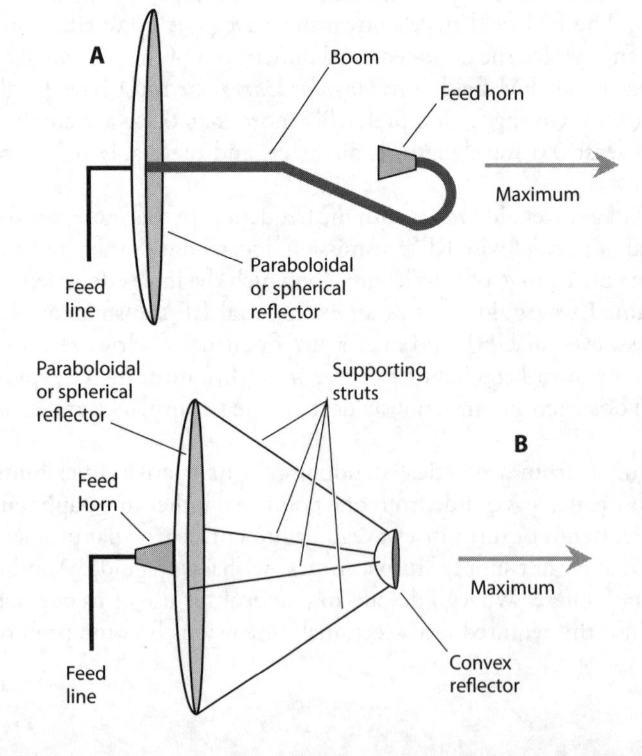

35-11 Dish antennas with conventional feed (A) and Cassegrain feed (B).

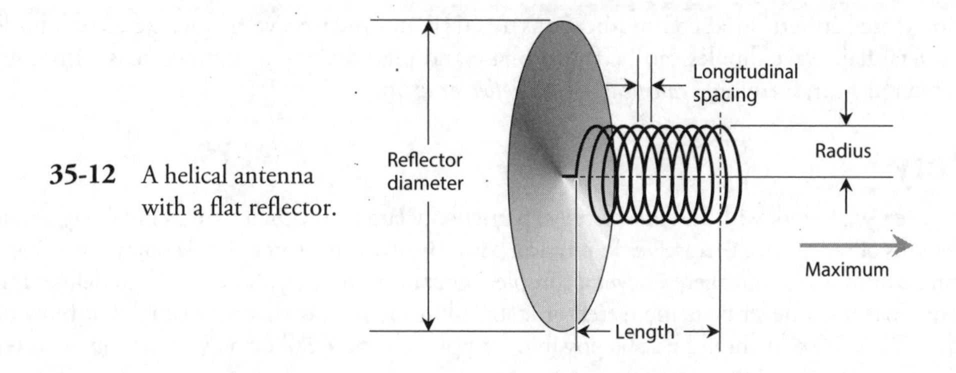

35-12 A helical antenna with a flat reflector.

As we increase the diameter of the dish reflector in wavelengths, the gain, the f/b ratio, and the f/s ratios all increase, and the width of the main lobe decreases, making the antenna more sharply unidirectional. A dish antenna must measure at least several wavelengths in diameter for proper operation. The reflecting element can consist of sheet metal, a screen, or a wire mesh. If a screen or mesh is used, the spacing between the wires must be a small fraction of a wavelength. At microwave frequencies, large dish antennas can have forward gain figures that exceed 35 dBd.

Helical

A *helical antenna* is a high-gain, unidirectional antenna that transmits and receives EM waves with circular polarization. Figure 35-12 illustrates the construction of a typical helical antenna. The reflector diameter should be at least 0.8 wavelength at the lowest operating frequency. The radius of the helix should be approximately 0.17 wavelength at the center of the intended operating frequency range. The longitudinal spacing between helix turns should be approximately 0.25 wavelength in the center of the operating frequency range. The entire helix should measure at least a full wavelength from end to end at the lowest operating frequency. When properly designed, this type of antenna can provide about 15 dBd forward gain. Helical antennas, like dish antennas, are used primarily at UHF and microwave frequencies.

Corner Reflector

Figure 35-13 illustrates a *corner reflector* with a half-wave open-dipole driven element. This design provides some power gain over a half-wave dipole. The reflector is made of wire mesh, screen, or sheet metal. The *flare angle* of the reflecting element equals approximately 90°. Corner reflectors

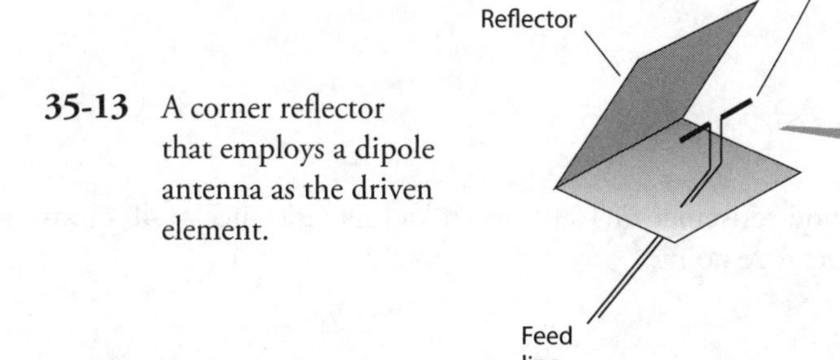

35-13 A corner reflector that employs a dipole antenna as the driven element.

are widely used in terrestrial communications at UHF and microwave frequencies. For additional gain, several half-wave dipoles can be fed in phase and placed along a common axis with a single, elongated reflector, forming a *collinear corner-reflector array.*

Safety

All engineers who work with antenna systems, particularly large arrays or antennas involving the use of long lengths of wire, place themselves in physical peril. By observing some simple safety guidelines, the risk can be minimized, but there's never a complete guarantee of safety. Some basic guidelines follow.

Antennas must never be maneuvered or installed in such a way that they can fall or blow down on power lines. Also, it should not be possible for power lines to fall or blow down on an antenna, even in the event of a violent storm.

Wireless equipment having outdoor antennas should not be used during thundershowers, or when lightning exists anywhere in the vicinity. Antenna construction and maintenance should never be undertaken when lightning is visible or thunder can be heard, even if a storm appears far away. Ideally, antennas should be disconnected from electronic equipment, and connected to a substantial earth ground, at all times when the equipment is not in use.

Tower and antenna climbing constitutes a job for professionals only. No inexperienced person should ever attempt to climb any antenna support structure.

Indoor transmitting antennas can expose operating personnel to EM field energy. The extent of the hazard, if any, posed by such exposure has not been firmly established. However, sufficient concern exists among some experts to warrant checking the latest publications on the topic.

Caution!

For complete information on antenna safety matters, you should consult a professional antenna engineer or a comprehensive reference devoted to antenna design and construction. You should also consult and heed all electrical and building codes for your city, state, or province.

Quiz

Refer to the text in this chapter if necessary. A good score is at least 18 correct. Answers are in the back of the book.

1. An antenna has a radiation resistance of 900 ohms and a loss resistance of 100 ohms. What's its efficiency to the nearest percentage point?
   (a) 66 percent
   (b) 75 percent
   (c) 90 percent
   (d) 93 percent

2. An antenna has a radiation resistance of 10 ohms and a loss resistance of 40 ohms. What's its efficiency to the nearest percentage point?
   (a) 20 percent
   (b) 25 percent

(c) 33 percent
(d) 67 percent

3. A circular loop antenna that measures exactly one wavelength in circumference, and that lies entirely in a single plane, radiates best
   (a) in all possible directions; it's isotropic.
   (b) in the plane perpendicular to the loop axis.
   (c) along the loop axis in both directions.
   (d) along the loop axis in one direction.

4. How long is a wavelength in free space at 7.05 MHz? Round your answer to three significant figures.
   (a) 279 ft
   (b) 140 ft
   (c) 69.8 ft
   (d) 34.9 ft

5. How long should you cut *either side* of a straight, center-fed, half-wavelength open dipole wire antenna for operation at a fundamental frequency of 7.05 MHz, taking into account the typical velocity factor for thin wire? Round your answer to three significant figures.
   (a) 33.1 ft
   (b) 66.2 ft
   (c) 132 ft
   (d) 265 ft

6. The feed-point impedance of a half-wave, center-fed, two-wire folded dipole is a pure resistance of
   (a) eight times that of a half-wave open dipole made from a single wire.
   (b) four times that of a half-wave open dipole made from a single wire.
   (c) twice that of a half-wave open dipole made from a single wire.
   (d) the same as that of a half-wave open dipole made from a single wire.

7. Even if an antenna system has a lot of loss resistance, it can work efficiently if you design it to have an extremely
   (a) low reactance.
   (b) high reactance.
   (c) low radiation resistance.
   (d) high radiation resistance.

8. How tall is a ground-mounted, quarter-wavelength vertical antenna designed for 10 MHz? Assume that the tubing used for the radiating element has a velocity factor of 92 percent. Express your answer to two significant figures.
   (a) 6.9 ft
   (b) 14 ft
   (c) 23 ft
   (d) 45 ft

9. You want to transmit radio signals with a base-fed, inductively loaded, short vertical antenna over perfectly conducting ground. The inductor comprises a low-loss coil. You feed the antenna with low-loss 50-ohm coaxial cable. Your radio is designed to work into a nonreactive, 50-ohm load. You'll get the best results if you
   (a) make the antenna less than ⅛ wavelength tall.
   (b) use thick copper tubing as the radiating element.
   (c) install a wide-range antenna tuner at the feed point.
   (d) feed the antenna directly with the cable.

10. In terms of physical size, waveguides are practical at all frequencies
   (a) below 3 MHz.
   (b) below 300 MHz.
   (c) above 3 MHz.
   (d) above 300 MHz.

11. The feed-point impedance of a Zepp antenna is a pure,
   (a) high resistance.
   (b) low resistance.
   (c) high capacitive reactance.
   (d) high inductive reactance.

12. In a three-element Yagi, the director is half-wave resonant at
   (a) the same frequency as that of the driven element.
   (b) twice that of the driven element.
   (c) a slightly lower frequency than that of the driven element.
   (d) a slightly higher frequency than that of the driven element.

13. To get reasonable efficiency with a half-wavelength vertical antenna mounted over ground with fair conductivity (using a wide-range antenna tuner at the antenna's base to match the radiator to the feed line), you should
   (a) make the radiating element a thick, solid metal rod.
   (b) use a ground rod and a few radials (you won't need many).
   (c) bury at least 100 radials measuring a half wavelength or more.
   (d) salt the earth near the antenna to improve the ground conductivity.

14. Imagine a ground-plane antenna measuring ¼ wavelength tall, with four radials each ¼ wavelength long and drooping at 45°. You elevate the feed point at least ¼ wavelength above the surface and let the radials double as guy wires. This antenna has a purely resistive feed-point impedance of approximately
   (a) 37 ohms.
   (b) 50 ohms.
   (c) 73 ohms.
   (d) 100 ohms.

15. When the total loss resistance in an antenna equals exactly half the feed-point radiation resistance, the antenna's efficiency is
    (a) 33 percent.
    (b) 50 percent.
    (c) 67 percent.
    (d) 75 percent.

16. In a seven-element Yagi, the longest element is the
    (a) driven element.
    (b) first director.
    (c) last director.
    (d) reflector.

17. A small loopstick receiving antenna is most sensitive to signals arriving from
    (a) all possible directions; it's isotropic.
    (b) directions in the plane perpendicular to the loop axis.
    (c) along the loop axis in both directions.
    (d) along the loop axis in one direction.

18. Which of the following transmission-line types is ideally suited for use with a half-wave folded dipole?
    (a) Parallel-wire line
    (b) Waveguide
    (c) Coaxial cable
    (d) Any of the above; it doesn't matter

19. Which of the following antennas works best for satellite TV?
    (a) Folded dipole
    (b) Ground plane
    (c) Loopstick
    (d) Dish

20. Which of the following antenna types can provide significant power gain over a half-wave dipole?
    (a) Yagi
    (b) Dish
    (c) Helical
    (d) All of the above

Test: Part 4

DO NOT REFER TO THE TEXT WHEN TAKING THIS TEST. A GOOD SCORE IS AT LEAST 37 CORRECT. Answers are in the back of the book. It's best to have a friend check your score the first time, so you won't memorize the answers if you want to take the test again.

1. In most microcontrollers, Random Access Memory (RAM) is used to store
   (a) programs only.
   (b) programs and variables.
   (c) data only.
   (d) programs and constant values.
   (e) nonvolatile data.
2. Figure Test 4-1 illustrates a pressure sensor. What's the material marked X?
   (a) Quartz crystal
   (b) Dielectric foam
   (c) Conductive foam
   (d) Solid dielectric
   (e) Semiconducting foam

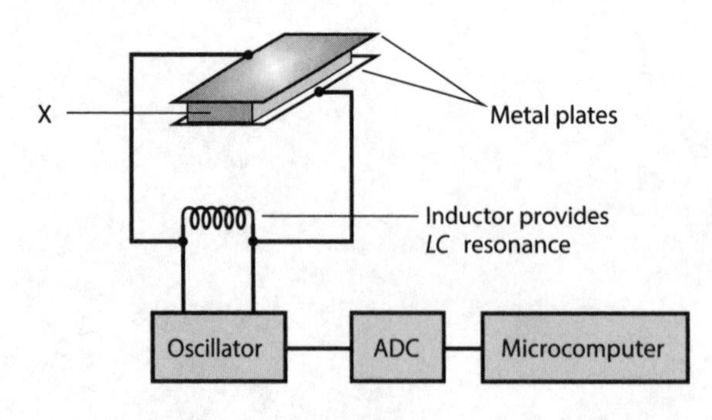

Test 4-1 Illustration for Part 4 Test Question 2.

3. You'll notice considerable latency in a geostationary-satellite communications link because of
   (a) the polar orbit that the satellite follows.
   (b) the satellite's high altitude.
   (c) complicated signal processing in the satellite repeater.
   (d) handoffs among multiple satellites.
   (e) All of the above

4. A stepper motor shaft rotates
   (a) continuously at a speed that you can adjust.
   (b) continuously at a speed that you can't adjust.
   (c) only when it receives signals at certain frequencies.
   (d) in increments, rather than continuously.
   (e) back and forth, reversing at a time interval that you can adjust.

5. Figure Test 4-2 is a block diagram of an audio
   (a) waveform analyzer.
   (b) frequency synthesizer.
   (c) frequency separator.
   (d) graphic equalizer.
   (e) timbre adjuster.

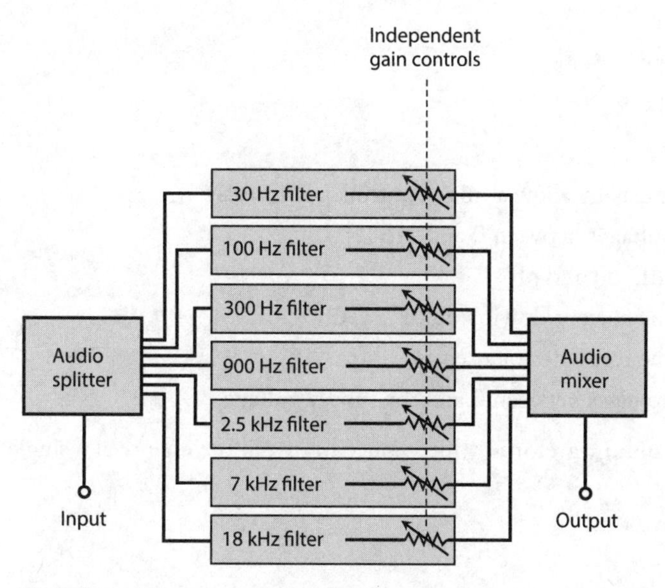

Test 4-2 Illustration for Part 4 Test Question 5.

6. What's the technical term for a device that determines the distance to an object by transmitting sound pulses toward it and measuring the echo time?
   (a) Acoustar
   (b) Sonar
   (c) Radar
   (d) Sonic ranging
   (e) Echo ranging
7. Which of the following statements best describes the role of a CPU?
   (a) Fetch and execute program instructions
   (b) Regulate the voltage of the microcontroller
   (c) Store program code
   (d) Store variables
   (e) Convert analog readings into digital values
8. Which of the following statements about the Arduino is false?
   (a) You can only program an Arduino using a Windows computer.
   (b) A USB socket can supply power to an Arduino.
   (c) The Arduino needs a computer to program it.
   (d) The Arduino Uno can accept plug-in "shields" that fit onto its GPIO pins.
   (e) The Arduino Uno has a 16-MHz clock.
9. Which of the following is considered a soft recording medium for audio use?
   (a) Compact disk
   (b) Flash drive
   (c) Analog tape
   (d) Digital tape
   (e) Vinyl disk
10. Analog inputs usually allow a microcontroller to directly
    (a) measure voltages between 0 and 10V.
    (b) turn an LED on and off.
    (c) measure the current flowing out of a GPIO pin.
    (d) measure the resistance of a sensor.
    (e) measure voltages between 0 and the supply voltage.
11. The simplest audio waveform, which concentrates all the energy at a single frequency, is a
    (a) square wave.
    (b) ramp.

(c) sinusoid.
(d) triangular wave.
(e) harmonic.

12. In a graded-index optical fiber,
   (a) the refractive index is lowest along the central axis, and abruptly increases at a certain distance from that axis.
   (b) the refractive index is highest along the central axis, and abruptly decreases at a certain distance from that axis.
   (c) the refractive index is lowest along the central axis, and steadily increases as you move outward.
   (d) the refractive index is highest along the central axis, and steadily decreases as you move outward.
   (e) the refractive index is constant throughout.

13. Figure Test 4-3 is a functional diagram of
   (a) an ordinary infrared-emitting diode (IRED).
   (b) an ordinary light-emitting diode (LED).
   (c) a vertical-cavity surface-emitting laser (VCSEL).
   (d) a photodiode.
   (e) a quantum diode.

14. Every Arduino sketch must contain
   (a) at least 10 lines of code.
   (b) only the function "setup."
   (c) only the function "loop."
   (d) both the functions "setup" and "loop."
   (e) global variables.

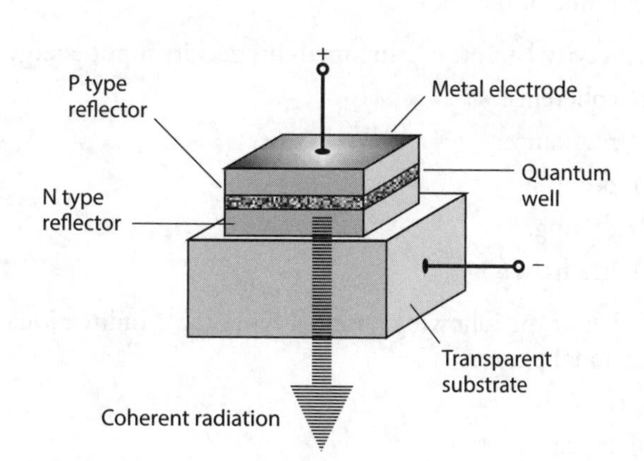

Test 4-3 Illustration for Part 4 Test Question 13.

15. Visible light from a source that radiates at a single wavelength, and in which all of the wavefronts line up with each other, is called
    (a) coherent.
    (b) phase-coincidental.
    (c) sinusoidal.
    (d) polychromatic.
    (e) broadband.

16. A half-wavelength vertical antenna, fed at the base and mounted on the surface, radiates efficiently even if the ground conductivity is not very good. That's because the feed-point impedance is
    (a) a pure capacitive reactance.
    (b) a pure inductive reactance.
    (c) a pure and high resistance.
    (d) a pure and low resistance.
    (e) zero.

17. An eight-bit analog input will provide a digital numeric value between
    (a) 1 and 256.
    (b) 0 and 1023.
    (c) 1 and 1024.
    (d) 0 and 100.
    (e) 0 and 255.

18. In an Arduino sketch, the word "const"
    (a) must be placed before every variable declaration.
    (b) will reduce the amount of flash storage needed for a program if used with any variable whose value never changes.
    (c) indicates that a variable is consistent.
    (d) should be used only in a function definition.
    (e) None of the above

19. In a cavity laser, the lasing medium gets its input energy by means of a process called
    (a) coherence.
    (b) resonance.
    (c) pumping.
    (d) driving.
    (e) irradiation.

20. Which of the following antenna types has a bidirectional radiation pattern, as opposed to a unidirectional pattern?
    (a) Dish
    (b) Horn
    (c) Helical

(d) Broadside array
(e) Corner reflector

21. A visible-light laser has a vivid, intense hue ("color") because
(a) the beam is extremely narrow.
(b) the light waves come out in many different phases.
(c) the output energy is spread over a wide range of wavelengths.
(d) the output energy is concentrated at a single wavelength.
(e) its beam contains a tremendous amount of energy.

22. Which of the following statements about PWM is false?
(a) PWM stands for Pulse Width Modulation.
(b) PWM control of an LED results in the LED turning on and off.
(c) PWM outputs are true analog outputs.
(d) With a PWM duty cycle of 50%, the PWM output will be on half the time.
(e) The frequency of PWM pulses is normally configurable.

23. What's the technical term for the device shown in Fig. Test4-4?
(a) Booster
(b) Repeater
(c) Remodulator
(d) Reamplifier
(e) Extender

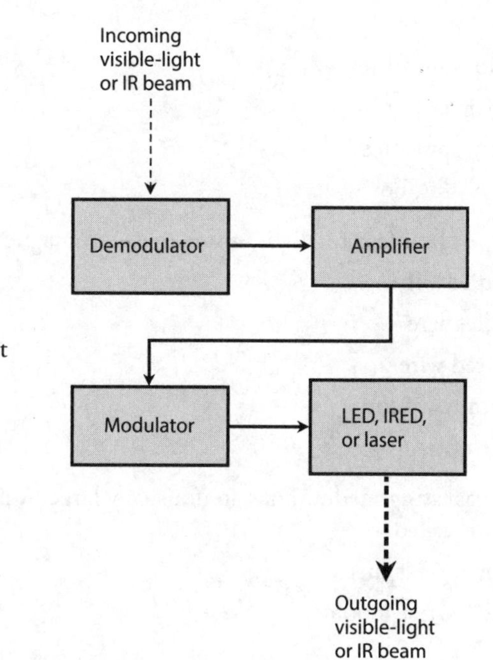

Test 4-4 Illustration for Part 4 Test Question 23.

24. The lasing medium in a ruby laser comprises
    (a) yttrium and garnet.
    (b) aluminum oxide and chromium.
    (c) gallium and arsenic.
    (d) pure elemental germanium.
    (e) indium dioxide.

25. Which of the following statements about "setup" and "loop" is false?
    (a) "Setup" will run when the Arduino is powered up.
    (b) "Loop" will run repeatedly when the Arduino is powered up.
    (c) Pressing the Reset button will result in "setup" being run.
    (d) After the "loop" function has run, the Arduino checks for serial programming updates.
    (e) "Setup" can contain only one line of code.

26. Which of the following antennas is a parasitic array?
    (a) Helical
    (b) Broadside
    (c) End-fire
    (d) Dish
    (e) Quad

27. Which of the following devices might cause electromagnetic interference (EMI) to a hi-fi sound system?
    (a) Computer
    (b) Radio transmitter
    (c) Electric power lines
    (d) Home appliances
    (e) Any of the above

28. What type of cable is totally immune to electromagnetic interference (EMI)?
    (a) Bundled wire
    (b) Parallel wire
    (c) Twisted wire
    (d) Helically wound
    (e) Fiber-optic

29. When a gas lasing medium has an unusually large number of electrons in high-energy shells, the condition is called
    (a) an energy surplus.
    (b) excess carrier mobility.

(c) an energy imbalance.
(d) a population inversion.
(e) negative charge surplus.

30. What's the maximum positive number that can be stored in a variable of type "int" in Arduino C?
(a) 255
(b) 127
(c) 2,147,483,647
(d) 65,535
(e) 32,767

31. In a hi-fi stereo sound system, you might find any, or all, of the following components except one. Which one?
(a) Amplifiers
(b) Loudspeakers
(c) AM/FM tuner
(d) Compact-disk (CD) player
(e) AM/FM transmitter

32. What type of antenna transmits and receives radio waves with circular polarization?
(a) Helical
(b) Broadside
(c) End-fire
(d) Dish
(e) Quad

33. At the feed point, a single-turn, circular-loop antenna a half wavelength in circumference exhibits a
(a) high inductive reactance.
(b) high capacitive reactance.
(c) low radiation resistance.
(d) high radiation resistance.
(e) short circuit.

34. In a hi-fi stereo sound system, which of the following controls allows you to adjust the relative loudness of the left channel versus the right channel?
(a) Tone
(b) Node
(c) Ratio
(d) Harmonic
(e) Balance

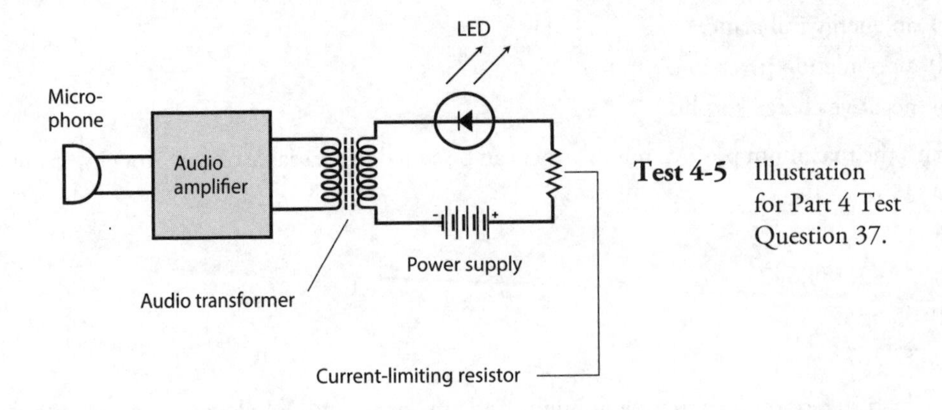

Test 4-5 Illustration for Part 4 Test Question 37.

35. An external pull-up resistor on a digital input will
    (a) stop the switch contacts from "bouncing."
    (b) allow a switch to be connected using a longer lead than would otherwise be possible.
    (c) keep it in a low logic state.
    (d) allow it to function as an output.
    (e) allow it to function as an analog input.

36. The waveform of a note at Middle C from a clarinet differs from the waveform of a note at Middle C from a trumpet. That's why they don't sound alike. It reflects a difference in
    (a) tone.
    (b) pitch.
    (c) timbre.
    (d) wavelength.
    (e) propagation speed.

37. Figure Test 4-5 is a block diagram of a
    (a) cellular repeater.
    (b) ham radio transmitter.
    (c) modulated-light transmitter.
    (d) device that interferes with cell phones.
    (e) circuit for testing LEDs and IREDs.

38. Piezoelectric transducers are well suited for use in
    (a) ultrasonic intrusion detectors.
    (b) hi-fi stereo equipment.
    (c) radio antenna systems.
    (d) navigation systems.
    (e) pressure sensors.

39. The Arduino Uno clock frequency is
    (a) 4 MHz.
    (b) 8 MHz.
    (c) 16 MHz.
    (d) 20 MHz.
    (e) 80 MHz.

40. If you design a modulated-light communications system for line-of-sight use in clear air at sea level, keep in mind that the most absorption (per kilometer) occurs for the color
    (a) red.
    (b) orange.
    (c) yellow.
    (d) green.
    (e) blue.

41. At a fixed location, the distance to the radar horizon depends on the
    (a) height of the antenna.
    (b) contour of the surrounding terrain.
    (c) transmitter output power.
    (d) weather conditions in the vicinity.
    (e) All of the above

42. What purpose can the small loop shown in Fig. Test 4-6 effectively serve?
    (a) Radio direction finding (RDF) below 20 MHz
    (b) Transmitting below 100 kHz
    (c) An isotropic receiving antenna at any frequency
    (d) The tuned circuit in a microwave oscillator
    (e) Transmitting or receiving at any frequency

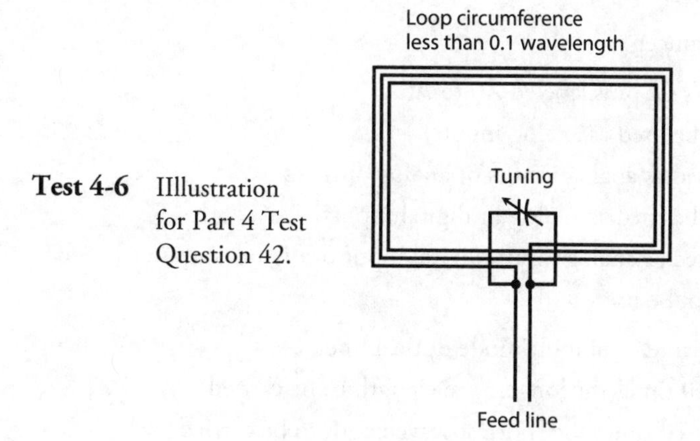

Test 4-6 Illlustration for Part 4 Test Question 42.

43. With respect to hi-fi sound systems, what's the technical term for the ratio of the maximum power output to the minimum power output that the system can produce while maintaining acceptable performance with low distortion?

    (a) Dynamic range
    (b) Loudness ratio
    (c) Power range
    (d) Distortion coefficient
    (e) Audio output ratio

44. A microcontroller's digital outputs can be directly connected to

    (a) high-power loads such as motors.
    (b) switches and sensors.
    (c) computer monitors.
    (d) low-current loads such as LEDs.
    (e) USB peripherals such as printers.

45. A half-wavelength straight wire fed at one end with a quarter-wave section of parallel-wire transmission line is

    (a) a zepp antenna.
    (b) an open dipole antenna.
    (c) a longwire antenna.
    (d) an isotropic antenna.
    (e) a quarter-dipole antenna.

46. Which of the following devices comprises a coil-and-magnet assembly that converts varying electric currents into sound waves?

    (a) Acoustic converter
    (b) Acoustic repeater
    (c) RF transducer
    (d) Loudspeaker
    (e) DC transducer

47. The Arduino's six pins labeled A0 to A5

    (a) can only be used as analog inputs.
    (b) can be used as analog inputs or analog outputs.
    (c) can only be used as analog or digital inputs.
    (d) can be used as analog input, digital input or digital output.
    (e) should not be used.

48. What's the diameter of multimode optical fiber?

    (a) At least 10 times the longest wavelength to be carried
    (b) Less than 10 times the shortest wavelength to be carried

(c) At least 10 nanometers (nm)
(d) Less than 10 nm
(e) It doesn't matter

49. Which of the following systems operates at the shortest wavelength?
(a) Radar
(b) Global Positioning System (GPS)
(c) Photoelectric proximity sensor
(d) Loran
(e) FM broadcast station

50. A quantum number expresses
(a) the amount of energy in a light beam.
(b) the wavelength of a photon.
(c) the total number of electrons in an atom.
(d) an electron's energy level in an atom.
(e) the extent to which energy waves are coherent.

Final Exam

DO NOT REFER TO THE TEXT WHEN TAKING THIS TEST. A GOOD SCORE IS AT LEAST 75 CORRECT. Answers are in the back of the book. It's best to have a friend check your score the first time, so you won't memorize the answers if you want to take the test again.

1. Complete the following statement with a single word that makes it true. "When you connect multiple inductors and/or capacitors in parallel, their ________ values all add up to give you the net ________."
   (a) reactance
   (b) impedance
   (c) resistance
   (d) susceptance
   (e) conductance

2. The atomic weight of an atom approximately equals the number of
   (a) neutrons in it.
   (b) protons in it.
   (c) neutrons plus the number of protons in it.
   (d) electrons in it.
   (e) neutrons plus the number of electrons in it.

3. What's the technical term for the length of time between a specific point in an AC cycle and the same point in the next cycle?
   (a) Frequency
   (b) Cycle time
   (c) Period
   (d) Duration
   (e) Wavelength

4. Figure Exam-1 shows a device that detects the presence of electrical
   (a) current.
   (b) resistance.
   (c) power.
   (d) energy.
   (e) charge.

5. You should *never* expect to see an autotransformer used
   (a) as a loopstick antenna.
   (b) for RF impedance matching.
   (c) when isolation between windings is important.
   (d) to step an RF voltage down by a small factor.
   (e) to step an RF voltage up by a small factor.

6. When you express a point in the RX_L quarter-plane as a vector, you give that point a unique
   (a) magnitude and direction.
   (b) combination of resistance and inductance.
   (c) combination of resistance and frequency.
   (d) pure resistance.
   (e) pure reactance.

7. You have a string of holiday ornament bulbs. They're all connected in series. One of the bulbs burns out. What happens?
   (a) The current in every other bulb increases.
   (b) The voltage across every other bulb increases.
   (c) The power consumed by every other bulb increases.
   (d) The current in, voltage across, and power consumed by every other bulb all increase.
   (e) All the other bulbs go out, leaving the whole string dark.

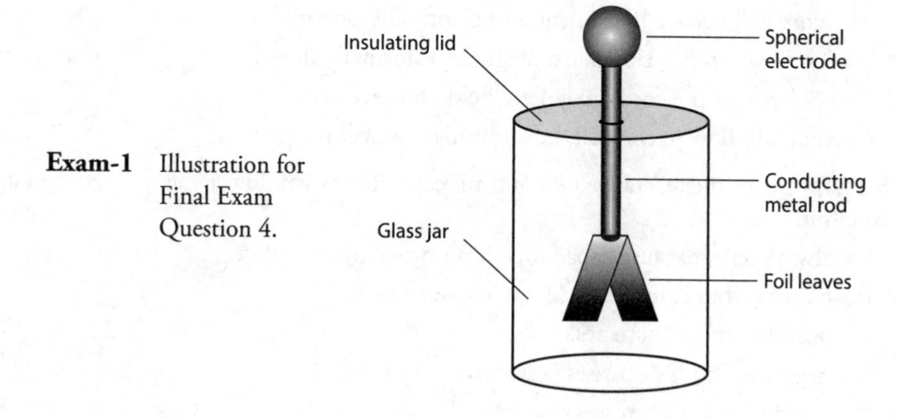

Exam-1 Illustration for Final Exam Question 4.

8. A complex number
   (a) is the same thing as an imaginary number.
   (b) comprises a real number plus or minus an imaginary number.
   (c) is a one-dimensional quantity.
   (d) quantifies the characteristic impedance of a transmission line.
   (e) can't quantify anything in the real world.

9. Power that exists only within the reactive part of an AC circuit is sometimes called
   (a) theoretical power.
   (b) apparent power.
   (c) true power.
   (d) false power.
   (e) imaginary power.

10. Roughly how much current, passing through your heart, does it take to kill you?
    (a) 0.1 mA
    (b) 1 mA
    (c) 10 mA
    (d) 100 mA
    (e) Any of the above

11. An inductor works by storing and releasing energy in the form of
    (a) an electric field.
    (b) a magnetic field.
    (c) electric charge.
    (d) pure reactance.
    (e) complex impedance.

12. What will happen if you connect a battery to the "current source" terminals of the device shown in Figure Exam-2?
    (a) The battery voltage will increase.
    (b) The core will behave like a magnet (it normally doesn't).
    (c) The core will stop acting like a magnet (it normally does).
    (d) The polarity of the core's magnetic field will reverse.
    (e) Current will flow in the coil, but nothing else will happen.

13. What happens to the reactance of a 100-pF capacitor as you lower the frequency of an AC signal through it?
    (a) It's always zero, because capacitors don't have any reactance.
    (b) It's negative, but it increases (gets closer to zero).
    (c) It's positive, and it increases.
    (d) It's negative, and it decreases (gets farther from zero).
    (e) It's positive, but it decreases.

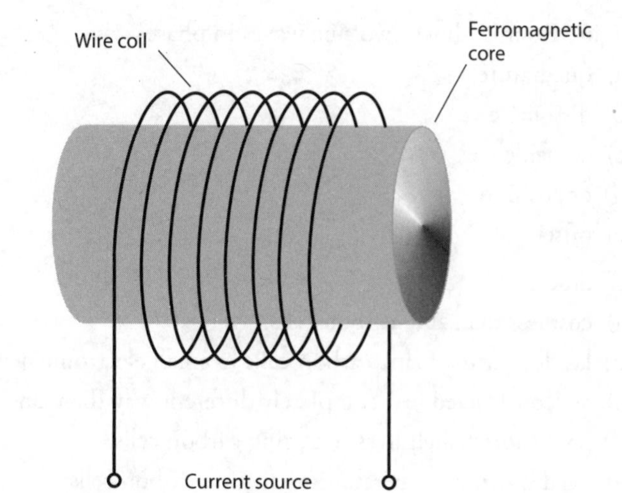

Exam-2 Illustration for Final Exam Question 12.

14. The Arduino Pro Mini differs from the Arduino Uno in that
    (a) it has no analog inputs.
    (b) it has no built-in USB interface.
    (c) it is a larger board for professional use.
    (d) it costs more than the Arduino Uno.
    (e) None of the above

15. In a bipolar transistor, the term *zero bias* means that two of the three electrodes are at the same voltage. Which electrodes?
    (a) Emitter and base
    (b) Base and gate
    (c) Emitter and collector
    (d) Source and drain
    (e) Emitter and gate

16. When used for audio peak clipping, a diode voltage limiter
    (a) has high gain and sensitivity.
    (b) often distorts the sound.
    (c) can break into oscillation.
    (d) needs a high-voltage battery.
    (e) drains a lot of current from the battery.

17. Bleeder resistors
    (a) should be connected in parallel with power-supply filter capacitors.
    (b) should be connected in series with power-supply filter capacitors.
    (c) can keep power-supply filter capacitors from burning out.
    (d) can be connected across batteries to smooth out the voltage.
    (e) can minimize the risk of electrocution from solar cells.

18. Figure Exam-3 shows two sine waves in phase
    (a) quadrature.
    (b) dissonance.
    (c) coincidence.
    (d) opposition.
    (e) offset.

19. Alkaline cells
    (a) cost less than zinc-carbon cells.
    (b) last longer than zinc-carbon cells in small electronic devices.
    (c) are constructed in a completely different way than zinc-carbon cells.
    (d) have shorter shelf lives than zinc-carbon cells.
    (e) need warmer temperatures than zinc-carbon cells.

20. Compared with circuits built up from discrete components, integrated circuits (ICs) offer all of the following advantages except one. Which one?
    (a) Better reliability
    (b) Reduced size
    (c) Lower current demand
    (d) Higher inductance values
    (e) Higher speed

21. Which of the following actions could damage a microcontroller?
    (a) Connecting a 5-V digital output to a 3.3-V input on a second microcontroller.
    (b) Connecting a 3.3-V digital output to a 5-V input on a second microcontroller.
    (c) Connecting a 1.5-V cell between GND and a digital input of a microcontroller.

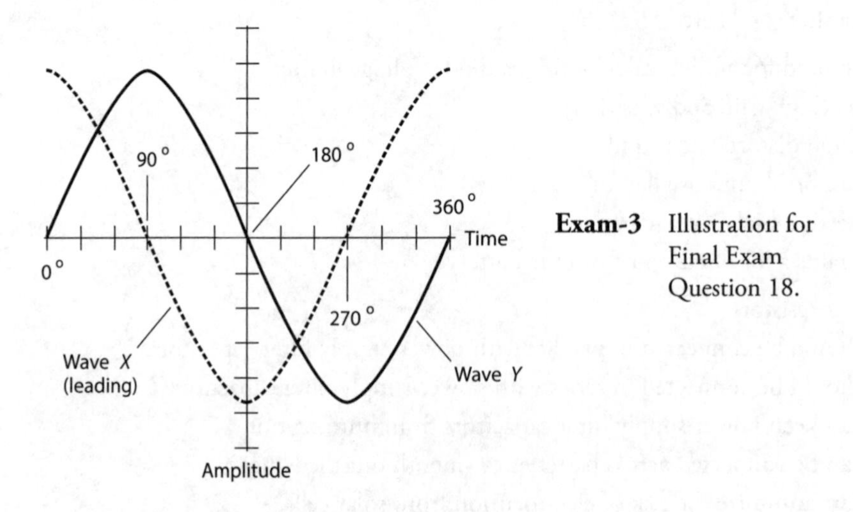

Exam-3 Illustration for Final Exam Question 18.

(d) Connecting a digital output pin through a 1000-ohm resistor to another digital output pin on the same microcontroller.
(e) Connecting a digital output pin directly to a digital input pin on the same microcontroller.

22. You'll find a cold cathode in a
(a) gas-filled electron tube.
(b) multigrid electron tube.
(c) bipolar transistor.
(d) junction field-effect transistor (JFET).
(e) metal-oxide-semiconductor field-effect transistor (MOSFET).

23. A key advantage of JFETs over bipolar transistors is the fact that JFETs
(a) last longer.
(b) can handle more power.
(c) generate less internal noise.
(d) have lower input impedance.
(e) can take more physical abuse.

24. Table Exam-1 shows the digital inputs and output for

Table Exam-1. Truth Table for Final Exam Question 24

Input X	Input Y	Output
0	0	1
0	1	1
1	0	1
1	1	0

(a) an OR gate.
(b) a NOR gate.
(c) an AND gate.
(d) a NAND gate.
(e) an XOR gate.

25. When you want to calculate a circuit's *power* gain (or loss) in decibels, the input and output impedances don't matter
(a) if their reactance components are both inductive or both capacitive.
(b) if both are purely reactive.
(c) if both are purely resistive.
(d) if the input has constant amplitude.
(e) under any circumstances.

26. A well-designed power supply has rectifier diodes whose peak-inverse-voltage (PIV) ratings
    (a) equal the positive or negative peak AC input voltage.
    (b) significantly exceed the positive or negative peak AC input voltage.
    (c) equal the peak-to-peak AC voltage across the transformer primary.
    (d) ensure that they constantly remain in a state of avalanche breakdown.
    (e) are less than their own forward breakover voltages.

27. A 12-V battery drives 777 mA of current through a resistor. Remembering the rules of rounding and significant figures, how should you express the resistance?
    (a) 15 ohms
    (b) 15.4 ohms
    (c) 9.3 ohms
    (d) 9.32 ohms
    (e) 65 ohms

28. Which of the following things can a displacement transducer do?
    (a) Measure the peak-to-peak amplitude of an AC wave
    (b) Convert sound waves to radio signals
    (c) Convert infrared, visible, or ultraviolet radiation to DC
    (d) Convert an electrical signal to mechanical rotation
    (e) Measure the frequency of an irregular wave

29. In an amplitude-modulated (AM) signal, the carrier wave conveys no information itself, but it nevertheless consumes
    (a) 10 percent of the power.
    (b) 25 percent of the power.
    (c) 33 percent of the power.
    (d) 50 percent of the power.
    (e) more than 50 percent of the power.

30. Every time a polar-orbiting satellite goes around the earth once, the earth rotates beneath it. The rotation of the earth rarely coincides with the orbital period of a satellite, so successive satellite apogees occur over different points on the surface. This phenomenon is called
    (a) satellite shift.
    (b) orbital displacement.
    (c) apogee drift.
    (d) earth-satellite differential.
    (e) nothing; it has no technical name.

31. If 1 km of wire has a conductance of 0.6 S, then 3 km of the same wire has a conductance of
    (a) 1.8 S.
    (b) 1.2 S.

(c) 0.2 S.
(d) 0.3 S.
(e) 0.6 S.

32. A half-wave, center-fed dipole antenna has *gain* over
(a) an isotropic antenna.
(b) a Yagi antenna.
(c) a quad antenna.
(d) a dish antenna.
(e) a longwire antenna.

33. What's the technical name for a laser that produces brief, powerful radiation bursts at regular intervals?
(a) Burst laser
(b) Intermittent laser
(c) Pulsed laser
(d) Digital laser
(e) Flash laser

34. A capacitor works by storing and releasing energy in the form of
(a) an electric field.
(b) a magnetic field.
(c) electric charge.
(d) pure reactance.
(e) complex impedance.

35. One gigabyte equals *exactly*
(a) 10^3 bytes.
(b) 10^6 bytes.
(c) 10^9 bytes.
(d) 10^{12} bytes.
(e) None of the above

36. Suppose that a battery, connected to an unchanging load, delivers constant current for a while and then "dies fast." In technical terms, that battery has a
(a) flat ampere-hour characteristic.
(b) flat energy-loss curve.
(c) flat discharge curve.
(d) nonlinear discharge contour.
(e) linear discharge contour.

37. Which of the following antenna types is designed mainly for use at ultra-high and/or microwave frequencies?

(a) Horn
(b) Dish
(c) Helical
(d) Corner reflector
(e) All of the above

38. If you want to build a bipolar-transistor RF power amplifier with the highest efficiency possible, and if you intend to use it for FM only, you should bias the transistor

(a) beyond saturation.
(b) at saturation.
(c) midway between cutoff and saturation.
(d) at cutoff.
(e) beyond cutoff.

39. You connect a circuit whose output has a 1000-ohm, reactance-free impedance to the primary of a transformer. As a result, the secondary exhibits a 10-ohm, reactance-free impedance. What's the transformer's primary-to-secondary turns ratio?

(a) 100:1
(b) 10:1
(c) 1:10
(d) 1:100
(e) You need more information to answer this question.

40. If you connect a 1.00 V DC electrochemical cell across a 2.00-ohm resistor, how much current will flow through the resistor?

(a) 125 mA
(b) 250 mA
(c) 500 mA
(c) 707 mA
(d) 1.41 A
(e) 2.00 A

41. Refer to Fig. Exam-4, which shows the impedance vectors for two hypothetical circuits. The diagram suggests that if you connect these two circuits in series, you'll get a third, more complicated circuit that contains

(a) pure resistance without reactance.
(b) some resistance with a little bit of capacitive reactance.
(c) some resistance with a little bit of inductive reactance.
(d) pure capacitive reactance.
(e) pure inductive reactance.

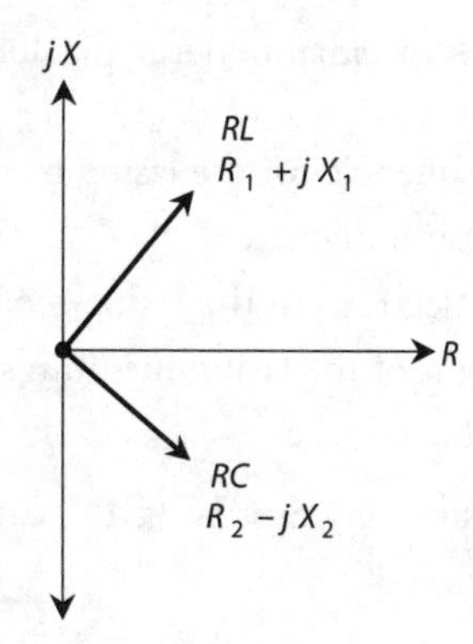

Exam-4 Illustration for Final Exam Question 41.

42. Imagine a perfect sine wave without any DC component. You can shift the phase of that wave by a certain number of degrees and end up with the same wave "upside-down." How many degrees?
    (a) 45°
    (b) 90°
    (c) 180°
    (d) 270°
    (e) 360°

43. In a perfect sine wave with no DC component, the peak-to-peak amplitude is
    (a) half the positive peak amplitude.
    (b) about 0.707 times the positive peak amplitude.
    (c) the same as the positive peak amplitude.
    (d) about 1.414 times the positive peak amplitude.
    (e) twice the positive peak amplitude.

44. If the inductive reactance and the resistance have the same ohmic value in an *RL* circuit, then the phase angle is
    (a) 0°.
    (b) 45°.
    (c) 90°.
    (d) 30°.
    (e) 60°.

45. In a microcontroller, a 10-bit analog input gets digitized, changing it to a number in the range
    (a) 0 to 256.
    (b) 0 to 1023.
    (c) 0 to 4096.
    (d) 0 to 255.
    (e) 0 to 100.

46. The wavelengths of radiant energy produced by a cavity laser depend on the cavity's physical length and also on the
    (a) resonant frequencies of the lasing medium's atoms.
    (b) cavity's physical diameter.
    (c) number of electrons in the lasing medium's atoms.
    (d) atomic weight of the lasing medium's atoms.
    (e) All of the above

47. The Global Positioning System (GPS) can locate a point in
    (a) one dimension.
    (b) two dimensions.
    (c) three dimensions.
    (d) four dimensions.
    (e) five dimensions.

48. Which of the following characteristics is an advantage of a rheostat over a potentiometer in some situations?
    (a) Rheostats can handle higher frequencies.
    (b) Rheostats can be more precisely adjusted.
    (c) Rheostats can handle more current.
    (d) Rheostats work better with DC than with 60-Hz utility AC.
    (e) All of the above

49. In a PNP transistor, you'd normally set the DC collector voltage
    (a) the same as the base voltage.
    (b) more positive than the emitter voltage.
    (c) more negative than the emitter voltage.
    (d) the same as the emitter voltage.
    (e) at zero (ground potential).

50. If you want an electric power delivery system to work its best, then
    (a) the transmission line's characteristic impedance must be as low as possible.
    (b) the transmission line's characteristic impedance must be as high as possible.
    (c) the load impedance must comprise a pure reactance equal to the characteristic impedance of the transmission line.
    (d) the load impedance must comprise a pure resistance equal to the characteristic impedance of the transmission line.
    (e) the ground system must have as little reactance as possible.

51. With a permeability-tuned solenoid-coil inductor, moving more of the core into the coil
    (a) increases the inductance.
    (b) does not change the inductance.
    (c) reduces the inductance.

(d) increases the frequency.
(e) reduces the reactance.

52. Arduino shields are
(a) protective devices that prevent damage to GPIO pins.
(b) programs for the device.
(c) a collection of Arduino software.
(d) boards that you can add to the Arduino to provide extra hardware features.
(e) None of the above

53. A power supply can incorporate a circuit that applies a reduced AC voltage to the transformer for a couple of seconds immediately after power-up, and then delivers the full voltage once the filter capacitors have charged completely. This precaution minimizes the risk of damage to the
(a) transformer core.
(b) filter capacitors.
(c) rectifier diodes.
(d) voltage regulator.
(e) bleeder resistors.

54. If the real-number part of a complex-number impedance is zero and the imaginary part is nonzero (positive or negative), then the number denotes
(a) pure reactance.
(b) pure capacitance but not inductance.
(c) pure inductance but not capacitance.
(d) a short circuit.
(e) an open circuit.

55. In Fig. Exam-5, the small black dots represent electrons and the larger white dots represent holes. This drawing portrays a semiconductor diode in a state of
(a) forward bias below the forward breakover voltage.
(b) forward bias at or beyond the forward breakover voltage.
(c) reverse bias below the avalanche voltage.
(d) reverse bias at or beyond the avalanche voltage.
(e) zero bias.

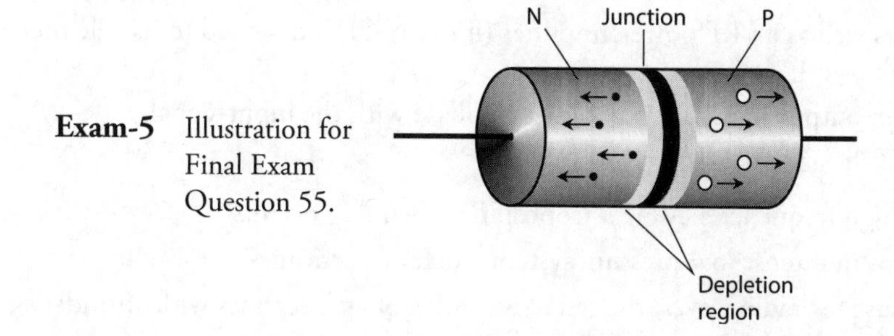

Exam-5 Illustration for Final Exam Question 55.

56. Imagine a circuit that has finite, nonzero resistance but no reactance. You send an AC signal through it. What's the phase angle?

(a) It depends on the signal frequency
(b) 0°
(c) 45°
(d) 90°
(e) 180°

57. Which of the following devices, if any, converts 150-kHz AC to high-pitched sound waves that most people can hear?

(a) Woofer
(b) Mixer
(c) Tweeter
(d) Inverter
(e) None of the above

58. Fill in the blanks in the following statement with a single word that makes it true. "In either a series circuit or a parallel circuit that operates from a battery, the sum of the ________s in each component always equals the total ________ that the circuit demands from the battery."

(a) amperage
(b) voltage
(c) wattage
(d) charge
(e) Any of the above

59. What type of IC *always* needs precautions to protect it against electrostatic discharges?

(a) Metal oxide semiconductor (MOS)
(b) Random-access memory (RAM)
(c) Transistor-transistor logic (TTL)
(d) Large-scale integration (LSI)
(e) Emitter-coupled logic (ECL)

60. Figure Exam-6 shows a common-gate amplifier that uses a JFET. A circuit of this type usually

(a) has less gain than a common-source amplifier.
(b) is less likely to break into oscillation than a common-source amplifier.
(c) works well as an RF power amplifier (if the JFET is designed to handle moderate or high power).
(d) has an output signal that coincides in phase with the input signal.
(e) All of the above

61. At very high frequencies (VHF), tropospheric bending occurs

(a) except in intense low-pressure systems such as hurricanes.
(b) because for radio waves, the refractive index of air decreases with altitude.
(c) only during geomagnetic storms caused by unusual activity on the sun.

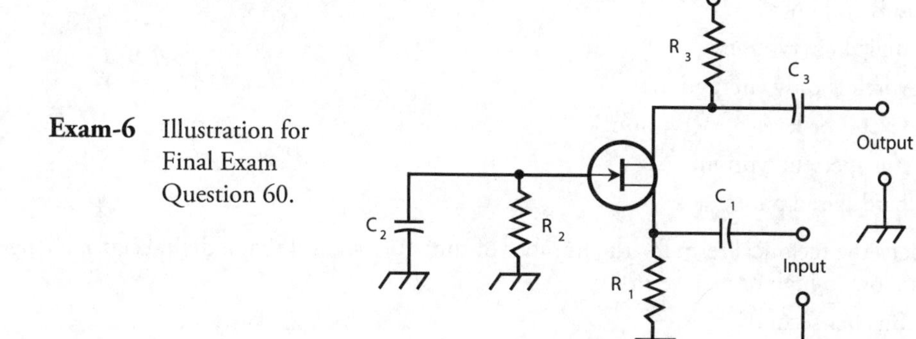

Exam-6 Illustration for Final Exam Question 60.

(d) because air gets more prone to ionization as the altitude increases.
(e) at no time; it's a widespread misconception.

62. In general, a voltmeter should have
(a) the highest possible internal resistance.
(b) the lowest possible internal resistance.
(c) the highest possible sensitivity.
(d) the ability to withstand the highest possible current.
(e) the ability to dissipate the highest possible amount of power.

63. In a phase-locked loop (PLL) circuit, the output stays at the same frequency as the reference-oscillator frequency thanks to a phase
(a) divider.
(b) comparator.
(c) multiplier.
(d) stabilizer.
(e) splitter.

64. Which of the following types of diode might you use in a power-supply voltage regulator?
(a) Rectifier
(b) PIN
(c) Zener
(d) Varactor
(e) Gunn

65. If you connect five 50-pF capacitors in parallel, you get a net capacitance of
(a) 250 pF.
(b) 125 pF.
(c) 50 pF.
(d) 25 pF.
(e) 10 pF.

66. If you have security concerns about using a cordless phone system in your home, the best thing to do is employ
    - (a) digital encryption.
    - (b) wireless-only encryption.
    - (c) level-2 or level-3 encryption.
    - (d) mil-spec encryption.
    - (e) hard-wired phone sets.

67. What's the technical term for the number of times per second that a digital signal changes state (from low to high or vice-versa)?
    - (a) Bits per second
    - (b) Digital frequency
    - (c) Baud rate
    - (d) Signal shift rate
    - (e) Triggering rate

68. Doppler radar can measure or estimate the
    - (a) frequency of a radio signal.
    - (b) wind speed in a tornado.
    - (c) distance between two ships at sea.
    - (d) depth of the ocean at a specific location.
    - (e) intensity of the lightning in a thundershower.

69. A ferromagnetic material
    - (a) concentrates magnetic lines of flux that pass through it.
    - (b) increases the magnetomotive force around a current-carrying wire.
    - (c) causes the current in a wire to increase.
    - (d) causes the current in a wire to decrease.
    - (e) increases the number of ampere-turns in a coil of wire.

70. Figure Exam-7 is a schematic diagram of a
    - (a) grounded-emitter circuit.
    - (b) grounded-gate circuit.
    - (c) grounded-collector circuit.
    - (d) grounded-base circuit.
    - (e) None of the above

71. Sixty degrees of phase represents
    - (a) ⅙ of a cycle.
    - (b) ¼ of a cycle.
    - (c) ⅓ of a cycle.
    - (d) ½ of a cycle.
    - (e) ⅔ of a cycle.

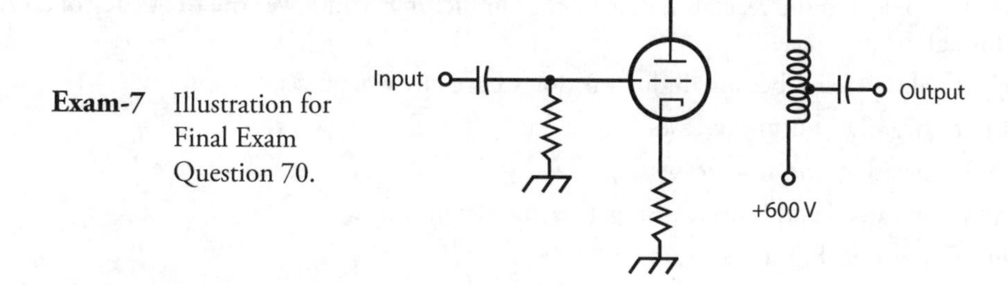

Exam-7 Illustration for Final Exam Question 70.

72. In cellular telephone systems, the term *handoff* refers to the process, as a mobile phone set moves around, of
    (a) substituting one phone set for another.
    (b) switching from one repeater to another.
    (c) switching from one network to another.
    (d) switching from the unencrypted mode to an encrypted mode.
    (e) switching from one level of encryption to another.

73. The Arduino Uno has two microcontroller chips because
    (a) that way, it can perform twice as fast.
    (b) one of the microcontrollers is dedicated to providing a USB interface.
    (c) one of the microcontrollers provides a video interface.
    (d) one microcontroller provides access to the GPIO pins and the other one performs the processing.
    (e) that way, if one processor fails the other one can take over.

74. In a pulsed laser, the ratio of the peak output power to the average output power varies in *inverse* proportion to the
    (a) duty cycle.
    (b) frequency.
    (c) beam radius.
    (d) beam divergence.
    (e) wavelength.

75. In a certain advanced form of radio communication, the transmitter carrier frequency varies in a controlled manner, independent of the signal modulation. The receiver is programmed to follow the transmitter frequency from instant to instant, so it "thinks" that the signal frequency remains constant. What's the technical name for this mode?
    (a) Spread spectrum
    (b) Variable frequency
    (c) Synchronized frequency
    (d) Coherent carrier
    (e) Fluctuating carrier

76. What happens to the reactance of a 10-mH inductor as you lower the frequency of an AC signal through it?

(a) It's always zero, because inductors don't have any reactance.
(b) It's negative, but it increases (gets closer to zero).
(c) It's positive, and it increases.
(d) It's negative, and it decreases (gets farther from zero).
(e) It's positive, but it decreases.

77. You have a package of fixed resistors. The manufacturer quotes their value as 56.0 ohms, plus or minus 10.0 percent. You measure the resistance of each component with an ohmmeter. Which of the following readings indicates a reject?

(a) 50.0 ohms
(b) 51.0 ohms
(c) 52.0 ohms
(d) 59.7 ohms
(e) 61.1 ohms

78. You might use a tunnel diode as a

(a) high-speed switch.
(b) voltage limiter.
(c) variable capacitor.
(d) high-voltage rectifier.
(e) low-power oscillator.

79. Which of the following things can have resonant properties?

(a) Piezoelectric crystals
(b) Antennas
(c) Sections of transmission line
(d) *LC* circuits
(e) All of the above

80. The characteristic impedance of a parallel-wire RF transmission line depends on the

(a) standing-wave ratio (SWR).
(b) length of the line.
(c) voltage between the wires.
(d) frequency of the signal.
(e) spacing between the wires.

81. If you want to exercise good engineering practice, you should make a series-parallel network of resistors using

(a) the highest-wattage resistors you have on hand at the moment.
(b) resistors that are all as nearly identical as possible.

(c) a series combination of parallel-connected resistors.
(d) a parallel combination of series-connected resistors.
(e) whatever you have on hand at the moment, as long as it works.

82. If two AC sine waves have identical frequency and phase, and if both of them lack DC components, then when you combine them you get
(a) a sine wave whose amplitude equals the difference between the amplitudes of the original waves.
(b) a sine wave whose amplitude equals the sum of the amplitudes of the original waves.
(c) a square wave whose amplitude equals the difference between the amplitudes of the original waves.
(d) a square wave whose amplitude equals the sum of the amplitudes of the original waves.
(e) no wave at all, because the original waves cancel each other out.

83. Which of the following statements about a microcontroller's GPIO pins is false?
(a) A GPIO pin can change mode from an input to an output while a program is running.
(b) A GPIO pin, when used as an output, can directly drive an LED with a suitable current-limiting resistor.
(c) GPIO pins often have a second function, such as a serial interface pin.
(d) GPIO pins normally operate at 3.3 V or 5 V.
(e) GPIO pins are fixed as either inputs or outputs during manufacture.

84. Which of the following statements applies to an emitter-follower circuit?
(a) The output signal is in phase opposition relative to the input signal.
(b) You apply the input signal between the emitter and ground.
(c) It can operate as a stable RF oscillator.
(d) The gain increases as the frequency increases.
(e) It can match a high impedance to a low impedance.

85. If you connect five 50-pF capacitors in series, you get a net capacitance of
(a) 250 pF.
(b) 125 pF.
(c) 50 pF.
(d) 25 pF.
(e) 10 pF.

86. You can increase the current-delivering capacity of a solar panel by
(a) connecting multiple solar cells in series when you build it.
(b) connecting an alkaline battery in series with it.
(c) connecting multiple solar cells in parallel when you build it.
(d) connecting a lead-acid battery in series with it.
(e) charging it from a wall outlet before using it.

87. The *minority* carriers in P type semiconductor material are
(a) atomic nuclei.
(b) protons.
(c) neutrons.
(d) electrons.
(e) holes.

88. A coil wound inside a pot-core shell
(a) can carry more current than it could without the shell.
(b) allows you to get a large inductance in a small space.
(c) gets more efficient as you increase the frequency.
(d) can be easily adjusted to vary the inductance.
(e) has no reactance, unlike other coil types.

89. Movement of holes in a semiconductor material
(a) is the same thing as electron movement in the same direction.
(b) can occur only if the current is high enough.
(c) constitutes an electric current.
(d) prevents the material from conducting current.
(e) makes the material act like a perfect conductor.

90. You're building a low-power transformer and you need a winding inductance of 7 H. Interwinding capacitance does not concern you. For best results, you should use
(a) an air core.
(b) a solenoid core.
(c) a toroid core.
(d) a pot core.
(e) an E core.

91. A serious, competent audiophile would *never* use a VU meter to measure or infer
(a) sound intensity.
(b) sound decibels.
(c) the power output of an audio amplifier.
(d) the efficiency of an audio amplifier.
(e) the likelihood of distortion on sound peaks.

92. Which of the following units expresses magnetic field strength at a specific location?
(a) Gilbert
(b) Ampere-turn
(c) Weber
(d) Maxwell
(e) Gauss

93. In the antenna shown in Fig. Exam-8, the object marked X is a
    (a) phase coordinator.
    (b) synchronizer.
    (c) parasitic element.
    (d) reflector.
    (e) signal coupler.

94. Each point in the RX_C quarter-plane corresponds to a unique
    (a) resistance.
    (b) capacitance.
    (c) frequency.
    (d) combination of resistance and capacitance.
    (e) combination of resistance and capacitive reactance.

95. Which of the following circuits produces an AC output signal that opposes the input signal in phase?
    (a) Common-gate
    (b) Common-drain
    (c) Common-source
    (d) Common-base
    (e) Common-collector

96. A vector pointing downward and toward the right ("southeast") in the resistance-reactance (RX) half-plane portrays
    (a) pure resistance.
    (b) resistance and inductive reactance.
    (c) resistance and capacitive reactance.
    (d) pure inductive reactance.
    (e) pure capacitive reactance.

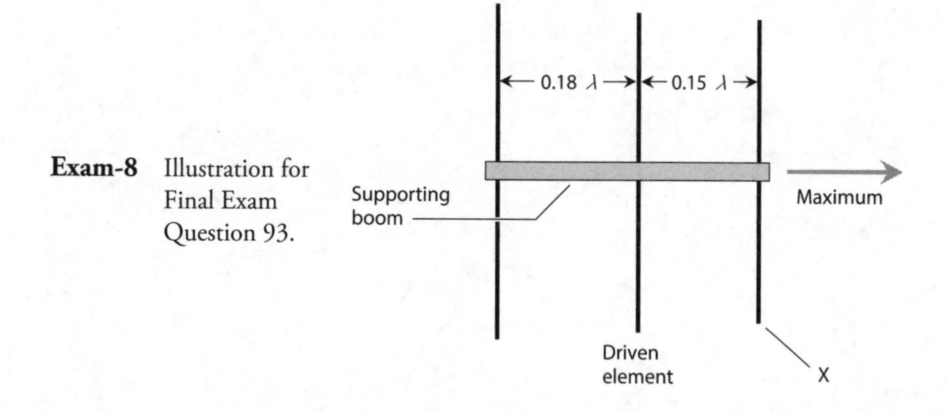

Exam-8 Illustration for Final Exam Question 93.

97. When the direct and reflected sound waves from a point source (say, a trumpet's horn) arrive at a certain point in phase opposition, then a listener at that point is

(a) at an antinode.
(b) in a dead zone.
(c) at the threshold of hearing.
(d) at a nonlinear point.
(e) in an opposition zone.

98. You connect a voltage divider across a battery. The circuit comprises several resistors in series. What is the highest voltage output you can get from it?

(a) A small fraction of the battery voltage
(b) About half the battery voltage
(c) The battery voltage exactly
(d) The battery voltage divided by the number of resistors
(e) The battery voltage times the number of resistors

99. If a power-supply fuse has a spring inside along with a short wire or strip, it's probably a

(a) quick-break fuse.
(b) high-current fuse.
(c) low-voltage fuse.
(d) DC fuse.
(e) slow-blow fuse.

100. What type of IC combines many signals into one?

(a) Differentiator
(b) Multiplexer
(c) Integrator
(d) Comparator
(e) Operational amplifier

A
APPENDIX

Answers to Quizzes, Tests, and Final Exam

Chapter 1

1. a	2. d	3. d	4. a	5. a
6. d	7. d	8. b	9. c	10. a
11. b	12. b	13. a	14. b	15. d
16. a	17. c	18. d	19. b	20. d

Chapter 2

1. c	2. c	3. a	4. c	5. b
6. a	7. b	8. c	9. c	10. d
11. c	12. b	13. a	14. b	15. a
16. c	17. d	18. c	19. c	20. a

Chapter 3

1. a	2. b	3. d	4. a	5. c
6. d	7. a	8. a	9. b	10. c
11. b	12. d	13. d	14. a	15. c
16. b	17. c	18. a	19. d	20. c

Chapter 4

1. a	2. c	3. c	4. b	5. d
6. a	7. d	8. b	9. a	10. c
11. b	12. d	13. c	14. a	15. b
16. c	17. d	18. a	19. b	20. c

Chapter 5

1. a	2. c	3. b	4. d	5. a
6. a	7. d	8. c	9. a	10. a
11. a	12. d	13. d	14. c	15. b
16. a	17. d	18. d	19. d	20. a

Chapter 6

1. d	2. d	3. d	4. c	5. b
6. c	7. b	8. c	9. d	10. a
11. b	12. b	13. b	14. a	15. a
16. c	17. d	18. a	19. c	20. a

Chapter 7

1. d	2. b	3. c	4. c	5. a
6. a	7. b	8. d	9. b	10. a
11. c	12. a	13. b	14. a	15. b
16. c	17. c	18. b	19. c	20. a

Chapter 8

1. a	2. d	3. c	4. d	5. d
6. b	7. a	8. a	9. b	10. a
11. b	12. d	13. c	14. c	15. a
16. b	17. b	18. d	19. d	20. a

Test: Part 1

1. c	2. d	3. c	4. d	5. a
6. e	7. e	8. a	9. e	10. d
11. a	12. c	13. c	14. e	15. e
16. b	17. c	18. b	19. e	20. a
21. b	22. d	23. c	24. d	25. c
26. b	27. e	28. a	29. d	30. a
31. a	32. d	33. e	34. e	35. d
36. a	37. a	38. e	39. a	40. b
41. c	42. b	43. e	44. c	45. a
46. a	47. b	48. a	49. e	50. b

Chapter 9

1. c	2. b	3. d	4. c	5. b
6. a	7. c	8. c	9. c	10. b
11. a	12. a	13. c	14. d	15. c
16. d	17. c	18. d	19. c	20. c

Chapter 10

1. d	2. a	3. b	4. a	5. c
6. a	7. d	8. a	9. b	10. b
11. c	12. a	13. b	14. c	15. d
16. b	17. b	18. c	19. c	20. c

Chapter 11

1. a	2. a	3. c	4. b	5. d
6. c	7. a	8. d	9. c	10. c
11. b	12. c	13. d	14. d	15. a
16. c	17. d	18. b	19. c	20. b

Chapter 12

1. c	2. a	3. d	4. a	5. b
6. d	7. a	8. c	9. b	10. b
11. a	12. a	13. b	14. c	15. b
16. a	17. c	18. c	19. c	20. d

Chapter 13

1. d	2. c	3. d	4. c	5. b
6. a	7. b	8. c	9. b	10. a
11. b	12. d	13. c	14. b	15. c
16. d	17. c	18. c	19. a	20. d

Chapter 14

1. d	2. c	3. b	4. c	5. c
6. a	7. b	8. c	9. b	10. a
11. b	12. a	13. d	14. a	15. d
16. d	17. a	18. b	19. c	20. b

Chapter 15

1. c	2. b	3. c	4. a	5. a
6. c	7. b	8. d	9. c	10. a
11. c	12. d	13. b	14. b	15. a
16. d	17. a	18. d	19. c	20. d

Chapter 16

1. c	2. d	3. c	4. c	5. d
6. b	7. d	8. c	9. a	10. b
11. a	12. c	13. d	14. b	15. d
16. a	17. c	18. d	19. b	20. a

Chapter 17

1. a	2. d	3. c	4. b	5. b
6. a	7. d	8. c	9. b	10. c
11. c	12. c	13. d	14. a	15. c
16. b	17. b	18. a	19. c	20. d

Chapter 18

1. b	2. a	3. d	4. a	5. c
6. b	7 c	8. d	9. a	10. b
11. c	12. c	13. b	14. d	15. a
16. d	17. c	18. b	19. a	20. c

Test: Part 2

1. d	2. a	3. b	4. d	5. e
6. a	7. c	8. c	9. a	10. c
11. a	12. d	13. d	14. a	15. c
16. d	17. a	18. c	19. e	20. a
21. b	22. d	23. c	24. b	25. e
26. a	27. a	28. a	29. a	30. d
31. b	32. e	33. c	34. b	35. b
36. e	37. d	38. a	39. c	40. b
41. d	42. e	43. c	44. e	45. a
46. d	47. e	48. c	49. b	50. e

Chapter 19

1. a	2. c	3. c	4. d	5. b
6. b	7. c	8. b	9. c	10. a
11. a	12. d	13. a	14. b	15. b
16. a	17. a	18. d	19. d	20. b

Chapter 20

1. d	2. b	3. d	4. a	5. a
6. c	7. c	8. b	9. d	10. a
11. b	12. c	13. d	14. a	15. a
16. c	17. b	18. d	19. c	20. b

Chapter 21

1. d	2. c	3. a	4. c	5. a
6. b	7. d	8. b	9. c	10. a
11. c	12. a	13. b	14. c	15. a
16. c	17. c	18. d	19. c	20. b

Chapter 22

1. b	2. d	3. b	4. a	5. c
6. b	7. b	8. c	9. d	10. a
11. d	12. c	13. a	14. d	15. d
16. a	17. d	18. c	19. a	20. b

Chapter 23

1. b	2. c	3. b	4. c	5. a
6. d	7. a	8. d	9. a	10. b
11. d	12. d	13. c	14. a	15. d
16. b	17. b	18. c	19. a	20. b

Chapter 24

1. d	2. b	3. b	4. c	5. a
6. a	7. b	8. d	9. d	10. b
11. c	12. d	13. a	14. a	15. c
16. a	17. c	18. a	19. c	20. b

Chapter 25

1. c	2. d	3. a	4. d	5. c
6. b	7. d	8. c	9. d	10. b
11. a	12. c	13. b	14. d	15. b
16. c	17. b	18. c	19. d	20. d

Chapter 26

1. d	2. b	3. a	4. a	5. c
6. c	7. a	8. a	9. a	10. d
11. b	12. a	13. c	14. d	15. a
16. b	17. a	18. d	19. a	20. a

Chapter 27

1. b	2. a	3. d	4. c	5. c
6. b	7. d	8. b	9. d	10. b
11. c	12. b	13. d	14. c	15. b
16. a	17. d	18. b	19. c	20. a

Chapter 28

1. c	2. b	3. b	4. d	5. b
6. c	7. c	8. d	9. c	10. a
11. d	12. a	13. d	14. c	15. b
16. d	17. b	18. c	19. b	20. a

Test: Part 3

1. b	2. a	3. a	4. e	5. a
6. c	7. d	8. d	9. c	10. d
11. b	12. c	13. e	14. b	15. d
16. b	17. a	18. d	19. c	20. e
21. a	22. e	23. e	24. b	25. d
26. a	27. a	28. e	29. a	30. a
31. b	32. e	33. a	34. a	35. a
36. c	37. d	38. a	39. c	40. d
41. a	42. d	43. b	44. b	45. c
46. d	47. c	48. a	49. b	50. d

Chapter 29

1. b	2. a	3. c	4. b	5. b
6. c	7. c	8. b	9. d	10. d
11. d	12. d	13. b	14. a	15. d
16. b	17. a	18. a	19. c	20. c

Chapter 30

1. b	2. d	3. c	4. a	5. d
6. d	7. d	8. d	9. b	10. a
11. b	12. a	13. d	14. c	15. c
16. b	17. d	18. a	19. d	20. c

Chapter 31

1. d	2. a	3. d	4. c	5. a
6. d	7. d	8. b	9. d	10. b
11. b	12. c	13. a	14. a	15. b
16. b	17. a	18. d	19. c	20. b

Chapter 32

1. b	2. b	3. b	4. a	5. c
6. c	7. a	8. d	9. d	10. b
11. a	12. a	13. a	14. a	15. c
16. a	17. d	18. c	19. b	20. c

Chapter 33

1. a	2. a	3. c	4. d	5. d
6. a	7. c	8. b	9. c	10. a
11. c	12. b	13. b	14. d	15. c
16. d	17. a	18. b	19. c	20. c

Chapter 34

1. b	2. d	3. b	4. d	5. b
6. a	7. c	8. c	9. d	10. a
11. a	12. d	13. c	14. b	15. a
16. a	17. c	18. b	19. a	20. b

Chapter 35

1. c	2. a	3. c	4. b	5. a
6. b	7. a	8. c	9. c	10. d
11. a	12. d	13. b	14. b	15. c
16. d	17. b	18. a	19. d	20. d

Test: Part 4

1. c	2. b	3. b	4. d	5. d
6. b	7. a	8. a	9. b	10. e
11. c	12. d	13. c	14. d	15. a
16. c	17. e	18. b	19. c	20. d
21. d	22. c	23. b	24. b	25. e
26. e	27. e	28. e	29. d	30. e
31. e	32. a	33. d	34. e	35. b
36. c	37. c	38. a	39. c	40. e
41. e	42. a	43. a	44. d	45. a
46. d	47. d	48. a	49. c	50. d

Final Exam

1. d	2. c	3. c	4. e	5. c
6. a	7. e	8. b	9. e	10. d
11. b	12. b	13. d	14. b	15. a
16. b	17. a	18. a	19. b	20. d
21. a	22. a	23. c	24. d	25. c
26. b	27. a	28. d	29. e	30. c
31. c	32. a	33. c	34. a	35. e
36. c	37. e	38. e	39. b	40. c
41. c	42. c	43. e	44. b	45. b
46. a	47. c	48. c	49. c	50. d
51. a	52. d	53. c	54. a	55. c
56. b	57. e	58. c	59. a	60. e
61. b	62. a	63. b	64. c	65. a
66. e	67. c	68. b	69. a	70. e
71. a	72. b	73. b	74. a	75. a
76. e	77. a	78. e	79. e	80. e
81. b	82. b	83. e	84. e	85. e
86. c	87. d	88. b	89. c	90. d
91. d	92. e	93. c	94. e	95. c
96. c	97. b	98. c	99. e	100. b

B
APPENDIX
Schematic Symbols

ammeter

A

amplifier, general

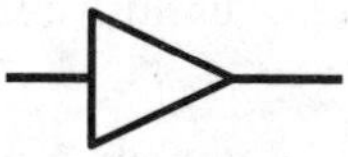

amplifier, inverting

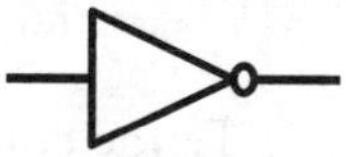

amplifier, operational

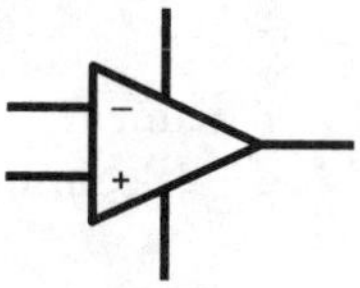

−
+

AND gate

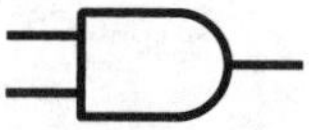

antenna, balanced

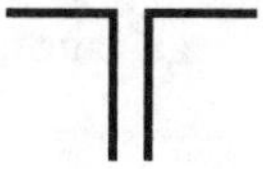

antenna, general

antenna, loop

antenna, loop, multiturn

battery, electrochemical

- +

capacitor, feedthrough

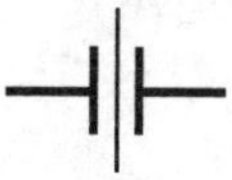

capacitor, fixed

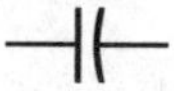

capacitor, variable

capacitor, variable, split-rotor

capacitor, variable, split-stator

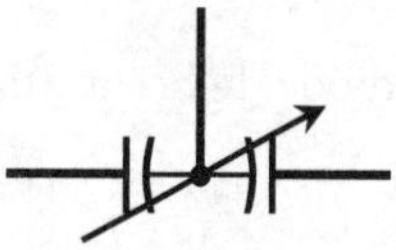

cathode, electron-tube, cold

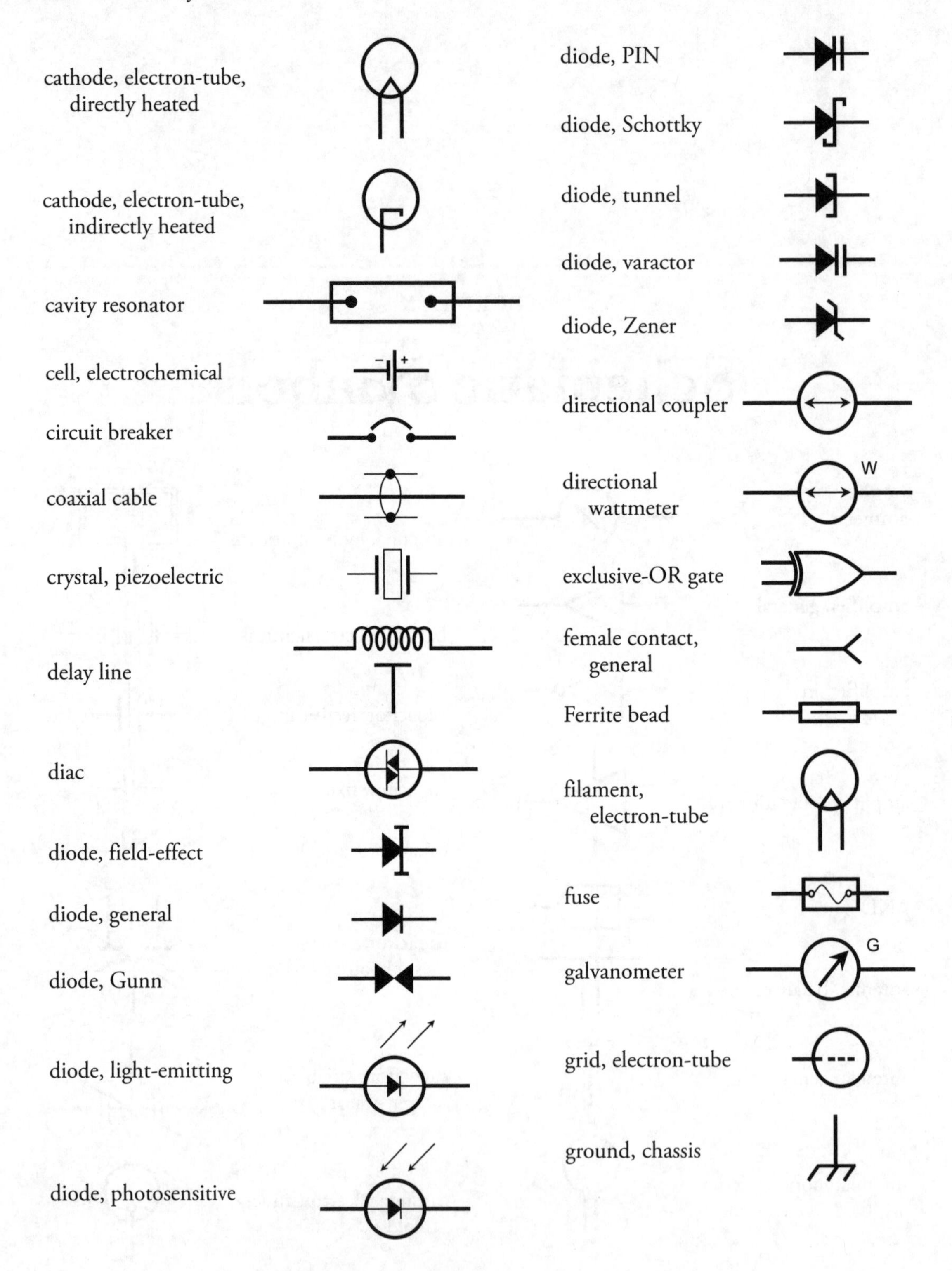
cathode, electron-tube, directly heated
cathode, electron-tube, indirectly heated
cavity resonator
cell, electrochemical
–
+
circuit breaker
coaxial cable
crystal, piezoelectric
delay line
diac
diode, field-effect
diode, general
diode, Gunn
diode, light-emitting
diode, photosensitive
diode, PIN
diode, Schottky
diode, tunnel
diode, varactor
diode, Zener
directional coupler
directional wattmeter
W
exclusive-OR gate
female contact, general
Ferrite bead
filament, electron-tube
fuse
galvanometer
G
grid, electron-tube
ground, chassis

ground, earth

handset

headset, double

headset, single

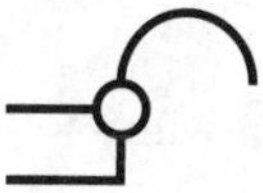

headset, stereo

L R

inductor, air core

inductor, air core, bifilar

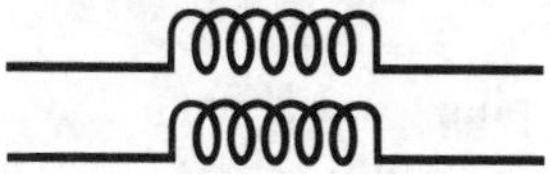

inductor, air core, tapped

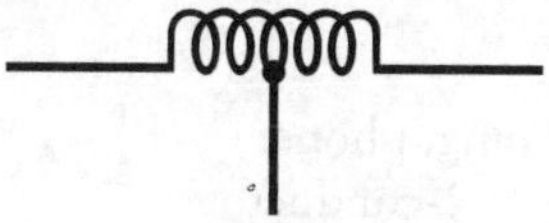

inductor, air core, variable

inductor, iron core

inductor, iron core, bifilar

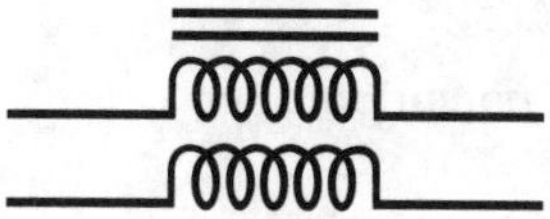

inductor, iron core, tapped

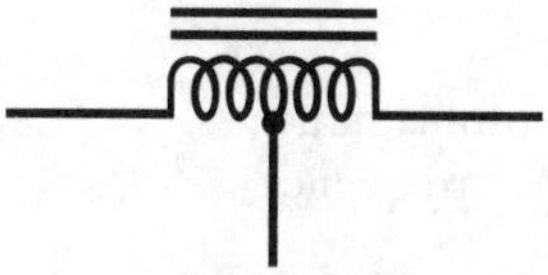

inductor, iron core, variable

inductor, powdered-iron core

inductor, powdered-iron core, bifilar

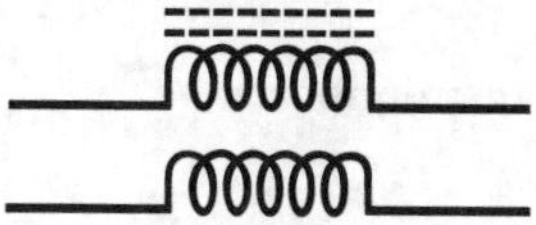

inductor, powdered-iron core, tapped

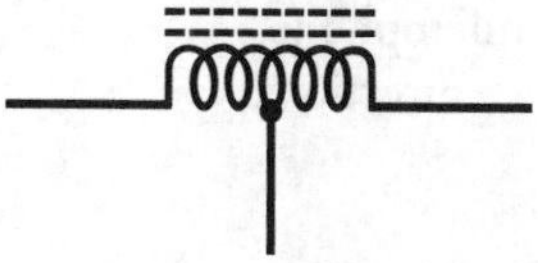

inductor, powdered-iron core, variable

integrated circuit, general

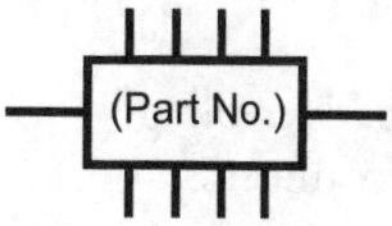

(Part No.)

jack, coaxial or phono

jack, phone, 2-conductor

jack, phone, 3-conductor

key, telegraph

lamp, incandescent

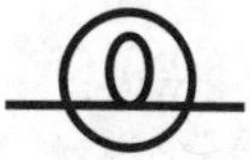

lamp, neon

male contact, general

meter, general

microammeter

μA

microphone

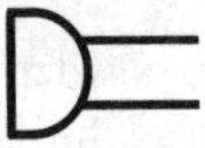

microphone, directional

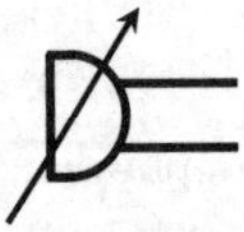

milliammeter

mA

NAND gate

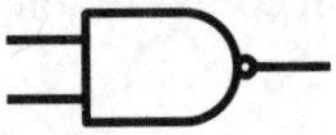

negative voltage connection

−

NOR gate

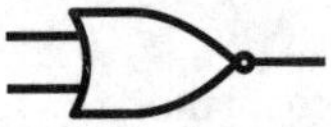

NOT gate

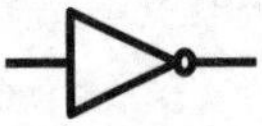

optoisolator

OR gate

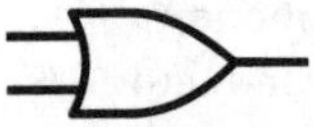

outlet, 2-wire, nonpolarized

outlet, 2-wire, polarized

outlet, 3-wire

outlet, 234-volt

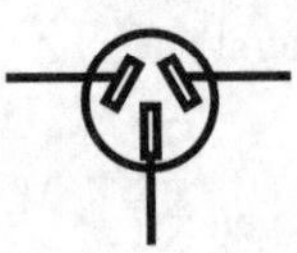

plate, electron-tube

plug, 2-wire, nonpolarized

plug, 2-wire, polarized

plug, 3-wire

plug, 234-volt

plug, coaxial or phono

plug, phone, 2-conductor

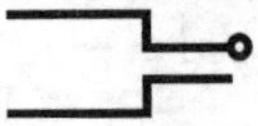

plug, phone, 3-conductor

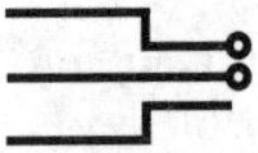

positive voltage connection

+

potentiometer

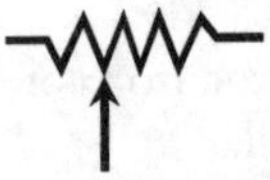

probe, radio-frequency

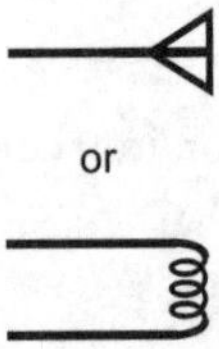

or

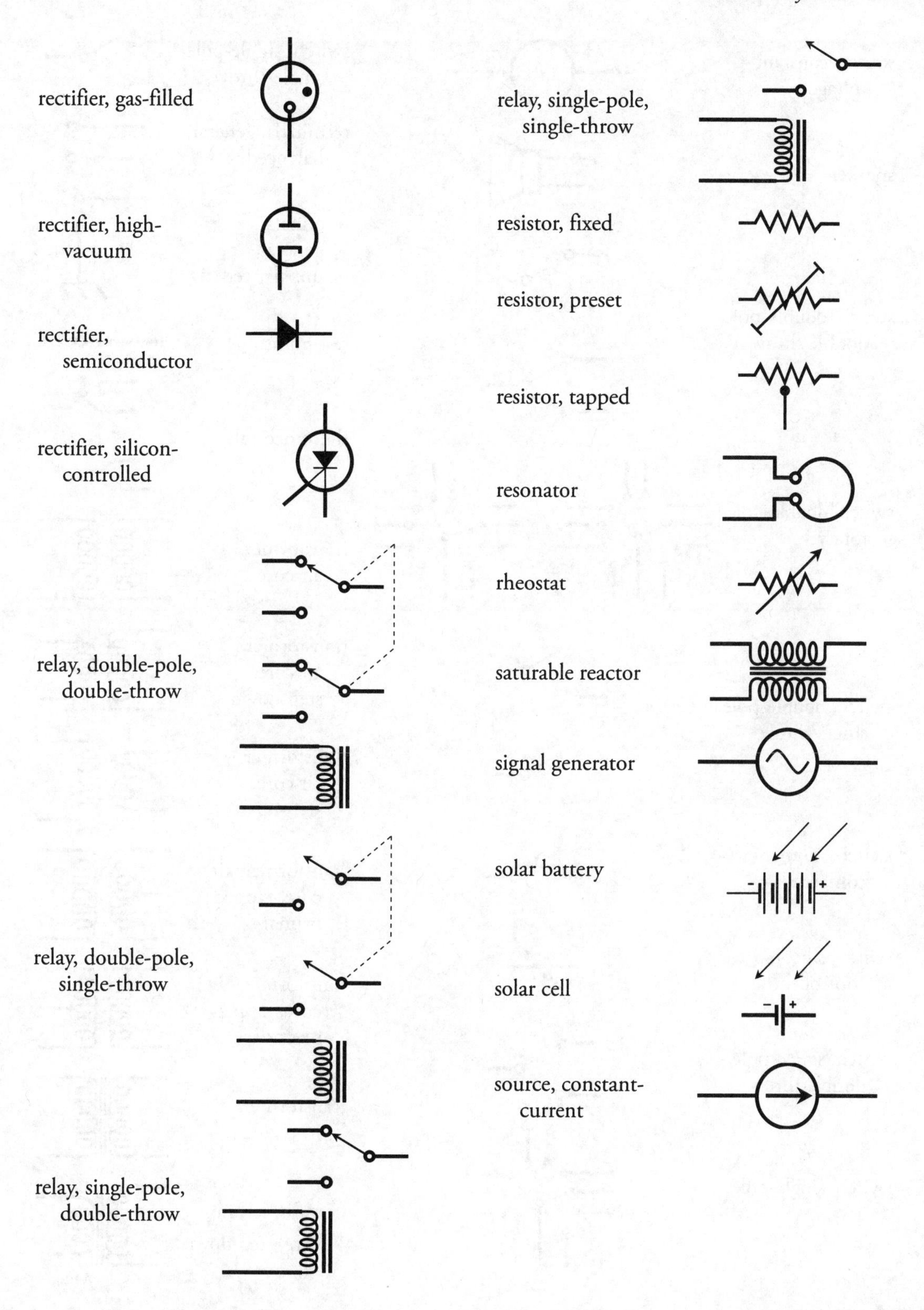

rectifier, gas-filled
rectifier, high-vacuum
rectifier, semiconductor
rectifier, silicon-controlled
relay, double-pole, double-throw
relay, double-pole, single-throw
relay, single-pole, double-throw
relay, single-pole, single-throw
resistor, fixed
resistor, preset
resistor, tapped
resonator
rheostat
saturable reactor
signal generator
solar battery
−
+
solar cell
−
+
source, constant-current

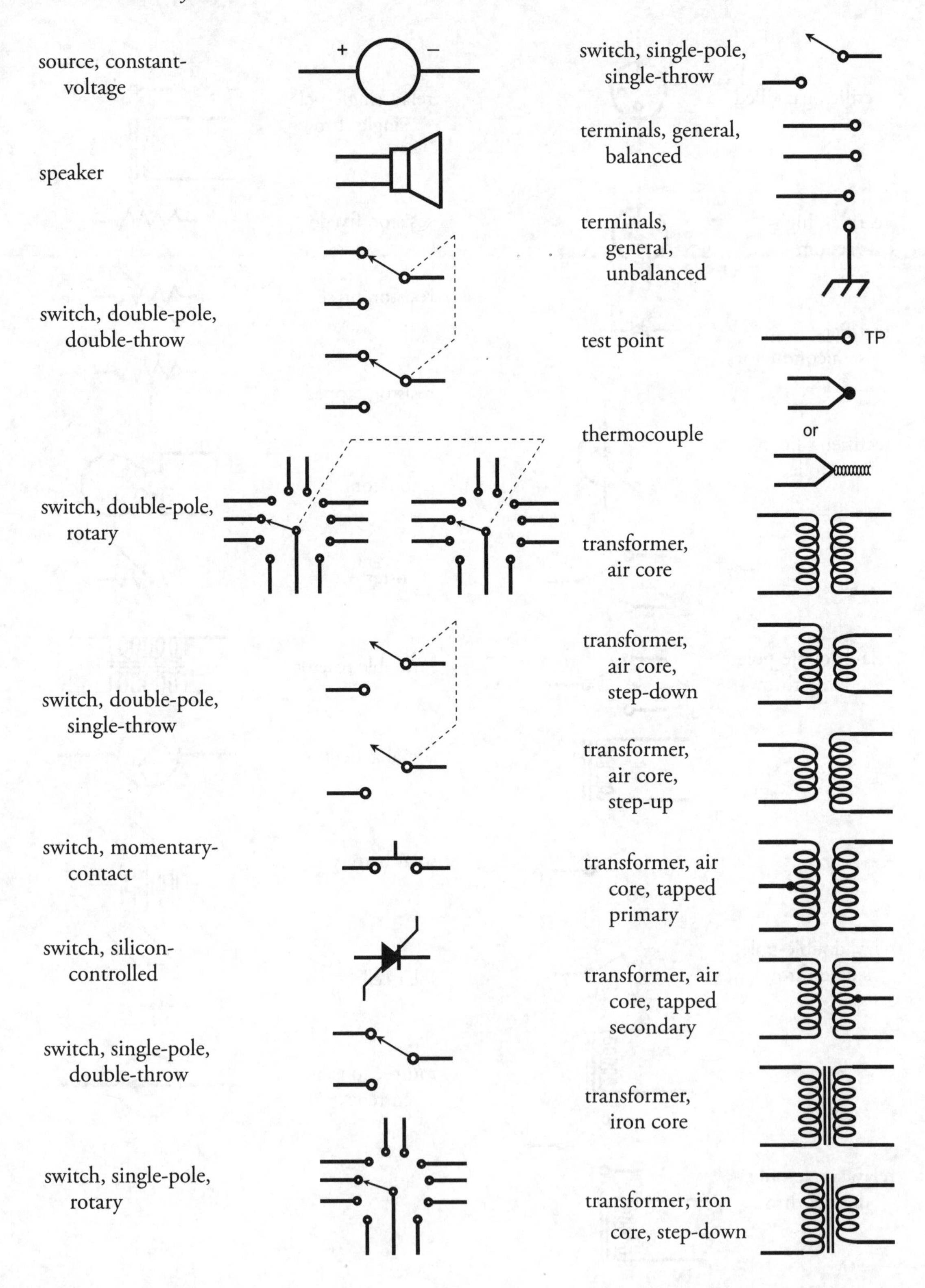
source, constant-voltage
+
–
speaker
switch, double-pole, double-throw
switch, double-pole, rotary
switch, double-pole, single-throw
switch, momentary-contact
switch, silicon-controlled
switch, single-pole, double-throw
switch, single-pole, rotary
switch, single-pole, single-throw
terminals, general, balanced
terminals, general, unbalanced
test point
TP
thermocouple
or
transformer, air core
transformer, air core, step-down
transformer, air core, step-up
transformer, air core, tapped primary
transformer, air core, tapped secondary
transformer, iron core
transformer, iron core, step-down

transformer, iron core, step-up

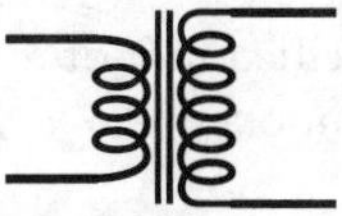

transformer, iron core, tapped primary

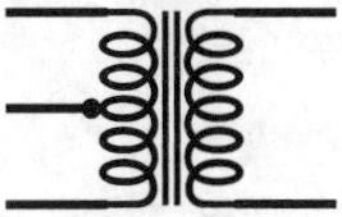

transformer, iron core, tapped secondary

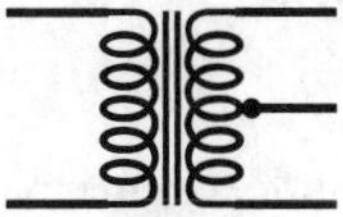

transformer, powdered-iron core

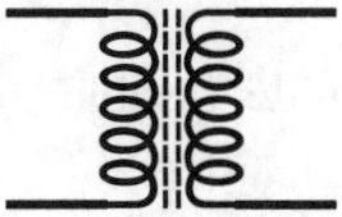

transformer, powdered-iron core, step-down

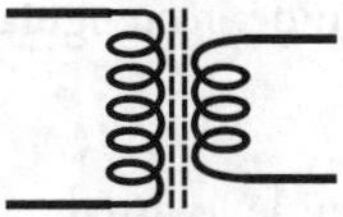

transformer, powdered-iron core, step-up

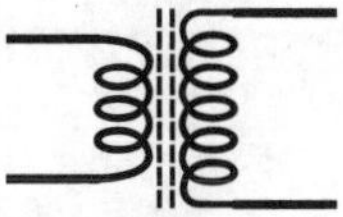

transformer, powdered-iron core, tapped primary

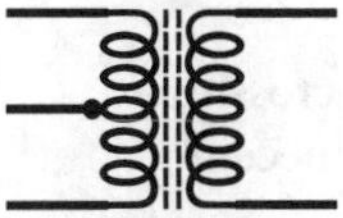

transformer, powdered-iron core, tapped secondary

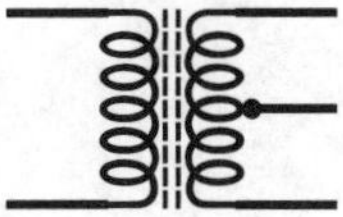

transistor, bipolar, NPN

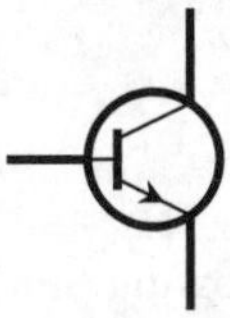

transistor, bipolar, PNP

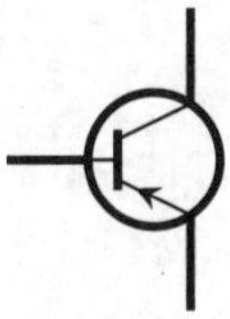

transistor, field-effect, N-channel

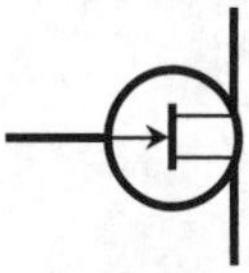

transistor, field-effect, P-channel

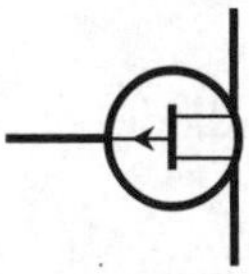

transistor, MOS field-effect, N-channel

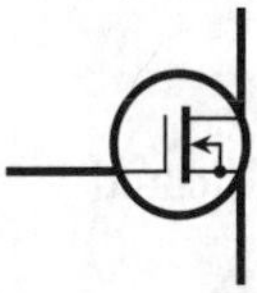

transistor, MOS field-effect, P-channel

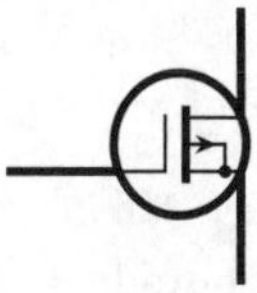

transistor, photosensitive, NPN

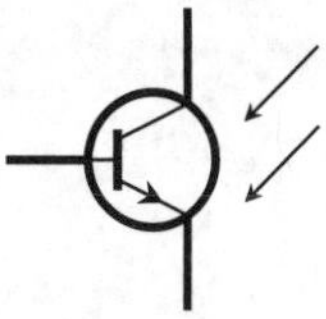

transistor, photosensitive, PNP

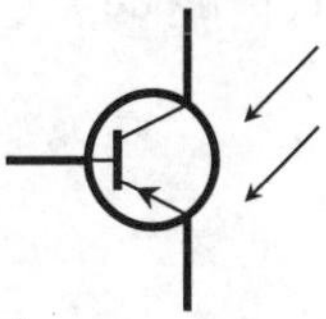

transistor, photosensitive, field-effect, N-channel

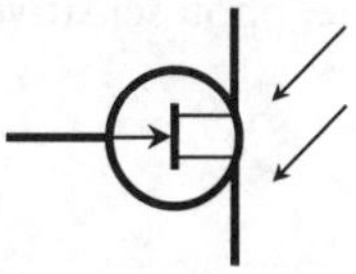

transistor, photosensitive, field-effect, P-channel

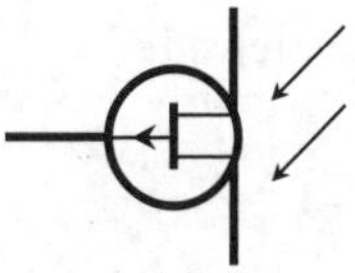

transistor, unijunction

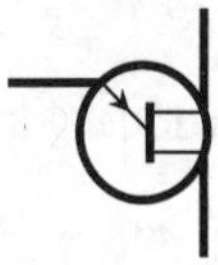

triac

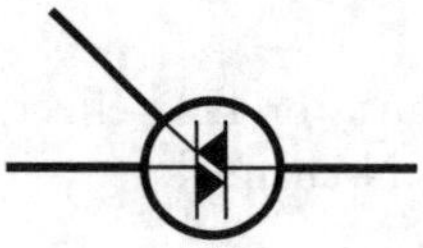

tube, diode

tube, heptode

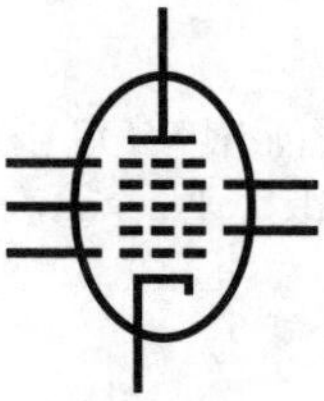

tube, hexode

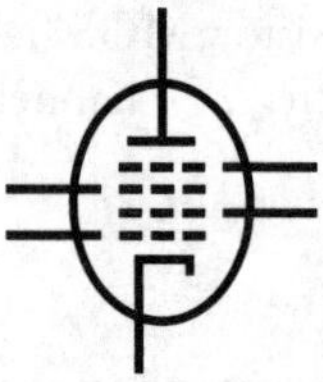

tube, pentode

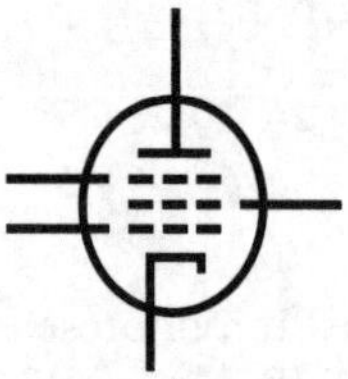

tube, photosensitive

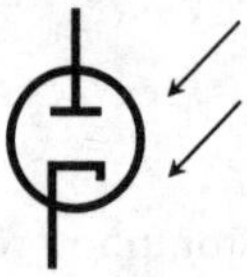

tube, tetrode

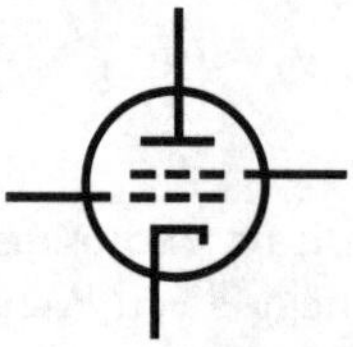

tube, triode

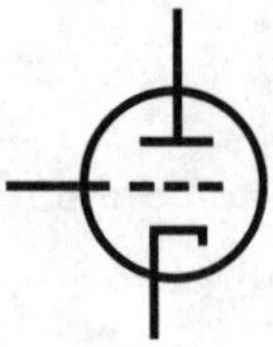

unspecified unit or component

voltmeter

V

wattmeter

W

waveguide, circular

waveguide, flexible

waveguide, rectangular

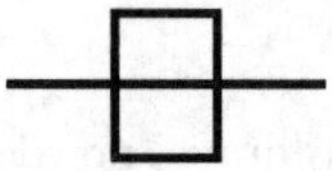

waveguide, twisted

wires, crossing, connected

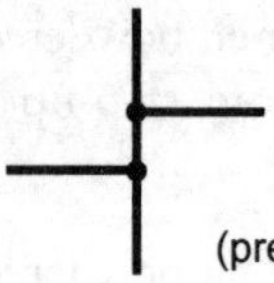

(preferred)

or

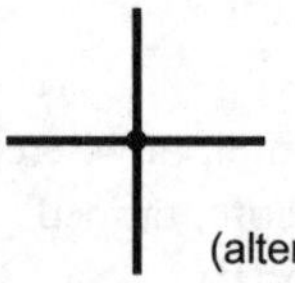

(alternative)

wires, crossing, not connected

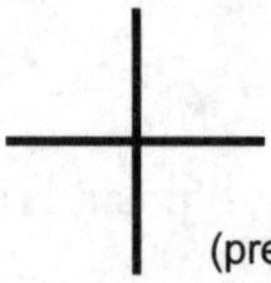

(preferred)

or

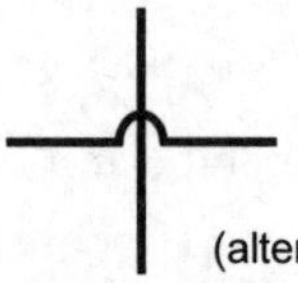

(alternative)

Suggested Additional Readings

- American Radio Relay League, Inc. *The ARRL Handbook for Radio Communications.* Newington, CT: ARRL, revised annually.
- Frenzel, Louis E., Jr., *Electronics Explained.* Burlington, MA: Newnes/Elsevier, 2010.
- Gerrish, Howard, *Electricity and Electronics.* Tinley Park, IL: Goodheart-Wilcox Co., 2008.
- Gibilisco, Stan, *Electricity Demystified*, 2nd ed. New York: McGraw-Hill, 2011.
- Gibilisco, Stan, *Electronics Demystified*, 2nd ed. New York: McGraw-Hill, 2011.
- Gibilisco, Stan, *Ham and Shortwave Radio for the Electronics Hobbyist.* New York: McGraw-Hill, 2014.
- Gibilisco, Stan, *Technical Math Demystified.* New York: McGraw-Hill, 2006.
- Gussow, Milton, *Schaum's Easy Outline of Basic Electricity Revised.* New York: McGraw-Hill, 2011.
- Horn, Delton, *Basic Electronics Theory with Experiments and Projects*, 4th ed. New York: McGraw-Hill, 1994.
- Kybett, Harry, *All New Electronics Self-Teaching Guide*, 3rd ed. Hoboken, NJ: John Wiley & Sons, Inc., 2008.
- Kybett, Harry and Boysen, Earl, *Complete Electronics Self-Teaching Guide with Projects*, 4th Ed. Hoboken, NJ: John Wiley & Sons, Inc., 2012.
- Miller, Rex and Miller, Mark, *Electronics the Easy Way*, 4th ed. Hauppauge, NY: Barron's Educational Series, 2002.
- Mims, Forrest M., *Getting Started in Electronics.* Niles, IL: Master Publishing, 2003.
- Monk, Simon, *Hacking Electronics.* New York: McGraw-Hill, 2013.
- Monk, Simon, *Programming Arduino: Getting Started with Sketches.* New York: McGraw-Hill, 2011.

- Monk, Simon, *Programming Arduino Next Steps: Going Further with Sketches*. New York: McGraw-Hill, 2013.
- Monk, Simon, *Programming the Raspberry Pi*, 2nd ed. New York: McGraw-Hill, 2015.
- Morrison, Ralph, *Electricity: A Self-Teaching Guide*, 3rd ed. Hoboken, NJ: John Wiley & Sons, Inc., 2003.
- Santiago, John, *Circuit Analysis for Dummies*. Hoboken, NJ: John Wiley & Sons, Inc., 2013.
- Schertz, Paul and Monk, Simon, *Practical Electronics for Inventors*, 3rd ed. New York: McGraw-Hill, 2013.
- Shamieh, Cathleen, *Electronics for Dummies*, 3rd ed. Hoboken, NJ: John Wiley & Sons, Inc., 2015.
- Slone, G. Randy, *TAB Electronics Guide to Understanding Electricity and Electronics*, 2nd ed. New York: McGraw-Hill, 2000.
- Van Valkenburg, Mac and Middleton, Wendy, *Reference Data for Engineers: Radio, Electronics, Computers and Communications*, 9th ed. Burlington, MA: Elsevier/Newnes, 2001.
- Veley, Victor, *The Benchtop Electronics Handbook*. New York: McGraw-Hill, 1998.

Index

A

B

C

D

F

G

H

I

M

N

O

P

Q

S

T

U

V

W

X

Y

Z